MAGNETOELECTRIC COMPOSITES

WOODHEAD PUBLISHING SERIES IN ELECTRONIC AND OPTICAL MATERIALS

MAGNETO-ELECTRIC COMPOSITES

MATERIALS, STRUCTURES, AND APPLICATIONS

SECOND EDITION

Edited by

GOPALAN SRINIVASAN
Oakland University, Rochester, MI, United States

SHASHANK PRIYA
Department of Materials Science and Engineering, The Pennsylvania State University, University Park, PA, United States; Chemical Engineering and Materials Science, University of Minnesota, Minneapolis, MN, United States

NIAN SUN
Northeastern University, Boston, MA, United States

ELSEVIER

WP WOODHEAD PUBLISHING
An imprint of Elsevier

Woodhead Publishing is an imprint of Elsevier
50 Hampshire Street, 5th Floor, Cambridge, MA 02139, United States
125 London Wall, London EC2Y 5AS, United Kingdom

ISBN: 978-0-443-18606-6

For information on all Woodhead Publishing publications visit our website at https://www.elsevier.com/books-and-journals

Publisher: Matthew Deans
Acquisitions Editor: Stephen Jones
Editorial Project Manager: Namrata Aggarwal
Production Project Manager: Erragounta Saibabu Rao
Cover Designer: Vicky Pearson Esser

Typeset by TNQ Tech

Contents

4. Layered multiferroic laminates

Gopalan Srinivasan, Zhaoqiang Chu, Mohammad Javad Pourhosseini Asl, Zhonghui Yu and Shuxiang Dong

5. Recent advances in ferroic, multiferroic, and magnetoelectrics materials and phenomena

Sumanta Kumar Karan, Abbey Knoepfel, Su Chul Yang, Aman Nanda, Deepam Maurya, Yongke Yan and Shashank Priya

6. Multiferroic nanostructures

Peng Zhou, Ying Liu, Gopalan Srinivasan, J.F. Li and Dwight Viehland

7. Magnetoelectric energy harvesting from kHz to GHz

R. Sriramdas, Sumanta Kumar Karan, M. Sanghadasa, Mirza Bichurin, O. Sokolov, Shashank Priya, Nian Sun, Zhaoqiang Chu, MohammadJavad PourhosseiniAsl and Shuxiang Dong

8. Magnetoelectric current/magnetic field sensor

Y. Zhou, Shashank Priya, Yisi Liu, Xianfeng Liang, Cunzheng Dong and Nian Sun

9. Magnetoelectric sensors based on magnetoelectric MEMS

Dirk Meyners, Jeffrey McCord and Eckhard Quandt

10. Magnetoelectric NEMS sensors

Bin Luo, Cunzheng Dong and Nian Sun

11. Magnetoelectric antennas for VLF

Yisi Liu, Cunzheng Dong, Yifan He and Nian Sun

12. Magnetoelectric antennas for RF applications

Bin Luo and Nian Sun

Contributors

Mohammad Javad Pourhosseini Asl   Peking University, Beijing, China

Mirza Bichurin   Politechnic Institute, Novgorod State University, Veliky Novgorod, Russia

Zhaoqiang Chu   Peking University, Beijing, China

Shuxiang Dong   Peking University, Beijing, China

Cunzheng Dong   Northeastern University, Boston, MA, United States

Yifan He   Northeastern University, Boston, MA, United States

S. Hosur   School of Electrical Engineering and Computer Science, The Pennsylvania State University, University Park, PA, United States

Rui Huang   Northeastern University, Boston, MA, United States

Sumanta Kumar Karan   Department of Materials Science and Engineering, The Pennsylvania State University, University Park, PA, United States; Center for Clean Energy Engineering, University of Connecticut, Storrs, CT, United States

M. Kiani   School of Electrical Engineering and Computer Science, The Pennsylvania State University, University Park, PA, United States

Abbey Knoepfel   Department of Materials Science and Engineering, The Pennsylvania State University, University Park, PA, United States

J.F. Li   Virginia Tech, Blacksburg, VA, United States

Xianfeng Liang   Northeastern University, Boston, MA, United States

Yisi Liu   Northeastern University, Boston, MA, United States

Ying Liu   Oakland University, Rochester, MI, United States

Xiaxin Liu   Northeastern University, Boston, MA, United States

Bin Luo   Northeastern University, Boston, MA, United States

Deepam Maurya   Bio-inspired Materials and Devices Laboratory (BMDL), Center for Energy Harvesting Materials and Systems (CEHMS), Blacksburg, VA, United States

Jeffrey McCord   Kiel University, Kiel, Germany

Dirk Meyners   Kiel University, Kiel, Germany

Aman Nanda   Department of Materials Science and Engineering, The Pennsylvania State University, University Park, PA, United States

Shashank Priya   Department of Materials Science and Engineering, The Pennsylvania State University, University Park, PA, United States; Chemical Engineering and Materials Science, University of Minnesota, Minneapolis, MN, United States; Virginia Tech, Blacksburg, VA, United States

Eckhard Quandt Kiel University, Kiel, Germany

M. Sanghadasa U.S. Army Combat Capabilities Development Command Aviation & Missile Center, Redstone Arsenal, AL, United States

S. Sharma Department of Materials Science and Engineering, The Pennsylvania State University, University Park, PA, United States

O. Sokolov Politechnic Institute, Novgorod State University, Veliky Novgorod, Russia

Gopalan Srinivasan Oakland University, Rochester, MI, United States

R. Sriramdas Department of Materials Science and Engineering, The Pennsylvania State University, University Park, PA, United States; Department of Mechanical Engineering, Amrita School of Engineering, Coimbatore, Amrita Vishwa Vidyapeetham, Coimbatore, Tamil Nadu, India

Nian Sun Northeastern University, Boston, MA, United States

Dwight Viehland Virginia Tech, Blacksburg, VA, United States

Alexandria Ruth Will-Cole Sandia National Laboratory, Albuquerque, NM, United States

Yongke Yan Department of Materials Science and Engineering, The Pennsylvania State University, University Park, PA, United States

Su Chul Yang Department of Chemical Engineering, Dong-A University, Busan, South Korea

Zhonghui Yu Peking University, Beijing, China

Peng Zhou Oakland University, Rochester, MI, United States

Y. Zhou Virginia Tech, Blacksburg, VA, United States

CHAPTER

1

Theory of magnetoelectric phenomena in composites

Mirza Bichurin, and O. Sokolov

Politechnic Institute, Novgorod State University, Veliky Novgorod, Russia

1.1 Introduction

The magnetoelectric (ME) effect in composites consists of the induction of electric polarization in an applied external magnetic field or, conversely, in magnetization by an external electric field, and is the result of the interaction of the electric, magnetic, and elastic subsystems of the composite. In connection with the fact that many theoretical and experimental studies are currently devoted to ME composites, it is of interest to give a summary of pioneering works in the theory and application of the ME effect. Fundamental are the works of Landau, Lifshitz, Dzyaloshinsky, Astrov, and Rado et al. (Astrov, 1961; Dzyaloshinskii, 1960; Folen et al., 1961; Landau & Lifshitz, 1980) on the discovery of the ME effect. The first experimental works in single crystals have shown that the magnitude of the effect is insufficient for practical applications (Magnetoelectric Interaction Phenomena in Crystals, 1975) and the interest, after the works of Van Suchtelen et al. (Van den Boomgard et al., 1974; Van Run et al., 1974; VanSuchtelen, 1972), has shifted toward bulk and layered composites. In the same years, work was carried out in Russia on the creation and study of the properties of single-phase and heterophase materials, called ferroelectromagnets (Ferroelectromagnet Substances, 1990; Smolenskii & Chupis, 1982). In subsequent years, heterophase ferroelectromagnets based on ferroelectric and ferrite began to be called bulk and layered magnetoelectric composites. Further, it is necessary to note the works (Bichurin, Petrov, & Srinivasan, 2003; Harshe et al., 1993a, 1993b; Lopatin et al., 1994; Nan, 1994; Ryu et al., 2002), in which the theory and experiment of the ME effect in the low-frequency range were presented. Since the main purpose of the research was to increase the ME effect, the next

https://doi.org/10.1016/B978-0-443-18606-6.00009-5

important step was made in the work where it was proposed to investigate the ME effect in the electromechanical resonance (EMR) range (Bichurin, Filippov, et al., 2003), because the energy exchange between the magnetic and electrical component in the composite occurs through mechanical oscillations. This led to an increase in the effect by one to two orders of magnitude and made it possible to proceed to the solution of practical problems. As for the microwave range, here from the very beginning of the research it was clear that the use of an electric field instead of a magnetic field for tuning of parameters should give a big benefit in terms of the speed and efficiency of the devices (Bichurin, 1994; Bichurin & Petrov, 1988). One of the drawbacks of ME devices is the need for a bias field, which is most often created by permanent magnets. This disadvantage was corrected in a patent (Bichurin et al., 2009) by using gradient magnetostrictive-piezoelectric structures, since in this case it is possible to obtain an internal bias field and an electric field for the polarization of the composite without external additional devices. The performed experiments (Bichurin et al., 2020; Mandal et al., 2011) fully confirmed the calculated ratios. After this brief historical introduction, we present the useful information on the ME effect contained in the Proceedings of the MEIPIC conference and a number of reviews and books (Magnetoelectric Interaction Phenomena in Crystals, 1975; Proceedings of the 2nd International, 1993; Proceedings of the 3rd International, 1997; Proceedings of the Fourth Conference, 2002; Proceedings of the Fifth Conference, 2004; Fiebig, 2005; Eerenstein et al., 2006; Ramesh & Spaldin, 2007; Priya et al., 2007; Bichurin et al., 2007; Nan et al., 2008; Zhai et al., 2008; Vaz et al., 2010; Srinivasan, 2010; Wang et al., 2010; Ma et al., 2011; Tatarenko & Bichurin, 2012; Magnetoelectricity in Composites, 2012; Sun & Srinivasan, 2012; Burdin et al., 2013; Wang et al., 2014; Bichurin & Petrov, 2014; Composite Magnetoelectrics, 2015; Palneedi et al., 2016; Song et al., 2018; Chu et al., 2018; Leung et al., 2018; Carman & Sun, 2018; Viehland et al., 2018; Lin et al., 2018; Bichurin et al., 2019; Spaldin & Ramesh, 2019; Yang et al., 2020; Bichurin, Petrov, et al., 2021; Gao et al., 2021; Wang et al., 2021; Spetzler et al., 2021; Liang, Chen, & Sun, 2021; Liang, Matyushov, et al., 2021). There are many works devoted to the calculation of individual EMR regimes of the low-frequency ME effect. But the results of the calculations are scattered among different sources. Therefore, authors of the chapter decided to consider from a unified standpoint the theory of the direct ME effect in composites in the low-frequency range, including all modes of EMR regime: longitudinal, bending, longitudinal-shear, and torsional ones. In order to compare the calculated and experimental results, data are presented for symmetric/three-layer and asymmetric/two-layer structures based on Metglas and various piezoelectrics: PZT, lithium niobate, and gallium arsenide. Note that the quasi-static regime, as a special case, is also considered and the accuracy of the formulas for the resonant frequencies of various EMR modes is also discussed. Considering that we have previously published the main relationships for the ME effect in the FMR region for layered and bulk structures, here we have presented new manifestations of the ME effect in the FMR region for nanofilms of ferromagnetic metals (FM): nickel and permalloy on various piezoelectric substrates: Quartz, PZT and PMN-PT for different bias field orientations. Particular attention is paid considering the "substrate effect". In the case of the microwave ME effect in the PZT-YIG-GGG structure, an assessment is made of the influence of the substrate effect on the shift of the FMR line in an electric field.

1.2 Direct low-frequency ME effect in composites

The paper (Bichurin et al., 2022) considers the physical foundations of the application of the linear ME effect for devices in the low-frequency range, including the EMR regime. The main theoretical expressions for the ME voltage coefficients in the case of a symmetric and asymmetric composite structure in the quasi-static and resonant modes are given. The area of EMR considered here includes longitudinal, bending, longitudinal shear, and torsional modes. Explanations are given for finding the main resonant frequencies of the modes under study.

Here, we give a relatively brief summary of the main results of the consideration of the longitudinal, bending, longitudinal-shear and torsional modes of the low-frequency ME effect in layered composites in the EMR regime and the quasi-static case.

1.2.1 Longitudinal modes

First, we consider the propagation of a longitudinal elastic wave along the length of the composite. Let us consider small mechanical vibrations in a composite under the influence of a small external alternating magnetic field $h_1(\tau)$ in the presence of a constant magnetic field H_0. The electric field arising during the direct ME effect is directed perpendicular to the plane of the composite, since the electrodes from which the ME voltage is removed are located on the upper and lower faces of the piezoelectric. The X axis will be directed along the long side of the composite, the Z axis will be directed perpendicular to the composite plane, and the Y axis will be directed along the width of the composite, respectively. The time dependence of the alternating magnetic field is harmonic $h_1(\tau) \sim e^{i\omega\tau}$.

1.2.1.1 Symmetrical structures

We will consider the ME composite with the form of a thin narrow plate, in which the thickness and width are much less than the length. Layers of the magnetostrictive phase of the same thickness are located above and below the piezoelectric layer of the composite. The magnetic fields are directed along the length of the composite.

The transverse component of the electric displacement vector:

$$D_3 = \varepsilon\varepsilon_0 E_3 + d_{31}{}^pT_1, \tag{1.1}$$

and the longitudinal component of the mechanical stress of the piezoelectric phase:

$${}^pT_1 = \frac{1}{{}^ps_{11}}{}^pS_1 - \frac{d_{31}}{{}^ps_{11}}E_3, \tag{1.2}$$

where, E_3 is the electric field in a piezoelectric.

The longitudinal component of the stress tensor of the magnetostrictive phase:

$${}^mT_1 = {}^mY^{Bm}S_1 - \bar{q}_{11}h_1, \tag{1.3}$$

where h_1 is the external alternating magnetic field away from the ME composite,

$$\bar{q}_{11} = {}^mY^Bq_{11}, \tag{1.4}$$

Young's modulus of the magnetostrictive phase at constant magnetic induction:

$${}^mY^B = \frac{{}^mY}{1 - {}^mK_{11}^2}. \tag{1.5}$$

The square of the coefficient of magnetomechanical coupling is:

$${}^mK_{11}^2 = \frac{{}^mYq_{11}^2}{\mu\mu_0}, \tag{1.6}$$

The longitudinal component of the composite strain tensor, considering the ideal mechanical connection between the phases:

$${}^pS_1 = {}^mS_1 = S_1 = \frac{\partial U_x}{\partial x}, \tag{1.7}$$

where, U_x is the longitudinal displacement.

The longitudinal component of the composite stress tensor is:

$$T_1 = {}^mv^mT_1 + {}^pv^pT_1 = c_{11}S_1 - {}^mv\bar{q}_{11}h_1 - {}^pv\frac{d_{31}}{{}^ps_{11}}E_3, \tag{1.8}$$

where, volume fractions of the piezoelectric and magnetostrictive phases are:

$$\begin{aligned} {}^{p}v &= \frac{{}^{p}t}{{}^{p}t + 2{}^{m}t} \\ {}^{m}v &= \frac{2{}^{m}t}{{}^{p}t + 2{}^{m}t} \end{aligned}, \tag{1.9}$$

effective composite stiffness coefficient:

$$c_{11} = \frac{{}^{p}v}{{}^{p}s_{11}} + {}^{m}v\,{}^{m}Y^{B}, \tag{1.10}$$

composite effective density:

$$\rho = {}^{p}v\,{}^{p}\rho + {}^{m}v\,{}^{m}\rho. \tag{1.11}$$

Solution of the motion equation for longitudinal deformations $\rho\frac{\partial^2 U_x}{\partial t^2} = \frac{\partial T_1}{\partial x}$ is obtained as:

$$U_x = A\cos(kx) + B\sin(kx), \tag{1.12}$$

where A, B are unknown constants,

and wave number is:

$$k = \sqrt{\frac{\rho}{c_{11}}}\omega. \tag{1.13}$$

To find the ME voltage coefficient, we use the boundary conditions for free fixing of both ends of the composite:

$$\begin{aligned} T_1|_{x=-\frac{l}{2}} &= 0 \\ T_1|_{x=\frac{l}{2}} &= 0 \end{aligned}. \tag{1.14}$$

Solving the system of Eq. (1.14), we find the values of the constants A,B.

The transverse component of the electric field vector can be found from the condition that the electric induction flux through the interface between the upper layer of the magnetostrictive phase and the piezoelectric is equal to zero:

$$\int_{-\frac{l}{2}}^{\frac{l}{2}} D_3 dx = 0. \tag{1.15}$$

Expressing E_3 from Eq. (1.15), we find the ME voltage coefficient, taking into account the fact that the electric field exists only in the piezoelectric phase:

$$\alpha_E = \frac{E_3{}^p\nu}{h_1} = -\frac{{}^m\nu{}^p\nu\bar{q}_{11}d_{31}{}^ps_{11}\tan(\eta)}{\varepsilon\varepsilon_0{}^ps_{11}^2c_{11}\eta + d_{31}^2\left[{}^p\nu\tan(\eta) - c_{11}{}^ps_{11}\eta\right]}, \tag{1.16}$$

where, $\eta = \frac{kl}{2}$.

Below in Fig. 1.2 the dependence of the ME voltage coefficient is shown on the frequency of the alternating magnetic field for 2 cases, when PZT and a cut of lithium niobate y + 128 degrees are taken as the piezoelectric phase. Metglas is taken as the magnetostrictive phase. For this and subsequent calculations, the following Metglas thicknesses were taken ${}^mt = 29\ \mu m$ and for piezoelectric ${}^pt = 0.5\ mm$, length of the ME composite $l = 10\ mm$. To take into account energy losses during EMR, we set $\omega = 2\pi\left(1 + \frac{1}{Q}i\right)f$, where Q is the quality factor of the resonance. The value of the quality factor for longitudinal and bending modes Q = 130.

The maximum value of the ME voltage coefficient, ceteris paribus, for LN cut y + 128 degrees is greater than for PZT, since it has a larger ratio $\frac{d_{31}}{\varepsilon}$.

Fundamental resonant frequency for the longitudinal oscillation mode of the ME effect in the composite in Fig. 1.1, both ends of which are freely fixed:

$$f_r = \frac{1}{2l}\sqrt{\frac{c_{11}}{\rho}}. \tag{1.17}$$

1.2.1.2 *Asymmetric structures*

An asymmetric ME structure differs from a symmetrical one in that the layer of the magnetostrictive phase is only on one side of the piezoelectric.

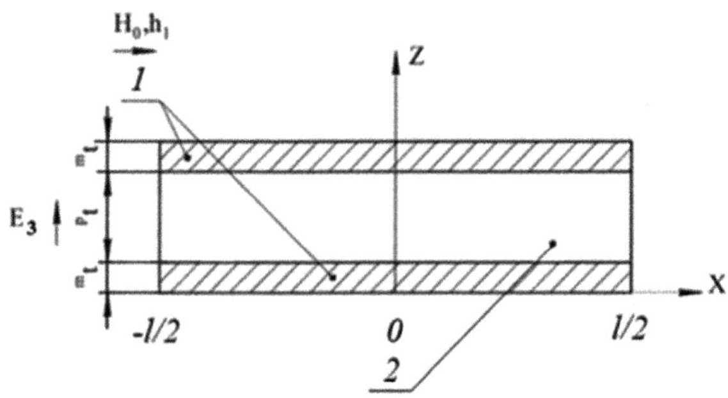

FIGURE 1.1 The structure of a symmetric ME composite for calculate of the longitudinal mode. 1 is magnetostrictive phases. 2 is piezoelectric phase. ${}^pt$ and ${}^mt$ are thicknesses of the piezoelectric and magnetostrictive phases.

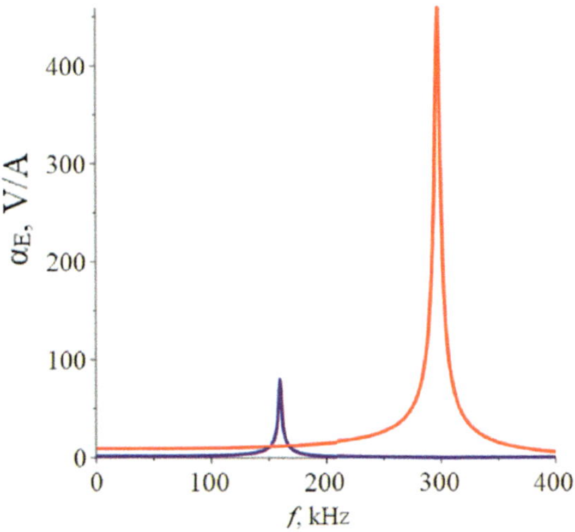

FIGURE 1.2 Estimated dependence of the ME voltage coefficient on the frequency of an alternating magnetic field applied to a symmetric ME composite. Blue line is for PZT layer. Red line is for LN cut y + 128 degrees layer.

For an asymmetric ME structure in the longitudinal mode, the ME voltage coefficient is calculated according to Section 1.2.1.1, provided that factor 2 should be removed before ${}^{m}t$ in Eq. (1.9).

1.2.2 Bending mode

The use of the ME effect for longitudinal vibration modes is limited by high resonant frequencies (about hundreds of kHz for a sample length and width of several millimeters). Also, at high resonant frequencies of ME composites, eddy current losses in the magnetostrictive phase of the composite can be high for transition metals and rare earth metal alloys, such as Terfenol-D, which leads to inefficient transfer of mechanical vibration energy from the magnetostrictive phase to the piezoelectric one during the direct ME effect. To reduce resonant frequencies, it is possible to increase the size of ME composites, but in this case, the use of ME composites in small-sized electronic devices will be impossible. An alternative way to implement the ideal ME coupling of the composite phases is to use the bending modes of the ME effect (Bichurin et al., 2019). It is assumed in the calculations that the resonant vibration frequency of the ME composite in the bending mode will be much lower compared to the longitudinal modes of the ME effect. Recent studies have shown a giant ME effect in bending modes in several layered ME structures.

We consider small mechanical bending vibrations in an asymmetric ME composite in Fig. 1.3, since in a symmetric composite, this mode of the

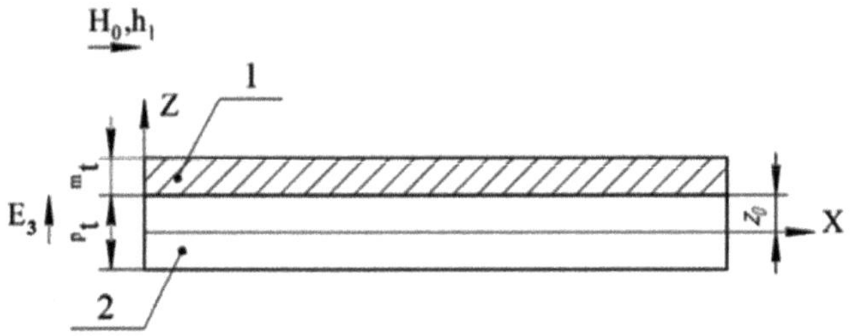

FIGURE 1.3 The structure of an asymmetric magnetoelectric composite for calculating the bending mode. 1 is magnetostrictive phases. 2 is piezoelectric phase. ${}^{p}t$ and ${}^{m}t$ are thicknesses of the piezoelectric and magnetostrictive phases. z_0 is the position of the boundary between the piezoelectric and magnetostrictive phases relative to the neutral line.

ME effect does not arise. We assume that the composite has the form of a thin narrow plate, the thickness and width of which are much less than the length. The orientations of the magnetic and electric fields are the same as those for the longitudinal mode of the ME effect.

The full thickness of the composite:

$$t = {}^{p}t + {}^{m}t. \tag{1.18}$$

Volume fractions of the piezoelectric and magnetostrictive phases are:

$$\begin{aligned} {}^{p}v &= \frac{{}^{p}t}{t}, \\ {}^{m}v &= \frac{{}^{m}t}{t}. \end{aligned} \tag{1.19}$$

The longitudinal component of the strain tensor of the composite, considering the ideal mechanical connection between the phases:

$${}^{m}S_1 = {}^{p}S_1 = S_1 = -z\frac{\partial^2 w}{\partial x^2}, \tag{1.20}$$

where, w is transverse displacement.

The longitudinal components of the stress tensor of the piezoelectric and magnetostrictive phases of the composite and the transverse component of the electric field of the piezoelectric phase:

$${}^{p}T_1 = c_{11}^{D} S_1 - h_{31} D_3, \tag{1.21}$$

$${}^{m}T_1 = -z\,{}^{m}Y^{B}\frac{\partial^2 w}{\partial x^2} - \overline{q}_{11} h_1, \tag{1.22}$$

$$E_3 = -h_{31} S_1 + \beta_{33}^{S} D_3, \tag{1.23}$$

Where:

$$c_{11}^{D} = \left({}^{p}s_{11}^{E} - \frac{d_{31}^{2}}{\varepsilon_{33}^{T}\varepsilon_{0}}\right)^{-1}$$
$$h_{31} = \frac{c_{11}^{D}d_{31}}{\varepsilon_{33}^{T}\varepsilon_{0}} \quad . \tag{1.24}$$
$$\beta_{33}^{S} = \frac{1 + h_{31}d_{31}}{\varepsilon_{33}^{T}\varepsilon_{0}}$$

Bending moment:

$$M = \int_{z_0 - {}^{p}t}^{z_0} bz^{p}T_1 dz + \int_{z_0}^{z_0 + {}^{m}t} bz^{m}T_1 dz = \quad , \tag{1.25}$$
$$= -b\frac{\partial^2 w}{\partial x^2}D - b^{p}t^2\langle h_{31}\rangle D_3 - b^{m}t^2\langle q_{11}\rangle h_1$$

where, b is composite width,

$$\langle h_{31}\rangle = \frac{1}{{}^{p}t^2}\int_{z_0 - {}^{p}t}^{z_0} zh_{31}dz = \frac{2z_0 - {}^{p}t}{2^{p}t}h_{31}$$
$$. \tag{1.26}$$
$$\langle q_{11}\rangle = \frac{1}{{}^{m}t^2}\int_{z_0}^{z_0 + {}^{m}t} z\overline{q}_{11}dz = \frac{\overline{q}_{11}(2z_0 + {}^{m}t)}{2^{m}t}$$

Full cylindrical stiffness of the composite:

$$D = \frac{1}{3}\left[c_{11}^{D}\,{}^{p}t\left({}^{p}t^2 - 3^{p}tz_0 + 3z_0^2\right) + {}^{m}Y^{Bm}t\left({}^{m}t^2 + 3^{m}tz_0 + 3z_0^2\right)\right]. \tag{1.27}$$

Electrical voltage in a piezoelectric:

$$U = \int_{z_0 - {}^{p}t}^{z_0} E_3 dz = {}^{p}t^2\langle h_{31}\rangle\frac{\partial^2 w}{\partial x^2} + {}^{p}t\beta_{33}^{S}D_3. \tag{1.28}$$

Expressing from Eq. (1.28) the electrical displacement in the piezoelectric D_3 and substituting into Eq. (1.25), we obtain:

$$M = -bt^3\langle c_{11}\rangle\frac{\partial^2 w}{\partial x^2} - \frac{b^{p}t\langle h_{31}\rangle}{\beta_{33}^{S}}U - b^{m}t^2\langle q_{11}\rangle h_1, \tag{1.29}$$

where:

$$\langle c_{11}\rangle = \frac{1}{t^3}\left(D - \frac{{}^p t^3 \langle h_{31}\rangle^2}{\beta_{33}^S}\right). \tag{1.30}$$

The position of the boundary between the piezoelectric and magnetostrictive phases relative to the neutral line z_0 is determined from the minimum condition $\langle c_{11}\rangle$:

$$z_0 = \frac{\left(c_{11}^{D\,p}t^2 - {}^m Y^{B\,m}t^2\right)\beta_{33}^S - h_{31}^2\,{}^p t^2}{2\left({}^m Y^{B\,m}t + c_{11}^{D\,p}t\right)\beta_{33}^S - h_{31}^2\,{}^p t}. \tag{1.31}$$

General solution of the flexural vibration equation $\rho bt \frac{\partial w^2}{\partial \tau^2} = \frac{\partial M}{\partial x^2}$ is obtained as:

$$w = C_1 \cosh(kx) + C_2 \sinh(kx) + C_3 \cos(kx) + C_4 \sin(kx), \tag{1.32}$$

where, C_1, C_2, C_3, C_4 are unknown constants,

The wave number:

$$k = \left(\frac{\rho}{t^2\langle c_{11}\rangle}\omega^2\right)^{\frac{1}{4}}, \tag{1.33}$$

Integrating Eq. (1.28) over x considering the open circuit condition $\int_0^l D_3 dx = 0$, we obtain:

$$\begin{aligned} Ul &= {}^p t^2 \langle h_{31}\rangle \frac{\partial w}{\partial x}\Big|_0^l = \\ &= {}^p t^2 \langle h_{31}\rangle k[C_1 r_2 + C_2(r_1 - 1) - C_3 r_4 + C_4(r_3 - 1)] \end{aligned}, \tag{1.34}$$

where:

$$\begin{aligned} r_1 &= \cosh(kl) \\ r_2 &= \sinh(kl) \\ r_3 &= \cos(kl) \\ r_4 &= \sin(kl) \end{aligned}. \tag{1.35}$$

The boundary conditions for free ends of the beam are given by:

$$\begin{aligned} &\frac{\partial M}{\partial x}(0) = \frac{\partial M}{\partial x}(l) = 0, \\ &M(0) = M(l) = 0. \end{aligned} \tag{1.36}$$

Solving the system of Eqs, (1.34) and (1.36), we can find the values of the constants $C_1 - C_4$ and voltage U in the piezoelectric phase of the composite.

Then the ME voltage coefficient:

$$\alpha_E = \frac{U}{h_1 t} = \frac{2^m t^{2p} t^2 \langle q_{11} \rangle \langle h_{31} \rangle \beta_{33}^S (r_1 r_4 + r_2 r_3 - r_2 - r_4)}{t\left[\langle c_{11} \rangle k l t^3 \beta_{33}^S (1 - r_1 r_3) - 2^p t^3 \langle h_{31} \rangle^2 (r_1 r_4 + r_2 r_3 - r_2 - r_4)\right]}. \tag{1.37}$$

The fundamental resonant frequency for the bending mode of oscillations of the ME effect in the composite in Fig. 1.3, both ends of which are freely fixed:

$$f_r = \frac{\chi^2 t}{2\pi l^2}\sqrt{\frac{\langle c_{11} \rangle}{\rho}}. \tag{1.38}$$

$$\chi = 4,73$$

Below in Fig. 1.4 the estimated dependence of the ME voltage coefficient is shown on the frequency of the alternating magnetic field f, kHz for the same ME structures as in section 1.2.1.1.

The maximum value of the ME voltage coefficient for LN cut y + 128 degrees is greater than for PZT, for the same reason as for the longitudinal mode.

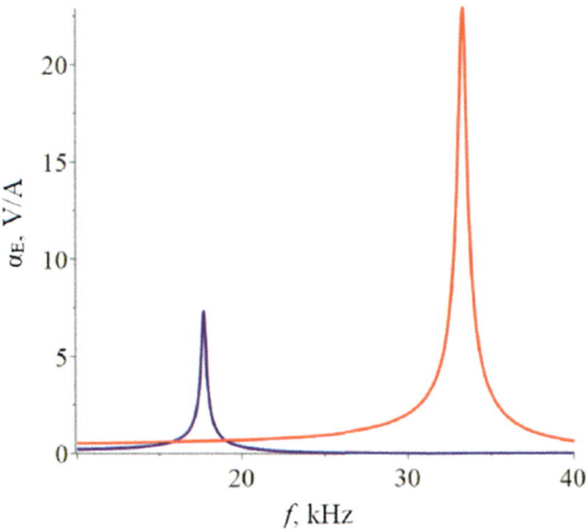

FIGURE 1.4 Dependence of the ME voltage coefficient on the frequency of an alternating magnetic field applied to an asymmetric ME composite. Bending mode of the ME effect. Blue line is for PZT layer. Red line is for LN cut y + 128 degrees layer.

In practice, in excitation of the bending mode of the ME effect, cantilever fastening of the ME composite is often used, when one end is rigidly clamped and the other is free. The calculation for this case differs only in the boundary conditions for the pinched end, and the final expression for the ME voltage coefficient has the form

$$\alpha_E = -\frac{{}^{m}t^{2p}t^{2}\langle q_{11}\rangle\langle h_{31}\rangle\beta_{33}^{S}(r_1r_4 + r_2r_3)}{\left[\langle c_{11}\rangle klt^{3}\beta_{33}^{S}(1 + r_1r_3) + {}^{p}t^{3}\langle h_{31}\rangle^{2}(r_1r_4 + r_2r_3)\right]t}. \quad (1.39)$$

The equation for the fundamental resonant frequency in this case coincides with Eq. (1.38), but $\chi = 1.875$. Thus, the main resonant frequency of the bending mode with cantilevered fixing is much less than with free one. In this case, the maximum values of the ME voltage coefficient are only slightly less than in the case of free composite ends.

1.2.3 Longitudinal-shear modes

1.2.3.1 Symmetrical structures

We consider small longitudinal shear mechanical vibrations (Bichurin & Petrov, 2014) in an ME composite in Fig. 1.5. The constant bias field is directed along the width of the composite, the alternating magnetic field is directed along the length of the composite. For the longitudinal shear mode the equations for calculating the ME stress coefficient are the same as for the longitudinal mode, but the mechanical stress tensors are replaced ${}^{m}T_1, {}^{p}T_1$ on ${}^{m}T_6, {}^{p}T_6$, piezoelectric and piezomagnetic modulus d_{31}, q_{11} on d_{36}, q_{16}, compliance factor of the piezoelectric phase ${}^{p}s_{11}$ on ${}^{p}s_{66}$, Young's modulus of magnetostrictive phases ${}^{m}Y$ on torsion module ${}^{m}G$, effective coefficient of composite stiffness c_{11} on c_{66}.

In Fig. 1.5 the notations are the same as in Fig. 1.1.

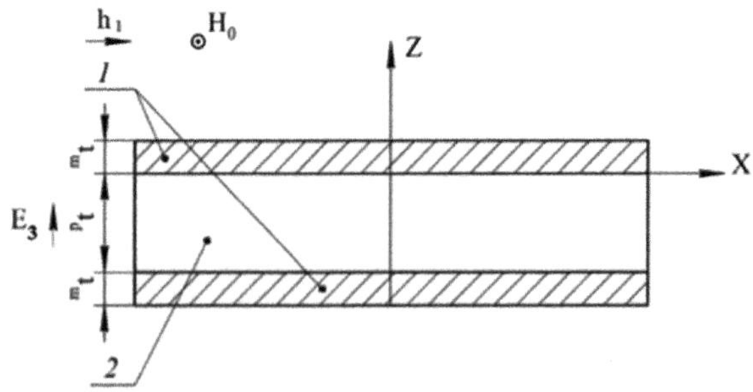

FIGURE 1.5 The structure of a symmetric ME composite for calculate of the longitudinal shear mode.

The component of the strain tensor for longitudinal-shear vibrations at considering the ideal mechanical connection between the phases:

$$ {}^{m}S_6 = {}^{p}S_6 = S_6 = \frac{\partial U_x}{\partial y} + \frac{\partial U_y}{\partial x} = \frac{\partial U_y}{\partial x}. \tag{1.40} $$

The wave number:

$$ k = \sqrt{\frac{\rho}{c_{66}}}\omega. \tag{1.41} $$

Then the ME voltage coefficient, considering the fact that the electric field exists only in the piezoelectric phase for free fixing of the composite, will be equal to:

$$ \alpha_E = \frac{E_3{}^{p}\nu}{h_1} = -\frac{{}^{m}\nu^{p}\nu\bar{q}_{16}d_{36}{}^{p}s_{66}\tan(\eta)}{\varepsilon\varepsilon_0{}^{p}s_{66}^{2}c_{66}\eta + d_{36}^{2}\left[{}^{p}\nu\tan(\eta) - c_{66}{}^{p}s_{66}\eta\right]}, \tag{1.42} $$

where $\eta = \frac{kl}{2}$.

Below in Fig. 1.6 the dependence of the ME voltage coefficient is shown on the frequency of the alternating magnetic field. The following previously unspecified material parameters of Metglas were used in the calculation ${}^{m}G = 3.85 \cdot 10^{10}\ Pa$, $q_{16} = 1.0 \cdot 10^{-9}\ m/A$ and the same thickness as before; material parameters and thickness of GaAs: ${}^{p}\rho = 5320\ kg/m^3, {}^{p}G = 5.94 \cdot 10^{10}\ Pa, \varepsilon = 12.9, d_{36} = -2.69 \cdot 10^{-12}\ m/V, {}^{p}t = 2 \cdot 10^{-4}\ m$. Length of the ME composite $l = 2.3 \cdot 10^{-2}$ m, width is $b = 3 \cdot 10^{-4}$ m. The value of the quality factor of the EMR for ME composite with GaAs is $Q = 300$.

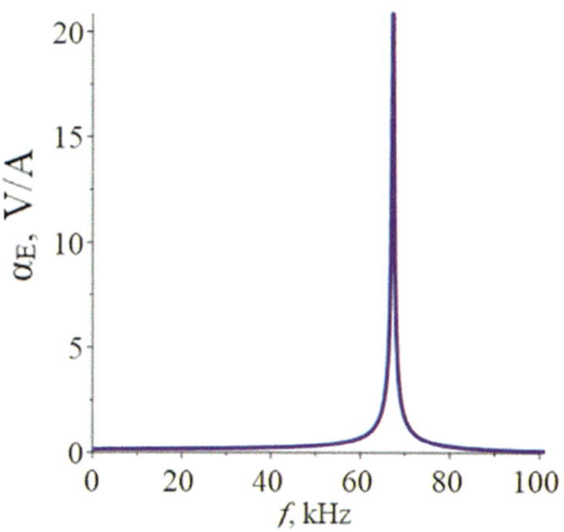

FIGURE 1.6 Dependence of the ME voltage coefficient on the frequency of an alternating magnetic field applied to a symmetric ME composite with GaAs layer for longitudinal-shear mode.

Fundamental resonant frequency for the longitudinal-shear oscillation mode of the magnetoelectric effect in the composite in Fig. 1.5, both ends of which are freely fixed:

$$f_r = \frac{1}{2l}\sqrt{\frac{c_{66}}{\rho}}. \tag{1.43}$$

1.2.3.2 *Asymmetric structure*

For an asymmetric ME structure in the longitudinal shear mode, the ME voltage coefficient is calculated according to Section 1.2.3.2, considering the coefficient 2 should be removed before ${}^{m}t$ in Eq. (1.9). Orientations of alternating magnetic and electric fields, as well as a constant magnetic field correspond to the orientations in Fig. 1.5.

1.2.4 Torsional mode

We consider small mechanical torsional vibrations (Bichurin, Sokolov, & Lobekin, 2021) in an asymmetric ME composite in Fig. 1.7, since in a symmetric composite, the torsional mode of the ME effect does not arise. The orientations of the magnetic and electric fields are the same as for the longitudinal shear mode. With this orientation of the constant and alternating magnetic fields, torsional vibrations occur about the X axis directed along the length of the composite.

The total thickness t of the ME composite is determined from Eq. (1.18).

Shear components of the strain tensor are:

$$\begin{aligned} S_5 &= y\frac{\partial\theta}{\partial x} \\ S_6 &= -z\frac{\partial\theta}{\partial x}, \end{aligned} \tag{1.44}$$

where, θ is the twist angle.

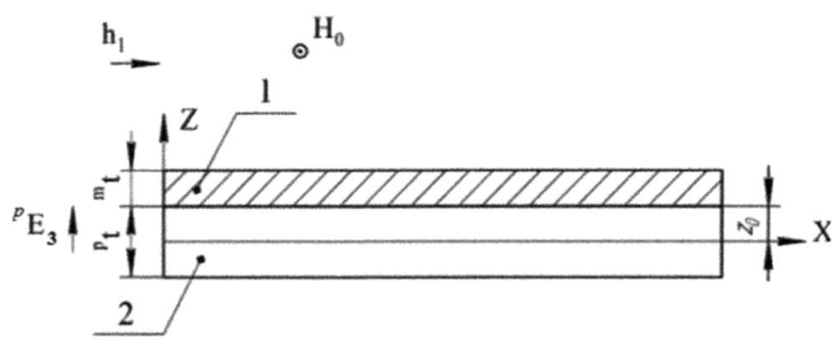

FIGURE 1.7 The position of the interface between the piezoelectric and magnetostrictive phases relative to the axis of rotation of the two-layer composite. 1 is magnetostrictive phases, 2 is piezoelectric phase. ${}^{p}t$ and ${}^{m}t$ are thicknesses of the piezoelectric and magnetostrictive phases.

Tangent components of the piezoelectric stress tensor and electrical displacement are:

$$ {}^pT_5 = {}^pGS_5 = {}^pGy\frac{\partial\theta}{\partial x}, \tag{1.45} $$

$$ {}^pT_6 = -{}^pG^Dz\frac{\partial\theta}{\partial y} - h_{36}D_3, \tag{1.46} $$

$$ D_3 = -d_{36}{}^pGz\frac{\partial\theta}{\partial x} + \left(\varepsilon\varepsilon_0 - {}^pGd_{36}^2\right){}^pE_3, \tag{1.47} $$

where, ${}^pE_3$ is an electric field in a piezoelectric.

$$ \begin{aligned} {}^pG^D &= \frac{\varepsilon\varepsilon_0{}^pG}{\varepsilon\varepsilon_0 - {}^pGd_{36}^2} \\ h_{36} &= \frac{d_{36}{}^pG}{\varepsilon\varepsilon_0 - {}^pGd_{36}^2} \end{aligned}. \tag{1.48} $$

Tangent components of the magnetostrictive phases stress tensor are:

$$ \begin{aligned} {}^mT_5 &= {}^mGS_5 = {}^mGy\frac{\partial\theta}{\partial x} \\ {}^mT_6 &= {}^mG(S_6 - q_{16}h_1) = -{}^mGz\frac{\partial\theta}{\partial x} - \bar{q}_{16}h_1 \end{aligned}, \tag{1.49} $$

Where:

$$ \bar{q}_{16} = {}^mGq_{16}. \tag{1.50} $$

The torque is:

$$ \begin{aligned} M &= \int_{-\frac{b}{2}}^{\frac{b}{2}} dy \int_{z_0-{}^pt}^{z_0} \left(y^pT_5 - z^pT_6\right)dz + \int_{-\frac{b}{2}}^{\frac{b}{2}} dy \int_{z_0}^{z_0+{}^mt} \left(y^mT_5 - z^mT_6\right)dz = , \\ &= K\frac{\partial\theta}{\partial x} + b^p t^2\langle h_{36}\rangle D_3 + b^m t^2\langle q_{16}\rangle h_1 \end{aligned} \tag{1.51} $$

Where:

$$ \begin{aligned} K &= {}^pK + {}^mK \\ {}^pK &= \frac{1}{3}{}^pG^D\left(z_0^3 - \left(z_0 - {}^pt\right)^3\right)b + \frac{1}{12}{}^pG^ptb^3 \\ {}^mK &= {}^mG\left(\frac{1}{3}\left(\left(z_0 + {}^mt\right)^3 - z_0^3\right)b + \frac{1}{12}{}^mtb^3\right) \end{aligned}, \tag{1.52} $$

$$\langle h_{36}\rangle = \frac{1}{{}^{p}t^{2}} \int\limits_{z_0-{}^{p}t}^{z_0} z h_{36} dz = \frac{h_{36}\left(2z_0 - {}^{p}t\right)}{2{}^{p}t}$$
$$\langle q_{16}\rangle = \frac{1}{{}^{m}t^{2}} \int\limits_{z_0}^{z_0+{}^{m}t} z \bar{q}_{16} dz = \frac{\bar{q}_{16}\left(2z_0 + {}^{m}t\right)}{2{}^{m}t} \quad (1.53)$$

Expressing from Eq. (1.47) ${}^{p}E_3$, we get the electrical voltage on the piezoelectric:

$$U = \int\limits_{z_0-{}^{p}t}^{z_0} {}^{p}E_3 dz = {}^{p}t^{2}\langle h_{36}\rangle \frac{\partial\theta}{\partial x} + {}^{p}t\beta_{33}^{S} D_3, \quad (1.54)$$

Where:

$$\beta_{33}^{S} = \frac{1}{\varepsilon\varepsilon_0 - {}^{p}G d_{36}^{2}}. \quad (1.55)$$

Expressing from Eq. (1.54) electrical displacement D_3 in piezoelectric and substituting in Eq. (1.51), we get:

$$M = -bt^{3}\langle G\rangle \frac{\partial\theta}{\partial x} - \frac{b^{p}t\langle h_{36}\rangle}{\langle\beta_{33}^{S}\rangle} U + b^{m}t^{2}\langle q_{16}\rangle h_1, \quad (1.56)$$

Where:

$$\langle G\rangle = \frac{1}{bt^{3}}\left(K - \frac{b^{p}t^{3}\langle h_{36}\rangle^{2}}{\beta_{33}^{S}}\right). \quad (1.57)$$

The position of the interface between the piezoelectric and magnetostrictive phases relative to the axis of rotation of the composite beam z_0 is determined from the condition of the minimum effective shear modulus of the sample $\langle G\rangle$:

$$z_0 = \frac{{}^{p}G^{Dp}t^{2}\beta_{33}^{S} - {}^{m}G^{m}t^{2}\beta_{33}^{S} - h_{36}^{2}\,{}^{p}t^{2}}{2\left({}^{m}G^{m}t\beta_{33}^{S} + {}^{p}G^{Dp}t\beta_{33}^{S} - h_{36}^{2}\,{}^{p}t\right)}. \quad (1.58)$$

Solution of the motion equation for torsional vibrations $J\frac{\partial\theta^2}{\partial\tau^2} = \frac{\partial M}{\partial x}$ is obtained as:

$$\theta = A\cos(kx) + B\sin(kx), \quad (1.59)$$

where, A, B are unknown constants,

The wave number:

$$k = \omega\sqrt{\frac{J}{bt^3\langle G\rangle}}. \tag{1.60}$$

The moment of inertia of the sample per unit width:

$$J = {}^p\rho^p I + {}^m\rho^m I, \tag{1.61}$$

The polar moment of the piezoelectric and magnetostrictive phases:

$$\begin{aligned} {}^pI &= \frac{1}{3}\left(z_0^3 - \left(z_0 - {}^pt\right)^3\right)b + \frac{1}{12}{}^ptb^3 \\ {}^mI &= \frac{1}{3}\left(\left(z_0 + {}^mt\right)^3 - z_0^3\right)b + \frac{1}{12}{}^mtb^3 \end{aligned}. \tag{1.62}$$

Boundary conditions for a free sample are:

$$\begin{aligned} M\left(\frac{l}{2}\right) &= 0 \\ M\left(-\frac{l}{2}\right) &= 0 \end{aligned}. \tag{1.63}$$

Integrating Eq. (1.54) over x, considering the open circuit condition $\int_{-\frac{l}{2}}^{\frac{l}{2}} D_3 dx = 0$ and combining with boundary conditions, we obtain a linear system of three nonhomogeneous algebraic equations in three unknowns A, B, U.

Solving this system, we find the voltage on the piezoelectric U and then the ME voltage coefficient:

$$\alpha_E = \frac{U}{th_1} = \frac{2^p t^{2m} t^2 \langle h_{36}\rangle\langle q_{16}\rangle\beta_{33}^S \tan\eta}{t\left(klt^3\langle G\rangle\beta_{33}^S + 2\langle h_{36}\rangle^{2p} t^3 \tan\eta\right)}, \tag{1.64}$$

where $\eta = \frac{kl}{2}$.

In the calculation, the same material parameters and dimensions of the ME composite were used for the longitudinal shear mode. The results of calculation are presented in Fig. 1.8.

Fundamental resonant frequency for the torsional mode in the composite in Fig. 1.7, both ends of which are freely fixed:

$$f_r = \frac{1}{2l}\sqrt{\frac{bt^3\langle G\rangle}{J}}. \tag{1.65}$$

FIGURE 1.8 Dependence of the ME voltage coefficient on the frequency of an alternating magnetic field applied to asymmetric ME composite with GaAs layer for torsional mode.

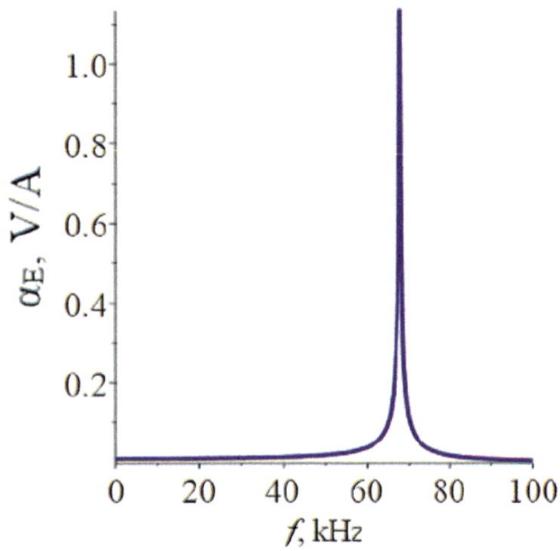

1.2.5 Accuracy of resonance formulas

In this section, the accuracy of the formulas for the fundamental frequencies of the above modes of the ME effect in the EMR region is considered. To determine the accuracy of the formula for the resonant frequency of the longitudinal vibration mode of an ME structure with freely fixed ends, let us consider Eq. (1.16). Obviously, the resonant frequency should vanish the denominator of this expression. But this denominator consists of the main term and a term proportional to the dimensionless quantity $\frac{d_{31}^2}{\varepsilon\varepsilon_0{}^p s_{11}}$. If this dimensionless quantity is small compared to unity, then the corresponding term in the denominator can be neglected, and then Eq. (1.17) will be obtained for the fundamental resonant frequency for the planar mode of the ME effect. The accuracy of the formula for the resonant frequency of the bending mode of oscillations of an ME structure with the same pinning condition as for the longitudinal mode is determined similarly. In Eq. (1.37), the denominator consists of the main term and the second term, which is proportional to the dimensionless quantity $\frac{\langle h_{31}\rangle^{2p}t^3}{t^3\langle c_{11}\rangle\beta_{33}^S}$. If this dimensionless quantity is small compared to unity, then the corresponding term in the denominator can be neglected, and then Eq. (1.38) is obtained for the bending mode of the ME effect. For the longitudinal shear and torsional modes of the ME effect, the accuracy of the formulas for fundamental resonant frequencies is determined from the same considerations with similar conditions $\frac{d_{36}^2}{{}^p s_{66}\varepsilon\varepsilon_0} \ll 1$ and $\frac{\langle h_{36}\rangle^{2p}t^3}{t^3\langle G\rangle\beta_{33}^S} \ll 1$, respectively.

1.2.6 Quasi-static regime

The magnitude of the ME coefficients for quasi-static regime is determined by the values of the piezoelectric and piezomagnetic coefficients for the individual phases of the composite, the values of the compliance and stiffness coefficients of each phase, as well as the efficiency of transfer of mechanical stresses between the individual phases of the composite.

1.2.6.1 Quasi-static regime of longitudinal and longitudinal-shear modes for symmetric ME composites

Assuming in Eqs. (1.16) and (1.42) the frequency f is equal to zero, we obtain the equations for the ME voltage coefficient in the quasi-static regime of the longitudinal and longitudinal-shear modes:

$$\alpha_E = \frac{{}^m\nu^p\nu\bar{q}_{11}d_{31}}{{}^m\nu^m Y^B d_{31}^2 - \varepsilon\varepsilon_0{}^p s_{11}c_{11}}, \quad (1.66)$$

$$\alpha_E = \frac{{}^m\nu^p\nu\bar{q}_{16}d_{36}}{{}^m\nu^m G d_{36}^2 - \varepsilon\varepsilon_0{}^p s_{66}c_{66}}, \quad (1.67)$$

and present it in the form in Fig. 1.9.

It can be seen from Fig. 1.9, that for each symmetric ME structure in the quasi-static regime, there is an optimal ratio of volume fractions between the phases of the composite, at which the ME voltage coefficient reaches a maximum.

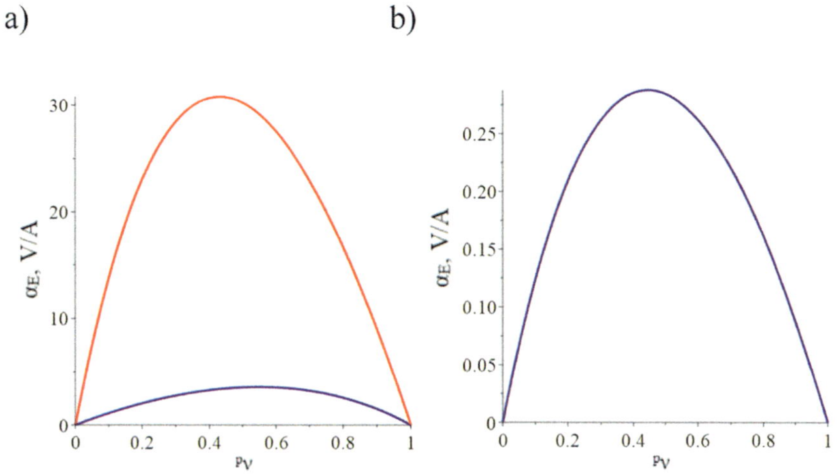

FIGURE 1.9 Dependence of the ME voltage coefficient on the volume fraction of the piezoelectric phase of a symmetric ME composite in a quasi-static regime. (a) Longitudinal mode. Blue line is for PZT layer. Red line is for LN cut y + 128 degrees layer. (b) Longitudinal shear mode.

1.2.6.2 Quasi-static regime for asymmetric ME composites

In an asymmetric composite, in the quasi-static regime of the ME effect and the orientation of a constant magnetic field directed along the length of the composite, both longitudinal and bending vibration modes arise simultaneously.

For the quasi-static regime, the motion equation of the bending vibrations of the ME composite has the form:

$$\frac{\partial T_1}{\partial x} = 0. \tag{1.68}$$

This means that T_1 must not depend on x. It is obvious that S_1 must not depend on x either. Since both the longitudinal and bending modes are excited in the asymmetric ME structure in the quasi-static regime:

$$S_1 = A + zB. \tag{1.69}$$

The first condition for the static equilibrium of an ME composite is that the total longitudinal force is equal to zero:

$$\int_{z_0-{}^pt}^{z_0} {}^pT_1 dz + \int_{z_0}^{z_0+{}^mt} {}^mT_1 dz = 0. \tag{1.70}$$

The second condition for the static equilibrium of the ME composite is the zero total moment is given by:

$$\int_{z_0-{}^pt}^{z_0} z{}^pT_1 dz + \int_{z_0}^{z_0+{}^mt} z{}^mT_1 dz = 0. \tag{1.71}$$

Eqs. (1.70) and (1.71) form a linear inhomogeneous system of two equations with two unknowns A, B. Solving them, we obtain A, B.

Substituting Eq. (1.69) in Eq. (1.23), and taking into account that due to the open circuit condition $D_3 = 0$, we get:

$$E_3 = -h_{31}(A + zB). \tag{1.72}$$

Then, the voltage across the piezoelectric is:

$$U = \int_{z_0-{}^pt}^{z_0} -h_{31}(A + zB)dz = -h_{31}\left[A{}^pt + \frac{1}{2}B{}^pt\left(2z_0 - {}^pt\right)\right]. \tag{1.73}$$

Substituting the values found from the system of equations A, B in Eq. (1.73), find the voltage on the piezoelectric and get the final expression for the ME voltage coefficient:

$$\alpha_E = -\frac{{}^{m}t^{p}t\left(c_{11}^{D}\,{}^{p}t^{3} + {}^{m}Y^{B\,m}t^{3}\right)\cdot}{\left[\left(c_{11}^{D}\right)^{2\,p}t^{4} + 4c_{11}^{D}\,{}^{m}Y^{B\,p}t^{m}t^{3} + 6c_{11}^{D}\,{}^{m}Y^{B\,p}t^{2\,m}t^{2} +\right.} \cdot \frac{\cdot h_{31}\bar{q}_{11}}{\left. +4c_{11}^{D}\,{}^{m}Y^{B\,m}t^{p}t^{3} + \left({}^{m}Y^{B}\right)^{2\,m}t^{4}\right]\left({}^{m}t + {}^{p}t\right)} \tag{1.74}$$

Below in Fig. 1.10 the dependence of the ME voltage coefficient is shown on the volume fraction of the piezoelectric for 2 cases, when PZT and a cut of lithium niobate y + 128 degrees are taken as the piezoelectric phase. Metglas is taken as the magnetostrictive phase.

In the presence of a bending mode, the longitudinal mechanical strain associated with it has different signs above and below the neutral plane. This causes such a specific behavior of the dependence of the ME voltage coefficient on the piezoelectric volume fraction.

In an asymmetric composite in the quasi-static regime of the ME effect with the orientation of a constant magnetic field H_0 along the width of the composite longitudinal-shear and torsional modes arise simultaneously. In the quasi-static regime, there are no vibrations in the direction of the length of the composite. It means that S_5 and S_6 must not depend on x.

Since both the longitudinal shear and torsional modes are excited in the asymmetric ME structure under the quasi-static regime:

$$\begin{aligned} S_5 &= yB \\ S_6 &= A - zB \end{aligned} \tag{1.75}$$

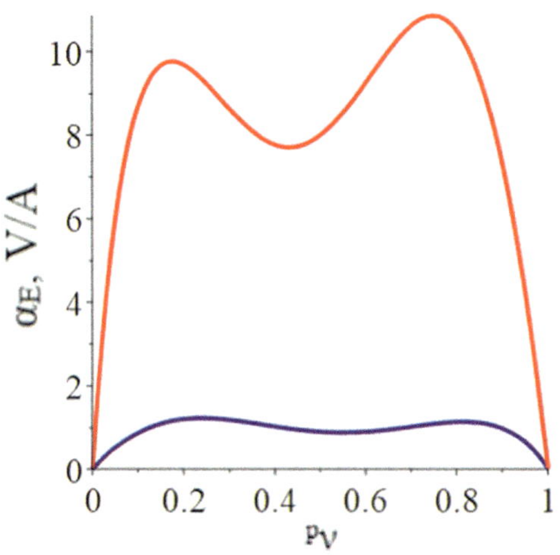

FIGURE 1.10 Dependence of the ME voltage coefficient on the volume fraction of the piezoelectric phase of an asymmetric ME composite when the constant magnetic field H_0 is oriented along the length of the composite. Quasi-static regime of the ME effect. Blue line is for PZT layer. Red line is for LN cut y + 128 degrees layer.

The first condition for the static equilibrium of an ME composite is that the total tangential force on the area perpendicular to the x axis along the y axis is equal to zero:

$$\int_{z_0-{}^pt}^{z_0} {}^pT_6 dz + \int_{z_0}^{z_0+{}^mt} {}^mT_6 dz = 0. \tag{1.76}$$

The second condition for the static equilibrium of the ME composite is the zero torque and is given by:

$$\int_{-\frac{b}{2}}^{\frac{b}{2}} dy \int_{z_0-{}^pt}^{z_0} (y^pT_5 - z^pT_6)dz + \int_{-\frac{b}{2}}^{\frac{b}{2}} dy \int_{z_0}^{z_0+{}^mt} (y^mT_5 - z^mT_6)dz = 0. \tag{1.77}$$

Eqs. (1.76) and (1.77) form a linear inhomogeneous system of two equations with two unknowns A, B. Solving them, we obtain A, B.

Taking into account Eq. (1.75) and the open circuit condition, we obtain the electric field in the piezoelectric phase:

$${}^pE_3 = -h_{36}(A - zB). \tag{1.78}$$

Then, electrical voltage in piezoelectric:

$$U = \int_{z_0-{}^pt}^{z_0} -h_{36}(A - zB)dz = -h_{36}\left[A^pt + \frac{1}{2}B^pt(2z_0 - {}^pt)\right]. \tag{1.79}$$

Substituting the values found from the system of equations A, B in Eq. (1.79), find the voltage on the piezoelectric and get the final expression for the ME voltage coefficient:

$$\alpha_E = \frac{{}^pt^mt^mGq_{16}h_{36}\left[{}^pG^{Dp}t^3 + {}^mG^mt^3 + b^2\left({}^pG^pt + {}^mG^mt\right)\right]}{\left[{}^mG^{2m}t^2\left(b^2 + {}^mt^2\right) + {}^pG^{Dp}t^2\left({}^pG^{Dp}t^2 + 6^mt^{2m}G + 4^mt^mG^pt + b^{2p}G\right) + {}^mG^pt^mt\left(b^{2p}G + 4^pG^{Dm}t^2 + {}^pG^Db^2\right)\right]\left({}^pt + {}^mt\right)}, \tag{1.80}$$

and present it in the form in Fig. 1.11.

As we noted earlier, the magnitude of the ME stress coefficient for the torsional mode is an order of magnitude smaller than that for the longitudinal shear one. Therefore, the torsional mode does not make a very significant contribution in the quasi-static regime, and the ME voltage

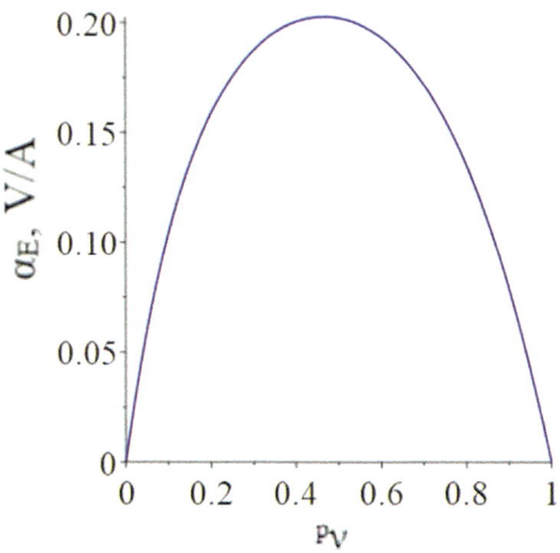

FIGURE 1.11 Dependence of the ME voltage coefficient on the piezoelectric volume fraction for an asymmetric Metglas/GaAs ME structure in the quasi-static regime of torsional and longitudinal-shear modes.

coefficient in the quasi-static regime is mainly determined by the contribution of the longitudinal shear mode. Therefore, the general nature of the dependence of the ME voltage coefficient on the volume fraction of the piezoelectric remains the same as in Fig. 1.9b.

1.3 Converse low frequency ME effect

This section presents an example for calculating the converse ME coefficient of a free symmetric ME structure in the longitudinal mode and in the case of a quasi-static regime.

In the case of the converse ME effect (Bichurin et al., 2012), an external alternating electric field causes deformation of the piezoelectric layer. This deformation, due to the mechanical coupling of the phases of the ME composite, is transferred to the magnetostrictive phase and leads to the appearance of a variable magnetic induction in it. Thus, it is possible after connecting an alternating electrical voltage to a piezoelectric to measure the average amplitude of the alternating magnetic induction by the test coil wound on composite. The converse ME coefficient can be calculated by dividing the variable magnetic induction averaged over the length of the sample in the magnetostrictive phase by the value of the applied electric field to the piezoelectric phase of the ME composite.

We consider an ME composite in the form of a thin narrow plate, whose thickness and width are much less than the length. Layers of the magnetostrictive phase of the same thickness are located above and below the piezoelectric layer of the composite in Fig. 1.12. An alternating electric

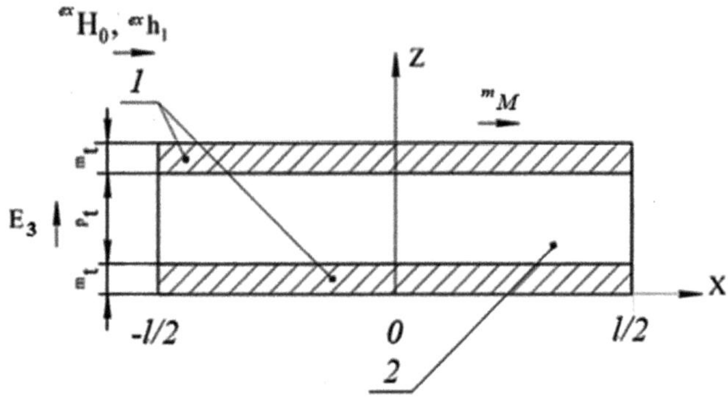

FIGURE 1.12 Structure of a symmetric ME composite for calculating the longitudinal mode of the inverse ME effect. 1 is magnetostrictive phases, 2 is piezoelectric phase. ${}^{p}t$ and ${}^{m}t$ are the thicknesses of the piezoelectric and magnetostrictive phases.

field E_3 applied to the piezoelectric phase of the composite is directed along the Z axis.

There is no alternating magnetic field in outer space: ${}^{ex}h_1 = 0$. According to the condition of the continuity of the tangential components of the magnetic field strength when passing through the boundary of the magnetostrictive phase, we obtain for the variable strength of the magnetic field in magnetostriction phase: ${}^{ex}h_1 = h_1 = 0$.

The expressions for the volume fraction of the piezoelectric and magnetostrictive phases and for the longitudinal components of the strain tensors of the magnetostrictive and piezoelectric phases of the composite are determined according to Eqs. (1.9) and (1.7), respectively.

Longitudinal components of mechanical stresses of piezoelectric and magnetostrictive phases and magnetic induction in magnetostrictive phases for $h_1 = 0$:

$$ {}^{p}T_1 = \frac{{}^{p}S_1}{{}^{p}s_{11}} - \frac{d_{31}E_3}{{}^{p}s_{11}}, \tag{1.81} $$

$$ {}^{m}T_1 = {}^{m}Y {}^{m}S_1, \tag{1.82} $$

$$ {}^{m}B_1 = q_{11} {}^{m}T_1, \tag{1.83} $$

where, E_3 is the alternating electric field applied to the piezoelectric phase of the magnetoelectric structure.

The longitudinal component of the composite stress tensor is:

$$ T_1 = {}^{m}\nu {}^{m}T_1 + {}^{p}\nu {}^{p}T_1 = c_{11}S_1 - {}^{p}\nu \frac{d_{31}}{{}^{p}s_{11}} E_3. \tag{1.84} $$

Effective composite stiffness coefficient c_{11} is determined according to Eq. (1.10). For determining A, B we use the boundary conditions for free fixing of both ends of the ME composite in Eq. (1.14). Then the alternating magnetic induction that occurs in the magnetostrictive phase, with the inverse ME effect:

$$^{m}B_1 = \frac{q_{11}{}^{m}Y^{p}\nu d_{31}E_3\cos(kx)}{^{p}s_{11}c_{11}\cos(\eta)}, \tag{1.85}$$

where $\eta = \frac{kl}{2}$.

The alternating magnetic induction averaged over the length of the sample in the magnetostrictive phase with the inverse ME effect:

$$\overline{B}_{1m} = \frac{\int_{-\frac{l}{2}}^{\frac{l}{2}} B_{1m}(x)dx}{l} = \frac{q_{11}Y_m\nu_p d_{31}E_3 tg(\eta)}{s_{11p}c_{11}\eta}, \tag{1.86}$$

Converse ME coefficient:

$$\alpha_{inv} = \frac{^{m}\overline{B}_1}{E_3} = \frac{q_{11}{}^{m}Y^{p}\nu d_{31}tg(\eta)}{^{p}s_{11}c_{11}\eta}. \tag{1.87}$$

The geometric dimensions and materials of ME composites for calculating the converse ME coefficient will be the same as those for calculating the direct ME voltage coefficient for the longitudinal mode.

To take into account energy losses during EMR, we set $\omega = 2\pi\left(1 + \frac{1}{Q}i\right)f$, where Q is the quality factor of the resonance. The value of the quality factor for longitudinal mode of Q = 130. The results of calculating the converse ME coefficient according to Eq. (1.87) are presented in Fig. 1.13.

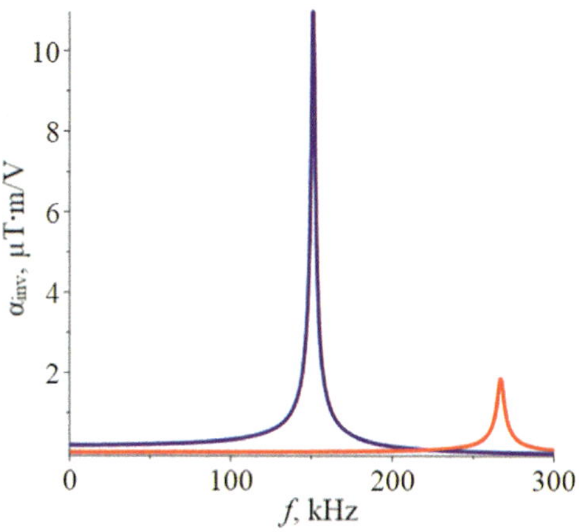

FIGURE 1.13 Dependence of the converse ME coefficient on the frequency of an alternating electric field applied to the piezoelectric phase of a symmetrical ME composite. Longitudinal mode of the ME effect. Blue line is for PZT layer. Red line is for LN cut y + 128 degrees layer.

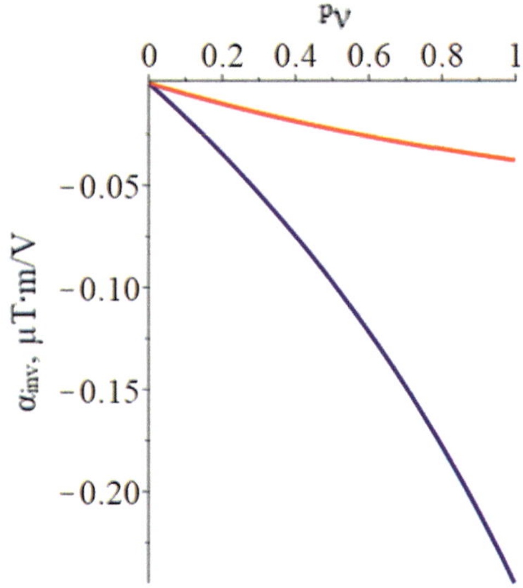

FIGURE 1.14 Dependence of the converse ME coefficient for a symmetric ME composite on the piezoelectric volume fraction. Blue line is for PZT layer. Red line is for LN cut y + 128 degrees layer.

The piezoelectric coefficient d_{31} of PZT is greater than that of LN cut y + 128 degrees. Therefore, the value of the converse ME coefficient for PZT is also greater than for LN cut y + 128 degrees.

Assuming in Eq. (1.87) the frequency f is equal to zero, we obtain the formula for converse ME coefficient in the quasi-static regime:

$$\alpha_{inv} = \frac{q_{11}{}^{m}Y^{p}\nu d_{31}}{p_{\nu} + m_{\nu}{}^{m}Y^{B}p s_{11}}, \tag{1.88}$$

and present it in the form in Fig. 1.14.

For the above reasons, the value of the converse ME coefficient in the quasi-static mode for PZT is greater than for LN cut y + 128 degrees.

1.4 Microwave ME effect

Along with the direct and converse low-frequency ME effects, the microwave ME effect is of great interest for practical applications. It lies in the fact that, in the ME structure, when an electric field is applied to the piezoelectric phase, the FMR line shifts in the magnetic phase. It should be noted that previous studies were carried out, as a rule, with the bias field oriented along the crystallographic axes. Subsequent works, in which the angular dependences of the ME effect were studied, showed that for a number of orientations of the bias field, the ME effect is much larger compared to the effect measured along the principal axes (Nan et al., 2014).

The reasons for this can be related to both two-magnon scattering (Srivastava et al., 1999; Xue et al., 2017) and the accumulation of charges at the interface (Duan et al., 2008). An increase in the ME effect when the bias field is oriented along directions different from the main ones gives hope for practical application, although it should be noted that the FMR line width also changes in this case (Hou et al., 2018). Understanding the origin and quantitative calculation of the microwave ME effect is important in designing and optimizing the ME microwave devices. In this section, the microwave ME effect is considered for ME structures in which ferromagnetic metals and such a well-known ferrite as yttrium iron garnet (YIG) are used as the magnetostrictive phase (Bichurin et al., 2023; Tatarenko et al., 2019). First, a simplified model is considered that does not take into account bending vibrations and the presence of a substrate. At the end of the section, the influence of the substrate on the microwave ME effect is considered.

1.4.1 Ferromagnetic metal-piezoelectric structures

Consider two-layer ME structure consisting of magnetic and piezoelectric phases in Fig. 1.15.

In Fig. 1.15 an ME structure consisting of 1 (magnet) and 2 (piezoelectric) with the directions of the constant bias and alternating magnetic fields and the external electric one is shown. We will consider a ferromagnetic metal as a magnet. The sample is oriented so that the piezoelectric thickness corresponds to the crystallographic direction [011] in the case of PMN-PT. Also, in the case of PMN-PT, the directions of the x and y axes coincide with the crystallographic directions [100] and [01-1], respectively.

Let us consider a thin nickel film to which a bias field is applied along the film plane. We will assume that the value of this field is sufficiently large,

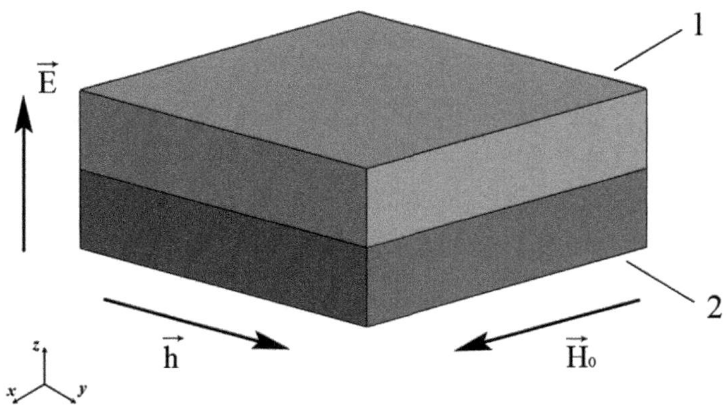

FIGURE 1.15 Two-layer ME structure.

and the nickel film is uniformly magnetized to saturation. Axis 1 (x) is directed along $\vec{H}_0$, then it has components $(H_0, 0, 0)$, and the equilibrium magnetization has components $(M_0, 0, 0)$. In this case, the 1 (x) axis is directed along the easy magnetization axis of the nickel film.

The motion equation of magnetization in a thin nickel film under the action of a high-frequency magnetic field $\vec{h}$ in the presence of a constant magnetic field $\vec{H}_0$ under the condition $h \ll H_0$ and at considering the dissipation:

$$\frac{\partial \vec{M}}{\partial t} = \gamma \left[\vec{M} \times \frac{\partial W}{\mu_0 \partial \vec{M}} \right] + i\omega\alpha \frac{\left[\vec{M}_0 \times \vec{m} \right]}{M_0}, \tag{1.89}$$

Using Eq. (1.89), the imaginary parts of the complex components of the high-frequency magnetic susceptibility tensor χ''_{22}, χ''_{33} were found.

Demagnetizing factors related to sample shape:

$$\begin{aligned} N_{11}^F &= N_{22}^F = 0 \\ N_{33}^F &= 1 \end{aligned}, \tag{1.90}$$

and demagnetizing factors associated with magnetic uniaxial anisotropy have the form:

$$\begin{aligned} N_{11}^a &= -\frac{H_a}{M_0} \\ N_{22}^a &= N_{33}^a = 0 \end{aligned} \tag{1.91}$$

The resonant values of the constant bias field H_0 for χ''_{22}, χ''_{33} , which were numerically found in Maple for the frequency $f = 10$ GHz, coincide with each other and with the value found from the resonance condition to within 1 A/m:

$$\omega^2 = \omega_0^2, \tag{1.92}$$

Where:

$$\omega_0^2 = \gamma^2 \left(H_0 + \left\{ \sum N_{22}^i - \sum N_{11}^i \right\} M_0 \right) \left(H_0 + \left\{ \sum N_{33}^i - \sum N_{11}^i \right\} M_0 \right). \tag{1.93}$$

We find the resonant value of the constant bias field H_0, as

$$H_0 = \frac{\gamma M_0 \left(2 \sum N_{11}^i - \sum N_{22}^i - \sum N_{33}^i \right) + \sqrt{\gamma^2 M_0^2 \left(\sum N_{22}^i - \sum N_{33}^i \right)^2 + 4\omega^2}}{2\gamma}. \tag{1.94}$$

Substituting Eqs. (1.90) and (1.91) into Eq. (1.94), we obtain $H_0 = 148696 \frac{A}{m} \approx 1869$ Э for the resonant frequency $f = 10$ GHz. Similarly, we find the resonant value of the constant bias field H_0 for a thin 50H permalloy film, which equals to $H_0 = 63011 \frac{A}{m} \approx 792$ Oe for the resonant frequency $f = 10$ GHz.

FMR line shift in a two-layer ME structure of Ferromagnetic Metal/Piezoelectric.

Let us find the demagnetizing factors for the Ni/Piezoelectric structure associated with the action of a constant electric field. Let us consider the case when the constant electric field is directed along the three (z) axis.

Consider the mechanical balance of the structure. The boundary conditions for a free sample in the form of a rectangular parallelepiped, the base of which is a square, are equal to:

$$^mT_3 = {}^pT_3 = 0, \tag{1.95}$$

$$\begin{aligned} {}^mt^mT_1 + {}^pt^pT_1 &= 0, \\ {}^mt^mT_2 + {}^pt^pT_2 &= 0. \end{aligned} \tag{1.96}$$

Taking into account Eq. (1.95), the constitutive equations for nickel:

$$\begin{aligned} S_1 &= {}^ms_{11}{}^mT_1 + {}^ms_{12}{}^mT_2, \\ S_2 &= {}^ms_{12}{}^mT_1 + {}^ms_{11}{}^mT_2 \end{aligned} \tag{1.97}$$

and for piezoelectric:

$$\begin{aligned} S_1 &= {}^ps_{11}{}^pT_1 + {}^ps_{12}{}^pT_2 + d_{31}E, \\ S_2 &= {}^ps_{12}{}^pT_1 + {}^ps_{22}{}^pT_2 + d_{32}E. \end{aligned} \tag{1.98}$$

Solving Eqs. (1.96)–(1.98) together, we find:

$$\begin{aligned} {}^mT_1 &= \frac{(d_{31}{}^ms_{11}{}^pt + d_{31}{}^ps_{22}{}^mt - d_{32}{}^ms_{12}{}^pt - d_{32}{}^ps_{12}{}^mt)^ptE}{{}^ms_{11}^2{}^pt^2 + {}^ms_{11}{}^ps_{11}{}^mt^pt + {}^ms_{11}{}^ps_{22}{}^mt^pt - {}^ms_{12}^2{}^pt^2 - 2{}^ms_{12}{}^ps_{12}{}^mt^pt + {}^ps_{11}{}^ps_{22}{}^mt^2 - {}^ps_{12}^2{}^mt^2} \\ {}^mT_2 &= \frac{(d_{32}{}^ms_{11}{}^pt + d_{32}{}^ps_{11}{}^mt - d_{31}{}^ms_{12}{}^pt - d_{31}{}^ps_{12}{}^mt)^ptE}{{}^ms_{11}^2{}^pt^2 + {}^ms_{11}{}^ps_{11}{}^mt^pt + {}^ms_{11}{}^ps_{22}{}^mt^pt - {}^ms_{12}^2{}^pt^2 - 2{}^ms_{12}{}^ps_{12}{}^mt^pt + {}^ps_{11}{}^ps_{22}{}^mt^2 - {}^ps_{12}^2{}^mt^2} \end{aligned} \tag{1.99}$$

Part of the free energy of nickel associated with the presence of stresses arising from the action of an electric field on a piezoelectric is:

$$\Delta W_{ME} = -\frac{3\lambda_{100}}{2M_0^2}\left({}^mT_1M_1^2 + {}^mT_2M_2^2 + {}^mT_3M_3^2\right). \tag{1.100}$$

From Eq. (1.100) we find the effective demagnetizing factors arising from the action of the electric field $\overrightarrow{E}$:

$$N_{11}^{E} = -\frac{3\lambda_{100}}{\mu_0 M_0^2} {}^{m}T_1,$$
$$N_{22}^{E} = -\frac{3\lambda_{100}}{\mu_0 M_0^2} {}^{m}T_2, \tag{1.101}$$
$$N_{33}^{E} = 0.$$

We write the resonance condition without taking into account the effective demagnetizing factors arising from the action of the electric field $\overrightarrow{E}$:

$$\omega^2 = \gamma^2\left(H_0 + \left\{\sum_{i\neq E} N_{22}^{i} - \sum_{i\neq E} N_{11}^{i}\right\} M_0\right)\left(H_0 + \left\{\sum_{i\neq E} N_{33}^{i} - \sum_{i\neq E} N_{11}^{i}\right\} M_0\right). \tag{1.102}$$

Next, we write the resonance condition, taking into account the small correction associated with small effective demagnetizing factors arising from the action of the electric field $\overrightarrow{E}$:

$$\omega^2 = \gamma^2\left(H_0 + \delta H_E + \left\{\sum_{i\neq E} N_{22}^{i} - \sum_{i\neq E} N_{11}^{i} + N_{22}^{E} - N_{11}^{E}\right\} M_0\right)\cdot$$
$$\cdot\left(H_0 + \delta H_E + \left\{\sum_{i\neq E} N_{33}^{i} - \sum_{i\neq E} N_{11}^{i} + N_{33}^{E} - N_{11}^{E}\right\} M_0\right) \tag{1.103}$$

Let us subtract Eq. (1.102) from Eq. (1.103), retaining in the resulting equation only terms of the first order of smallness in $\delta H_E, N_{11}^{E}, N_{22}^{E}, N_{33}^{E}$:

$$M_0 Q_2\left(N_{22}^{E} - N_{11}^{E}\right) + M_0 Q_3\left(N_{33}^{E} - N_{11}^{E}\right) + Q_1 \delta H_E = 0, \tag{1.104}$$

Where:

$$Q_1 = 2H_0 + \left\{\sum_{i\neq E} N_{22}^{i} + \sum_{i\neq E} N_{33}^{i} - 2\sum_{i\neq E} N_{11}^{i}\right\} M_0$$
$$Q_2 = H_0 + \left\{\sum_{i\neq E} N_{33}^{i} - \sum_{i\neq E} N_{11}^{i}\right\} M_0 \quad . \tag{1.105}$$
$$Q_3 = H_0 + \left\{\sum_{i\neq E} N_{22}^{i} - \sum_{i\neq E} N_{11}^{i}\right\} M_0$$

Now we find the shift of the resonance line, as:

$$\delta H_E = \frac{M_0\left[Q_2\left(N_{11}^E - N_{22}^E\right) + Q_3\left(N_{11}^E - N_{33}^E\right)\right]}{Q_1}. \tag{1.106}$$

Substitute Eqs. (1.90) and (1.91) into Eq. (1.105):

$$\begin{aligned} Q_1 &= 2H_0 + M_0 + 2H_a \\ Q_2 &= H_0 + M_0 + H_a \\ Q_3 &= H_0 + H_a \end{aligned} \quad . \tag{1.107}$$

Then, from Eq. (1.106) we find:

$$\delta H_E = \frac{-3\lambda_{100}\left(Q_2\left\{{}^mT_1 - {}^mT_2\right\} + Q_3{}^mT_1\right)}{\mu_0 M_0 Q_1}. \tag{1.108}$$

Eq. (1.108) is also valid for the NiFe/Piezoelectric structure. The results of calculating the dependence of the FMR line shift on the electric field in Ferromagnetic Metals/piezoelectric structures are shown in Fig. 1.16. The material parameters that were used in the calculations are given in Table 1.1. Thickness of ferromagnetic metal ${}^mt = 5 \cdot 10^{-8} m$ and piezoelectric ${}^pt = 1 \cdot 10^{-4} m$.

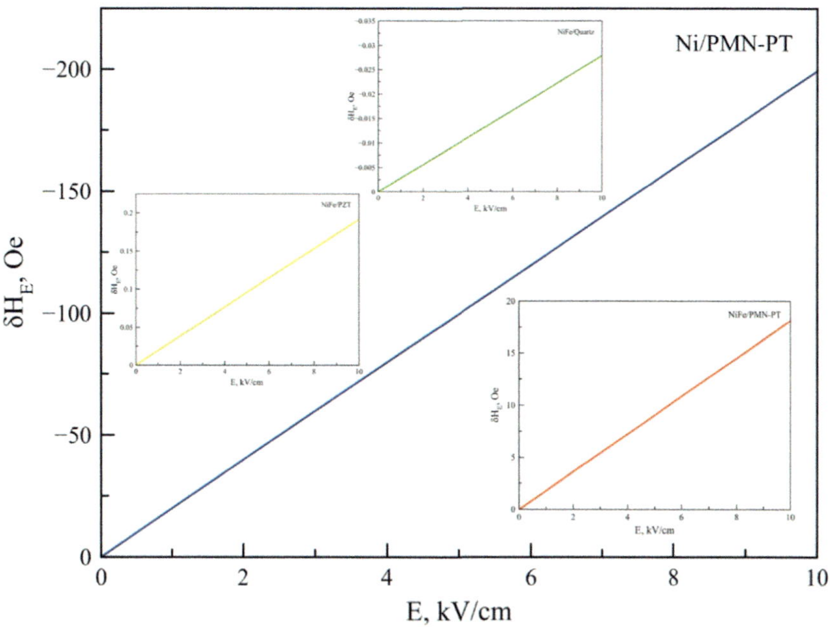

FIGURE 1.16 Dependence of the FMR resonance line shift on the electric field.

TABLE 1.1 Material parameters of magnets and piezoelectrics.

Material	Ni	NiFe	YIG cut (001)	0.67 PMN-0.33 PT (PMN-PT) cut (011)	Quartz X-cut	PZT	Langatate X-cut
s_{11} (10^{-12} m^2/N)	20	6.67	4.8	69	12.8	15.3	5.27
s_{12} (10^{-12} m^2/N)	−7	−1.93	−1.4	−33.4	−1.22	−5	−1.84
s_{22} (10^{-12} m^2/N)	–	–	–	22.9	9.6	15.3	9.13
M_0 (A/m)	262698	1230158	142143	–	–	–	–
γ (m/C)	236803	220000	221000	–	–	–	–
λ_{100} (10^{-6})	−35	4.6	−1.4	–	–	–	–
H_a (A/m)	15900	0	−3343	–	–	–	–
d_{31} (10^{-12} m/V)	–	–	–	−940	0	−175	0
d_{32} (10^{-12} m/V)	–	–	–	475	−2.29	−175	−6.54

In Fig. 1.16 the dependence of the FMR line shift on the electric field is shown for the Ni/PMN-PT, NiFe/PMN-PT, NiFe/PZT, and NiFe/Quartz structures when the constant bias field is directed in the plane of the nickel film along the 1 (x) axis. As shown in Eq. (1.108), the shift of the FMR line depends rather intricately on the corresponding components of the magnetostrictive phase stress tensor ${}^{m}T_1, {}^{m}T_2$, which, in turn, also depend not quite trivially on the corresponding components of the piezoelectric tensor d_{31}, d_{32}. Thus, the sign of the resonant line shift, depending on the total action of all the above factors, can be either positive or negative. As can be seen from the graphs in Fig. 1.16, the largest shift of the FMR line is observed in the structure where PMN-PT is used as the piezoelectric phase, and the smallest shift is observed in the structure where the quartz X-cut is used as the piezoelectric. This result is expected and understandable. As can be seen from Table 1.1, PMN-PT has the largest piezo coefficients of the considered piezoelectrics. the quartz X-cut, on the contrary, has the smallest piezo coefficients.

The results obtained show that in order to clearly observe the resonance line shifts in an electric field, it is necessary to have narrow FMR lines for magnetic phases of ME composite.

1.4.2 Ferrite-piezoelectric structures

Fig. 1.15 shows the ME structure, and in this section, we will consider ferrite as a magnet.

In Bichurin et al. (2019), a method was developed for calculating the microwave ME effect in two-layer ME structures in which YIG ferrite was used as the magnetostrictive phase. Here we present this method of calculation, having supplemented the set of piezoelectrics used there with several more for completeness of the picture, and having drawn the appropriate conclusions and recommendations on the choice of piezoelectrics for observing the maximum microwave ME effect.

Let us consider a thin YIG plate to which a constant bias field $\vec{H}_0$ is applied along the plane of the plate. We will assume that the value of this field is sufficiently large and the YIG plate is uniformly magnetized to saturation. Axis 2 (y) is directed along $\vec{H}_0$, then it has components $(0, H_0, 0)$, and the equilibrium magnetization has components $(0, M_0, 0)$. Axis 1 (x) is directed along the edge of the YIG plate, which coincides with the crystallographic direction [100]. The motion equation of magnetization in a thin YIG plate under the action of $\vec{h}$ in the presence of $\vec{H}_0$ under the condition $h \ll H_0$, taking into account dissipation, is given in section 1.4.1 (Eq. 1.89). Using Eq. (1.89), we found the imaginary parts of the complex components of the high-frequency susceptibility χ''_{11}, χ''_{33} that are of interest to us.

Free energy density related to magnetic crystalline cubic anisotropy:

$$W_K = \frac{K_1}{M_0^4}\left(M_x^2 M_y^2 + M_y^2 M_z^2 + M_z^2 M_x^2\right). \quad (1.109)$$

Demagnetizing factors related to sample shape is equal Eq. (1.90):

Demagnetizing factors associated with magnetic crystalline cubic anisotropy:

$$N_{11}^a = N_{33}^a = -\frac{H_a}{M_0}, N_{22}^a = 0. \quad (1.110)$$

The resonant values of the constant bias field H_0 for χ_{11}'', χ_{33}'', which were numerically found in Maple for the frequency $f = 10$ GHz, coincide with each other with an accuracy of 1 A/m and with the value found from the resonance condition in Eq. (1.92), where

$$\omega_0^2 = \gamma^2\left(H_0 + \left\{\sum N_{11}^i - \sum N_{22}^i\right\}M_0\right)\left(H_0 + \left\{\sum N_{33}^i - \sum N_{22}^i\right\}M_0\right). \quad (1.111)$$

We find the resonant value of the constant bias field H_0:

$$H_0 = \frac{\gamma M_0\left(2\sum N_{22}^i - \sum N_{11}^i - \sum N_{33}^i\right) + \sqrt{\gamma^2 M_0^2\left(\sum N_{11}^i - \sum N_{33}^i\right)^2 + 4\omega^2}}{2\gamma}. \quad (1.112)$$

Substituting Eq. (1.90) and Eq. (1.110) into Eq. (1.112), we obtain $H_0 = 218501 \frac{A}{m} \approx 2746\ Oe$ for the resonant frequency $f = 10$ GHz.

Let us now consider the case when the constant bias field is directly perpendicular to the plane of the YIG plate. We will assume that the value of this field is sufficiently large and the YIG plate is uniformly magnetized to saturation. Axis 3 (z) is directed along $\overrightarrow{H_0}$, then it has components $(0, 0, H_0)$, and the equilibrium magnetization has components $(0, 0, M_0)$. Axis 1 (x) is directed along the edge of the YIG plate, which coincides with the crystallographic direction [100].

For this case, the resonant value of the constant bias field H_0:

$$H_0 = \frac{\gamma M_0\left(2\sum N_{33}^i - \sum N_{11}^i - \sum N_{22}^i\right) + \sqrt{\gamma^2 M_0^2\left(\sum N_{11}^i - \sum N_{22}^i\right)^2 + 4\omega^2}}{2\gamma}. \quad (1.113)$$

From Eq. (1.113) we obtain $H_0 = 429649 \frac{A}{m} \approx 5399\ Oe$ for the resonant frequency $f = 10$ GHz.

FMR line shift in a two-layer magnetoelectric structure of YIG/Piezoelectric.

The effective demagnetization factors for the YIG/Piezoelectric structure, associated with the action of a constant electric field that is directed along the three (z) axis, are similar to the demagnetization factors given in Section 1.4.1 in Eq. (1.101).

We write the resonance condition without taking into account the effective demagnetizing factors arising from the action of the electric field $\vec{E}$, for the case when the constant magnetic field $\vec{H}_0$ is directed along the axis 2 (y):

$$\omega^2 = \gamma^2\left(H_0 + \left\{\sum_{i\neq E} N_{11}^i - \sum_{i\neq E} N_{22}^i\right\}M_0\right)\left(H_0 + \left\{\sum_{i\neq E} N_{33}^i - \sum_{i\neq E} N_{22}^i\right\}M_0\right). \tag{1.114}$$

Next, we write the resonance condition, taking into account the small correction ${}^1\delta H_E$ associated with small effective demagnetizing factors arising from the action of the electric field $\vec{E}$:

$$\begin{aligned}\omega^2 = \gamma^2&\left(H_0 + {}^1\delta H_E + \left\{\sum_{i\neq E} N_{11}^i - \sum_{i\neq E} N_{22}^i + N_{11}^E - N_{22}^E\right\}M_0\right)\cdot\\ &\cdot\left(H_0 + {}^1\delta H_E + \left\{\sum_{i\neq E} N_{33}^i - \sum_{i\neq E} N_{22}^i + N_{33}^E - N_{22}^E\right\}M_0\right)\end{aligned} \tag{1.115}$$

Let us subtract Eq. (1.114) from Eq. (1.115), retaining in the resulting equation only terms of the first order of smallness in ${}^1\delta H_E, N_{11}^E, N_{22}^E, N_{33}^E$:

$$M_0Q_1(N_{11}^E - N_{22}^E) + M_0Q_3(N_{33}^E - N_{22}^E) + Q_2\,{}^1\delta H_E = 0, \tag{1.116}$$

Where:

$$\begin{aligned}Q_1 &= H_0 + \left\{\sum_{i\neq E} N_{33}^i - \sum_{i\neq E} N_{22}^i\right\}M_0\\ Q_2 &= 2H_0 + \left\{\sum_{i\neq E} N_{11}^i + \sum_{i\neq E} N_{33}^i - 2\sum_{i\neq E} N_{22}^i\right\}M_0.\\ Q_3 &= H_0 + \left\{\sum_{i\neq E} N_{11}^i - \sum_{i\neq E} N_{22}^i\right\}M_0\end{aligned} \tag{1.117}$$

Now we find the resonance line shift:

$${}^1\delta H_E = \frac{M_0\left[Q_1(N_{22}^E - N_{11}^E) + Q_3(N_{22}^E - N_{33}^E)\right]}{Q_2}. \tag{1.118}$$

Substitute Eqs. (1.90) and (1.110) into Eq. (1.117):

$$\begin{aligned} Q_1 &= H_0 - H_a + M_0 \\ Q_2 &= 2(H_0 - H_a) + M_0 \,. \\ Q_3 &= H_0 - H_a \end{aligned} \tag{1.119}$$

Then, from Eq. (1.118) we find:

$$^{1}\delta H_E = \frac{3\lambda_{100}\left(Q_1\{^{m}T_1 - {}^{m}T_2\} - Q_3{}^{m}T_2\right)}{\mu_0 M_0 Q_2}. \tag{1.120}$$

The case when the magnetizing field is directed perpendicular to the YIG plate is considered similarly. It turns out the following equation:

$$^{2}\delta H_E = \frac{3\lambda_{100}\left[^{m}T_1 + {}^{m}T_2\right]}{2\mu_0 M_0}. \tag{1.121}$$

The material parameters used in the calculation are given in section 1.4.1. in Table 1.1 ferrite and piezoelectric thicknesses are $^{m}t = 2 \cdot 10^{-5} m$, $^{p}t = 5 \cdot 10^{-4} m$, respectively. The results of calculating the dependence of the FMR line shift on the electric field in YIG/piezoelectric structures are shown in Fig. 1.17.

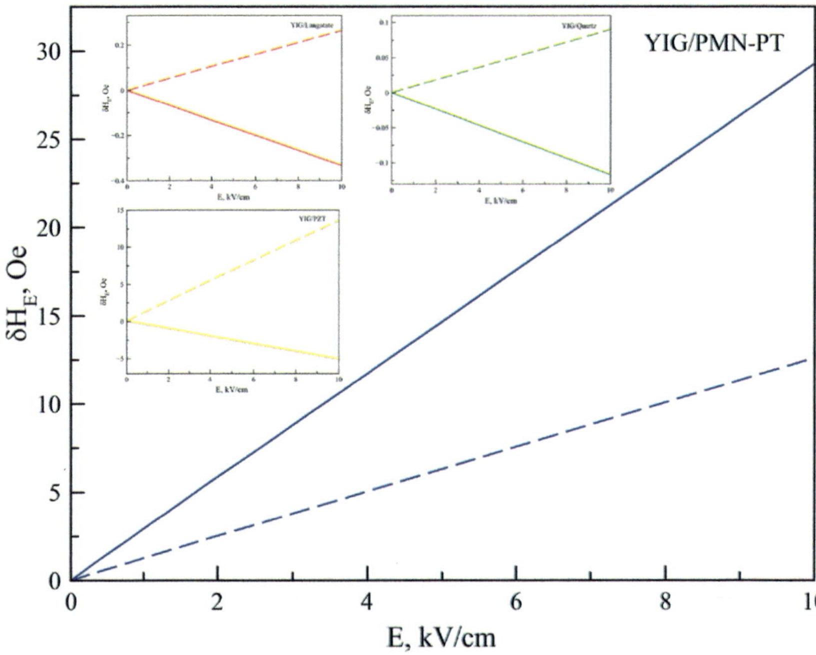

FIGURE 1.17 Dependence of the FMR line shift on an electric field in YIG/piezoelectric structures. Solid line is constant magnetic field along axis 2 (y), dashed line is constant magnetic field along axis 3 (z).

In Fig. 1.17 the dependences of the FMR line shift on an electric field in the YIG/PMN-PT, YIG/PZT, YIG/Quartz, YIG/langatate structures are shown. The solid lines show the dependences for the constant bias field $\vec{H}_0$ directed in the plane of the YIG plate along the 2 (y) axis; the dotted lines are for the constant bias field directed perpendicular to the YIG plate along axis 3 (z). Among the cases considered, the largest FMR line shift is for PMN-PT, when a constant bias field is directed in the plane of the YIG plate along the 2(y) axis. This is due both to the fact that PMN-PT has the highest piezoelectric coefficients and to the fact that the side edges of the (011) crystal of the PMN-PT cut coincide with the [100] and [01−1] crystallographic directions. In this case, the value of $(d_{32} - d_{31})$, which is proportional to the FMR line shift, has the maximum value. The decrease in the shift of the FMR line for PZT is due to the fact that its piezoelectric coefficients are smaller than those of PMN-PT. In addition, due to the symmetry of the PZT, the values of d_{31} and d_{32} for it coincide. Since the piezoelectric coefficients of quartz and langatate are approximately two orders of magnitude lower than those of PZT and PMN-PT, the FMR line shift in an electric field is relatively small for these materials.

As a result, since the width of the FMR line for YIG is about 1 Oe, it seems possible to experimentally observe the shift of the FMR line in an electric field in the considered ME structures.

1.4.3 "Substrate effect" in the microwave ME effect range

In practice, a thin epitaxial layer of the YIG magnetostrictive ferrite phase is formed on a gallium gadolinium garnet (GGG) substrate. In this paragraph, we take into account in our theoretical model the effect of the GGG substrate on the FMR resonance line shift for ME piezoelectric/YIG/GGG structures and compare these results with those obtained earlier, when the need for the substrate was not taken into account.

Let us consider a thin YIG film to which a constant bias field $\vec{H}_0$ is applied along the plane of the plate in Fig. 1.18. We will assume that the value of this field is sufficiently large and the thin YIG film is uniformly magnetized to saturation. Axis 2 (y) is directed along $\vec{H}_0$, then $\vec{H}_0$ has components $(0, H_0, 0)$, and the equilibrium magnetization has components $(0, M_0, 0)$. Let us direct the axis 1 (x) along the edge of the GGG plate, which coincides with the crystallographic direction of the YIG [100].

The values of the resonant bias field H_0 for a thin YIG film for two orientations of the bias field are the same as in section 1.4.2.

Let us find the demagnetizing factors for the PMN-PT/YIG/GGG structure associated with the action of a constant electric field. Let us consider the case when the constant electric field is directed along the three (z) axis. Since both longitudinal and bending modes are excited

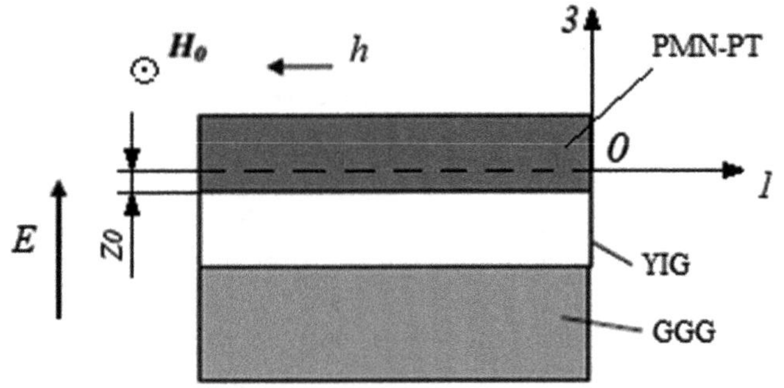

FIGURE 1.18 Structure of the ME composite PMN-PT/YIG/GGG.

along both X and Y axes in an asymmetric ME structure, the components of the strain tensor along axes 1 and 2 have the form:

$$S_1 = A_1 + zB_1, \tag{1.122}$$

$$S_2 = A_2 + zB_2, \tag{1.123}$$

where A_1, A_2 and B_1, B_2 are unknown constants associated with the planar and bending modes, respectively.

Consider the mechanical equilibrium of the structure. Boundary conditions for a free sample in the form of a rectangular parallelepiped:

for stress tensors of the ME composite along axis 3 have the form:

$$^{m}T_3 = {^{p}T_3} = 0, \tag{1.124}$$

for longitudinal forces in the ME composite along axes 1 and 2:

$$\begin{matrix} F_1 = 0 \\ F_2 = 0 \end{matrix}, \tag{1.125}$$

for bending moments in the ME composite along axes 1 and 2:

$$\begin{matrix} M_1 = 0 \\ M_2 = 0 \end{matrix}. \tag{1.126}$$

Taking into account Eq. (1.124), the constitutive equations for YIG are:

$$\begin{matrix} S_1 = {^{m}s_{11}}{^{m}T_1} + {^{m}s_{12}}{^{m}T_2} \\ S_2 = {^{m}s_{12}}{^{m}T_1} + {^{m}s_{11}}{^{m}T_2} \end{matrix}, \tag{1.127}$$

for piezoelectric:

$$\begin{matrix} S_1 = {^{p}s_{11}}{^{p}T_1} + {^{p}s_{12}}{^{p}T_2} + d_{31}E \\ S_2 = {^{p}s_{12}}{^{p}T_1} + {^{p}s_{22}}{^{p}T_2} + d_{32}E \end{matrix}, \tag{1.128}$$

for substrate:

$$\begin{aligned} S_1 &= {}^s s_{11}{}^s T_1 + {}^s s_{12}{}^s T_2 \\ S_2 &= {}^s s_{12}{}^s T_1 + {}^s s_{11}{}^s T_2 \end{aligned}. \tag{1.129}$$

From Eqs. (1.127)–(1.129) we express the necessary stress components:

$$\begin{aligned} {}^m T_1 &= {}^m c_{11} S_1 + {}^m c_{12} S_2 \\ {}^m T_2 &= {}^m c_{11} S_2 + {}^m c_{12} S_1 \\ {}^p T_1 &= {}^p c_{11}^E S_1 + {}^p c_{12}^E S_2 + h_{31} E \\ {}^p T_2 &= {}^p c_{22}^E S_2 + {}^p c_{12}^E S_1 + h_{32} E \\ {}^s T_1 &= {}^s c_{11} S_1 + {}^s c_{12} S_2 \\ {}^s T_2 &= {}^s c_{11} S_2 + {}^s c_{12} S_1 \end{aligned}, \tag{1.130}$$

where ${}^m c_{11}, {}^m c_{12}$; ${}^p c_{11}^E, {}^p c_{12}^E, {}^p c_{22}^E, {}^s c_{11}, {}^s c_{12}$ are the effective stiffness coefficients of the magnetic, piezoelectric and substrate, respectively; h_{31}, h_{32} are piezo coefficients.

We determine the longitudinal forces as:

$$\begin{aligned} F_1 &= \int_{-z_0 - {}^m t - {}^s t}^{-z_0 - {}^m t} {}^s T_1 dz + \int_{-z_0 - {}^m t}^{-z_0} {}^m T_1 dz + \int_{-z_0}^{{}^p t - z_0} {}^p T_1 dz \\ F_2 &= \int_{-z_0 - {}^m t - {}^s t}^{-z_0 - {}^m t} {}^s T_2 dz + \int_{-z_0 - {}^m t}^{-z_0} {}^m T_2 dz + \int_{-z_0}^{{}^p t - z_0} {}^p T_2 dz \end{aligned}, \tag{1.131}$$

and the bending moments as:

$$\begin{aligned} M_1 &= \int_{-z_0 - {}^m t - {}^s t}^{-z_0 - {}^m t} z^s T_1 dz + \int_{-z_0 - {}^m t}^{-z_0} z^m T_1 dz + \int_{-z_0}^{{}^p t - z_0} z^p T_1 dz \\ M_2 &= \int_{-z_0 - {}^m t - {}^s t}^{-z_0 - {}^m t} z^s T_2 dz + \int_{-z_0 - {}^m t}^{-z_0} z^m T_2 dz + \int_{-z_0}^{{}^p t - z_0} z^p T_2 dz \end{aligned}. \tag{1.132}$$

Substituting the necessary components of the stresses and strain tensors and integrating in Eqs. (1.131) and (1.132), we group the resulting factors in front of the unknown constants A_1, A_2, B_1, B_2 and the electric field strength E. To eliminate the dependence of the longitudinal forces along the axes 1 and 2 on the bending deformations along the axes 2 and 1, respectively, we equate to zero the resulting multiplier in front of the unknown constant B_2 for the longitudinal force F_1, it is also the multiplier in front of the unknown constant B_1 for the longitudinal force F_2:

$$\frac{{}^{s}c_{12}\left[\left(-z_0-{}^{m}t-{}^{s}t\right)^2-\left(-z_0-{}^{m}t\right)^2\right]}{2}+\frac{{}^{m}c_{12}\left[\left(-z_0-{}^{m}t\right)^2-z_0^2\right]}{2}$$
$$+\frac{{}^{p}c_{12}^{E}\left[z_0^2-\left({}^{p}t-z_0\right)^2\right]}{2}=0. \tag{1.133}$$

In such a case, for bending moments, the factor in front of the unknown constant A_2 for the bending moment M_1 will also be equal to zero, and the factor in front of the unknown constant A_2 for the bending moment M_1 will be equal to zero. But even now the solution of the system of four linear inhomogeneous equations for four unknowns A_1, A_2, B_1, B_2, composed of the equilibrium conditions in Eqs. (1.125) and (1.126), looks too cumbersome. Therefore, we do not present it here but use it only in further calculations to obtain the final result. From Eq. (1.133) we find the distance from the neutral line of the composite to the interface between the YIG and the piezoelectric:

$$z_0=\frac{{}^{p}t^{2p}c_{12}^{E}-{}^{m}t^{2m}c_{12}-{}^{s}t^{s}c_{12}\left(2^{m}t+{}^{s}t\right)}{2\left({}^{p}t^{p}c_{12}^{E}+{}^{m}t^{m}c_{12}+{}^{s}t^{s}c_{12}\right)}. \tag{1.134}$$

Since the stress components of the magnetostrictive phase will depend on, but the thickness of the magnetostrictive phase is small, then we can take for z the coordinate of this phase middle as:

$$z=-\frac{{}^{m}t}{2}-z_0. \tag{1.135}$$

We substitute the found constants A_1, B_1, A_2, B_2 into Eqs. (1.122) and (1.123), then we substitute the obtained strain tensor components along axes 1 and 2 into the YIG stress components ${}^{m}T_1, {}^{m}T_2$ from Eq. (1.130).

Further, the YIG stress components found in this way ${}^{m}T_1, {}^{m}T_2$, in which the influence of the substrate and bending vibrations are now taken into account, are substituted into Eq. (1.120), which gives the resonance line shift at a constant magnetic field directed along the plane of the YIG thin film. Similarly, to calculate the resonance line shift for the case when the magnetizing field is directly perpendicular to the thin film, the found YIG stress components ${}^{m}T_1, {}^{m}T_2$ are substituted into Eq. (1.121). These calculations were used to plot the dependence of the shift of the resonance line on the magnitude of the applied electric field to the ME composite of piezoelectric/YIG/GGG for two cases of magnetizing field orientation in Fig. 1.19. For ease of comparison, in Fig. 1.19 the corresponding graphs for the ME composite of piezoelectric/YIG are shown. At calculating the

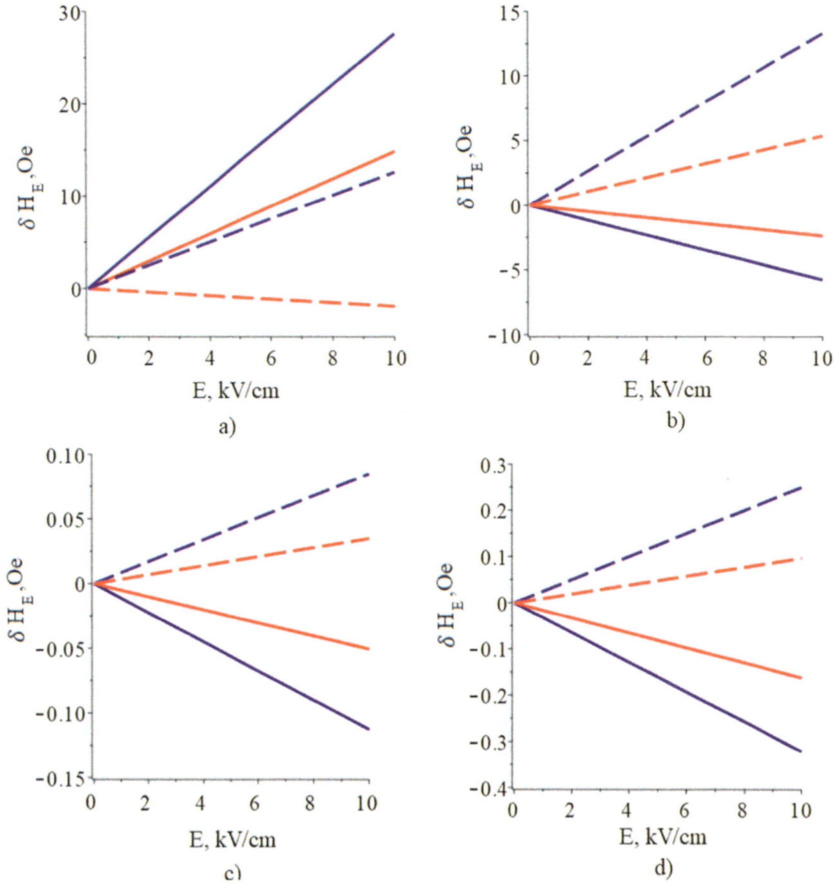

FIGURE 1.19 Dependence of the FMR line shift on the electric field in the structures: (a) YIG/PMN-PT is blue line, PMN-PT/YIG/GGG is red line, (b) YIG/PZT is blue line, PZT/YIG/GGG is red line, (c) YIG/Quartz is blue line, Quartz/YIG/GGG is red line, (d) YIG/langatate is blue line, langatate/YIG/GGG is red line solid line is constant magnetic field directed in the plate YIG along axis 2 (y), dashed line is constant magnetic field is directed perpendicular to the YIG plate along axis 3 (z).

resonance line shift, the material parameters of YIG with (001) cut, (011) cut 0.67 PMN-0.33 PT (PMN-PT), PZT, quartz X-cut, and langatate X-cut, presented in Table 1.1 of Section 1.4.2, were used.

Parameters of GGG: ${}^{s}s_{11} = 4,4\cdot10^{-12}\ \frac{m^2}{N}, {}^{s}s_{12} = -1,2\cdot10^{-12}\ \frac{m^2}{N}, {}^{s}t = 5\cdot 10^{-4}\ m$. Below are graphs of the dependence of the resonance line shift on the applied electric field to the ME composite of piezoelectric/YIG/GGG.

Based on the results of calculations, it was determined that for the piezoelectric/YIG/GGG structures, the FMR line shift is approximately two times less than that for the YIG/piezoelectric structures for both types of orientation of the constant magnetic field. Also, for the PMN-PT/YIG structure, taking into account the GGG substrate, a change in the sign of the shift of the FMR resonance line for a constant magnetic field, which is perpendicular to the YIG plate, is observed. The results of the calculation confirm the "clamping effect" that occurs in the presence of a substrate. This effect is due to the fact that mechanical deformations that occur in a piezoelectric under the action of an electric field, in the presence of a substrate, are transferred not only to the magnetostrictive phase, but also to the substrate. This weakens the mechanical stresses arising in the YIG, compared with the case without the substrate, and leads to decrease of the FMR line shift. It should be noted that bending strains differ from planar strains in that the strain tensor components above and below the neutral plane have different signs. Therefore, at taking into account the substrate in our model, we simultaneously included the bending deformations for the PMN-PT/YIG/GGG structure, and the sign of the FMR line shift changed compared to the PMN-PT/YIG structure.

1.5 Conclusions

The introduction provides a brief historical description of a number of important decisions related to the development of the physics and technics of ME composites, as well as a list of books and reviews that can help young researchers get acquainted with this new field of functional electronics. The theory of low-frequency direct ME effect in symmetric and asymmetric magnetostrictive-piezoelectric structures in longitudinal and bending, as well as longitudinal-shear and torsional modes was considered. Expressions are obtained for the ME voltage coefficients in the quasi-static and EMR modes. Additionally, for the EMR mode, approximate formulas for the main resonant frequencies were obtained and their accuracy was investigated. A comparison of the obtained theoretical results with known data from the literature and experiment for the GaAs-Metglas and $LiNbO_3$-Metglas structures showed satisfactory agreement. In the development of our previous calculations, here we considered the ME effect in the FMR region for FM-piezoelectric structures. Nickel and permalloy were used as the FM, while the piezoelectric was represented by Quartz, PZT, and PMN-PT. In contrast to the previous results, the calculation of the ME effect in the YIG-piezoelectric and YIG-GGG-piezoelectric structures made it possible to analyze the "substrate effect" and evaluate the effect of FMR line shift in an electric field.

An analysis of the studies carried out and the results obtained allows us to determine that the future directions should include the following.

1.5.1 Theoretical and experimental research of "substrate effect" and "mismatch effect"

Here, it is important to carry out purposeful calculations for experimentally observed effects in order to be able to compare the results obtained. In addition, one should keep in mind the possibility of depositing magnetic micro- and nanofilms directly onto a piezoelectric layer, which can play the role of a substrate (Lou et al., 2009).

1.5.2 ME effect in the magnetoacoustic resonance range

Great interest is the experimental study of the ME effect in the region of shear and torsional modes in the EMR range. Due to the fact that these modes are of higher frequency, it becomes possible to proceed to the study of the ME effect in the region of magnetoacoustic resonance (MAR) (Bichurin et al., 2019). Preliminary estimates show the possibility of obtaining giant ME effects in the MAR region, but the difficulty of observing this effect lies in the need to match the frequency of one of the EMR modes with the FMR frequency in the structure under study. Since the frequency ranges of EMR and FMR differ significantly, it seems that to observe this effect, the FMR frequency should be lowered by choosing a magnetic component with a low saturation magnetization, and the EMR frequency should be increased by reducing the thickness of the piezoelectric layer and using EMR harmonics.

1.5.3 Microwave ME effect

To improve the efficiency of the microwave ME effect, it is necessary to use magnetic films with a narrow magnetic resonance line (FMR linewidth less than 10 Oe), low values of the resonant magnetic field and the control electric one. The production of magnetic films with narrow resonance lines is a serious technological problem that must be solved in the future. An example in this regard is a YIG film on a GGG substrate, which has an FMR linewidth of less than 1 Oe. Since the magnitude of the bias field is related to the demagnetizing factors and the saturation magnetization, then, obviously, it is necessary to choose the tangential orientation of the bias field and the magnetic film with the minimum saturation magnetization. As for the control voltage, estimates show that a voltage of 1 V can provide an FMR line shift of several line widths for a piezoelectric film thickness of up to 1 μm (Tatarenko & Bichurin, 2012).

References

Astrov, D. N. (1961). Magnetoelectric effect in chromium oxide. *Soviet Physics Journal of Experimental and Theoretical Physics, 13*, 729.

Bichurin, M. I. (1994). Magnetoelectrics in microwave range. *Ferroelectrics, 161*, 53.

Bichurin, M. I., Filippov, D. A., Petrov, V. M., Laletsin, V. M., Paddubnaya, N. N., & Srinivasan, G. (2003). Resonance magnetoelectric effects in layered magnetostrictive-piezoelectric composites. *Physical Review B: Condensed Matter, 68*, Article 132408.

Bichurin, M. I., & Petrov, V. M. (1988). Magnetic resonance in layered ferrite-ferroelectric structures. *Soviet Physics Journal of Experimental and Theoretical Physics, 58*, 2277.

Bichurin, M., & Petrov, V. (2014). Modeling of magnetoelectric effects in composites. *Springer Series in Materials Science, 201*, 108p.

Bichurin, M. I., Petrov, R. V., Leontiev, V. S., Sokolov, O. V., Turutin, A. V., Kuts, V. V., Kubasov, I. V., Kislyuk, A. M., Temirov, A. A., Malinkovich, M. D., & Parkhomenko, Y. N. (2020). Self-biased bidomain LiNbO3/Ni/Metglas magnetoelectric current sensor. *Sensors, 20*, 7142. https://doi.org/10.3390/s20247142

Bichurin, M. I., Petrov, V. M., & Petrov, R. V. (2012). Direct and inverse magnetoelectric effect in layered composites in electromechanical resonance range: A review. *Journal of Magnetism and Magnetic Materials, 324*, 3548.

Bichurin, M. I., Petrov, V. M., Petrov, R. V., & Tatarenko, A. S. (2019). *Magnetoelectric composites*. Singapore: Pan Stanford Publishing Pte. Ltd., 280p.

Bichurin, M. I., Petrov, V. M., & Semenov, G. A. (2009). Magnetoelectric material for components of radioelectronic devices. *Patent of Russian Federation, № 2363074.*

Bichurin, M., Petrov, R., Sokolov, O., Leontiev, V., Kuts, V., Kiselev, D., & Wang, Y. (2021). Magnetoelectric magnetic field sensors: A review. *Sensors, 21*, 6232.

Bichurin, M. I., Petrov, V. M., & Srinivasan, G. (2003). Theory of low-frequency magnetoelectric coupling in magnetostrictive-piezoelectric bilayers. *Physical Review B: Condensed Matter, 68*, Article 054402.

Bichurin, M., Sokolov, O., Ivanov, S. V., Ivasheva, E., Leontiev, V., Lobekin, V., & Semenov, G. (2023). Modeling the composites for magnetoelectric microwave devices. *Sensors, 23*, 1780.

Bichurin, M., Sokolov, O., Ivanov, S. V., Leontiev, V., Petrov, D., Semenov, G., & Lobekin, V. (2022). Physics of composites for low-frequency magnetoelectric devices. *Sensors, 22*, 4818.

Bichurin, M. I., Sokolov, O. V., & Lobekin, V. N. (2021). Torsion mode of the magnetoelectric effect in a Metglas/GaAs. *IEEE Magnetics Letters, 13*, 1.

Bichurin, M. I., Viehland, D., & Srinivasan, G. (2007). Magnetoelectric interactions in ferromagnetic—Piezoelectric layered structures: Phenomena and devices. *Journal of Electroceramics, 19*, 243.

Burdin, D. A., Chashin, D. V., Ekonomov, N. A., Fetisov, Y. K., Fetisov, L. Y., Gollapudi, S., & Srinivasan, G. (2013). Resonance mixing of alternating current magnetic fields in a multiferroic composite. *Journal of Applied Physics, 113*, 033902.

Carman, G. P., & Sun, N. X. (2018). Strain-mediated magnetoelectrics: Turning science fiction into reality. *MRS Bulletin, 43*, 822.

Chu, Z., PourhosseiniAsl, M., & Dong, S. (2018). Review of multi-layered magnetoelectric composite materials and devices applications. *Journal of Physics D Applied Physics, 51*, Article 243001.

Eds. G. Srinivasan, S. Priya, N.X. Sun. *Composite magnetoelectrics* (2015). Cambridge, UK: Woodhead Pulishing, 381p.

Duan, C.-G., et al. (2008). Surface magnetoelectric effect in ferromagnetic metal films. *Physical Review Letters, 101*, Article 137201.

Dzyaloshinskii, I. E. (1960). On the magneto-electrical effect in antiferromagnets. *Soviet Physics Journal of Experimental and Theoretical Physics, 10*, 628.

Eerenstein, W., Mathur, N. D., & Scott, J. F. (2006). Multiferroic and magnetoelectric materials. *Nature, 442*, 759.

Eds. Yu. N. Venevtsev, V.N. Lyubimov. *Ferroelectromagnet Substances* (1990). Moscow: Science, 184p. (in Russian).

Fiebig, M. (2005). Revival of the magnetoelectric effect. *Journal of Physics D Applied Physics, 38*, R1.

Folen, V. J., Rado, G. T., & Stalder, E. W. (1961). Anysotropy of the magnetoelectric effect in Cr2O3. *Physical Review Letters, 6*, 607.

Gao, J., Jiang, Z., Zhang, S., Mao, Z., Shen, Y., & Chu, Z. (2021). Review of magnetoelectric sensors. *Actuators, 10*, 109.

Harshe, G., Dougherty, J. O., & Newnham, R. E. (1993a). Theoretical modelling of multilayer magnetoelectric composites. *International Journal of Applied Electromagnetics in Materials, 4*, 145.

Harshe, G., Dougherty, J. P., & Newnham, R. E. (1993b). Theoretical modelling of 3-0, 0-3 magnetoelectric composites. *International Journal of Applied Electromagnetics in Materials, 4*, 161.

Hou, W., Zhou, Z., Xue, X., Guan, M., Hu, Z., & Liu, M. (2018). Voltage control of two-magnon scattering in multiferroic layers for tunable magnetoelectric devices. *IEEE Transactions on Magnetics.* https://doi.org/10.1109/TMAG.2018.2850064

Landau, L. D., & Lifshitz, E. M. (1980). *Statistical physics* (3rd ed.). Pergamon, 562p.

Leung, C. M., Li, J., Viehland, D., & Zhuang, X. (2018). A review on applications of magnetoelectric composites: From heterostructural uncooled magnetic sensors, energy harvesters to highly efficient power converters. *Journal of Physics D Applied Physics, 51*, Article 263002.

Liang, X., Chen, H., & Sun, N. X. (2021). Magnetoelectric materials and devices. *APL Materials, 9*, Article 041114.

Liang, X., Matyushov, A., Hayes, P., Schell, V., Dong, C., Chen, H., He, Y., Will-Cole, A., Quandt, E., Martins, P., McCord, J., Medarde, M., Lanceros-Méndez, S., van Dijken, S., & Sun, N. X. (2021). Roadmap on magnetoelectric materials and devices. *IEEE Transactions on Magnetics, 57*(8), 1.

Lin, H., Page, M. R., McConney, M., Jones, J., Howe, B., & Sun, N. X. (2018). Integrated magnetoelectric devices: Filters, pico-Tesla magnetometers, and ultracompact acoustic antennas. *MRS Bulletin, 43*, 841.

Lopatin, S., Lopatina, I., & Lisnevskaya, I. (1994). Magnetoelectric PZT/Ferrite composite materials. *Ferroelectrics, 162*, 63.

Lou, J., Liu, M., Reed, D., Ren, Y., & Sun, N. X. (2009). Giant electric field tuning of magnetism in novel multiferroic FeGaB/lead zinc niobate–lead titanate (PZN-PT) heterostructures. *Advances in Materials, 21*, 4711.

Ma, J., Hu, J., Li, Z., & Nan, C. W. (2011). Recent progress in multiferroic magnetoelectric composites: From bulk to thin films. *Advanced Materials, 23*, 1062.

Eds. A. J Freeman, H. Schmid. *Magnetoelectric interaction phenomena in crystals (MEIPIC-I, Seattle, USA, 1973)* (1975). London. N.-J. Paris: Gordon and Breach, 228p.

Eds. M. I. Bichurin and D. Viehland. *Magnetoelectricity in composites* (2012). Singapore: Pan Stanford Publishing, 286p.

Mandal, S. K., Sreenivasulu, G., Petrov, V. M., & Srinivasan, G. (2011). Magnetization-graded multiferroic composite and magnetoelectric effects at zero bias. *Physical Review B: Condensed Matter, 84*, Article 014432.

Nan, C.-W. (1994). Magnetoelectric effect in composites of piezoelectric and piezomagnetic phases. *Physical Review B: Condensed Matter, 50*, 6082.

Nan, C.-W., Bichurin, M. I., Dong, S., Viehland, D., & Srinivasan, G. (2008). Multiferroic magnetoelectric composites: Historical perspectives, status, and future directions. *Journal of Applied Physics, 103*, Article 031101.

Nan, T., Zhou, Z., Liu, M., et al. (2014). Quantification of strain and charge co-mediated magnetoelectric coupling on ultra-thin Permalloy/PMN-PT interface. *Scientific Reports, 4*, 3688.

Palneedi, H., Annapureddy, V., Priya, S., & Ryu, J. (2016). Status and perspectives of multiferroic magnetoelectric composite materials and applications. *Actuators, 5*, 9.

Priya, S., Islam, R., Dong, S., & Viehland, D. (2007). Resent advancements in magnetoelectric particulate and laminate composites. *Journal of Electroceramics, 19*, 149.

Proceedings of the 2nd international conference on magnetoelectnc interaction phenomena in crystals (MEIPIC-2, ascona). Eds. Schmid H., Janner A., Grimmer H., Rivera J.-P., and Ye Z.-G *Ferroelectrics, 161–162*, (1993), 748p.

Proceedings of the 3rd international conference on magnetoelectric interaction phenomena in crystals (MEIPIC-3, novgorod). Ed. M.I Bichurin *Ferroelectrics, 204*, (1997), 356p.

Eds. M. Fiebig, V.V. Eremenko, and I.E. Chupis. *Proceedings of the Fifth conference on magnetoelectric interaction phenomena in crystals (MEIPIC-5, Sudak)* (2004). Kiuwer Academic Publishers, NATO Sciences Series, 334p.

Proceedings of the Fourth conference on magnetoelectric interaction phenomena in crystals (MEIPIC-4, Veliky Novgorod). Ed. M.I. Bichurin *Ferroelectrics, 279–280*, (2002), 386p.

Ramesh, R., & Spaldin, N. A. (2007). Multiferroics: Progress and prospects in thin films. *Nature Materials, 6*, 21.

Ryu, J., Priya, S., Uchino, K., & Kim, H.-E. (2002). Magnetoelectric effect in composites of magnetostrictive and piezoelectric materials. *Journal of Electroceramics, 8*, 107.

Smolenskii, G. A., & Chupis, I. E. (1982). Ferroelectromagnets. *Soviet Physics - Uspekhi, 25*, 475–490.

Song, H.-C., et al. (2018). Broadband dual phase energy harvester: Vibration and magnetic field. *Applied Energy, 225*, 1132.

Spaldin, N. A., & Ramesh, R. (2019). Advances in magnetoelectric multiferroics. *Nature Materials, 18*, 203.

Spetzler, B., Bald, C., Durdaut, P., Reermann, J., Kirchhof, C., Teplyuk, A., Meyners, D., Quandt, E., Höft, M., Schmidt, G., & Faupel, F. (2021). Exchange biased delta-E efect enables the detection of low frequency pT magnetic felds with simultaneous localization. *Scientific Reports, 11*, 5269.

Srinivasan, G. (2010). Magnetoelectric composites. *Annual Review of Materials Researsh, 40*, 153.

Srivastava, A. K., Hurben, M. J., Wittenauer, M. A., et al. (1999). Angle dependence of the ferromagnetic resonance linewidth and two magnon losses in pulsed laser deposited films of yttrium iron garnet, MnZn ferrite, and NiZn ferrite. *Journal of Applied Physics, 85*, 7838.

Sun, N. X., & Srinivasan, G. (2012). Voltage control of magnetism in multiferroic heterostructures and devices. *Spin, 2*, Article 1240004, 1.

Tatarenko, A. S., & Bichurin, M. I. (2012). Microwave magnetoelectric devices. *Advances in Condensed Matter Physics, 2012*, Article 286562.

Tatarenko, A., Sokolov, O., Ivanov, S., Bichurin, M., & Wang, Y. (2019). Microwave magnetoelectric effect in structures based on ferromagnetic metals. In *ITM web conference* (Vol 30).

Van den Boomgard, J., et al. (1974). An in situ grown eutectic magnetoelectric composite materials: Part I. *Journal of Materials Science, 9*, 1705.

Van Run, A. M. J. G., et al. (1974). An in situ grown eutectic magnetoelectric composite materials: Part II. *Journal of Materials Science, 9*, 1710.

Van Suchtelen, J. (1972). Product properties: A new application of composite materials. *Philips Research Reports, 27*, 28.

Vaz, C. A. F., Hoffman, J., Ahn, C., & Ramesh, R. (2010). Magnetoelectric coupling effects in multiferroic complex oxide composite structures. *Advanced Materials, 22*, 2900.

Viehland, D., Wuttig, M., McCord, J., & Quandt, E. (2018). Magnetoelectric magnetic field sensors. *MRS Bulletin, 43*, 834.

Wang, J., Chen, A., Li, P., & Zhang, S. (2021). Magnetoelectric memory based on ferromagnetic/ferroelectric multiferroic heterostructure. *Materials, 14*, 4623.

Wang, Y., Hu, J., Lin, Y., & Nan, C.-W. (2010). Multiferroic magnetoelectric composite nanostructures. *NPG Asia Materials, 2*, 61.

Wang, Y., Li, J., & Viehland, D. (2014). Magnetoelectrics for magnetic sensor applications: Status, challenges and perspectives. *Materials Today, 6*, 269.

Xue, X., Zhou, Z., Dong, G., et al. (2017). Discovery of enhanced magnetoelectric coupling through electric field control of two-magnon scattering within distorted nanostructures. *ACS Nano, 11*, 9286.

Yang, F., Du, L., Yu, H., & Huang, P. (2020). Magnetic and electric energy harvesting technologies in power grids: A review. *Sensors, 20*, 1496.

Zhai, J., Xing, Z., Dong, S., Li, J., & Viehland, D. (2008). Magnetoelectric laminate composites: An overview. *Journal of the American Ceramic Society, 91*, 351.

CHAPTER

2

Magnetic, piezoelectric, and magnetoelectric materials and phenomena

Dirk Meyners[1], Bin Luo[2], Jeffrey McCord[1], Eckhard Quandt[1], and Nian Sun[2]

[1] Kiel University, Kiel, Germany; [2] Northeastern University, Boston, MA, United States

2.1 Introduction

As mentioned in Chapter 1, different mechanisms have been reported to lead to ME coupling, among which strain-mediated ME coupling has been the strongest (Vaz, 2012). Strain-mediated ME laminates comprising piezoelectric and magnetostrictive materials have demonstrated much higher ME coupling strength than single-phase multiferroic materials (Chu, PourhosseiniAsl et al., 2018; Luo, Will-Cole et al., 2024), such as $BiFeO_3$ (Spaldin & Ramesh, 2019). For bulk laminate composites, the theoretical ME coefficient α_{ME}^{L-L} for the longitudinal–longitudinal (*L*-*L*) mode is given by Eq. (2.1), where s_{33}^{E} and s_{33}^{H} are the elastic compliances of the piezoelectric and magnetostrictive layers, respectively, k_{33} is the electromechanical coupling coefficient of the piezoelectric layer, n is the ratio of magnetic layer thickness to the total thickness of the ME laminate, and $d_{33,m}$ and $g_{33,p}$ are the longitudinal piezomagnetic and piezoelectric coefficients, respectively (Dong, Zhai et al., 2006).

$$\alpha_{ME}^{L-L} = \frac{n d_{33,m} g_{33,p}}{n s_{33}^{E}\left(1 - {k_{33}}^{2}\right) + (1-n) s_{33}^{H}}, \tag{2.1}$$

https://doi.org/10.1016/B978-0-443-18606-6.00008-3

Clearly, the ME coupling coefficient is dependent on the electromechanical and magnetomechanical properties of the constitutive piezoelectric and ferromagnetic layers.

High ME coupling has been achieved in magnetostrictive materials, such as $(Fe_{80}Ga_{20})_{88}B_{12}$ or ferrite films (Liu, Obi et al., 2009) grown on ferroelectric substrates, namely $Pb(Zn_{1/3}Nb_{2/3})O_3-PbTiO_3$ (PZN-PT) or $Pb(Mg_{1/3}Nb_{2/3})O_3-PbTiO_3$ (PMN-PT) (Liu, Obi et al., 2010; Lou, Liu et al., 2009; Nan, Zhou et al., 2014; Wang, Gao et al., 2018; Yang, Zhao et al., 2009; Zhou, Li et al., 2014). However, the large E-fields required to tune the magnetization are not feasible for most working devices, e.g., 8 kV/cm translating to 400 V for 0.5 mm thick ferroelectric slab is required to fully manipulate the magnetization (Yang, Zhao et al., 2009). Additionally, the bulk, single-crystal ferroelectric substrates are not easily integrated onto Si substrates due to CMOS processing incompatibility. However, thin-film ME heterostructures have been integrated with CMOS-compatible processes on Si using piezoelectric film, namely AlN, ZnO, $Al_{0.73}Sc_{0.27}N$, coupled to a magnetostrictive film, such as $(Fe_{80}Ga_{20})_{88}B_{12}$ or $(Fe_{90}Co_{10})_{78}Si_{12}B_{10}$, for ME devices for both communication and sensing applications (Fichtner, Wolff et al., 2019; Li, Matyushov et al., 2017; Liang, Chen et al., 2020; Nan, Lin et al., 2017; Parsapour, Pashchenko et al., 2018; Su, Niekiel et al., 2020). For strain-mediated ME composites, it is critical to select piezoelectric materials with a high piezoelectric coefficient. While perovskite ferroelectrics can exhibit high piezoelectric coefficients (Piazza, Felmetsger et al., 2012), they are not CMOS compatible due to the requirement of epitaxial growth; therefore, AlN- and Sc-doped AlN are the premier candidates for use in ME composites (Piazza, Felmetsger et al., 2012; Tasnadi, Alling et al., 2010). In addition, fabrication methods have been developed to integrate ferroelectric $LiNbO_3$ on Si wafers via an ion slicing and film transfer approach (Rabiei & Steier, 2005; Mayrhofer, Persson et al., 2016, Churaev et al., 2023), thus enabling $LiNbO_3$ deployment in future ME-based IoT devices.

There is a clear rationale for incorporating magnetostrictive thin-film systems into magnetoelectric (ME) composites. Thin-film systems can be constructed as multilayers, wherein magnetic couplings are employed to modify pivotal properties. Considering the important application as a magnetic field sensor, it is shown how the zero field operation and the directionality in the response of a vector field sensor are achieved. The various sensor noise mechanisms are identified, with a particular emphasis on those of magnetic origin. Finally, measures in the magnetic design of the multilayers are evaluated with the objective of resolving the antagonism between high sensor sensitivity and low sensor noise.

2.2 Piezoelectric/electrostrictive/ferroelectric material for magnetoelectric coupling

Among piezoelectric/electrostrictive materials, lead zirconate titanate (PZT), lead magnesium niobate–lead titanate (PMN-PT), lead zinc niobate–lead titanate (PZN-PT), and aluminum nitride (AlN) have been widely used in integrated ME devices to convert electric excitation to strain. Table 2.1 shows a cross comparison of the most commonly used piezoelectric/electrostrictive materials for integrated ME applications (He, Luo et al., 2021). PZT is a polycrystalline ceramic ferroelectric material with a high electromechanical coupling coefficient and an effective piezoelectric coefficient of −200 to −300 pC/N (Berlincourt, Krueger et al., 2003) at morphotropic phase boundary (MPB) composition corresponding to Zr/Ti ratios of about 58/42. The PZT piezoelectric ceramic also has the advantages of low-cost and easier shaping process (Erhart, Půlpán et al., 2017). The main drawback of PZT materials is that the contamination during the process is often unavoidable due to the lead element in it, which greatly degrades its application in ME devices (Tu, Chu et al., 2019). Moreover, the relatively high acoustic loss of the PZT film at high frequency range prevents it from achieving an efficient ME coupling. In Table 2.1, PZT materials are classified into hard and soft PZTs, where the hard PZT materials usually have smaller piezoelectric coefficient with higher energy efficiency. Depending on the circumstance of the specific applications, various types of PZT materials are commercially available. Different from the PZT film, aluminum nitride (AlN) is a textured polycrystalline nonferroelectric material with an advantage of lower loss tangent and compatibility with CMOS technology. Another unique advantage of AlN is its high electric Curie temperature, which is desirable for electronics applications in high temperature and inferior heat dissipation environment. However, the moderate piezoelectric coefficient of about −2 pC/N makes it unsuitable for ME devices requiring high ME coupling (Dubois & Muralt, 1999; Sanchez-Rojas, Hernando et al., 2008; Tsubouchi, Sugai et al., 1981). Compared with PZT and AlN, lead magnesium niobate–lead titanate (PMN-PT) and lead zinc niobate–lead titanate (PZN-PT), as single crystal ferroelectric materials, show larger piezoelectric coefficients and lower loss tangents. The piezoelectric performance of these materials highly depends on their composition and crystal orientation. The composition here means the element ratio of rhombohedral phase (PZN or PMN) and tetragonal phase (PT). The PZN–PT composition that induces MPB with (001) crystal cut can give a d_{33} piezoelectric coefficient as high as 2800 pC/N (Lim, Rajan et al., 2007). For the PMN-PT film, a 31% PT and

TABLE 2.1 Comparison of commonly used ferro/piezoelectric/electrostrictive materials for ME applications (He, Luo et al., 2021).

	Soft PZT	**Hard PZT**	**PMN-PT**	**PZN-PT**	**AlN**	**AlScN**	**$BiFeO_3$-based ceramics**
d_{31} (pC/N)	−175∼-274	−58∼-140	−1800		−2		
d_{32} (pC/N)			900	−3800			
d_{33} (pC/N)	400–593	149–320	2000	2800	3.5–4	24.6–27.6	42
T_c (℃)	195–360	300–330	145	160	>2000	>600	810–830
ε_r	1850–2870	325–1375	∼6000	∼6000	∼10	9–24	<200
Tan δ	1.25–2.5	0.2–0.4	0.0042	0.0047	0.0005	0.004–0.006	0.005–0.55
References	Erhart, Půlpán et al. (2017)	Erhart, Půlpán et al. (2017)	(Han, Yan et al., 2005; Yang & Zu, 2016)	(Lim, Rajan et al., 2007; Yang & Zu, 2016)	(Dubois & Muralt, 1999; Ling, Xiaoning et al., 2017; Sanchez-Rojas, Hernando et al., 2008; Tsubouchi, Sugai et al., 1981)	(Akiyama, Kamohara et al., 2009; Fichtner, Wolff et al., 2019; Mayrhofer, Persson et al., 2016; Olsson, Tang et al., 2020; Parsapour, Pashchenko et al., 2018; Zhang, Wu et al., 2020)	(Chauhan, Kumar et al., 2012; Chen, Wang et al., 2012; Heron, Schlom et al., 2014; Lou, Liu et al., 2009; Mishra, Pradhan et al., 2008)

optimized cut give a d_{31} of about −1800 pC/N and a d_{33} of about 2000 pC/N (Han, Yan et al., 2005). Although the aforementioned two materials deliver superior piezoelectric performance, the difficulties of composition control during the deposition of these materials may induce the lack of performance repeatability. The grain growth of these single crystal materials is also challenging. Two major deposition methods include flux method and Bridgman method (Uchino, 2017).

2.3 Magnetostrictive thin film materials for magnetoelectrics

The approach to ME implements with composite materials expands the possibilities for the selection of the magnetic phase under the condition of material compatibility. The selection should be based on the fundamental equation for the ME effect size (Nan et al., 2008). The ME voltage coefficient for an open-circuited ME device is given by:

$$\alpha_{ME} = \frac{\partial E}{\partial H} = \frac{\partial E}{\partial \sigma}\frac{\partial \sigma}{\partial \lambda}\frac{\partial \lambda}{\partial H}, \tag{2.2}$$

with the piezoelectric voltage coefficient $\partial E/\partial \sigma$ and the mechanical coupling coefficient $\partial \sigma/\partial \lambda$. Generally, the magnetostrictive strain λ is conditional on the material's magnetization M and thus magnetic field dependent. Accordingly, the piezomagnetic coefficient $\partial \lambda/\partial H$ depends on the magnetic susceptibility $\chi = \partial m/\partial H$:

$$\frac{\partial \lambda(m)}{\partial H} = \frac{\partial \lambda(m)}{\partial m}\frac{\partial m}{\partial H} = \frac{\partial \lambda(m)}{\partial m}\chi \tag{2.3}$$

Here, the notation of reduced magnetization is used with $m = M/M_S$ and the material's saturation magnetization M_S. In accordance with Eq. (2.3), the magnetic phase should manifest both, a large magnetostriction and a large susceptibility, which motivate the integration of soft magnetic amorphous foils in bulk laminated composites (Table 2.2). The magnetostrictive material in the ME composites should have high saturation magnetostriction λ_s and high piezomagnetic coefficient dλ/dH, near unity magnetomechanical coupling factor, low hysteresis loss, low eddy current damping, and low Gilbert damping, or more generally, narrow ferromagnetic resonance linewidth. A high magnetomechanical coupling factor close to unity indicates an efficient coupling between the magnetic and mechanical energies (Dong, Li et al., 2018). Depending on the specific applications, other magnetic and elastic properties of the materials may also be considered. The low magnetic loss tangent, low FMR bandwidth, and low coercive field are essential for high-frequency applications. Some particular properties

TABLE 2.2 Soft magnetic amorphous materials used in bulk laminated and thin-film ME composites.

Bulk laminated ME composites	
Magnetostrictive phase	**References**
FeBSiC (Metglas Inc., Conway, SC)	Dong et al. (2006)
$Fe_{74.4}Co_{21.6}Si_{0.5}B_{3.3}Mn_{0.1}C_{0.1}$ (Vacuumschmelze GmbH & Co. KG, Germany)	Wang et al. (2012)
Metglas alloy 2605SA1 (Metglas Inc., Conway, SC)	Zhai, Dong, et al. (2006)
Vitrovac 7600F (Vacuumschmelze GmbH & Co. KG, Germany)	Li et al. (2011)
2826 MB type foil ($Fe_{0.4-0.5}Ni_{0.4-0.5}Mo_{0.05-0.1}B_{0.01-0.05}$) (Hitachi Metals Europe GmbH)	Turutin et al. (2018)
Thin film ME composites	
Magnetostrictive layer material	**References**
FeGaB	Pettiford et al. (2008)
FeCoSiB	Greve et al. (2010)
FeGaC	Liang et al. (2019)
CoFeC	Wang et al. (2019)

like ΔE effect, where the Young's modules of materials vary with the applied magnetic field, are also needed for specific applications including sensors.

A cross-comparison of commonly used piezomagnetic/magnetostrictive materials for ME composite is given in Table 2.3 including Terfenol-D ($Tb_{0.7}Dy_{0.3}Fe_2$), Galfenol ($Fe_{81}Ga_{19}$), FeGa, FeCoSiB, FeGaB, FeBSiC (Metglas), and FeGaC (He, Luo et al., 2021). Owing to the largest magnetostriction of Terfenol-D, the Terfenol-D/PZN-PT composite with giant electric field induces effective magnetic anisotropy field of 3500 Oe, corresponding to an ME coefficient of 580 Oe cm/kV (Liu, Li et al., 2012). The ME coefficient is almost 4 times that in Fe_3O_4/PZN-PT (Liu, Obi et al., 2009). Even though Terfenol-D exhibits giant saturation magnetostriction of 1600 ppm, it is lossy and has a high saturation field of several kOe, which induces a relatively small piezomagnetic coefficient. Compared with Terfenol-D, binary FeGa alloys show a high saturation magnetostriction constant of 400 ppm for single crystals (Galfenol [$Fe_{81}Ga_{19}$]) and 275 ppm for directional solidified polycrystalline alloys (FeGa), a low saturation

TABLE 2.3 Comparison of commonly used piezomagnetic/magnetostrictive materials for ME applications (He, Luo et al., 2021).

	Terfenol-D	Metglas	FeGa	FeGaB	FeCoSiB	FeGaC
λ_s (ppm)	1600	30	400	70	158	81.2
$(d\lambda/dH)_{max}$ ($\times10^{-6}$/Oe)	2.4	3.8	1.5	7		9.71
μ_r	10	45000	73 at 50 kHz	~400	~145 at 600 MHz	
T_c (C)	650	395	675			674
References	Engdahl and Mayergoyz (2000)	Chu, Shi et al. (2017)	(Clark, Wun-Fogle et al., 2002; Srisukhumbowornchai & Guruswamy, 2001)	(Dong, Li et al., 2018; Lou, Insignares et al., 2007)[	(Hrkac, Lage et al., 2014; Schell, Müller et al., 2020; Xiao, Han et al., 2007)	Liang, Dong et al. (2019)

field of about 100 Oe and a high saturation magnetization of 18 kG. However, the FeGa films have a large FMR linewidth of 450–600 Oe at X band and strong loss at microwave frequencies (Butera, Gómez et al., 2005). The shortcoming indicates that the FeGa alloys are far from being an ideal magnetic material to be incorporated into RF/microwave ME devices. Boron (B), as a well-known metalloid element, has been added into the FeGa and FeCo films to improve their magnetic properties (Lou, Insignares et al., 2007; Dong, Li et al., 2018; Liang, Dong et al., 2019). It has been experimentally demonstrated that the addition of B in FeCo films contributes to an excellent soft magnetization and microwave properties by refining the grain size and diminishing magnetocrystalline anisotropy (Kim, Kim et al., 2004; Platt, Minor et al., 2001). Doping $Fe_{80}Ga_{20}$ films with boron or carbon causes a crystallographic phase transition from the body centered cubic (BCC) phase to an amorphous structure, leading to excellent soft magnetic properties (<2 Oe coercive field), high magnetostriction (78 ppm), near unity magnetomechanical coupling factor (0.84), and low ferromagnetic resonance linewidth in *X*-band (<20 Oe linewidth) in the BCC-amorphous phase boundary region (Dong, Li et al., 2018; Liang, Dong et al., 2019; Lou, 2010; Lou, Insignares et al., 2007). Moreover, the addition of boron in the composition of FeCoSiB and FeGaB thin films contributes to the demonstration of giant ME coefficients and low loss tangents in ME devices RF/microwave devices. FeGaB material exhibits excellent performance on magnetic softness, saturation magnetization, saturation magnetostriction, piezomagnetic coefficient, ΔE effect, and magnetomechanical coupling factor and has been regarded as an ideal piezomagnetic film material for ME devices (Lou, Insignares et al., 2007). As indicated in (Lou, Insignares et al., 2007), with the addition of B rising from 0% to 21% in FeGa, the coercivity H_c decreased from about 100 Oe to 1 Oe and the effective anisotropy field H_k dropped from 120 Oe to 15 Oe due to the material phase transition (MPB). The reduced coercivity and anisotropy significantly enhance magnetic softness. Additionally, as shown in Fig. 2.1a, the B content–induced phase transition results in a lower threshold magnetic field to generate a magnetostrictive response. The maximum magnetostriction of 70 ppm achieved at 12% B content was three times that of the binary FeGa films. The FMR linewidth of FeGaB films also dropped dramatically from 700 Oe for binary FeGa films to 16 Oe at 21% B content, as shown in Fig. 2.1b. The maximum piezomagnetic coefficient of the FeGaB films has been reported as about 7 ppm Oe^{-1}, which is much higher than multiple magnetostrictive materials, including Terfenol-D (Tb-Dy-Fe), Galfenol (Fe–Ga), and Metglas (FeBSiC). The comparison was made in (Lou, Liu et al., 2009) and graphically shown in Fig. 2.2. The maximum piezomagnetic

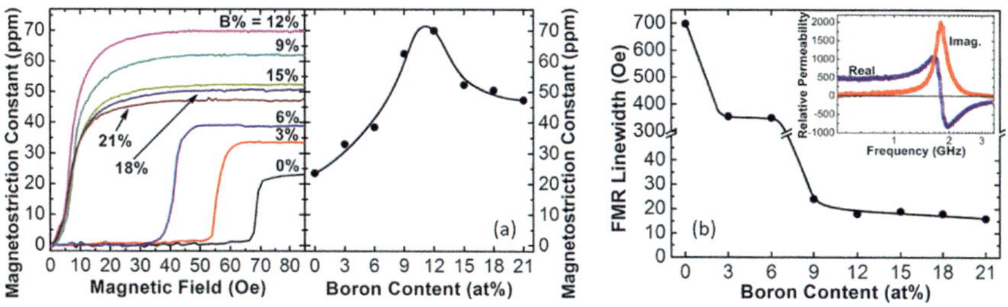

FIGURE 2.1 (a) Boron content–dependent magnetostriction property of FeGaB thin-film material. (b) Boron content–dependent FMR linewidth of FeGaB thin-film material (Lou, Insignares et al., 2007)

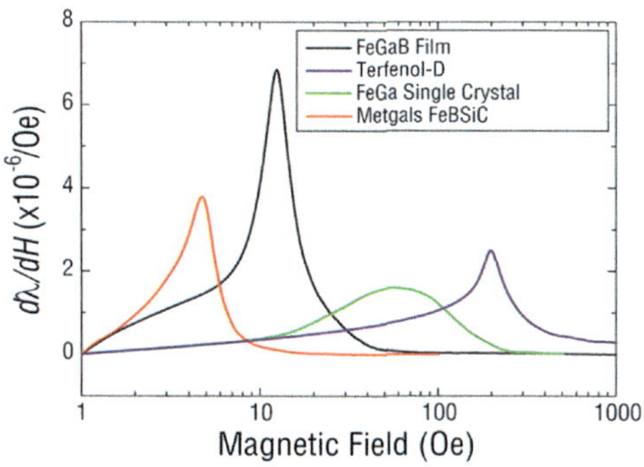

FIGURE 2.2 Piezomagnetic coefficients (dλ/dH) of different types of magnetostrictive alloys (Lou, Liu et al., 2009).

coefficient can be further enhanced to 12 ppm Oe^{-1} after annealing due to the effective release of residual stress inside the film and the reduction of total anisotropy energy (Dong, Li et al., 2018). More recently, Bayesian optimization has been shown as an efficient machine learning method for elucidating structure–property relationships in ferromagnetic materials for ME-based MEMS applications (Will-Cole et al., 2022). Novel growth approaches utilizing molecular beam epitaxial growth of $Fe_{1-x}Ga_x$ films on PMN-PT have expanded the possible composition range for the disordered BCC (A2) phase space, achieving an ultra-high magnetostriction at a Ga concentration of $x = 30\%$ with a diminishing stiffness, thus limiting their applications (Meisenheimer, Steinhardt et al., 2021). Underlayers such as NiFe, Cu, Cr, or Ti seed layer (<5 nm) prior to magnetic film deposition were demonstrated to promote soft magnetic properties in Fe–Ga alloys (Acosta, Fitzell et al., 2020).

Aside from the magnetomechanical properties and static magnetic behavior, the magnetic loss is a key property, particularly for RF ME NEMS. Magnetic loss can be understood through characterizing the ferromagnetic resonance linewidth and/or the Gilbert damping parameter (Gilbert, 2004). Yttrium iron garnet ($Y_3Fe_5O_{12}$, YIG) films are the pinnacle for low magnetic losses with an intrinsic magnetic damping parameter as low as $\simeq 3 \times 10^{-5}$, with the lowest reported magnetic damping parameter for the thin film regime at $\simeq 5.2 \times 10^{-5}$ (Spencer 1959; Ding, Liu et al., 2020). This low magnetic damping behavior enables long spin wave decay lengths more than 100 μm (Maendl, Stasinopoulos et al., 2017; Yu, Kelly et al., 2014). A key challenge for YIG integration in ME devices is the lattice-matching requirement for epitaxy. Most YIG films are grown on $Gd_3Ga_5O_{12}$ (GGG) since GGG has the same garnet structure as YIG with a lattice constant of a = 12.383 Å and a 0.06% lattice mismatch between YIG and GGG (Liu, Jin et al., 2018; Makino, Nakamura et al., 1976). While the GGG substrate allows for heteroepitaxial growth of YIG films, the GGG substrate can serve as a source of contamination in sputtered YIG films (Jermain, Aradhya et al., 2017; Shigematsu, Ando et al., 2016). While liquid-phase epitaxy-grown YIG films have shown minimal to no contamination from the GGG substrate, liquid phase epitaxy is not as back end of line (BEOL) compatible as sputtering (Will-Cole, Hart et al., 2023). Recently, a buffer layer of diamagnetic $Y_3Sc_{2.5}Al_{2.5}O_{12}$ thin film on GGG enabled the growth of YIG films with off-axis sputtering with suppressed Gd-contamination to enable decent Gilbert damping at few-K temperatures (Guo, McCullian et al., 2022). Despite recent work suppressing Gd contamination from the GGG substrate, there is a need for more investigation into the integration of YIG with CMOS-compatible substrates such as silicon. Ion slicing has been successfully executed on YIG and other oxide films such as ferroelectric BaTiO3, LiNbO3 around 20 years ago (Levy, Osgood et al., 1998; Izuhara, Gheorma et al., 2003; Rabiei & Steier, 2005, Churaev et al., 2023). However, the integration of YIG film on the piezoelectric layer has not been demonstrated, and this is likely a viable path forward for efficient and high *Q* ME-based IoT devices.

Remarkable ME coupling effects were demonstrated with, e.g., $\alpha_{ME} = 22\ \mathrm{V}/(\mathrm{cmOe})$ and $\alpha_{ME} = 61.5\ \mathrm{V}/(\mathrm{cmOe})$ using PZT (Dong et al., 2006) or PMN-PT (Wang et al., 2012) as piezoelectric phase, respectively. For thin-film composites, sputter-deposited, soft magnetic amorphous layers have been proposed, and a ME coefficient $\alpha_{ME} = 3.1\ \mathrm{V}/(\mathrm{cmOe})$ has been validated for a FeCoSiB and AlN composite (Greve et al., 2010). Exploiting voltage-enhancing schemes based on mechanical resonances, record values $\alpha_{ME} \approx 7000\ \mathrm{V}/(\mathrm{cmOe})$ and $\alpha_{ME} = 20000\ \mathrm{V}/(\mathrm{cmOe})$ have been reported for bulk laminate (LASER-treated FeBSi (Chu et al., 2017)) and thin-film composites (FeCoSiB (Kirchhof et al., 2013)).

From the magnetization and field dependence of the piezomagnetic coefficient, it follows directly that there is an operating point with optimal magnetic bias for maximum ME response. Moreover, in Eq. (2.2), the coefficients are tensor quantities with space dependent entries. In particular, according to Eq. (2.3), a spatial variation in magnetization results in a locally varying magnetostrictive response. The performance of any ME device will therefore be influenced not only by the choice of material but also by the specific magnetization configuration present. Control of the magnetization behavior is therefore desired. This aspect provides the incentive to fabricate ME composites by thin-film technology. Aside from the potential to device miniaturization (Onuta et al., 2011), which is essential for the use in dense sensor arrays and medical implants, e.g., thin-film technology enables the fabrication of complex multilayer structures. These multilayers in turn can be designed with the aim of tailoring the operating point or also the magnetization behavior. A disadvantage of the thin-film approach is that the interaction volume and thus the total magnetoelectrically converted energy are reduced compared to ME composites assembled with bulk materials.

2.4 Magnetostrictive multilayers for magnetoelectrics—Exchange bias for zero field operation

The above given ME coefficients were achieved in the presence of magnetic field bias less than or equal to 10 *Oe*. In laboratory setups, these holding fields can be easily realized by external magnetic field sources. In demonstrator setups for technical applications, permanent magnet arrays are often used as, e.g., in (Gao et al., 2012; Li et al., 2010; Ren et al., 2018). As demonstrated in (Giang et al., 2012), coils can also be used for excitation with alternating fields. The external field then takes over the role of the DC bias. For some applications, however, the requirement of a magnetic bias is disadvantageous. The additional magnetic field sources require installation space. In arrays with composites arranged in different spatial directions, care must also be taken to ensure that the stray fields do not affect adjacent composites (Li et al., 2012). Both requirements limit the potential for miniaturization of ME assemblies. In magnetic field sensor applications, the situation is further complicated by the fact that the fields can influence the object to be measured or potentially worsen the noise level of the sensor (Salzer et al., 2018).

Under the collective term "self-bias" ME composite, various approaches have been published aiming at replacing the external holding field with internal fields or couplings. A comprehensive overview of the approaches is provided in the review article (Zhou et al., 2016). The authors classify self-biased ME composites into the categories of (a) functionally graded ferromagnetic based (Mandal et al., 2010; Petrov & Srinivasan, 2008), (b) exchange bias mediated (Lage et al., 2012), (c) magnetostriction hysteresis based (Srinivasan et al., 2002), (d) built-in stress mediated (Kumar et al., 2021), and (e) nonlinear composites (Jahns et al., 2012; Kamentsev et al., 2006). For bulk laminates, the working principles are based on the use of hysteretic effects in the ferromagnetic phase (Zhang et al., 2014; Zhou et al., 2012), the incorporation of hard magnetic layers (Li et al., 2013), the Villari effect (Kumar et al., 2021; Villari, 1865), and the application of AC fields instead of a DC bias (Kamentsev et al., 2006).

In hysteretic composites, the desired magnetization state can be set by an initialization field to then operate the ME composite in remanence without bias (Laletin et al., 2012). Hysteretic ME composites can, however, change their performance if external magnetic fields that are present in the meantime change the magnetization state to an altered remanent one. Re-initialization is then required, so that such hysteretic composites generally also require an integrated field source. In this context, it is important to point out a special feature of the approach to q-graded materials presented by Laletin et al., in 2012 (Laletin et al., 2012). By combining two materials with different piezomagnetic coefficients (NZFO and Ni), the authors have succeeded in fabricating a composite which is characterized by a plateau in the ME characteristic and exhibits an almost unchanged ME coefficient for DC magnetic fields in the range of about ±100 Oe (Fig. 2.3).

In a straightforward way, thin-film technology facilitates the formation of the magnetic phase from individual layers of different materials into

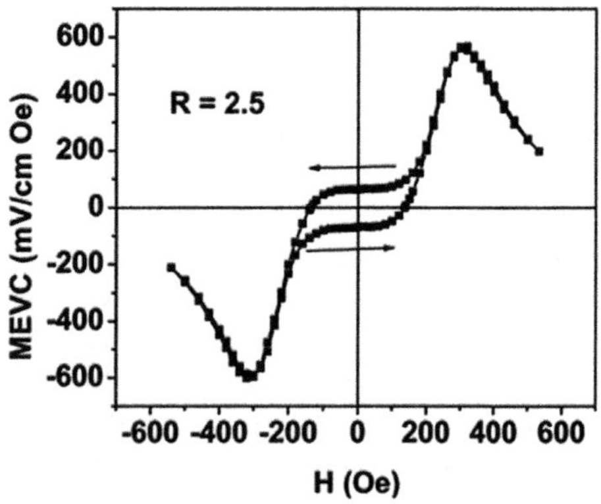

FIGURE 2.3 Magnetoelectric voltage coefficient versus magnetic field measured for an alternating magnetic field with 20 Hz frequency. The bulk laminate is made out of a trilayer from nickel zinc ferrite, lead zirconate titanate, and nickel. The ferrite-to-Ni thickness ratio R and the direction of H variation (*arrows*) are indicated. *From Laletin et al., 2012.*

complex multilayers. By this, a substantial increase in the degrees of freedom can be used to advantage in the design of the magnetic phase. Properties such as saturation magnetization, magnetostriction, susceptibility, anisotropy, piezomagnetic coefficient at zero field, and layer thickness can be adjusted layer by layer. The thin-film approach not only enables the use of magnetic coupling effects acting across the structure edges, but also couplings mediated at the atomic level across the interfaces of the layers. A good starting point for exploring the possibilities offered by this extended design freedom is to adapt well-known concepts from other thin-film applications such as read heads, flux concentrators, and magnetoresistive sensors (Hempstead et al., 1978). In particular, much research and development have been devoted to the exchange coupling between a ferromagnetic and an antiferromagnetic material, also known as exchange bias. It was first observed in cobalt particles with oxide shell (Meiklejohn & Bean, 1957) and was subsequently considered for diverse applications such as perpendicular recording media (Ohkoshi et al., 1985), magnetic sensors for read heads (Tsang, 1984) and automotive applications (Giebeler et al., 2001), magnetic random access memories (Parkin et al., 1999), and biosensors (Ehresmann et al., 2015). A theoretical concept for using the exchange bias effect in a ME device, precisely a ME read head, was first proposed by Vopsaroiu in 2011 (Vopsaroiu et al., 2007). The concept is to subject the magnetostrictive material to unidirectional anisotropy to achieve a prealignment of magnetization just as with an external bias (Lage et al.).

However, the prealignment of the magnetization should be neither perpendicular nor parallel to the external field to be transduced by the ME composite. A perpendicular configuration leads to a symmetric magnetostriction curve $\lambda(H)$ with vanishing slope at zero field (Lage et al., 2012). With parallel alignment, no torque acts on the magnetization and the piezomagnetic response to e.g., a small AC field consequently remains likewise vanishingly small. The angle enclosed by the unidirectional anisotropy and the external field must therefore be adjusted to fulfill $0° < \varphi < 90°$ condition.

Fig. 2.4 outlines the magnetic behavior for mutlilayers of antiferromagnetic $Mn_{70}Ir_{30}$ and $Fe_{50}Co_{50}$. The layer stack was fabricated by sputter deposition and includes 40 MnIr/FeCo layers, where a single layer of FeCo had a thickness of 20 nm. The layers were deposited on 300 μm thick silicon substrates, which were diced into 2 mm × 20 mm bending beams for the magnetostriction measurements based on the principle of Klokholm (1976). Fig. 2.4a shows the magnetization curves for different angles φ ranging from 18° to 72°. The shift of the curves along the field axis ($-H_{EB,long}$) as a consequence of the exchange bias coupling is clearly visible. The reduction of the shift with increasing angle φ is the result of

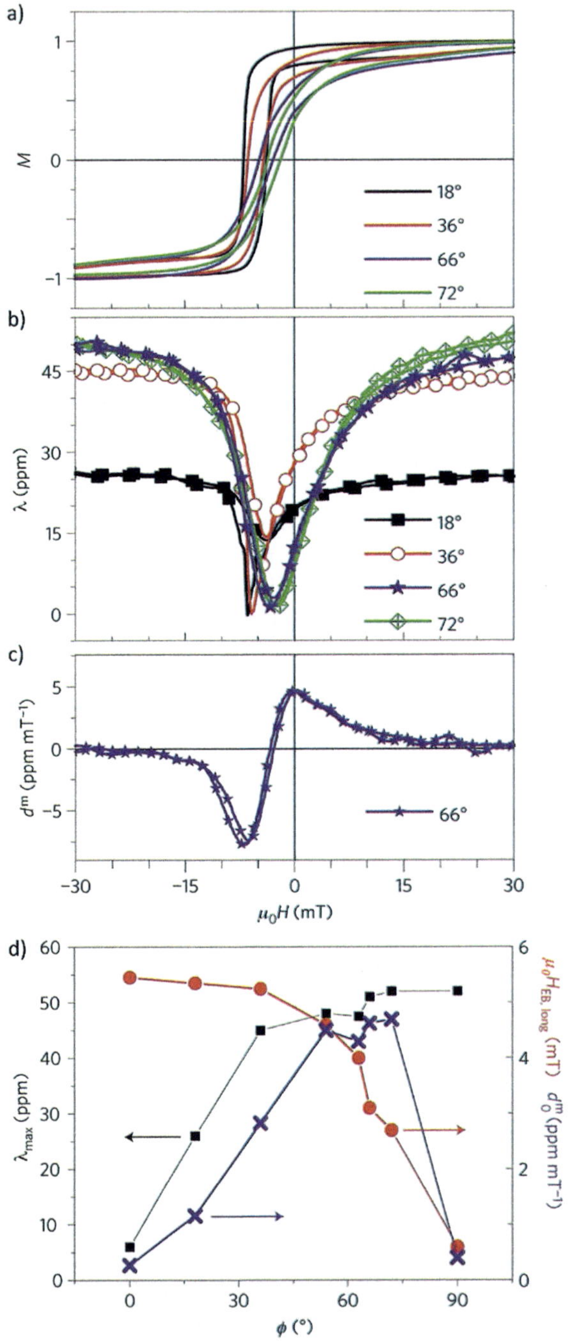

FIGURE 2.4 Normalized magnetization curves (a) and corresponding magnetostriction curves (b) of MnIr/FeCo multilayer stacks on cantilever samples for selected values of the inclination angle φ. (c) Piezomagnetic coefficient versus magnetic field for $\varphi = 66°$. (d) Maximum magnetostrictive strain λ_{max} obtained for magnetic fields along the longitudinal cantilever axis, $H_{EB,long}$ and piezomagnetic coefficient d_0^m for zero magnetic field as a function of φ. *From Lage et al., 2012.*

the projection of the exchange bias field onto the measurement direction (Fig. 2.4d, red circles) (Hu et al., 2002). In addition to the loop shift, the slope of the magnetization curves nearby zero magnetization also reduces with increasing angle φ as expected for angled exchange biased systems (Camarero et al., 2005). Of more relevance for the application in an ME composite are the changes that can be derived from the magnetostriction curves (Fig. 2.4b) and, in particular, for their slopes at vanishing magnetic field bias. For the specific MnIr/FeCo system studied here, a maximum piezomagnetic coefficient of 4.6 ppm/mT is obtained at zero field for $\varphi = 66^\circ$ (Fig. 2.4c). Using such magnetostrictive multilayer and 2 μm thick sputter-deposited AlN, a composite was fabricated. To prove the concept, the composite was subjected to an AC field with 1 Oe amplitude and the ME voltage was recorded while varying the magnetic bias. Consequently, it showed a local maximum of the ME coefficient $\alpha_{ME} = 21$ V/cmOe under zero bias condition (Lage et al., 2012).

Since the exchange bias coupling is an interfacial effect, it follows directly that the exchange bias field is reciprocal to the thickness of the magnetostrictive layer (Nogués & Schuller, 1999). Accordingly, the interfacial character of the coupling sets an upper limit for the magnetostrictive layer thickness t_{FM} in the range of few 100 nm, where a loop shift is still observed (Lage et al., 2012). However, the $1/t_{FM}$ dependence can also be used to tune the coupling strength to compromise between stable magnetization alignment and still high magnetic susceptibility. The limitation in individual magnetostrictive layer thickness implies that multilayers must be fabricated, as in the example above, to increase the magnetic interaction volume in the composite. To ensure that the growth conditions and the closely related magnetic layer properties do not vary across the stack, the fabrication processes need to be closely monitored. In particular, growth-induced layer stresses, roughness, and morphology in the magnetostrictive multilayers are of major concern.

Lage et al. (2012) incorporated a repetitive seed layer of Ta 7/Cu 3 [nm] into the magnetostrictive multilayer to induce a {111} textured growth of the MnIr promoting the exchange bias effect (Suzuki & Ouchi, 2004; van Driel et al., 2000; Wiśniowski et al., 2006). However, in the stack with a total FeCo thickness of about 1 μm, cross-sectional transmission electron microscopy (TEM) revealed that there is a decrease in the degree of texture in the layers due to grain coarsening with increasing distance from the substrate/multilayer interface. Thus, in this example, a technologically conditioned upper limit for the total thickness of the sputter-deposited multilayers was set. In a consecutive step, the concept was transferred to AlN based composites with amorphous FeCoSiB layers which avoids grain coarsening. A further increase of the ME coefficient in the first bending mode and at zero field to $\alpha_{ME} = 96.7$ V /(cmOe) was achieved. Applied as a magnetic field sensor, a limit of detection LOD =

$10\ \mathrm{pT}/\mathrm{Hz}^{1/2}$ was demonstrated for resonant magnetic fields (Lage et al., 2012).

In order to achieve a well-defined exchange coupling across the entire interface without local variation in strength as well as direction, the magnetic layers are often subjected to a thermal treatment in a magnetic field after deposition (Nogués & Schuller, 1999; O'Grady et al., 2010). The treatment aims to thermally activate the antiferromagnetic grains by heating above the blocking temperature, so that subsequent cooling in a magnetic field can imprint a magnetic order in the ferromagnet/antiferromagnet system, which then leads to a reproducible shift in the ferromagnet's hysteresis. At the same time, for maximum ME effect, substantial crystallization of the amorphous ferromagnetic layers must be avoided (Yang et al., 2011). For the MnIr/FeCoSiB system, high resolution TEM studies on as-deposited films revealed that a thin crystalline layer forms at the interface, which extends into the FeCoSiB layer with 10 nm wide and 50 nm high polycrystalline filaments (Hrkac et al., 2014). Annealing at 250 °C for 30 minutes did not lead to structural changes, but to the desired unidirectional magnetic anisotropy. In contrast, the treatment at a temperature $T = 350\ ^\circ\mathrm{C}$ caused an increase in the crystalline volume fraction, which illustrates the responsivity of the layer system to the annealing temperature. Since the exchange bias coupling in such ferromagnet/antiferromagnet systems is initialized by the magnetization of the ferromagnet (Miltényi et al., 1999), an alternative to temperature treatment and subsequent field cooling is to grow the layers in the presence of a magnetic field (see, e.g., Lin et al., 1995). In such procedure, the ferromagnetic layer is first deposited and magnetized by the field, so that the growing antiferromagnet can also acquire an ordered state under the influence of the energy influx of the incoming film forming particles. A viable ferromagnetic seed layer for MnIr is NiFe (Soh et al., 2015), which can be inserted in the magnetostrictive multilayer of an ME composite between the amorphous ferromagnet and MnIr layers (Schell et al., 2023).

Due to their high sensitivity at zero magnetic field bias, composites with exchange biased multilayers can be easily arranged in dense arrays of ME devices. The general feasibility of such ME assembly was demonstrated on the basis of a two-dimensional vector magnetometer (Lage et al., 2013). The ME vector magnetometer was set up with two AlN-coated Si cantilevers rotated 90° to each other in the measurement plane. Serving as the magnetostrictive phase, a 10-fold repeating sequence of Ta 5/Cu 3/$Mn_{70}Ir_{30}$ 8/$Fe_{70.2}Co_{7.8}Si_{12}B_{10}$ 100 [nm] was used on the 3 mm× 25 mm beams. By field cooling, the unidirectional anisotropy was imprinted at an angle of 45° to the long flexural axis. Fig. 2.5 illustrates the sensor concept in a simplified single spin picture. The different response of the two sensors to an AC test field can be taken to determine the field

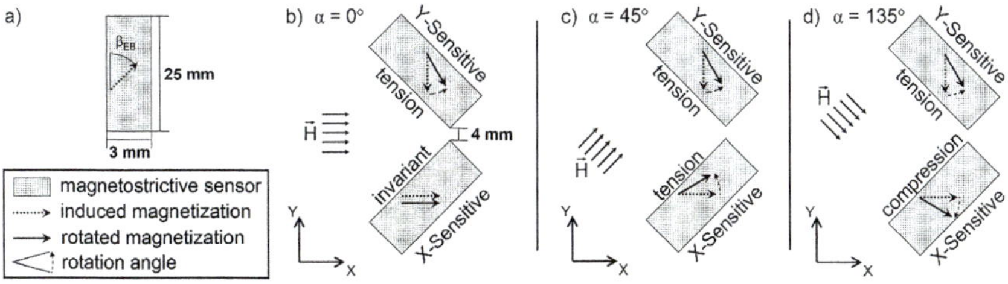

FIGURE 2.5 Concept of a two-dimensional ME vector magnetometer. (a) Sensing element designed as ME cantilever exhibiting a unidirectional anisotropy canted by $\beta_{EB} = 45°$ relating to the long lever axis. (b) Two sensing elements under the influence of an external field H. The magnetic field angle α indicates the alignment of the magnetic field to the x-axis of the sensor assembly. Due to the angled assembly, only the upper element responses. (c–d) Due to the symmetry, the ME voltage magnitude is the same for both elements for particular field angles α. The responses are either in-phase (c) or out-of-phase (d). *From Lage et al., 2013.*

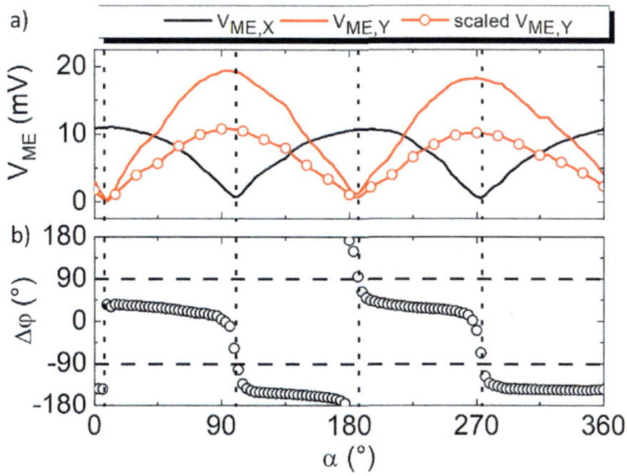

FIGURE 2.6 ME voltage of two sensing elements for an AC field of 1 Oe magnitude and at a frequency of 335 Hz. The 3 mm × 25 mm sized ME cantilevers are arranged as depicted in Fig. 2.5. (a) The signal amplitudes follow the absolute values of sine functions of the AC field angle α. To provide equal signal amplitudes, the y-component is deduced by scaling factor of 1.8. (b) Phase difference $\Delta\varphi$ between the two sensor element signals as a function of α. *From Lage et al., 2013.*

angle (Fig. 2.5b). However, the ME voltage magnitudes are always positive, so the inverse tangent of the ratio of both results in calculated angles between 0° and 90°. In order to be able to discriminate symmetrical cases as in Fig. 2.5c and d, the phase difference $\Delta\varphi$ between the two sensor signals had to be evaluated in parallel to the ME voltage amplitudes (Fig. 2.6). Finally, the discrimination of the cases with $\Delta\varphi \in [-90°; +90°]$ or

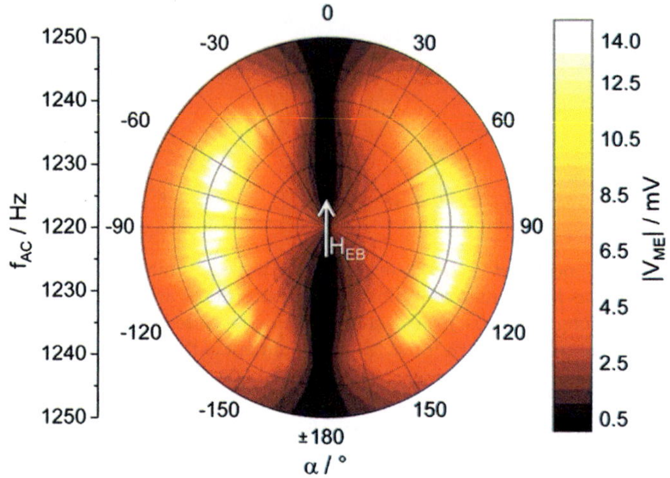

FIGURE 2.7 Voltage response at zero field of an exchange biased ME sensor. The angle α denotes the angle between unidirectional anisotropy and the direction of the applied AC magnetic field. The long cantilever axis of the ME sensor and the unidirectional anisotropy form an angle of 45°. *From Jahns et al., 2013*

$\Delta\varphi \notin [-90°; +90°]$ enabled the full range of magnetic field angles $\alpha = 0°$ to $\alpha = 180°$ for AC signals and demonstrated the successful detection of a two-dimensional magnetic vector.

A prerequisite for the successful implementation of the vector magnetometer was the directionality in the magnetic response of the individual sensor elements (Fig. 2.7). Due to the prealignment of magnetization in the exchange-biased layers, the magnetostrictive response to a small alternating magnetic field H_{AC} vanishes when the field is aligned in this energetically favored direction ($\alpha = 0°$ in Fig. 2.7). In this configuration, no torque acts on the magnetization and $\partial m/\partial H$ equals zero (cf. Eq. 2.3). Accordingly, perpendicular to the unidirectional anisotropy, the ME response is maximum. The measurements confirm that the ME voltage is approximately proportional to the vector product of H_{AC} and M and follows a sinusoidal function of the angle α. This simple concept presumes, however, that the magnetization can be described similar to a single vector, i.e., that it is decently homogeneous and not decomposed into domains.

A systematic study to investigate the critical thickness of FeCoSiB above which the concept for self-biased ME composites by exchange bias cannot be successfully implemented has been presented in (Lage et al., 2014). In this study, multiple AlN cantilevers, designed and fabricated in the same way as described above and in (Jahns et al., 2013), were coated with N layers of Ta/Cu/MnIr/FeCoSiB. The thickness t_{FM} of the individual FeCoSiB layers was chosen in such a way that $Nt_{FM} = 1$ μm was satisfied. The diagram of Fig. 2.8a graphs the ME coefficient measured in the first bending resonance at zero field as a function of the thickness t_{FM}. In addition, the reciprocal of the total anisotropy field $H_{K,tot}$ is shown. In the case of parallel alignment, $H_{K,tot}$ can be considered to be a combination of the contributions of the uniaxial magnetic anisotropy and the unidirectional exchange bias anisotropy. For small thicknesses, the ME coefficient initially follows the trend of increasing magnetic susceptibility, explained here by the decrease in the anisotropy field $H_{K,tot}$. At $t_{FM} =$ 333 nm, the coefficient reaches a maximum with 430 V/cmOe, and $H_{K,tot}$ remains almost unchanged compared to its magnitude for $t_{FM} = 200$ nm. Then, α_{ME} declines again with even greater film thicknesses. For these composites with high t_{FM}, the maximum ME coefficient is observed at nonzero magnetic field bias. To gain insight into the thickness-dependent behavior, the top magnetic layer was examined by magneto-optical Kerr effect (MOKE) microscopy (Fig. 2.8b–g). In magnetic multilayers without exchange bias coupling (Fig. 2.8b), demagnetization leads to a multidomain state. In contrast, the multilayers with $t_{FM} = 50$ nm and 100 nm remain in a single-domain state caused by the dominant exchange bias coupling. The systems with 200 nm single-layer thickness also exhibit a homogeneous magnetization in the center but form domains at the edge that are oriented approximately antiparallel to the unidirectional anisotropy. The demagnetizing field exceeds the exchange bias field in these edge regions, so that the domain formation is energetically favored. Above $t_{FM} = 200$ nm, multidomain states form that extend over the entire region of the magnetic layers. The unequal distribution of magnetic domains observed for the 333 nm FeCoSiB thickness can be attributed to the net effect of exchange bias coupling. The occurrence of curved magnetic domain walls in Fig. 2.8 indicates that the effective anisotropies are not fully aligned and subject to local variation.

In summary, the experiments reveal that the zero bias ME coefficient for the studied geometry of composites no longer follows the $1/H_{K,tot}$ dependence from layer thicknesses above 200 nm. The magnetization collapses into multidomain states and the self-bias concept by exchange bias fails.

In general, the occurrence of multidomain states is relevant from more fundamental aspects, as abrupt magnetization changes become likely. In particular, such jumps in magnetization can significantly limit the achievable performance of magnetic field sensors, a typical device application of ME composites. Therefore, the following section focuses on the role of magnetic domain formation in magnetic thin-film systems for ME sensors and the trade-off between large magnetic susceptibility and domain control.

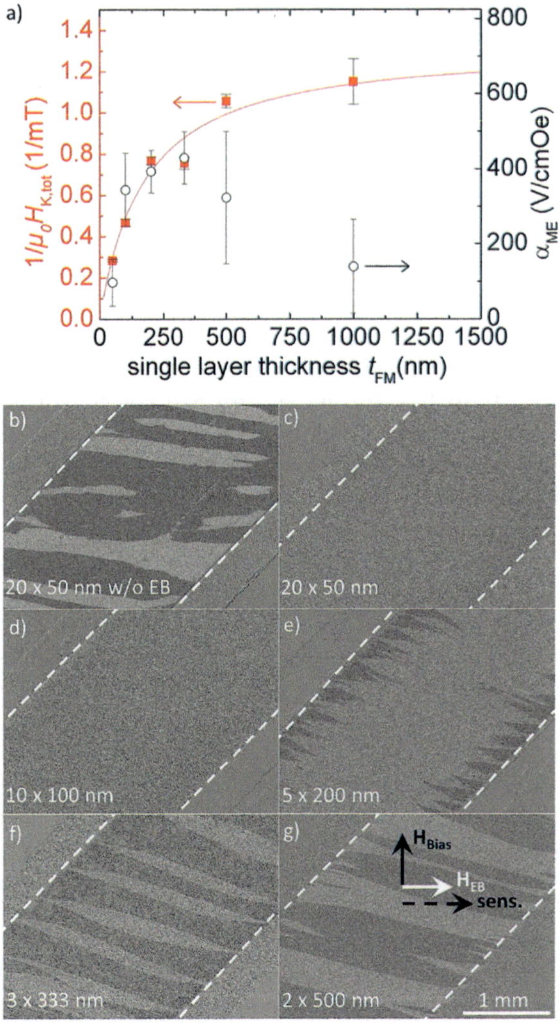

FIGURE 2.8 (a) Inverse of the total anisotropy field and ME voltage coefficient as a function of the thickness of the individual single ferromagnetic layer. (b–g) Top-layer domain configurations at zero magnetic field of magnetostrictive multilayers with different layer thicknesses. Prior to the imaging, a degaussing alternating field was applied perpendicular to the magnetic anisotropy to provoke a magnetic equilibrium state. All samples exhibit exchange bias coupling except for the sample in (b). The number of ferromagnetic layers N and the ferromagnetic layer thicknesses t_{FM} are indicated. The orientation of H_{EB} and the magneto-optical sensitivity direction (sens.) are shown in (f). An optional magnetic bias H_{Bias} applied during ME characterization is also displayed. *All subfigures (a–g) are from Lage et al., 2014.*

2.4.1 The antagonism of sensitivity versus noise—Magnetostrictive multilayers for magnetic domain control

The utilization of ME composites as magnetic field sensors is of great significance and drives enormous research efforts of various groups worldwide. The interest primarily emerged from the potentially very high sensitivity and the possibility of passive operation (Zhai, Xing, et al., 2006).

As highly sensitive magnetic field sensor, in addition to the high ME coupling effect (Eq. 2.2 in section *magnetostrictive thin film materials for magnetoelectrics*), further requirements arise depending on the application scenario such as low power consumption, small available space, high bandwidth, low cross-sensitivity to environmental changes, high linearity of sensor response, and low manufacturing costs. The decisive factor for successful practical application, however, is that the detection limit achieved in the frequency range of interest is sufficient to sense the desired magnetic field signals. The limit of detection (LOD), i.e., the smallest detectable field, is determined by the quotient of noise level and sensitivity. For better comparability, the LOD is usually set at the level of the magnetic field for which the signal-to-noise ratio is 1 at a measurement bandwidth of 1 Hz. In the development of magnetic field sensors for the highest level of sensitivity, the reduction of sensor noise is just as important as the increase in sensitivity. This section identifies the main noise sources of ME sensor systems in general, followed by a detailed discussion of the contribution of magnetic noise and strategies to reduce it in thin-film ME sensors used in direct detection mode or with magnetic frequency conversion. The focus is on loss mechanisms, as losses are related to stochastic changes in the sensor state and sensor noise via the fluctuation–dissipation theorem (Landau & Lifshitz, 1980). More complex, piezoelectrically modulated sensors such as magnetic surface acoustic wave sensors or converse ME sensors are discussed as high bandwidth sensor concepts in Chapter 9, *Magnetoelectric sensors based on magnetoelectric MEMS*.

Although an ME sensor system consists of a small number of components, a thorough noise analysis is not trivial and must take into account the mode of operation. Principally, a distinction is necessary between extrinsic and intrinsic sources of noise (Wang et al., 2014). The extrinsic sources include changes in ambient temperature, magnetic and electric stray fields, mechanical, and acoustic waves. Various kinds of sensors are subjected to these extrinsic influences and a number of established measures such as shielding and gradiometric arrangement of sensor units, and more specific measures such as mode separation, and tuning fork-like sensor assemblies have been applied to ME sensors (Jahns et al., 2013;

Salzer et al., 2016; Shen et al., 2013; Spetzler et al., 2021; Xing et al., 2009). Intrinsic noise sources of an ME sensor system are both in the ME sensor itself and in the readout electronics. Any kind of electronics with resistors is subjected to thermal fluctuation in the movement of charge carriers resulting in Johnson–Nyquist noise (Johnson, 1928; Nyquist, 1928). Other important electronic contributions are the shot noise (Schottky, 1918) that occurs when the discrete charge carriers flow across potential barriers and the flicker ($1/f$) noise that is due to fluctuations in the charge carrier density and mobility and is ubiquitous in MOS structures (Hung et al., 1990). The relevance of electronic noise for ME magnetic field sensors was realized at an early stage and modeled. An analysis for ME cantilever sensors has been provided by Wang et al. (Bichurin et al., 2021).

In an equivalent circuit diagram, the ME composite is to be considered as a combination of a charge source, a leakage resistance, and a capacitor. Consequently, the ME sensing element itself is afflicted with dielectric loss noise and leakage resistance noise. Fig. 2.9a illustrates an example of an equivalent circuit with the different noise contributions for the so-called direct detection of an alternating magnetic field by a resonantly tuned ME cantilever senor (Krantz & Gerken, 2020). This direct sensing concept is attractive due to its simplicity and the very good detection limits in the low $\mathrm{pt}/\mathrm{Hz}^{1/2}$ range for thin-film sensors (Piorra et al., 2013) and below

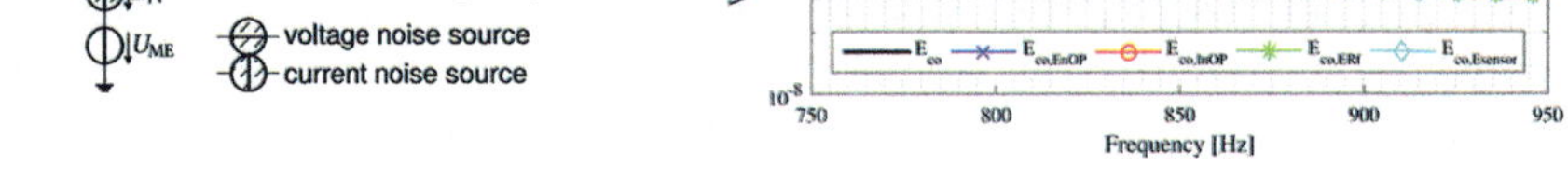

FIGURE 2.9 (a) Equivalent circuit diagram for the direct detection mode of an ME cantilever sensor system. On the readout electronics side, noise contributions from the operational amplifier (E_n: voltage noise, I_n: current noise), the feedback resistor (E_F) and the cabling (E_C) must be considered. On the ME sensor side, the contributions E_{ME} from the piezoelectric phase are considered as well as thermally induced noise E_{TV} from the mechanical oscillating circuit which is resonantly tuned to the desired AC magnetic field. (b) Calculated contributions to the overall voltage noise density E_{co}. The labels denote the voltage and current noise of the operational amplifier $E_{co,EnOP}$ and $E_{co,InOP}$, the thermal electrical noise of a feedback resistor $E_{co,ERf}$, and $E_{co,Esensor}$ the superimposed thermal noise contribution from the piezoelectric and the thermal–mechanical noise from the ME cantilever. At mechanical resonance, the thermal–mechanical contribution dominates $E_{co,Esensor}$ as it is amplified by the mechanical resonator. *(a) From Krantz & Gerken, 2020. (b) From Durdaut et al., 2017.*

for bulk-laminated ME composites (Zhai, Xing, et al., 2006). Using the equivalent circuit model and taking into account the transfer functions of the individual components, the noise voltage density U_{SC} at the output of the charge amplifier can be calculated.

On the ME sensor side, in addition to the noise contributions from the piezoelectric phase, thermally induced noise from the mechanical oscillating circuit is also taken into account (E_{TV} in Fig. 2.9a). Thermal–mechanical noise occurs in any micromechanical system (Djuric, 2000) and, in the context of miniature cantilevers, is a well-known issue from, e.g., atomic force microscopy (Albrecht et al., 1991; Butt & Jaschke, 1995). It appears because a mechanical resonator in thermal equilibrium gives off energy to its surroundings through losses. Due to the random movement of the gas molecules surrounding it, the energy transfer between cantilever and the gas medium is also random. As a result, a frequency-dependent deflection noise is observed which is converted via the piezoelectric to voltage noise. In addition, further intrinsic losses arise within the oscillating structure (Saulson, 1990) including thermoelastic friction (Prabhakar & Vengallatore, 2007), support (Hao et al., 2003), mounting (Lübbe et al., 2010), and surface losses (Sandberg et al., 2005). However, viscoelastic air damping (Chen et al., 1994) is typically the largest contributor to thermal–mechanical noise, especially when ME sensors are operated at resonance and at atmospheric pressure (Kirchhof et al., 2013). Even for centimeter-sized thin-film cantilever sensors with free-standing cantilever dimensions of 23 mm × 2.2 mm × 350 μm, the relevance of thermal-mechanical noise has been demonstrated (Fig. 2.9) (Durdaut et al., 2017). For the ME sensor investigated, there with a magnetostrictive layer stack consisting of 5-times repetitions of Ta 5/Cu 3/MnIr 8/FeCoSiB 200 [nm] and due to the amplification by mechanical resonance, the thermal–mechanical noise was dominant over thermal–electric noise of the piezoelectric and read-out electronics. Hence, thermal–mechanical noise has been identified as the contribution limiting the achievable LOD of such resonant ME cantilevers operated in the first bending mode (Durdaut et al., 2017; Krantz & Gerken, 2020). Thermally induced magnetization noise, another possible contribution to ME sensor noise, is also not limiting because of the dominant thermal–mechanical noise.

When operating the sensors, though, it should be considered that hysteretic properties of the piezomagnetic phase are transferred to the ME response. In particular, the sensor sensitivity does depend on the specific magnetic domain state of the magnetostrictive phase. Urs et al. investigated an ME thin-film sensor with a 4 μm thick single layer of FeCoSiB (Urs et al., 2014). The authors report a factor of two reduction in ME coefficient for equal magnetic bias but different field history. Thus, to achieve optimum sensor performance, an appropriate sensor initialization procedure is required to set the magnetic domain state in the FeCoSiB layer. By means of MOKE investigations, Urs et al. showed that magnetization reversal in the laterally extended sensor (20 mm × 2 mm) is influenced by local and edge effects. Although the magnetic anisotropy could be well defined and aligned by magnetic field cooling inside the sensor, the magnetization reversal for fields perpendicular to this anisotropy is not rotation dominated. Instead, transitions between different magnetic domain states with closure domains and spike domains are observed in the course of a magnetic field loop. Following the authors' explanations, one reason for the complex magnetization behavior is stress relaxation at the edge of the magnetostrictive layer, which rotates the effective anisotropy by 90° and aligns it with the structural edge (McCord, 2004).

The possibility that magnetoelastic effects can negatively influence the reproducibility of the response of magnetic sensors not only applies to ME composites but has also been demonstrated, for example, for magnetoresistive read heads with shields made of materials with different magnetostrictive strengths (Nakamoto et al., 1999). Consistently, in 1984, Narishige et al. realized a connection between read heads reproducibility, magnetic domain formation, and the sign of magnetostriction (Narishige et al., 1984).

Aside from the potentially hysteretic ME response and the LOD limitation by thermal–mechanical noise, the concept of direct detection by means of ME cantilever sensors also faces challenges in terms of bandwidth, which renders it unsuitable for a number of applications. The narrow bandwidth can be attributed to the fact that the high sensitivity resulting from resonance amplification is only attainable for fields within a frequency range determined by the resonance frequency and the quality factor of the mechanical resonator. This limitation is partially overcome by frequency conversion techniques in which the ME sensors are actively driven by an electrical or magnetic field to temporarily modulate the transfer function (Gillette et al., 2011; Hayes et al., 2016; Jahns et al., 2012). Demodulation can then be used to extract in particular low-frequency magnetic field signals from the sensor output. In the case of magnetic frequency conversion (MFC), a strong alternating magnetic field is used to drive the ME sensor along its nonlinear magnetostriction curve. The sensor thus acts as a frequency mixer, and in the additional presence of the field to be measured, sum and difference signals are generated. By selecting a specific excitation field frequency, these frequencies can be brought into resonance with the cantilever, again increasing sensitivity with the quality factor. However, such dynamic excitation potentially introduces new sources of noise that must be taken into account because, in general, further dynamic losses occur in the hysteretic magnetostrictive material under excitation.

Generally, the losses occurring in a soft magnetic material can be divided into quasi-static hysteresis loss, classical eddy current loss, and excess loss (Bertotti, 1988; Lin et al., 2004). Accordingly, the term excess loss covers all dynamic loss mechanisms except the eddy currents. MFC has been extensively investigated for the detection of low frequency signals in the range of a few Hertz to a few hundred Hertz, as this is the expected frequency range of challenging applications such as biomagnetism (Romani et al., 1982). In this frequency range, hysteresis losses predominate, but dynamic loss mechanisms may become relevant depending on the frequency of the modulation field. For their phenomenological description, the complex magnetic permeability $\mu = \mu' + i\mu''$ and the magnetic loss factor $\tan\delta_m = \frac{\mu''}{\mu'}$ are considered (Goodenough, 2002). In a field loop with driving frequency $2\pi f = \omega$, the hysteresis losses can then be determined by the integral over the magnetostrictive layer volume V (Lee & Romalis, 2008):

$$P_{hyst} = \int_V \frac{1}{2}\omega\mu'' H^2 dv. \tag{2.4}$$

A characteristic of magnetic losses is that their intensity depends on the specific domain states that are formed during the excitation. Furthermore, energy dissipation occurs in the transitions between different states due to irreversible processes such as domain wall nucleation and annihilation (Brailsford, 1970). Therefore, the quality factor of the modulated ME sensor is influenced by a magnetic bias and is lower than in the saturated magnetic state. For the same reason, it also depends on the excitation intensity. Consistently, magnetic bias–dependent quality factors have been reported for electrically modulated ME sensors (Durdaut et al., 2020). Furthermore, because the detection limit is determined by loss induced noise over sensitivity, the modulation field H_{mod} for the highest sensitivity of magnetically modulated ME sensors is usually not the same as that for the best LOD (see the example below in Fig. 2.10) (Urs et al., 2020).

A correlation between noise level and amplitude of an AC magnetic field is not a phenomenon limited to ME sensors and has been experimentally confirmed, e.g., for magnetoresistive elements by Jagielinski et al. (1986). A significant reduction of the noise level of magnetoresistive sensors can be achieved by selecting a suitable amplitude of the alternating field and by additionally applying a DC magnetic field bias (see also Decker and Tsang (1980)). Jagielinski et al. explained the noise level reduction by the circumstance that the DC field leads to a domain-free state and thus Barkhausen noise is avoided. Decker et al. find by Bitter pattern studies on 10 μm to 100 μm Permalloy structures that buckling

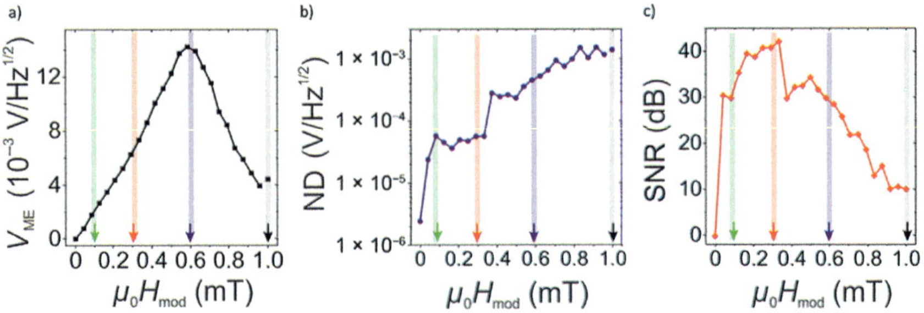

FIGURE 2.10 Properties of an ME cantilever modulated by a magnetic field H_{mod} of various amplitudes. The frequency of the field to be measured $\mu_0H_{mea} = 10^{-3}$mT was set to $f_{mea} = 10$ Hz. (a) The ME signal amplitude at 10 Hz reaches a maximum at $\mu_0H_{mod} =$ 0.6 mT (marked by the blue arrow in all diagrams). (b) The voltage noise density *ND* develops in a step-wise manner. After an initial rise, *ND* reaches a plateau at $\mu_0H_{mod} =$ 0.08 mT (green arrow) (c) resulting signal-to-noise ratio with maximum magnitude at $\mu_0H_{mod} = 0.333$ mT (red arrow), which is clearly below the modulation field amplitude with maximum sensitivity. At a magnetic field of 1 mT, the onset of magnetic saturation is reached (black arrow). *From Urs et al., 2020.*

domain patterns are prevented by a DC field, and therefore, the MR response becomes low noise.

For effective frequency conversion in ME sensors, the amplitude of the modulation field is often set in the order of magnitude of the magnetic anisotropy field to fully exploit the nonlinearity of the magnetostriction curve. This makes H_{mod} several orders of magnitude larger than the desired magnetic signal. The magnetically modulated ME sensor is therefore a predestinated model for the study of magnetically induced noise contributions. For example, Urs et al. (2020) used in-situ quasi-static and time-resolved MOKE microscopy combined with measurements of the noise voltage density, to identify specific domain processes that cause the occurrence of different category groups of modulation. These category groups differ in modulation field amplitude, level of noise voltage density, and achievable detection limit (see Fig. 2.10). In agreement with their investigations on resonant ME cantilevers (Lage et al., 2014), they find for modulated sensors with a 1 μm thick FeCoSiB layer that external magnetic fields lead via domain wall nucleation and annihilation to transitions between domain states characterized by very different magnetic domain periodicity. While large amplitudes of the modulation field result in high sensitivity, they also lead to hysteresis losses and a high noise voltage density due to the nonreversible magnetic processes. The optimal modulation field with respect to the signal-to noise ratio (SNR) and the LOD must therefore be chosen as large as possible on the one hand but on the other hand, must also be kept so small that transitions between domain states with different periodicities do not occur. Quintessence of the study is that magnetic interactions with structural defects are not a necessary condition for significant magnetic noise contributions. In the present case, the authors conclude, it is the aforementioned micromagnetic reorganization processes that are induced at high modulation fields and limit sensor performance. The obvious solution, then, is to control the magnetization state such that magnetization reversal can only happen through a reversible rotational process. This challenge also exists for other sensor concepts such as fluxgates and giant magnetoimpedance (GMI) sensors, whose operation requires alternating fields to excite the magnetization (Aschenbrenner & Goubau, 1936; Mohri et al., 1993; Primdahl, 1979). Accordingly, switching of Barkhausen volumes and domain wall displacements in the magnetization are reported to contribute to magnetically induced noise of parallel fluxgate magnetometers (Chu et al., 2017; Tejedor et al., 1994; van Bree et al., 1974). Consequently, Mattia Butta classifies the principle of the always saturated core to avoid magnetic noise contributions as essential to the concept of fundamental mode orthogonal fluxgates (Butta, 2017; Sasada, 2002). GMI sensors are also commonly operated with a DC bias to reduce sensor noise by eliminating complex domain states (Dolabdjian & Ménard, 2017).

Any solution to domain control in ME sensors must overcome the dilemma that approaches to stabilizing the magnetization state are likely to be accompanied by a reduction in magnetic susceptibility and sensor sensitivity (see eq. (2.2) and eq. (2.3) in section *Magnetostrictive thin film materials for magnetoelectrics*). Here, the approach of ME thin-film sensors again allows the use of magnetic exchange couplings but now to control the magnetization reversal by the internal couplings. The advantage of internal bias has already been explained in the previous section *Magnetostrictive multilayers for magnetoelectrics—exchange bias for zero field operation* and is less important for MFC sensors due to the external modulating AC field. However, to a certain extent, the layer concepts developed for MFC sensors can be applied to electrically modulated sensors, which opens up the possibility of low-noise magnetic field sensors without external magnetic supporting fields.

In this context and based on the findings for zero field operation (s. *Magnetostrictive multilayers for magnetoelectrics—exchange bias for zero field operation*), Röbisch et al. compare the MFC performance of two types of ME sensors in cantilever design that differ in the implementation of the magnetostrictive phase (Röbisch et al., 2015). One sensor has a single 3 μm thick FeCoSiB layer, the other a multilayer consisting of five repetitions of Ta 5/Cu 3/MnIr 8/FeCoSiB 200 [nm]. Both sensors feature AlN as piezoelectric material and are fabricated on 300 μm thick silicon to cantilevers with dimensions of 25.2 mm × 2.2 mm.

The diagrams in Fig. 2.11 summarize the main results of the sensor comparison with respect to sensitivity and detection limit. The sensor

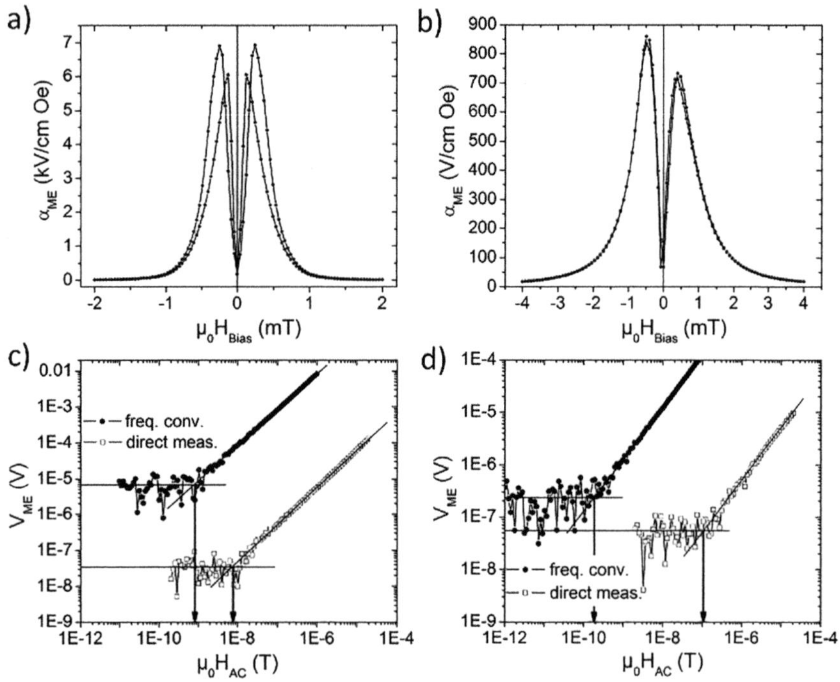

FIGURE 2.11 Magnetoelectric characterization of two thin-film composite sensors with either a single FeCoSiB layer (a, c) or an exchange biased multilayer (b, d). The diagrams in (a) and (b) display the ME coefficients in response to a resonant AC magnetic field with 1×10^{-3}Oe amplitude and as a function of magnetic bias. The graphs in (c) and (d) are plots of the ME voltage versus AC magnetic field with a frequency $f_{AC} = 10$ Hz. The ME voltage was measured by a lock-in amplifier with signal integration times of 3 s and 1 s for the direct and the frequency conversion measurements, respectively. The modulation field amplitude and magnetic bias were individually tuned for the single FeCoSiB and multilayer sensors. The arrows point to the minimum detectable AC field determined by linear fits of the ME voltage and averaged noise level. *From Röbisch et al., 2015.*

with single layer reaches its maximum resonant ME coefficient with $\alpha_{ME} = 6.9$ kV/cmOe at a magnetic field bias of ±2.4 Oe (Fig. 2.11a). At this operating point, the detection limit for resonant magnetic fields with a frequency of $f_{res} = 876$ Hz is low and approximately $1\ \mathrm{pT}/\mathrm{Hz}^{1/2}$ (not shown). However, the ME response is subject to hysteresis, and for low-frequency fields, the detection limit deteriorates considerably due to the lack of resonance effect and increased noise ($LOD \approx 10\ \mathrm{nT}/\mathrm{Hz}^{1/2}$ for 10 Hz and a signal integration time of 3 s). By modulating the sensor with a field $H_{mod} = 0.96$ Oe and $f_{mod} = f_{res} - 10$ Hz, the sensitivity can be substantially increased by three orders of magnitude. At the same time, however, the noise voltage density grows by about 2 orders of magnitude due to the magnetization excitation and the associated effects described above, so that the improvement of the LOD remains very limited (Fig. 2.11c). The exchange bias coupling in the ME sensor with the magnetic multilayer now alters the magnetization behavior of the magnetostrictive phase. The hysteresis in the ME response is significantly reduced (Fig. 2.11b), and MOKE imaging of the remanent magnetization state reveals closure spike domains confined to the marginal area of the sensor (Röbisch et al., 2015). It can therefore be assumed that the magnetization inside the sensor responds to the modulation field with a coherent rotation. Consistently, the noise increase under modulation remains moderate (Fig. 2.11d), and the detection limit improves by more than two orders of magnitude to an LOD approximately equal to $180\,\mathrm{pT}/\mathrm{Hz}^{1/2}$ at 10 Hz. Compared to the sensor with a single layer of FeCoSiB, the improvement is only about a factor of 5. This is due to the fact that the overall thickness of the FeCoSiB is reduced by a factor of 3, and the effective anisotropy field of the exchange-biased sensor is about twice as large, which illustrates the importance of the dilemma between magnetic stabilization and susceptibility reduction outlined above. The most straightforward approach is to increase the overall thickness of the magnetostrictive component of the magnetic phase by repeating the exchange biased layer sequence. Table 2.4 lists the results of such an experiment, in which the LODs of otherwise identical sensors with appropriately varied magnetostrictive phases were measured.

First it should be noted that the observed 800-fold degradation of the detection limit of the sensor with single FeCoSiB layer when changing from resonant fields to fields with 10 Hz frequency cannot be explained by the conversion losses alone but is also due to the increase in the noise voltage density caused by the modulation field and the periodic magnetization reversal (cf. Fig. 2.11c).

A second observation is that, as expected, the LOD of the exchange bias sensors in resonance improves approximately linearly with increasing total FeCoSiB layer thickness. However, it is noteworthy that the opposite trend is observed for the frequency conversion of the 10 Hz signals. For example, the LOD deteriorates from $180\,\mathrm{pT}/\mathrm{Hz}^{1/2}$ to $540\,\mathrm{pT}/\mathrm{Hz}^{1/2}$ when the number of magnetic layers is quadruplicated. The upscaling of the preferentially parallel magnetized layers inevitably leads to a strengthening of the demagnetizing field, which apparently results in magnetic domain formation and corresponding increased domain activity and thus a deterioration of the detection limit. Using fluxmetric magnetization loop tracer measurements on similar sensors, it was found that during a remagnetization process about one third of the magnetic

TABLE 2.4 Limit of detection at selected signal frequencies for thin-film ME cantilever sensors.

Operational mode	Magnetostrictive phase	Total FeCoSiB layer thickness	H_{bias}	Signal frequency	LOD
Resonance	Single FeCoSiB layer	3 μm	2.4 Oe	≈ 1 kHz	$1\ pT/Hz^{1/2}$
Resonance	5 × { Ta/Cu/MnIr/ FeCoSiB 200 nm}	1 μm	5 Oe	≈ 1 kHz	$10\ pT/Hz^{1/2}$
Resonance	10 × { Ta/Cu/MnIr/ FeCoSiB 200 nm}	2 μm	5 Oe	≈ 1 kHz	$6\ pT/Hz^{1/2}$
Resonance	20 × { Ta/Cu/MnIr/ FeCoSiB 200 nm}	4 μm	5 Oe	≈ 1 kHz	$1.5\ pT/Hz^{1/2}$
Frequency conversion	Single FeCoSiB layer	3 μm	N/A	10 Hz	$800\ pT/Hz^{1/2}$
Frequency conversion	5 × { Ta/Cu/MnIr/ FeCoSiB 200 nm}	1 μm	N/A	10 Hz	$180\ pT/Hz^{1/2}$
Frequency conversion	10 × { Ta/Cu/MnIr/ FeCoSiB 200 nm}	2 μm	N/A	10 Hz	$450\ pT/Hz^{1/2}$
Frequency conversion	20 × { Ta/Cu/MnIr/ FeCoSiB 200 nm}	4 μm	N/A	10 Hz	$540\ pT/Hz^{1/2}$

From Röbisch et al., 2017.

domains align antiparallel to the exchange bias and a multidomain state is formed (Jovičević Klug et al., 2019). The concept of parallel exchange bias multilayers meets its limits here. Upscaling to increase sensitivity while maintaining low noise can only succeed if an increase in the demagnetizing field is prevented.

One possibility is to bring the adjacent magnetic layers via exchange bias coupling not in parallel but into a well-ordered antiparallel state (Jovičević Klug et al., 2019). The magnetic flux then closes across the edges, and the demagnetizing field and the stray field energy are dramatically reduced. Upscaling is facilitated since it is irrelevant for the stray field energy per magnetic layer how many stray field compensated layer pairs are stacked on top of each other. Under the control of in-situ MOKE observations, Klug et al. established a temperature treatment that allows such a simultaneous adjustment of single-domain magnetization in the individual layers and layer-by-layer antiparallel magnetization. Key features of the procedure are that the layer stack containing the antiferromagnetic material MnIr is heated above its blocking temperature, integrally demagnetized by an alternating field, and then cooled in the field-free state. Furthermore, the multilayer is adjusted to the temperature treatment and composed of 20 pairs, which are slightly asymmetric in FeCoSiB thickness with Ta 5/Cu 3/MnIr 8/FeCoSiB 90/Ta 5/Cu 3/MnIr 8/FeCoSiB 110 [nm]. This asymmetry leads to layer-wise changing Zeeman energies that promote layer-by-layer antiparallel alignment and simultaneous integral demagnetization of the stack. If the sputter deposition system offers the necessary flexibility, the same magnetic alignment can also be achieved during deposition by applying opposing magnetic fields during multilayer growth. The positive effect of the stray field compensation can be seen in the layer-wise single-domain state, the achieved control of the magnetization behavior with dominant rotation as well as the associated magnetic noise level, which is an order of magnitude lower than for a comparable sensor with parallel exchange bias at the same modulation field amplitude (Jovičević Klug et al., 2019). However, due to the stabilizing effect of the magnetostatic coupling, the ME coefficient of the antiparallel sensor is reduced by a factor of 3. In the experiment, approximately equal detection limits of $120\,\mathrm{pT}/\mathrm{Hz}^{1/2}$ at 10 Hz are found for both sensor types (parallel and antiparallel), indicating the presence of a common noise source. A detailed analysis of the noise sources in the MFC measurements shows that the source of the modulation field must be considered (Salzer et al., 2018). By using a filter in the field source circuit, the pump noise was mitigated, and the 10 Hz detection limit of the sensor with antiparallel exchange bias was improved to $60\,\mathrm{pT}/\mathrm{Hz}^{1/2}$. Noise modeling further demonstrates that this achieved detection limit is very close to the limit given by thermal–mechanical noise at $40\ \mathrm{pT}/\mathrm{Hz}^{1/2}$ (Salzer et al., 2018).

In summary, the current state of the art is that magnetic noise contributions do not limit the detection limit for both resonant and MFC thin-film sensors. However, the implemented magnetic layer system must be matched to the sensor design and excitation. In the case of magnetically modulated sensors, a layer-by-layer single-domain state could be established by using magnetic multilayers and exploiting magnetic couplings across interfaces and edges. The magnetization process during modulation is therefore dominated by magnetization rotation and hysteretic losses are strongly reduced. Further improvement of the sensor performance is possible by increasing the effective piezomagnetic coefficient, increasing the magnetic volume fraction and reducing nonmagnetic noise contributions such as thermal–mechanical oscillations.

References

Acosta, A., et al. (2020). Underlayer effect on the soft magnetic, high frequency, and magnetostrictive properties of FeGa thin films. *Journal of Applied Physics, 128*(1).

Akiyama, M., et al. (2009). Enhancement of piezoelectric response in scandium aluminum nitride alloy thin films prepared by dual reactive cosputtering. *Advanced Materials, 21*(5), 593–596.

Albrecht, T. R., Grütter, P., Horne, D., & Rugar, D. (1991). Frequency modulation detection using high- Q cantilevers for enhanced force microscope sensitivity. *Journal of Applied Physics, 69*(2), 668–673.

Aschenbrenner, H., & Goubau, G. (1936). Eine Anordnung zur Registrierung rascher magnetischer Störungen. *Hochfrequenztechnik und Elektroakustik, 47*(6), 177–181.

Berlincourt, D., et al. (2003). Technical publication TP-226 properties of piezoelectricity ceramics. *Morgan electro ceramics, 12*.

Bertotti, G. (1988). General properties of power losses in soft ferromagnetic materials. *IEEE Transactions on Magnetics, 24*(1), 621–630.

Bichurin, M., Petrov, R., Sokolov, O., Leontiev, V., Kuts, V., Kiselev, D., et al. (2021). Magnetoelectric magnetic field sensors: A review. *Sensors, 21*(18), 6232.

Brailsford, F. (1970). Domain-wall energy and hysteresis loss in ferromagnetic material. *Proc. Inst. Electr. Eng. UK, 117*(5), 1052.

Butera, A., et al. (2005). Growth and magnetic characterization of epitaxial $Fe_{81}Ga_{19}/MgO$ (100) thin films. *Journal of Applied Physics, 98*(3).

Butt, H.-J., & Jaschke, M. (1995). Calculation of thermal noise in atomic force microscopy. *Nanotechnology, 6*(1), 1–7.

Butta, M. (2017). Orthogonal fluxgate magnetometers. In A. Grosz, M. J. Haji-Sheikh, & S. C. Mukhopadhyay (Eds.), *High sensitivity magnetometers* (pp. 63–102). Switzerland: Springer.

Camarero, J., Sort, J., Hoffmann, A., García-Martín, J. M., Dieny, B., Miranda, R., et al. (2005). Origin of the asymmetric magnetization reversal behavior in exchange-biased systems: Competing anisotropies. *Physical Review Letters, 95*(5), Article 57204.

Chauhan, S., et al. (2012). Multiferroic, magnetoelectric and optical properties of Mn doped $BiFeO_3$ nanoparticles. *Solid State Communications, 152*(6), 525–529.

Chen, X., et al. (2012). Structure, ferroelectric and piezoelectric properties of multiferroic Bi0. 875Sm0. 125FeO3 ceramics. *Journal of Alloys and Compounds, 541*, 173–176.

Chen, G. Y., Warmack, R. J., Thundat, T., Allison, D. P., & Huang, A. (1994). Resonance response of scanning force microscopy cantilevers. *Review of Scientific Instruments, 65*(8), 2532–2537.

Chu, Z., et al. (2018). Review of multi-layered magnetoelectric composite materials and devices applications. *Journal of Physics D: Applied Physics, 51*(24), Article 243001.

Chu, Z., Shi, H., Shi, W., Liu, G., Wu, J., Yang, J., et al. (2017). Enhanced resonance magnetoelectric coupling in (1-1) connectivity composites. *Advanced Materials, 29*(19), Article 1606022.

Churaev, M., Wang, R. N., Riedhauser, A., et al. (2023). A heterogeneously integrated lithium niobate-on-silicon nitride photonic platform. *Nature Communications, 14*.

Clark, A. E., et al. (2002). Magnetostrictive properties of Galfenol alloys under compressive stress. *Materials Transactions, 43*(5), 881–886.

Decker, S., & Tsang, C. (1980). Magnetoresistive response of small permalloy features. *IEEE Transactions on Magnetics, 16*(5), 643–645.

Ding, J. J., et al. (2020). Sputtering growth of low-damping yttrium-iron-garnet thin films. *Ieee Magnetics Letters, 11*.

Djuric, Z. (2000). Noise sources in microelectromechanical systems. In *2000 22nd International Conference on Microelectronics. Proceedings (Cat. No. 00TH8400)* (Vol. 1, pp. 85–96). IEEE.

Dolabdjian, C., & Ménard, D. (2017). Giant magneto-impedance (GMI) magnetometers. In A. Grosz, M. J. Haji-Sheikh, & S. C. Mukhopadhyay (Eds.), *High sensitivity magnetometers* (pp. 103–126). Switzerland: Springer.

Dong, C. Z., et al. (2018). Characterization of magnetomechanical properties in FeGaB thin films. *Applied Physics Letters, 113*(26).

Dong, S., Zhai, J., Li, J., & Viehland, D. (2006). Near-ideal magnetoelectricity in high-permeability magnetostrictive/piezofiber laminates with a (2-1) connectivity. *Applied Physics Letters, 89*(25), Article 252904.

Dubois, M.-A., & Muralt, P. (1999). Properties of aluminum nitride thin films for piezoelectric transducers and microwave filter applications. *Applied Physics Letters, 74*(20), 3032–3034.

Durdaut, P., Rubiola, E., Friedt, J.-M., Muller, C., Spetzler, B., Kirchhof, C., et al. (2020). Fundamental noise limits and sensitivity of piezoelectrically driven magnetoelastic cantilevers. *Journal of Microelectromechanical Systems, 29*(5), 1347–1361.

Durdaut, P., Salzer, S., Reermann, J., Robisch, V., Hayes, P., Piorra, A., et al. (2017). Thermal-mechanical noise in resonant thin-film magnetoelectric sensors. *IEEE Sensors Journal, 17*(8), 2338–2348.

Ehresmann, A., Koch, I., & Holzinger, D. (2015). Manipulation of superparamagnetic beads on patterned exchange-bias layer systems for biosensing applications. *Sensors, 15*(11), 28854–28888.

Engdahl, G., & Mayergoyz, I. D. (2000). *Handbook of giant magnetostrictive materials*. Elsevier.

Erhart, J., et al. (2017). *Piezoelectric ceramic resonators*. Springer.

Fichtner, S., et al. (2019). AlScN: A III-V semiconductor based ferroelectric. *Journal of Applied Physics, 125*(11).

Gao, J., Wang, Y., Li, M., Shen, Y., Li, J., & Viehland, D. (2012). Quasi-static (f<10−2Hz) frequency response of magnetoelectric composites based magnetic sensor. *Materials Letters, 85*, 84–87.

Giang, D. T. H., Duc, P. A., Ngoc, N. T., Hien, N. T., & Duc, N. H. (2012). Spatial angular positioning device with three-dimensional magnetoelectric sensors. *Review of Scientific Instruments, 83*(9), Article 95006.

Giebeler, C., Adelerhof, D. J., Kuiper, A., van Zon, J., Oelgeschläger, D., & Schulz, G. (2001). Robust GMR sensors for angle detection and rotation speed sensing. *Sensors and Actuators A: Physical, 91*(1–2), 16–20.

Gilbert, T. L. (2004). A phenomenological theory of damping in ferromagnetic materials. *IEEE Transactions on Magnetics, 40*(6), 3443–3449.

Gillette, S. M., Geiler, A. L., Gray, D., Viehland, D., Vittoria, C., & Harris, V. G. (2011). Improved sensitivity and noise in magneto-electric magnetic field sensors by use of modulated AC magnetostriction. *IEEE Magn. Lett., 2*, Article 2500104.

Goodenough, J. B. (2002). Summary of losses in magnetic materials. *IEEE Transactions on Magnetics, 38*(5), 3398–3408.

Greve, H., Woltermann, E., Quenzer, H.-J., Wagner, B., & Quandt, E. (2010). Giant magnetoelectric coefficients in (Fe90Co10)78Si12B10-AlN thin film composites. *Applied Physics Letters, 96*(18), Article 182501.

Guo, S., et al. (2022). Low damping at few-K temperatures in $Y_3Fe_5O_{12}$ epitaxial films isolated from Gd3Ga5O12 substrate using a diamagnetic $Y_3Sc_{2.5}Al_{2.5}O_{12}$ spacer. *Journal of Magnetism and Magnetic Materials, 562*, Article 169795.

Han, P., et al. (2005). Cut directions for the optimization of piezoelectric coefficients of lead magnesium niobate–lead titanate ferroelectric crystals. *Applied Physics Letters, 86*(5).

Hao, Z., Erbil, A., & Ayazi, F. (2003). An analytical model for support loss in micromachined beam resonators with in-plane flexural vibrations. *Sensors and Actuators A: Physical, 109*(1–2), 156–164.

Hayes, P., Salzer, S., Reermann, J., Yarar, E., Röbisch, V., Piorra, A., et al. (2016). Electrically modulated magnetoelectric sensors. *Applied Physics Letters, 108*(18), Article 182902.

He, Y., et al. (2021). Integrated magnetics and magnetoelectrics for sensing, power, RF, and microwave electronics. *IEEE Journal of Microwaves, 1*(4), 908–929.

Hempstead, R., Krongelb, S., & Thompson, D. (1978). Unidirectional anisotropy in nickel-iron films by exchange coupling with antiferromagnetic films. *IEEE Transactions on Magnetics, 14*(5), 521–523.

Heron, J. T., et al. (2014). Electric field control of magnetism using $BiFeO_3$-based heterostructures. *Applied Physics Reviews, 1*(2).

Hrkac, V., Lage, E., Köppel, G., Strobel, J., McCord, J., Quandt, E., et al. (2014). Amorphous FeCoSiB for exchange bias coupled and decoupled magnetoelectric multilayer systems: Real-structure and magnetic properties. *Journal of Applied Physics, 116*(13), Article 134302.

Hu, J., Jin, G., & Ma, Y. (2002). Thickness and angular dependencies of exchange bias in ferromagnetic/antiferromagnetic bilayers. *Journal of Applied Physics, 92*(2), 1009–1013.

Hung, K. K., Ko, P. K., Hu, C., & Cheng, Y. C. (1990). A unified model for the flicker noise in metal-oxide-semiconductor field-effect transistors. *IEEE Transactions on Electron Devices, 37*(3), 654–665.

Izuhara, T., et al. (2003). Single-crystal barium titanate thin films by ion slicing. *Applied Physics Letters, 82*, 616–618.

Jagielinski, T., Doyle, W., & Smith, N. (1986). The measurement of Barkhausen noise in magnetoresistive elements using the magnetoresistive susceptibility method. *IEEE Transactions on Magnetics, 22*(5), 680–682.

Jahns, R., Greve, H., Woltermann, E., Quandt, E., & Knöchel, R. (2012). Sensitivity enhancement of magnetoelectric sensors through frequency-conversion. *Sensors and Actuators A: Physical, 183*, 16–21.

Jahns, R., Piorra, A., Lage, E., Kirchhof, C., Meyners, D., Gugat, J. L., et al. (2013). Giant magnetoelectric effect in thin-film composites. *Journal of the American Ceramic Society, 96*(6), 1673–1681.

Jermain, C. L., et al. (2017). Increased low-temperature damping in yttrium iron garnet thin films. *Physical Review B, 95*(17).

Johnson, J. B. (1928). Thermal agitation of electricity in conductors. *Physics Reviews, 32*(1), 97–109.

Jovičević Klug, M., Thormählen, L., Röbisch, V., Toxværd, S. D., Höft, M., Knöchel, R., et al. (2019). Antiparallel exchange biased multilayers for low magnetic noise magnetic field sensors. *Applied Physics Letters, 114*(19), Article 192410.

Kamentsev, K. E., Fetisov, Y. K., & Srinivasan, G. (2006). Low-frequency nonlinear magnetoelectric effects in a ferrite-piezoelectric multilayer. *Applied Physics Letters, 89*(14), Article 142510.

Kim, I., et al. (2004). Effects of boron contents on magnetic properties of Fe-Co-B thin films. *IEEE Transactions on Magnetics, 40*(4), 2706–2708.

Kirchhof, C., Krantz, M., Teliban, I., Jahns, R., Marauska, S., Wagner, B., et al. (2013). Giant magnetoelectric effect in vacuum. *Applied Physics Letters, 102*(23), Article 232905.

Klokholm, E. (1976). The measurement of magnetostriction in ferromagnetic thin films. *IEEE Transactions on Magnetics, 12*(6), 819–821.

Krantz, M. C., & Gerken, M. (2020). Limit of thermal-vibration noise in magnetic field detection with magnetoelectric-composite cantilevers. *Physical Review Applied, 13*(5), Article 054047.

Kumar, A., Arout Chelvane, J., & Arockiarajan, A. (2021). Enhanced self-biased magnetoelectric response in novel distributed disc structure Ni/PZT composite. *Materials Letters, 305*, Article 130834.

Lübbe, J., Tröger, L., Torbrügge, S., Bechstein, R., Richter, C., Kühnle, A., et al. (2010). Achieving high effective Q -factors in ultra-high vacuum dynamic force microscopy. *Measurement Science and Technology, 21*(12), Article 125501.

Lage, E., Kirchhof, C., Hrkac, V., Kienle, L., Jahns, R., Knöchel, R., et al. (2012). Exchange biasing of magnetoelectric composites. *Nature Mater, 11*(6), 523–529.

Lage, E., Meyners, D., & Quandt, E. (2012). *Magnetostrictive layer system*. WO2012/175567 A1.

Lage, E., Urs, N. O., Röbisch, V., Teliban, I., Knöchel, R., Meyners, D., et al. (2014). Magnetic domain control and voltage response of exchange biased magnetoelectric composites. *Applied Physics Letters, 104*(13), Article 132405.

Lage, E., Woltering, F., Quandt, E., & Meyners, D. (2013). Exchange biased magnetoelectric composites for vector field magnetometers. *Journal of Applied Physics, 113*(17), Article 17C725.

Laletin, U., Sreenivasulu, G., Petrov, V. M., Garg, T., Kulkarni, A. R., Venkataramani, N., et al. (2012). Hysteresis and remanence in magnetoelectric effects in functionally graded magnetostrictive-piezoelectric layered composites. *Physical Review B: Condensed Matter, 85*(10), Article 104404.

Landau, L. D., & Lifshitz, E. M. (1980). *Statistical physics*. Oxford: Pergamon Press.

Lee, S.-K., & Romalis, M. V. (2008). Calculation of magnetic field noise from high-permeability magnetic shields and conducting objects with simple geometry. *Journal of Applied Physics, 103*(8), Article 84904.

Levy, M., et al. (1998). Crystal ion slicing of single-crystal magnetic garnet films. *Journal of Applied Physics, 83*(11), 6759–6761.

Li, M., Berry, D., Das, J., Gray, D., Li, J., & Viehland, D. (2011). Enhanced sensitivity and reduced noise floor in magnetoelectric laminate sensors by an improved lamination process. *Journal of the American Ceramic Society, 94*(11), 3738–3741.

Li, M. H., et al. (2017). Ultra-sensitive NEMS magnetoelectric sensor for picotesla DC magnetic field detection. *Applied Physics Letters, 110*(14).

Li, M., Wang, Y., Gao, J., Gray, D., Li, J., & Viehland, D. (2012). Dependence of magnetic field sensitivity of a magnetoelectric laminate sensor pair on separation distance: Effect of mutual inductance. *Journal of Applied Physics, 111*(3), Article 33923.

Li, M., Wang, Z., Wang, Y., Li, J., & Viehland, D. (2013). Giant magnetoelectric effect in self-biased laminates under zero magnetic field. *Applied Physics Letters, 102*(8), Article 82404.

Li, P., Wen, Y., Liu, P., Li, X., & Jia, C. (2010). A magnetoelectric energy harvester and management circuit for wireless sensor network. *Sensors and Actuators A: Physical, 157*(1), 100–106.

Liang, X., Dong, C., Celestin, S. J., Wang, X., Chen, H., Ziemer, K. S., et al. (2019). Soft magnetism, magnetostriction, and microwave properties of Fe-Ga-C alloy films. *IEEE Magn. Lett., 10*, 1–5.

Liang, X. F., et al. (2020). Mechanically driven SMR-based MEMS magnetoelectric antennas. In *2020 ieee international symposium on antennas and propagation and north American radio science meeting* (pp. 661–662).

Lim, L. C., et al. (2007). Characterization of flux-grown PZN-PT single crystals for high-performance piezo devices. *IEEE Transactions on Ultrasonics, Ferroelectrics, and Frequency Control, 54*(12), 2474–2478.

Lin, T., Tsang, C., Fontana, R. E., & Howard, J. K. (1995). Exchange-coupled Ni-Fe/Fe-Mn, Ni-Fe/Ni-Mn and NiO/Ni-Fe films for stabilization of magnetoresistive sensors. *IEEE Transactions on Magnetics, 31*(6), 2585–2590.

Lin, D., Zhou, P., Fu, W. N., Badics, Z., & Cendes, Z. J. (2004). A dynamic core loss model for soft ferromagnetic and power ferrite materials in transient finite element analysis. *IEEE Transactions on Magnetics, 40*(2), 1318–1321.

Ling, W., et al. (2017). Frequency-dependent complex magnetic permeability and magnetic losses of Fe-Ga alloy. In *2017 20th international conference on electrical machines and systems (ICEMS)*. IEEE.

Liu, M., et al. (2009). Giant electric field tuning of magnetic properties in multiferroic ferrite/ferroelectric heterostructures. *Advanced Functional Materials, 19*(11), 1826–1831.

Liu, M., et al. (2010). Electrical tuning of magnetism in Fe_3O_4/PZN-PT multiferroic heterostructures derived by reactive magnetron sputtering. *Journal of Applied Physics, 107*(7).

Liu, M., et al. (2012). Electrically induced enormous magnetic anisotropy in Terfenol-D/lead zinc niobate-lead titanate multiferroic heterostructures. *Journal of Applied Physics, 112*(6).

Liu, M. C., et al. (2018). Atomic-scale structure and chemistry of YIG/GGG. *AIP Advances, 8*(8).

Lou, J. (2010). Electrostatically tunable microwave multiferroic heterostructures with novel magnetic materials. In *P. D. Northeastern Univeristy.*

Lou, J., et al. (2007). Soft magnetism, magnetostriction, and microwave properties of FeGaB thin films. *Applied Physics Letters, 91*(18).

Lou, J., et al. (2009). Giant electric field tuning of magnetism in novel multiferroic FeGaB/lead zinc niobate-lead titanate (PZN-PT) heterostructures. *Advanced Materials, 21*(46), 4711–4715.

Luo, B., et al. (2024). Magnetoelectric microelectromechanical and nanoelectromechanical systems for the IoT. In *Nature Reviews Electrical Engineering* (pp. 1–18).

Maendl, S., et al. (2017). Spin waves with large decay length and few 100 nm wavelengths in thin yttrium iron garnet grown at the wafer scale. *Applied Physics Letters, 111*(1).

Makino, H., et al. (1976). Lattice parameter variations in czochralski grown gadolinium gallium garnet single crystals. *Japanese Journal of Applied Physics, 15*(3), 415–419.

Mandal, S. K., Sreenivasulu, G., Petrov, V. M., & Srinivasan, G. (2010). Flexural deformation in a compositionally stepped ferrite and magnetoelectric effects in a composite with piezoelectrics. *Applied Physics Letters, 96*(19), Article 192502.

Mayrhofer, P., et al. (2016). Properties of $Sc_xAl_{1-x}N$ ($x = 0.27$) thin films on sapphire and silicon substrates upon high temperature loading. *Microsystem Technologies, 22*, 1679–1689.

McCord, J. (2004). Irregular domain patterns in structured magnetic thick films. *Journal of Applied Physics, 95*(11), 6855–6857.

Meiklejohn, W. H., & Bean, C. P. (1957). New magnetic anisotropy. *Physics Reviews, 105*(3), 904–913.

Meisenheimer, P. B., et al. (2021). Engineering new limits to magnetostriction through metastability in iron-gallium alloys. *Nature Communications, 12*(1).

Miltényi, P., Gierlings, M., Bamming, M., May, U., Güntherodt, G., Nogués, J., et al. (1999). Tuning exchange bias. *Applied Physics Letters, 75*(15), 2304–2306.

Mishra, R., et al. (2008). Effect of yttrium on improvement of dielectric properties and magnetic switching behavior in $BiFeO_3$. *Journal of Physics: Condensed Matter, 20*(4), Article 045218.

Mohri, K., Kawashima, K., Kohzawa, T., & Yoshida, H. (1993). Magneto-inductive element. *IEEE Transactions on Magnetics, 29*(2), 1245–1248.

Nakamoto, K., Narumi, S., Kawabe, T., Kobayashi, T., & Fukui, H. (1999). Shield-related signal instability in magnetoresistive heads. *Journal of Applied Physics, 85*(8), 5846–5848.

Nan, C.-W., Bichurin, M. I., Dong, S., Viehland, D., & Srinivasan, G. (2008). Multiferroic magnetoelectric composites: Historical perspective, status, and future directions. *Journal of Applied Physics, 103*(3).

Nan, T. X., et al. (2014). Quantification of strain and charge co-mediated magnetoelectric coupling on ultra-thin Permalloy/PMN-PT interface. *Scientific Reports, 4.*

Nan, T., et al. (2017). Acoustically actuated ultra-compact NEMS magnetoelectric antennas. *Nature Communications, 8.*

Narishige, S., Hanazono, M., Takagi, M., & Kuwatsuka, S. (1984). Measurements of magnetic characteristics of thin film heads using magneto-optical method. *IEEE Transactions on Magnetics, 20*(5), 848–850.

Nogués, J., & Schuller, I. K. (1999). Exchange bias. *Journal of Magnetism and Magnetic Materials, 192*(2), 203–232.

Nyquist, H. (1928). Thermal agitation of electric charge in conductors. *Physics Reviews, 32*(1), 110–113.

O'Grady, K., Fernandez-Outon, L. E., & Vallejo-Fernandez, G. (2010). A new paradigm for exchange bias in polycrystalline thin films. *Journal of Magnetism and Magnetic Materials, 322*(8), 883–899.

Ohkoshi, M., Tamari, K., Harada, M., Honda, S., & Kusuda, T. (1985). Microstructure and exchange anisotropy of Co-CoO sputtered films with perpendicular magnetization. *IEEE Translation Journal on Magnetics in Japan, 1*(1), 37–38.

Olsson, R. H., et al. (2020). Doping of aluminum nitride and the impact on thin film piezoelectric and ferroelectric device performance. In *2020 IEEE custom integrated circuits conference (CICC)*. IEEE.

Onuta, T.-D., Wang, Y., Long, C. J., & Takeuchi, I. (2011). Energy harvesting properties of all-thin-film multiferroic cantilevers. *Applied Physics Letters, 99*(20), Article 203506.

Parkin, S. S. P., Roche, K. P., Samant, M. G., Rice, P. M., Beyers, R. B., Scheuerlein, R. E., et al. (1999). Exchange-biased magnetic tunnel junctions and application to nonvolatile magnetic random access memory (invited). *Journal of Applied Physics, 85*(8), 5828–5833.

Parsapour, F., et al. (2018). Material constants extraction for AlScN thin films using a dual mode baw resonator. In *2018 IEEE micro electro mechanical systems (MEMS)*. IEEE.

Petrov, V. M., & Srinivasan, G. (2008). Enhancement of magnetoelectric coupling in functionally graded ferroelectric and ferromagnetic bilayers. *Physical Review B: Condensed Matter, 78*(18), Article 184421.

Pettiford, C., Lou, J., Russell, L., & Sun, N. X. (2008). Strong magnetoelectric coupling at microwave frequencies in metallic magnetic film/lead zirconate titanate multiferroic composites. *Applied Physics Letters, 92*(12), Article 122506.

Piazza, G., et al. (2012). Piezoelectric aluminum nitride thin films for microelectromechanical systems. *MRS Bulletin, 37*(11), 1051–1061.

Piorra, A., Jahns, R., Teliban, I., Gugat, J. L., Gerken, M., Knöchel, R., et al. (2013). Magnetoelectric thin film composites with interdigital electrodes. *Applied Physics Letters, 103*(3), Article 32902.

Platt, C. L., et al. (2001). Magnetic and structural properties of FeCoB thin films. *IEEE Transactions on Magnetics, 37*(4), 2302–2304.

Prabhakar, S., & Vengallatore, S. (2007). Thermoelastic damping in bilayered micromechanical beam resonators. *Journal of Micromechanics and Microengineering, 17*(3), 532–538.

Primdahl, F. (1979). The fluxgate magnetometer. *Journal of Physics E: Scientific Instruments, 12*(4), 241.

Röbisch, V., Salzer, S., Urs, N. O., Reermann, J., Yarar, E., Piorra, A., et al. (2017). Pushing the detection limit of thin film magnetoelectric heterostructures. *Journal of Materials Research, 32*(6), 1009–1019.

Röbisch, V., Yarar, E., Urs, N. O., Teliban, I., Knöchel, R., McCord, J., et al. (2015). Exchange biased magnetoelectric composites for magnetic field sensor application by frequency conversion. *Journal of Applied Physics, 117*(17), 17B513.

Rabiei, P., & Steier, W. H. (2005). Lithium niobate ridge waveguides and modulators fabricated using smart guide. *Applied Physics Letters, 86*(16).

Ren, Y., Ouyang, J., Wang, W., Wu, X., Yang, X., Zhang, Y., et al. (2018). Rotating magnetoelectric sensor for DC magnetic field measurement. *IEEE Transactions on Magnetics, 54*(11), 1–3.

Romani, G. L., Williamson, S. J., & Kaufman, L. (1982). Biomagnetic instrumentation. *Review of Scientific Instruments, 53*(12), 1815–1845.

Salzer, S., Jahns, R., Piorra, A., Teliban, I., Reermann, J., Höft, M., et al. (2016). Tuning fork for noise suppression in magnetoelectric sensors. *Sensors and Actuators A: Physical, 237*, 91–95.

Salzer, S., Röbisch, V., Klug, M., Durdaut, P., McCord, J., Meyners, D., et al. (2018). Noise limits in thin-film magnetoelectric sensors with magnetic frequency conversion. *IEEE Sensors Journal, 18*(2), 596–604.

Sanchez-Rojas, J.-L., et al. (2008). Advanced determination of piezoelectric properties of AlN thin films on silicon substrates. In *2008 IEEE ultrasonics symposium*. IEEE.

Sandberg, R., Mølhave, K., Boisen, A., & Svendsen, W. (2005). Effect of gold coating on the Q -factor of a resonant cantilever. *Journal of Micromechanics and Microengineering, 15*(12), 2249–2253.

Sasada, I. (2002). Orthogonal fluxgate mechanism operated with dc biased excitation. *Journal of Applied Physics, 91*(10), 7789.

Saulson, P. R. (1990). Thermal noise in mechanical experiments. *Physical Review D – Particles and Fields, 42*(8), 2437–2445.

Schell, V., et al. (2020). Magnetic anisotropy controlled FeCoSiB thin films for surface acoustic wave magnetic field sensors. *Applied Physics Letters, 116*(7).

Schell, V., Spetzler, E., Wolff, N., Bumke, L., Kienle, L., McCord, J., et al. (2023). Exchange biased surface acoustic wave magnetic field sensors. *Scientific Reports, 13*(1), 8446.

Schottky, W. (1918). Über spontane Stromschwankungen in verschiedenen Elektrizitätsleitern. *Annalen der Physik, 362*(23), 541–567.

Shen, Y., McLaughlin, K., Gao, J., Li, M., Li, J., & Viehland, D. (2013). Effective optimization of magnetic noise for a Metglas/$Pb(Zr,Ti)O_3$ magnetoelectric sensor array in an open environment. *Materials Letters, 91*, 307–310.

Shigematsu, E., et al. (2016). Significant reduction in spin pumping efficiency in a platinum/yttrium iron garnet bilayer at low temperature. *Applied Physics Express, 9*(5), Article 053002.

Soh, W. T., Yeow, Y., Zhong, X., & Ong, C. K. (2015). Inverse spin Hall effect of antiferromagnetic MnIr in exchange biased NiFe/MnIr films. *Journal of Physics D Applied Physics, 48*(34), Article 345002.

Spaldin, N. A., & Ramesh, R. (2019). Advances in magnetoelectric multiferroics. *Nature Materials, 18*(3), 203–212.

Spencer, E. G., & Clogston, A. M. (1959). Low-temperature lin-width maximum in yttrium iron garnet. *Physical Review Letters, 3.*

Spetzler, B., Bald, C., Durdaut, P., Reermann, J., Kirchhof, C., Teplyuk, A., et al. (2021). Exchange biased delta-E effect enables the detection of low frequency pT magnetic fields with simultaneous localization. *Scientific Reports, 11*(1), 5269.

Srinivasan, G., Rasmussen, E. T., Levin, B. J., & Hayes, R. (2002). Magnetoelectric effects in bilayers and multilayers of magnetostrictive and piezoelectric perovskite oxides. *Physical Review B: Condensed Matter, 65*(13), Article 134402.

Srisukhumbowornchai, N., & Guruswamy, S. (2001). Large magnetostriction in directionally solidified FeGa and FeGaAl alloys. *Journal of Applied Physics, 90*(11), 5680–5688.

Su, J. X., et al. (2020). AlScN-based MEMS magnetoelectric sensor. *Applied Physics Letters, 117*(13).

Suzuki, T., & Ouchi, K. (2004). Crystal orientation of Mn–Ir antiferromagnetic films for exchange-biased soft-magnetic underlayer in perpendicular recording media. *IEEE Transactions on Magnetics, 40*(4), 2347–2349.

Tasnadi, F., et al. (2010). Origin of the anomalous piezoelectric response in wurtzite $Sc_xAl_{1-x}N$ alloys. *Physical Review Letters, 104*(13).

Tejedor, M., Hernando, B., & Sánchez, M. L. (1994). Reversible permeability for perpendicularly superposed induction in metallic glasses for fluxgate sensors. *Journal of Magnetism and Magnetic Materials, 133*(1–3), 338–341.

Tsang, C. (1984). Magnetics of small magnetoresistive sensors (invited). *Journal of Applied Physics, 55*(6), 2226–2231.

Tsubouchi, K., et al. (1981). AlN material constants evaluation and SAW properties on AlN/Al_2O_3 and AlN/Si. In *1981 ultrasonics symposium*. IEEE.

Tu, C., et al. (2019). Mechanical-resonance-enhanced thin-film magnetoelectric heterostructures for magnetometers, mechanical antennas, tunable RF inductors, and filters. *Materials, 12*(14), 2259.

Turutin, A. V., Vidal, J. V., Kubasov, I. V., Kislyuk, A. M., Malinkovich, M. D., Parkhomenko, Y. N., et al. (2018). Magnetoelectric metglas/bidomain y + 140°-cut lithium niobate composite for sensing fT magnetic fields. *Applied Physics Letters, 112*(26), Article 262906.

Uchino, K. (2017). *Advanced piezoelectric materials: Science and technology.* Woodhead Publishing.

Urs, N. O., Golubeva, E., Röbisch, V., Toxvaerd, S., Deldar, S., Knöchel, R., et al. (2020). Direct link between specific magnetic domain activities and magnetic noise in modulated magnetoelectric sensors. *Physical Review Applied, 13*(2), Article 24018.

Urs, N. O., Teliban, I., Piorra, A., Knöchel, R., Quandt, E., & McCord, J. (2014). Origin of hysteretic magnetoelastic behavior in magnetoelectric 2-2 composites. *Applied Physics Letters, 105*(20), Article 202406.

van Bree, J. L. M. J., Poulis, J. A., & Hooge, F. N. (1974). Barkhausen noise in fluxgate magnetometers. *Applied Scientific Research, 29*(1), 59–68.

van Driel, J., Boer, FR de, Lenssen, K.-M. H., & Coehoorn, R. (2000). Exchange biasing by Ir19Mn81: Dependence on temperature, microstructure and antiferromagnetic layer thickness. *Journal of Applied Physics, 88*(2), 975–982.

Vaz, C. A. (2012). Electric field control of magnetism in multiferroic heterostructures. *Journal of Physics: Condensed Matter, 24*(33), Article 333201.

Villari, E. (1865). Ueber die Aenderungen des magnetischen Moments, welche der Zug und das Hindurchleiten eines galvanischen Stroms in einem Stabe von Stahl oder Eisen hervorbringen. *Annalen der Physik, 202*(9), 87–122.

Vopsaroiu, M., Blackburn, J., & Cain, M. G. (2007). A new magnetic recording read head technology based on the magneto-electric effect. *Journal of Physics D Applied Physics, 40*(17), 5027–5033.

Wang, J., Dong, C., Wei, Y., Lin, X., Athey, B., Chen, Y., et al. (2019). Magnetostriction, soft magnetism, and microwave properties in Co–Fe–C alloy films. *Physical Review Applied, 12*(3), Article 034011.

Wang, X., et al. (2018). Size-dependent magnetic properties of FeGaB/Al_2O_3 multilayer micro-islands. *Physics Letters A, 382*(23), 1505–1508.

Wang, Y., Gao, J., Li, M., Hasanyan, D., Shen, Y., Li, J., et al. (2012). Ultralow equivalent magnetic noise in a magnetoelectric Metglas/Mn-doped $Pb(Mg_{1/3}Nb_{2/3})O_3$-$PbTiO_3$ heterostructure. *Applied Physics Letters, 101*(2), Article 22903.

Wang, Y. J., Gao, J. Q., Li, M. H., Shen, Y., Hasanyan, D., Li, J. F., et al. (2014). A review on equivalent magnetic noise of magnetoelectric laminate sensors. *Philosophical transactions. Series A, Mathematical, physical, and engineering sciences, 372*(2009), Article 20120455.

Wiśniowski, P., Stobiecki, T., Kanak, J., Reiss, G., & Brückl, H. (2006). Influence of buffer layer texture on magnetic and electrical properties of IrMn spin valve magnetic tunnel junctions. *Journal of Applied Physics, 100*(1), Article 13906.

Will-Cole, A. R., et al. (2022). Application of bayesian optimization and regression analysis to ferromagnetic materials development. *IEEE Transactions on Magnetics, 58*.

Will-Cole, A. R., et al. (2023). Negligible magnetic losses at low temperatures in liquid phase epitaxy grown $Y_3F_5O_{12}$ films. *Physical Review Materials, 7*(5), Article 054411.

Xiao, F.-Q., et al. (2007). Out-of-plane magnetic anisotropy and microwave permeability of magnetoelastic FeCoSiB amorphous thin films. *Journal of Electronic Science and Technology, 5*(3), 226–229.

Xing, Z., Zhai, J., Li, J., & Viehland, D. (2009). Investigation of external noise and its rejection in magnetoelectric sensor design. *Journal of Applied Physics, 106*(2), Article 24512.

Yang, J. J., et al. (2009). Electric field manipulation of magnetization at room temperature in multiferroic $CoFe_2O_4/Pb(Mg_{1/3}Nb_{2/3})_{(0.7)}Ti_{0.3}O_3$ heterostructures. *Applied Physics Letters, 94*(21).

Yang, Y., Gao, J., Wang, Z., Li, M., Li, J.-F., Das, J., et al. (2011). Effect of heat treatment on the properties of Metglas foils, and laminated magnetoelectric composites made thereof. *Materials Research Bulletin, 46*(2), 266–270.

Yang, Z., & Zu, J. (2016). Comparison of PZN-PT, PMN-PT single crystals and PZT ceramic for vibration energy harvesting. *Energy Conversion and Management, 122*, 321–329.

Yu, H. M., et al. (2014). Magnetic thin-film insulator with ultra-low spin wave damping for coherent nanomagnonics. *Scientific Reports, 4*.

Zhai, J., Dong, S., Xing, Z., Li, J., & Viehland, D. (2006). Giant magnetoelectric effect in Metglas/polyvinylidene-fluoride laminates. *Applied Physics Letters, 89*(8), Article 83507.

Zhai, J., Xing, Z., Dong, S., Li, J., & Viehland, D. (2006). Detection of pico-Tesla magnetic fields using magneto-electric sensors at room temperature. *Applied Physics Letters, 88*(6).

Zhang, Z., et al. (2020). Coupling of magneto-strictive FeGa film with single-crystal diamond MEMS resonator for high-reliability magnetic sensing at high temperatures. *Materials Research Letters, 8*(5), 180–186.

Zhang, J., Li, P., Wen, Y., He, W., Yang, A., Wang, D., et al. (2014). Giant self-biased converse magnetoelectric effect in multiferroic heterostructure with single-phase magnetostrictive materials. *Applied Physics Letters, 105*(17), Article 172408.

Zhou, Y., Chul Yang, S., Apo, D. J., Maurya, D., & Priya, S. (2012). Tunable self-biased magnetoelectric response in homogenous laminates. *Applied Physics Letters, 101*(23), Article 232905.

Zhou, W. P., et al. (2014). Electric field manipulation of magnetic and transport properties in $SrRuO_3/Pb(Mg_{1/3}Nb_{2/3})O_{-3}$-$PbTiO_3$ heterostructure. *Scientific Reports, 4*.

Zhou, Y., Maurya, D., Yan, Y., Srinivasan, G., Quandt, E., & Priya, S. (2016). Self-biased magnetoelectric composites: An overview and future perspectives. *Energy Harvesting and Systems, 3*(1), 1–42.

CHAPTER

3

Magnetoelectric characterization techniques

Gopalan Srinivasan[1], *Bin Luo*[2], *Cunzheng Dong*[2], *Rui Huang*[2], *and Nian Sun*[2]

[1] **Oakland University, Rochester, MI, United States;** [2] **Northeastern University, Boston, MA, United States**

3.1 Introduction

Measurement techniques on magnetic field and electric field responses of the multiferroic composites are considered here. A wide variety of techniques for frequencies of excitation fields ranging from DC to 110 GHz could be used (Nan et al., 2008). There are two broad categories of ME studies: (i) sample response to magnetic fields, termed direct ME effect (DME) and (ii) response to an electric field, called converse ME effect (CME). Several techniques have been reported for studies on both direct and converse ME coupling (Vopson et al., 2017). Investigations on the strengths of DME include low-frequency ME voltage coefficient and at the bending and longitudinal acoustic resonance modes, nonlinear ME effects for large ac magnetic fields, static magnetic field induced polarization, and magneto-dielectric effects. For CME, one could measure low-frequency ME effects by applying an ac magnetic field and measuring the induced magnetic flux in a coil wound around the composite, static electric field E induced magnetization or anisotropy field, E-tuning of inductance or ferromagnetic resonance, and magneto-optic Kerr effect under an applied electric field (Vopson et al., 2017). Scanning probe microscopy is the preferred techniques for ME measurements on nanoparticles, *1D* structures such as nanotubes and core-shell wires, and *2D* structures such as nanobilayers and nanopillars in a host matrix (Viehland et al., 2018). These techniques are discussed in the following sections.

Magnetoelectric Composites
https://doi.org/10.1016/B978-0-443-18606-6.00014-9

The integration of magnetostrictive thin films with piezoelectric materials (magnetoelectric heterostructure) into nanoelectromechanical systems (NEMS) technology has achieved great advantage in magnetic field detection and RF communication systems (Nan et al., 2013; Tianxiang Nan et al., 2017; Li et al., 2017; al., , 2003; Lou et al., 2009), owing to their high sensitivity and high coupling efficiency with a small size and low power consumption. The characterization of saturation magnetostriction λ_s, one of the significant figures of merit of magnetostrictive thin films, is also discussed.

3.2 Direct ME effects

3.2.1 Low-frequency ME effects

A parameter of importance for the DME is the magnetoelectric voltage coefficient (MEVC), i.e., the induced polarization resulting from an applied ac magnetic field. The induced polarization δP is related to the ac magnetic field δH by the expression, $\delta P = \alpha\ \delta H$, where α is the second rank ME-susceptibility tensor and is expressed in s/m in SI units (in Gaussian units $\alpha = 4\pi\ \delta P/\delta H$ is dimensionless). One generally determines $\alpha = \delta P/\delta H$ by subjecting the sample to a bias field H and an ac field δH and measuring δP or the electric field δE. The ME voltage coefficient (MEVC) $\alpha_E = \delta E/\delta H$ (expressed in V/A or V/cm Oe) is related to α by $\alpha = \varepsilon_0\ \varepsilon_r\ \alpha_E$, where ε_r is the relative permittivity of the material. For such measurements it is necessary to first pole the ferroelectric phase in the composite in an electric field. In addition to the ac magnetic field δH, a DC bias H is essential for measurements under a large magnetostriction in the magnetic phase so that the strength of ME coupling is enhanced. The need for DC bias is illustrated in Fig. 3.1 in which schematic variation of

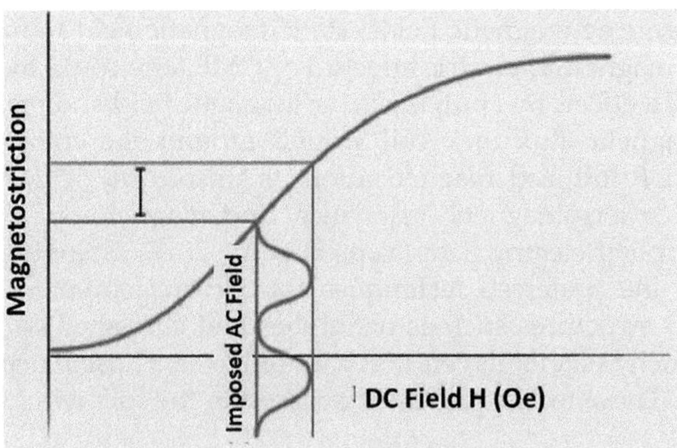

FIGURE 3.1 AC magnetostriction under DC bias H and AC magnetic field. position sensor.

magnetostriction λ of the ferromagnet is shown as a function of H. For $H = 0$, the mechanical deformation under the ac magnetic field (or AC magnetostriction) is rather small and $\alpha_E \approx 0$. With a bias field, however, the AC magnetostriction can be substantially enhanced. It is also clear from Fig. 3.1 that $\alpha_E \approx 0$ when the magnetostriction attains saturation.

Fig. 3.2 shows a typical set up for the measurement of MEVC (Vopson et al., 2017). The sample is subjected to a bias field H (with the use of an electromagnet) and an ac field δH produced by a pair of Helmholtz coils. The coils must not be wound (or mounted) on the pole pieces of the magnet. The sample must be shielded from picking up noise by placing it inside a metal box.

It is necessary to use a 3-terminal network for measurement of the differential voltage across the sample. Lock-in-detection is essential for accurate determination of MEVC. One can measure MEVC as a function of the bias field H, frequency and amplitude of δH, applied field directions, and temperature. Representative MEVC dependence on H, f, and temperature for a multilayer composite of nickel ferrite and PZT are shown in Fig. 3.3 (Srinivasan et al., 2001). The MEVC is relatively high when H and δH are parallel to each other and to the sample plane compared to out-of-plane magnetic fields due to demagnetization associated with out-of-plane fields.

3.2.2 ME effects at acoustic resonance modes (1 kHz–2 MHz)

A powerful tool for investigations on the nature of ME interactions is measurements at electromechanical resonances that could occur, depending on the sample dimensions, over a few kHz to several MHz

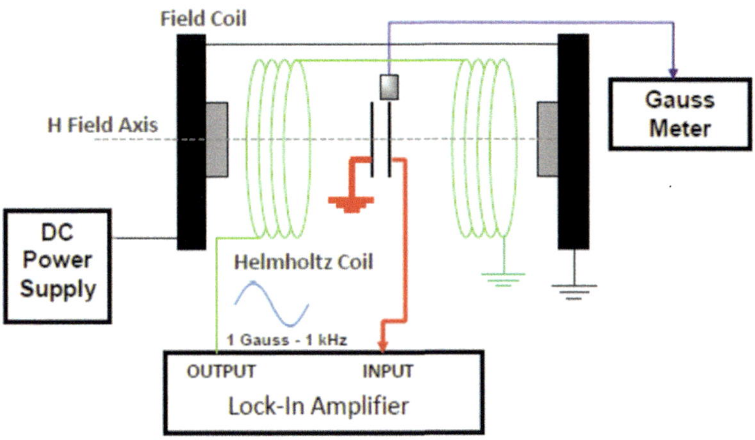

FIGURE 3.2 A typical set up for low-frequency magneto-electric voltage coefficient.

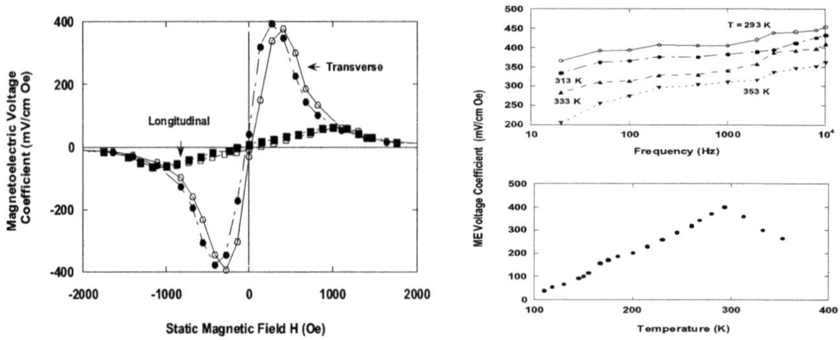

FIGURE 3.3 Representative data showing MEVC versus H and temperature dependence (Srinivasan et al., 2001).

(Nan et al., 2008; Vopson et al., 2017). The nature of coupling is the standard effect, i.e., induced polarization due to AC magnetic field, but the frequency of the AC field is tuned to bending, longitudinal, or thickness electromechanical resonance modes. A pulsed magnetic field technique for studies on frequency characteristics of ME coefficients was reported in Ref (Fetisov et al., 2004). As shown in Fig. 3.4 the method is based on the excitation of a composite with magnetic field pulses, followed by the measurement of induced ME voltage across the sample and Fourier analysis of the signal.

3.2.3 Nonlinear ME effects

Most studies on ferromagnetic-ferroelectric composites were devoted to investigation of ME interactions under a low ac magnetic field amplitude for which the voltage response is a linear function of the ac field. But the ferromagnetic materials are characterized by nonlinear magnetostriction λ versus field dependence (as in Fig. 3.1)

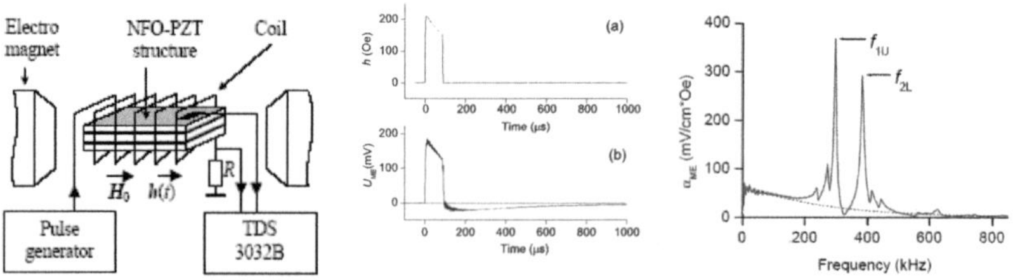

FIGURE 3.4 Schematics of pulsed-magnetic field ME measurement system, sample response to a pulsed magnetic field, and fourier transform showing peaks in MEVC at acoustic resonance modes (Fetisov et al., 2004).

and ferroelectric materials are characterized by nonlinear dependence of the piezoelectric coefficient d on E. Thus, under certain conditions and for rather high amplitude of the ac fields, a nonlinear ME response is expected in the composites. Many reports in the recent past have dealt with nonlinear phenomena in ME composites. Frequency doubling and generation of harmonics were observed in composites such as layered ferrite-PZT (Burdin et al., 2013). Fig. 3.5 illustrates a specific example for linear and nonlinear ME effects. When a bias field H and an ac field of 2.8 Oe are applied to a composite the mechanical deformation occurs at the excitation frequency (Burdin et al., 2013). A linear ME response characterized by a resonance ME coupling at $f = f_r = 700$ Hz as in Fig. 3.5a was reported under this low ac field amplitude. But when $H = 0$ and for the same ac field of 10 Oe, the ac magnetostriction occurs at twice the excitation frequency. This frequency doubling leads to appearance of the peak in Fig. 3.5b for an excitation frequency $f = f_r/2$. It was also reported that nonlinearity of the magnetostriction results in doubling of the frequency and generation of voltages with sum and difference frequencies when ac signals of two different frequencies are applied as in Fig. 3.5c (Burdin et al., 2013).

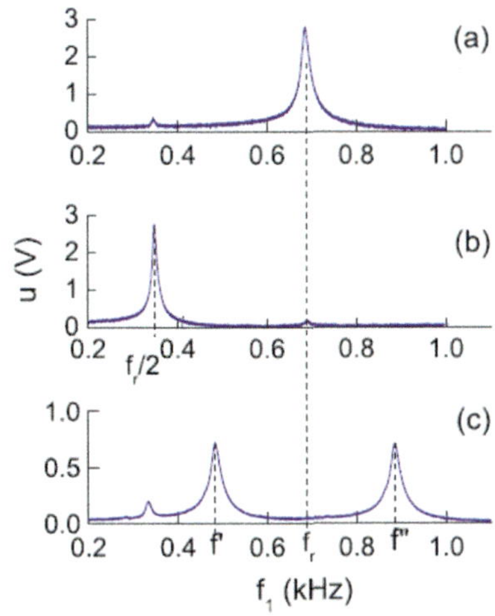

FIGURE 3.5 Measured dependences of the generated voltage u versus frequency: (a) For linear ME effect showing a peak at the bending resonance, for bias field $H = 18$ Oe, and ac field $h = 2.8$ Oe; (b) non-linear ME effect showing frequency doubling for $H = 0$, $h = 14$ Oe. The peak occurs when the excitation occurs at half the bending mode frequency; (c) for nonlinear field mixing showing peaks at sum and difference of two ac fields, $H = 0$, $f_1 = 700$ Hz, $h_1 = 10$ Oe, $f_2 = 0.2$ kHz, $h_2 = 10$ Oe (Burdin et al., 2013).

3.2.4 Static magnetic field induced ferroelectric polarization

The strength of DME could also be measured in terms of the polarization induced by a static magnetic field H. A ferroelectric tester capable of measuring P versus E is used. Samples are placed in a measurement cell and data on ferroelectric hysteresis data is obtained for a series of in-plane or out-of-plane H. Several composites are reported to show a change in the remnant polarization P_r and the coercive field E_c under H. The strength of ME coupling is then defined by the fractional change in the remnant polarization, $\Delta P_r/P_r\ (H = 0) = [P_r(H) - P(H = 0)]/P_r\ (H = 0)$, where ΔP_r is the change in P_r. Fig. 3.6 show P versus E and $\Delta P_r/P_r\ (H = 0)$ versus H for discs of core-shell fibers of nickel ferrite and PZT. A related measurement procedure of importance is positive-up-negative-down (PUND) measurements under H. In PUND measurements, a series of five voltage pulses are used for switchable polarization measurements in a composite as described in Ref (Sreenivasulu et al., 2014a).

3.2.5 Magneto-dielectric effects (MDE)

The magneto-dielectric effect (MDE) involves studies on the influence of a static magnetic field on the dielectric constant of a material (Sreenivasulu et al., 2014b). It is a powerful tool for investigations on the nature on direct-ME effects in composite multiferroics. A variety of measurements are possible: (i) magnetic field induced variation in the permittivity ε over a wide frequency range, from a few hundred Hz to 110 GHz, (ii) ε versus H under dielectric resonance in the composite, and (iii) dielectric resonance frequency f_r versus H. For MDE at low-frequencies, one could

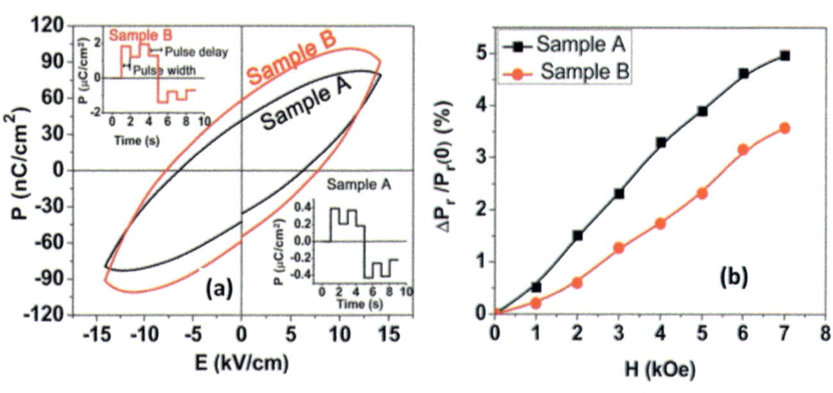

FIGURE 3.6 (a) Polarization versus electric field E for discs containing nanofibers with PZT core and nickel ferrite shell (sample A) and nickel ferrite core and PZT shell (sample B).(b) Fractional change in remnant polarization as a function of H for the fiber discs (Sreenivasulu et al., 2014a).

perform magneto-capacitance measurements on the composite. MDE over 1–110 GHz usually requires the use of a vector network analyzer for measurements of reflected and transmitted power in a composite. Measurement procedure involves placing a sample of appropriate shape inside a rectangular waveguide or coaxial airline, resonance cavity, or between horn antennas. Waveguide fixtures are band-limited but operate at higher frequencies than coaxial airline. Composites must be rectangular in shape and completely fill the cross-section of a waveguide without gaps at the fixture walls. Coaxial airlines require samples of specific shape and dimensions as well. During the measurements a bias static magnetic field is applied either parallel or perpendicular to the sample plane. Fig. 3.7a shows real part of the relative permittivity ε_r' versus f for a series of H for a composite of core-shell nanoparticles of nickel ferrite and barium titanate (Sreenivasulu et.al., 2014b). These data were obtained for the sample placed in a waveguide. Fig. 3.7a also shows the fractional change in the permittivity as a function of H.

Investigations on MDE at dielectric resonance in the composites are of interest since the resonance frequency is a function of ε_r. Thus, any H-induced changes in ε_r' are directly reflected as a change in resonance frequency. One may measure (i) ε_r' versus f as a function of H in the region of dielectric resonance and (ii) resonance frequency f_r versus H. It is essential that an appropriate sample size for dielectric resonance over the specific frequency range is chosen. Typical data on ε_r' versus f for a series of H over the range 16–24 GHz are shown in Fig. 3.7b. A resonance is clearly seen in the data and is due to dielectric resonance in the sample of core-shell nanocomposite. With the application of H, a general decrease in ε_r' is observed except for a narrow frequency range extending from 19 to 20 GHz over which ε_r' is found to increase with increasing H. The fractional change in the dielectric constant defined as $\Delta\varepsilon r'/\varepsilon r'(0) = [\varepsilon r'(H)\text{-}\ \varepsilon r'(H=0)]/\varepsilon r'(H=0)$ for $H = 4$ kOe is shown in

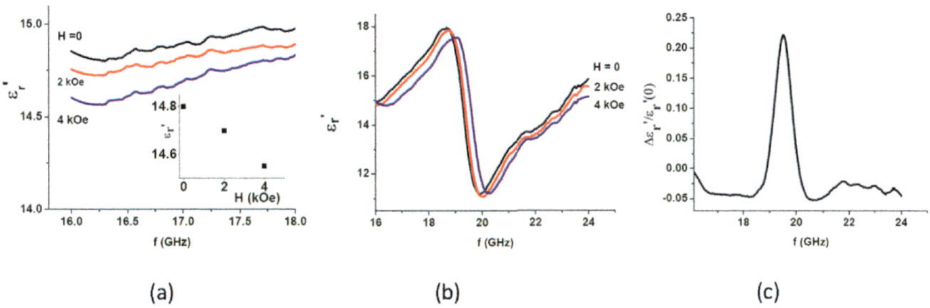

FIGURE 3.7 Magneto-dielectric effects measured on a disc of core-shell nanoparticles of barium titanate and nickel ferrite for (a) away from dielectric resonance, (b) at dielectric resonance, and (c) fractional change in ε_r' for the data in (b) (Sreenivasulu et al., 2014b).

Fig. 3.7. For frequencies away from the resonance region one measures 1–5% decrease in ε_r', while at resonance, the dielectric constant increases by 23%. Thus, a giant magneto-dielectric effect in the composite with a 28% net change in permittivity is evident in the data. A change in the resonance frequency with H is also evident from the data in Fig. 3.7. The shift in f_r with H could be attributed to the change in the permittivity that arises due to strain mediated ME coupling in the composite (Sreenivasulu et al., 2014b).

3.3 Converse ME effects

3.3.1 Low-frequency CME

In the case of converse ME effect, an external AC electric field E produces a deformation of piezoelectric layers resulting in a change in the magnetic parameters of the composite (Fetisov et al., 2007). Thus, as a pickup could wound on a composite, one could apply an AC voltage across the piezoelectric and measure the resulting change in the magnetic induction B by measuring the induced voltage in the coil. The strength of the ME coupling is measured in terms of $\alpha_E = B/E$. Fig. 3.8 shows a schematic view of the set up. A trilayers composite of PZT-Ni-PZT is shown with a coil wound around the sample for measuring the induced B. A DC bias field H is applied to the sample. An AC voltage U_0 applied to PZT results in an induced U_{in} in the coil. The AC voltages generated in the coil could be measured as functions of frequency and magnitude of input ac fields and for different values and orientations of bias magnetic field H. Fig. 3.8 shows typical U_{in} versus H for in-plane and out-of-plane bias magnetic fields (Fetisov et al., 2007).

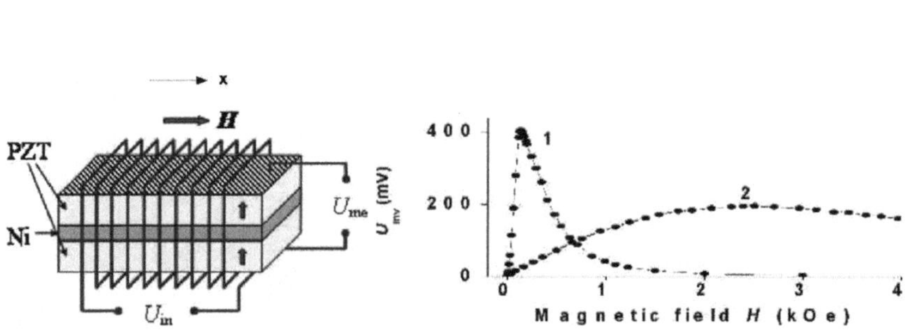

FIGURE 3.8 Low-frequency converse ME effects in trilayers of Ni and PZT. An ac voltage is applied across the trilayers and the resulting induced voltage due to CME is measured with the pickup coil. The CME voltage is shown as a function of static field H for field parallel (1) and perpendicular (2) to the sample plane (Fetisov et al., 2007).

3.3.2 Electric field induced magnetization

The nature of *static* E-tuning of magnetic parameters of composites can be studied by electric-field induced changes in the magnetic hysteresis loops with the use of a vibrating sample magnetometer (VSM), an SQUID magnetometer, or similar magnetization measurement systems. Fig. 3.9 shows E-dependence of M versus H for a composite of NFO/PZN-PT. Data are for H along the [100] and [110] axes of PZN-PT (Li et al., 2011). Both orientations display an E-induced easy magnetization process. A measure of the strength of the ME interactions is the fractional change in the remnant magnetization defined by $\Delta M/M(E = 0) = [M(E)\text{-}M(E = 0)]/M(E = 0)$ reaches 15% for $E = 8$ kV/cm applied across PZN-PT.

3.3.3 Static electric field tuning of inductance

A ferromagnetic–ferroelectric composite could potentially form the core of a high frequency inductor. Such an inductor could be tuned with a DC electric field E. The strength of the ME effect could be expressed in terms of the ratio of change in the inductance to E (Lou et al., 2009). Electrostatically tunable ME inductors with two layers of Metglas and one layer of PZT is shown in Fig. 3.10. Measured inductance L versus f for a series of E is also shown in figure. A decrease in L with increasing E is evident from the data (Lou et al., 2009).

3.3.4 Electrostatic tuning of ferromagnetic resonance (FMR)

Ferromagnetic resonance is ideal for studies on CME effects. There are two types of ME interactions at FMR: (i) coupling between microwave

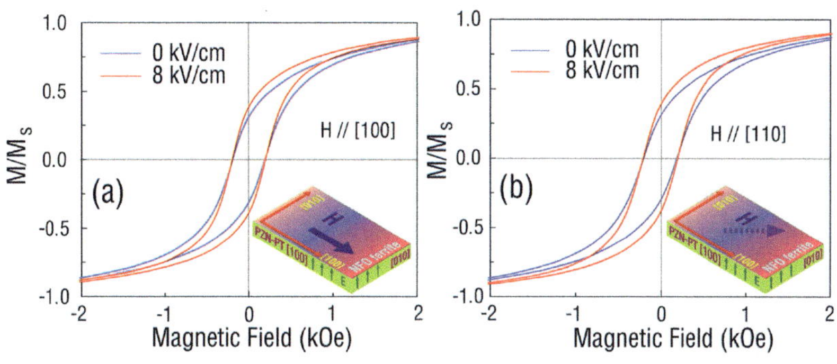

FIGURE 3.9 Magnetic hysteresis under static electric field for studies on CME. Figures show the influence of E on M versus H for a film of nickel ferrite deposited on PZN-PT substrates. H is applied either parallel to [100] or [110] axis of PZN-PT (Li et al., 2011).

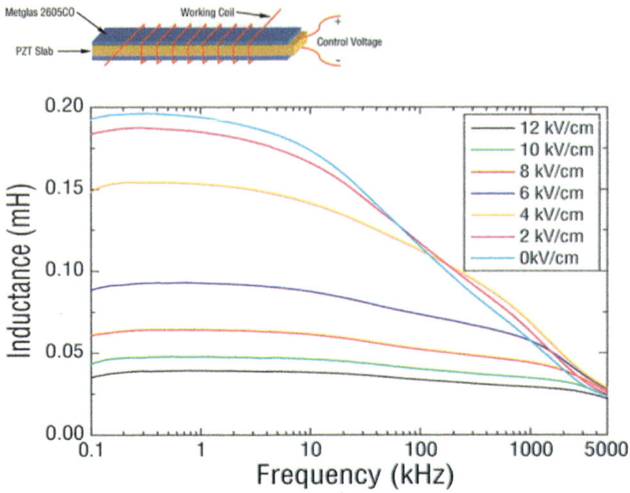

FIGURE 3.10 Inductance L measured for a coil wound around a Metglas-PZT composite. L is measured as a function of frequency for a series of static electric field applied across PZT (Lou et al., 2009).

magnetic field and DC electric field and (ii) between microwave magnetic and electric fields (Nan et al., 2008). Only the first effect is considered here. The technique is based on the fact that magnetic resonance frequency is strain dependent and that the magnitude of this strain dependence is determined by the piezoelectric and magnetoelastic constants. The E induced mechanical strain manifests as an internal magnetic field in the ferromagnetic layer, leading to a shift δH_E in the FMR field. The shift δH_E is strongly influenced by, among other factors, the sample magnetization, magnetostriction λ and piezoelectric coupling d. Information on the nature of ME coupling could therefore be obtained from δH_E versus E data. Such measurements of course could only be performed on composites in which the ferromagnetic resonance has a low FMR linewidth.

The FMR measurements could be done with a (i) cavity, (ii) stripline or (iii) shorted waveguides. Measurements with cavities are done using a modified electron spin resonance spectrometer. A TE_{102} reflection type cavity with *h*oles (1 mm diameter) at the center of the cavity bottom or at one-quarter wavelength from the bottom on the narrow side was used in the work reported in Ref.13. The holes facilitate placing the sample *outside* the cavity so that electric field can be applied to the piezoelectric layer. One then measures FMR profiles as a function of E as illustrated in Fig. 3.11. Data on field shifts δH_E (or shift in FMR frequency δf) are obtained for a series of E and the strength of ME coupling $A = \delta H_E/E$ $(=\delta f/E)$ is estimated from the slope of δH_E versus E data (Li et al., 2011).

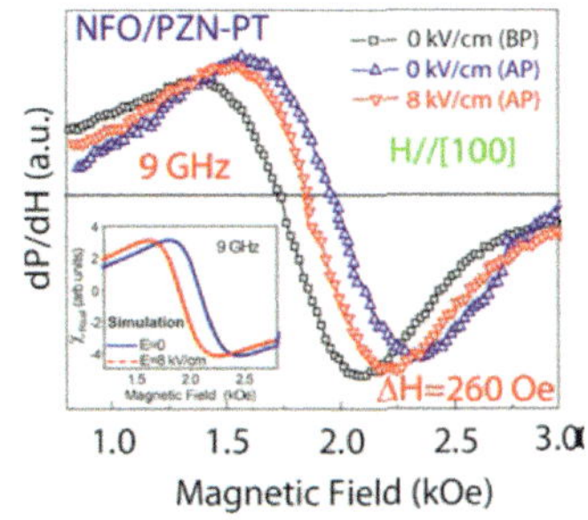

FIGURE 3.11 Profiles of power absorbed versus static field H showing FMR in an NFO/PZN-PT bilayer. Notice the shift in the profiles with the application of E to PMN-PT (Li et al., 2011).

Resonant cavity techniques are appropriate for CME studies at a fixed frequency. Measurements over a wide frequency band are possible with a stripline transducer or a coplanar waveguide. The schematics of the set up with a microstripline structure are shown in Fig. 3.12 (Fetisov et al., 2006). A composite of yttrium iron garnet and PZT is shown on a microstrip transducer on an alumina substrate. The structure is placed between the poles of an electromagnet so that a bias magnetic field H could be applied parallel or perpendicular to the sample plane. A DC electrical field is generated across PZT by applying a voltage. Microwave measurements are carried out using a vector network analyzer. Low input power on the order of 1 mW is chosen to prevent heating of the sample due to power absorption at FMR. Profiles of the scattering matrix parameter S_{11} versus f as in Fig. 3.12 are recorded for a series of H and E. For $E = 0$, the spectra contained a well-defined FMR absorption peak at ~5 GHz with a 3 dB linewidth $\Delta f = 3.4$ MHz. With the application of $E = 10$ kV/cm, the FMR peak shifted to a higher frequency by $\delta f_1 = 16$ MHz. When the polarity of E was reversed by reversing the voltage applied to PZT, the peak downshifted by $\delta f_1 = -18$ MHz. A similar electric field tuning of the

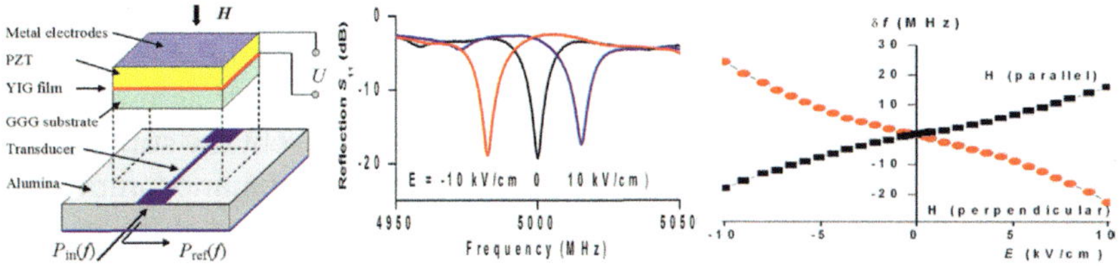

FIGURE 3.12 A microstripline transducer for wide-band CME studies by E-tuning of FMR. Profiles showing FMR profiles are for E = 0 and ± 10 kV/cm. FMR frequency shift versus E data are for static filed H parallel or perpendicular to sample plane (Fetisov et al., 2006).

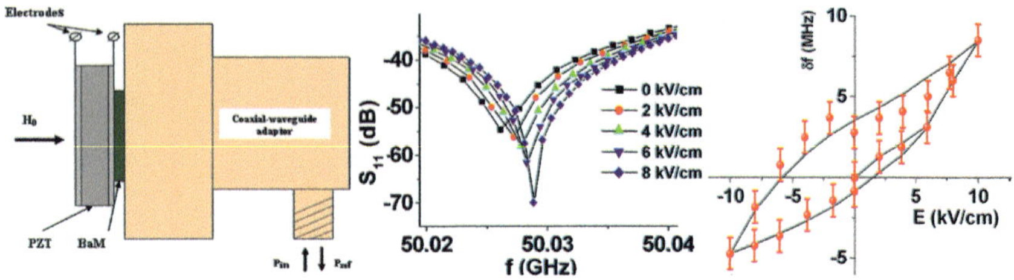

FIGURE 3.13 A shorted waveguide set up for CME studies by E-tuning of FMR, FMR profiles for a series of E for a bilayer of barium ferrite-PZT, and FMR frequency shift versus E for the bilayer (Srinivasan et al., 2006).

absorption peak could be observed for *H* perpendicular to the bilayer plane as shown in the figure. However, the shifts in FMR are usually higher in magnitude than for in-plane *H* (Fetisov et al., 2006).

Since stripline measurements in general are limited to frequencies below 50 GHz, it is necessary to use resonance cavities or waveguide shorts for *E*-tuning of FMR for CME studies at higher frequencies. The measurement cell for such studies on a bilayer of barium ferrite and PZT is shown in Fig. 3.13 and consists of a coaxial-to- (WR-17) waveguide adapter (Carignan et al., 2011). A bilayer of M-type hexagonal barium ferrite (BaM) bonded to silvered surface of PZT is shown. The sample when placed at the open end of the waveguide the silver electrode at the BaM-PZT interface acted as a classical short load of the waveguide. A vector network analyzer is then used for data on S_{11} versus f for a series of *H* and for *E*. Typical profiles of FMR under *E* and shift in FMR frequency δf versus *E* are also shown in Fig. 3.13 (Srinivasan et al., 2006).

3.4 Scanning probe microscopy techniques for ME effects in nanocomposites

There have been considerable efforts in recent years on ME effects in nanostructured composites. Several groups have reported on the synthesis of composites of core-shell nanoparticles, nanopillars in a host matrix, nanobilayers, coaxial tubes in a template, and core-shell nanowires. Some of these nanocomposites have been further self-assembled or magnetic and/or electric field assisted assembly to obtain superstructures. One in general anticipates a strong ME coupling in nanocomposites due to a large surface-to-volume ratio. Studies on the nature of ME coupling in these nanocomposites would require innovative techniques. The preferred tool so far has been scanning probe microscopy. Piezo force microscopy (PFM) under an applied magnetic field is used for studies on

DME and magnetic force microscopy (MFM) in an applied voltage for CME (Beguhn et al., 2012; Carignan et al., 2011; Liu et al., 2020). A few representative examples on the utility of these techniques for studies on bilayers, nanopillars, and nanotubes and wires are discussed here.

For DME in nanobilayers and core-shell fibers, several groups used PFM under H. Piezo force microscopy was used to monitor the ferroelectric domain characteristics and switching under the influence of a magnetic field (Caruntu et al., 2012). Data on PFM phase and amplitude and piezoelectric displacement were used for quantitative estimates of ME coupling strength. Fig. 3.14 shows results of PFM measurements under a magnetic field for a nanobilayer of PTO and NFO (Caruntu et al., 2012). The PFM phase images show bright and dark spots corresponding, respectively, regions with polarization upward and downward. The image in Fig. 3.14a is for $H = -600$ Oe with primarily bright regions and few dark regions. When the polarity of H was switched and increased to 600 Oe, the dark regions grew at the expense of bright regions. These images indicate reversal of polarization in PTO with a magnetic field due to magnetostriction of NFO.

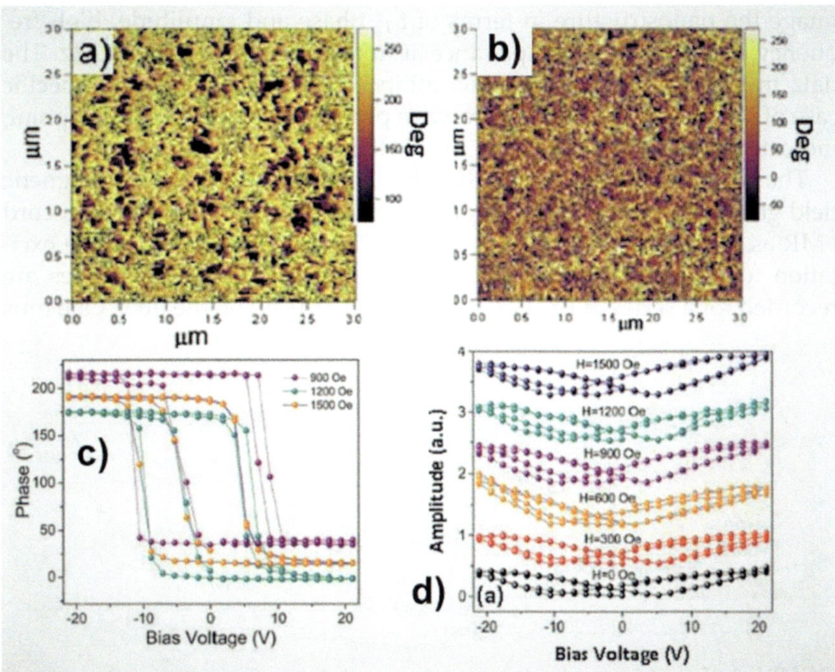

FIGURE 3.14 The PFM phase images for (a) H −600 Oe and (b) H = +600 Oe for a nanobilayer sample of PTO and NFO. The phase angle versus V and amplitude versus V for a series of H values are shown in (c) and (d) respectively (Caruntu et al., 2012).

Fig. 3.14c shows phase hysteresis data for PTO–NFO film as a function of *H*. With increasing value of *H*, the data shows significant change in the phase angle as well as 180 degree switch in the phase angle. The presence of 180 degree domains in PTO is also confirmed by the butterfly-like variation of phase amplitude with voltage in Fig. 3.14d. As H is increased, the linear portion of the variation in the amplitude becomes steeper. Overall, the results in Fig. 3.14 is evidence for utility of PFM measurements under *H* for characterizing the DME interactions in nanocomposites.

Similarly, magnetic force microscopy under a static electric field is useful for information on the nature of CME (Beguhn et al., 2012; Berry et al., 1975). A strong CME in a bilayer of BTO-CFO was inferred from changes in MFM images attributed to reversal in in-plane magnetization under an electric field (Liu et al., 2020). CME studies by FMR were reported on array of core-shell ferrite-ferroelectric nanotubes grown in an alumina template (Beguhn et al., 2012). A recent report of importance in this regard is the studies on CME in a core-shell nanofiber of NFO and PZT by performing FMR with a scanning microwave microscope (SMM) in which the probe is a source of microwave powered by a vector network analyzer (Liu et al., 2020; Viehland et al., 2018). With the SMM, one can image the nanostructure in terms of S_{11} phase and amplitude, high frequency capacitance and impedance at frequencies from 1 to 26 GHz. The data in Fig. 3.15 demonstrate the utility of the techniques for a specific case of NFO-PZT fiber. The microwave probe is located at a specific point, indicated by a dot in Fig. 3.15a, for FMR profiles.

The amplitude and phase of S_{11} is then measured as the static magnetic field generated by an electromagnet is scanned from 0 to 2 kOe to record FMR as in Fig. 3.15b. Following this measurement, the microwave excitation for a specific microwave excitation frequency, FMR profiles are recorded for a series of DC voltage applied to the tip on the fiber. One thus

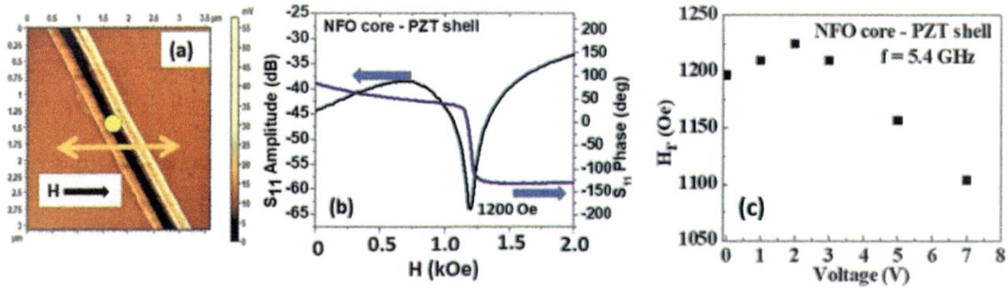

FIGURE 3.15 (a) S_{11} amplitude image of a core shell nanofiber with nickel ferrite core and PZT shell. (b) Profiles of S_{11} amplitude and phase versus static magnetic field *H* showing FMR at 8.25 GHz in the coaxial fiber. (c) Resonance field Hr versus bias voltage applied to a fiber of NFO core and PZT shell (Liu et al., 2020).

obtains data on the variation in the resonance field Hr as a function of V to estimate the strength of CME.

3.5 Magnetostriction characterization of thin film

The realization of strong ME coupling (α) requires a large piezomagnetic coefficient $d_{33,m}$ (i.e., $d\lambda/dH$) in the magnetic phase, that is, a soft magnetostrictive material with a high magnetostriction constant (λ) at a low saturation field (H). Unlike bulk materials, it is usually difficult to characterize these magnetomechanical properties in thin films since measuring their strain change would not be possible using the strain gauge method due to the loading effect. Few data were available for novel thin film materials. Several techniques have been developed to characterize magnetostriction and ΔE effect of thin films, as shown in Table 3.1. Some techniques have limitations on the instrumentation setup and sample mounting. For instance, the laser beam technique requires a large space area on the table to setup the sophisticated optical system including He–Ne laser diodes, adjustable mirrors, and position sensors (Tam et al., 1989). For the nanoindentation method (Shima et al., 1999), there is a minimum penetration depth of the thin film to determine the Young's modulus and magnetostriction. Therefore, for a thin film thickness less than 60 nm, it would be impossible to use this method, and the samples would be damaged after characterization.

In 2018, Dong et al. developed a simple, compact, and sensitive system to measure the saturation magnetostriction, piezomagnetic coefficient, ΔE effect, and magnetomechanical coupling factor by using a noncontact optical technique (Dong et al., 2018).

Both the magnetostriction tester and the ΔE tester share a similar structure and can be assembled with very few parts. High vibrational detection sensitivity for the magnetostriction measurement and high frequency shift sensitivity for the ΔE effect characterization has been achieved by the system. In the magnetostriction tester, an MTI-2000 fiber-optic sensor was utilized to detect the deflection of the sample tip due to strain changes. An adjustable AC driving magnetic field ranging from 0 to 300 Oe was provided by two pairs of mutually perpendicular set Helmholtz coils as shown in Fig. 3.16a, which provide enough space for various sample dimensions to be confined in a homogeneous field region. An equivalent amplitude and 90 degrees phase shift current in the two pairs of Helmholtz coil were configured to generate a rotating magnetic field, which would induce domain rotation inside the magnetostrictive thin

TABLE 3.1 Existing techniques on thin film magnetostriction and ΔE effect measurement.

Measurement technique	Detectivity	Strength and weakness
Permeance meter (λ) (Kawakami et al., 1983)	0.05 ppm	Difficult to apply stress on thin films, only λ_s can be measured
Cantilever capacitance (λ) (Klokholm, 1976)	0.1 ppm	Accuracy would be easily affected by air current and ambient temperature
Laser beam (λ) (Tam et al., 1989)	0.1 ppm	Require a large space area for the setup and sophisticated optical system
Piezoelectric ceramic (λ) (Arai et al., 1989)	–	Magnetic and nonmagnetic films are difficult to be perfectly aligned
Nano-indentation (λ) (Shima et al., 1999)	0.02 ppm	Film Young's modulus can be determined and minimum thickness requirement
Magnetostriction tester using an MTI-2000 fiber-optic sensor (λ) (Dong et al., 2018)	0.015 ppm	Noncontact, temperature independent, compact size with few parts to setup
Vibrating reed (ΔE) (Berry et al., 1975)	0.5 Hz	Operate at low frequency and large frequency interval
Magnetostriction tester using an MTI-2000 fiber-optic sensor (ΔE) (Dong et al., 2018)	0.01 Hz	Share a similar structure with the magnetostriction tester and high frequency resolution

films, as shown in Fig. 3.16d. As it is shown in Fig. 3.16c, the deflection due to the strain change of magnetostriction can be expressed as

$$d = \lambda \frac{3t_f l^2}{t_s^2} \frac{E_f(1 - v_s)}{E_s\left(1 + v_f\right)} \tag{3.1}$$

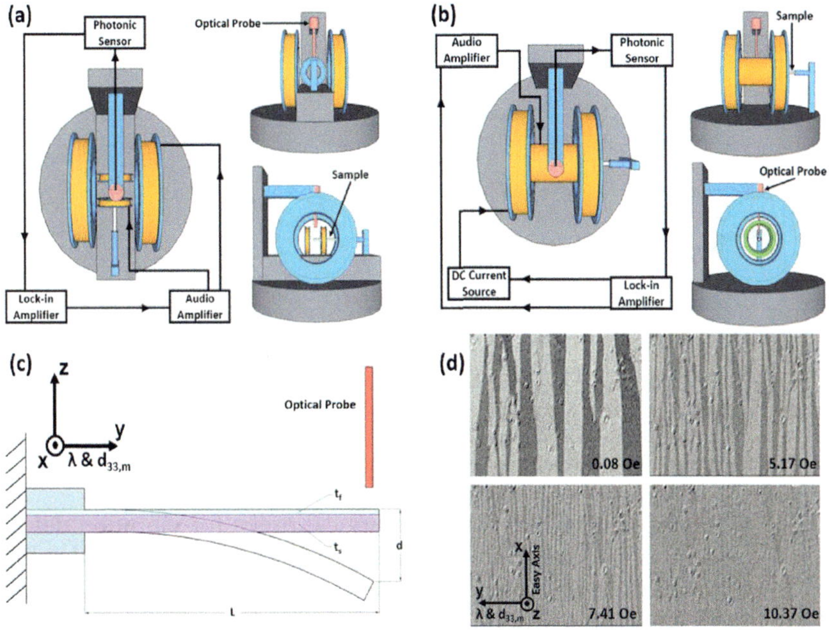

FIGURE 3.16 3D schematic of the home-made (a) magnetostriction tester and (b) ΔE tester in three-view. (c) Illustration of the detection of displacement by the optical probe when the sample was clamped at one end.

where, t_f and t_s are the thickness of the film and the substrate, respectively, l is the distance between the clamping edge and the probe location, and λ is the magnetostriction constant, and E_f and E_s and ν_f and ν_s represent Young's moduli and Poisson's ratios of the film and the substrate, respectively. The displacement between the sample tip and the optical probe can be detected and converted to an AC voltage signal, which was recorded using an SR-830 lock-in amplifier. Thereby, the magnetostriction constant and piezomagnetic coefficient can be demonstrated. Since the vibration signal is dependent upon the dimension and resonant frequency of the cantilever, to make the measurement accurate and repeatable, a standard sample substrate dimension of $35 \times 3.5\ \text{mm} \times 0.15\ \text{mm}$ has always been used and the AC driving frequency was set to 25 Hz, which was off resonance of the cantilever to avoid mechanical amplification. After that, the system was calibrated with the reported value of typical magnetostrictive thin films of Ni, Co, FeGa, and Permalloy. The accuracy and repeatability of the system have been confirmed by the measurement results of typical different magnetostrictive thin films, as shown in Fig. 3.17a.

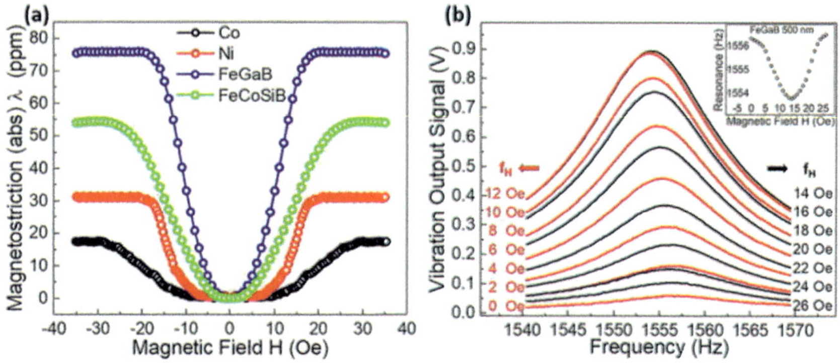

FIGURE 3.17 (a) Measurement results of the magnetostriction constant in the absolute value of different magnetic thin films. (b) Vibration output signal near resonance under different magnetic fields of a 500 nm-thick FeGaB thin film sample. The inset shows the resonant frequency shift with magnetic field variation due to the ΔE effect.

The ΔE tester also utilizes the same optical probe for vibration detection. Fig. 3.16b shows the schematic of the ΔE tester. A 0.1 Oe AC driving field with the frequency sweep range around 1555 Hz was provided by a solenoid, which would induce a constant vibration at each sample's second mechanical resonant frequency. Outside the solenoid, one pair of Helmholtz coil was set to provide the DC bias field, which would induce magnetostrictive strain change and thereby the change of Young's modulus. To obtain the maximum strain change and therefore the strongest vibration signal, both the field directions of the solenoid and Helmholtz coil were set perpendicular to the easy axis of samples. According to the modification of Euler-Bernoulli beam theorem, the ΔE effect can be expressed as (Beguhn et al., 2012)

$$\Delta E_f = \frac{E_s d_s}{3 d_f}\left[\left(\frac{f_H}{f_0}\right)^2 - 1\right] \tag{3.2}$$

where d_s and d_f and E_s and E_f are the thicknesses and Young's moduli of the silicon substrate and the film, respectively, f_H the resonant frequency at various DC bias fields, and f_0 the resonant frequency in the demagnetizing state. By detecting the shift of resonant frequency, the change of Young's modulus could be determined. Fig. 3.17b shows the resonant frequency shift along with a changing bias magnetic field from 0 to 26 Oe of a 500 nm-thick FeGaB multilayer thin film. Due to the ΔE effect, as the bias magnetic field increases from 0 to 12 Oe, the resonant frequency decreased to a minimum and the vibration amplitude increased to its maximum. As the bias field continues to increase from 14 to 26 Oe, the resonant frequency shifts back to its maximum and the vibration amplitude decreased to its minimum. The high sensitivity and simplicity of this system made it a good candidate tool for the characterization of magnetomechanical properties in magnetostrictive thin films, helping develop advanced magnetoelectric devices.

Summary

This chapter is intended to provide a broad overview of the measurement techniques for characterizing the strain mediated ME interactions in ferromagnetic-ferroelectric composites. Techniques for the measurements of both direct- and converse ME effects are considered in this section. Further details on the techniques are provided in the references cited in this chapter.

References

al, P. S.e. (2003). Magnetically tunable SAW-resonator. In *IEEE international frequency control symposium and PDA exhibition jointly with the 17th European frequency and time forum*. IEEE.

Arai, K. I., et al. (1989). Measurement of thin film's magnetostriction with piezoelectric ceramic substrates. *IEEE Transactions on Magnetics, 25*(5), 4201–4203.

Beguhn, S., et al. (2012). A new highly sensitive broadband ferromagnetic resonance measurement system with lock-in detection. *Journal of Applied Physics, 111*(7), 07A503-507A503-503.

Berry, B. S., & Pritchet, W. C. (1975). Vibrating reed internal friction apparatus for films and foils. *IBM Journal of Research and Development, 19*, 334–343.

Burdin, D. A., et al. (2013). Resonance mixing of alternating current magnetic fields in a multiferroic composite. *Journal of Applied Physics, 113*(3).

Carignan, L.-P., et al. (2011). Ferromagnetic nanowire metamaterials: Theory and applications: Special issue on radio-frequency nanoelectronics. *IEEE Transactions on Microwave Theory and Techniques, 59*(10), 2568–2586.

Caruntu, G., et al. (2012). Probing the local strain-mediated magnetoelectric coupling in multiferroic nanocomposites by magnetic field-assisted piezoresponse force microscopy. *Nanoscale, 4*(10), 3218–3227.

Dong, C., et al. (2018). Characterization of magnetomechanical properties in FeGaB thin films. *Applied Physics Letters, 113*(26).

Fetisov, Y. K., et al. (2004). Wide-band magnetoelectric characterization of a ferrite-piezoelectric multilayer using a pulsed magnetic field. *Solid State Communications, 132*(1), 13–17.

Fetisov, Y. K., et al. (2007). Inverse magnetoelectric effects in a ferromagnetic–piezoelectric layered structure. *Journal of Materials Research, 22*(8), 2074–2080.

Fetisov, Y. K., & Srinivasan, G. (2006). Electric field tuning characteristics of a ferrite-piezoelectric microwave resonator. *Applied Physics Letters, 88*(14).

Kawakami, K., et al. (1983). A high frequency permeance meter for anisotropic films and its application in the determination of magneto striction constants. *IEEE Transactions on Magnetics, 19*(5), 2154–2156.

Klokholm, E. (1976). The measurement of magnetostriction in ferromagnetic thin films. *IEEE Transactions on Magnetics, 12*(6), 819–821.

Li, N., et al. (2011). Electrostatic tuning of ferromagnetic resonance and magnetoelectric interactions in ferrite-piezoelectric heterostructures grown by chemical vapor deposition. *Applied Physics Letters, 99*(19). 192502-192502-192503.

Li, M., et al. (2017). Highly sensitive DC magnetic field sensor based on nonlinear ME effect. *IEEE sensors letters, 1*(6), 1–4.

Liu, Y., et al. (2020). Converse magneto-electric effects in a core-shell multiferroic nanofiber by electric field tuning of ferromagnetic resonance. *Scientific Reports, 10*(1), 20170-20170.

Lou, J., et al. (2009). Electrostatically tunable magnetoelectric inductors with large inductance tunability. *Applied Physics Letters, 94*(11), 112508-112508-112503.

Nan, C.-W., et al. (2008). Multiferroic magnetoelectric composites: Historical perspective, status,and future directions. *Journal of Applied Physics, 103*(3), 031101-031101-031135.

Nan, T., et al. (2013). Self-biased 215MHz magnetoelectric NEMS resonator for ultra-sensitive DC magnetic field detection. *Scientific Reports, 3*(1), 1985.

Shima, T., & Fujimori, H. (1999). An accurate measurement of magnetostriction of thin films by using nano-indentation system. *IEEE Transactions on Magnetics, 35*(5), 3832–3834.

Srinivasan, G., et al. (2001). Magnetoelectric bilayer and multilayer structures of magnetostrictive and piezoelectric oxides. *Physical Review B: Condensed Matter, 64*(21).

Srinivasan, G., et al. (2006). Millimeter-wave magnetoelectric effects in bilayers of barium hexaferrite and lead zirconate titanate. *Applied Physics Letters, 89*(15). 152508-152508-152503.

Sreenivasulu, G., et al. (2014a). Magnetic field assisted self-assembly of ferrite-ferroelectric core-shell nanofibers and studies on magneto-electric interactions. *Applied Physics Letters, 104*(5).

Sreenivasulu, G., et al. (2014b). Controlled self-assembly of multiferroic core-shell nanoparticles exhibiting strong magneto-electric effects. *Applied Physics Letters, 104*(5).

Tam, A. C., & Schroeder, H. (1989). A new high-precision optical technique to measure magnetostriction of a thin magnetic film deposited on a substrate. *IEEE Transactions on Magnetics, 25*(3), 2629–2638.

Tianxiang Nan, H. L., Gao, Y., Matyushov, A., et al. (2017). Acoustically actuated ultra-compact NEMS magnetoelectric antennas. *Nature Communications, 8*.

Viehland, D., et al. (2018). Tutorial: Product properties in multiferroic nanocomposites. *Journal of Applied Physics, 124*(6).

Vopson, M. M., et al. (2017). Measurement techniques of the magneto-electric coupling in multiferroics. *Materials, 10*(8), 963.

CHAPTER

4

Layered multiferroic laminates*

Gopalan Srinivasan[1], *Zhaoqiang Chu*[2], *Mohammad Javad Pourhosseini Asl*[2], *Zhonghui Yu*[2], *and Shuxiang Dong*[2]

[1] **Oakland University, Rochester, MI, United States;** [2] **Peking University, Beijing, China**

Part A

4.1 Equivalent circuit analysis of magnetoelectric laminated composites

The equivalent circuit method offers a comprehensive understanding of both the static and dynamic analysis of magnetoelectric (ME) laminates. It encompasses electromechanical resonance and off-resonance analysis. In this chapter, an analytical method employing magnetoelectric equivalent circuits is applied to laminate composites consisting of magnetostrictive and piezoelectric layers. The method encompasses four distinct modes of operation: longitudinal magnetized and transverse polarized (L—T), longitudinal magnetized and polarized (L—L), transverse magnetized and polarized (T—T), and circumferentially magnetized and circumferentially polarized (C—C). These modes are illustrated in Fig. A.1a—d (Srinivasan et al., 2001; Dong et al., 2004; Shuxiang Dong et al., 2006; Ce-Wen Nan et al., 2008).

* This chapter consists of two parts. Part A is on models for the analysis of magnetoelectric laminates and Part B is on experiments on ME effects in thick film laminate composites. References are provided at the end of each section.

https://doi.org/10.1016/B978-0-443-18606-6.00015-0

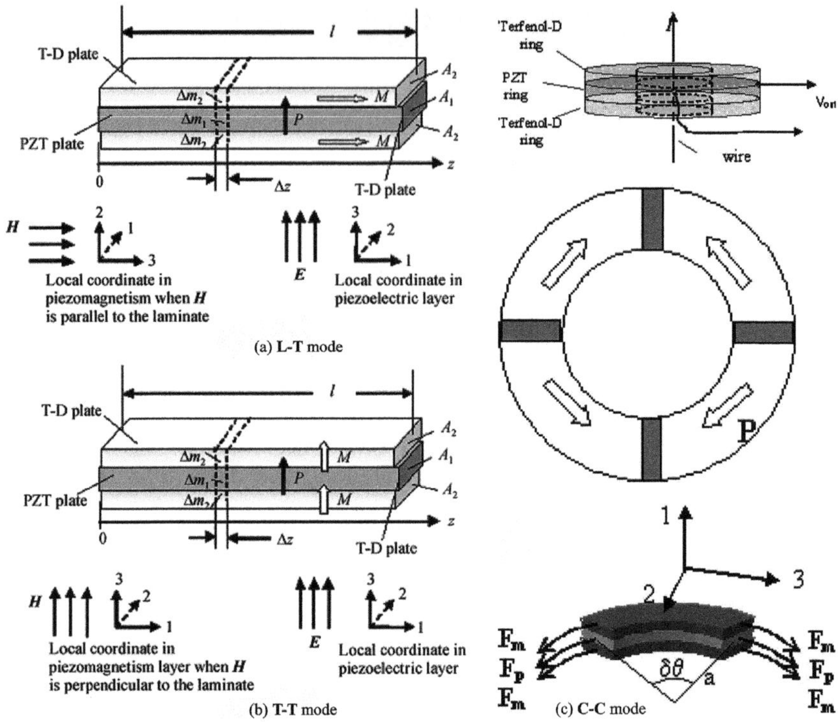

FIGURE A.1 Magnetostrictive/piezoelectric laminate composites (a) laminated L—T mode, (b) laminated T—T mode, (c) laminated C—C ring mode. In piezoelectric or magnetic phase, the local coordinate axis "3" is assumed to be defined by the polarization or magnetization direction (Dong et al., 2004; Dong & Zhai, 2008).

Based on laminated ME composites, Dong et al. proposed an equivalent circuit approach, which is convincing for dynamic modeling of ME coupling effect. The approach builds upon the magnetostrictive and piezoelectric constitutive equations and accounts for the mutual coupling between the magnetostrictive and piezoelectric layers through elastic interaction involving strain *(S)* and stress *(T)*. This interaction is governed by an equation of motion, which is further stimulated by a magnetic field (*H*) (Dong et al., 2003; Shuxiang Dong, 2003; Shuxiang Dong and Viehland, 2003).

According to the equivalent circuit method, the magnetoelectric (ME) coupling in ME laminates exhibits a strong dependence on their working modes. Specifically, the ME voltage coefficients in L—T or L—L modes are considerably higher than those in the T—T mode. Furthermore, the method predicts that there exists an optimal thickness ratio between the piezoelectric and magnetostrictive layers, which maximizes the ME coefficients and resonance effects. Additionally, it is observed that the ME voltage coefficient is approximately Q_m times higher when driven at resonance. It is important to note that these predictions have been validated through experimental results (Dong & Zhai, 2008).

4.1.1 Multilayered longitudinal vibration modes

It is assumed that the magnetostrictive/piezoelectric laminates have long-type configurations, where the piezoelectric layer is polarized either along its thickness or length directions and is subjected to stress from two magnetostrictive layers along their length (i.e., principal strain) direction. Consequently, the piezoelectric constitutive equations for thickness and length polarization are provided below.

Thickness poling case (Shuxiang Dong, 2003; Shuxiang Dong & Viehland, 2003, Shuxiang Dong & Viehland, 2006):

$$S_{1p} = s_{11}^{E} T_{1p} + d_{31,p} E_3, \quad D_3 = d_{31,p} T_{1p} + \varepsilon_{33}^{T} E_3, \tag{A.1}$$

Length polarization case,

$$S_{3p} = s_{33}^{D} T_{3p} + g_{33,p} D_3, \quad E_3 = -g_{33,p} T_{3p} + \beta_{33}^{T} D_3, \tag{A.2}$$

where D_3 is the electric displacement, ε_{33}^{T} and β_{33}^{T} are the dielectric permittivity and impermeability under constant stress T, s_{11}^{E} and s_{33}^{D} are the elastic compliances of the piezoelectric material under constant electric field E or constant electric displacement D, $d_{31,p}$ and $g_{33,p}$ are the transverse piezoelectric constant and longitudinal piezoelectric voltage constants, and T_{1p}, T_{3p}, and S_{1p} S_{3p} are the stress and strain of the piezoelectric layer imposed by the magnetostrictive layers. When H is applied parallel to the longitudinal axis of the laminate, a longitudinal (33) strain is excited. The piezomagnetic constitutive equations for the (33) longitudinal mode are:

$$S_{3m} = s_{33}^{H} T_{3m} + d_{33,m} H_3, \quad B_3 = d_{33,m} T_{3m} + \mu_{33}^{T} H_3, \tag{A.3}$$

where B_3 is the magnetization along the length direction, μ_{33}^{T} is the permeability under constant stress, s_{33}^{H} is the elastic compliance of the magnetostrictive layer under constant H, $d_{33,m}$ is the longitudinal magnetostrictive constant, and T_{3m} and S_{3m} are the stress and strain in the longitudinal direction of the magnetostrictive layers imposed on the piezoelectric layer. The provided constitutive equations are linear relationships that do not consider loss components. It is well known that both piezoelectric and magnetostrictive materials exhibit significant nonlinearities, particularly when subjected to resonance drive.

Assuming harmonic motion, along a given direction z, it will be supposed that three small mass units, Δm_i, in the laminate have the same displacement $u(z)$. This follows from Fig. A.1, by assuming that the layers in the laminate act only in a coupled manner. Following Newton's Second Law, we then have an equation of motion to couple the piezoelectric and piezomagnetic constitutive Eqs. (A.1)–(A.3) (Shuxiang Dong, 2003; Shuxiang Dong & Viehland, 2003; Shuxiang Dong & Viehland, 2006; Ce-Wen Nan et al., 2008):

$$\bar{\rho}\frac{\partial^2 u}{\partial t^2} = n\frac{\partial T_{i,m}}{\partial z} + (1-n)\frac{\partial T_{1,p}}{\partial z}, (0 < n < 1, \mathrm{i} = 1 \text{ or } 3), \quad (A.4)$$

where $n = A_m/A_{\text{lam}} = t_m/t_{\text{lam}}$, is a geometric factor, and $\bar{\rho} = \frac{\rho_p A_p + \rho_m A_m}{A_{\text{lam}}}$ is the average mass density of the laminate, $t_m = t_{m1} + t_{m2}$ is the total thickness of the magnetic phase layers, A_p and A_m are the cross-sectional areas of the magnetic and piezoelectric phase layers respectively, and ρ_p and ρ_m are the mass densities of the piezoelectric and magnetostrictive layers. For a given laminate width w_{lam} and thickness t_{lam}, the total cross-sectional area of the laminate is $A_{\text{lam}} = t_{\text{lam}}\, w_{\text{lam}}$, and the total thickness is the sum of the layer thickness $t_{\text{lam}} = t_p + t_m$, where t_p and t_m are the thickness of piezoelectric and magnetostrictive layers, respectively (Shuxiang Dong & Viehland, 2003, Shuxiang Dong & Viehland, 2006; Ce-Wen Nan et al., 2008).

By combining piezoelectric and piezomagnetic constitutive Eqs. (A.1)–(A.3), solutions to the equation of motion (8.4) can be derived for longitudinally magnetized *(M)* and transversely or longitudinally poled *(P)* ME modes, named as L–T or L–L modes. Correspondingly, the two magneto–(elasto)–electric (or ME) equivalent circuits for L–T and L–L modes under free boundary conditions can be derived, as shown in Fig. A.2a and b. In these figures, an applied magnetic field *(H)* acts as a magnetic-induced "mechanical voltage" ($\varphi_m H_3$) and then induces a "mechanical current" ($\dot{u}$) via the magnetoelastic effect with a coupling factor (φ_m). In turn, $\varphi_m H_3$ results in an electrical voltage *(V)*, and $\dot{u}$ results in a real electric current (I_p) across the piezoelectric layer, due to electromechanical coupling. A transformer with a turn-ratio of φ_p can then be used to represent the electromechanical coupling. In the circuits shown in Fig. A.2, Z_1 and Z_2 represent the characteristic mechanical impedances of the composite, and C_0 denotes the clamped capacitance of the piezoelectric plate.

4.1.2 ME voltage coefficients at low frequency

Following Fig. A.2a and b, open-circuit conditions have been supposed from above, where the current (I_p) from the piezoelectric layer is zero. Thus, the capacitive load (C_0) (and $-C_0$) can be moved to the main circuit loop. Applying Ohm's law to the mechanical loop, the following ME field

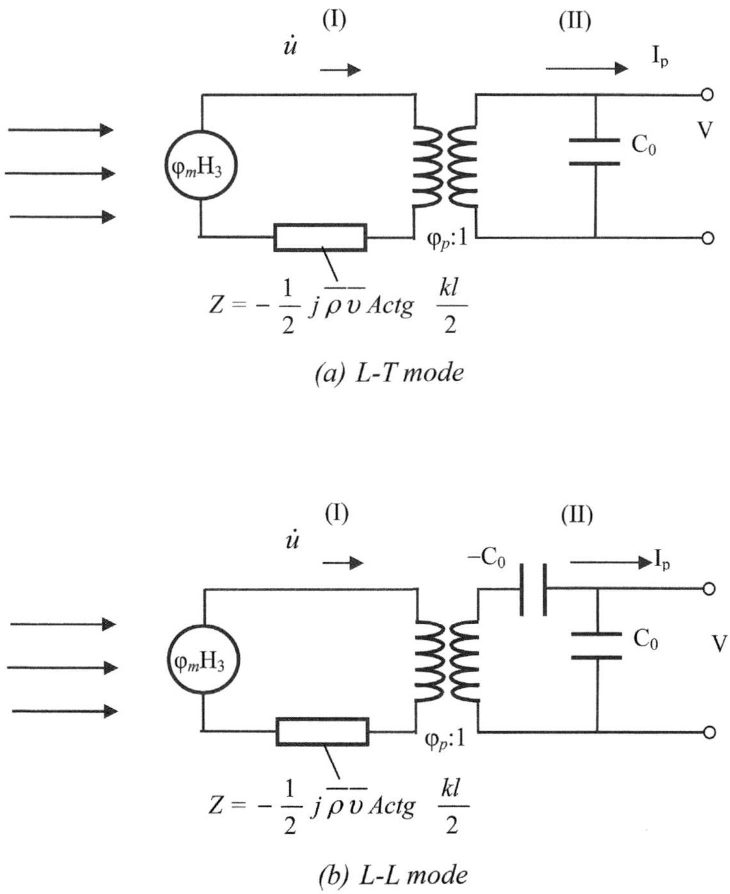

FIGURE A.2 Magneto-elastic-electric bi-effect equivalent circuits for (a) L−T and (b) L−L mode (Priya et al., 2007; Ce-Wen Nan et al., 2008).

coefficients at low-frequency for L−T and L−L modes can be directly derived as magnetoelectric voltage coefficient in the L−T mode. Under a small ac magnetic field H_{ac} applied along the longitudinal axis of the laminate, the L−T mode ME voltage coefficient is (Shuxiang Dong & Viehland; Shuxiang Dong, 2003; Shuxiang Dong & Viehland, 2003):

$$\left|\frac{\mathrm{d}E}{\mathrm{d}H_3}\right|_{(\mathrm{L-T})} = \frac{nd_{33,m}g_{31,p}}{ns_{11}^E\left(1-k_{31,p}^2\right)+(1-n)s_{33}^H}, \tag{A.5}$$

$$\left|\frac{\mathrm{d}E}{\mathrm{d}H_3}\right|_{(\mathrm{T-T})} = \frac{nd_{31,m}g_{31,p}}{ns_{11}^E\left(1-k_{31,p}^2\right)+(1-n)s_{11}^H}, \tag{A.6}$$

These are the formulas for the ME field coefficients predicted by the equivalent method. The values dE/dH_3 are proportional to the longitudinal piezomagnetic constant $(d_{33,m})$ and to the transverse or longitudinal piezoelectric voltage constants $(g_{31,p})$ or $(g_{33,p})$. In addition, the ME voltage coefficient is strongly related to the thickness ratio (n) of the Magnetostrictive layers and to the elastic compliances of both phases. Also, please note that these predicted ME electric field coefficients from Eqs. (A.1) and (A.2) are much higher than measured the values in our prior reports therefore, a rectifying factor (β) was introduced to modify these formulas (Shuxiang Dong & Viehland; Shuxiang Dong, 2003; Shuxiang Dong & Viehland2003; Shuxiang Dong et al., 2003; Dong et al., 2006).

This formula provides a quantitative calculation for the ME voltage coefficient using the equivalent circuit method. The value of the ME coefficient is directly proportional to the transverse piezoelectric voltage constant $(g_{31,p})$ and the longitudinal piezomagnetic constant $(d_{33,m})$. Additionally, it is important to note that a larger thickness (t_{lam}) will result in a higher ME coefficient. This information is crucial for the design and optimization of ME laminates.

In Fig. A.3a, the calculated ME voltage coefficients for the L−T (longitudinally magnetized to transversely poled) and T−T (transversely poled to transversely poled) modes are depicted as a function of n. These coefficients were determined using Eqs. (A.5) and (A.6). A maximum value for $|dV/dH_3|_{L-T}$ of ~2.9V/(Oe-cm) was found for one L−T laminate composite of Terfenol-D and PZT-5 at n~0.67. The maximum value for $|dV/dH_3^{\circ}|_{(L-T)}$ of ~0.7V/(Oe-cm) can be seen at n~0.68 for the T−T mode. Clearly, the L−T coefficient $|dV/dH_3|_{L-T}$ is much larger than the T−T one (Dong & Zhai, 2008).

4.1.3 ME coefficients at resonance frequency

The piezomagnetic and piezoelectric layers in the laminate are coupled through strain $S(z)$ and stress $T(z)$. When an applied magnetic field H is directed along the length of the magnetostrictive layer, it induces forced oscillation in the piezoelectric layer along the longitudinal axis, resulting in a voltage generated between the end and middle electrodes. At the first longitudinal resonance frequency, this laminate functions as a half-wavelength $(\lambda/2)$ ME resonator. The middle position of the laminate represents a node line where the vibration velocity (mechanical current) is zero. Assuming symmetric vibration of the laminate and symmetric polarization of the piezoelectric layer about the node line, the induced ME voltages (α_{ME}) across both piezoelectric elements in the layer are equivalent. Furthermore, the magneto-elastic-electric equivalent circuits for both parts of the laminate are identical. These circuits can be derived by

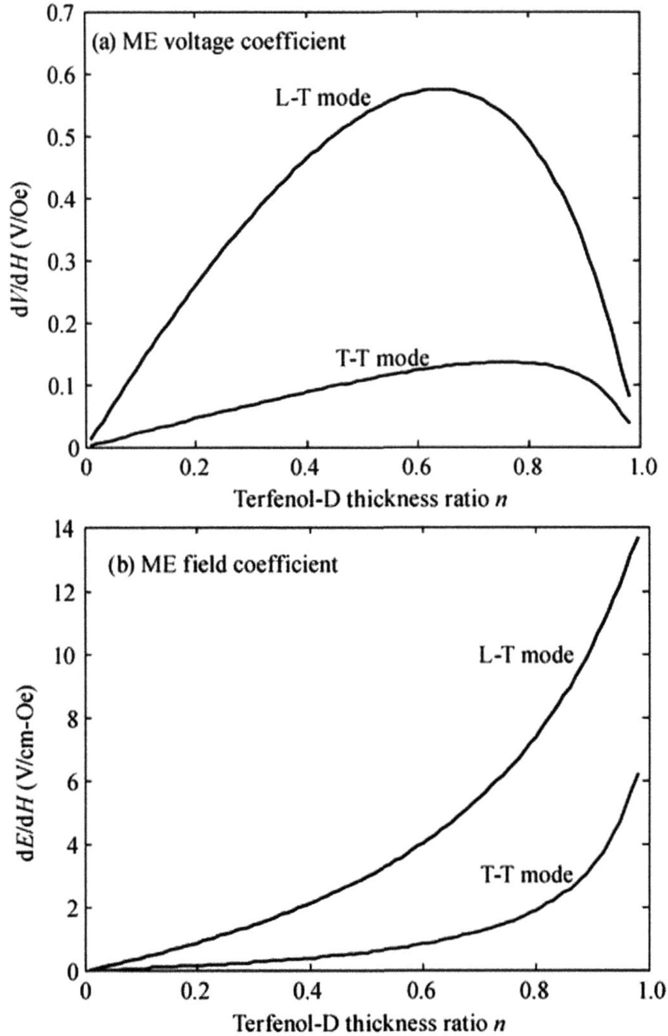

FIGURE A.3 Calculated ME voltage and field coefficients for (a) L–T and (b) T–T Terfenol-D/PZT-5 laminates (Dong & Zhai, 2008).

utilizing the piezoelectric and piezomagnetic constitutive equations, applying Newton's second law of motion to the laminate, and subsequently determining analogous electrical parameters.

Fig. A.4b shows the magneto-elasto-electric equivalent circuit in the region around the resonance frequency (ω_S). This circuit is applicable to either section of the laminate about the nodal line. It is useful for understanding the ME coupling of the laminate under resonance drive

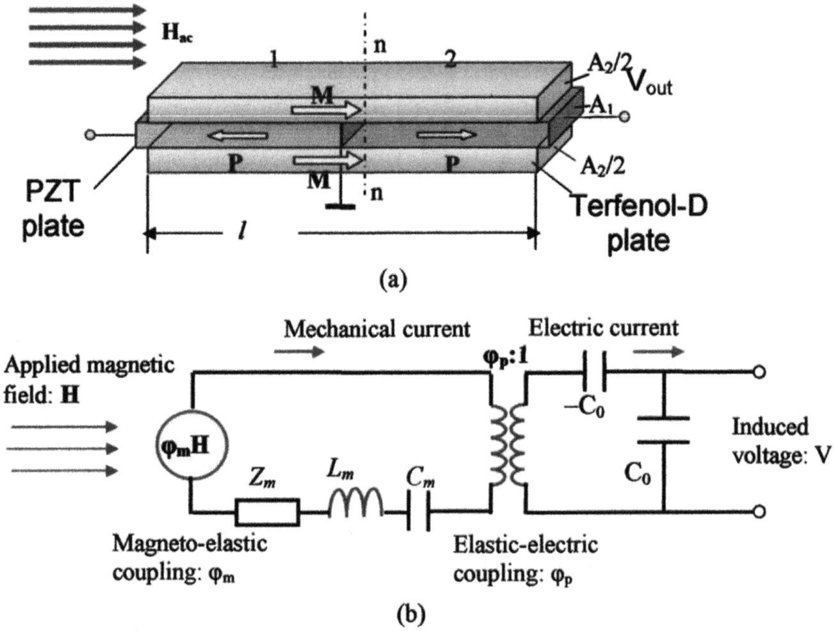

FIGURE A.4 (a) Configuration of (L−L) mode magnetostrictive/piezoelectric laminate composite. Arrows M and P designate the magnetization and polarization directions, respectively. (b) Magnetoelastoelectric equivalent circuit of (L−L) ME laminate resonator, where $R_m = \pi\ Z_0/4Q_m$, $L_m = \pi\ Z_0/4\omega_s$, and $L_m = 1/\omega_s^2\ L_m$ are the motional mechanical impedance, inductance, and capacitance, respectively; $\varphi_m = A_1 d_{33,m}/s_{33}^H$ is the magnetoelastic coupling factor; $\underline{\varphi}_p = 2A_1 g_{33,p}/l s_{33}^D \beta_{33}$ is the elastoelectric coupling factor; and $C_0 = 2A_1/l\ \rho_{33}$ is the clamped capacitance of the piezoelectric layer (Dong & Zhai, 2008; Shuxiang Dong & Viehland, 2003; Shuxiang Dong & Viehland, 2006).

conditions. An applied magnetic field *(H)* acts as a "mechanical voltage" (i.e., $\varphi_m H$) via a coupling factor (φ_m). This excites a "mechanical current" through the magnetoelastic effect. In turn, this results in an electrical voltage *(V)* across the two ends of the piezoelectric layer due to electromechanical coupling. A transformer with a turn-ratio of φ_p can be used to represent the electromechanical coupling in the circuit.

In Fig. A.4b, the magneto-elasto-electric equivalent circuit is illustrated, focusing on the region around the resonance frequency *(ω_s)*. This circuit is applicable to either section of the laminate around the nodal line and aids in understanding the ME coupling of the laminate under resonance drive conditions. When an applied magnetic field *(H)* is present, it acts as a "mechanical voltage" ($\varphi_m H$) through a coupling factor (φ_m). This mechanical voltage excites a "mechanical current" through the magnetoelastic effect. As a result of electromechanical coupling, an electrical voltage *(V)* is generated across the two ends of the piezoelectric layer. The electromechanical coupling in the circuit can be represented by a transformer with a turn-ratio of φ_p. Using this equivalent circuit, the ME voltage coefficient ($\alpha_{ME}(\omega)$) can be derived as a function of the AC magnetic-field frequency for the (L–L) ME laminate resonator (Dong & Zhai, 2008; Shuxiang Dong & Viehland, 2003; Shuxiang Dong & Viehland, 2006).

$$\alpha_{ME}(\omega) = \frac{\partial V_{ME}}{\partial H_3} = \left| \beta \frac{\varphi_p^2}{j\omega C_0} \frac{\varphi_m}{R_m + j\omega L_m + \dfrac{1}{j\omega C_m}} \right|, \tag{A.7}$$

where R_m, L_m, and C_m are motional resistance, inductance, and capacitance of the laminate, respectively, and C_0 is the clamped capacitance. At a resonance frequency of $\omega_s = (1/L_m C_m)^{1/2}$,$\alpha_{ME}(\omega)$ reaches a maximum value of:

$$\alpha_{ME,\omega_s} = \beta \frac{4Q_m \varphi_m \varphi_p^2}{n Z_0 \omega_s C_0}, \tag{A.8}$$

where $\beta <= 1$ is a factor related to applied magnetic field bias (H_{dc}) and at the optimum H_{dc} it equals 1, Q_m is the effective mechanical quality factor of the laminate composite including contributions from the Terfenol-D and piezoelectric layers, and the bonding between the layers, and Z_0 is the characteristic mechanical impedance of the laminate. Eq. (A.8) predicts at resonance that α_{ME} is $8Q_m/\pi^2$ higher than at subresonant frequencies.

Using the materials parameters for Terfenol-D and PZT-8, α_{ME} was calculated as a function of the frequency, as shown in Fig. A.5a. Using the experimental thickness ratio for the Terfenol-D layers of $n = 0.76$, and assuming $Q_m = 50$, the maximum value of α_{ME} was predicted to be 10.96V/Oe at a resonance frequency of $f_s = 20.03$ kHz ($\omega_s = 2\pi f_s$).

The ME voltage across the end and middle (ground) electrodes of the PZT layer in the laminate was measured using a lock-in amplifier method. The measurements were conducted while maintaining a DC magnetic bias of $H_{dc} = 200$ Oe, as it is known that the maximum inflection point of the quadratic strain-magnetic-field (ε–E) response occurs near this bias level.

A pair of Helmholtz coils was used to generate a small H_{ac} by applying an input AC current (I_{coil}). The frequency of H_{ac} was varied within the bandwidth of 1–40 kHz. The induced ME voltage of the (L–L) laminate exhibited a maximum value near the resonance frequency $f_s = 19.96$ kHz, as shown in Fig. A.5b. This observation is consistent with the predictions of Eq. (A.8).

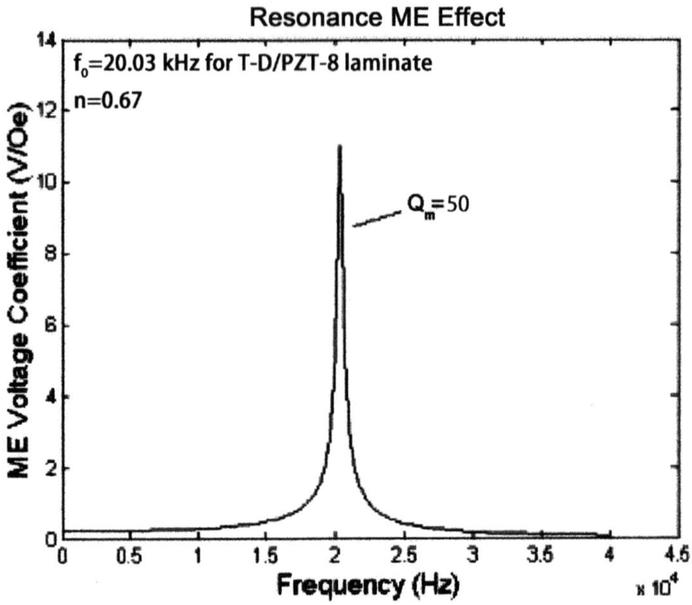

(a)

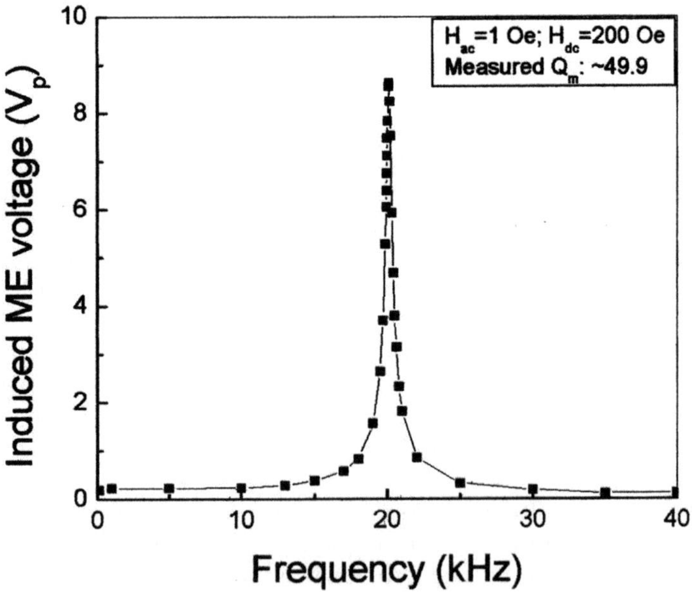

(b)

FIGURE A.5 Induced magnetoelectric voltage α_{ME} as a function of frequency: (a) calculated ME voltage using above mentioned equations and (b) measured one using a lock-in amplifier method (Shuxiang Dong, 2003).

The maximum value of α_{ME} was determined to be 8.7 V using a drive field of $H_{ac} = 1$ Oe, a drive frequency of 19.96 kHz, and a magnetic bias of $H_{dc} = 200$ Oe. Consequently, the ME voltage coefficient $\alpha_{ME}(\omega_s)$ at resonance was found to be 8.7 V/Oe.

Furthermore, the ME voltage induced in the L–T mode ME composite can be enhanced by a factor of 8 Q_m/π^2 at electromechanical resonance. Correspondingly, the ME coupling coefficient at resonance can be expressed as follows (Chu et al., 2017):

$$\left|\frac{dE}{dH_3}\right|_{(L-T)} = \frac{8Q_m}{\pi^2}\frac{nd_{33,m}g_{32,p}}{ns_{22}^E\left(1-k_{32,p}^2\right)+(1-n)s_{33}^H}\frac{S_{coil}}{w_m t_m \cdot \mu_r}, \tag{A.9}$$

where n is the piezomagnetic-phase thickness ratio that determines the optimum t_m for a given piezoelectric phase, $d_{33,m}$ and $g_{32,p}$ are the longitudinal piezomagnetic and transverse piezoelectric voltage coefficients, S_{22}^E and S_{33}^H are the elastic compliances of the piezoelectric and piezomagnetic phases, and $k_{32,p}$ is the electromechanical coupling coefficient of the piezoelectric phase.

4.2 Bilayer bending vibration mode

By introducing the concepts of mechanical voltage and mechanical current, a bending mode ME coupling equivalent circuit is developed, which proves to be useful in predicting the ME coefficients in the vicinity of the resonance frequency. Fig. A.6a–c illustrate the local coordinate system, schematic diagram, and equivalent circuit of the laminated bilayer magnetoelectric composite. In this equivalent circuit diagram, the coupling factors γ_m and γ_p represent the magnetoelastic and elastoelectric (electromechanical) transform ratios, respectively. When an applied magnetic field (H_3) is present, it generates a mechanical voltage, $\gamma_m H_3$, through the magnetoelastic effect with a coupling factor γ_m. Consequently, it induces a motional current, which leads to an electric voltage (V) across the piezoelectric layer due to electromechanical coupling. The electromechanical coupling in the circuit can be represented by a transformer with a turn-ratio of γ_p.

Under resonant drive, the impedance (Z_m) depicted in Fig. A.6b can be expressed in the form of a serial RLC circuit, as illustrated in Fig. A.6c. In the circuit, L_m is motional inductance $\left[L_m = 1/(2i)(dZ_m/d\omega)|_{\omega=2\pi f_1} = 0.0231a^3\left(\rho_p h_p + \rho_m h_m\right)/b\right]$; C_m is motional capacitance $\left[C_m = 1/(4\pi^2 f_1{}^2 L_m)\right]$. Because of energy dissipation, a motional resistance R_m is introduced $\left[R_m = (2\pi f_1 L_m)/Q_m\right]$, where Q_m is

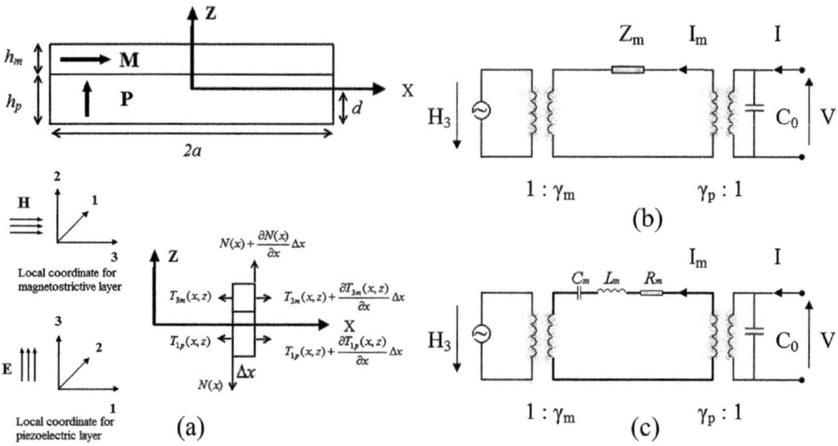

FIGURE A.6 (a) Local coordinate and schematic diagram of the laminated bilayer magnetoelectric composite and stress analysis in the differential element. (b) Magneto-elastic-electric equivalent circuit for bending mode of laminated bilayer magnetoelectric composites. (c) RLC equivalent circuit under resonance drive, where L_m is motional inductance $\left[L_m = 1/(2i)(dZ_m/d\omega)|_{\omega=2\pi f_1} = 0.0231a^3\left(\rho_p h_p + \rho_m h_m\right)/b\right]$, C_m is motional capacitance $\left[C_m = 1/\left(4\pi^2 f_1{}^2 L_m\right)\right]$, and R_m is motional resistance $[R_m = (2\pi f_1 L_m)/Q_m]$ (Dong, 2009).

the mechanical quality factor of the laminated bilayer composite]. Note that by assuming complex elastic coefficients in constitutive equations, the mechanical quality factor can be found. The resonance bilayer ME voltage coefficient $|dV/dH_3|$ as a function of the ac magnetic field at the resonance frequency, $\omega = \pi_1 = 2\pi f_1$, ME voltage coefficient reaches a maximum value of (Dong, 2009):

$$\left|\frac{dV}{dH_3}\right|_{\omega 1} \approx \frac{\gamma_m \gamma_p}{\omega_1 C_0 \left|R_m + i\omega_1 L_m + \dfrac{1}{i\omega_1 C_m}\right|} = \frac{Q_m \gamma_m \gamma_p}{\omega_1^2 L_m C_0}, \qquad \text{(A.10)}$$
$$= Q_m \gamma_m \gamma_p C_m / C_0.$$

It is well-established that the frequency plays a significant role in influencing the ME coupling in laminated composites. When the laminated composite operates in resonance mode, the ME effect can be greatly enhanced, often resulting in an ME voltage output several orders of magnitude higher than that of off-resonance ME laminates. According to the given equation, at resonance, the ME coefficient is directly proportional to the mechanical quality factor.

To analyze the effect of material parameters and dimensional parameters on the bilayer ME coefficient, Eq. (A.10) is rewritten as (Dong, 2009):

$$\begin{aligned}\left|\frac{dV}{dH_3}\right|_{\omega 1} &= Q_m\gamma_m\gamma_p C_m/C_0 \\ &= \frac{1}{s^D_{11p}} h_p Q_m d_{33m} g_{31p} G\left(h_m/h_p, s^H_{33m}/s^D_{11p}\right)\end{aligned}, \tag{A.11}$$

It shows that the ME coefficient is also proportional to the piezomagnetic constant d_{33m}, piezoelectric constant g_{31p}, and geometric factor $G(h_m/h_p, s^H_{33m}/s^D_{11p})$ (which is a function of the thickness ratio h_m/h_p and elastic compliance ratio s^H_{33m}/s^D_{11p}.

4.3 C–C radial vibration mode

Traditionally, studies on magnetostrictive and piezoelectric laminates have focused on simple geometries such as discs, squares, or rectangles. These laminates are typically operated in modes such as transverse-transverse (T–T), longitudinal-transverse (L–T), or longitudinal-longitudinal (L–L), where the magnetic field (*H*) and electric field (*E*) are oriented in specific directions relative to the principle strain direction of the laminate. These modes are suitable for detecting magnetic fields with a constant direction.

However, in many practical scenarios, there is a need to detect alternating current (AC) rotating or vortex magnetic fields generated by current-carrying wires, as found in power-integrated circuits and superconducting films. Therefore, it becomes necessary to develop a magnetostrictive-piezoelectric (ME) laminate specifically designed for detecting vortex magnetic fields.

Fig. A.1c illustrates our ring-type laminate design specifically tailored for this purpose. It comprises two Terfenol-D rings magnetized in the circumferential direction and one PZT ring with m segments polarized in the same circumferential direction. The PZT ring is sandwiched between the two Terfenol-D rings. This laminated ring operates in a circumferential-circumferential (C–C) mode relative to its principal vibration direction. When a vortex magnetic field is applied to the laminated ring, it induces a continuous magnetic flux in the Terfenol-D rings, intensifying the principle strain in the circumferential direction (Dong et al., 2004; Shuxiang Dong & Viehland, 2004).

In the C–C mode, the piezomagnetic and piezoelectric layers are coupled through strain *S(z)* and stress *T(z)*. Applying an AC vortex magnetic field (H_{ac}) along the circumferential direction of the magnetostrictive rings causes the piezoelectric ring to undergo forced oscillation in the same direction. This excites a radial-symmetric vibration mode in the piezoelectric ring, generating a voltage across each segment of the piezoelectric ring through the longitudinal piezoelectric constant ($d_{33,p}$). Assuming the laminated ring is thin, the transverse (31) vibration mode (width or thickness) can be neglected compared to its circumferential principal strain. Consequently, the voltage induced by the magnetic field is a result of the C–C mode effect.

Expressions for the ME voltage coefficients were derived from the equivalent circuit depicted in Fig. A.7a. Under open-circuit conditions,

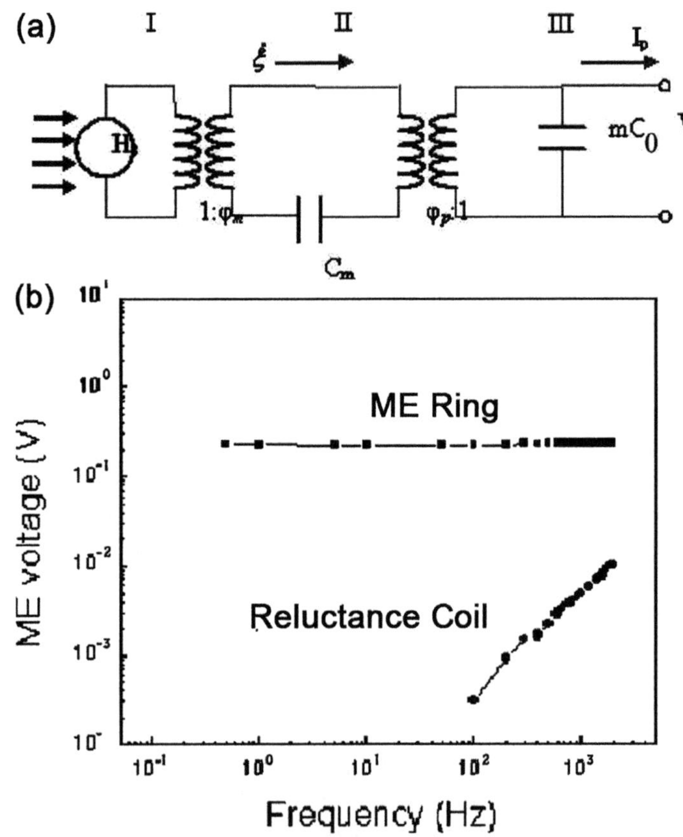

FIGURE A.7 (a) Magnetoelasto-electric bieffect equivalent circuit for the C–C mode magnetostrictive/piezoelectric laminated ring under a vortex magnetic field. (b) ME voltage responses of our C–C laminated ring and a reluctance coil as a function of the drive frequency of an applied vortex magnetic field for $10^{-1} < f < 10^{3}$ Hz (Dong et al., 2004; Shuxiang Dong & Viehland, 2004).

the current *(I)* flowing through the piezoelectric layer is zero. Consequently, the capacitive load (mC_0) can be relocated to the mechanical section of the circuit. Assuming that the ratio of the induced voltage *(V)* to the magnetic exciting voltage $(\varphi_m H_3)$ is proportional to the ratio of their impedances, according to Ohm's law, we obtain the following expressions (Dong et al., 2004):

$$\left|\frac{V}{\varphi_m H_3}\right| = \beta \left| \frac{\dfrac{\varphi_p{}^2}{j\omega m C_0}}{Z_m + \dfrac{\varphi_p{}^2}{j\omega m C_0}} \right|, \tag{A.12}$$

where $\beta = 1(V/N)$ is a factor related to the dc magnetic bias applied to the magnetostrictive layers. For $\omega << \omega_0$, where ω_0 is the free angular resonance frequency of the laminated ring, the mechanical losses can be neglected. Thus, the value of the mechanical impedance (Z_m) is $\sim 1/j\omega C_m$. The ME voltage coefficient under a small magnetic field excitation can then be approximated in differential form as (Dong et al., 2004):

$$\left|\frac{dV}{dH_3}\right| = \beta \frac{\varphi_p^2 \varphi_m}{\dfrac{mC_0}{C_m} + \varphi_p^2}, \tag{A.13}$$

Eq. (A.11b) shows that the ME voltage coefficient is dependent upon φ_m of the Terfenol-D rings and φ_p of the PZT one. Indeed, high magnetoelastic and electromechanical couplings are essential factors that contribute to the design of laminated rings with a high ME voltage coefficient. These couplings determine the efficiency of the conversion between magnetic and electric energy in the composite structure. By optimizing both the magnetoelastic and electromechanical couplings, the laminated ring can achieve a higher ME voltage coefficient. This coefficient represents the ability of the device to generate a larger electrical output in response to a given magnetic input, making it a crucial consideration in the design of laminated rings for ME applications. Eq. (A.11b) can be expressed as a function of the geometric factor *(n)*, as (Dong et al., 2004):

$$\left|\frac{dV}{dH_3}\right|_{(C-C)} = \beta \frac{2\pi n(1-n) d_{33,m} d_{33,p}^2 A_{lam^\circ}}{n\varepsilon_{33}^S \left(s_{33}^E\right)^2 + (1-n) s_{33}^H \left(s_{33}^E \varepsilon_{33}^S + d_{33,p}^2\right)}, \tag{A.14}$$

where $\varepsilon^S{}_{33}=(1-k^2{}_{33,p})$. The value of $|dV/dH3|_{(C-C)}$ is directly proportional to the $d_{33,p}$ and $d_{33,m}$ constants; also, a larger value of A_{lam° results in a higher value of $|dV/dH_3|_{(C-C)}{}^\circ$ as more magnetic flux is captured.

In addition, the ME voltage coefficient is strongly dependent on the thickness ratio of the Terfenol-D layers, n.

A toroidal-type variable reluctance coil, consisting of 100 turns, was fabricated and utilized for the detection of an AC vortex magnetic field. Fig. A.7b illustrates the induced voltage as a function of the drive frequency for both the reluctance coil sensor and the C–C mode laminated ring. It can be observed that the induced voltage for the reluctance coil is significantly smaller (approximately 0.1–0.01 times) compared to the C–C mode laminated ring.

Increasing the number of coils in the reluctance coil may potentially result in higher magnetic sensitivity. However, it was discovered that the induced voltage for the reluctance coil is strongly dependent on the frequency, whereas the C–C laminated ring remains frequency-independent within the range of $10^{-1} < f < 10^3$ Hz. For instance, at a frequency of 10^2 Hz, the rate of flux change passing through the coil is sufficiently small, causing the signal detected by the coil to become indistinguishable from the noise floor. Another alternative for a vortex sensor is a Hall device, but its sensitivity is relatively low. Experimental measurements have confirmed that the C–C mode laminated ring exhibits an ME voltage coefficient approximately 3–5 times higher than the L–T or L–L mode laminated rings, which aligns with theoretical predictions, although the measured values are lower than the predicted values, suggesting potential for improvement. The results indicate that Terfenol-D- and PZT-laminated rings possess a high sensitivity of less than 10^{-8} T to a vortex magnetic field. With optimization of materials and assembly methods, the sensitivity limit could potentially be enhanced to the range of 10^{-11} to 10^{-12} T (Dong et al., 2004).

4.4 Power conversion efficiency of ME gyration

Indeed, power conversion efficiency in ME devices is a crucial factor that determines their overall performance. It is primarily influenced by two fundamental energy conversion processes: (1) Magnetic energy to mechanical vibration energy conversion: This conversion process occurs through the magnetostrictive effect in the ferromagnetic phase of the ME composite. The applied magnetic field induces changes in the magnetic properties of the ferromagnetic material, resulting in mechanical strain and vibration. To enhance power conversion efficiency, it is important to select optimum ferromagnetic materials with high magnetostrictive coefficients and favorable magnetic properties. Additionally, the geometric structure and design of the ME composite should be optimized to maximize the magnetostrictive response and mechanical energy transfer.

(2) Mechanical energy to electrical energy conversion (and vice versa): This conversion process involves the piezoelectric effect in the piezoelectric phase of the ME composite. Mechanical strain or vibration applied to the piezoelectric material generates electric charges, which can be converted into electrical energy. Conversely, an applied electric field can induce mechanical strain or vibration in the piezoelectric material. To enhance power conversion efficiency, it is crucial to utilize high-quality piezoelectric materials with high piezoelectric coefficients and excellent electromechanical coupling properties. Optimal geometric structures and configurations should also be employed to maximize the piezoelectric response and facilitate efficient energy conversion.

In both cases, high coupling factors (k), which represent the effectiveness of energy transfer between different phases, and high mechanical quality factors (Q_m), which characterize the energy storage and dissipation capabilities of the ME composite, are essential for enhancing power conversion efficiency. These factors are influenced by the material properties, composite structure, and resonance behavior of the ME device (Shuxiang Dong et al., 2006; Leung et al., 2016, 2017a, 2017b; Zhuang et al., 2019).

The mechanical quality factor $(Q_{m,\text{eff}})$ of an ME laminate composite is indeed influenced by the mechanical quality factors of the individual phases, namely, $Q_{m,m}$ in the magnetostrictive phase and $Q_{m,p}$ in the piezoelectric phase, as well as the thickness ratio (n) between them. The relationship can be expressed as $Q_{m,\text{eff}} = (n/Q_{m,m}+(1\text{-}n)/Q_{m,p})^{-1}$. In the case of the ferromagnetic material Metglas, which exhibits a high permeability (over 45,000) and a high resonant magneto-elastic (M-E) coupling efficiency of 97%, it possesses a strong magnetomechanical conversion ability. This means that it is capable of efficiently converting magnetic energy to mechanical vibration energy.

To highlight the importance of the mechanical quality factor (Q_m) in the L–T mode magnetoelectric composite, three different types of ME laminated composites were fabricated: (i) sample #1, the amorphous Metglas/Mn-doped (hard) PMN-PZT, (ii) sample #2, nanocrystalline Metglas/undoped (soft) PMN-PT-laminated composites, and (iii) sample #3, the nanocrystalline Metglas/Mn-doped (hard) PMN-PZT single crystal one (PourhosseiniAsl et al., 2020).

The schematic and equivalent circuit of the coil-ME gyrator operating in the L–T mode under a free vibration condition is shown in Fig. A.8a and b, respectively. In this circuit, R_m represents the motional mechanical resistance, C_m represents the motional mechanical capacitance, and L_m represents the motional mechanical inductance. These elements collectively represent the mechanical response of the system. The working principle of the ME gyrator is based on the combination effects of Faraday's electromagnetic induction and ME coupling. When an input AC

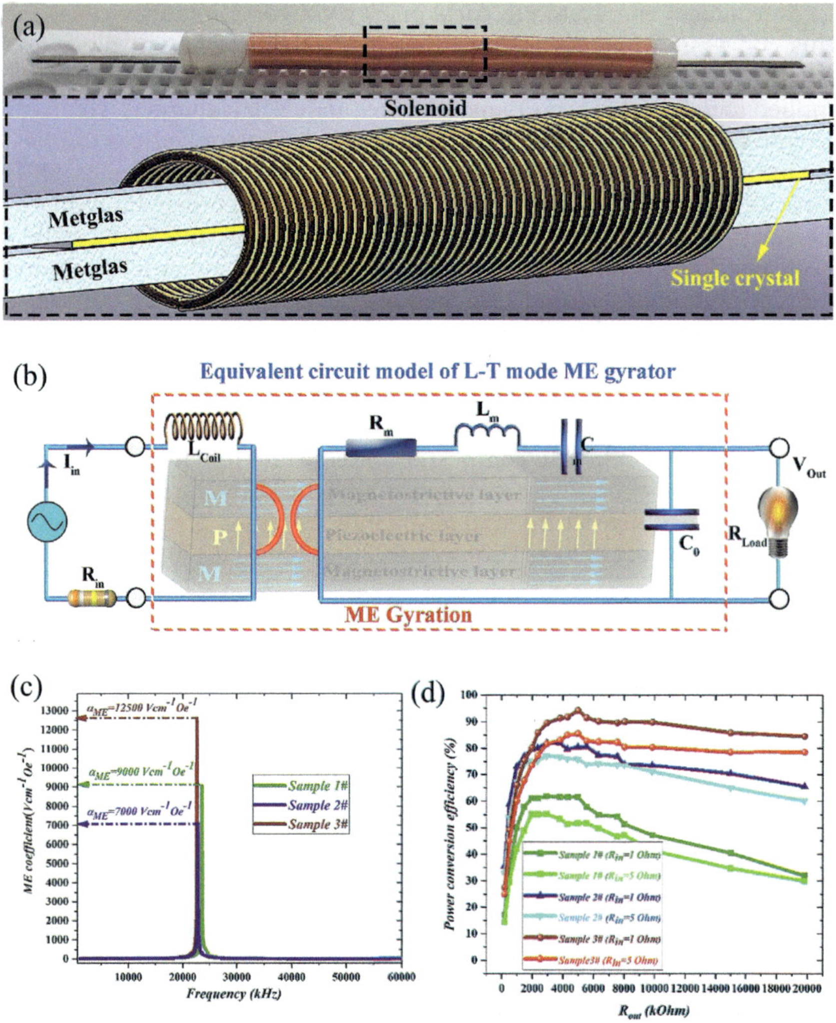

FIGURE A.8 (a) Schematic of the magnetoelectric structure. (b) Equivalent circuit model of the L-T mode coil-ME gyrator. (c) Measured dependences of the generated ME voltage coefficient versus frequency: Sample #1: Heat-treated Metglas/soft PMN-PT laminated, sample 2#: Untreated Metglas/hard PMN-PZT and sample 3#: Heat-treated Metglas/hard PMN-PZT, and (d) power efficiency of gyrators. In measurement, a small resistance (R_{in}) was connected in series for measuring the input current (I_{in}) at the input port, and the input voltage (V_{in}) applied to the ME gyrator was calculated by the output voltage of an ac power amplifier and voltage drop across R_{in}. Note that different input (R_{in}) has slightly effect on power conversion efficiency (η). The real input power (P_{in}) was calculated by $P_{in} = V_{in} \times I_{in} \times \cos\theta$ in the input port, a varying resistor (R_{out}) was connected to the output port as a resistance load, and the real output power (P_{out}) was calculated by $P_{out} = \frac{V_{out}^2}{R_{out,}}$. The power conversion efficiency η can be estimated by $\eta = (\frac{P_{out}}{P_{in}}) \times 100$ (PourhosseiniAsl et al., 2020).

current is applied to the wrapped coil, it generates an oscillating AC magnetic field (H_{ac}). This magnetic field interacts with the ferromagnetic phase in the ME composite, inducing an amplified voltage output from the piezoelectric phase via the ME effect.

Fig. A.7c illustrates the measured ME coefficient (α_{ME}), which quantifies the magneto-elasto-electric (ME) coupling, as a function of frequency for three different samples: #1, #2, and #3. These samples were subjected to an optimized DC bias field. Notably, sample #3 operating in the longitudinal vibration mode at the resonance frequency of approximately 23 kHz exhibits a remarkably high ME coefficient of 12,500 V/Oe cm. This suggests a strong magneto-elasto-electric coupling in this sample. The high quality factors (Q_m) in both the ferromagnetic and ferroelectric phases contribute to a larger effective $Q_{m,\mathrm{eff}}$ in the ME laminated composite, thereby enhancing its electromechanical resonance. By considering the 3-dB bandwidth at the resonance frequency ($Q_{m,\mathrm{eff}} = f_r/\Delta f_{r(3\text{-dB})}$) and the measured data shown in Fig. A.8c, the estimated effective $Q_{m,\mathrm{eff}}$ values for ME samples #1, #2, and #3 are 390, 320, and 570, respectively. Sample #3, with the largest $Q_{m,\mathrm{eff}}$, exhibits the highest ME coupling coefficient. This can be attributed to the reduced mechanical loss due to nanocrystallization in the ferromagnetic phase and low dielectric loss in the hard ferroelectric phase. Overall, the results demonstrate that the optimization of the mechanical quality factors and material properties in both the ferromagnetic and ferroelectric phases plays a crucial role in achieving high ME coupling coefficients and enhancing the performance of ME-laminated composites.

The maximum efficiency (η_{Max}) for a ME gyrator working at its resonance frequency can be estimated as the ratio of the output power (P_{out}) to the input power (P_{in}). Assuming the eddy current loss in multilayered ME composites can be neglected, the maximum efficiency (η_{Max}) for a ME gyrator working at its resonance frequency can be estimated as (Shuxiang Dong et al., 2006; PourhosseiniAsl et al., 2020):

$$\eta_{\mathrm{Max}} = \frac{\varphi_P^2}{\varphi_P^2 + \dfrac{\pi Z_0 C_0 \omega_S}{Q_{m,\mathrm{eff}}}}, \tag{A.15}$$

where $\varphi_P \left(= \frac{w d_{32,p}}{S_{22}^E} = k_{32,p}\, w \sqrt{\frac{\varepsilon_{33}^T}{S_{22}^E}} \right)$ is the electromechanical coupling factor; Z_0 is the motional mechanical impedance; C_0 is the electrical capacitance; ω_S is the resonance angular frequency; and $Q_{m,\mathrm{eff}}$ is the mechanical quality factor of the ME-laminated composite. Clearly, both high $k_{32,p}$ and φ_p in piezoelectric phase as well as a high $Q_{m,\mathrm{eff}}$ of ME composite will result in an enhancement in power conversion efficiency (η_{Max}) of ME gyrator.

In Fig. A.8d, the power conversion efficiency (η) of coil-ME samples #1-#3 is presented. It is observed that coil-ME sample #3, which consists of a nanocrystalline Metglas/hard Mn-doped PMN-PZT composite, exhibits an ultra-high efficiency of approximately 95% at a load resistance (R_{out}) of 5 kOhm. On the other hand, coil-ME samples #1 and #2 demonstrate lower efficiencies, with approximately 61.5% at $R_{out} = 3$ kOhm and 82% at $R_{out} = 5$ kOhm, respectively.

These results clearly highlight the significance of a high mechanical quality factor ($Q_{m,eff}$) in the ME composite for enhancing the power conversion efficiency (η) of the coil-ME system. The higher $Q_{m,eff}$ in sample #3 enables it to achieve a remarkably high efficiency compared to the other samples. It should be noted that power conversion efficiency can be influenced by various factors, including losses, impedance matching, and nonidealities in the ME system. Therefore, optimizing the mechanical quality factor and addressing other relevant factors are crucial for achieving higher efficiency in ME devices.

References

Ce-Wen Nan, M. I. B., Dong, S., & Viehland, D. (2008). Multiferroic magnetoelectric composites: Historical perspective, status, and future directions. *Journal of Applied Physics, 103*.

Chu, Z., et al. (2017). Enhanced resonance magnetoelectric coupling in (1-1) connectivity composites. *Advanced Materials, 29*(19).

Dong, M. G. A. S. (2009). A resonance-bending mode magnetoelectric coupling equivalent circuit. *Ieee Transactions on Ultrasonics , Ferroelectrics, and Frequency Control, 56*, 11.

Dong, S., et al. (2003). Giant magnetoelectric effect in laminate composites. *Philosophical Magazine Letters, 83*(12), 769–773.

Dong, S., et al. (2004). Circumferentially magnetized and circumferentially polarized magnetostrictive/piezoelectric laminated rings. *Journal of Applied Physics, 96*(6), 3382–3387.

Dong, S., et al. (2006). Magnetoelectric coupling, efficiency, and voltage gain effect in piezoelectric-piezomagnetic laminate composites. *Journal of Materials Science, 41*(1), 97–106.

Dong, S., & Zhai, J. (2008). Equivalent circuit method for static and dynamic analysis of magnetoelectric laminated composites. *Science Bulletin, 53*(14), 2113–2123.

Leung, C. M., et al. (2016). Power conversion efficiency and resistance tunability in coil-magnetoelectric gyrators. *Applied Physics Letters, 109*(20).

Leung, C. M., et al. (2017a). Highly efficient solid state magnetoelectric gyrators. *Applied Physics Letters, 111*(12).

Leung, C. M., et al. (2017b). Importance of composite parameters in enhanced power conversion efficiency of Terfenol-D/PZT magnetoelectric gyrators. *Applied Physics Letters, 110*(11).

PourhosseiniAsl, M., et al. (2020). Versatile power and energy conversion of magnetoelectric composite materials with high efficiency via electromechanical resonance. *Nano Energy, 70*.

Priya, S., et al. (2007). Recent advancements in magnetoelectric particulate and laminate composites. *Journal of Electroceramics, 19*(1), 149–166.

Shuxiang Dong, J. (2003). Characterization of magnetoelectric laminate composites operated in longitudinal-transverse and transverse– transverse modes. *Journal of Applied Physics, 95*, 5.

Shuxiang Dong, J.-F. L., & Viehland, D. Giant magneto-electric effect in laminate composites. Ieee Transactions on Ultrasonics, Ferroelectrics, and Frequency Control, 50, 10.

Shuxiang Dong, J. C., Li, J. F., & Viehland, D. (2003). Enhanced magnetoelectric effects in laminate composites of Terfenol-D/Pb(Zr,Ti)O_3 under resonant drive. *Applied Physics Letters, 83*, 23.

Shuxiang Dong, J. Z., Li, J. F., & Viehland, D. (2006). Magnetoelectric gyration effect in $Tb_{1-x}Dy_xFe_{2-y}$/Pb(Zr,Ti)O_3 laminated composites at the electromechanical resonance. *Applied Physics Letters, 89*.

Shuxiang Dong, J.-F. L., & Viehland, D. (2003). Longitudinal and transverse magnetoelectric voltage coefficients of magnetostrictive/piezoelectric laminate composite: Theory. *Ieee Transactions on Ultrasonics , Ferroelectrics, and Frequency Control, 50*, 10.

Shuxiang Dong, J. F. L., & Viehland, D. (2004). Voltage gain effect in a ring-type magnetoelectric laminate. *Applied Physics Letters, 84*.

Shuxiang Dong, J.-F. L., & Viehland, D. (2006). Magnetoelectric coupling, efficiency, and voltage gain effect in piezoelectric-piezomagnetic laminate composites. *Journal of Materials Science, 41*, 97–106.

Srinivasan, G., et al. (2001). Magnetoelectric bilayer and multilayer structures of magnetostrictive and piezoelectric oxides. *Physical Review B, 64*(21).

Zhuang, X., et al. (2019). Power conversion efficiency and equivalent input loss factor in magnetoelectric gyrators. *IEEE Transactions on Industrial Electronics, 66*(4), 2499–2505.

Part B

4.5 Experiments on ME effects in thick film laminate composites

Theory/model magnetic oxides including spinel ferrites, hexagonal ferrites, or lanthanum manganites for the magnetic phase and lead zirconate titanate (PZT), barium titanate (BTO), or lead magnesium niobate-lead titanate (PMN-PT) for the ferroelectric phase are considered (Nan et al., 2008; Cho & Priya, 2011; Wang et al., 2014). Composites with 3D transitions metals such as Ni, Co, Fe, and alloys such as Terfenol-D, galfenol, permendur, or Metglas are also discussed (Srinivasan, 2010; Ma et al., 2011; Lawes & Srinivasan, 2011). Efforts on composites with piezoelectric AlN, lanthanum-gallium-tantalate (langatate), and quartz are also addressed (Sreenivasulu, Petrov, et al., 2012; Sreenivasulu, Fetisov, et al., 2012). Early works on ME composites dealt with bulk composites prepared by mixing and sintering ferrite and barium titanate powders, whereas later works have mainly focused on layered composites and nanostructures (Gheevarughese et al., 2007; Mandal et al., 2010, 2011; Petrov & Srinivasan, 2008; Sun & Srinivasan, 2012; Wang et al., 2011). The strongest ME coupling is expected in a layered structure due to (i) low leakage currents and (ii) ease of poling to align the electric dipoles and strengthen the piezoelectric effect. Thick film–layered samples are synthesized by laminating and sintering of the two phases made by tape casting or by bonding single crystal or polycrystalline platelets. Efforts on nanocomposites are not part of this chapter and are discussed in Chapter 6.

Several techniques were employed for studies on direct and converse ME coupling in these composites (Nan et al., 2008). A giant low-frequency direct-ME (DME) effect and resonance enhancement of the coupling at bending or electromechanical resonance were reported for a variety of systems (Srinivasan, 2010; Wang et al., 2010). Ferromagnetic resonance (FMR) is a powerful tool for studies on converse-ME (CME) effects since the electric field E applied to the composite manifests as frequency or field shift in the FMR profiles and the strength of CME coupling could be obtained from the data on the shifts as a function of E. Strong CME effects in ferrite–ferroelectric composites were reported with the use of FMR studies (Sun & Srinivasan, 2012; Zavislyak et al., 2013). Models were developed for low-frequency and resonance DME effects and for CME at FMR (Nan et al., 2008). Another key area of focus in this chapter is the linear ME effects that arise due to nonlinearity in the magnetostriction or piezoelectric coupling coefficients in the composites. Significant developments in the application aspects of metals/oxide layered composites include magnetic sensors and E-tunable ferrite resonator, filter, phase shifter, and miniature antennas (Tatarenko & Srinivasan, 2011; Sun & Srinivasan, 2012). We conclude this chapter with a brief discussion on recent advances on the use of ME laminates for a variety of applications such as sensors, energy harvesting, gyrators, and antennas. Extensive details on the application aspects of laminates as well as nanocomposites are provided in the second part of this book.

4.5.1 Ferromagnetic—ferroelectric composites

The materials of choice for the oxide ME composites are ferrimagnetic spinels ferrites, M-, Y-, Z-type hexagonal ferrites, or yttrium iron garnet (YIG) and ferroelectric PZT, PMN-PT, or BTO due to their robust ferroic response at room temperature and high chemical, thermal, and mechanical stability. The key parameters for the choice of ferrite for strong DME coupling are the magnetostriction λ and the piezomagnetic coefficient $q = d\lambda/dH$ and the permeability μ that is a measure of the magnetic field confinement. The strength of DME coupling is critically dependent on the q-value. The strength of CME is directly proportional to λ. For measurements on CME by FMR, both the FMR line-width and the magnetic anisotropy field are important parameters. Nickel ferrite and Ni-Zn ferrites have high values of λ, q, and μ for achieving composites with strong DME (Srinivasan, 2010). For strong CME and for E-tunable FMR devices such as resonators and filters, one prefers a ferrite with high λ, low high-frequency losses, and an anisotropy field facilitates FMR in the frequency range of interest for nominal bias magnetic fields (Sun & Srinivasan, 2012). Spinel ferrites with cubic anisotropy of 50–500 Oe and YIG are used in composites for strong CME and device applications over range 1–20 GHz. Y-type hexagonal ferrites $Ba_2Zn_2Fe_{12}O_{22}$ (Zn_2Y) with easy plane anisotropy of 1–10 kOe are preferred for 20–40 GHz. M-type hexagonal ferrites with uniaxial anisotropy, $BaFe_{12}O_{19}$ (BaM), SrM and Al-substituted hexaferrites, and substituted ferrites including (Ba,Al) M and (Sr,Al) M are ideal for the frequency range 40–110 GHz (Mathe et al., 2008; Pullar, 2012).

Both PZT and PMN-PT are ferroelectrics of importance for use in the composite. For strong CME, one desires a high value of the piezoelectric coupling coefficient d (Cross, 1993), whereas for strong DME, it is necessary to have a high value for d/ε (ε is the dielectric constant) (Nan et al., 2008). The ferroelectric nature of PZT, $Pb(Zr_{1-x}Ti_x)O_3$, is critically dependent on the Zr/Ti ratio. Much interest is focused on compositions close to the morphotropic phase boundary (MPB) for Ti ratio of ~47% because of their large piezoelectric constants $d_{33} = 200–600$ pm/V (Cross, 1993).

Similarly PMN-PT, (1-x) $Pb(Mg_{1/3}Nb_{2/3})O_3$ – x $PbTiO_3$, which is a relaxor ferroelectric shows an MPB around 33%Ti and $d_{33} = 1000–2500$ pm/V (Cross, 1993). Piezoelectrics such as ZnO, AlN, quartz, and langatate with very high values of d/ε have been used in composites showing strong DME for use as magnetic sensors (Sreenivasulu, Petrov, et al., 2012; Sreenivasulu, Laletin, et al., 2012). Synthesis of layered thick film composites of ferrite-PZT, lanthanum calcium manganite (LCMO)-PZT, and lanthanum strontium manganite (LSMO)-PZT involved laminating and sintering thick films of the two phases made by tape-casting or epoxy bonding of platelets of single crystal or polycrystalline oxides (Srinivasan et al, 2002; Srinivasan, 2010; Zhang et al., 2009). Thin film composites are made by techniques including PLD, MBE, and CVD (Nan et al., 2008; Li et al., 2011).

Composites with ferromagnetic metals/alloys with high values of λ and q are of interest for accomplishing very high ME coupling strengths for useful technologies. The FM layer serves the dual role of ferromagnetic phase and electrodes for the structure. Studies on composites with 3d-transition metals (Ni, Fe, Co), permendur (an alloy of Fe-Co-V), Metglas, Galfenol (alloy of Fe and Ga), and Terfenol-D reveal strong ME coupling (Laletin et al., 2005; Sreenivasulu, Petrov, et al., 2012; Srinivasan et al., 2005; Wang et al., 2011). Among the 3d-transition metals, Ni has the highest λ and q, whereas Fe has the lowest λ (Srinivasan et al., 2005). Composites with permendur and Galfenol have high ME coupling strength, whereas those with Metglas with high q are reported to have the strongest DME coupling (Greve et al., 2010; Sreenivasulu, Laletin, et al., 2012;

Wu et al., 2013). Terfenol-D has one of highest λ, but the piezomagnetic coefficient q is rather small compared to Metglas or Galfenol that results in a weak DME in composites with ferroelectrics (Hockel et al., 2011; Park et al., 2010; Yao, Hou, Dong, & Li, 2011). Some recent studies on composites involved the use of both ferromagnetic alloys and ferrimagnetic oxides (Yang et al., 2011).

4.5.2 Direct magnetoelectric effects

The electrical response of a composite to an applied magnetic field is termed direct ME effect. Measurement techniques for DME include (i) subjecting the sample to an ac magnetic field δH and a bias magnetic field H and measure the ME voltage coefficient, (ii) magneto-dielectric effect involving the measurement of ε as a function of bias field H, and (iii) change in polarization P with H. This section focuses on studies of DME in layered composites.

4.5.2.1 *Low-frequency ME effect*

This measurement is on electric field response of composites that are first polarized in an electric field E and then subjected to an ac magnetic field δH and a bias field H. The resulting ac voltage δV is measured across the sample thickness. The ME voltage coefficient MEVC $= \delta E/\delta H = \delta V/(t\ \delta H)$, where t is the thickness of ferroelectric (or piezoelectric) layer, is measured as a function of bias field H and frequency f of δH and for a series of magnetic field orientation. Assuming the sample in the (1,2) plane, the MEVC is usually measured for two conditions: (i) for H and δH parallel to each other and to the sample plane (1,2) and δE measured along direction 3, and (ii) all the three fields parallel to each other and perpendicular to sample plane. Fig. B.1 shows representative data on H dependence of MEVC measured at 300 K for layered ferrite-PZT composites (Nan et al., 2008; Srinivasan, 2010). Fig. B.1a is for a multilayer of nickel ferrite (NFO)-PZT prepared by laminating and sintering 10–40 μm thick ferrite and PZT prepared by tape casting (Srinivasan et al., 2001). Data are for (i) δH_1 and H_1 parallel to each other and to the sample plane and (ii) δH_3 and H_3 parallel to each other and perpendicular to the sample plane and $f = 100$ Hz. A general increase in MEVC is observed with H to a peak value, followed by a rapid drop. According to a model that assumes a one-dimensional composite with length much higher than the width and thickness, the MEVC for in-plane magnetic fields along direction *1* and E measured along direction *3* is given by (Sreenivasulu, Petrov, et al., 2012)

$$\text{MEVC} = \frac{\delta E_3}{\delta H_1} = \frac{{}^m q_{11}\, {}^p d_{13}/{}^p\varepsilon_{33}}{r\, {}^m s_{11} + {}^p s_{11}} \tag{B.1}$$

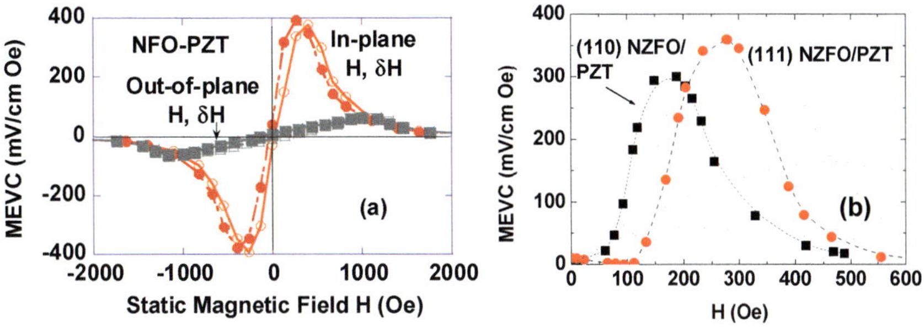

FIGURE B.1 Low frequency direct magnetoelectric (DME) effects in ferrite-PZT composites. The ME voltage coefficient (MEVC) is shown as a function of bias field H. (a) Multilayers of nickel ferrite (NFO)-PZT: Data are for H and ac field δH (at 100 Hz) parallel to each other and for increasing (filled symbols) and decreasing (open symbols) H. Results are for the fields applied either parallel or perpendicular to the sample plane. (b) ME voltage coefficient versus H for epoxy bonded laminates of (110) and (111) nickel zinc ferrite-PZT for in-plane magnetic fields (Srinivasan et al., 2001; Srinivasan, 2010).

where ${}^{p}s_{11}$, ${}^{m}s_{11}$, ${}^{p}d_{13}$, and ${}^{m}q_{11}$ are compliance, and piezoelectric/piezomagnetic coupling coefficients for ferroelectric and piezomagnetic layers, respectively, ${}^{p}\varepsilon_{33}$ is the permittivity of piezoelectric layer, and $r = {}^{p}t/{}^{m}t$ with ${}^{p}t$ and ${}^{m}t$ denoting the thicknesses of piezoelectric and piezomagnetic layers. The coefficients are directly proportional to the piezomagnetic coupling $q = d\lambda/dH$, and the H-dependence in Fig. B.1a tracks the slope of λ versus H. Saturation of λ at high field leads to MEVC of zero. One measures an order of magnitude higher MEVC for in-plane magnetic fields compared to out-of-plane orientation for magnetic fields, and this is due large demagnetizing fields (and therefore small piezomagnetic coupling q) for out-of-plane magnetic fields. Similar studies were reported on bilayers of single crystal ferrites bonded to ferroelectrics. Representative MEVC data for such laminates for in-plane magnetic fields are shown in Fig. B.1b. The results are for PZT and (110) $Ni_{0.9}\ Zn_{0.1}\ Fe_2\ O_4$ (NZFO) or (111) NZFO (Gheevarughese et al., 2007). Although there is dependence of MEVC on the orientation of magnetic fields relative to crystallographic direction due the directional dependence of λ, the overall MEVC is comparable to polycrystalline NFO-PZT data in Fig. B.1a. Similar low-frequency ME coupling was investigated in bilayers of a variety of laminates with single crystalline or polycrystalline ferrites and PZT or single crystal lead magnesium niobate-lead titanate (PMN-PT) and yttrium iron garnet (YIG) (Nan et al., 2008; Srinivasan, 2010).

It is clear from the data in Fig. B.1 that the ME response vanishes when magnetostriction saturates at high bias magnetic fields. But strong ME effects for H as high as 18 kOe were reported in bilayers of PZT and single-crystal hexagonal M-, Y-, or Z-type ferrites and is due to the fact that saturation of λ requires in-plane bias fields higher than the uniaxial magnetocrystalline anisotropy fields of 10–40 kOe for some of the hexagonal ferrites (Mathe et al., 2008). Representative results in Fig. B.2 for (Sr,Al)M or Zn_2Y and PZT show a weak coupling in bilayers with M-ferrites compared to a much stronger coupling for samples with Zn_2Y. In bilayers with (Sr,Al)M, MEVC was reported to increase with increasing Al-substitution. Theoretical estimates of MEVC were in good agreement with the data in Fig. B.3 (Mathe et al., 2008).

The results in Figs. B.1 and B.2 are for composites with ferrimagnetic oxides. Similar studies were reported for composites with ferromagnetic oxides. Lanthanum manganites with divalent substitutions such as Ba, Ca, or Sr show metallic type electrical conduction and ferromagnetism

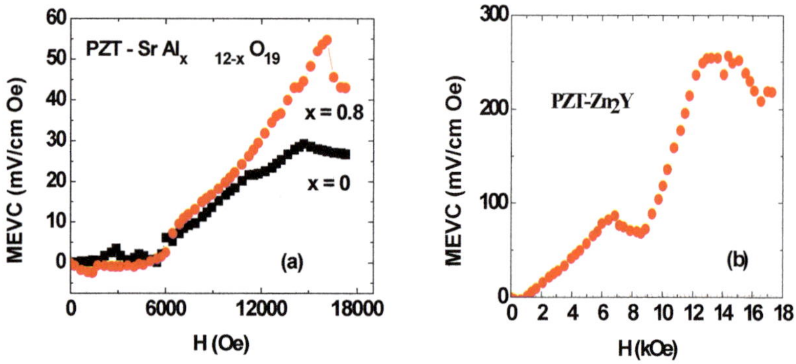

FIGURE B.2 Bias field *H* dependence of MEVC for in-plane magnetic fields for bilayers of PZT and single crystal (a) M-type Al-substituted strontium hexaferrite (Sr,Al)M and (b) Y-type hexagonal ferrite (Mathe et al., 2008).

FIGURE B.3 MEVC versus H data for in-plane magnetic fields and as a function of temperature for a bilayer of single crystal $La_{0.7}Sr_{0.3}MnO_3$ (LSMO) and PZT (Zhang et al., 2009).

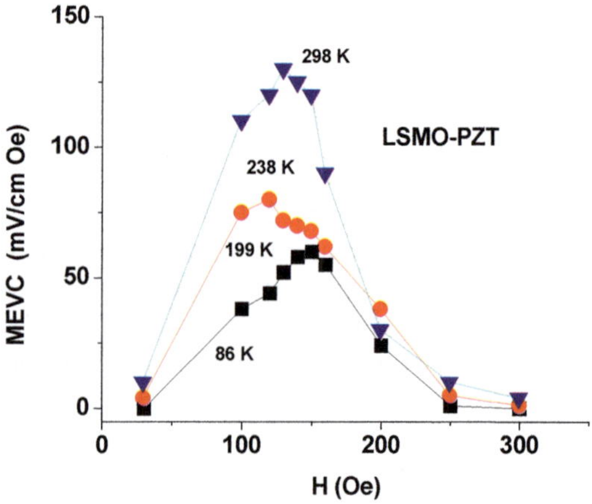

due to double exchange interactions. The oxides are of interest for studies on ME coupling in layered composites because of high λ and low resistivity. Bilayers and multilayers of polycrystalline and single crystal $La_{0.7}Sr_{0.3}MnO_3$ (LSMO)-PZT and $La_{0.7}Ca_{0.3}MnO_3$ (LCMO)-PZT were studied (Srinivasan et al., 2002; Zhang et al., 2009). The coupling was stronger in LSMO-PZT than in LCMO-PZT. Representative results in Fig. B.3 on MEVC versus *H* for in-plane magnetic fields for bilayers with PZT and single crystal LSMO indicate strong ME interactions. Measurements on temperature dependence of MEVC showed a decrease in MEVC with decreasing temperature.

Although laminates of PZT with ferrites and manganates show strong strain mediated coupling, the highest MEVC, however, were reported for composites of PZT or PMN-PT with ferromagnetic metals or alloys. Most of the studies to-date focused on samples of ferroelectrics with 3D transition metal oxides, permendur, Metglas, galfenol, FeGaB, or Terfenol-D (Greve et al., 2010; Hockel et al., 2011; Laletin et al., 2005; Park et al., 2010; Wu et al., 2013; Yao et al., 2011). Among the 3d-transition metals, Ni has the highest λ and q, whereas Fe has the lowest λ. At low frequencies, the highest MEVC was measured for Ni-PZT-Ni trilayers, and the weakest ME response was measured in samples with Co (Laletin et al., 2005). Samples of Fe-PZT-Fe showed unique characteristics such as zero-crossing and sign reversal in MEVC versus *H*. One of the best values for low-frequency MEVC, ~52 V/cm Oe, was measured in samples of Metglas and a piezofiber and was attributed to high q value for Metglas and excellent magnetic field confinement field due to high permeability (Wang et al., 2011). The ME response of several ferromagnetic metal-/alloy-based composites were found to be nonlinear under high ac field strengths (Fetisov et al., 2018). Another alloy-based system of interest showing high MEVC has been the high-q, high-permeability FeCoSiB alloy and piezoelectric AlN (Greve et al., 2010). The system, in particular, shows excellent resonance ME coupling as discussed later. Galfenol, an alloy of $Fe_{1-x}Ga_x$, has high magnetostriction. Compositions with $x = 0.15-0.3$ have saturation magnetostriction as high as $\lambda_S \sim 400 \cdot 10^{-6}$ in a relatively low bias magnetic fields and, therefore, a high q. Composites of galfenol-PZT are found to show strong ME coupling (Fetisov, 2009; Wu et al., 2013). Another alloy-based system of importance is composites with Terfenol-D with MEVC comparable to all-oxide composites. Even though Terfenol-D has high magnetostriction, the q value is rather small. Ferromagnetic alloy-based composites show excellent ME response at mechanical resonance as discussed in the section that follows.

4.5.2.2 ME coupling at bending and electromechanical resonances

It is of interest to investigate the dependence of the MEVC on the frequency f of the ac magnetic field. A bilayer of ferrite-PZT, for example, undergoes a bending deformation in a magnetic field due to its inherent asymmetry. One therefore anticipates a resonance enhancement in the MEVC when f is tuned to the bending resonance frequency (Nan et al., 2008; Srinivasan, 2010). Such bending resonances, however, will be absent in a symmetric ferrite-PZT-ferrite trilayer. Similarly, a resonance enhancement is also expected at electromechanical resonance corresponding to radial, longitudinal, or thickness acoustic modes of the bilayer or a symmetric trilayers. Several models have been proposed for ME coupling at bending modes, but most involve numerical solution of equations. For longitudinal acoustic mode, under $1-d$ approximation, one obtains the following expressions for the resonance frequency (Sreenivasulu, Petrov, et al., 2012):

$$f = \frac{1}{2L}\sqrt{\frac{{}^{p}s_{11} + r^{m}s_{11}}{{}^{p}s_{11}{}^{m}s_{11}\left(r^{p}\rho + {}^{m}\rho\right)}}, \tag{B.2}$$

where ρ is the density, and the peak ME voltage coefficient at the axial mode frequency is given by

$$\frac{\delta E_3}{\delta H_1} = \frac{8Q}{\pi^2}\frac{{}^{m}q_{11}{}^{p}d_{13}/{}^{p}\varepsilon_{33}}{r^{m}s_{11} + {}^{p}s_{11}}, \tag{B.3}$$

where Q_a is the quality factor for the EMR resonance.

Representative data for MEVC at bending and EMR are shown in Fig. B.4 for an epoxy-bonded bilayer of (111) $Ni_{0.8}$ $Zn_{0.2}$ Fe_2 O_4 and PZT (Gheevarughese et al., 2007). The profile of MEVC versus f shows a peak

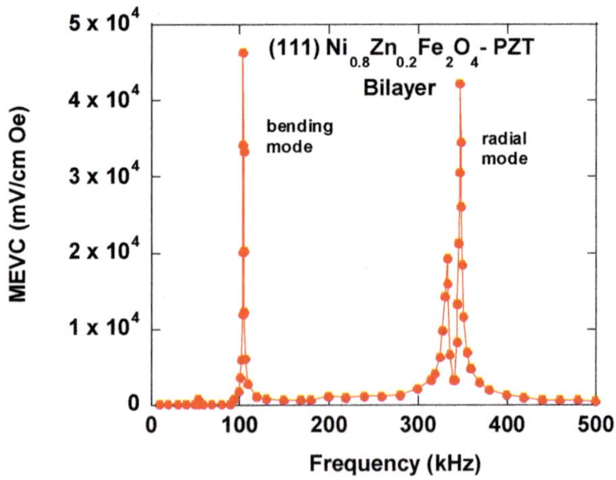

FIGURE B.4 Frequency dependence of MEVC for a bilayer of NZFO and PZT. The peaks correspond to bending and radial acoustic resonances in the sample (Gheevarughese et al., 2007).

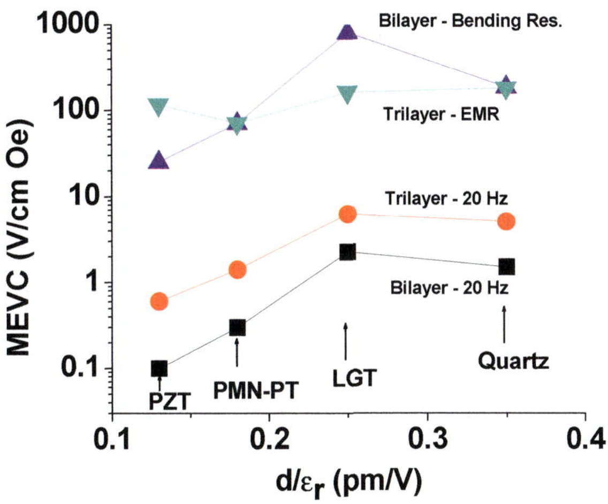

FIGURE B.5 Low-frequency and resonance MEVC as a function of the ratio of the piezoelectric coupling coefficient d to the relative permittivity ε_r for bilayers and trilayers of permendur and ferroelectric PZT or PMN-PT and piezoelectric LGT or quartz (Sreenivasulu, Petrov, et al., 2012).

MEVC of 48 V/cm Oe due to the bending resonance at ~110 kHz and 45 V/cm Oe due to radial acoustic mode at ~350 kHz. The secondary peak at 330 kHz could be higher harmonics of the bending mode.

Studies on trilayer composites made of circular discs of 3d-ferromagnetic metals and PZT revealed EMR at frequencies in the range 100 kHz-3 MHz for radial and thickness modes. The corresponding MEVC was in the range 4–90 V/cm Oe (Laletin et al., 2005). A recent study compared the low-frequency and resonance ME effects in bilayers and trilayers of composites with permendur and ferroelectric PZT and PMN-PT and piezoelectric langatate and quartz (Sreenivasulu, Petrov, et al., 2012). Fig. B.5 shows the resonance value of MEVC as a function of the ratio of the piezoelectric coupling coefficient d to the relative permittivity ε_r. It is clear from the data in Fig. B.5 that the MEVC scales with the ratio d/ε_r. But the highest resonance MEVC to-date, 20 kV/cm Oe was reported for AlN–FeCoSiB for measurements under vacuum that reduces damping of bending resonance that is present in air (Greve et al., 2010).

4.5.2.3 DME coupling in functionally graded composites

Magneto-electric effects in composites with functionally graded ferroics are of fundamental and technological importance. Since the MEVC is proportional to the product of the piezomagnetic coefficient q and piezoelectric coefficient d, models were developed for ME coupling in layered composites with graded q and d. The theory predicts stronger ME interactions in graded composites compared to homogeneous systems and strong ME coupling for zero external magnetic bias field (Petrov & Srinivasan, 2008).

The low-frequency ME interactions in functionally graded layered composites have been studied in several systems. Results for two such systems are provided in Fig. B.6. Fig. B.6(i) shows the low-frequency MEVC versus H data for bilayers of PZT and a functionally graded ferromagnetic layer with two ferrites. The results are for graded ferrite layer with discs of NFO and $Ni_{0.7}$ $Zn_{0.3}$ Fe_2O_4 that were stacked to achieve grading of q (Mandal et al., 2010).

A strong ME coupling arising out of grading and an ME response in the absence of an external bias magnetic field H are observed as shown in Fig. B.6(i). The maximum MEVC in Fig. B.6(i) was a factor two higher than for a bilayer of NFO–PZT and was attributed to grading in q that counteracts the flexural deformation due to asymmetry associated with bilayers (Mandal et al., 2010). Another observation of significance in Fig. B.6(i) is the large MEVC for $H = 0$ for the graded systems and its dependence on the grading scheme. The zero-bias MEVC is due to a torque which arises from interaction between in-plane ac induction δM produced by the applied as magnetic field δH and a grading induced internal magnetic field H_{int} in the ferrite. Compositional grading of NZFO

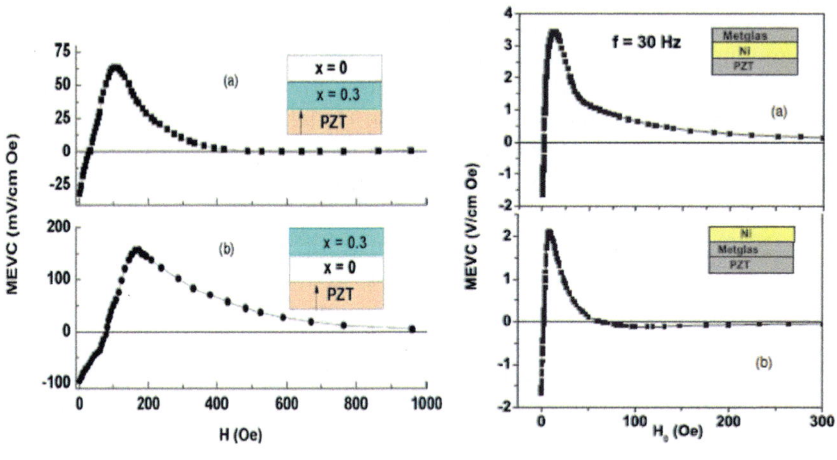

(i) (ii)

FIGURE B.6 Low-frequency ME effects in functionally (magnetization) graded composites. (i) MEVC versus H data at room temperature for a sample of PZT-nickel ferrite-nickel zinc ferrite. Notice the large MEVC for zero bias field and its dependence on the grading scheme dependence on the grading scheme. (ii) Similar data for samples of PZT and graded ferromagnetic layer with Ni and Metglas (Mandal et al., 2010, 2011).

was shown to result in an internal field of magnitude $H_{\text{int}} \approx 30$ Oe in the grading direction.

Fig. B.6(ii) shows results on the observation of a similar zero-bias ($H = 0$) ME effect in a bilayer of PZT and a ferromagnetic layer in which the magnetization is graded with the use of Ni and Metglas. One also observes a dependence of the zero-bias MEVC on the grading scheme (Mandal et al., 2011). Similar studies of significance are the use of nickel zinc ferrite and Ni for functional grading (Yang et al., 2010). A zero-bias MEVC was measured in the system and was attributed to an induced bias magnetic field due to the difference in the magnetic susceptibility and coercivity for Ni and the ferrite. Another study of importance is the exchange biasing of a composite to achieve zero-bias MEVC (Lage et al., 2012). In this study, the exchange bias field acting on the ferromagnetic layer was controlled by controlling the layer thickness so that the piezomagnetic coefficient was maximum for zero external bias magnetic field.

4.5.3 Converse ME effects

The magnetic response of a composite to an applied electric field is termed converse ME effect, and the primary tools for studies on CME are measurements of magnetization M versus H under a static electric field E, tuning of ferromagnetic resonance (FMR) in E, and electric field control of magnetization or ferromagnetic domain switching by magnetic force microscopy. In this section, we discuss these tools for CME studies with two representative examples: M versus H for NFO/PZN-PT and FMR in a bilayer of single crystal Y-type hexagonal ferrite and PMN-PT (Tatarenko & Srinivasan, 2011; Li et al., 2011).

The converse ME effect was reported for direct liquid injection-chemical vapor deposition (DLI-CVD) grown NFO films on (001)-single crystal lead zinc niobate-lead titanate (PZN-PT) or PMN-PT substrates. Models predicted a strong CME in such heterostructures because of high values of λ and d and since such lattice-matched heterostructures are free of any foreign medium at the interface, as in epoxy bonded bilayers, or even in polycrystalline films deposited directly on piezoelectric substrates. The nature of CME was studied by electric-field induced changes in the magnetic hysteresis loops, which were carried out using a vibrating sample magnetometer (VSM). Fig. B.7 shows E-dependence of M versus H for H applied along the **[100] and [110]** directions of NFO/PZN-PT, respectively. Both of them display an E-induced easy magnetization process.

$$\Delta M / M(E=0) = [M(E) - M(E=0)] / M(E=0) \tag{B.4}$$

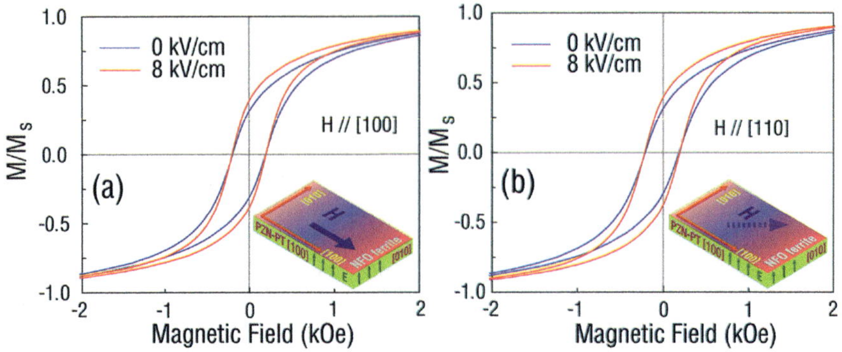

FIGURE B.7 Magnetization versus *H* data for an epitaxial thin film of NFO on PZN-PT substrate for electric field $E = 0$ and 8 kV/cm. The converse ME effect results in an induced magnetization in NFO (Li et al., 2011).

The fractional change in the remnant magnetization defined as reaches 15% with $E = 8$ kV/cm applied across PZN-PT. Assuming $4\pi M_s = 3.2$ kG and for $M(E = 0)/M_s = 0.3$ based on hysteresis data in Fig. B.7, an ME coupling coefficient of $A = \Delta M/E = 18$ G cm/kV is estimated and found to be in very good agreement with *A* obtained from static *E*-field tuning FMR (Li et al., 2011).

Another widely used technique for investigations on CME is static electric field tuning of FMR. An electric field *E* applied to the composite produces a mechanical deformation in PZT that results in an induced anisotropy in the ferrite, leading to a shift in the resonance field or frequency. The very first study was on bulk samples of 90% YIG-10% PZT and the ME coupling was found to be quite weak because of low PZT concentration (Srinivasan, 2010). But in bulk samples with any higher PZT, FMR line broadening masked electric field effect on the resonance field. Such a line broadening, however, could be easily eliminated in layered samples. Models were developed for the phenomenon in bilayers. The *E*-induced shift δH_E in the resonance field is determined by the magnetostriction λ and piezoelectric coupling coefficient *d*. When *E* and *H* are directed along a symmetry axis, <100> for example, perpendicular to the (*1,2*) plane of a bilayer, one obtains (Srinivasan, 2010),

$$\delta H_E = \frac{3\lambda_{100} d_{31} E_3}{M_0\left[\left({}^p s_{11} + {}^p s_{12}\right)(1-v) + \left({}^m s_{11} + {}^m s_{12}\right)v\right]}, \tag{B.5}$$

where M_0 is the saturation magnetization and v is the volume fraction for the piezoelectric phase. Since the measurement accuracy depends very much on the FMR line width, the techniques in general are appropriate for bilayers consisting of single crystals platelets or epitaxial films of low-loss ferrites such as YIG, lithium ferrite, nickel ferrite, and hexagonal ferrites

(Sun & Srinivasan, 2012; Tatarenko & Srinivasan, 2011). *E*-tuning of FMR has also been utilized for CME in composites with metallic ferromagnetic alloys such as FeGaB films (Sun & Srinivasan, 2012).

Next, we discuss a representative example for *E*-tuning of FMR in the case of a laminate composite with hexagonal ferrite and a ferroelectric. Microwave magnetoelectric (ME) effects over 8–25 GHz were studied in bilayers of single crystal Y-type hexagonal ferrite $Ba_2Zn_2Fe_{12}O_{22}$ (Zn_2Y) and polycrystalline PZT or single crystal PMN-PT (Tatarenko et al., 2012). The bilayers were made by bonding the ferrite and piezoelectric platelets.

The strength of ME interactions $A = \delta H/E$ (or $\delta f/H$) was measured from data on resonance field shift δH or frequency shift δf in an electric field *E*. Studies on ME interactions were performed on the ferrite/PMN-PT bilayer on a microstripline transducer as shown in Fig. B.8. A static magnetic field *H* was applied parallel to the sample plane. Profiles of the scattering parameter S_{11} versus *f* in Fig. B.8 show absorption corresponding to FMR with a 3 dB frequency-width $\Delta f = 70$ MHz ($\Delta H = 25$ Oe). Data were obtained on frequency shift in FMR when a static field *E* was applied to PZT or PMN-PT for the estimation of ME coefficient *A*. For epoxy bonded Zn_2Y-PZT, the measured $A = 1.2$ MHz cm/kV, and it is enhanced by almost an order of magnitude to $A \approx 10$ MHz cm/kV for eutectic-bonded Zn_2Y/PMN-PT. The order of magnitude difference in *A*-value for samples with PZT and

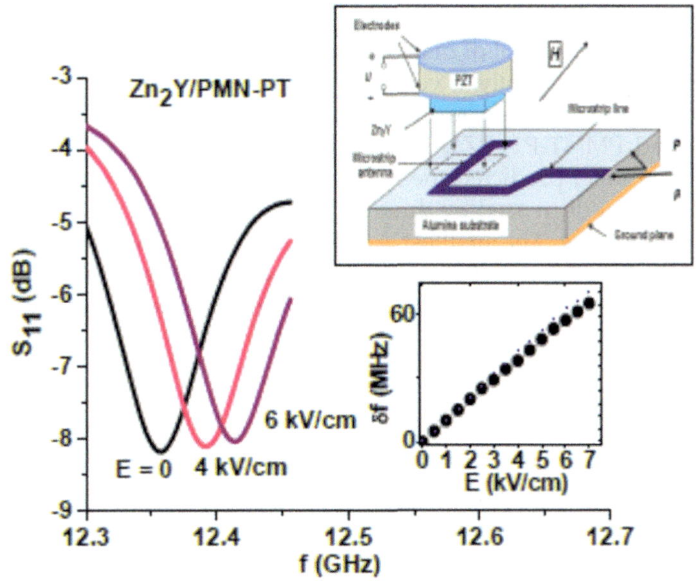

FIGURE B.8 Data showing tuning of ferromagnetic resonance in a bilayer of single crystal Y-type hexagonal ferrite and PMN-PT. The insets show the stripline device used for the measurements and shift in FMR frequency versus *E* (Tatarenko et al., 2012).

PMN-PT is primarily due to the difference in the value of the piezoelectric coupling constant $d \approx 400$ pC/N for PZT and $d \approx 2500$ pC/N for PMN-PT (Tatarenko et al., 2012). The measured A for Zn_2Y-PZT is comparable to values reported for YIG-PZT as expected since magnetic parameters for Zn_2Y ($\lambda = 1.8$ ppm, $4\pi M = 2.8$ kG) are close to that of YIG ($\lambda = 2$ ppm, $4\pi M = 1.78$ kG (Sun & Srinivasan, 2012; Tatarenko & Srinivasan, 2011).

Similar studies on Fe_3O_4/PZN-PT demonstrated strong CME (Sun & Srinivasan, 2012). The electric field was found to induce a magnetic anisotropy field as high as 400–600 Oe, depending on the field orientation, and an ME coefficient $A = 68$–100 Oe cm/kV. In another study on layered samples of Al-substituted nickel zinc ferrite and PZT, E-tuning of FMR was used to determine the strength of CME (Li et al., 2013). A related technique, that is, ac electric field switching of magnetic induction in a ferrite–ferroelectric composite was employed for studies on CME (Yang et al., 2011).

Although one generally expects a large FMR line-width in composites of ferromagnetic alloys-ferroelectrics, studies on composites with thin films of FeGa revealed a line-width of less than 60 Oe (frequency half-width of 350 MHz) (Liu et al., 2013). A giant E-tuning of FMR, by 1200 Oe (equivalent to 5.3 GHz) and an ME coefficient $A = 880$ MHz cm/kV, was achieved for the FeGa-based composite. Studies on CME in ferromagnetic metals or alloys and ferroelectric composites involved resonant CME in FeGa-PZT-FeGa trilayers (Wu et al., 2013), magnetic domain switching in FeGa-BTO and FeGa-PZT (Sun & Srinivasan, 2012), and electric field control of magnetization in Ni–ferroelectric composites (Kim et al., 2013). Finally, a very recent model for CME considered primarily composites of ferromagnetic metals (Fe, Ni, Co) or alloys (NiFe, FeGaB) and PMNT-PT, PZN-PT, or piezoelectric quartz (Bichurin et al., 2023). The model predicted the highest A-value of 90 Oe cm/kV for Fe/PZN-PT, and it decreased by three orders of magnitude when PZN-PT is replaced by quartz. The smallest A-value of 5 Oe cm/kV for estimated for NiFe/PZN-PT. The model when applied to composites with YIG films on nonmagnetic or nonferroelectric substrates, A value decreased by a factor of two.

4.5.4 Nonlinear ME effects

The strain-mediated coupling phenomena discussed so far in ferromagnetic–ferroelectric composites were on linear ME effects and the measurements of the interaction coefficient involved the application of a small amplitude ac magnetic field δH and measuring the induced voltage δV. One, in general, finds linear variation in the induced voltage with the ac magnetic field amplitude. For very large excitation field amplitudes, however, nonlinearity of the magnetic and magnetostrictive parameters with the amplitude of δH results in a rich variety of ME phenomena and are discussed in this section (Fetisov et al., 2018). The nonlinear ME (NLME) effects were first reported in a multilayer of NFO and PZT (Kamentsev et al., 2006). The sample contained alternate layers of nickel zinc ferrite (NZFO) and PZT of thickness 18 μm each. The ME measurements were done without any bias magnetic field and for applied ac magnetic fields $H(t) = H \cos(2\pi f_0 t)$ at frequencies $f_0 = 1$ mHz–10 Hz and amplitudes $H = 200$ Oe and 1 kOe. Fig. B.9 shows profiles of the ac field $H(t)$ at $f_0 = 0.03$ Hz and ME voltage $V(t)$ and its frequency spectra obtained by fast Fourier transform for the above two ac field amplitudes (Kamentsev et al., 2006). For $H_0 = 200$ Oe, one observes in Fig. B.9 doubling of the frequency for the voltage $V(t)$ due to the fact that the piezomagnetic coefficient q is positive for both positive and negative half of the harmonic ac magnetic fields. In addition to the frequency doubling, the frequency spectra for $V(t)$ show a few even harmonics. When H_0 is increased to 1 kOe, the profile for $V(t)$ shows significant distortion of the wave form and also negative values for its amplitude.

Its frequency spectra show several even harmonics with slowly decreasing amplitude in the low frequency region and both odd and even harmonics with a much more rapid decrease in the amplitude in the high frequency region. The harmonics arise due to nonlinearity of the magnetostriction at high magnetic fields and with increasing H leads to further distortion in $V(t)$ and the presence of multiple harmonics. The negative polarity of the ME voltage at high fields was attributed to finite electrical conductivity of the composite (Kamentsev et al., 2006). The

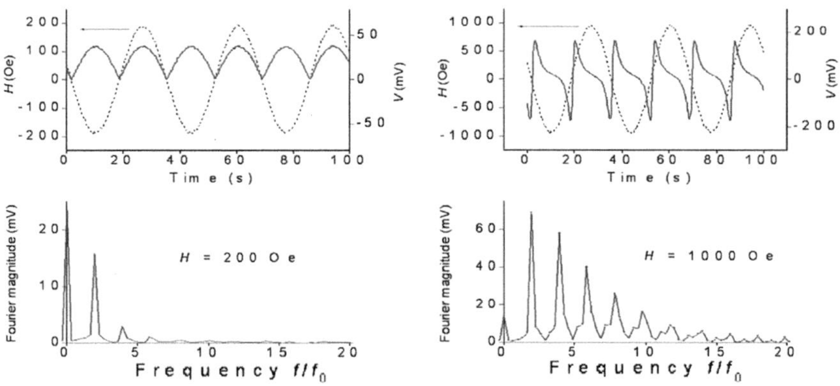

FIGURE B.9 Nonlinear ME effects in a multilayer of $Ni_{0.8}Zn_{0.2}Fe_2O_4$, (NZFO) and PZT. Plots on the left show applied magnetic field $H(t)$ (dashed lines) and ME voltage $V(t)$ (solid lines) and the corresponding frequency spectra for the ME signals. The ac field frequency $f_0 = 0.03$ Hz and amplitudes of $H_0 = 200$ Oe (left) and 1000 Oe (right) (Kamentsev et al., 2006).

NLME effects associated orientation of magnetic fields and due to nonlinearity in composite magnetization and were reported recently (Burdin et al., 2021).

The NLME phenomenon could be utilized to enhance the ME coupling under resonance conditions such as the excitation of bending and longitudinal modes in a laminate and was demonstrated in a composite of ferromagnetic FeBSiC alloy and PZT (Fetisov et al., 2018). Fig. B.10 shows data on frequency dependence of the ME coefficient in the composite for two different conditions: (a) when a bias field $H_0 = 18$ Oe and ac field $h = 2.8$ Oe are applied and (b) for $H_o = 0$ and $h = 14$ Oe. In Fig. B.10a, one observes resonance enhancement of the ME coupling at the bending mode frequency $f_r = 750$ Hz. A small peak in the ME voltage is also observed at $f_r/2 = 350$ Hz and can be attributed to the frequency doubling discussed earlier. The NLME in this case of small ac magnetic field strength is rather weak. When the ac field h is increased to 14 Oe and for zero DC bias field, the profile in Fig. B.10b shows an intense peak at $f = f_r/2$ and is due to strong NLME coupling and frequency doubling in the

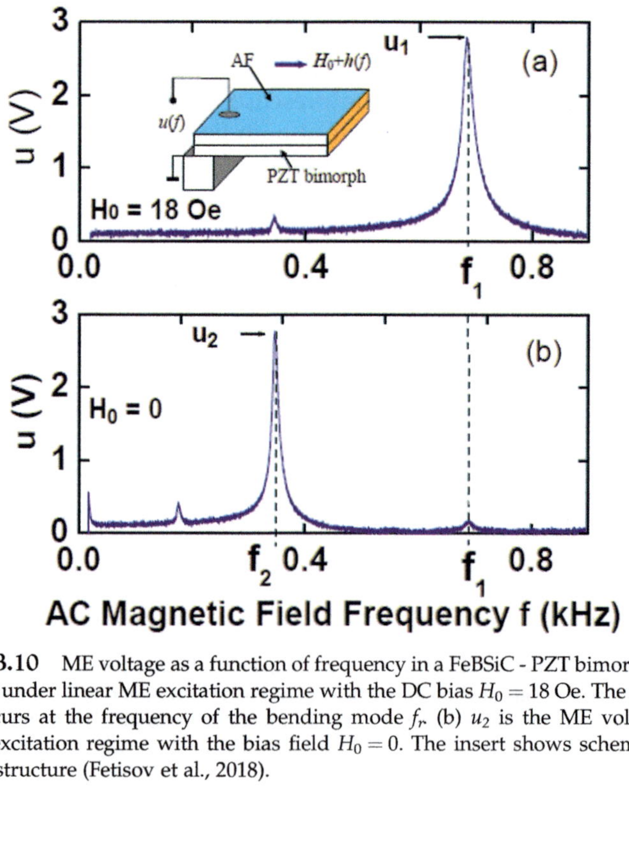

FIGURE B.10 ME voltage as a function of frequency in a FeBSiC - PZT bimorph. (a) u_1 is the voltage under linear ME excitation regime with the DC bias $H_0 = 18$ Oe. The peak in the voltage occurs at the frequency of the bending mode f_r. (b) u_2 is the ME voltage under nonlinear excitation regime with the bias field $H_0 = 0$. The insert shows schematic of the composite structure (Fetisov et al., 2018).

composite. It was found that the effect was even stronger when PZT in the composite was replaced by a PZT bimorph. A theory for NLME under resonance was also discussed in this work (Fetisov et al., 2018). A similar NLME effect under the converse ME effects in the same composite was reported under longitudinal mode excited by an ac electric field.

Finally, recently reported nonlinear ME phenomenon leading to parametric excitation of subharmonics in a laminate composite is considered (Burdin et al., 2020). A disc-shaped bilayer consisting of ferromagnetic FeBSiC and PZT as in Fig. B.11 was used in the study. The disc was subjected to an in-plane DC magnetic bias field $H = 0-100$ Oe. A pair of coils as shown in Fig. B.11 was used to either generate an ac magnetic field or as a pickup coil to measure the voltage under converse ME effect. Two different schemes, corresponding to direct or converse ME effects were used in the study. In the first scheme, an ac magnetic field $h = 0-5$ Oe at $f = 9-12$ kHz was applied parallel to H and the voltage V across PZT layer was measured and its frequency spectrum was obtained by fast Fourier transform. As the magnitude of h was increased a peak appeared in the frequency spectrum of V as seen in Fig. B.11a at a frequency f_r corresponding an acoustic resonance mode in the composite. With further increase in h, subharmonic peaks at $f_r/2$ as in Fig. B.11b appeared in the frequency spectrum. In the second scheme, the sample was subjected to an ac electric field $E = 0-500$ V/cm at 9–12 kHz across

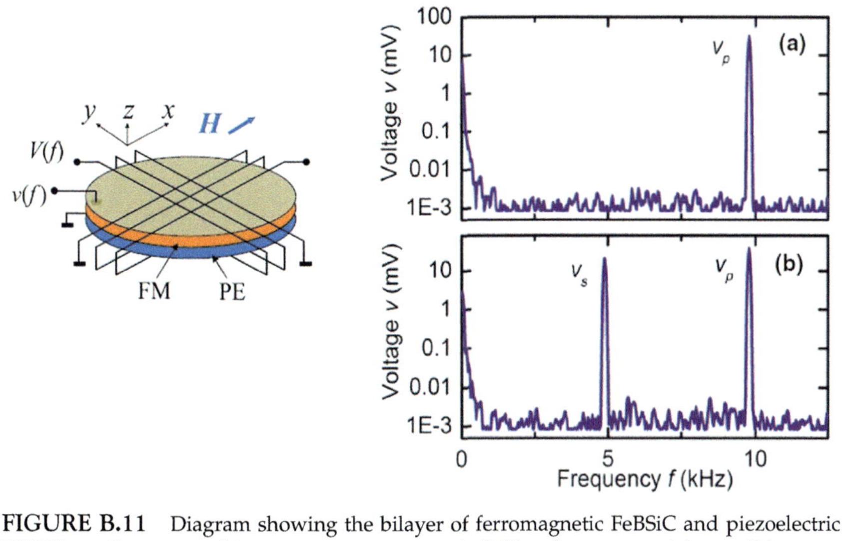

FIGURE B.11 Diagram showing the bilayer of ferromagnetic FeBSiC and piezoelectric PZT. The coils were used to generate an ac magnetic field or for use as a pickup coil for measurement under CME. The fast Fourier transforms show (a) peak in the ME voltage at acoustic resonance at ~ 10 kHz due to DME and (b) parametric excitations at half the acoustic resonance mode for DME above a threshold ac magnetic field (Burdin et al., 2020).

the PZT layer, and the voltage V generated in the pickup coil due to CME effect was measured. Once again, a peak appeared in the frequency spectrum of V at the acoustic mode frequency and subharmonics were seen when the E exceeded a threshold value. The parametric excitation of the subharmonics was attributed to magneto-acoustic nonlinearity in FeBSiC layer. A model for the phenomenon was also discussed. The nonlinear ME effects were utilized for wideband ac magnetometers and DC magnetic field sensors (Burdin et al., 2016; Fetisov et al., 2014).

4.5.5 Recent advances: ME laminates for gyrators

Efforts in the past decade mainly focused on avenues for enhancing the strengths of ME interactions in laminate composites made of high permeability and high magnetostriction ferromagnetic alloys such as Metglas and galfenol and ferroelectric PZT, PMN-PT, and PVDF (Wang et al., 2015). Several review articles provide details on these studies (Palneedi et al., 2016). The primary thrust of the research on ME laminates, however, transitioned to device applications in recent years. Specific applications include pico-Tesla magnetic sensors (Ge et al., 2022), energy harvesters (He & Liu, 2022; Narita & Fox, 2018; Swain et al., 2020), very-low-frequency (VLF) transmitting and receiving antennas (Niu & Ren, 2022), RF, and microwave and millimeter wave devices (Sun & Srinivasan, 2012), and gyrators (Zhang et al. 2019; Zhuang et al., 2018). The second part of this book deals with most of these device applications. Here we discuss a specific device application: ME gyrators.

Gyrators made of ME laminates are unique circuit elements capable of converting a current-to-voltage (or vice versa) (Leung et al., 2017). In general, they are symmetric trilayers with a ferroelectric layer placed between two ferromagnetic layers as in Fig. B.12 or a ferromagnetic layer

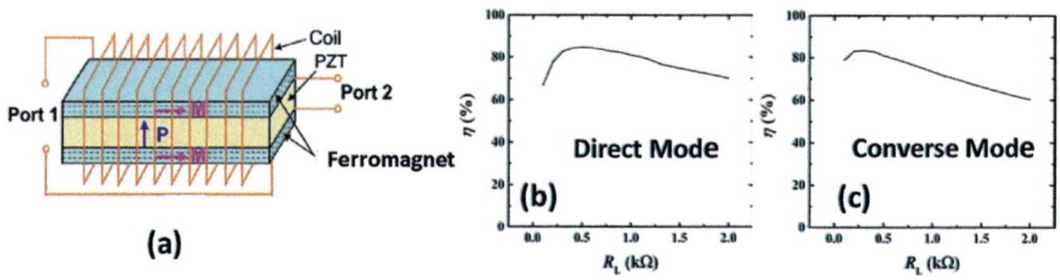

FIGURE B.12 (a) Schematic showing a gyrator with a symmetric trilayer ME composite and a solenoid. The gyrator is capable of converting a current in the solenoid is converted to a voltage across PZT through DME or a voltage across PZT is converted into a current in the solenoid by CME. (b) Power conversion efficiency η as a function of the load resistance R_L under DME in a gyrator made of nickel zinc ferrite and PZT. (c) η versus R_L under CME in a gyrator.

between two ferroelectric platelets and are placed inside a solenoid (Leung et al., 2017). A bias magnetic field is applied parallel and along the length of the laminate plane. The conversion of an AC current in the coil to a voltage across the ferroelectric layer thus occurs due to the direct ME effect. A similar conversion of a voltage applied across the ferroelectric layer to a current in the pickup coil can be realized due to the converse ME effect. Symmetric trilayers are preferred for the gyrator since they have a much higher ME coupling compared to bilayers. For efficient conversion of current to voltage, the AC current in the coil is applied at a frequency corresponding to acoustic resonance modes of the composite.

Power conversion efficiency is a key performance metrics for the gyrators. The primarily losses in the gyrators are the resistive loss in the solenoid and losses at the mechanical resonance in the composite. According to a model for the gyrator, the power efficiency η is proportional to the product of magneto-mechanical coupling factor and the quality factor for the mechanical resonance (Zhuang et al., 2018). Fig. B.12 shows η as a function of the load resistance R_L in a gyrator with nickel zinc ferrite and PZT and operating under DME and CME. The maximum power efficiency (PE) in both cases exceeds 85% (Leung et al., 2017). Efficiency as much as 92% is predicted for a composite with Metglas and hard PZT (Zhuang et al., 2018).

There were several studies on avenues for enhancing the PE in gyrators including the use of ferrites with a variety substitution such as Mn, Y, and Sm in order to modify magnetic parameters, thereby enhancing the strength ME interactions. A bias magnetic field is essential for optimizing the ME coupling to achieve the maximum PE in a gyrator. There were several attempts to eliminate the need for a bias magnetic field with the use of self-biased ME composites (Zhang et al., 2019). These include the use of substitutions to increase the remnant magnetization in the ferrite that will act as a built-in bias or functionally grade ferrites to induce the necessary bias field. A recent study reported the elimination of the solenoid with the use of ferrite layers that were metalized that essentially acted as a solenoid (Fetisov & Chashin, 2022).

4.5.6 Conclusions

There have been considerable interests in recent years on mechanical strain–mediated coupling in layered composites of ferromagnetic and ferroelectric phases. The nature of the ME coupling is studied by the sample response to a magnetic field or an electric field. Ferromagnetic metal or oxide and ferroelectric- or piezoelectric-based composites show evidence for strong direct-ME coupling at low frequencies and further enhancement in the coupling strength under mechanical resonance conditions. A strong converse–ME effect was also evident from electric field switching of magnetization, magnetic domain switching, variation in inductance, or E-tuning of FMR. Several applications, including magnetic sensors, high-frequency inductors, memory devices, and high-frequency signal processing devices have been proposed and demonstrated. Most of the studies discussed in this chapter are on thick film—layered composites or bonded bilayers. Ultra-thin-film multiferroic heterostructures have not been studied in detail so far. Difficulties in preparing high-quality heterostructures have impeded the studies on these composites. It is therefore necessary to overcome the challenges in depositing epitaxial films of the composite structures. It is also necessary to explore in detail the ME coupling in nanocomposites.

References

Bichurin, M., Sokolov, O., Ivanov, S., Ivasheva, E., Leontiev, V., Lobekin, V., & Semenov, G. (2023). Modeling the composites for magnetoelectric microwave devices. *Sensors, 23*(4), 1780.

Burdin, D. A., Chashin, D. V., Ekonomov, N. A., Fetisov, Y. K., & Stashkevich, A. A. (2016). High-sensitivity dc field magnetometer using nonlinear resonance magnetoelectric effect. *Journal of Magnetism and Magnetic Materials, 405*, 244–248.

Burdin, D. A., Chashin, D. V., Ekonomov, N. A., Preobrazhenskii, V. L., Gordeev, S. N., & Fetisov, Y. K. (2020). Parametric generation of subharmonics in a composite multiferroic resonator. *Physical Review Applied, 13*(5), Article 054070.

Burdin, D., Ekonomov, N. A., Vopson, M. M., & Fetisov, Y. K. (2021). Enhancement of the nonlinear magnetoelectric effect in a ferromagnet-piezoelectric heterostructure due to nonlinearity of magnetization. *Applied Physics Letters, 118*(13), Article 132901.

Cho, K. H., & Priya, S. (2011). Direct and converse effect in magnetoelectric laminate composites. *Applied Physics Letters, 98*(23), Article 232904.

Cross, L. E. (1993). Ferroelectric ceramics: Tailoring properties for specific applications. In N. Setter (Ed.), *Ferroelectric ceramics*. Basel: Birkhäuser.

Fetisov, Y. K., Burdin, D., Chashin, D., & Ekonomov, N. (2014). High-sensitivity wideband magnetic field sensor using nonlinear resonance magnetoelectric effect. *IEEE Sensors Journal, 14*(7), 2252–2256.

Fetisov, Y. K., & Chashin, D. V. (2022). Magnetoelectric coil-free voltage transformer based on monolithic ferrite-piezoelectric heterostructure. *Sensors and Actuators A: Physical, 344*, Article 113737.

Fetisov, L. Y., Chashin, D., et al. (2018). Nonlinear converse magnetoelectric effects in a ferromagnetic-piezoelectric bilayer. *Applied Physics Letters, 113*(21), Article 212903.

Ge, B., Zhang, J., et al. (2022). Ultra-low anisotropy magnetoelectric sensor in ferrite/piezoelectric toroidal composites. *Journal of Magnetism and Magnetic Materials, 564*, Article 170115.

Gheevarughese, V., Laletsin, U., Petrov, V. M., Srinivasan, G., & Fedotov, N. A. (2007). Low-frequency and resonance magnetoelectric effects in lead zirconate titanate and single-crystal nickel zinc ferrite bilayers. *Journal of Materials Research, 22*(8), 2130–2135.

Greve, H., Woltermann, E., Jahns, R., Marauska, S., Wagner, B., Knöchel, R., … Quandt, E. (2010). Low damping resonant magnetoelectric sensors. *Applied Physics Letters, 97*(15), Article 152503.

He, W., & Liu, S. (2022). A magnetoelectric energy harvester for low-frequency vibrations and human walking. *Journal of Magnetism and Magnetic Materials, 542*, Article 168609.

Hockel, J. L., Wu, T., & Carman, G. P. (2011). Voltage bias influence on the converse magnetoelectric effect of PZT/terfenol-D/PZT laminates. *Journal of Applied Physics, 109*(6), Article 064106.

Kamentsev, K. E., Fetisov, Y. K., & Srinivasan, G. (2006). Low-frequency nonlinear magnetoelectric effects in a ferrite-piezoelectric multilayer. *Applied Physics Letters, 89*(14), Article 142510.

Kim, H. K., Schelhas, L. T., Keller, S., Hockel, J. L., Tolbert, S. H., & Carman, G. P. (2013). Magnetoelectric control of superparamagnetism. *Nano Letters, 13*(3), 884–888.

Lage, E., Kirchhof, C., Hrkac, V., Kienle, L., Jahns, R., Knöchel, R., … Meyners, D. (2012). Exchange biasing of magnetoelectric composites. *Nature Materials, 11*(6), 523–529.

Laletin, V. M., Paddubnaya, N., Srinivasan, G., De Vreugd, C. P., Bichurin, M. I., Petrov, V. M., & Filippov, D. A. (2005). Frequency and field dependence of magnetoelectric interactions in layered ferromagnetic transition metal-piezoelectric lead zirconate titanate. *Applied Physics Letters, 87*(22), Article 222507.

Lawes, G., & Srinivasan, G. (2011). Introduction to magnetoelectric coupling and multiferroic films. *Journal of Physics D: Applied Physics, 44*(24), Article 243001.

Leung, C. M., Zhuang, X., Friedrichs, D., Li, J., Erickson, R. W., Laletin, V., … Viehland, D. (2017). Highly efficient solid state magnetoelectric gyrators. *Applied Physics Letters, 111*(12), Article 122904.

Li, N., Liu, M., Zhou, Z., Sun, N. X., et al. (2011). Electrostatic tuning of ferromagnetic resonance and magnetoelectric interactions in ferrite-piezoelectric heterostructures grown by chemical vapor deposition. *Applied Physics Letters, 99*, Article 192502.

Li, M., Zhou, Z., Liu, M., Lou, J., Oates, D. E., Dionne, G. F., … Sun, N. X. (2013). Novel NiZnAl-ferrites and strong magnetoelectric coupling in NiZnAl-ferrite/PZT multiferroic heterostructures. *Journal of Physics D: Applied Physics, 46*(27), Article 275001.

Liu, M., Zhou, Z., Nan, T., Howe, B. M., Brown, G. J., & Sun, N. X. (2013). Voltage tuning of ferromagnetic resonance with bistable magnetization switching in energy-efficient magnetoelectric composites. *Advanced Materials, 25*(10), 1435–1439.

Nan, C. W., Bichurin, M. I., Dong, S., Viehland, D., & Srinivasan, G. (2008). Multiferroic magnetoelectric composites: Historical perspective, status, and future directions. *Journal of Applied Physics, 103*(3), Article 031101.

Narita, F., & Fox, M. (2018). A review on piezoelectric, magnetostrictive, and magnetoelectric materials and device technologies for energy harvesting applications. *Advanced Engineering Materials, 20*(5), Article 1700743.

Niu, Y., & Ren, H. (2022). Transceiving signals by mechanical resonance: A miniaturized standalone low frequency (lf) magnetoelectric mechanical antenna pair with integrated DC magnetic bias. *IEEE Sensors Journal, 22*(14), 14008–14017.

Ma, J., Hu, J., Li, Z., & Nan, C. W. (2011). Recent progress in multiferroic magnetoelectric composites: From bulk to thin films. *Advanced Materials, 23*(9), 1062–1087.

Mandal, S. K., Sreenivasulu, G., Petrov, V. M., & Srinivasan, G. (2010). Flexural deformation in a compositionally stepped ferrite and magnetoelectric effects in a composite with piezoelectrics. *Applied Physics Letters, 96*(19), Article 192502.

Mandal, S. K., Sreenivasulu, G., Petrov, V. M., & Srinivasan, G. (2011). Magnetization-graded multiferroic composite and magnetoelectric effects at zero bias. *Physical Review B: Condensed Matter and Materials Physics, 84*(1), Article 014432.

Mathe, V. L., Srinivasan, G., & Balbashov, A. M. (2008). Magnetoelectric effects in bilayers of lead zirconate titanate and single crystal hexaferrites. *Applied Physics Letters, 92*(12), Article 122505.

Palneedi, H., Annapureddy, V., Priya, S., & Ryu, J. (March 2016). Status and perspectives of multiferroic magnetoelectric composite materials and applications. *Actuators, 5*(1), 9.

Park, C. S., Cho, K. H., Arat, M. A., Evey, J., & Priya, S. (2010). High magnetic field sensitivity in Pb (Zr, Ti) O3–Pb (Mg1/3Nb2/3) O3 single crystal/Terfenol-D/Metglas magnetoelectric laminate composites. *Journal of Applied Physics, 107*(9), Article 094109.

Petrov, V. M., & Srinivasan, G. (2008). Enhancement of magnetoelectric coupling in functionally graded ferroelectric and ferromagnetic bilayers. *Physical Review B: Condensed Matter and Materials Physics, 78*(18), Article 184421.

Pullar, R. C. (2012). Hexagonal ferrites: A review of the synthesis, properties and applications of hexaferrite ceramics. *Progress in Materials Science, 57*(7), 1191–1334.

Sreenivasulu, G., Petrov, V. M., Fetisov, L. Y., Fetisov, Y. K., & Srinivasan, G. (2012). Magnetoelectric interactions in layered composites of piezoelectric quartz and magnetostrictive alloys. *Physical Review B: Condensed Matter and Materials Physics, 86*(21), Article 214405.

Sreenivasulu, G., Fetisov, L. Y., Fetisov, Y. K., & Srinivasan, G. (2012). Piezoelectric single crystal langatate and ferromagnetic composites: Studies on low-frequency and resonance magnetoelectric effects. *Applied Physics Letters, 100*(5).

Sreenivasulu, G., Laletin, U., Petrov, V. M., Petrov, V. V., & Srinivasan, G. (2012). A permendur-piezoelectric multiferroic composite for low-noise ultrasensitive magnetic field sensors. *Applied Physics Letters, 100*(17), Article 173506.

Srinivasan, G., De Vreugd, C. P., Laletin, V. M., Paddubnaya, N., Bichurin, M. I., Petrov, V. M., & Filippov, D. A. (2005). Resonant magnetoelectric coupling in trilayers of ferromagnetic alloys and piezoelectric lead zirconate titanate: The influence of bias magnetic field. *Physical Review B: Condensed Matter and Materials Physics, 71*(18), Article 184423.

Srinivasan, G., Rasmussen, E. T., Gallegos, J., Srinivasan, R., Bokhan, Y. I., & Laletin, V. M. (2001). Magnetoelectric bilayer and multilayer structures of magnetostrictive and piezoelectric oxides. *Physical Review B, 64*(21), Article 214408.

Srinivasan, G., Rasmussen, E. T., Levin, B. J., & Hayes, R. (2002). Magnetoelectric effects in bilayers and multilayers of magnetostrictive and piezoelectric perovskite oxides. *Physical Review B, 65*(13), Article 134402.

Srinivasan, G. (2010). Magnetoelectric composites. *Annual Review of Materials Research, 40*(1), 153–178.

Sun, N. X., & Srinivasan, G. (September 2012). Voltage control of magnetism in multiferroic heterostructures and devices. *Spin, 2*(03), Article 1240004.

Swain, A. B., Kumar, S., Subramanian, V., & Murugavel, P. (2020). Engineering resonance modes for enhanced magnetoelectric coupling in bilayer laminate composites for energy harvesting applications. *Physical Review Applied, 13*(2), Article 024026.

Tatarenko, A. S., Murthy, D. V. B., & Srinivasan, G. (2012). Hexagonal ferrite-piezoelectric composites for dual magnetic and electric field tunable 8–25 GHz microstripline resonators and phase shifters. *Microwave and Optical Technology Letters, 54*(5), 1215–1218.

Tatarenko, A. S., & Srinivasan, G. (2011). A strain engineered voltage tunable millimeter-wave ferrite phase shifter. *Microwave and Optical Technology Letters, 53*(2), 261–264.

Wang, Y., Gray, D., Berry, D., Gao, J., Li, M., Li, J., & Viehland, D. (2011). An extremely low equivalent magnetic noise magnetoelectric sensor. *Advanced Materials, 23*(35), 4111.

Wang, Y., Hu, J., Lin, Y., & Nan, C. W. (2010). Multiferroic magnetoelectric composite nanostructures. *NPG Asia Materials, 2*(2), 61–68.

Wang, Y., Li, J., & Viehland, D. (2015). A differential magnetoelectric heterostructure: Internal noise reduction and external noise cancellation. *Journal of Applied Physics, 118*(21), Article 214103.

Wang, Z., Zhang, Y., Viswan, R., Li, Y., Luo, H., Li, J., & Viehland, D. (2014). Electrical and thermal control of magnetic coercive field in ferromagnetic/ferroelectric heterostructures. *Physical Review B, 89*(3), Article 035118.

Wu, G., Nan, T., Zhang, R., Zhang, N., Li, S., & Sun, N. X. (2013). Inequivalence of direct and converse magnetoelectric coupling at electromechanical resonance. *Applied Physics Letters, 103*(18), Article 182905.

Yang, S. C., Park, C. S., Cho, K. H., & Priya, S. (2010). Self-biased magnetoelectric response in three-phase laminates. *Journal of Applied Physics, 108*(9), Article 093706.

Yang, S., Cho, K. H., Park, C. S., & Priya, S. (2011). Self-biased converse magnetoelectric effect. *Applied Physics Letters, 99*(20).

Yao, Y. P., Hou, Y., Dong, S. N., & Li, X. G. (2011). Giant magnetodielectric effect in Terfenol-D/PZT magnetoelectric laminate composite. *Journal of Applied Physics, 110*(1), Article 014508.

Zavislyak, I. V., Sohatsky, V. P., Popov, M. A., & Srinivasan, G. (2013). Electric-field-induced reorientation and flip in domain magnetization and light diffraction in an yttrium-iron-garnet/lead-zirconate-titanate bilayer. *Physical Review B: Condensed Matter and Materials Physics, 87*(13), Article 134417.

Zhang, N., Srinivasan, G., & Balbashov, A. M. (2009). Low-frequency magnetoelectric interactions in single crystal and polycrystalline bilayers of lanthanum strontium manganite and lead zirconate titanate. *Journal of Materials Science, 44*, 5120–5126.

Zhang, J., Zhu, W., Chen, D., Qu, H., et al. (2019). Magnetoelectric effects and power conversion efficiencies in gyrators with compositionally-graded ferrites and piezoelectrics. *Journal of Magnetism and Magnetic Materials, 473*, 131–135.

Zhuang, X., Leung, C. M., Li, J., Srinivasan, G., & Viehland, D. (2018). Power conversion efficiency and equivalent input loss factor in magnetoelectric gyrators. *IEEE Transactions on Industrial Electronics, 66*(4), 2499–2505.

CHAPTER

5

Recent advances in ferroic, multiferroic, and magnetoelectrics materials and phenomena

Sumanta Kumar Karan[1,5], *Abbey Knoepfel*[1], *Su Chul Yang*[2], *Aman Nanda*[1], *Deepam Maurya*[3], *Yongke Yan*[1], *and Shashank Priya*[1,4]

[1] Department of Materials Science and Engineering, The Pennsylvania State University, University Park, PA, United States; [2] Department of Chemical Engineering, Dong-A University, Busan, South Korea; [3] Bio-inspired Materials and Devices Laboratory (BMDL), Center for Energy Harvesting Materials and Systems (CEHMS), Blacksburg, VA, United States; [4] Chemical Engineering and Materials Science, University of Minnesota, Minneapolis, MN, United States; [5] Center for Clean Energy Engineering, University of Connecticut, Storrs, CT, United States

5.1 Introduction

The realization of a material with simultaneous presence of strong coupling between electric and magnetic order, termed as "multiferroic magnetoelectrics," would be a milestone for modern electronics and multifunctional materials (Eerenstein et al., 2006). It will open the gateway for very high-density memory storage media using both magnetic and electric polarization and the possibility of electrically reading or writing magnetic memory devices (and vice versa) (Nan et al., 2008). These applications have already prompted a surge of activity in magnetoelectronics and spintronics where phenomenon such as carrier effects in

Magnetoelectric Composites
https://doi.org/10.1016/B978-0-443-18606-6.00012-5

magnetic semiconductors and high-correlation effects in colossal magnetoresistive compounds are being investigated. Investigations have revealed the presence of both ferroelectricity and magnetism in number of materials such as perovskite type $BiFeO_3$, $BiMnO_3$, the boracite family, $BaMF_4$ compounds (M, divalent transition metal ions), hexagonal $RMnO_3$ (R, rare earths), and the rare earth molybdates, but none seem to provide large coupling between them (Boomgaard & Born, 1978; Dai et al., 2004a; Fischer et al., 1972; Hornreich, 1969; Kadomtseva et al., 2004; Ruette et al., 2004; Singh, 1996; Srinivas et al., 2004; Suryanarayana, 1994; Van den Boomgaard et al., 1974; Van Suchtelen, 1972; Wang et al., 2003). Recently, rare earth manganites such as $TbMnO_3$, $DyMnO_3$, and $TbMn_2O_5$ have been reported to exhibit reproducible electric polarization under magnetic fields; however, the magnitude of the magnetoelectric (ME) coefficient (unit of V/cm.Oe) is quite small (Ederer & Spaldin, 2004; Fiebig, 2005; Fiebig et al., 2002; Hur et al., 2004; Kimura et al., 2003; Lottermoser et al., 2004; Srinivas et al., 2004; Van Aken et al., 2004). Single-phase materials exhibiting ME effects should show two coupled-transitions: one from ferroelectric to paraelectric states, and another from ferro/ferri/antiferro-magnetic to paramagnetic states: the ME effect then arises due to coupling between the magnetic and polar sublattices. Unfortunately, all the known single-phase materials suffer from drawback that the ME effect is considerably weak even at low temperatures.

Several researchers have explored the magnitude of ME effect in single-phase materials and have presented alternative ways to overcome the fundamental problem related to filling of electronic states. Recent investigations on single-phase multiferroics have revealed that the origin of the ME effect is often associated with a particular exchange mechanism for various families of compounds, such as orbital ordering, Jahn–Teller distortion, super/double exchange, and/or geometric magnetic frustration have been cited as the origin source of ME effects. Two prominent family of Mn-based compounds: $TbMnO_3$, $DyMnO_3$, and $TbMn_2O_5$ exhibit coupling due to frustrated spin resulting in magnetoelastically induced lattice modulations, while buckling of the layered MnO_5 polyhedra accompanied by displacements of the Y ions results in finite ME coupling in $YMnO_3$. Detailed theoretical calculations have been conducted on these materials, but none seem to indicate the possibility of shifting the transition temperature to room temperature.

There are two approaches that are promising for room-temperature multiferroic MEs. One based on solid solutions exhibiting morphotropic phase boundary (MPB) or polymorphic phase boundary (PPB), and the other based on composite structure that utilizes product property. It has been shown both experimentally and theoretically that the strain-mediated ME coupling in magnetostrictive-piezoelectric layered composite or laminate is several orders of magnitude larger as compared to single-phase materials. Direct ME (DME) and converse ME (CME) effect in the laminate composites consisting of piezoelectric layer (P-layer) and magnetostrictive layer (M-layer) can be represented by following schematics (Cho & Priya, 2011; Hosur et al., 2021; Karan et al., 2024):

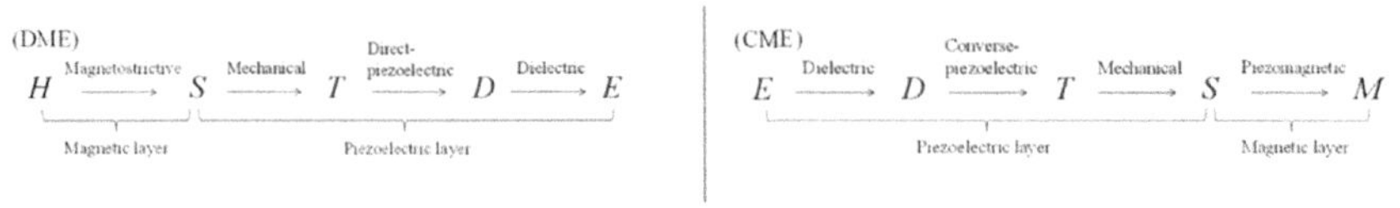

where H is magnetic field, S is mechanical strain, T is mechanical stress, D is electric displacement, E is (Cho et al., 2010) electric field, and M is magnetization. In the off-resonance condition, the ME coefficient in the composites is determined from Eq. (5.1), using the piezoelectric voltage coefficient (Eq. 5.2):(Cho et al., 2010)

$$\alpha_{E,31} = \frac{dE_3}{dH_1} = \frac{nq_{11}g_{31}^*}{nS_{11}^E\left(1 - {k_{31}}^2\right) + (1-n)S_{11}^H} \tag{5.1}$$

$${g_{31}}^* = g_{31}\frac{1}{\left(\tan\delta' + \tan\theta' + \dfrac{C - C_f}{C_f}\right)} \tag{5.2}$$

H_1 is the longitudinally applied magnetic field along 1 axis, E_3 is the generated electric field perpendicular to H_1 direction (3 axis), n is the volume fraction of the magnetostrictive layer ($n = V_{magnetic}/(V_{piezoelectric} + V_{magnetic})$; V is the volume), q_{11} is the piezomagnetic coefficient, g_{31} is the piezoelectric voltage coefficient, k_{31} is the electromechanical coupling coefficient, piezoelectric loss is shown as $\tan\theta'$, dielectric loss is shown as $\tan\delta'$, C is the capacitance at a given frequency, C_f is capacitance at 1 kHz, and S_{11}^E and S_{11}^H are the elastic compliances of the piezoelectric and magnetostrictive layers, respectively.

Considerable progress has been made on both these points as described in this chapter. Extensive investigations have been conducted on the BFO–$BaTiO_3$ (BTO) solid solutions that show the presence of MPB as shown in Fig. 5.1 (Yang et al., 2013). The enhanced magnetic properties with decreasing BFO mole fraction from 0.775 to 0.750 were attributed to active spin modulation of ordered Fe–O–Fe. It was found that the substitution of large Ba^{2+} and Ti^{4+} ions on the A and B sites, respectively, enlarges the distortion of bond angle of Fe–O–Fe to release the spiral magnetic modulation (Yang et al., 2013). $BiFeO_3$ (BFO) consists of a cycloidal spatially modulated spin structure in conjunction with an antiferromagnetic order that deteriorates the macroscopic magnetization. The atomic pair distribution function (PDF) for pure BFO ($x = 1.0$) was found to

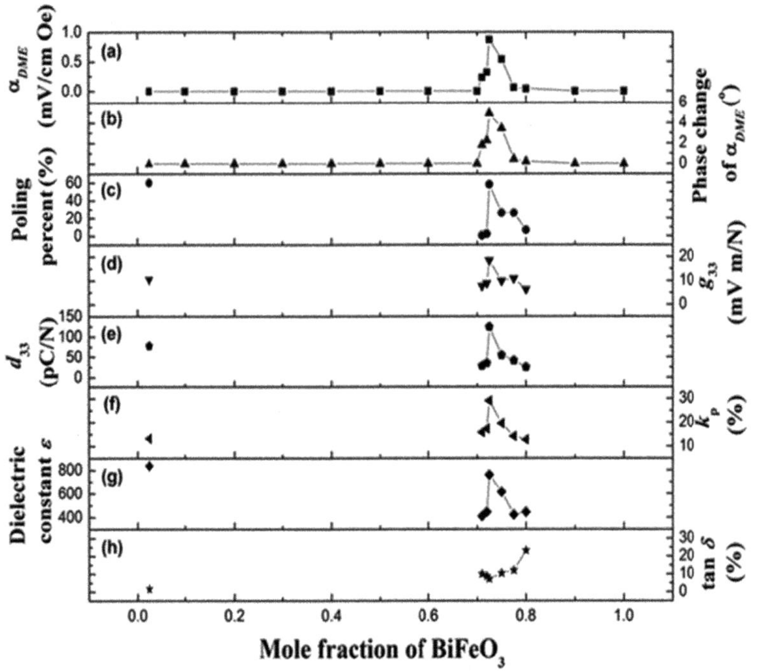

FIGURE 5.1 Variation of magnetoelectric, piezoelectric, and dielectric properties as a function of BFO mole fraction for BTO–xBFO ceramics with x = 0.025–1.0; (a) DME coefficient (α_{DME}), (b) phase change of α_{DME}, (c) poling percent, (d) piezoelectric voltage constant (g_{33}), (e) piezoelectric charge constant (d_{33}), (f) radial mode electromechanical coupling factor (k_p), (g) dielectric constant (ε), and (h) tangent loss factor (tan δ). *Reproduced with permission from Yang et al. (2013).*

fit with a rhombohedral structure model (S.G. R3c) with lattice parameters in hexagonal setting (a = 5.583 Å, c = 13.835 Å and gamma = 120 degrees). The PDFs for BT–xBFO (x = 0.725) were found fit at lower r values with a structure model featuring monoclinic structure (S.G. Cm) with parameters (a = 5.499 Å, b = 5.631 Å, c = 3.974 Å, and beta = 91 degrees). At higher r values, a rhombohedral model (S.G. R3c) (a = 5.558 Å and c = 13.834 Å and gamma = 120 degrees) provide better fit.

Thus, this material was found to be rhombohedral on average with local monoclinic distortions/symmetry. These results are highly exciting toward the realization of a room temperature single-phase multiferroic ME material.

In the single-phase ME materials, the upper limit for ME susceptibility can be given as (Fiebig, 2005; Ideshita et al., 1978):

$$\alpha_{ij}{}^2 < \left(\kappa_{ii}^e \chi_{jj}^m\right) \tag{5.3}$$

where κ^e and χ^m are the electric and magnetic susceptibilities respectively. A similar relationship can be derived based on the thermodynamic considerations given as: (O'dell, 1963)

$$\alpha_{ij} < \left(\varepsilon_{ii}\mu_{jj}\right)^{1/2} \tag{5.4}$$

where ε and μ are the electric permittivity and magnetic permeability, respectively. In terms of energy functions, the ME coupling coefficient has been defined as: (Eerenstein et al., 2006)

$$k_{ME} = \frac{U_{mutual}}{\left(U_{electric} \cdot U_{magnetic}\right)^{1/2}} \tag{5.5}$$

where U_{mutual} $(= ½\alpha_{ij}E_iH_j, \textit{where } H_j \textit{ is the magnetic field})$ is the mutual energy, $U_{electric}$ $(= ½\varepsilon_{ii}E_iE_i)$ is the electric energy, and $U_{magnetic}$ $(= ½\mu_{jj}H_jH_j)$ is the magnetic energy. In order to evaluate the efficiency of energy conversion from magnetic to electric forms, or vice-versa, a working definition of the coupling coefficient can be written as: (Sahoo et al., 2015)

$$k_{me}^2 = k_{ij,piezo}^2 \, k_{ij,magnetic}^2 \tag{5.6}$$

where $k_{ij,\ \mathrm{piezo}}$ is the coupling coefficient of the piezoelectric phase, and $k_{ij,\ magnetic}$ is the coupling coefficient of the magnetic phase.

As mentioned earlier, single-phase materials suffer from the drawback that the ME effect is extremely small at room temperature. For example, the highest ME coefficient has been reported for antiferromagnetic Cr_2O_3 crystals, which is 0.74 mV/cm·Oe at a Neel temperature of 34°C (Kita et al., 1979a, 1979b). On the other hand, multiphase ME composites exhibit much higher magnitude of ME coefficient at room temperature (Zhai et al., 2008). Composite geometry can be designed based on the connectivity. The connectivity in general can be given by the ratio $\frac{(n+3)!}{3!n!}$, where n is number of phases. For two-phase system, there are 10 types of connectivity. For three- and four-phase systems, there are 20 and 35 types of connectivity. Fig. 5.2 shows three types of connectivity for multiphase ME composites (Bichurin et al., 2011; Pradhan et al., 2020). The 3-0 type particulate ME composites were constructed by embedding one-phase particles in a matrix of another phase and found to exhibit coupling coefficient of several 10 ~ 100 mV/cm·Oe at room temperature. The ME coefficient in 3-0 type particulate composites can reach up to several thousand mV/cm·Oe at resonance frequency, but this magnitude is still below the theoretical magnitude due to the problems related to the

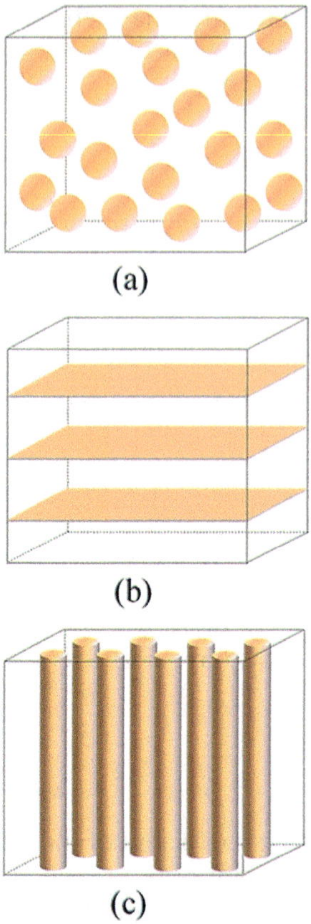

FIGURE 5.2 Schematic diagram for connectivity of multiphase ME composites: (a) 3-0 type particulate structure, (b) two to two type laminate structure, and (c) one to three type cylinder-matrix structure (Pradhan et al., 2020).

interdiffusion between piezoelectric and magnetostrictive phases during high-temperature sintering, thermal expansion mismatch between the two phases, and low resistivity. In order to improve the resistivity in ME composites, two to two type ME laminates have been synthesized by using piezoelectric and magnetostrictive layers and shown to exhibit ME coefficient of several thousand mV/cm·Oe at room temperature due to reduction of the leakage problem (Bichurin et al., 2011). The one to three type ME composites were investigated for high ME coefficient by exploiting contribution from both piezoelectric coefficient d_{33} and piezomagnetic coefficient q_{11}. The PZT/Terfenol-D composites were reported with ME coefficient of 500 mV/cm·Oe at off-resonance condition and 18,200 mV/cm·Oe at a resonance condition by Ma et al. (2007). Laminate composites are generally fabricated by bonding magnetostrictive and piezoelectric layers using silver epoxy followed by annealing at lower temperature of ~100°C (Palneedi et al., 2016). The most common geometry for laminate composites is "sandwich" structure where the piezoelectric layer is arranged between two magnetostrictive ones. Under applied magnetic field, the strain in magnetostrictive layer is transferred to the piezoelectric layer through bonding layer resulting in generation of electrical charge.

In 2–2 ME laminate composites, there can be four possible modes as shown in Fig. 5.3 (Pradhan et al., 2020). In L-L mode, both the applied H field and the measured E field are parallel to the principle vibration mode; in T-L mode, the applied H field is perpendicular to the induced E field, which is measured parallel to the principle mode; in L-T mode, the applied H field is parallel, and the induced E field is measured perpendicular to the principle mode; and in T-T mode, the applied H field and induced E are perpendicular to the principle mode (Priya et al., 2007).

ME coefficient for two to two type laminate composites has been modeled by Srinivasan et al. (2001). The transverse ME coefficient for the laminate composite was given as:

$$\alpha_{E31} = \frac{\delta E_3}{\delta H_1} = \frac{-2d^p_{31} q^m_{11} v^m}{\left(s^m_{11} + s^m_{12}\right)\varepsilon^{T,P}_{33} v^p + \left(s^p_{11} + s^p_{12}\right)\varepsilon^{T,P}_{33} v^m - 2\left(d^p_{31}\right)^2 v_m} \tag{5.7}$$

The longitudinal ME coefficient for the laminate composite was given as:

$$\alpha_{E33} = \frac{\delta E_3}{\delta H_3} = \frac{-2d^p_{31} q^m_{13} v^m}{\left(s^m_{11} + s^m_{12}\right)\varepsilon^{T,P}_{33} v^p + \left(s^p_{11} + s^p_{12}\right)\varepsilon^{T,P}_{33} v^m - 2\left(d^p_{31}\right)^2 v_m} \tag{5.8}$$

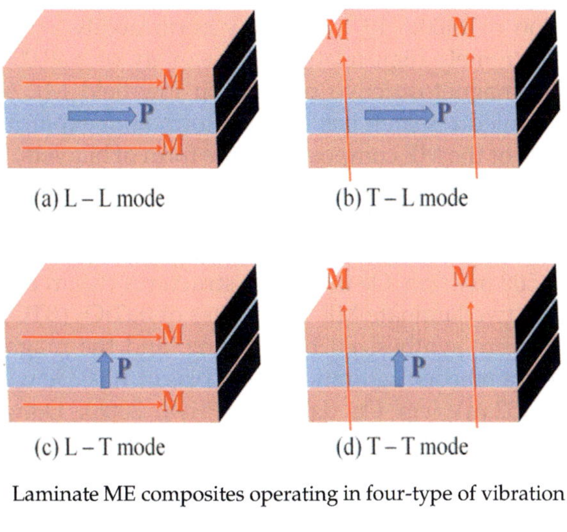

FIGURE 5.3 Laminate ME composites operating in four-type of vibration modes (Zhang et al., 2018).

where d_{31}^{p} is piezoelectric coefficient, v^m and v^p are volumes of magnetostrictive and piezoelectric phases, s_{11}^{p}, s_{12}^{p} are elastic compliances for piezoelectric phase, s_{11}^{m}, s_{12}^{m} are elastic compliances for magnetostrictive phase, q_{11} is piezomagnetic coefficient of the magnetostrictive phase, and $\varepsilon_{33}^{T,P}$ is the permittivity of the piezoelectric phase.

In the case of laminate ME composites with two to two connectivity, detailed research has been conducted in the form of bilayers, trilayers, and multilayers of piezoelectric and magnetostrictive materials. Table 5.1 shows the ME properties of two to two type laminate ME composites. The push-pull mode ME laminates consisting of PZT, Terfenol D, and Metglas were found to exhibit ME coefficient of ~20,000 mV/cm·Oe with high magnetic sensitivity in resonance condition (Dong, Zhai, Bai, Li, & Viehland, 2005). Partial texturing of piezoelectric phase in two to two laminate composites was found to show 67% improvement of ME coefficient compared with randomly oriented laminate (Islam & Priya, 2009).

One of the problems related to the design of magnetic field sensors is the need for DC bias magnets. Additional magnets make the device bulky and require insulation. Thus, there is a critical need for the design of self-biased ME laminates. In this section, a discussion is included on the magnitude of DC bias required for the two to two type laminates. Two to two type laminate ME composites can be categorized by the type of magnetostrictive material used in the design: (i) ferrite, (ii) Terfenol-D, and (iii) Metglas. Each of these ME laminate has different ME behavior due to difference in coercive field (H_c), permeability (μ), and magnetization (M). In Table 5.2, varying ferrite oxides are available to co-sinter with piezoelectric matrix and were found to exhibit different magnetostrictions due to their varying magnetic properties (Priya et al., 2007). Terfenol-D in Table 5.3 has the largest magnetostriction of 1400×10^{-6} with low permeability, while Metglas in Table 5.4 has extremely high relative permeability of >40,000 and thus is quite attractive magnetostrictive material in low-field ME sensors (Zhai et al., 2008). Detailed research has been performed in exploring the two to two connectivity of the ME composites in the form of bilayers, trilayers, and multilayers of piezoelectric and magnetostrictive materials (Zhai et al., 2008). Islam et al. have synthesized PZT-NCZF bilayer and NCZF-PZT-NCZF trilayer using cofiring technique and found a high ME coefficient on the order of ~500 mV/cm·Oe in trilayer structures (Islam & Priya, 2008c). Lead-free two to two laminate ME composites of $BaTiO_3$ (BTO)/($Ni_{0.8}Zn_{0.2}$) Fe_2O_4 (NZFO) were synthesized by co-firing, and the magnitude of ME voltage coefficient reached saturation at DC bias of 1000 Oe with maximum magnitude of 152 mV/cm Oe (Islam & Priya, 2006). Dong et al. have studied several multilayered ME configurations by using materials PZT, Terfenol D, and Metglas and have reported quite high ME coefficients on the order of 10 V/cm·Oe in resonance condition (Dong et al., 2003a, 2003b, 2004a, 2004b, 2005b).

TABLE 5.1 Magnetoelectric laminate composites.

Compositions	DC bias/ frequency	dE/dH (mV/ cm·Oe)	References
$Pb(Zr,Ti)O_3/TbDyFe_2$	4200 Oe/ 1 kHz	4680	Ryu et al. (2001a)
$Pb(Zr_{0.5}Ti_{0.5})O_3/Tb_{1-x}Dy_xFe_2$	4000 Oe/ 1 kHz	5900	Ryu et al. (2002)
$Pb(Mg_{1/3}Nb_{2/3})O_3-PbTiO_3/Tb_{1-x}Dy_xFe_2$	400 Oe/ 1 kHz	4300	Dong et al. (2004c)
$BaTiO_3/Tb_{1-x}Dy_xFe_2$	350 Oe/ 1 kHz	2100	Liu et al. (2004)
$Ni/Pb(Zr_{0.5}Ti_{0.5})O_3/Ni$	80 Oe/ 1 kHz	450	Laletin et al. (2005)
$BaTiO_3/Ni_{0.8}Zn_{0.2}Fe_2O_4$	1000 Oe/ 1 kHz	152	Islam and Priya (2006)
Terfenol-D/$Pb(Zr,TiO)_3$/μ-metal	240 Oe/ 1 kHz	>1400	Dong et al. (2006)
Fe-doped $BaTiO_3/Tb_{1-x}Dy_xFe_{2-y}$	350 Oe/ 100 Hz	2100	Zhang, Fan, et al. (2007)
$0.9Pb(Zr_{0.52}Ti_{0.48})O_3-0.1Pb(Zn_{1/3}Nb_{2/3})O_3/Ni_{0.6}Cu_{0.2}Zn_{0.2}Fe_2O_4$	400 Oe/ 1 kHz	468	Islam and Priya (2008b)
$BaTi_{0.99}Cr_{0.01}O_3/Tb_{1-x}Dy_xFe_{2-y}$	355 Oe/ 100 Hz	2753	Ke et al. (2009)
$Pb(Zr,Ti)O_3-Pb(Mg_{1/3}Nb_{2/3})O_3$ single crystal/Terfenol-D/ Metglas	1011 Oe/ 1 kHz	5150	Park et al. (2010)
Mn-doped $Na_{0.5}Bi_{0.5}TiO_3-BaTiO_3$ single crystal/Terfenol-D	400 Oe/ 1 kHz	1320	Zhang, Xia, et al. (2006)
$0.675Pb(Mg_{1/3}Nb_{2/3})O_3$-$0.325PbTiO_3$ with 1 vol.% $BaTiO_3$/Metglas	10 Oe/ 1 kHz	4410	Wang et al. (2008)
$Fe_{64}Co_{17}Si_7B_{12}$/PVDF	17.5 Oe/ 172 kHz	16,500	Lasheras et al. (2015)
Metglas (fiber)/PMN-PT (fiber)/Metglas (fiber)	2 Oe/1 kHz	29,310	Chu et al. (2017)
Terfenol-D/PVDF/Terfenol-D	2300/1 kHz	7930	Lu et al. (2015)
$Pb(Zr_{0.52}Ti_{0.48})O_3/CoFe_2O_4$ (bulk ceramic)	600 Oe/ 1 kHz	155	Palneedi et al. (2018)

Continued

TABLE 5.1 Magnetoelectric laminate composites.—cont'd

Compositions	DC bias/ frequency	dE/dH (mV/ cm·Oe)	References
NCZF/textured 0.675PMN-0.325PT/NCZF	—/1 kHz	1200	Yan, Zhou, and Priya (2013)
Terfenol-D/P(VDF-TrFE)	150 Oe/—	38	Brito-Pereira et al. (2017)

TABLE 5.2 Magnetoelectric laminate based on ferrite.

Compositions	DC bias (Oe)	References
Pb(Zr, Ti)O_3/$NiFe_2O_4$–bilayer	100	Srinivasan et al. (2001)
Pb(Zr, Ti)O_3/$NiFe_2O_4$–multilayer	400	Srinivasan et al. (2001)
Pb(Zr, Ti)O_3/$Ni_{0.8}Zn_{0.2}Fe_2O_4$	100	Srinivasan et al. (2002)
Pb(Zr, Ti)O_3/$Co_{0.6}Zn_{0.4}Fe_2O_4$	450	Srinivasan, Rasmussen, and Hayes (2003)
Pb(Zr, Ti)O_3/$Ni_{0.8}Zn_{0.2}Fe_2O_4$	150	Srinivasan, Rasmussen, and Hayes (2003)
Pb(Zr, Ti)O_3/$Li_{0.35}Zn_{0.3}Fe_{2.35}O_4$	150	Srinivasan, Hayes, and Bichurin (2003)
Pb(Zr, Ti)O_3/$Ni_{0.93}Co_{0.02}Mn_{0.05}Fe_{1.95}O_4$	700	Babu et al. (2008)
$BaTiO_3$/$Ni_{0.8}Zn_{0.2}Fe_2O_4$	1000	Islam and Priya (2006)
Pb(Zr, Ti)O_3/$Ni_{0.8}Zn_{0.2}Fe_2O_4$	250	Zhang, Ke, et al. (2006)
$Ni_{0.6}Cu_{0.2}Zn_{0.2}Fe_2O_4$/ 0.9Pb($Zr_{0.52}Ti_{0.48}$)$O_3$–0.1Pb ($Zn_{1/3}Nb_{2/3}$)$O_3$/$Ni_{0.6}Cu_{0.2}Zn_{0.2}Fe_2O_4$	400	Islam, Ni, et al. (2008)
Pb($Zr_{0.52}Ti_{0.48}$)O_3/$CoFe_2O_4$ (bulk ceramic)	600	Palneedi et al. (2018)
PVDF/$Fe_{64}Co_{17}Si_7B_{12}$	17.5	Lasheras et al. (2015)

5.2 Lead-free piezoelectrics-ferrite composites and similar systems

As mentioned earlier, in composite systems, ME effect is combination of two types of materials property such as magnetostriction and piezoelectricity (Nan et al., 2008). On application of magnetic field,

TABLE 5.3 List of magnetoelectric laminate based on Terfenol-D.

Compositions	DC bias (Oe)	References
$Pb(Zr,Ti)O_3$/Terfenol-D	4200	Ryu et al. (2001a)
$Pb(Zr_{0.5}Ti_{0.5})O_3$/Terfenol-D	4000	Ryu et al. (2002)
$Pb(Zr,Ti)O_3$/Terfenol-D	200	Dong, Cheng, et al. (2003)
$Pb(Zr_{0.52}Ti_{0.48})O_3$/Terfenol-D	700	Wan et al. (2003)
$Pb(Mg_{1/3}Nb_{2/3})O_3-PbTiO_3$/Terfenol-D	400	Dong et al. (2004c)
$BaTiO_3$/Terfenol-D	350	Liu et al. (2004)
$Pb(Zr_{0.52}Ti_{0.48})O_3$/Terfenol-D	700	Wan, Li, et al. (2005)
$Pb(Mg_{1/3}Nb_{2/3})O_3-PbTiO_3$/Terfenol-D	450	Dong, Zhai, Bai, Li, and Viehland (2005)
Fe-doped $BaTiO_3$/Terfenol-D	350	Zhang, Fan, et al. (2007)
$Pb(Mg_{1/3}Nb_{2/3})O_3-PbTiO_3$/Terfenol-D	400	Wang et al. (2008)
$Pb(Mg_{1/3}Nb_{2/3})O_3-PbTiO_3/Tb_{0.3}Dy_{0.7}Fe_{1.92}$	400	Wang, Leung, Wang, et al. (2009)
$BaTi_{0.99}Cr_{0.01}O_3/Tb_{1-x}Dy_xFe_{2-y}$	355	Ke et al. (2009)
$Pb(Mg_{1/3}Nb_{2/3})O_3-PbTiO_3$/Terfenol-D	400	Wang, Leung, Or, et al. (2009)
$Pb(Zr,Ti)O_3$/Terfenol-D	300	Guo et al. (2010)
Mn-doped $Na_{0.5}Bi_{0.5}TiO_3-BaTiO_3$ single crystal/Terfenol-D	400	Zhang, Xia, et al. (2006)
Terfenol-D/PVDF/Terfenol-D	2300	Lu et al. (2015)
Terfenol-D/P(VDF-TrFE)	150	Brito-Pereira et al. (2017)
PZT/Terfenol-D	4	Grotel et al. (2023)

magnetostrictive phase produces strain which is transferred on to the piezoelectric phase that converts strain into electric charge. This conversion of applied magnetic field into electric field is termed as direct ME effect. On the other hand, under applied electric field, piezoelectric phase produces strain which is transferred on to the magnetostrictive phase that converts it into magnetic field, termed as converse ME effect. ME effect in 3-0 particulate ME composites can be illustrated by coupling between the piezoelectric matrix and magnetostrictive dispersant or vice-versa through elastic strain.

TABLE 5.4 List of magnetoelectric laminate based on Metglas.

Compositions	DC bias (Oe)	References
PVDF/Metglas	8	Zhai et al. (2006)
$Pb(Zr,Ti)O_3$ fiber/Metglas	<10	Wu et al. (2009)
PVDF/Metglas	52	Dong et al. (2009)
PVDF/Metglas	<10	Fang et al. (2009)
$Pb(Zn_{1/3}Nb_{2/3})_x(Zr_{0.5}Ti_{0.5})_{1-x}O_3$/Metglas	100–325	Park, Ahn, et al. (2009)
$0.675Pb(Mg_{1/3}Nb_{2/3})O_3$-$0.325PbTiO_3$ with 1 vol.% $BaTiO_3$/Metglas	10 Oe	Wang et al. (2008)
Metglas (fiber)/PMN-PT (fiber)/Metglas (fiber)	2 Oe	Chu et al. (2017)
$Pb(Zr,Ti)O_3$/Metglas	22 Oe	Palneedi et al. (2017)
PZT5/Metglas	5 Oe	Yu et al. (2020)

ME coefficient (α_E) for 3-0 type particulate composites can be represented by following equation given by Zubkov (1978b)

$$\begin{aligned}\alpha_E &= \left(\frac{dE}{dH}\right)_{comp} = \left(\frac{dS}{dH}\right)_{comp}\left(\frac{dE}{dS}\right)_{comp} \\ &= m_v \cdot \left(\frac{dS}{dH}\right)_{ferrite}(1-m_v)\cdot\left(\frac{dE}{dS}\right)_{piezoelectric}\end{aligned} \tag{5.9}$$

where m_v is the volume fraction of ferrite, (dE/dS) is the inverse of change in strain per unit magnetic field. Since $dE = g \cdot dX$ and $dS = dX/C$, where X is the stress, S is the strain, g is the piezoelectric voltage constant, and C is the stiffness, the equation can be demonstrated as (Zubkov, 1978a):

$$\alpha_{E33} = \left(\frac{dE}{dH}\right)_{comp} = m_v \cdot \left(\frac{dS}{dH}\right)_{ferrite}(1-m_v)\cdot(g_{33}\cdot C_{33})_{piezoelectric} \tag{5.10}$$

since $g_{33} = d_{33}/\varepsilon^T{}_{33}$, the ME coefficient for 3-0 type particulate composites directly depends on the charge coefficient d_{33} and the permittivity $\varepsilon^T{}_{33}$ of piezoelectric phase.

The ME coefficient of 3-0 particulate composites has been reported to be of the order of ~100 mV/cm·Oe as shown in Table 5.5, which is significantly higher than that of single-phase ME materials. However, particulate ME composites exhibit much lower ME coefficient than

TABLE 5.5 List of magnetoelectric particulate composites.

Compositions	DC Bias/ Frequency	dE/dH (mV/ cm·Oe)	References
0.6 $BaTiO_3$–0.4 $Ni_{0.97}Co_{0.03}Mn_{0.1}Fe_{1.90}O_4$	500 Oe/1 kHz	80	Boomgaard and Born (1978)
0.8 $Pb(Zr_{0.5}Ti_{0.5})O_3$–0.2 $NiCo_{0.02}Cu_{0.02}Mn_{0.1}Fe_{1.8}O_4$	1250 Oe/1 kHz	115	Ryu et al. (2001b)
0.41 $Pb(Zr_{0.5}Ti_{0.5})O_3$–0.59 $Ni_{1-x}ZnxFe_2O_4$	250 Oe/100 Hz	45	Srinivasan et al. (2004)
0.7 $Pb(Zr_{0.53}Ti_{0.47})O_3$–0.3 $CuFe_2O_4$	460 Oe/100 kHz	421	Dai et al. (2004b)
0.68 $Pb(Zr_{0.57}Ti_{0.43})O_3$–0.32 $NiFe_2O_4$	1000 Oe/1 kHz	80	Zhai et al. (2004)
0.7 $Sr_{0.5}Ba_{0.5}Nb_2O_6$–0.3 $Ni_{0.8}Zn_{0.2}Fe_2O_4$	1000 Oe/ 100 kHz	26	Li et al. (2006)
0.9 $Pb(Zr_{0.52}Ti_{0.48})O_3$–0.1 $NiFe_{1.9}Mn_{0.1}O_4$	600 Oe/1 kHz	90	Islam et al. (2007)
0.8 $Pb(Zr_{0.52}Ti_{0.48})O_3$–0.2 $NiFe_2O_4$	454 Oe/1 kHz	195	Islam, Bedekar, et al. (2008)
0.8 $Pb(Zr_{0.52}Ti_{0.48})O_3$–0.2 $Ni_{0.8}Zn_{0.2}Fe_2O_4$	1200 Oe/1 kHz	155	Islam and Priya (2008a)
0.82 (Martins and Lanceros-Méndez)–0.18 $NiMn_{0.1}Fe_{1.9}O_4$	175 Oe/1 kHz	150	Agrawal et al. (2009)
0.6 $Pb(Zr_{0.5}Ti_{0.5})O_3$–0.4 $CoFe_2O_4$	3000 Oe/1 kHz	70	Miclea et al. (2010)

theoretically predicted because of the problems related to low overall piezoelectric and magnetostrictive constants, interfacial defects, and low resistivity. Recently, investigations have focused on developing lead-free ME materials. Lead-free piezoelectric materials considered for ME laminates can be classified into three material groups (ABO_3): (i) $K_{0.5}Na_{0.5}NbO_3$ (KNN), (ii) $Na_{0.5}Bi_{0.5}TiO_3$ (NBT), and (iii) $BaTiO_3$ (BTO). Table 5.6 shows piezoelectric and electromechanical properties for compositions based on KNN, NBT, and BTO.

$BaTiO_3$–$CoFeO_4$ lead-free composites were the first ones investigated by Boomgaard et al., in 1978 and found to exhibit ME coefficient of 130 mV/cm·Oe. Since that time, few studies have been conducted on

TABLE 5.6 List of lead-free piezoelectric materials.

Compositions	d_{33} (pC/N)	k_p	T_c (°C)	$T_{O\text{-}T}/T_d$ (°C)	References
$(K_{0.5}Na_{0.5})NbO_3$	80	0.35	420	195	Egerton and Dillon (1959)
$(K_{0.465}Na_{0.465}Li_{0.07})NbO_3$	240	0.45	460	~20	Hollenstein et al. (2005)
$(K_{0.44}Na_{0.52}Li_{0.04})$ $(Nb_{0.84}Ta_{0.1}Sb_{0.06})O_3$	410	0.61	253	25	Lang et al. (2006)
0.95 $(K_{0.5}Na_{0.5})NbO_3$–0.05 $SrTiO_3$	200	0.37	277	27	Wang et al. (2006)
0.95 $(K_{0.5}Na_{0.5})NbO_3$–0.05 $LiTaO_3$	200	0.36	430	55	Guo et al. (2005)
0.94 $(K_{0.5}Na_{0.5})NbO_3$–0.06 $LiNbO_3$	235	0.42	460	70	Guo et al. (2004)
0.95 $(K_{0.5}Na_{0.5})NbO_3$–0.05 $LiSbO_3$	283	0.5	392	45	Zang et al. (2006)
0.7 $Bi_{0.5}Na_{0.5}TiO_3$–0.2 $Bi_{0.5}K_{0.5}TiO_3$–0.1 $Bi_{0.5}Li_{0.5}TiO_3$	216	0.401	350	160	Yuan et al. (2006)
x $(Na_{0.5}Bi_{0.5})TiO_3$–y $(K_{0.5}Bi_{0.5})TiO_3$–z $BaTiO_3$,	145	0.162	302	224	Zhang, Shrout, et al. (2007)
0.5 $Na_{0.5}Bi_{0.5}TiO_3$–0.5 $K_{0.5}Bi_{0.5}TiO_3$	150	0.22	320	210	Zhao et al. (2006)
$BaTiO_3$	190	0.36	115	0	Jaffe (1958)
0.95 $BaTiO_3$–0.05 $CaTiO_3$–Co	150	0.31	105	−45	Schofield and Brown (1957)
0.67Bi1.05 $(Fe_{0.97}Ga_{0.03})O_3$–0.33$BaTiO_3$	402	0.6	454	−/−	Lee et al. (2015)
$Bi_{0.5}(Na_{0.7}K_{0.2}Li_{0.1})_{0.5}TiO_3$	231	0.36	210	195	Lin et al. (2006)
0.96$(K_{0.5}Na_{0.5})$ $(Nb_{0.93}Sb_{0.07})$–(0.01)$CaZrO_3$-0.03(Bi, Ag)ZrO_3	680	0.5	176	R.T	Go et al. (2021)
(0.82) $Ba(Ti_{(1-y)}Sn_y)O_3$–0.18$(Ba_{(1-z)}Ca_z)TiO_3$ $0.10 \leq y \leq 0.12$ and $0 \leq z \leq 0.40$	730	0.7	75	31–41	Zhao et al. (2018)

lead-free ME composites as shown in Table 5.7 mainly because of the poor piezoelectric performance of lead-free materials. This problem still remains, but the emphasis on developing lead-free ME composites has increased due to environmental concerns.

Lead-free piezoelectric ceramics have gained much attention in the recent past due to their ability to form a complete substitutional solid solution, high ferroelectric–paraelectric phase transition temperature and low density. To compete with the lead counterpart, the compositions should manifest stable piezoelectric properties as a function of temperature, stress, and electric field. The base compositions for lead-free materials involve BFO($BiFeO_3$), BTO($BaTiO_3$), KNN($KNaNbO_3$), and BNT($BiNaTiO_3$)(Ahn, Maurya, et al., 2009; Go et al., 2021; Lee et al., 2015; Zhao et al., 2018). Among these BNT-BTO systems exhibit MPB similar to

TABLE 5.7 List of lead-free magnetoelectric composites.

Compositions	Type	DC Bias/ Frequency (kHz)	dE/dH (mV/ cm·Oe)	References
$BaTiO_3-CoFe_2O_4$	Particulate	270 Oe/160	2540	Ren et al. (2005)
0.7 $Sr_{0.5}Ba_{0.5}Nb_2O_6-0.3$ $Ni_{0.8}Zn_{0.2}Fe_2O_4$	Particulate	1–3 kOe/ 100	26.6	Li et al. (2006)
$BaTiO_3-CoFe_2O_4$	Particulate	2000 Oe/20	17.04	Nie et al. (2009)
$(K_{0.5}Na_{0.5})$ $NbO_3-LiSbO_3-CoFe_2O_4$	Particulate	2200 Oe/1	15	Zhou et al. (2009)
0.75 $BaZr_{0.08}Ti_{0.92}O_3-0.25$ $Co_{1.2-y}Mn_yFe_{1.8}O_4$, ($y = 0-0.4$)	Particulate	1500 Oe/1	2.5	Kambale et al. (2010)
$BaTiO_3/(Ni_{0.8}Zn_{0.2})$ Fe_2O_4	Laminate	1000 Oe/1	152	Islam and Priya (2006)
0.885 $Bi_{0.5}Na_{0.5}TiO_3-0.05$ $Bi_{0.5}K_{0.5}TiO_3-0.015$ $Bi_{0.5}Li_{0.5}TiO_3-0.05$ $BaTiO_3$/Terfenol-D	Laminate	600 Oe/131	2500	Lo et al. (2010)
Mn-doped $Na_{0.5}Bi_{0.5}TiO_3-BaTiO_3$ single crystal/ Terfenol-D	Laminate	400 Oe/1	1320	Zhang, Xia, et al. (2006)

PZT, while KNN system manifests PPT. Both MPB and PPB can be tuned by isovalent and aliovalent doping at both "A" and "B" sites and by forming ternary solid solutions with other lead-free perovskite systems (Wu et al., 2015; Xing et al., 2018). Piezoelectric and dielectric properties are the maximum at phase boundaries, so most of the advanced compositions are developed around MPB, and the PPB systems are either modified to form MPB or the ferroelectric–ferroelectric phase boundaries are engineered to converge at room temperature (Xing et al., 2018). Alterations to these compositions have the potential to improve d_{33}, K_p, g_{33}, etc. but might negatively impact the depolarizing and curie temperature resulting in low thermal stability (e.g., d_{33} or K_p v/s temperature). Compositions with improved piezoelectric properties exhibiting good thermal stability are suitable for multiple applications such as acoustic detectors, sensors, actuators, motors, etc.

Apart from phase, microstructural orientation has an effective role in enhancing d_{33} in randomly oriented compositions. Misoriented grains can affect the response of well-aligned domains, and the broader solid angle between the electric field and polar vector can result in ineffective polarization rotation resulting in clamped d_{33} and P_{sat} (Trolier-McKinstry et al., 2018). It can be achieved through TGG (templated grain growth) or RTGG (reactive templated grain growth) with a template acting as a seed crystal for growth. In lead-free systems, modified KNN and NBT systems are successfully textured around MPB, which yielded very high d_{33}. Complex NBT compositions like $(Na_{0.5}Bi_{0.5})_{0.935}Ba_{0.065}Ti_{0.978}(Fe_{0.5}Nb_{0.5})_{0.022}O_3$ are successfully textured along [001] direction with $NaNbO_3$(NN) templates with >95% texture degree (Liu et al., 2023). Similarly, $0.96(K_{0.5}Na_{0.5})(Nb_{0.965}Sb_{0.035})O_3$-$0.01CaZrO_3$-$0.03(Bi_{0.5}K_{0.5})HfO_3$ was also textured with NN templates and d_{33} enhanced by ~ 120% (Purusothaman et al., 2023).

5.2.1 Sintered KNNLS-NZF particulate composites

0.948 $K_{0.5}Na_{0.5}NbO_3$–0.052 $LiSbO_3$ (KNNLS) and $Ni_{0.8}Zn_{0.2}Fe_2O_4$ (NZF) ceramics have been synthesized through mixed oxide sintering method (Yang et al., 2011). The KNNLS powders were found to have the grain size <1 μm and exhibited perovskite structure as shown in Fig. 5.4a and b. The NZF powders had grain size range of <300 nm with spinel structure as shown in Fig. 5.4c and d. Fig. 5.5a–d shows the X-ray diffraction (XRD) patterns for $(1-x)$ KNNLS–x NZF ($x = 0.1$, 0.2, 0.3, and 0.4) composites sintered at varying temperatures (1040, 1050, 1060, and 1070°C) (Yang et al., 2011). The patterns clearly reflect that both phases of KNNLS and NZF stabilized respectively in perovskite and spinel structures without any trace of secondary phase in the composites. The mole fraction of NZF increased the intensity of spinel peaks increased, while the intensity of perovskite peaks decreased (Yang et al., 2013).

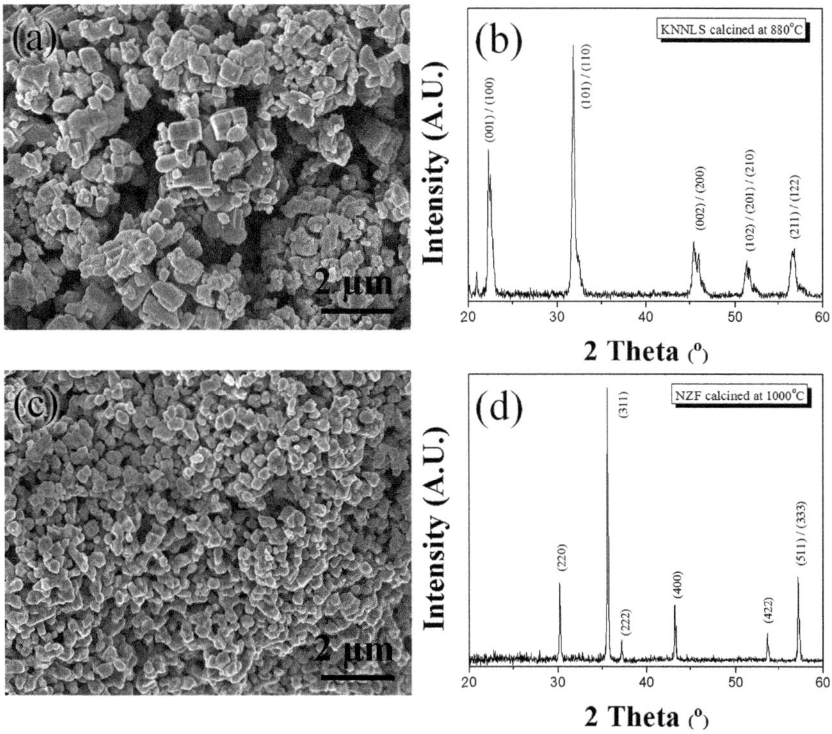

FIGURE 5.4 (a) SEM image and (b) XRD patterns of KNNLS calcined at 880°C. (c) SEM image and (d) XRD patterns of NZF calcined at 1000°C. *Taken with permission from Yang et al. (2011).*

Variations of longitudinal piezoelectric strain constant (d_{33}) and radial mode electromechanical coupling factor (k_p) with composition for all composites are shown in Fig. 5.6a and b. The magnitude of both d_{33} and k_p decreased with increasing NZF concentration and was found to maximize at the sintering temperatures of both 1050 and 1060°C regardless of x values (Yang et al., 2011). It can be noticed that there is critical sintering temperature of 1070°C above which the electromechanical properties decreased. Clearly 0.9 KNNLS–0.1 NZF composites exhibit variation of piezoelectric properties with sintering temperature. The changes occurring in other KNNLS–NZF composites are relatively small, and this could be associated to varying factors including poling percentage, grain size, resistivity, and space charge. Fig. 5.6c–d show that the magnitude of dielectric constant ($\varepsilon_r/\varepsilon_o$) decreased with increasing NZF concentration, and the magnitude of loss factor (tan δ) was found to minimize at the

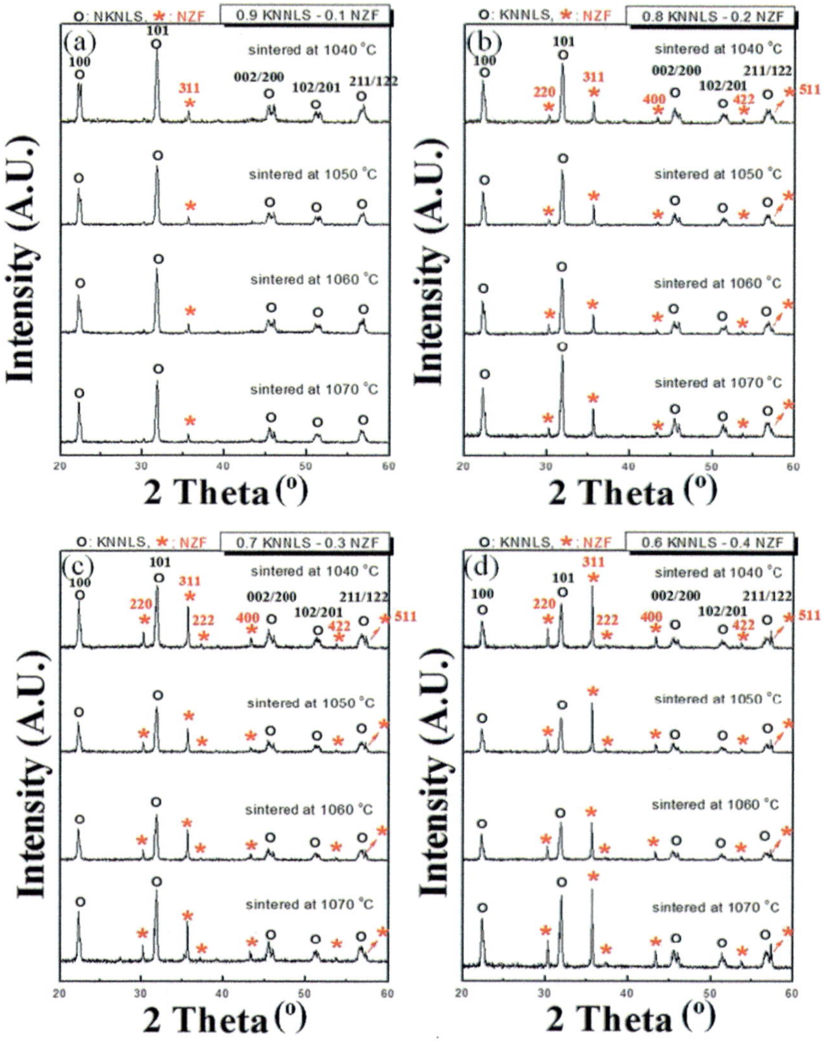

FIGURE 5.5 XRD patterns for (1−x) KNNLS−x NZF composites. (a) x = 0.1, (b) x = 0.2, (c) x = 0.3, and (d) x = 0.4. *Taken with permission from Yang et al. (2011).*

sintering temperature of 1050 and 1060°C (Yang et al., 2011). The low tan δ at the optimum sintering points resulted from the composite structures possessing high relative density of >95%. As the composites were sintered at 1040 and 1070°C, the relative density was found to decrease below 95% reflecting high porosity. Further, it is found in Fig. 5.6, the composites with high NZF concentration of $x = 0.4$ are still below the

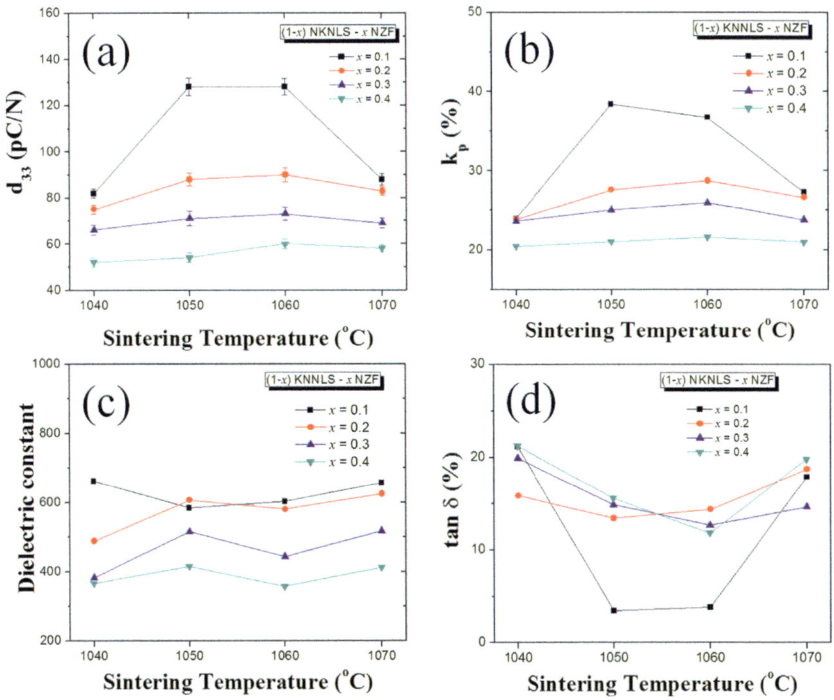

FIGURE 5.6 Piezoelectric and dielectric properties for (1−x) KNNLS−x NZF composites, (a) longitudinal piezoelectric constant, (b) radial mode coupling factor, (c) dielectric constant, and (d) tangent loss factor. *Taken with permission from Yang et al. (2011).*

percolation threshold as the dielectric and piezoelectric properties did not exhibit any significant drop (Yang et al., 2011).

Fig. 5.7a–d shows the magnetic hysteresis loops for the NZF particles and (1−x) KNNLS−x NZF ($x = 0.1, 0.2, 0.3$, and 0.4) composites measured by VSM at room temperature. The crushed NZF particles were found to exhibit saturation magnetization (M_s) of 63.90 emu/g as shown in Fig. 5.7a and b, which decreased for the sintered KNNLS-NZF composites (Yang et al., 2011). The magnitude of M_s for the composites was found to increase from 6.58 to 23.48 emu/g with increasing NZF concentration as shown in Fig. 5.7c–d (Yang et al., 2011). The remanent magnetization (M_r) for NZF particles was found to be 0.29 emu/g and the magnitude of M_r for the composites increased from 0.13 to 0.77 emu/g with increasing NZF concentration. The magnitude of coercive field (H_c) for KNNLS-NZF composites was found to be constant regardless of the change in composition (Yang et al., 2011).

In order to verify the coupling between magnetic polarization and mechanical strain for the (1−x) KNNLS−x NZF ($x = 0.1, 0.2, 0.3$, and 0.4)

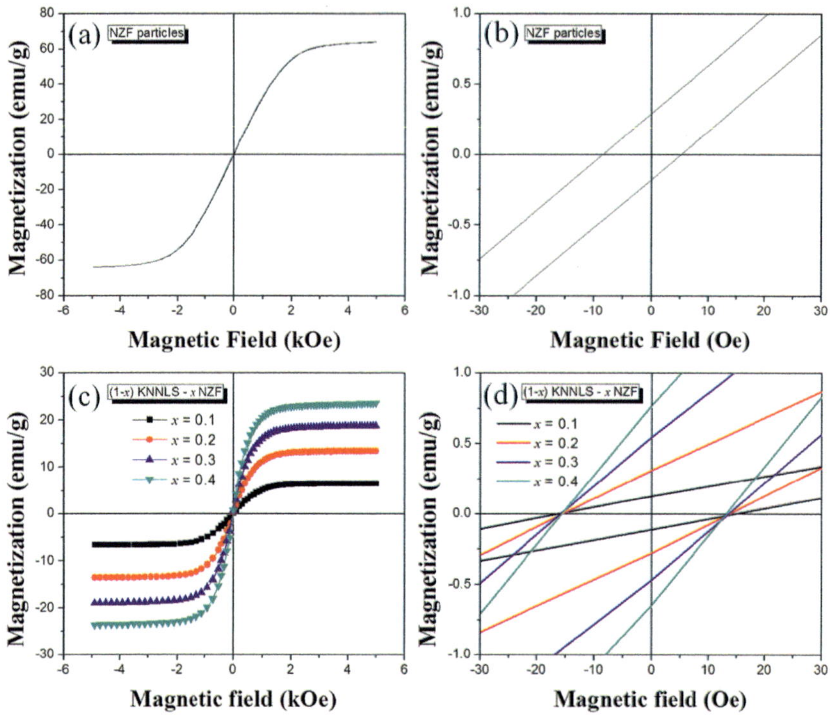

FIGURE 5.7 M–H curves for NZF particles and (1−x) KNNLS–x NZF composites. (a) Magnetization as a function of magnetic field for NZF particles, (b) amplified curve in the low magnetic field region shown in (a), (c) magnetization as a function of magnetic field for (1−x) KNNLS–x NZF composites, and (d) amplified curve in the low magnetic field region shown in (c). *Reprinted with permission from Yang et al. (2011).*

composites, magnetostriction (λ) and piezomagnetic coefficients (q) were determined in the range of 0–1000 Oe as shown in Fig. 5.8a–d. The parameters λ_{11} and λ_{12} correspond to in-plane parallel magnetostriction and in-plane perpendicular magnetostriction with respect to the applied magnetic field (H). With increasing NZF concentration, the magnitude of λ_{11} and λ_{12} was found to increase. The magnitude of λ_{11} and λ_{12} for composition $x = 0.4$ was found to be -8 ppm and 3 ppm (Yang et al., 2011). The in–plain piezomagnetic coefficients q_{11} and q_{12} were calculated by differentiating λ_{11} and λ_{12} with respect to magnetic field. As NZF concentration increases, the magnitude of q_{11} and q_{12} was found to increase and the peak shifted to low H (Yang et al., 2011). The position of peak in piezo-magnetic constant is an important characteristic in design of high-sensitivity magnetic field sensors as it corresponds with the peak observed in ME coefficient.

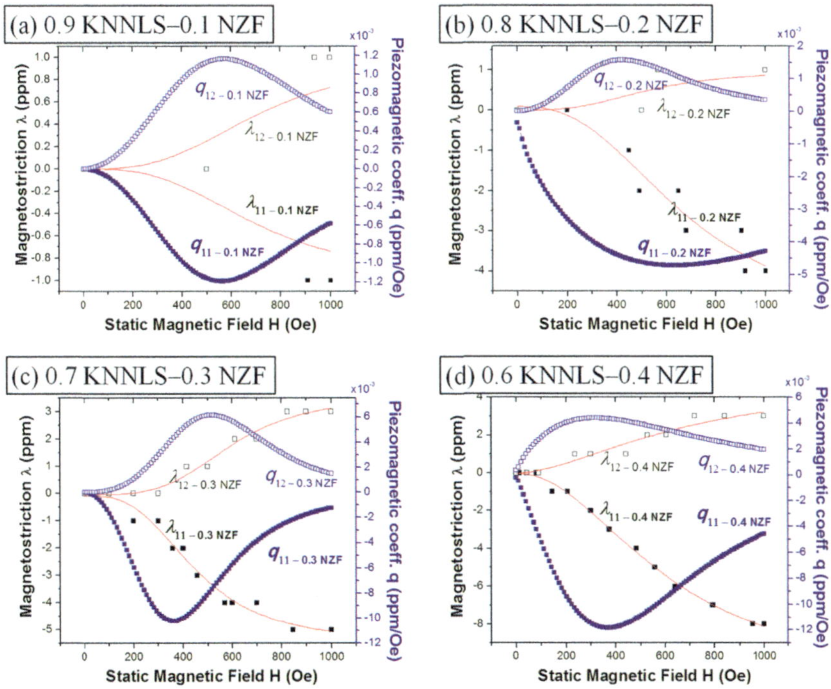

FIGURE 5.8 Magnetostrictions and piezomagnetic coefficients for (1−x) KNNLS−x NZF composites. (a) x = 0.1, (b) x = 0.2, (c) x = 0.3, and (d) x = 0.4. *Reprinted with permission from Yang et al. (2011).*

Fig. 5.9a–b show the variation of ME coefficient (α_E) and optimum DC bias (H_{bias}) for the (1−x) KNNLS−x NZF (x = 0.1, 0.2, 0.3, and 0.4) composites with sintering temperatures, measured at 1 kHz under H_{ac} = 1 Oe. With increasing NZF concentration, the magnitude of α_E was found to increase to 20.14 mV/cm·Oe under H_{bias} = 380 Oe at x = 0.3 and then slightly drop to 18.17 mV/cm·Oe under H_{bias} = 330 Oe at x = 0.4 (Yang et al., 2011). For all the compositions, the composites sintered at 1050 and 1060°C were found to exhibit maximum α_E due to their optimized piezoelectric and dielectric values as shown in Fig. 5.6. The magnitude of optimum H_{bias} decreased from 545 to 326 Oe with increasing NZF concentration. The reason for shift in H_{bias} is related to the difference in magnetic susceptibilities of the composites and variation in peak piezomagnetic coefficient as shown in Fig. 5.8 (Yang et al., 2011).

To form the island–matrix composite, following considerations were taken into account: (i) matrix had smaller particle size than that of island to achieve control of the surface area and (ii) sintering temperature of matrix and island was quite different to eliminate the elemental diffusion

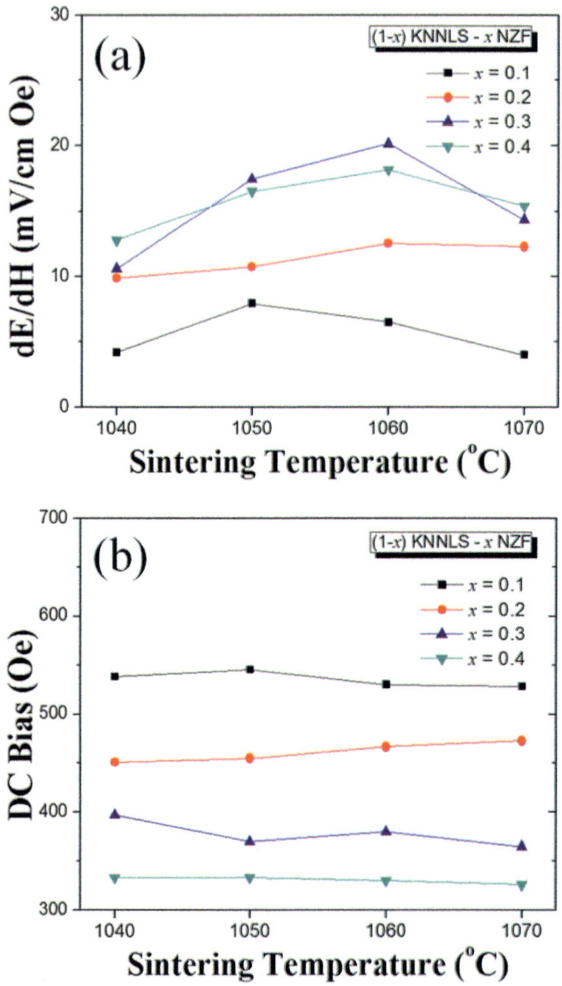

FIGURE 5.9 Variation of (a) ME coefficient and (b) H_{bias} for (1−x) KNNLS−x NZF composites with x = 0.1, 0.2, 0.3, and 0.4 as a function of sintering temperature. *Reprinted with permission from Yang et al. (2011).*

(Yang et al., 2011). Bordia et al. have described the sintering phenomenon with rigid inclusions in terms of transient stresses and strain rates (Bordia & Scherer, 1988). The analysis shows that volume fraction of the inclusion is an important parameter in determining the sintering of the composite. The KNNLS matrix normally has residual porosity due to the cubical shape of grains, but the KNNLS−NZF composites were found to exhibit high relative density of >95% (Ahn et al., 2010). This high density can be associated with NZF, which provides an interface for nucleation and growth. The densification rate of the composite is expressed as (Rahaman, 2017):

$$\frac{1}{\rho_m}\frac{d\rho_m}{dt}=\frac{1}{\rho_c}\frac{d\rho_c}{dt}\left(\frac{\rho_{co}}{\rho_{co}-v_{io}\rho_c}\right) \tag{5.11}$$

where ρ_c and ρ_m are the relative densities of composite and the matrix, and ρ_{co} and ν_{io} are initial values for density and volume fraction of inclusion. Three stages of sintering were identified in these systems. The first stage corresponds to rearrangement of random shape particles to form stack of plate-type particles. The second stage is characterized by rapid grain growth promoted by liquid phase. The liquid-phase formation was found to be related to Na_2O volatilization. In the final stage of sintering, densification slows and microstructural coarsening becomes the dominant process. The fraction control of liquid phase was an important factor to obtain a dense and uniform microstructure. Muffling was used to reduce the loss of the alkali ions during high-temperature sintering, and sintering temperature and time was used to control the volume fraction of the liquid phase (Ahn, Park, et al., 2009; Yang et al., 2011).

Fig. 5.10a–d shows the SEM micrographs for KNNLS–NZF composites with island–matrix microstructure sintered at 1060°C (Yang et al.,

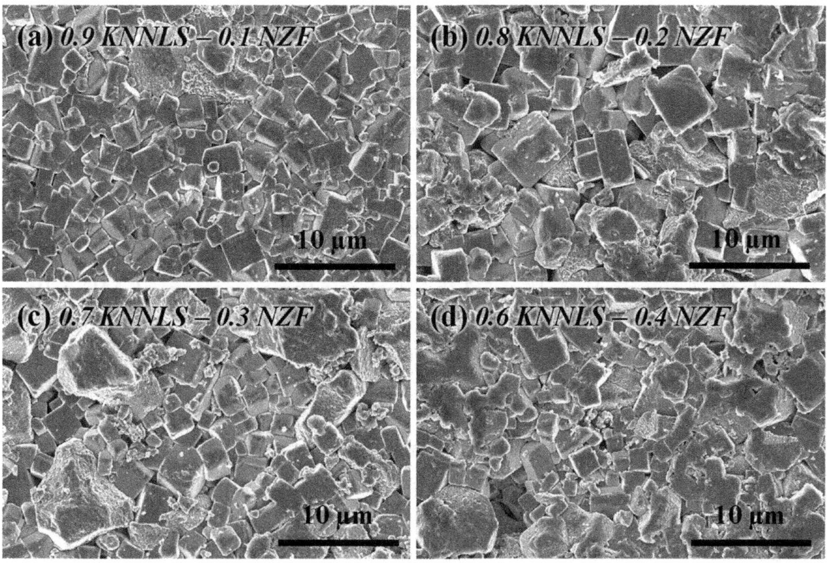

FIGURE 5.10 SEM micrographs for (1−x) KNNLS–x NZF composites sintered at 1060°C. (a) x = 0.1, (b) x = 0.2, (c) x = 0.3, and (d) x = 0.4. *Reprinted with permission from Ahn et al. (2010), Yang et al. (2011).*

2011). All the compositions exhibited dense microstructure with high density, which illustrates that growth occurred at the NZF interface leading to bridging of the residual porosity. The NZF islands were randomly dispersed in KNNLS matrix, which is highly beneficial for the poling of piezoelectric matrix. Two different grain structures can be noticed in the microstructure, cubical large grains corresponding to matrix and random-shaped small grains corresponding to island. There was no significant change observed in the size of the islands during sintering (Yang et al., 2011). From this microstructure, the connectivity of composites can be considered to be 3-0. The EDS elemental mapping images are shown in Fig. 5.11a–h, further confirm this conclusion. In this figure, the element Nb (white color in Fig. 5.11a–d) represents the KNNLS matrix, and Fe (white color in Fig. 5.11e–h) represents the NZF island. The NZF islands were isolated from each other for the composition $x = 0.1–0.3$, but the connectivity pattern changes at high mole fraction of $x = 0.4$.

5.2.2 Magnetoelectric effect of KNNLS-NZF/Ni laminates

Fig. 5.12 shows the schematic diagram for $(1-x)$ KNNLS$-x$ NZF/Ni/$(1-x)$ KNNLS$-x$ NZF ($x = 0.1–0.3$) laminates fabricated by embedding the Ni plate between KNNLS-NZF disks with same poling direction (Yang et al., 2011). The impedance and phase spectra for bimorph laminates were measured under a small applied AC electric field as shown in Fig. 5.13 (Yang et al., 2011). It was found that the radial mode laminates exhibited only resonance peaks over 200 kHz. Variation of α_E for the laminates was measured by applying H_{bias} with $H_{ac} = 1$ Oe at 1 kHz as shown in Fig. 5.14a. The laminates were found to exhibit typical ME behaviors with peak in ME coefficient and no-remanent α_E at zero DC bias. The magnitude of α_E for the laminates was found to increase from 161.5 to 261.3 mV/cm·Oe with decreasing NZF concentration (Yang et al., 2011). It can be shown that ME response in the laminates dominantly depends on piezoelectric properties. Fig. 5.14b–d show the variation of ME output voltage for the laminates measured by applying $H_{ac} = 1$ Oe with a step-change in DC magnetic field (Yang et al., 2011). At 1 kHz frequency, all laminates were able to measure step change of 1 μT DC magnetic field. The sensitivity data correspond well with the ME coefficient graph as the slope of the curve in Fig. 5.14a decreases with increasing NZF concentration (Yang et al., 2011). Low-field sensitivity is mainly governed by the magnitude of this slope and variation in piezomagnetic coefficient. The change in output voltage (V) with applied magnetic field (H) in off-resonance condition can be expressed as: $dV/dH = (dX/dH)\cdot g\cdot t$, where X is generated stress, g is piezoelectric voltage constant, and t is the

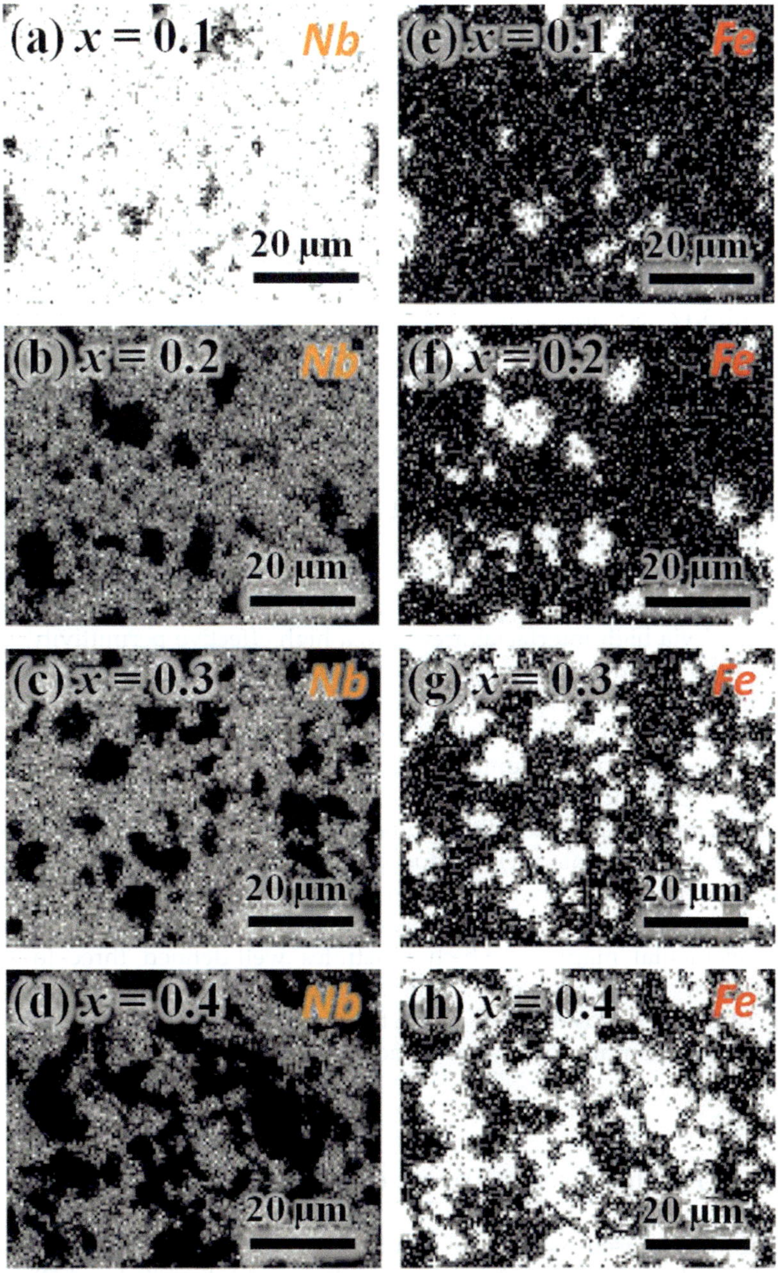

FIGURE 5.11 EDS elemental mapping images for (1−x) KNNLS−x NZF composites sintered at 1060°C. (a)−(d) SEM images for x = 0.1−0.4, (e)−(h) Nb mapping for x = 0.1−0.4, and (i)−(l) Fe mapping for x = 0.1−0.4. *Reprinted with permission from Yang et al. (2011).*

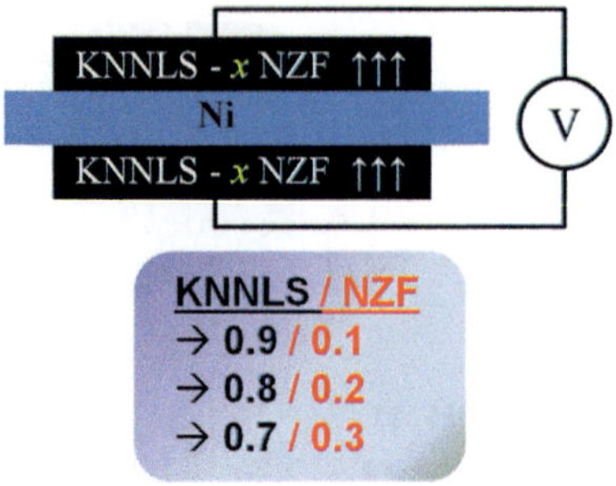

FIGURE 5.12 Schematic diagram for 2–2 ME laminate composites with variation of materials composition. (1−x) KNNLS–x NZF/Ni/(1−x) KNNLS–x NZF with x = 0.1, 0.2, and 0.3. *Reproduced with permission from Yang et al. (2011).*

thickness of the composite layer. The coefficient dX/dH can be shown to be proportional to q/s, where s is the elastic compliance of the NZF islands. Thus, the sensitivity will be dependent upon the mechanical impedance matching of the NZF phase with the KNNLS matrix and the composite layer with that of nickel.

Strong ME coupling requires: (i) large interphase elastic interactions, mediated via high interfacial areas, (ii) a high effective permittivity, and (iii) a high effective permeability. Recently, aerosol deposition (AD) process has been used to fabricate ME composite thick films, and one of the highest ME coefficients was reported (Park, Ryu, et al., 2009). Highly dense three to two nanocomposite thick films of $Pb(Zr,Ti)O_3$–$Pb(Zn,Nb)O_3$ (PZT-PZN) and $(Ni,Cu,Zn)Fe_2O_4$ (NCZF) with thickness >10 μm were synthesized on platinized silicon substrate at room temperature. High ME response was associated with reduced clamping effect, effective poling of piezoelectric matrix, and enhanced elastic coupling due to laminate structure. The XRD and STEM microstructure is shown in Fig. 5.15 along with elemental mapping which illustrates well-defined three to two connectivity (Park, Ryu, et al., 2009). In addition to improved ME characteristics, the deposition rate of ME films was exceptionally higher (over 1 μm/min) than any other conventional thin film process (Priya et al., 2009).

5.3 Self-biased textured cofired laminate systems

The widely used method for synthesis of ME laminate composites is bonding of magnetostrictive layer (such as Metglas, Terfenol-D) and piezoelectric layer (such as PZT, PMN-PT) together using epoxy resin. This fabrication method is relatively simpler, but the epoxy layer is much softer than both magnetostrictive alloy and ferroelectric ceramic, and

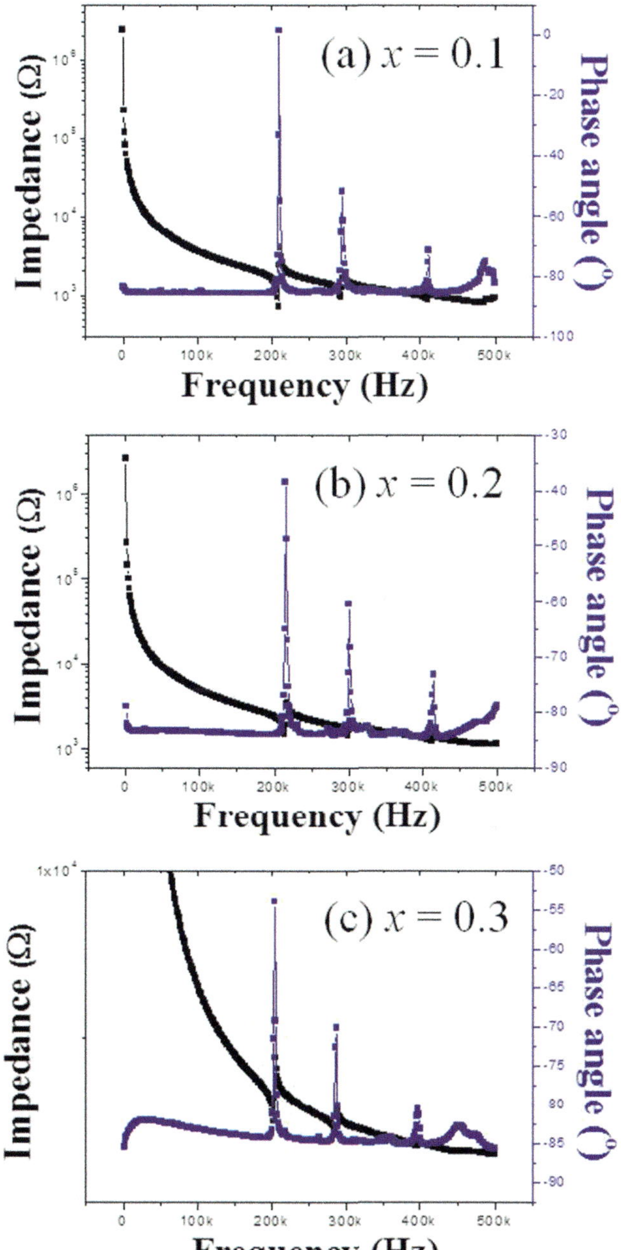

FIGURE 5.13 Impedance and phase spectra for radial mode configuration trilayer laminates with x = 0.1, 0.2, and 0.3 under AC electric field. *Reproduced with permission from Yang et al. (2011).*

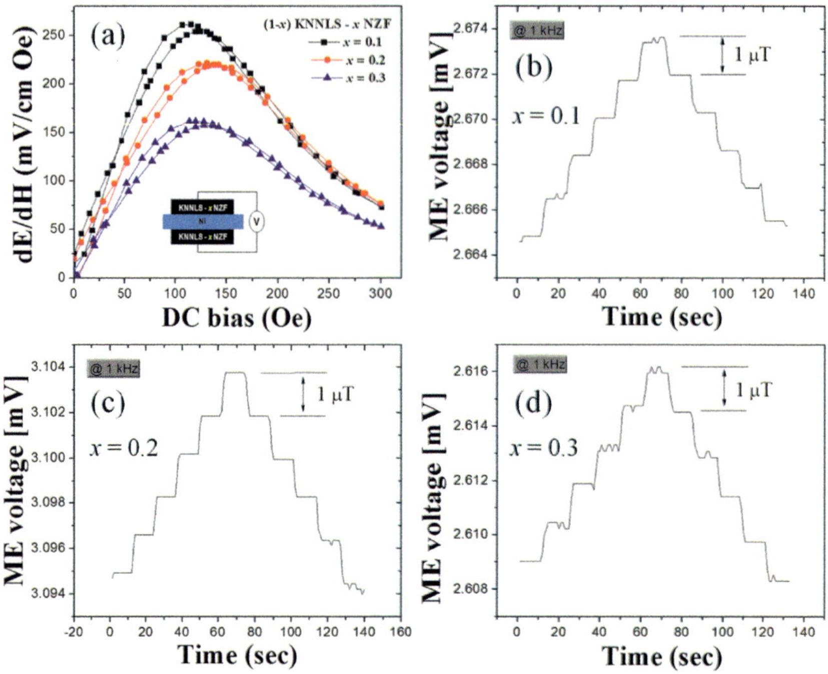

FIGURE 5.14 (a) ME coefficient and sensitivity limit of dc magnetic field for radial mode laminates. (b) x = 0.1, (c) x = 0.2, and (d) x = 0.3. *Reproduced with permission form Yang et al. (2011).*

thus, it will dampen the generated strain resulting in loss of efficiency. Co-firing of layered ME composite can avoid the interface weakening due to epoxy bonding and will also reduce the overall fabrication cost.

Magnetostrictive/piezoelectric/magnetostrictive (M/P/M) laminate with Ag inner electrodes has been cofired at 930°C for 4 has shown in (a). $Pb(Mg_{1/3}Nb_{2/3})O_3$-32.5$PbTiO_3$ [PMN-PT] and $(Ni_{0.6}Cu_{0.2}Zn_{0.2})Fe_2O_3$ [NCZF] were used as piezoelectric and magnetostrictive materials respectively. To improve the piezoelectric response, the PMN-PT piezoelectric materials were textured by using <001> $BaTiO_3$ (BTO) microcrystals synthesized by templated grain growth method (Yan, Zhou, & Priya, 2013).

Fig. 5.16b shows the cross-sectional scanning electron microscopy (SEM) images of co-fired NCZF/textured-PMN-PT/NCZF (abbreviated as C−N/T/N) laminate composite (Yan, Zhou, & Priya, 2013). There was no trace of delamination or micro-crack, and the thickness of textured PMN-PT piezoelectric layer and NCZF ME layer was similar. Energy-dispersive spectroscopy (EDS) line scanning analysis results in Fig. 5.16b show that the interfacial element diffusion and/or chemical

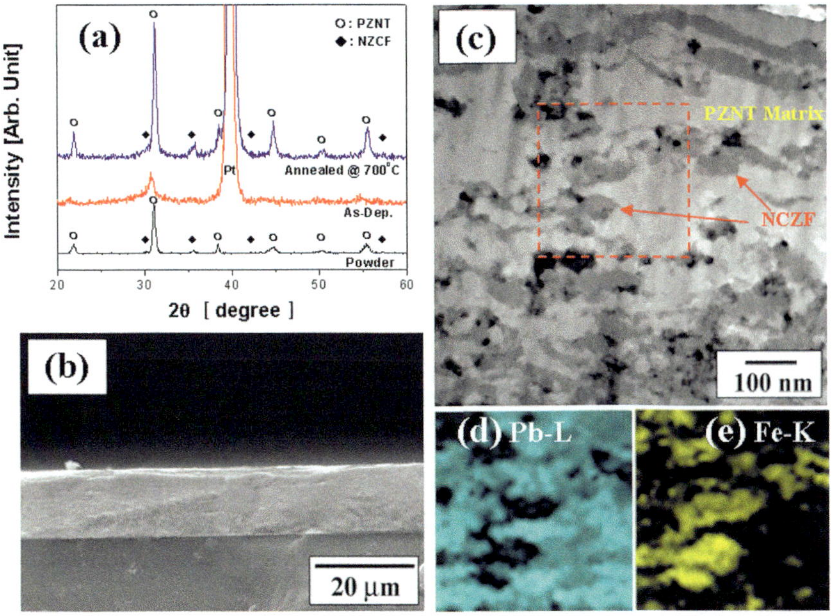

FIGURE 5.15 (a) XRD patterns of the powders and composite thick films, (b) cross-sectional SEM image, (c) STEM, (d) and (e) EDX mapping. *Reprinted with permission from Park, Ryu, et al. (2009).*

reactions occurring during the sintering process at high temperature were negligible, indicating the high chemical stability of NCZF, Ag, and PMN-PT. In a prior study conducted on NCZF/PZN-PZT system, the diffusion length for Cu was found to be in the range of 30 μm (Yan, Zhou, & Priya, 2013). EDS mapping images of C–N/T/N cross-sectional area (Fig. 5.16c) show the dense microstructure and sharp interface across different layers (Yan, Zhou, & Priya, 2013). The thickness of Ag electrode and PMN-PT piezoelectric layer was 10 and 140 μm, respectively. The fracture surface SEM images and corresponding planar X-ray diffraction (XRD) patterns of different layers are shown in Fig. 5.16d. The textured PMN-PT layer displays brick wall-like microstructure with well aligned BTO templates (black lines) inside. Increase of {00l} XRD peak intensity and decrease of other peaks clearly indicate the formation of texture. The texture degree calculated by Lotgering factor was found to be 66% (Yan, Zhou, & Priya, 2013). In TGG process, BTO template crystals are aligned in PMN–PT matrix powder by tape-casting process. PMN–PT matrix powders nucleate and grow on the aligned BTO template crystals during the sintering, thereby, yielding textured ceramics.

Fig. 5.17a shows the ME voltage coefficient (α_E) as a function of DC magnetic field at off-resonance frequency of 1 kHz (Yan, Zhou, & Priya,

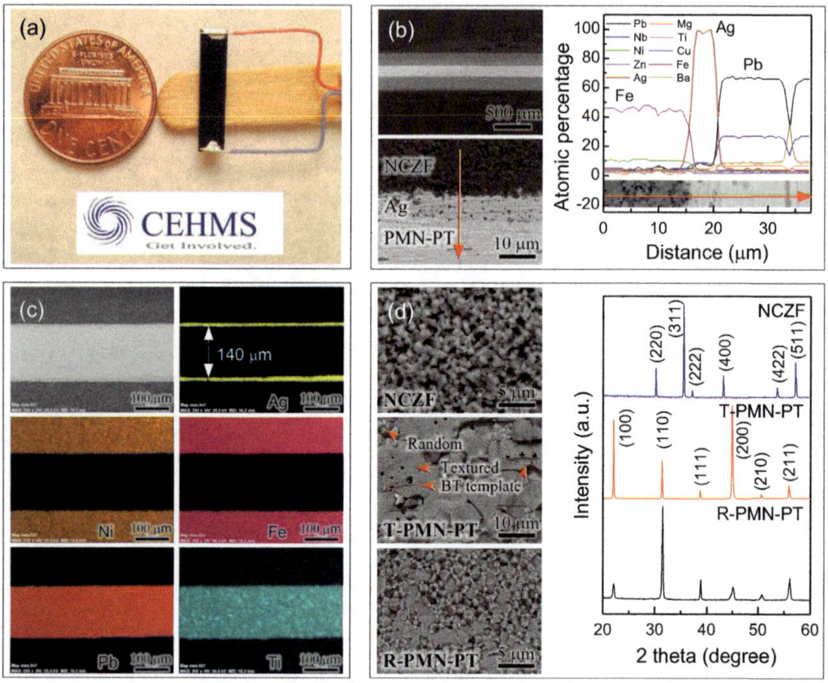

FIGURE 5.16 (a) Optical image of co-fired NCZF/T-PMN-PT/NCZF laminate; (b) polished cross-sectional SEM images and EDS/SEM element line scanning analysis; (c) element mapping of co-fired NCZF/T-PMN-PT/NCZF laminate; (d) fracture cross-sectional SEM images and planar XRD patterns of different layers. *Reprinted with permission from Yan, Zhou, and Priya (2013).*

2013). The maximum α_E of co-fired NCZF/random-PMN-PT/NCZF (abbreviated as C−N/R/N) is two times as that of epoxy bonded Metglas/random-PMN-PT/Metglas (abbreviated as B-M/R/M) laminate (Yan, Zhou, & Priya, 2013). Although the magnetostriction of Metglas ($\lambda = 40$ ppm) is twice that of NCZF ($\lambda \approx 20$ ppm), the strain transfer or coupling between magnetostrictive layer and piezoelectric layer by cofiring is much better than epoxy bonding. Via texturing, further improvement was achieved in co-fired textured sample (C−N/T/N). The large enhancement in <001> textured ceramics was related to the formation of "engineered domain state" which facilitates the rotation of <111> polarization toward the <001> direction. In Fig. 5.17a, it can also be observed that there is a strong hysteretic ME response in the co-fired composite. Considering to the relation $\alpha \propto q = d\lambda/dH_{DC}$, we estimated the effective λ behavior by integrating α with respect to the H_{DC} as shown in Fig. 5.18b (Yan, Zhou, & Priya, 2013). The epoxy bonded Metglas/PMN-PT/Metglas shows symmetrical λ-behavior with respect to the

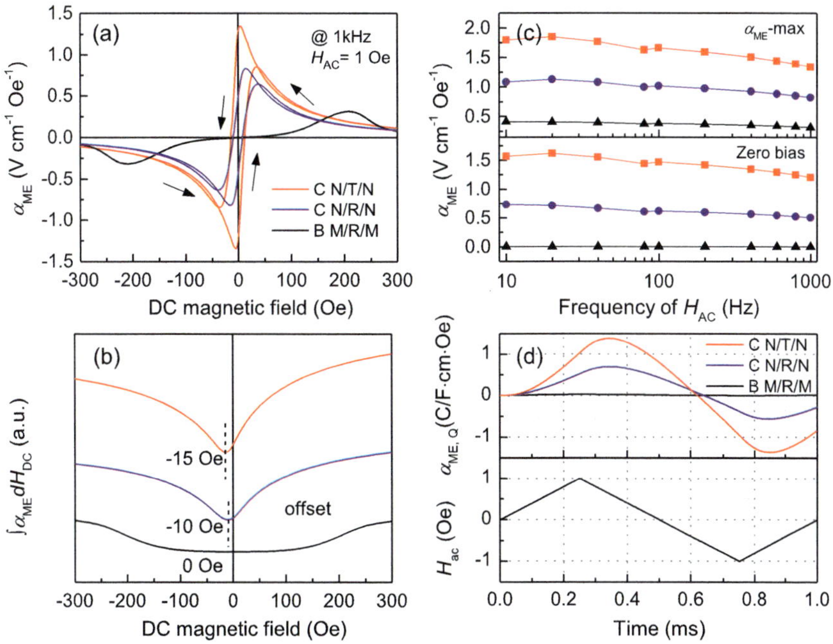

FIGURE 5.17 (a) ME voltage coefficients (α_{ME}) of co-fired NCZF/T-PMN-PT/NCZF (C–N/T/N), NCZF/R-PMN-PT/NCZF (C–N/R/N), and epoxy bonded Metglas/R-PMN-PT/Metglas (B-M/R/M) laminate; (b) Integral values of α_{ME} with respect to the DC magnetic field; (c) ME voltage coefficients (α_{ME}) as a function of the frequency of applied ac magnetic field (H_{AC}); (d) dynamic ME charge coefficients (α_{ME}) and applied ac magnetic field spectra as a function of time for C–N/T/N, C–N/R/N composite, and epoxy bonded Metglas/R-PMN-PT/Metglas (B-M/R/M) laminate, under zero-DC magnetic field. *Reprinted with permission from Yan, Zhou, and Priya (2013).*

H_{DC}, while the tendency of λ for co-fired composites was found to be asymmetric. The variation of α_E with frequency of applied H_{AC} is shown in Fig. 5.17c. It can be seen that the maximum α_E displays excellent stability. A giant ME voltage coefficient (>1200 mV cm^{-1} Oe^{-1}) at zero-bias was obtained in both C–N/R/N and C–N/T/N samples (Yan, Zhou, & Priya, 2013). The dynamic change in charge (Q) or polarization (P) of ME composite (Radiant: Precision Premier II, USA) measured under the condition of $H_{AC} = 1$ Oe at $f = 1$ kHz and $H_{DC} = 0$ Oe. ME charge coefficient $\alpha_{ME,Q}$ can be written as (Yan, Zhou, & Priya, 2013):

$$\alpha_{ME,Q} = \frac{dE}{dH} = \frac{dQ}{dH_{AC}} \times \frac{1}{Ct} = \frac{dP}{dH_{AC}} \times \frac{A}{Ct}, \quad (5.12)$$

where C, A, t are the capacitance, area and thickness of the piezoelectric layer, respectively. As shown in Fig. 5.17d, large ME charge coefficients, $\alpha_{ME,Q}$, under zero DC bias are observed which further confirms the

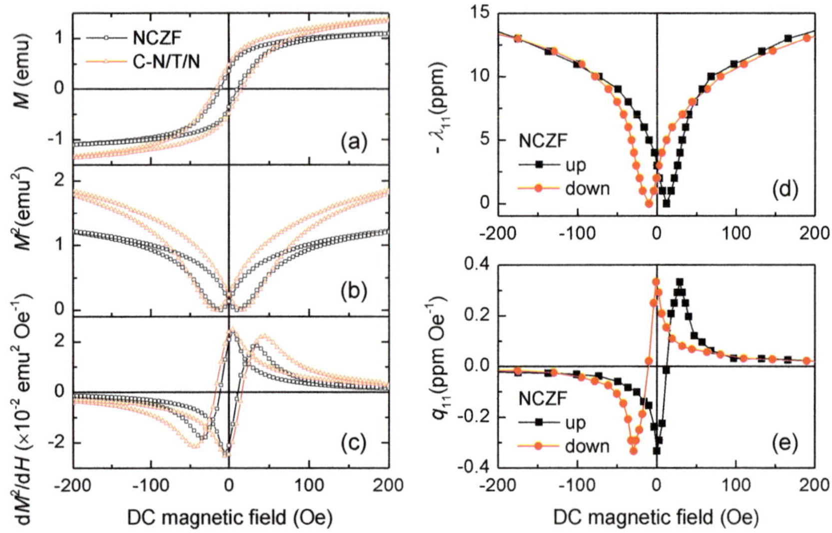

FIGURE 5.18 (a) Magnetization-magnetic field (M–H) loops, (b) square of magnetization as a function of magnetic field, and (c) differential of square magnetization-magnetic field [(dM^2/dH)-H] plot for pure NCZF (not co-fired) and C–N/T/N structure. DC magnetic field (H_{DC}) dependence of (d) the magnetostriction (λ) and (e) piezomagnetic coefficients (q) of NCZF (not co-fired). *Reprinted with permission from Yan, Zhou, and Priya (2013).*

presence of strong ME effect under zero bias.(Yan, Zhou, & Priya, 2013) The time delay of $\alpha_{ME,Q}$ related to H_{AC} is attributed to the interface between magnetostrictive and piezoelectric layer. Compared to the ME coefficient of 30 mV cm^{-1} Oe^{-1} obtained for NKNLS-NZF/Ni/NKNLS-NZF trilayer laminate (40′), ~100 mV cm^{-1} Oe^{-1} obtained for Ni-PZN-PZT bilayer laminate (12′), and ~400 mV cm^{-1} Oe^{-1} obtained for functionally graded Ni-NZFO-PZT laminate composites (3′) at zero-bias, these co-fired composite displayed extremely high ME response (Yan, Zhou, & Priya, 2013). A larger zero-bias ME coefficient of 1.65 V cm^{-1} Oe^{-1} at 100 Hz in a quasi-one dimensional ME sensor has been reported compared to cofired composite with less value (1.47 V cm^{-1} Oe^{-1}) (Chen et al., 2011). However, these co-fired multilayer composites based on LTCC have advantage of integration with other circuit components such as multilayer capacitor, resistor, inductor, and conductor to fabricate fully packaged electronic devices (Chen et al., 2011).

The ME voltage coefficient can be written as following:

$$\alpha_{ME} = \frac{\partial E}{\partial H} = \underbrace{\left|\frac{\partial E}{\partial D} \times \frac{\partial D}{\partial T} \times \frac{\partial T}{\partial S}\right|}_{P-layer} \times \underbrace{\left|\frac{\partial S}{\partial H}\right|}_{M-layer} = \left|\frac{d}{\varepsilon \times s}\right| \times q, \tag{5.13}$$

where E is the output electric field, H is the applied magnetic field, S is the mechanical strain, T is the mechanical stress, D is the electric displacement, d is the piezoelectric constant, ε is the dielectric constant, s is the elastic compliance of piezoelectric layer (P-layer), and q is piezomagnetic coefficient of magnetostriction layer (M-layer). Thus, the giant ME response of co-fired composite and self-biased hysteretic behavior can be explained by taking into account the piezoelectric and magnetostrictive characteristics separately. Since the parameters d, ε, and s of piezoelectric materials are independent of applied magnetic field (H), the self-biased should be related to the magnetostrictive materials. Fig. 5.18a shows the magnetization (M) of NCZF as a function of magnetic field in longitudinal direction by using vibrating sample magnetometer (VSM). The M-H loop of NCZF shows hysteresis (Yan, Zhou, & Priya, 2013). The ME behavior can be correlated with magnetostriction (λ) and magnetization (M) as: (Cahill et al., 1992; Yan, Zhou, & Priya, 2013)

$$\varphi \sim \frac{3\lambda\sigma}{\left(K + 2\pi M^2\right)} \text{ or } \lambda \propto M^2, \tag{5.14}$$

where φ is the angle of magnetic moments, K and σ are the anisotropy constant and stress respectively. M^2-H and $\partial M^2/\partial H$-H curves can be used to predict the nature of the magnetostriction coefficient (λ) and piezomagnetic coefficient ($q = d\lambda/dH$). To experimentally confirm the variation of q as a function of H_{DC}, in-plane magnetostriction coefficient (λ_{11}) was measured in parallel to applied H_{DC} by strain gauge method. The magnetostriction (λ) shows is asymmetric and displays "butterfly"-like characteristic (Fig. 5.18d), and the in-plane piezomagnetic coefficient ($q_{ij} = d\lambda_{ij}/dH$) shows similar behavior as $\partial M^2/\partial H$ and $\alpha_{ME,Q}$-H. Therefore, it can be concluded that the self-bias effect is mainly related to the hysteretic response of ferrite layer.

Table 5.8 lists the dielectric and ME properties of co-fired composites. Following conclusion can be drawn from this data: (1) the maximum α_E of NCZF/R-PMN-PT/NCZF was 2–3 times higher in magnitude (827 mV cm^{-1} Oe^{-1}) comparing to epoxy bonded Metglas/PMN-PT/

TABLE 5.8 Dielectric and magnetoelectric properties of co-fired layered ME composite (1 kHz) (Yan, Zhou, & Priya, 2013).

Sample	Electrode	ε_r	tanδ (%)	dE/dH (max) [mV cm^{-1} Oe^{-1}]	dE/dH (zero-bias) [mV cm^{-1} Oe^{-1}]
C–N/R/N	Co-fired	1336	1.1	827	510
C–N/T/N	Co-fired	1135	0.7	1346	1214

Metglas; (2) 63% increase in the magnitude of α_E (1346 mV cm^{-1} Oe^{-1}) are obtained via texturing; (3) textured composites show low loss due to "engineered domain configuration"; (4) the dielectric constant of textured sample was suppressed due to the introduce of low permittivity BTO template crystals; and (5) co-fired textured composites show large α_E under zero-bias (1214 mV cm^{-1} Oe^{-1}). Besides the increase in piezoelectric response, the lower losses of textured sample also play an important role in enhancing the α_E which can be understood by taking into account the modified piezoelectric voltage coefficient (g_{31}) given as (Palneedi et al., 2018):

$$g_{31}{}^{*} = g_{31}\frac{1}{\tan\delta' + \tan\theta' + \dfrac{C - C_f}{C_f}} \tag{5.15}$$

where $C{-}C_f/C_f$ is the rate of capacitance change, C is capacitance at a given frequency, C_f is capacitance at 1 kHz, and $\tan\delta'$ and $\tan\theta'$ are the intensive dielectric loss and piezoelectric loss respectively.

5.4 Textured microstructures—piezoelectric layer in composites

Since $Pb(Zr,Ti)O_3$ composition was discovered in the mid-1950s, many studies on developing high-performance piezoelectric materials have been conducted. As shown in Fig. 5.19a, commercialized piezoelectric ceramic compositions with high d_{33} value of 1000 pC/N or compositions with high g_{33} value of 40×10^{-3} V m N^{-1} can be found (Yan, Zhou, & Priya, 2013). However, compositions having high d_{33} usually show low g_{33} value, while high g_{33} compositions possess low d_{33}; thus, the achievement of high $d \cdot g$ coefficient from single piezoelectric composition is challenging. As seen in Fig. 5.19a, the $d_{33} \cdot g_{33}$ values of commercialized piezoelectric ceramics are confined below $16{,}500 \times 10^{-15}$ m^2 N^{-1} (Yan, Zhou, & Priya, 2013). This is a fundamental limitation imposed by the thermodynamics constraints. Using Landau–Devonshire theory, the energy due to the fluctuations in the polarization p and strain u can be written as: $f = \frac{1}{2\chi^u}p^2 + apu + \frac{1}{2}c^p u^2$, where c^u is the susceptibility under constant strain, a is the piezoelectric coefficient relating the polarization and strain, and c^p is the elastic stiffness under constant polarization. For simplicity, only one polarization and one strain component were taken into account. Thus, the derivative of polarization with respect to the applied stress X can be written as: $dp/dX = -as\chi + \textit{higher order terms}$, where s is the elastic compliance. This implies that any increase in the piezoelectric constant is always accompanied by the increase in

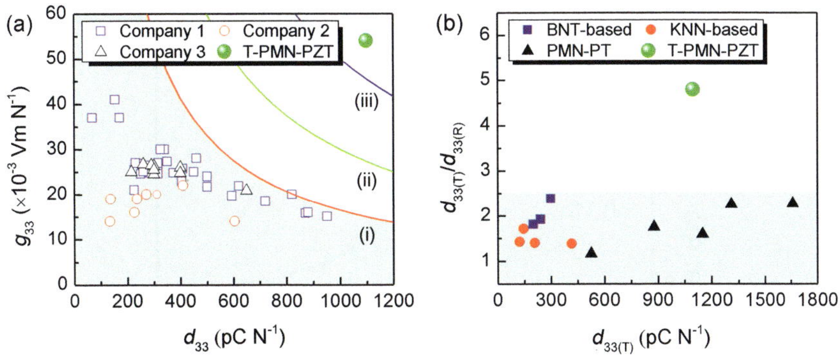

FIGURE 5.19 (a) Comparison of g_{33} and d_{33} values of various piezoelectric ceramics. Different colored lines (i), (ii), and (iii) indicate the plots of $g_{33} \cdot d_{33} = c$ function; (i) $c = 16{,}500 \times 10^{-15}\, m^2\, N^{-1}$, (ii) $c = 30{,}000 \times 10^{-15}\, m^2\, N^{-1}$, and (iii) $c = 50{,}000 \times 10^{-15}\, m^2\, N^{-1}$. (b) Comparison of $d_{33(T)}/d_{33(R)}$ ratio versus $d_{33(T)}$ of various textured piezoelectric ceramics. The $d_{33(T)}$ and $d_{33(R)}$ represent d_{33} of textured ceramic and that of randomly oriented counterpart, respectively. BNT and KNN represent $(Bi_{1-x}Na_x)TiO_3$ and $(K_{1-x}Na_x)NbO_3$ compositions, respectively. *Reprinted with permission from Yan, Zhou, and Priya (2013).*

transversal dielectric susceptibility, and thus, the product (*d.g*) is constant (Ishibashi & Iwata, 1999).

It is well known that <001> oriented relaxor-based piezoelectric single crystals such as $Pb(Mg_{1/3}Nb_{2/3})O_3–PbTiO_3$ (PMN-PT) and $Pb(Zn_{1/3}Nb_{2/3})O_3–PbTiO_3$ (PZN-PT) with rhombohedral structure near morphotropic phase boundary (MPB) exhibit d_{33} of >2000 pC/N and k_{33} of >92% (Park & Shrout, 1997). The piezoelectric response strongly depends on composition and orientation and shows largest near the MPB as illustrated in Fig. 5.20. These giant piezoelectric properties of <001> rhombohedral crystals are related to their piezoelectric anisotropy and "engineered domain state" that facilitate the rotation of <111> polarization (<111>: pseudocubic <111> directions of spontaneous polarization in rhombohedral perovskite) toward the <001> direction (<001>: pseudocubic <001> directions in rhombohedral perovskite) (Zhang & Li, 2012).

PZT- and relaxor-based ceramics with near MPB composition exhibit high dielectric and piezoelectric properties but low energy density due to reduced piezoelectric voltage coefficient. Defects and substituents have been used to influence the electromechanical properties of ceramics in poled conditions. Generally, acceptor doping atoms (Fe, Mn, Ni, Co) in the PZT-based ceramics results in (i) decreased dielectric constant and loss, (ii) lower elastic compliance, (iii) lower electromechanical coupling factor, and (iv) lower electromechanical losses. In contrast, donor doping atoms (La, Sb, Bi, and W) results in (i) increased dielectric constant and loss, (ii)

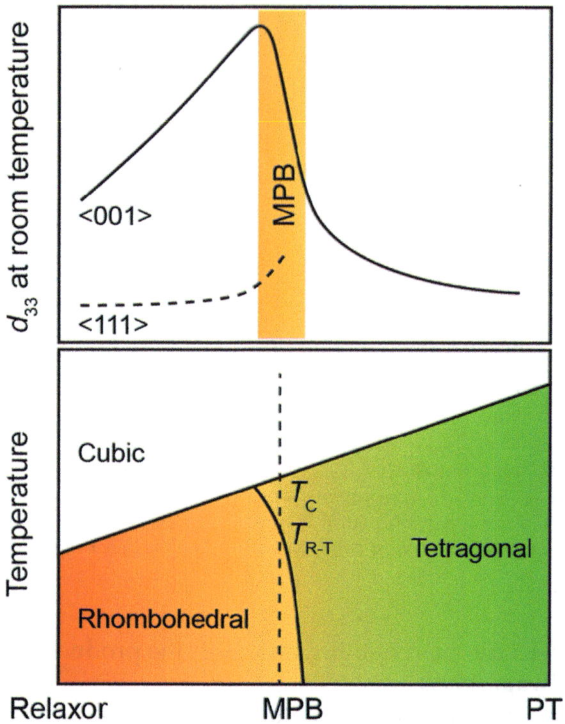

FIGURE 5.20 Schematic piezoelectric charge/strain coefficient, d_{33}, and phase diagram of PMN-PT solid solution (Park & Shrout, 1997; Priya et al., 2015).

increased elastic compliance, (iii) increased electromechanical coupling factor, and (iv) increased electromechanical losses. However, the success through these doping methods has been limited. Domain engineered relaxor piezoelectric single crystals have been grown to achieve high performance. However, low ferroelectric rhombohedral to tetragonal phase transition temperature ($T_{R\text{-}T} \sim 60–90°C$) and low coercive fields ($E_C \sim 2–3\ kV\ cm^{-1}$) limit the stability of electrical property in relaxor-based single crystals (Zhang & Li, 2012). Many attempts to grow PZT-based single crystals near MPB to achieve high and stable piezoelectric properties have been made but there is a fundamental challenge in accomplishing this task. PZT-based compositions cannot be readily grown in single crystal form because of incongruent melting behavior. For compositions near the morphotropic phase boundary (MPB), melting is incongruent and results in the precipitation of zirconia prior to PZT (Zhang & Li, 2012). Thus, any effort to grow MPB composition single crystal using liquid phase results in a mixture consisting of zirconia and Zr-deficient PZT grains. These considerations indicate the tremendous difficulty in synthesizing crystallographically oriented PZT materials.

A very different strategy for overcoming the fundamental limitations imposed by electrodynamics and thermodynamics is via designing a textured microstructure, which can achieve a giant magnitude of $d \cdot g$ coefficient with value of 59,000 × 10−15 m^2 N-1 (comparable to that of the single crystal counterpart and 359% higher than that of the best commercial compositions)(Yan, Cho, et al., 2013). The reason for realizing high $d \cdot g$ coefficient in textured ceramics is: (1) <001> texturing (grain orientation along the <001> crystallographic direction) of piezoelectric ceramic produces engineered domain state comparable to that of <001> single crystals, thus, resulting in high d values; (2) the existence of low ε templates produces a composite microstructure and suppresses the dielectric constant (ε) of piezoelectric ceramic.

In TGG process as illustrated in Fig. 5.21, template crystals are aligned in ceramic matrix powder by tape-casting process; then, matrix crystals nucleate and grow on the aligned template crystals during the sintering. To yield textured ceramics, the template crystals should possess a thin platelet morphology with {001} plane to facilitate their homogeneous alignment in the matrix and oriented epitaxial growth along the <001> direction. Rhombohedral composition given as $0.4Pb(Mg_{1/3}Nb_{2/3})O_3$-$0.25PbZrO_3$-$0.35PbTiO_3$ (PMN-PZT) was used as matrix because its <001> single crystal counterpart shows excellent piezoelectric properties ($d_{33} = 1530$ pC N^{-1}, $k_{33} = 0.93$) and much better thermal and electrical stabilities ($T_C = 211°C$ and $E_C = 4.5$ kV cm^{-1}) than those of PMN-PT or PZN-PT crystals ($T_C = 130–170°C$ and $E_C = 2–3$ kV cm^{-1}) (Priya et al., 2015; Zhang et al., 2007b, 2008). $BaTiO_3$ (BTO) was selected as template due to its chemical stability in PMN-containing ceramics and good lattice match with the PMN-PZT composition (lattice parameter: ~4.05 Å) (Cho & Priya, 2011; Messing et al., 2004). It was noted that the low relative permittivity of <001> BTO template crystal ($\varepsilon_{33}^T/\varepsilon_0 = 130$, ε_0: vacuum

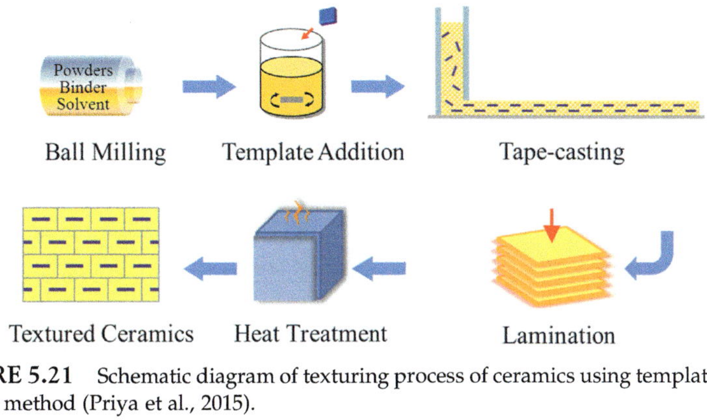

FIGURE 5.21 Schematic diagram of texturing process of ceramics using templated grain growth method (Priya et al., 2015).

permittivity) is very desirable to suppress the permittivity increase in <001> textured PMN-PZT because the TGG processed textured ceramic is the "composite" material composed of matrix and mono-dispersed template crystals (Wang et al., 1993).

BTO template crystals with desirable morphology were synthesized by topochemical microcrystal conversion method. As shown in Fig. 5.22a, first, $Bi_4Ti_3O_{12}$ platelets were synthesized by reacting Bi_2O_3 and TiO_2 powders in NaCl and KCl molten salts at 1050°C for 1 h (Gao et al., 2015). Next, $BaBi_4Ti_4O_{15}$ platelets were synthesized by reaction of $Bi_4Ti_3O_{12}$, TiO_2, and $BaCO_3$ in $BaCl_2$/KCl molten salts at 1050°C for 3 h. Finally, BTO platelets were obtained by topochemical reaction between $BaBi_4Ti_4O_{15}$ and $BaCO_3$ in KCl molten salt at 950°C for 3 h. In all these reactions, the weight ratio between reacting chemicals and salt(s) was maintained at 1:1. Fig. 5.22b shows the SEM image of the BTO seeds, which have length of 5–10 μm and thickness of 0.5–1 μm (Gao et al., 2015).

After embedding 5 vol% BTO template crystals, the PMN-PZT ceramic was textured at different sintering temperatures (Yan, Cho, et al., 2013). The microstructure evolution during the sintering is shown in Fig. 5.23a. At a sintering temperature of 600°C, the matrix PMN-PZT particles were observed to surround the BTO template without any growth in the matrix. The nucleation and growth on the surface of BTO template initiated around 800°C, and interface migration was observed from the sample sintered at 900°C. High <001> Lotgering factor of 90% with highly

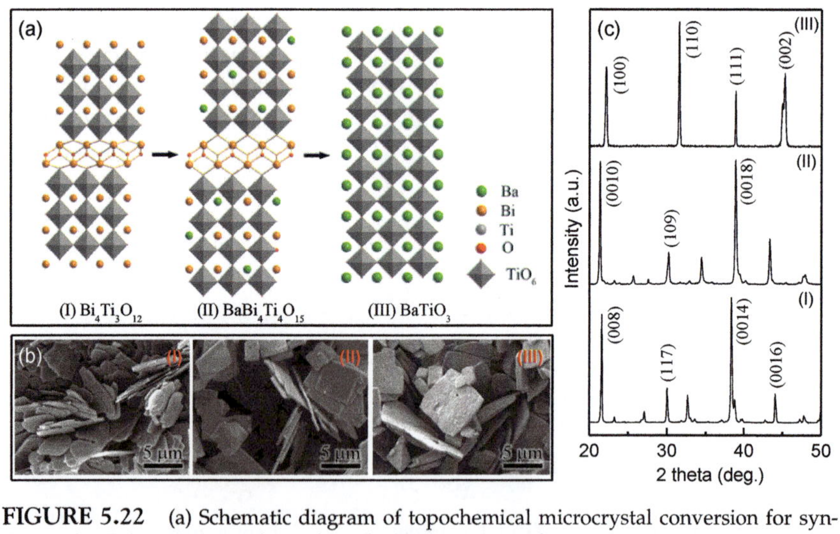

FIGURE 5.22 (a) Schematic diagram of topochemical microcrystal conversion for synthesis of $BaTiO_3$ microcrystal template. (b) SEM images and (c) XRD patterns of synthesized $Bi_4Ti_3O_{12}$ microcrystals, $BaBi_4Ti_4O_{15}$ microcrystals and $BaTiO_3$ microcrystals (Gao et al., 2015).

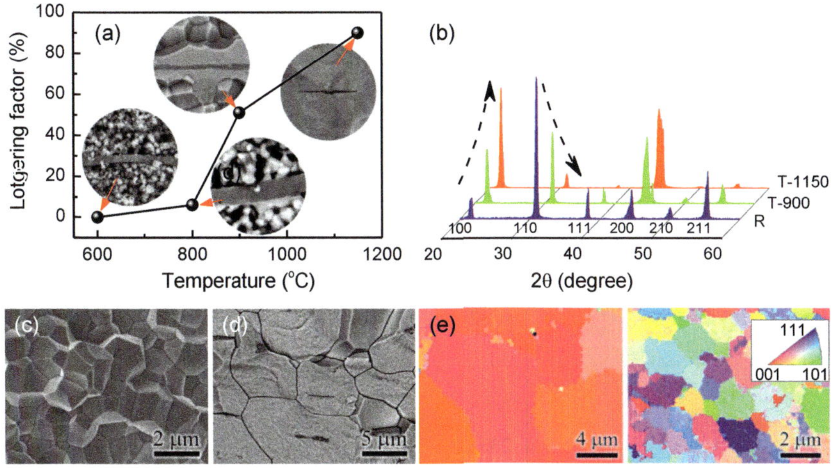

FIGURE 5.23 (a) <001> Lotgering factors of textured PMN-PZT samples sintered at various temperatures for 2 h. Inset SEM images show the evolution of matrix PMN-PZT crystal growth on BTO template crystal with variation of sintering temperature. (b) XRD patterns of random and textured PMN-PZT samples. R and T-1150 denote the random and textured sample sintered at 1150°C respectively, and T-900 denotes the textured sample sintered at 900°C. SEM images of (c) fractured random PMN-PZT and (d) fractured textured PMN-PZT; (e) EBSD inverse pole figure maps using stereographic projection of textured (left) and random (right) PMN-PZT ceramics. *Reprinted with permission from Yan, Cho, et al. (2013).*

textured microstructure was obtained from the specimen sintered at 1150°C (Yan, Cho, et al., 2013). The fine grain size of the matrix preserved the driving force required for growth until the larger grains above 1150°C started to impinge upon each other resulting in low energy boundaries. The textured development is evidenced by XRD patterns in Fig. 5.23b. It can be seen that the intensity of {001} peaks increased, while the other peaks intensity decreased with increasing sintering temperature. Compared to the randomly oriented ceramic in Fig. 5.23c, the highly textured sample in Fig. 5.23d clearly shows a brick wall-like structure developed with larger grain size. Furthermore, electron backscatter diffraction (EBSD) mapping images in Fig. 5.23e directly shows that all grains in textured sample were highly oriented along <001> direction unlike those in randomly oriented sample (Yan, Cho, et al., 2013).

Table 5.9 summarizes the piezoelectric and dielectric properties of randomly oriented ceramic (R-ceramic), <001> textured ceramic (T-ceramic), and <001> single crystal (S-crystal) of PMN-PZT composition. T-ceramic exhibited giant d_{33} of 1100 pC N^{-1} which is 4.8 times higher than that of R-ceramic (230 pC N^{-1}). This increasing ratio of d_{33} value between R- and T-ceramic (=4.8) was much higher than that of the other textured piezoelectric ceramics (usually less than 2.5) as shown in

TABLE 5.9 Piezoelectric and dielectric properties of PMN-PZT: randomly oriented ceramic (R-ceramic), <001> textured ceramic (T-ceramic) and <001> single crystal (S-crystal) (Yan, Cho, et al., 2013).

Properties	R-ceramic	T-ceramic	S-crystal[a]
d_{33} (pC N^{-1})	230	1100	1530
K	0.4 (k_p)	0.84 (k_p)	0.93 (k_{33})
$\varepsilon_{33}/\varepsilon_0$	915	2310	4850
g_{33} ($\times 10^{-3}$ V m N^{-1})	28.4	53.8	35.6
$d_{33} \cdot g_{33}$ ($\times 10^{-15}$ m^2 N^{-1})	6532	59,180	54,468
P_r (μC cm^{-2})	30	36	29
E_c (kV cm^{-1})	8.2	7.4	4.5
T_c (°C)	233	204	211

Priya et al. (2015).

Fig. 5.19b. Highly enhanced d_{33} of T-ceramic is also evidenced by strain versus electric field characteristic in Fig. 5.24. Compared to R-ceramic, T-ceramic exhibited a much higher strain value (0.24% at 20 kV cm^{-1}) proving the high strain coefficient and demonstrating its superior feasibility for actuator applications. Planar mode coupling coefficient (k_p) of T-ceramic was found to be more than two times higher than that of R-ceramic (Yan, Cho, et al., 2013). The most notable point is that via <001> texturing, the permittivity value was increased by only 2.5 times, while the d_{33} was increased by 4.8 times. This suppressed permittivity value of T-ceramic resulted in very high g_{33} of 53.8×10^{-3} V m N^{-1} (even higher than S-crystal value). High g_{33} of T-ceramic obtained by high d_{33} and suppressed permittivity yielded a giant transduction coefficient of $59{,}180 \times 10^{-15}$ m^2 N^{-1}, which is 9 times higher than that of R-ceramic and even higher than S-crystal value (Priya et al., 2015; Yan, Cho, et al., 2013).

Giant piezoelectric response of textured PMN-PZT ceramic can be understood by noticing the role of special microstructure where BTO templates are embedded in the matrix grains (Yan, Cho, et al., 2013). The BTO template inclusions have comparatively high polar anisotropy suppressing rotation of its polarization under applied electric field. This is contrary to the properties of the matrix phase, PMN-PZT, that being near morphotropic phase boundary (MPB), has a drastically reduced polar anisotropy. In this scenario, the PMN-PZT matrix can be regarded as "soft ferroelectric" under the constant electric and stress fields imposed on this matrix by BTO inclusions. These fields are generated by both fixed polarization of the BTO template inclusions and crystal lattice misfit

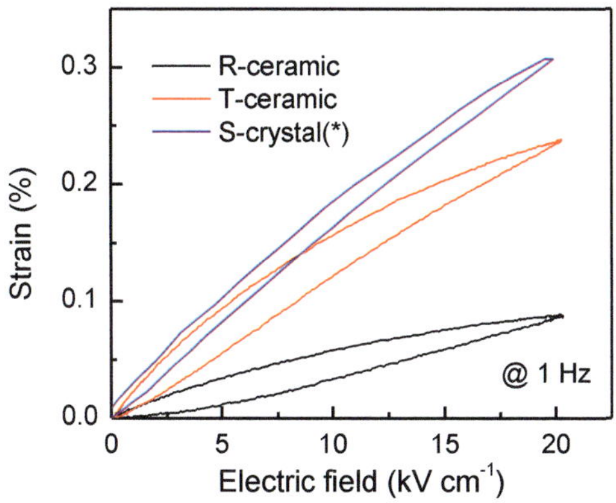

FIGURE 5.24 Strain versus electric field curves of randomly oriented ceramic (R-ceramic), <001> textured ceramic (T-ceramic) and <001> single crystal (S-crystal). *Reprinted with permission from Yan, Cho, et al. (2013).*

between the phases, which in turn depend on morphology, orientation relations with the matrix, and polarization and are determined by the processing. Given the low polar anisotropy of the PMN-PZT matrix, a direction of its energy-minimizing polarization is determined by a coupling with the BTO-generated fixed electric and stress fields imposed on the matrix. Application of external electric field produces a deviation of the polarization direction of the PMN-PZT matrix from its energy minimizing direction. This deviation produces the dielectric and strain response that are significant due to the "soft" character of the matrix (Priya et al., 2015; Yan, Cho, et al., 2013). A lifting of the applied electric field reverses the deviation and results in a recovery of the initial energy-minimizing polarization direction. This reversal caused by coupling with the BTO-generated constant field is essentially hysteresis-free. Therefore, the polar anisotropy of the PMN-PZT in this two-phase composite is determined by the interaction of the polarization of the PMN-PZT with the local static electric and stress fields generated by the BTO rather than by the intrinsic properties of the PMN-PZT. This coupling together with the low polar anisotropy of the PMN-PZT results in a combining of the maximum piezoelectric strain inherent to the PMN-PZT with the enhancement of blocking force caused by electrostatic and stress-induced interaction between the phases. The discovered processing regime providing a composite with very high texture of the PMN-PZT grains—this is a state that is close to the PMN-PZT single crystal with oriented BTO inclusions—is the important factor that results in high energy density (Priya et al., 2015).

The local switching characteristics across the templated region with the embedded BTO phase were studied using switching spectroscopy PFM technique (Yan, Cho, et al., 2013). Local hysteresis loops were collected across a 40×100 matrix, which contains the BTO phase as well as the

PMN-PZT matrix and switching parameters like imprint, switching polarization, and work of switching were mapped across the surface (Fig. 5.25) (Yan, Cho, et al., 2013). The hysteresis loops opened more in the matrix (area 2) in comparison to the BTO template (area 1), and the local switching behavior across the textured samples was in conformity with the macroscopic switching behavior. These results further confirmed that the texturing enhances the homogeneity of the polarization across the matrix, which thereby provides an additive effect promoting the low polar anisotropy. It also validates our hypothesis that the interaction between the polarization and local fields generated in the vicinity of BTO play an important role toward controlling the global response. Above discussed results demonstrate that <001> textured PMN-PZT ceramics can provide the high $d_{33} \cdot g_{33}$ coefficient.

Yan et al. demonstrated substantial piezoelectric voltage coefficients in grain-oriented modified $PbTiO_3$ materials, characterized by a high Curie temperature (T_c) of 364°C and an exceptionally large g_{33} value amounting to $115 \times 10^{-3}\,\mathrm{Vm\,N^{-1}}$, outperforming other established single-phase oxide materials (Yan et al., 2016). The exceptional piezoelectric voltage coefficient g_{33} originates from the maximized piezoelectric strain coefficient d_{33} and minimized dielectric permittivity ε_{33} within [001]-textured $PbTiO_3$ ceramics, where domain wall movements are notably absent.

Li et al. introduced a noteworthy development in the field of piezoceramics, presenting a viable alternative to single crystals with distinctive properties (Yang et al., 2021). Their work involved the implementation of a templated grain-growth method to create <001>-textured $Pb(In_{1/2}Nb_{1/2})O_3-Pb(Sc_{1/2}Nb_{1/2})O_3-PbTiO_3$ (PIN-PSN-PT) ceramics, enriched with a substantial amount of the refractory component Sc_2O_3 (Yang et al., 2021). This addition exhibited the capacity to enhance the rhombohedral-tetragonal phase transition temperatures (T_{rt}) within the system. Remarkably, these textured PIN-PSN-PT ceramics achieved an impressive high k_{33} value ranging from 85% to 89% and experienced a significant elevation in T_{rt}, reaching levels between 160°C and 200°C (Yang et al.,

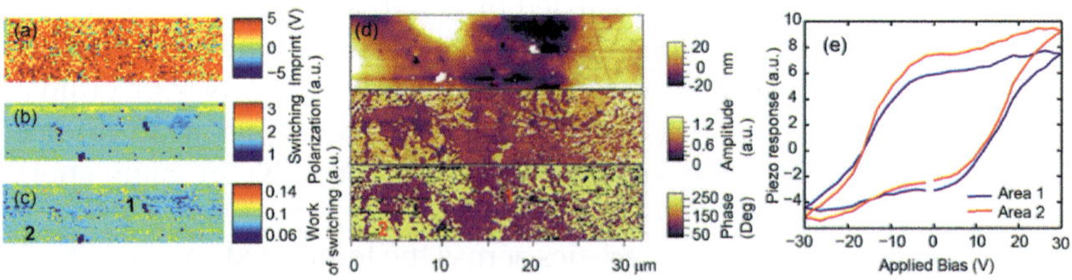

FIGURE 5.25 Analysis of local switching characteristics within the templated area of the PMN-PZT sample includes (a) imprint (horizontal offset), (b) polarization switching (loop height), (c) switching work (loop area), (d) piezoelectric response, and (e) the selection of specific local hysteresis loops. *Reprinted with permission from Yan, Cho, et al. (2013).*

2021). These advantages collectively position textured PIN-PSN-PT ceramics as a promising alternative to single crystals, thereby fostering the advancement of a wide array of cutting-edge piezoelectric devices. Yan et al. recently reported a high electromechanical coupling ($k > 0.9$) in <001> textured $Pb(Mg_{1/3}Nb_{2/3})O_3{-}Pb(Zr,Ti)O_3$ ceramics (Yan et al., 2022). The maximum value of k is observed when the electric field is oriented perpendicular to the polarization direction, whereas the minimum value of k is recorded when the electric field aligns parallel to the polarization (Yan et al., 2022). This observation implies that the k_{15} mode exhibits the highest k value among all modes in rhombohedral relaxor-PT single crystals. However, during the process of electrical poling, only k values with $\theta \in [0,90 \text{ degrees}]$ are achievable, and the highest k is attained when the electric field is aligned with the [001] direction. In other words, a single crystal that has been poled along the [001] direction can manifest the highest k_{33} value. In the case of polycrystalline ceramics, the situation differs because grain orientations are randomly distributed. Consequently, it is not possible to achieve the highest k_{33} value, as in the case of single crystals. Nevertheless, the use of [001]-textured polycrystals can aid in realizing the highest k_{33} within ceramics. In a [001]-textured polycrystal, all the grains are oriented along the [001] crystallographic axis, while the other crystallographic axes exhibit complete randomness (Yan et al., 2022). Leng et al. have engineered piezoelectric materials with high power capabilities by incorporating acceptor doping and orienting them in both ⟨001⟩ and ⟨111⟩ textured configurations (Leng, Yan, et al., 2023). Their work highlights the distinctive performance characteristics of these materials. Specifically, the Mn-doped ⟨001⟩ textured ceramics demonstrate a notably high *d*33 value of 725 pC N−1 while maintaining a moderate Qm value of 716. In contrast, the ⟨111⟩ textured ceramics exhibit lower *d*33 values (350 pC N−1) but demonstrate high Qm values (1495) (Leng, Yan, et al., 2023). This discrepancy arises from the interplay between soft and hard piezoelectric properties within the doped ⟨001⟩ textured ceramics. The unique combination of these properties allows the doped ⟨001⟩ textured ceramics to achieve a significantly increased vibration velocity, measuring approximately $0.90\,\text{m s}^{-1}$, which is 1.5 times higher than that of commercially available hard PZTs (Leng, Yan, et al., 2023). They also disclosed an innovative approach where they harnessed a synergistic design strategy that combines composition/phase selection, crystallographic texturing, defect engineering, and a water quenching technique (Leng, Wang, et al., 2023). This approach allowed them to create <001> textured ceramics with a composition of 2 mol% MnO_2-doped 0.19PIN-0.445PSN-0.365PT (Leng, Wang, et al., 2023). These ceramics exhibited remarkable figures of merit (FOM) values for $Q_m \times k_{31}^2$, ranging from 227 to 261, and showcased high $d_{33} \times g_{33}$ values (important for power generation) within the range of $28{-}35 \times 10^{-12}\,\text{m}^2\,\text{N}^{-1}$ (Leng, Wang, et al., 2023).

Additionally, these ceramics displayed low tan δ values between 0.3% and 0.39% and a high T_{r-t} (temperature range for the rhombohedral-tetragonal phase transition) of 140–190°C. These performance metrics far surpassed those of state-of-the-art piezoelectric materials. Furthermore, they introduced a novel room temperature poling technique known as water quenching (WQ), which enhanced the piezoelectric properties of the textured MnO_2-doped PIN-PSN-PT ceramics (Leng, Wang, et al., 2023).

Above discussion provided an overview of the texturing method for improving the properties of the piezoelectric layer. This is vital as the performance of ME composite (e.g., two to two composite made by bonding) is dependent on the magnitude of the piezoelectric coefficient. As new piezoelectric compositions are developed with higher piezoelectric coefficients, they can be utilized in the fabrication of the ME composite.

5.5 Magnetoelectric nanoparticles and their properties

Nanoparticles have gained significant attention across diverse application domains in recent years, particularly in scenarios where efficient motion within microchannels and heightened reactivity are desired (Hou et al., 2021; Rodzinski et al., 2016; Song, Listyawan, & Ryu, 2022; Wang et al., 2020). The small size with large surface area is exploited in such contexts. The benefits these nanoparticles offer have spurred research endeavors aimed at broadening their utility by integrating multifunctional attributes, especially within biomedical and environmental realms (Song, Listyawan, & Ryu, 2022). In medical applications, core–shell magnetic nanoparticles (MENPs) have garnered interest due to their compatibility with magnetic fields. MENPs are suitable for generation of an electric field within low magnetic permeability materials like the human body or ionic liquid solutions (Chen et al., 2017; Khizroev, 2019; Khizroev & Liang, 2019). It will be relevant to discuss here MENP properties and synthesis procedures.

At the nanoscale, core–shell MENPs exhibit a complex multiphase heterostructure, as depicted in Fig. 5.26a (Song, Listyawan, & Ryu, 2022). This structure manifests magnetostrictive and piezoelectric characteristics through elastic interfacial coupling between the core's magnetostrictive phase and the shell's piezoelectric phase under the influence of a magnetic field (Betal et al., 2016). The core typically comprises ferromagnetic materials with a magnetostrictive spinel structure (AB_2O_4) and substantial magnetic moments (Wang et al., 2020). Conversely, the shell consists of ferroelectric materials with a perovskite structure (ABO_3), showcasing

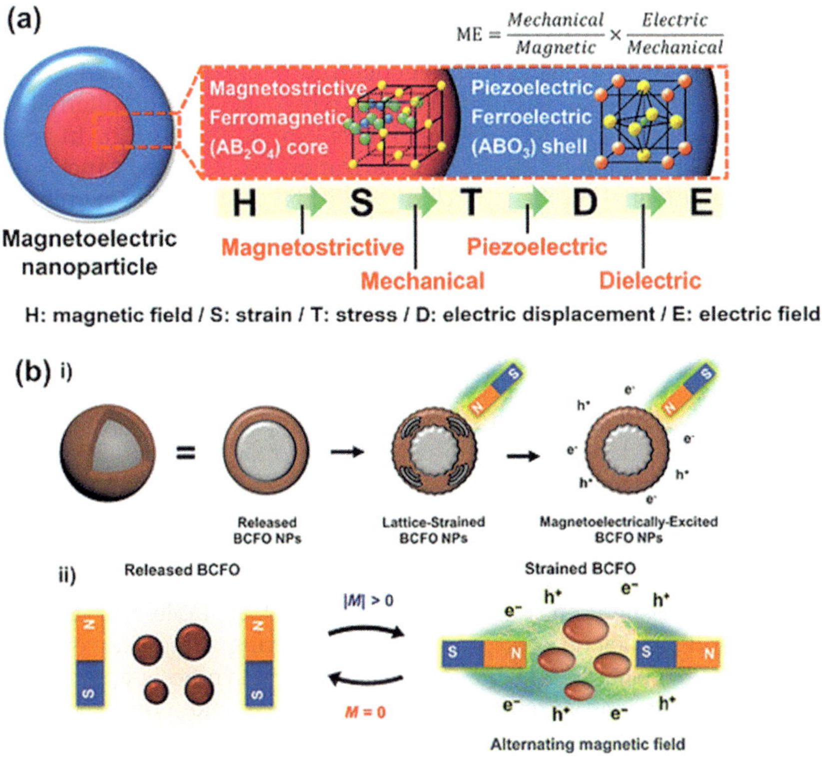

FIGURE 5.26 The magnetoelectric effect in MENPs. (a) The conceptual physics behind the transmission mechanism via the ME effect within MENPs, shows the interplay between the magnetostrictive core and the piezoelectric shell. (b), (i) presents a schematic illustration demonstrating the induction of the ME effect in MENPs under alternating magnetic field. (b), (ii), repetitive strain and release cycles within MENPs depends on magnetic moments of the applied magnetic field. *(B) (i) Reproduced with permission from Song, Kim, et al. (2022). (ii) Reproduced with permission from Jang and Park (2022).*

spontaneous electric polarization and piezoelectric attributes (Mhambi et al., 2021). Ferromagnetic candidates for the magnetostrictive core include Fe_3O_4 (Tanasă et al., 2019), $NiFe_2O_4$ (Ryu et al., 2006), and $CoFe_2O_4$, whereas perovskite materials like $PbTiO_3$, $BaTiO_3$ (Corral-Flores et al., 2010), and $BiFeO_3$ (Zhu et al., 2014) serve as options for the piezoelectric shell. If the core is piezoelectric and the shell is magnetostrictive, applying a magnetic field to the MENP partly shields the transmission of the electric field generated by the piezoelectric core. Consequently, this results in the direct transmission of electric energy to surrounding tissues or reactants. The successful combination of suitable materials is pivotal for achieving significant ME effects in MENPs.

The fundamental mechanism driving the ME coupling within MENPs is illustrated in Fig. 5.26b. Specifically, the asymmetry in dielectric properties within MENPs instigates the generation of an electric field, which propagates into the surrounding vicinity when subjected to alternating magnetic fields (Fig. 5.26bi) (Song, Listyawan, & Ryu, 2022). In most scenarios, the alternating electrical polarization leads to the induction of an electric potential on the surface of the MENPs. The application of an alternating magnetic field initiates vibrational strain within the magnetostrictive core, which is subsequently transmitted to the piezoelectric shell as mechanical stress. This transfer of stress occurs through the interfacial coupling between distinct phases of the core and shell (Pandey et al., 2020). Through the process of piezoelectric conversion, wherein stress is transformed into electric polarization, a continuous electric field is generated within the MENPs. This cycle of application and removal of the magnetic field perpetuates the generation of an electric field within the MENPs, as depicted in Fig. 5.26bii) (Song, Listyawan, & Ryu, 2022; Wang et al., 2020). Consequently, this phenomenon induces an electric field on the surface of the MENPs and concurrently triggers the generation of charge carriers—namely, electrons (e^-) and holes (h^+) within the surrounding tissue (Song, Listyawan, & Ryu, 2022). Table 5.10 summarizes several studies on core–shell MENPs, offering a detailed comparison of magnetoelectric voltage coefficient values, synthesis approaches, magnetoelectric measurement methodologies, and their diverse applications.

5.5.1 ME measurement procedure for core–shell magnetoelectric nanoparticles

Characterizing the ME coupling of MENPs holds paramount importance in generating accurate data that can used for distinct applications. The ME coupling, denoted by the ME voltage coefficient (αME), stems from the mechanical interfacial connection between the magnetostrictive core and the piezoelectric shell. This phenomenon can be mathematically represented as follows (Song, Kim, et al., 2022):

$$\alpha_{ME} = p \times \lambda = \frac{\Delta E}{\Delta S} \times \frac{\Delta S}{\Delta H} = \frac{\Delta E}{\Delta H} = \frac{\Delta V/t}{\Delta H} \left(Vcm^{-1}Oe^{-1}\right) \quad (5.16)$$

In Eq. (5.16), p and λ represent the piezoelectric and magnetostrictive coefficients, respectively, ΔE signifies the change in electric field resulting from the ME effect, ΔS corresponds to the change in strain originating from the magnetostrictive core, and ΔH denotes the magnitude of the external magnetic field being applied. The electric field is defined as the shift in voltage (ΔV) divided by the thickness (t) of the piezoelectric shell.

TABLE 5.10 Materials composition, synthesis approaches, ME voltage coefficients of magnetoelectric nanoparticles.

Material				
Magnetostrictive core	**Piezoelectric Shell**	**Synthesis method**	**ME voltage coefficient (V/cm.Oe)**	**References**
$CoFe_2O_4$ (CFO)	$BaTiO_3$ (BTO)	Hydrothermal/Sol–gel method	0.00813	Chaudhuri and Mandal (2015)
$CoFe_2O_4$ (CFO)	$BaTiO_3$ (BTO)	Sol–gel method	0.00918	Rao et al. (2017)
$CoFe_2O_4$ (CFO)	$BaTiO_3$ (BTO)	Sol–gel method	0.00000276	Kozielski et al. (2021)
$CoMnRFeO_4$ (CoMnRFe)	$BaTiO_3$ (BTO)	Sol–gel method	0.0249	Alfareed et al. (2022)
$CoFe_2O_4$ (CFO)	$BiFeO_3$ (BFO)	Sol–gel method	10–30	Jang and Park (2022)
$CoFe_2O_4$ (CFO)	$BiFeO_3$ (BFO)	Hydrothermal/Sol–gel method	405	Mushtaq, Chen, et al. (2019)
$CoFe_2O_4$ (CFO)	$BiFeO_3$ (BFO)	Hydrothermal method	1400	Mushtaq, Torlakcik, et al. (2019)
Fe_3O_4 (FO)	$BaTiO_3$ (BTO)	Hydrothermal/Sol–gel method	260	Zhang et al. (2021)
$CoFe_2O_4$ (CFO)	$BaTiO_3$ (BTO)	Sol–gel method	47	Song, Kim, et al. (2022)
$CoFe_2O_4$ (CFO)	$BiFeO_3$ (BFO)	Hydrothermal/Sol–gel method	325	Kim et al. (2022)
$CoFe_2O_4$ (CFO)	$BiFeO_3$ (BFO)	Hydrothermal/Sol–gel method	1700	Mushtaq et al. (2022)

By utilizing this definition of the ME voltage coefficient, accurate values can be derived from observing the output voltage signals generated by the MENPs in response to an applied magnetic field.

Accurate measurements of the ME properties for individual MENPs remain intricate due to their small size. Recently, a modified approach employing a piezoresponse force microscopy (PFM) probe system in the point I–V method has been applied to directly determine the ME voltage coefficient of an individual MENP (Fig. 5.27) (Jang & Park, 2022; Mushtaq, Torlakcik, et al., 2019; Song, Kim, et al., 2022). The measurement of the ME effect using an adapted PFM probe system relies on the converse piezoelectric phenomenon, converting the applied electrical energy into mechanical strain due to ME coupling. During the PFM measurement, the orientations of the MENP polarization, which change in response to the applied electric field, were contrasted with the piezoresponse hysteresis curves both in the presence and absence of an external magnetic field (Song, Listyawan, & Ryu, 2022). The switching of phase angles within the hysteresis loops, shifting between 0 and 180 degrees in the presence or absence of a magnetic field, suggests that the voltage signals obtained from MENP via PFM measurements can be attributed to the piezoelectric effect rather than an electrostatic response (Fig. 5.27ii).

The discernible asymmetric shifts in coercive voltages upon the application of an external magnetic field were obtained to calculate the ME voltage coefficient. This calculation is based on the difference between the positive (ΔV_c^+) and negative (ΔV_C^-) coercive voltage shifts, evident in the butterfly loops (S–E curve), both with and without a magnetic field, signifying the charge generation by the polarization of the piezoelectric shell within an individual MENP (Song, Kim, et al., 2022). Thus, the α_{ME} of an individual core–shell MENP can be determined using the following equation (Song, Kim, et al., 2022):

$$\alpha_{ME} = \frac{|\Delta V_c^+ - \Delta V_C^-|}{t \times \Delta H} \left(Vcm^{-1}Oe^{-1}\right) \tag{5.17}$$

Here, t represents the thickness of the piezoelectric shell enveloping the magnetostrictive core, while ΔH is the magnetic field administered to an individual MENP. Employing a PFM probe system for ME measurements

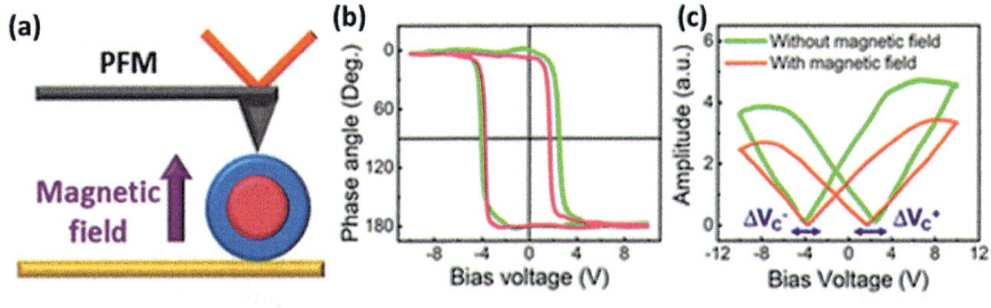

FIGURE 5.27 (i) Utilization of a piezoresponse force Microscopy (PFM) probe system to measure the ME effect of an individual MENP, (ii) depiction of hysteresis, and (iii) bipolar strain versus electric field (S–E) curves acquired from a solitary MENP, both in the presence and absence of a magnetic field. These curves serve as the basis for computing the ME voltage coefficient. *Reproduced with permission from Song, Kim, et al. (2022).*

presents an advantageous way to determine the αME of an individual MENP, accounting for size and morphology while circumventing polarization losses. Consequently, the exploration of ME measurements for a single MENP through probe systems utilizing PFM (Zhang et al., 2021) or scanning tunneling microscopy (STM) (Wang et al., 2020) is anticipated to garner increased interest and provide accelerated advancements in the future.

5.5.2 Synthesis procedure of magnetoelectric nanoparticles

MENPs can be synthesized via multiple different techniques, including hydrothermal, sol–gel, solvent evaporation, and solid-state reactions (Song, Listyawan, & Ryu, 2022). Among these, the sol–gel and hydrothermal methods stand out as the most prevalent approaches for creating core–shell structured MENPs with magnetostrictive and piezoelectric phases. The sol–gel process is particularly well-suited for controlling particle size due to its adaptable pH, temperature, and concentration parameters (Danks et al., 2016). This method involves generating a uniform precursor solution, wherein the precursor materials undergo ionization via hydrolysis. The solution, adjusted to the appropriate pH, is prepared by dissolving the precursor materials in a solvent and increasing the reaction temperature (Danks et al., 2016). After precursor solution evaporation, a dehydrated gel forms through alkoxide polycondensation. Subsequent calcination of the gel at elevated temperatures yields nanocrystalline materials (Song, Kim, et al., 2022). A notable instance of this approach is demonstrated by Song et al. who employed the sol–gel technique to synthesize core–shell $CoFe_2O_4$@$BaTiO_3$ (CFO@BTO) MENPs, featuring magnetostrictive $CoFe_2O_4$ cores encased in piezoelectric $BaTiO_3$ shells, as depicted in Fig. 5.28a (Song, Kim, et al., 2022). The coating of BTO shells onto the CFO cores involved dissolving $BaCO_3$ and Ti $(OCH(CH_3)_2)_4$ in a citric acid solution to produce Ba and Ti ions (Song, Kim, et al., 2022). Subsequently, CFO nanoparticles were introduced into the precursor solution containing Ba^{2+} and Ti^{4+}. Gelation around the CFO cores was induced through precursor solution evaporation, culminating in the thermal annealing production of 50 nm-sized MENPs exhibiting a BTO shell nanoarchitecture (Song, Kim, et al., 2022).

The hydrothermal method offers advantages such as heightened precursor solubility and rapid reactions at elevated temperatures and pressures within an autoclave, leading to reduced energy and cost requirements for the process (Hadjikhani et al., 2017; Mushtaq, Torlakcik, et al., 2019; Stimphil et al., 2017). In this approach, the precursor solution composed of starting materials undergoes hydrolysis and aging, resulting in the formation of an amorphous layer containing B-site ions in the

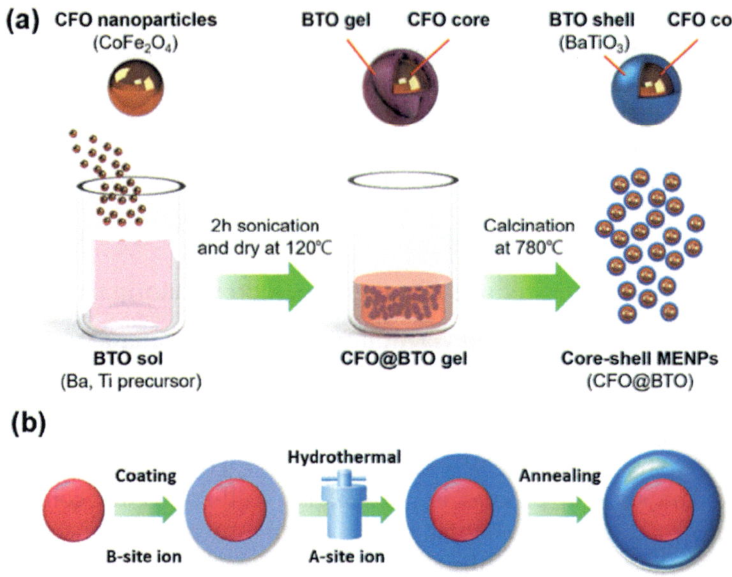

FIGURE 5.28 Commonly utilized techniques for crafting core–shell structured MENPs: (a) Production of $CoFe_2O_4$@$BaTiO_3$ (CFO@BTO) MENPs through the sol–gel approach. (b) Fabrication of core–shell MENPs via the hydrothermal method. *Reproduced with permission from Song, Kim, et al. (2022).*

perovskite structure (ABO_3) (depicted in Fig. 5.28b) (Lather et al., 2017). During the hydrothermal reaction in the autoclave, A-site ions are incorporated into the amorphous layer near the supercritical point of water. Rongzheng et al. demonstrated the efficacy of the hydrothermal synthesis method by selecting Fe_3O_4@$PbTiO_3$ core–shell particles (Liu et al., 2010). For the coating of Fe_3O_4 nanoparticles (synthesized via the hydrothermal approach) with $PbTiO_3$ shells, Ti^{4+} ions from the $Ti(SO_4)_2$ precursor formed a smooth layer on the Fe_3O_4 nanoparticles through an aging process. Subsequently, nanocrystalline $PbTiO_3$ shells were generated, involving the integration of A-site Pb^{2+} ions into the Ti hydroxide layer. This integration took place at relatively lower temperatures, facilitated by the high-pressure conditions within the autoclave. Additional annealing was conducted to enhance the crystallinity and density of the resulting core–shell Fe_3O_4@$PbTiO_3$ particles (Liu et al., 2010).

Furthermore, a solid-state reaction synthesis route has been employed to mechanically produce MENPs characterized with an irregular structure rather than a distinct core–shell arrangement (Lather et al., 2017). In this approach, piezoelectric BTO and magnetostrictive $Ni_{0.5}Co_{0.5}Fe_2O_4$ were milled and blended to synthesize BTO–NCF MENPs. The solid-state reaction hinges on high-temperature mechanical mixing and solid-state ion diffusion. Additionally, the solid-state reaction was utilized to synthesize single-grain BTO-$(Mn_{0.5}Zn_{0.5})Fe_2O$ MENPs featuring diverse geometrical configurations (Yang et al., 2012). Various other synthesis methods, including microemulsion-based, solvent evaporation, and sonochemical techniques, have been implemented to craft ferroelectric and ferromagnetic nanocrystals (Chhabra et al., 1995; Rondinone et al., 1999). Microemulsion-based synthesis triggers nanocrystal precipitation through chemical reactions involving inorganic salts, coupled with water removal. Using this method, MENPs containing CFO (Rondinone et al., 1999) and BFO (Chhabra et al., 1995) have been successfully synthesized. The solvent evaporation method is ideal for generating pure MENPs, capitalizing on the straightforward elimination of impurities from the solution during evaporation. For instance, $BiFeO_3$ was produced by drying metal ions blended with organic molecules, followed by solvent evaporation and calcination (Ghosh et al., 2005). In the case of sonochemical synthesis, nanoclusters coalesce due to the explosive collapse of bubbles resulting from acoustic cavitation. Additional calcination can yield perovskite nanocrystals, such as $SrTiO_3$ nanoparticles (Dutta & Tyagi, 2016).

Core–shell MENPs are progressively diversifying their scope of applications (Apu et al., 2022), encompassing fields like biomedicine for functions such as brain stimulation (Kujawska & Kaushik, 2023), cell regeneration (Song, Kim, et al., 2022), imaging (Khanal, 2016), drug delivery, and release as depicted in Fig. 5.29 (Abdelazim, 2017). To elaborate, the utilization of MENPs to create an electric field in the brain through an applied magnetic field holds potential for treating various ailments and expediting the regeneration (differentiation and proliferation) of neuronal cells. Furthermore, the electric field stemming from MENPs can facilitate the creation of channels in cell membranes, enabling the release of drugs loaded onto the expansive surface of the MENPs (Mhambi et al., 2021). Additionally, MENPs can be utilized for tissue imaging by detecting a self-induced magnetic field. This is achieved by leveraging the electric signals produced by the human body through the converse ME effect and subsequently mapping the MENPs' positions (Mushtaq, Chen, et al., 2019; Song, Listyawan, & Ryu, 2022). In the realm of electrocatalysts, the need for a substantial surface area and electrical imbalance is crucial to generate adequate surface charge capable of triggering ionic reactions within a solution. Consequently, MENPs present promising prospects as candidates for electrocatalytic applications (Mushtaq, Chen, et al., 2019). For more details about biomedical applications of ME nanoparticles, see chapter 14.

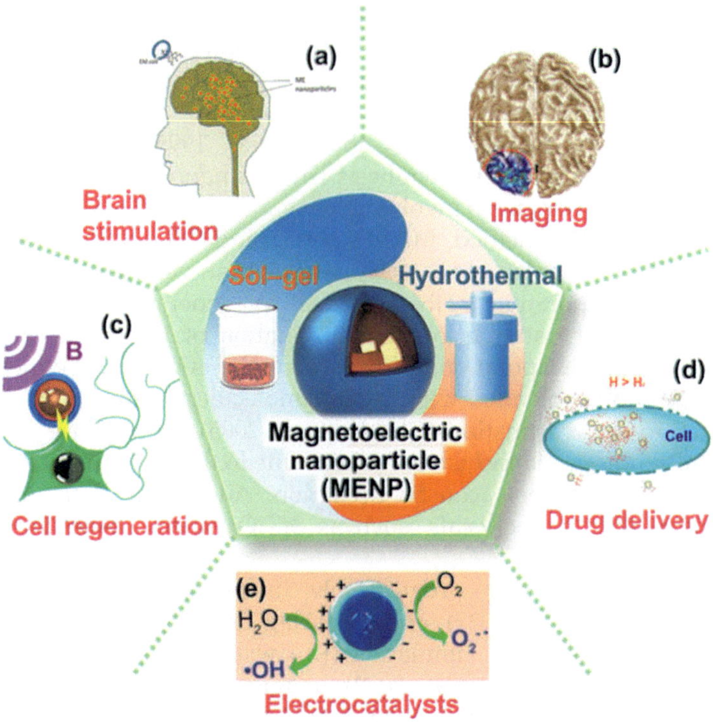

FIGURE 5.29 Different synthesis procedure and application of magnetoelectric nanoparticles from (Song, H. et al., 2022b): (a) Brain Stimulation. (b) Brain imaging. (c) Cell regeneration. (d) Drug delivery enabling nanoelectroporation. (e) Electrocatalysts for water purification. *(A) Reproduced with permission from Yue et al. (2012). Copyright 2012, PLoS ONE, (B) Reproduced with permission from Guduru et al. (2018). Copyright 2018, BMC Springer Nature, (C) Reproduced with permission from Song, Kim, et al. (2022). Copyright 2022, Royal Society of Chemistry, (D) Reproduced with permission from Guduru et al. (2013), Mushtaq, Chen, et al. (2019). Copyright 2013, Nature, (E) Reproduced with permission from Mushtaq, Chen, et al. (2019).*

5.6 Polymer-based magnetoelectric materials

While ceramic composites exhibit ME coefficients significantly higher than single-phase materials, they are vulnerable to fragility due to detrimental interface reactions, resulting in elevated electrical resistivity and dielectric losses exceeding 0.1 (Liu et al., 2009; Martins & Lanceros-Méndez, 2013; Nan, 1994). These limitations impede their integration into devices. Furthermore, ceramic composites could suffer from issues like high cost, high density, and brittleness, increasing the risk of operational failure. In contrast, polymer-based ME materials are promising alternatives to mitigate the challenges found in ceramic-based ME materials. Polymer-based ME materials can be readily produced using conventional low-temperature methods, shaping them into various forms like thin sheets or molded structures possessing enhanced mechanical properties (Scott, 2012).

There are three primary classifications of polymer-based ME composites: nanocomposites, polymers employed as binders, and laminated composites (Martins & Lanceros-Méndez, 2013). When creating particulate polymer-based ME nanocomposites, two specific polymers, polyurethane (PU) and poly(vinylidene fluoride-trifluoroethylene) (P(VDF-TrFE)), a PVDF copolymer, have been utilized. Much research has been conducted on 0-3 type ME polymer-based nanocomposites (Martins & Lanceros-Méndez, 2013). Piezoelectric polymers like PVDF and PU have found application in these ME nanocomposites due to their favorable piezoelectric and electrostrictive characteristics (Karan et al., 2015, 2016, 2019; Kepler & Anderson, 1978).

Electrostrictive PU-based ME composites have produced promising results. These composites exhibit a unique blend of both linear and quadratic ME responses in the PU film when filled. Notably, the linear ME effect achieved is on par with that of Cr_2O_3 single crystals, reaching up to 18 mV cm^{-1} Oe^{-1}. Moreover, there is a possibility of a linear magneto-elastoelectric interaction between the fillers and the polymer matrix, which is not induced by magnetostriction (Guyomar et al., 2008). In specific cases, linear voltage ME coefficients in PU/Fe_3O_4 and PU/nickel composites have been obtained at 11.4 and 6.0 mV cm^{-1} Oe^{-1}, respectively, under conditions of 7 Hz, 0 DC field, and 1 Oe AC field (Guyomar et al., 2008). The study suggested that the α value remains the same, even under increasing DC magnetic bias. This experimental observation strongly suggested that the magnetostrictive properties of the material have no influence on the PU/Fe_3O_4 and PU/Nickel ME composites. This empirical finding strongly implies that the magnetostrictive properties of the material do not exert any influence on the PU/Fe_3O_4 and PU/nickel sME composites (Jiles et al., 1988; Martins, Lasheras, et al., 2011).

Martins et al. incorporated ferrite nanoparticles, specifically $CoFe_2O_4$ and Ni0.5Zn0.5Fe_2O_4, into a polymer matrix composed of P(VDF-TrFE) instead of PVDF (Legrand, 1989; Martins, Moya, et al., 2011; Sencadas et al., 2004). This choice was made because of the copolymers of PVDF, with VDF molar percentages between 55 and 82, crystallize from the melt in the ferroelectric phase, a crucial factor in preparing PVDF-based ME nanocomposites (Legrand, 1989; Sencadas et al., 2004). The resulting P(VDF-TrFE)/ferrite nanocomposites exhibit a range of characteristics, including ferroelectric, piezoelectric, magnetic properties, and a direct ME effect, all dependent on the ferrite content. In the case of the P(VDF-TrFE)/$CoFe_2O_4$ nanocomposite, the resulting films display saturated hard magnetic properties, improved polarization, a notable piezoelectric response, and a maximum ME coupling coefficient (α_{33}) value of

41.3 mV cm^{-1} Oe^{-1}. Conversely, for P(VDF-TrFE)/$Ni0.5Zn0.5Fe_2O_4$ composites, α_{33} linearly increases with applied H_{DC}, reaching a maximum value of 1.35 mV cm^{-1} Oe^{-1} in samples containing 15 wt% ferrite (Martins, Moya, et al., 2011).

Unlike PU-based composites (Guyomar et al., 2009), the ME response of P(VDF-TrFE)/$CoFe_2O_4$ is noticeably affected by the magnetostriction of the ferrite nanoparticles. This is evident in the observation of an optimal H_{DC} magnetic field value, which leads to a distinct peak in the α value versus H_{DC} plot (Fig. 5.30) (Martins & Lanceros-Méndez, 2013).

The disparity between the in-plane and out-of-plane ME responses in P(VDF-TrFE)/$CoFe_2O_4$ nanocomposites is attributed to variations in the d_{33} and d_{31} P(VDF-TrFE) piezoelectric constants (Si et al., 2018; Wang et al., 1993). Zhang et al. conducted a study on the impact of $CoFe_2O_4$ nanoparticles on the morphology, ferroelectric, magnetic, and ME properties of P(VDF-TrFE)/$CoFe_2O_4$ nanocomposites. The concentration of ferrite nanoparticles significantly influences the ferroelectric and ME responses, with the α_{33} value reaching 40 mV cm^{-}1 Oe^{-1} (Zhang et al., 2009).

In contrast to the earlier section, polymer-as-binder composites do not employ the polymer as the piezoelectric component of the ME material. Instead, the polymer serves as a binding agent that holds the piezoelectric and magnetostrictive particles together, facilitating stress coupling between the piezoelectric matrix and the magnetostrictive fillers. The initial focus in this category involved investigating three-phase particulate composites consisting of Terfenol-D alloy, PZT, and PVDF (Fig. 5.31) (Nan et al., 2002).

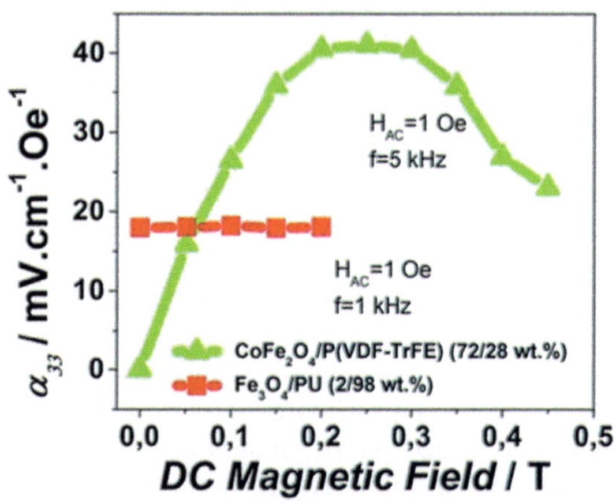

FIGURE 5.30 Comparison the characteristics of P(VDF-TrFE)/with those of PU/Fe_3O_4 magnetoelectric (ME) nanocomposites. *Reproduced with permission from Martins and Lanceros-Méndez (2013).*

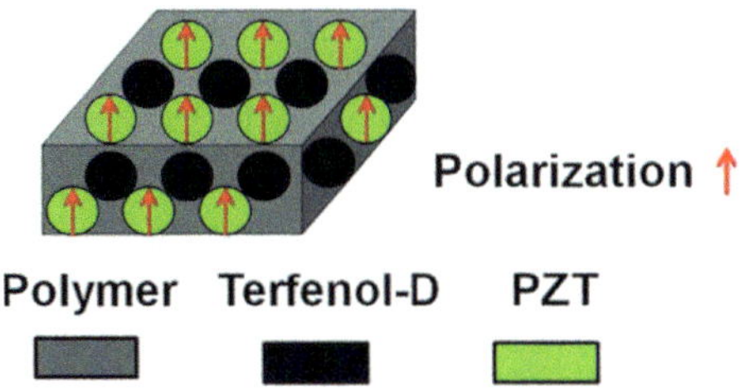

FIGURE 5.31 Illustrative depiction of the particulate composites comprising Terfenol-D/PZT/polymer based work. *Reproduced with permission from Shi et al. (2004).*

To obtain the ME response, a small volume fraction, f, Terfenol-D nanoparticles were included into a PZT/PVDF blend by blending technique (Shi et al., 2004). The resulting dielectric, piezoelectric, and ME properties reveal a percolation transition around f $\approx$ 0.12. When f remains below 0.07, the multifunctional composites exhibit piezoelectric as well as ME responses. However, as f falls within the range of 0.07–0.12, both piezoelectric and ME responses precipitously decline, vanishing at the percolation threshold (Martins & Lanceros-Méndez, 2013; Shi et al., 2004). Beyond this threshold, the composite takes on conductive properties and exclusively responds to magnetostriction. The highest recorded α_{33} value at 2 kOe reached 42 mV cm^{-1} Oe^{-1} at f = 0.06, less than half of the value obtained for PZT/ferrite ceramic composites (115 mV cm^{-1} Oe^{-1}) (Ryu et al., 2001b). As the ME response is basically determined by the $f_{\text{Terfenol-D}}$, the surfactants pretreatment of the Terfenol-D can change the ME coupling. Surfactants are commonly employed to enhance the dispersal and stability of nanoparticles in various matrices (Hilding et al., 2003; Pyun, 2007). In the context of PZT/ferrite ceramic composites, surfactants have been found to raise the percolation threshold.

In a previous section, the three-phase Terfenol-D/PZT/PVDF particulate composites had a relatively low allowable volume fraction of Terfenol-D, significantly constraining the ME response of these multifunctional nanocomposites. To overcome this limitation, an alternative class of ME materials has been developed, known as laminated composites. Employing a bilayer or multilayer configuration in ME composites offers distinct advantages compared to bulk nanocomposites. Notably, the issue of polarization loss in bulk composites, caused by leakage currents, can be effectively addressed in layered structures. These laminates allow for poling of the piezoelectric phase to augment ME coupling and also permit the adjustment of poling and applied field orientations to optimize ME coupling.

Laminated multifunctional (MF) composites have been developed, comprising a Terfenol-D/PVDF particulate composite layer sandwiched between two PZT/PVDF particulate layers, crafted through a hot-molding technique (Cai et al., 2003). Research findings indicate that when $f_{PVDF} \leq 0.3$, the low PVDF concentration results in suboptimal composite quality due to poor interphase connectivity, leading to diminished ME performance (Cai et al., 2003). The ME properties demonstrate improvement within the intermediate range of f_{PVDF} ($0.3 \leq f_{PVDF} \leq 0.5$). As f_{PVDF} surpasses 0.5 ($f_{PVDF} > 0.5$), the high concentration of inert PVDF leads to reduced dielectric, magnetostrictive, piezoelectric, and ME properties in the three-phase laminated Terfenol-D/PZT/PVDF composites. At 1 kHz, four kOe, and $f_{PVDF} = 0.5$, the maximum α_{33} value reaches 80 mV cm^{-1} Oe^{-1}. These laminated composites can reach ME values up to 3 V cm^{-1} Oe^{-1} at 100 kHz resonance frequency (Cai et al., 2004). The difference between longitudinal (α_{33}) and transverse (α_{31}) ME sensitivity, measuring three and 3.8 V cm^{-1} Oe^{-1}, respectively, is attributed to the anisotropy present in the laminated ME samples. Under high bias, magnetostriction saturates more quickly with in-plane bias than with out-of-plane bias, resulting in constant electric field in the PZT and decreases α_{31} by increasing magnetic bias (Cai et al., 2004). The PZT/PVDF particulate layer enclosed between two Terfenol-D/PVDF particulate composite layers shows maximum ME sensitivity (α_{33}) up to 300 mV cm^{-1} Oe^{-1} at > 50 kHz frequency and 6 V cm^{-1} Oe^{-1} at 80 kHz resonance frequency (Nan et al., 2005). The ME response of such composites is markedly influenced by the applied bias and the thickness ratio (t_p/L) between the Terfenol-D/PVDF layers and the PZT/PVDF layer. A bilayer disk, achieved by bonding a PZT disk to a Terfenol-D disk using low viscosity epoxy, exhibited three distinct resonance peaks at a bias of three kOe, the α_{33} values can reach 2.79 and 0.924 V cm^{-1} Oe^{-1} at 35 and 100 kHz, respectively (Shi et al., 2008).

Similarly, thin and flexible ME laminates (Fig. 5.32a) were also created using similar techniques but featured different magnetostrictive layers, that is, unimorph and three-layer sandwich configurations (Fig. 5.32c) (Zhai et al., 2006). Surprisingly, these laminates exhibited a remarkable ME response under an 8 Oe applied H_{DC}, which is 1/50th of the required applied H_{DC} for the previous ME laminates (Martins & Lanceros-Méndez, 2013). These compact magnetic field-sensitive ME

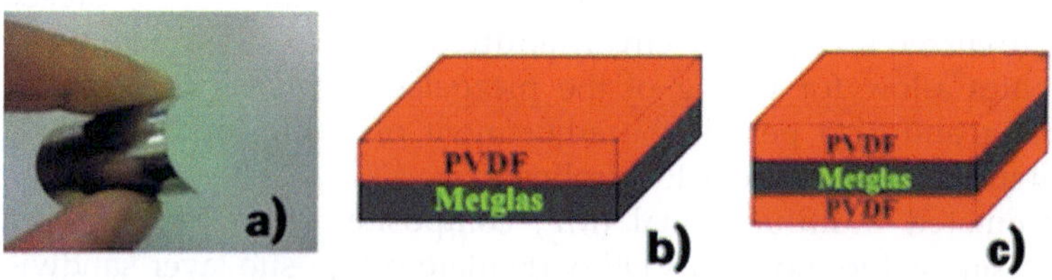

FIGURE 5.32 (a) Image of a pliable PVDF/Metglas unimorph laminate. (b) Configuration of the unimorph. (c) Configuration of the three-layer laminate. *Reproduced with permission from Martins and Lanceros-Méndez (2013).*

laminates demonstrated substantial ME voltage coefficients and displayed excellent sensitivity to slight fluctuations in both AC and DC magnetic fields. In general, the Metglas and PVDF layers were bonded together using epoxy to make a laminate for ME measurement. For the three-layer composites, α_{31} reached 238 V cm^{-1} Oe^{-1}, while the unimorph composites achieved an α_{31} of 310 V cm^{-1} Oe^{-1} operating at a 50 kHz longitudinal resonance frequency (Martins & Lanceros-Méndez, 2013). At lower frequencies, they reached a maximum value of 7.2 V cm^{-1} Oe^{-1} for both geometries. Despite Metglas's lower magnetostriction of 42 ppm, compared to Terfenol-D's, the effective piezomagnetic coefficient (d_{33m}) is notably larger for Metglas ($\sim 4 \times 10^{-6}$ Oe^{-1}), almost 3–4 times larger than that of Terfenol-D ($d_{33m} = 1.2 \times 10^{-6}$ Oe^{-1}), due to the smaller saturation magnetic field (Dong, Zhai, Bai, Li, & Viehland, 2005). The low H_{DC} is very essential to develop high-performance Metglas/PVDF composite to make them promising for practical applications.

Following the initial research on Metglas/PVDF laminate nanocomposites, several studies have been conducted toward exploring these promising ME materials. Notably, by leveraging the magnetic flux concentration effect associated with the sheet aspect ratio of Metglas, impressive values of α_{31} reached 21.46 V cm^{-1} Oe^{-1} in a PVDF/Metglas 2605SA1 laminate composite at 20 Hz frequency under a 3 Oe DC magnetic field (Martins & Lanceros-Méndez, 2013). As a summary of this section, the outcomes for the primary polymer-based ME materials are presented in Table 5.11, categorized by composite type.

5.7 Future outlook and summary of magnetoelectric materials and phenomena

Multiferroic ME materials continue to grow in interest due to potential applications in novel multifunctional devices, such as sensors, transducers, memories, and spintronics. Known single-phase materials exhibit a weak ME effect, even at low temperatures. There are two promising approaches for synthesizing room temperature multiferroic MEs. One utilizes solid solutions exhibiting MPB or PPB, and the other is based on composite structures that utilize the intrinsic material property. In this chapter, we first briefly reported the progress on ME solid solutions. $BiFeO_3$–$BaTiO_3$ (BTO) solid solutions show the presence of MPB ($BiFeO_3$ mole fraction from 0.750 to 0.775). The enhanced magnetic properties were attributed to active spin modulation of ordered Fe–O–Fe. These results are highly promising for the realization of a room temperature single-phase multiferroic ME material. We further presented recent progress on ME composites. To achieve high ME charge and voltage

TABLE 5.11 Comparison of the main characteristics of the developed polymer-based magnetoelectric materials (Martins & Lanceros-Méndez, 2013; Sasmal & Arockiarajan, 2023).

	Constitution	H_{DC} [Oe]	α [mV cm^{-1} Oe^{-1}]	$\alpha_{resonance}$ [mV cm^{-1} Oe^{-1}]	References
Nanocomposite	PE/Fe_3O_4	0	11.4	–	Guyomar et al. (2008)
	PE/Nickel	0	6	–	Guyomar et al. (2008)
	P(VDF-TrFE)/$Ni_{0.5}Zn_{0.5}Fe_2O_4$	5000	0.1	1.35	Martins, Moya, et al. (2011)
	P(VDF-TrFE)/$CoFe_2O_4$	2500	4.1	41.3	Martins, Lasheras, et al. (2011)
	P(VDF-TrFE)/$CoFe_2O_4$	2000	–	40	Zhang et al. (2009)
	PVDF-TrFE/$SmFeO_3$	1000		55	Ahlawat et al. (2019)
	PVDF-TrFE/$SmFeO_3$	1000		45	Ahlawat et al. (2018)
	PVDF/$YFeO_3$	8000		5.3	Sasmal et al. (2023)
	PVDF/$NiFe_2O_4$	3000		90	Mayeen et al. (2019)
	PVDF-TrFE/$La_{0.7}Sr_{0.3}MnO_3$	2500		155.7	Koner et al. (2021)
	PVDF-TrFE/$La_{0.7}Ba_{0.3}MnO_3$	3000		121.4	Koner et al. (2020)
	PVDF-TrFE/$CoFe_2O_4$	2500		40	Martins et al. (2014)
	PVDF-TrFE/$CoFe_2O_4$	2600		6.5	Martins et al. (2015)
	PVDF-TrFE/La_2CoMnO_6	5000		1.45	Bhadra et al. (2013)

Polymer as a binder composites	PVDF/Terfenol-D/PZT	2000	42	–	Nan et al. (2002)
	PEO/Terfenol-D/PZT	1400	1.3	–	Chau et al. (2009)
	Li^+-PEO/Terfenol-D/PZT		3.2	–	Chau et al. (2009)
	PMMA/Terfenol-D/PZT		4.8	–	Chau et al. (2009)
Laminate	PVDF/Terfenol-D/PZT	4000	80	3000	Cai et al. (2004)
	PVDF/Terfenol-D/PZT	4000	300	6000	Nan et al. (2005)
	Spurr epoxy/Terfenol-D/PZT	504	400	1100	Wan, Shen, and Wu (2005)
	PE/PVDF/Fe_3O_4	2000	753	–	Belouadah et al. (2011)
	VER/Terfenol-D/PZT	666	2700	–	Nersessian et al. (2004)
	PZT/Terfenol-D/epoxy	3000	1310	2790	Shi et al. (2008)
	Gd crystal/P(VDF-TrFE)/ silver conductive epoxy	200	500	–	Lu et al. (2010)
	PVDF/Metglas unimorph	8	7200	238,000	Zhai et al. (2006)
	PVDF/Metglas three-layer			310,000	
	PVDF/Metglas	8	21,460	–	Fang et al. (2009)
	PVDF/Metglas	3	400	–	Dong et al. (2009)
	PVDF-HPFP/Metglas	5	12,000	–	Lu et al. (2011)

Continued

TABLE 5.11 Comparison of the main characteristics of the developed polymer-based magnetoelectric materials (Martins & Lanceros-Méndez, 2013; Sasmal & Arockiarajan, 2023).—cont'd

	Constitution	H_{DC} [Oe]	α [mV cm^{-1} Oe^{-1}]	$\alpha_{resonance}$ [mV cm^{-1} Oe^{-1}]	References
	Cross-linked P(VDF-TrFE)/ Metglas 2605	4	17,700	383,000	Jin et al. (2011)
	PVDF/$Ni_{50}Mn_{29}Ga_{21}$	5100	1240	–	Zeng et al. (2010)
	PVDF-TrFE/Metglas	6		950,000	Kulkarni et al. (2014)
	PVDF/Metglas	5		240,000	Long et al. (2017)
	Metglas/PVDF/Metglas	4.75		190,000	Reis et al. (2016)
	Metglas/PVDF	5		176,000	Zhang et al. (2020)
	Metglas/PVDF/Metglas	7		150,000	Lasheras et al. (2015)
	PZT deposited on Ni foil	0		3.15	Palneedi et al. (2018)
	AlN layer grown on Ni foil	8.5		3.3	Nguyen et al. (2021)
	$BaTiO_3$ deposited on Metglas	20		0.055	Wang et al. (2011)

coefficients, a piezoelectric material with high d and g values is required. High-performance, cost-effective textured piezoelectric materials were developed with a piezoelectric response comparable to their single-crystal counterparts. Due to high d, k, and g values in textured ceramic, a 3.4 times improvement of the ME voltage coefficient was achieved. In addition to high piezoelectric properties of textured ceramics, functionally graded piezoelectric composites were presented, which exhibit unique overall functional response. Further, a strong coupling factor is necessary to achieve a high ME response. Here we presented cofiring approach for fabricating ME composites, which offer large interphase elastic interactions. Lastly, due to environmental concerns, recent developments in the area of lead-free piezoelectric and ME composites were discussed. Core–shell MENPs exhibit notable ME voltage coefficients, largely due to the strong interaction between the magnetostrictive core and piezoelectric shell, resulting in efficient phase-to-phase elastic interfacial coupling in ME composites. While several methods exist for characterizing the magnetoelectricity of individual MENPs, there remains a need for refining these techniques to ensure precise ME measurements. This precision is crucial for confirming the uniform characteristics and high-quality synthesis of MENPs. The realm of polymer-based ME materials is a rapidly increasing area of research, generating significant interest due to imminent applications. The achieved ME coefficient values, coupled with their adaptability to a wide range of magnetic field intensities, highlight the potential for polymer-based ME materials. Their flexibility, durability, and straightforward fabrication processes make them suitable for diverse applications, particularly within the domains of sensors and actuators.

References

Abdelazim, H. (2017). Magneto-electric nanocarriers for drug delivery: An overview. *Journal of Drug Delivery Science and Technology, 37*, 46–50.

Agrawal, S., Cheng, J., Guo, R., Bhalla, A. S., Islam, R. A., & Priya, S. (2009). Magnetoelectric properties of microwave sintered particulate composites. *Materials Letters, 63*, 2198–2200.

Ahlawat, A., Khan, A. A., Deshmukh, P., Tripathi, M., Shirolkar, M. M., Satapathy, S., Choudhary, R., & Karnal, A. (2019). A detailed study of magnetization and magnetoelectric effect in P (VDF-TrFe) based SmFeO 3 nanocomposites. *Journal of Materials Science: Materials in Electronics, 30*, 17765–17772.

Ahlawat, A., Satapathy, S., Shirolkar, M. M., Li, J., Khan, A. A., Deshmukh, P., Wang, H., Choudhary, R., & Karnal, A. (2018). Tunable magnetoelectric nonvolatile memory devices based on SmFeO3/P (VDF-TrFE) nanocomposite films. *ACS Applied Nano Materials, 1*, 3196–3203.

Ahn, C.-W., Maurya, D., Park, C.-S., Nahm, S., & Priya, S. (2009). A generalized rule for large piezoelectric response in perovskite oxide ceramics and its application for design of lead-free compositions. *Journal of Applied Physics, 105*.

Ahn, C. W., Park, C. S., Choi, C. H., Nahm, S., Yoo, M. J., Lee, H. G., & Priya, S. (2009). Sintering behavior of lead-free (K, Na) NbO3-basedpiezoelectric ceramics. *Journal of the American Ceramic Society, 92*, 2033–2038.

Ahn, C.-W., Park, C.-S., & Priya, S. (2010). Sintered composite for low temperature coefficient of piezoelectric property in knn based lead-free ceramics. *Functional Materials Letters, 3*, 35–39.

Alfareed, T. M., Slimani, Y., Almessiere, M. A., Shirsath, S. E., Hassan, M., Nawaz, M., Khan, F. A., Al-Suhaimi, E. A., & Baykal, A. (2022). Structure, magnetoelectric, and anticancer activities of core-shell Co0· 8Mn0. 2R0. 02Fe1· 98O4@ BaTiO3 nanocomposites (R= Ce, Eu, Tb, Tm, or Gd). *Ceramics International, 48*, 14640–14651.

Apu, E. H., Nafiujjaman, M., Sandeep, S., Makela, A. V., Khaleghi, A., Vainio, S., Contag, C. H., Li, J., Balasingham, I., & Kim, T. (2022). Biomedical applications of multifunctional magnetoelectric nanoparticles. *Materials Chemistry Frontiers, 6*, 1368–1390.

Babu, S. N., et al. (2008). Magnetoelectric properties in NCMF/PZT particulate and bulk laminate composites. *Journal of Physics D Applied Physics, 41*(1–6), Article 165407.

Belouadah, R., Guyomar, D., Guiffard, B., & Zhang, J.-W. (2011). Phase switching phenomenon in magnetoelectric laminate polymer composites: Experiments and modeling. *Physica B: Condensed Matter, 406*, 2821–2826.

Betal, S., Dutta, M., Cotica, L. F., Bhalla, A. S., & Guo, R. (2016). Control of crystalline characteristics of shell in core-shell magnetoelectric nanoparticles studied using HRTEM and holography. *Ferroelectrics, 503*, 68–76.

Bhadra, D., Masud, M. G., De, S., & Chaudhuri, B. (2013). Observation of large magnetodielectric and direct magnetoelectric behavior in LCMO/PVDF 0-3 nanocomposites. *Applied Physics Letters, 102*.

Bichurin, M., Petrov, V., Zakharov, A., Kovalenko, D., Yang, S. C., Maurya, D., Bedekar, V., & Priya, S. (2011). Magnetoelectric interactions in lead-based and lead-free composites. *Materials, 4*, 651–702.

Boomgaard, J., & Born, R. A. J. (1978). A sintered magnetoelectric composite material BaTiO3-Ni(Co, Mn)Fe2O4. *Journal of Materials Science, 13*, 1538–1548.

Bordia, R. K., & Scherer, G. W. (1988). On constrained sintering—III. Rigid inclusions. *Acta Metallurgica, 36*, 2411–2416.

Brito-Pereira, R., Ribeiro, C., Lanceros-Mendez, S., & Martins, P. (2017). Magnetoelectric response on Terfenol-D/P (VDF-TrFE) two-phase composites. *Composites Part B: Engineering, 120*, 97–102.

Cahill, D. G., Watson, S. K., & Pohl, R. O. (1992). Lower limit to the thermal conductivity of disordered crystals. *Physical Review B, 46*, 6131.

Cai, N., Nan, C.-W., Zhai, J., & Lin, Y. (2004). Large high-frequency magnetoelectric response in laminated composites of piezoelectric ceramics, rare-earth iron alloys and polymer. *Applied Physics Letters, 84*, 3516–3518.

Cai, N., Zhai, J., Nan, C.-W., Lin, Y., & Shi, Z. (2003). Dielectric, ferroelectric, magnetic, and magnetoelectric properties of multiferroic laminated composites. *Physical Review B, 68*, Article 224103.

Chau, K., Wong, Y., & Shin, F. (2009). Magnetoelectric effect of polymer electrolyte composites with Terfenol-D and lead zirconate titanate inclusions. *Applied Physics Letters, 94*.

Chaudhuri, A., & Mandal, K. (2015). Large magnetoelectric properties in CoFe2O4: BaTiO3 core–shell nanocomposites. *Journal of Magnetism and Magnetic Materials, 377*, 441–445.

Chen, R., Canales, A., & Anikeeva, P. (2017). Neural recording and modulation technologies. *Nature Reviews Materials, 2*, 1–16.

Chen, Y., Gillette, S. M., Fitchorov, T., Jiang, L., Hao, H., Li, J., Gao, X., Geiler, A., Vittoria, C., & Harris, V. (2011). Quasi-one-dimensional miniature multiferroic magnetic field sensor with high sensitivity at zero bias field. *Applied Physics Letters, 99*.

Chhabra, V., Lal, M., Maitra, A., & Ayyub, P. (1995). Nanophase BaFe12O19 synthesized from a nonaqueous microemulsion with Ba-and Fe-containing surfactants. *Journal of Materials Research, 10*, 2689–2692.

Cho, K.-H., Park, C.-S., & Priya, S. (2010). Effect of intensive and extensive loss factors on the dynamic response of magnetoelectric laminates. *Applied Physics Letters, 97*.

Cho, K.-H., & Priya, S. (2011). Direct and converse effect in magnetoelectric laminate composites. *Applied Physics Letters, 98*.

Chu, Z., Shi, H., Shi, W., Liu, G., Wu, J., Yang, J., & Dong, S. (2017). Enhanced resonance magnetoelectric coupling in (1-1) connectivity composites. *Advanced Materials, 29*, Article 1606022.

Corral-Flores, V., Bueno-Baques, D., & Ziolo, R. D. (2010). Synthesis and characterization of novel CoFe2O4–BaTiO3 multiferroic core–shell-type nanostructures. *Acta Materialia, 58*, 764–769.

Dai, Y., Bao, P., Zhu, J., Wan, J., Shen, H., & Liu, J. (2004a). Internal friction study on CuFe2O4/ PbZr0. 53Ti0. 47O3 composites. *Journal of Applied Physics, 96*, 5687–5690.

Dai, Y. R., Bao, P., Zhu, J. S., Wan, J. G., Shen, H. M., & Liu, J. M. (2004b). Internal friction study on CuFe2O4/PbZr0.53Ti0.47O3 composites. *Journal of Applied Physics, 96*, 5687–5690.

Danks, A. E., Hall, S. R., & Schnepp, Z. (2016). The evolution of 'sol–gel'chemistry as a technique for materials synthesis. *Materials Horizons, 3*, 91–112.

Dong, S. X., Cheng, J. R., Li, J. F., & Viehland, D. (2003). Enhanced magnetoelectric effects in laminate composites of Terfenol-D/Pb(Zr,Ti)O-3 under resonant drive. *Applied Physics Letters, 83*, 4812–4814.

Dong, S. X., Li, J. F., & Viehland, D. (2003). Longitudinal and transverse magnetoelectric voltage coefficients of magnetostrictive/piezoelectric laminate composite: Theory. *IEEE Transactions on Ultrasonics, Ferroelectrics, and Frequency Control, 50*, 1253–1261.

Dong, S. X., Li, J. F., & Viehland, D. (2004a). Circumferentially magnetized and circumferentially polarized magnetostrictive/piezoelectric laminated rings. *Journal of Applied Physics, 96*, 3382–3387.

Dong, S. X., Li, J. F., & Viehland, D. (2004b). Longitudinal and transverse magnetoelectric voltage coefficients of magnetostrictive/piezoelectric laminate composite: Experiments. *IEEE Transactions on Ultrasonics, Ferroelectrics, and Frequency Control, 51*, 794–799.

Dong, S. X., Li, J. F., Viehland, D., Cheng, J., & Cross, L. E. (2004c). A strong magnetoelectric voltage gain effect in magnetostrictive-piezoelectric composite. *Applied Physics Letters, 85*, 3534–3536.

Dong, X. W., Wang, B., Wang, K. F., Wan, J. G., & Liu, J. M. (2009). Ultra-sensitive detection of magnetic field and its direction using bilayer PVDF/Metglas laminate. *Sensors and Actuators A: Physical, 153*, 64–68.

Dong, S., Zhai, J., Bai, F., Li, J. F., & Viehland, D. (2005). Push-pull mode magnetostrictive/ piezoelectric laminate composite with an enhanced magnetoelectric voltage coefficient. *Applied Physics Letters, 87*(1–3), Article 062502.

Dong, S. X., Zhai, J. Y., Bai, F. M., Li, J. F., Viehland, D., & Lograsso, T. A. (2005). Magnetostrictive and magnetoelectric behavior of Fe-20 at. % Ga/Pb(Zr,Ti)O-3 laminates. *Journal of Applied Physics, 97*(1–6), Article 103902.

Dong, S. X., Zhai, J. Y., Li, J. F., & Viehland, D. (2006). Magnetoelectric effect in Terfenol-D/ Pb(Zr,TiO)(3)/mu-metal laminate composites. *Applied Physics Letters, 89*(1–3), Article 122903.

Dutta, D. P., & Tyagi, A. (2016). Weak room temperature ferromagnetism and ferroelectric behavior in sonochemically synthesized bismuth and iron codoped SrTiO3 nanoparticles. *Materials Letters, 164*, 368–371.

Ederer, C., & Spaldin, N. A. (2004). A new route to magnetic ferroelectrics. *Nature Materials, 3*, 849–851.

Eerenstein, W., Mathur, N., & Scott, J. F. (2006). Multiferroic and magnetoelectric materials. *Nature, 442*, 759–765.

Egerton, L., & Dillon, D. M. (1959). Piezoelectric and dielectric properties of ceramics in the system potassium—sodium niobate. *Journal of the American Ceramic Society, 42*, 438–442.

Fang, Z., Lu, S. G., Li, F., Datta, S., Zhang, Q. M., & El Tahchi, M. (2009). Enhancing the magnetoelectric response of Metglas/polyvinylidene fluoride laminates by exploiting the flux concentration effect. *Applied Physics Letters, 95*(1–3), Article 112903.

Fiebig, M. (2005). Revival of the magnetoelectric effect. *Journal of Physics D: Applied Physics, 38*, Article R123.

Fiebig, M., Lottermoser, T., Fröhlich, D., Goltsev, A. V., & Pisarev, R. V. (2002). Observation of coupled magnetic and electric domains. *Nature, 419*, 818–820.

Fischer, E., Gorodetsky, G., & Hornreich, R. (1972). A new family of magnetoelectric materials: A2M4O9 (A Ta, Nb; M Mn, Co). *Solid State Communications, 10*, 1127–1132.

Gao, T., Liao, J., Wang, J., Qiu, Y., Yang, Q., Zhang, M., Zhao, Y., Qin, L., Xue, H., & Xiong, Z. (2015). Highly oriented BaTiO 3 film self-assembled using an interfacial strategy and its application as a flexible piezoelectric generator for wind energy harvesting. *Journal of Materials Chemistry A, 3*, 9965–9971.

Ghosh, S., Dasgupta, S., Sen, A., & Sekhar Maiti, H. (2005). Low-temperature synthesis of nanosized bismuth ferrite by soft chemical route. *Journal of the American Ceramic Society, 88*, 1349–1352.

Go, S.-H., Kim, D.-S., Eum, J.-M., Shin, H.-S., Chae, S.-J., Kim, S.-W., Kim, E.-J., Woo, J.-U., & Nahm, S. (2021). Excellent piezoelectric properties of (K, Na)(Nb, Sb) O3-CaZrO3-(Bi, Ag) ZrO3 lead-free piezoceramics. *Journal of Alloys and Compounds, 889*, Article 161817.

Grotel, J., Pikula, T., & Mech, R. (2023). Application of the lock-in technique in magnetoelectric coupling measurements of the PZT/Terfenol-D composite. *Applied Sciences, 13*, 9543.

Guduru, R., Liang, P., Runowicz, C., Nair, M., Atluri, V., & Khizroev, S. (2013). Magnetoelectric nanoparticles to enable field-controlled high-specificity drug delivery to eradicate ovarian cancer cells. *Scientific Reports, 3*, 2953.

Guduru, R., Liang, P., Yousef, M., Horstmyer, J., & Khizroev, S. (2018). Mapping the Brain's electric fields with magnetoelectric nanoparticles. *Bioelectronic Medicine, 4*, 1–9.

Guo, Y. P., Kakimoto, K., & Ohsato, H. (2004). Phase transitional behavior and piezoelectric properties of (Na0.5K0.5)NbO3-LiNbO3 ceramics. *Applied Physics Letters, 85*, 4121–4123.

Guo, Y., Kakimoto, K.-I., & Ohsato, H. (2005). Na0.5K0.5)NbO3-LiTaO3 lead-free piezoelectric ceramics. *Materials Letters, 59*, 241–244.

Guo, Y. Y., Zhou, J. P., & Liu, P. (2010). Magnetoelectric characteristics around resonance frequency under magnetic field in Pb(Zr, Ti)O-3/Terfenol-D laminate composite. *Current Applied Physics, 10*, 1092–1095.

Guyomar, D., Guiffard, B., Belouadah, R., & Petit, L. (2008). Two-phase magnetoelectric nanopowder/polyurethane composites. *Journal of Applied Physics, 104*.

Guyomar, D., Matei, D., Guiffard, B., Le, Q., & Belouadah, R. (2009). Magnetoelectricity in polyurethane films loaded with different magnetic particles. *Materials Letters, 63*, 611–613.

Hadjikhani, A., Rodzinski, A., Wang, P., Nagesetti, A., Guduru, R., Liang, P., Runowicz, C., Shahbazmohamadi, S., & Khizroev, S. (2017). Biodistribution and clearance of magnetoelectric nanoparticles for nanomedical applications using energy dispersive spectroscopy. *Nanomedicine, 12*, 1801–1822.

Hilding, J., Grulke, E. A., George Zhang, Z., & Lockwood, F. (2003). Dispersion of carbon nanotubes in liquids. *Journal of Dispersion Science and Technology, 24*, 1–41.

Hollenstein, E., Davis, M., Damjanovic, D., & Setter, N. (2005). Piezoelectric properties of Li- and Ta-modified (K0.5Na0.5)NbO3 ceramics. *Applied Physics Letters, 87*(1–3), Article 182905.

Hornreich, R. (1969). The magnetoelectric effect: Some likely candidates. *Solid State Communications, 7*, 1081–1085.

Hosur, S., Sriramdas, R., Karan, S. K., Liu, N., Priya, S., & Kiani, M. (2021). A comprehensive study on magnetoelectric transducers for wireless power transfer using low-frequency magnetic fields. *IEEE Transactions on Biomedical Circuits and Systems, 15*, 1079–1092.

Hou, X., Zaks, T., Langer, R., & Dong, Y. (2021). Lipid nanoparticles for mRNA delivery. *Nature Reviews Materials, 6*, 1078–1094.

Hur, N., Park, S., Sharma, P. A., Ahn, J., Guha, S., & Cheong, S.-W. (2004). Electric polarization reversal and memory in a multiferroic material induced by magnetic fields. *Nature, 429*, 392–395.

Ideshita, T., Fukushima, Y., Honda, S., & Kusuda, T. (1978). *Detection of Walker oscillation by dynamic bubble collapse method in garnet material*. American Institute of Physics.

Ishibashi, Y., & Iwata, M. (1999). A theory of morphotropic phase boundary in solid-solution systems of perovskite-type oxide ferroelectrics. *Japanese Journal of Applied Physics, 38*, 800.

Islam, R. A., Bedekar, V., Poudyal, N., Liu, J. P., & Priya, S. (2008a). Magnetoelectric properties of core-shell particulate nanocomposites. *Journal of Applied Physics, 104*(1–4), Article 104111.

Islam, R. A., Jiang, J. C., Bai, F., Viehland, D., & Priya, S. (2007). Correlation between structural deformation and magnetoelectric response in (1-x) Pb(Zr0.52Ti0.48)O-3-xNiFe(1.9) Mn(0.1)O(4) particulate composites. *Applied Physics Letters, 91*(1–3), Article 162905.

Islam, R. A., Ni, Y., Khachaturyan, A. G., & Priya, S. (2008b). Giant magnetoelectric effect in sintered multilayered composite structures. *Journal of Applied Physics, 104*(1–5), Article 044103.

Islam, R. A., & Priya, S. (2006). Magnetoelectric properties of the lead-free cofired BaTiO3-(Ni0.8Zn0.2)Fe2O4 bilayer composite. *Applied Physics Letters, 89*(1–3), Article 152911.

Islam, R. A., & Priya, S. (2008a). Effect of piezoelectric grain size on magnetoelectric coefficient of Pb(Zr0.52Ti0.48)O-3-Ni0.8Zn0.2Fe2O4 particulate composites. *Journal of Materials Science, 43*, 3560–3568.

Islam, R. A., & Priya, S. (2008b). Large magnetoelectric coefficient in Co-fired Pb (Zr0.52Ti0.48)O-3-Pb (Zn1/3Nb2/3)O-3-Ni0.6Cu0.2Zn0.2Fe2O4 trilayer magnetoelectric composites. *Journal of Materials Science, 43*, 2072–2076.

Islam, R. A., & Priya, S. (2008c). Large magnetoelectric coefficient in Co-fired Pb (Zr(0.52) Ti(0.48))O(3)-Pb (Zn(1/3)Nb(2/3))O(3)-Ni(0.6)Cu(0.2)Zn(0.2)Fe(2)O(4) trilayer magnetoelectric composites. *Journal of Materials Science, 43*, 2072–2076.

Islam, R. A., & Priya, S. (2009). ME response of cofired trilayer magnetoelectric composites with partial texturing. *Journal of Materials Science, 44*, 5935–5938.

Jaffe, H. (1958). Piezoelectric ceramics. *Journal of the American Ceramic Society, 41*, 494–498.

Jang, J., & Park, C. B. (2022). Magnetoelectric dissociation of Alzheimer's β-amyloid aggregates. *Science Advances, 8*, Article eabn1675.

Jiles, D., Ostenson, J., Owen, C., & Chang, T. (1988). Barkhausen effect and discontinuous magnetostriction in Terfenol-D. *Journal of Applied Physics, 64*, 5417–5418.

Jin, J., Lu, S. G., Chanthad, C., Zhang, Q., Haque, M., & Wang, Q. (2011). Multiferroic polymer composites with greatly enhanced magnetoelectric effect under a low magnetic bias. *Advanced Materials, 23*, 3853–3858.

Kadomtseva, A. M., Zvezdin, A. K., Popov, Y. F., Pyatakov, A. P., & Vorob'ev, G. P. (2004). Space-time parity violation and magnetoelectric interactions in antiferromagnets. *Journal of Experimental and Theoretical Physics Letters, 79*, 571–581.

Kambale, R. C., Shaikh, P. A., Bhosale, C. H., Rajpure, K. Y., & Kolekar, Y. D. (2010). Studies on magnetic, dielectric and magnetoelectric behavior of (x) NiFe1.9Mn0.1O4 and (1-x) BaZr0.08Ti0.92O3 magnetoelectric composites. *Journal of Alloys and Compounds, 489*, 310–315.

Karan, S. K., Bera, R., Paria, S., Das, A. K., Maiti, S., Maitra, A., & Khatua, B. B. (2016). An approach to design highly durable piezoelectric nanogenerator based on self-poled PVDF/AlO-rGO flexible nanocomposite with high power density and energy conversion efficiency. *Advanced Energy Materials, 6*, Article 1601016.

Karan, S. K., Hosur, S., Kashani, Z., Leng, H., Vijay, A., Sriramdas, R., Wang, K., Poudel, B., Patterson, A. D., & Kiani, M. (2024). Magnetic field and ultrasound induced simultaneous wireless energy harvesting. *Energy and Environmental Science, 17*, 2129–2144.

Karan, S. K., Maiti, S., Agrawal, A. K., Das, A. K., Maitra, A., Paria, S., Bera, A., Bera, R., Halder, L., & Mishra, A. K. (2019). Designing high energy conversion efficient bio-inspired vitamin assisted single-structured based self-powered piezoelectric/wind/acoustic multi-energy harvester with remarkable power density. *Nano Energy, 59*, 169–183.

Karan, S. K., Mandal, D., & Khatua, B. B. (2015). Self-powered flexible Fe-doped RGO/PVDF nanocomposite: An excellent material for a piezoelectric energy harvester. *Nanoscale, 7*, 10655–10666.

Ke, W. L., Zhang, N., & Guo, Y. S. (2009). Magnetoelectric effect in laminate composites of Tb1-xDyxFe2-y and Cr doped BaTiO3. *Materials Research Innovations, 13*, 15–17.

Kepler, R., & Anderson, R. (1978). Piezoelectricity and pyroelectricity in polyvinylidene fluoride. *Journal of Applied Physics, 49*, 4490–4494.

Khanal, C. M. (2016). *Magnetoelectric nanoparticles: Paradigm shift in biomolecular diagnostics.*

Khizroev, S. (2019). Technobiology's enabler: The magnetoelectric nanoparticle. *Cold Spring Harbor Perspectives in Medicine, 9*, a034207.

Khizroev, S., & Liang, P. (2019). Engineering future medicines with magnetoelectric nanoparticles: Wirelessly controlled, targeted therapies. *IEEE Nanotechnology Magazine, 14*, 23–29.

Kim, D., Efe, I., Torlakcik, H., Terzopoulou, A., Veciana, A., Siringil, E., Mushtaq, F., Franco, C., Von Arx, D., & Sevim, S. (2022). Magnetoelectric effect in hydrogen harvesting: Magnetic field as a trigger of catalytic reactions. *Advanced Materials, 34*, Article 2110612.

Kimura, T., Goto, T., Shintani, H., Ishizaka, K., Arima, T.-H., & Tokura, Y. (2003). Magnetic control of ferroelectric polarization. *Nature, 426*, 55–58.

Kita, E., Siratori, K., & Tasaki, A. (1979). Electric shift in the anti-ferromagnetic resonance and the mechanism of the parallel magnetoelectric effect of CR2O3. *Journal of the Physical Society of Japan, 46*, 1033–1034.

Kita, E., Tasaki, A., & Siratori, K. (1979). Application of SQUID magnetometer to the measurement of magnetoelectric effect in CR2O3. *Japanese Journal of Applied Physics, 18*, 1361–1366.

Koner, S., Deshmukh, P., Ahlawat, A., Sagdeo, A., Singh, R., Karnal, A., & Satapathy, S. (2021). Effect of magnetic field on ferroelectric output voltage: A study on La0. 7Sr0. 3MnO3 (LSMO)/P (VDF-TrFE) flexible multiferroic nanocomposite films. *Journal of Materials Science: Materials in Electronics, 32*, 21780–21797.

Koner, S., Deshmukh, P., Khan, A. A., Ahlawat, A., Karnal, A., & Satapathy, S. (2020). Direction dependent strong magnetoelectric coupling in La0. 7Ba0. 3MnO3 embedded poly (vinylidenefluoride-co-trifluoroethylene) flexible nano-composite films at room temperature. *Scripta Materialia, 189*, 30–35.

Kozielski, K. L., Jahanshahi, A., Gilbert, H. B., Yu, Y., Erin, Ö., Francisco, D., Alosaimi, F., Temel, Y., & Sitti, M. (2021). Nonresonant powering of injectable nanoelectrodes enables wireless deep brain stimulation in freely moving mice. *Science Advances, 7*, Article eabc4189.

Kujawska, M., & Kaushik, A. (2023). Exploring magneto-electric nanoparticles (MENPs): A platform for implanted deep brain stimulation. *Neural Regeneration Research, 18*, 129.

Kulkarni, A., Meurisch, K., Teliban, I., Jahns, R., Strunskus, T., Piorra, A., Knöchel, R., & Faupel, F. (2014). Giant magnetoelectric effect at low frequencies in polymer-based thin film composites. *Applied Physics Letters, 104*.

Laletin, V. M., Paddubnaya, N., Srinivasan, G., De Vreugd, C. P., Bichurin, M. I., Petrov, V. M., & Filippov, D. A. (2005). Frequency and field dependence of magnetoelectric interactions in layered ferromagnetic transition metal-piezoelectric lead zirconate titanate. *Applied Physics Letters, 87*(1–3), Article 222507.

Lang, S. B., Zhu, W. Y., & Cross, L. E. (2006). Piezoelectric and pyroelectric properties of (K0.5Na0.5)(1-x)(Nb1-yTay)O-3 ceramics. *Ferroelectrics, 336*, 15–21.

Lasheras, A., Gutiérrez, J., Reis, S., Sousa, D., Silva, M., Martins, P., Lanceros-Mendez, S., Barandiarán, J., Shishkin, D., & Potapov, A. (2015). Energy harvesting device based on a metallic glass/PVDF magnetoelectric laminated composite. *Smart Materials and Structures, 24*, Article 065024.

Lather, S., Gupta, A., Dalal, J., Verma, V., Tripathi, R., & Ohlan, A. (2017). Effect of mechanical milling on structural, dielectric and magnetic properties of BaTiO3–Ni0. 5Co0. 5Fe2O4 multiferroic nanocomposites. *Ceramics International, 43*, 3246–3251.

Lee, M. H., Kim, D. J., Park, J. S., Kim, S. W., Song, T. K., Kim, M. H., Kim, W. J., Do, D., & Jeong, I. K. (2015). High-performance lead-free piezoceramics with high curie temperatures. *Advanced Materials, 27*, 6976–6982.

Legrand, J. (1989). Structure and ferroelectric properties of P (VDF-TrFE) copolymers. *Ferroelectrics, 91*, 303–317.

Leng, H., Wang, Y. U., Yan, Y., Karan, S. K., Wang, K., Li, X., Fanton, M., Fox, J. J., & Priya, S. (2023). Water quenched and acceptor-doped textured piezoelectric ceramics for off-resonance and on-resonance devices. *Small, 19*, Article 2204454.

Leng, H., Yan, Y., Li, X., Karan, S. K., Fanton, M., & Priya, S. (2023). High-power piezoelectric behavior of acceptor-doped< 001> and< 111> textured piezoelectric ceramics. *Journal of Materials Chemistry C, 11*, 2229–2240.

Li, Y. J., Chen, X. M., Lin, Y. Q., & Tang, Y. H. (2006). Magnetoelectric effect of Ni0.8Zn0.2-Fe2O4/Sr0.5Ba0.5Nb2O6 composites. *Journal of the European Ceramic Society, 26*, 2839–2844.

Lin, D., Xiao, D., Zhu, J., & Yu, P. (2006). Piezoelectric and ferroelectric properties of [Bi0.5 (Na1−x−yKxLiy) 0.5] TiO3 lead-free piezoelectric ceramics. *Applied Physics Letters, 88*.

Liu, K., Ma, W., Ben, N., Zhu, H., Xu, W., Gao, H., Samart, C., Kongparakul, S., Zhang, H., & Tan, H. (2023). Superior large-signal piezoelectric coefficient in textured NBT-based piezoceramics by template-induced structure and composition modulation. *ACS Applied Materials and Interfaces, 15*(22), 24595–24605.

Liu, G., Nan, C. W., Cai, N., & Lin, Y. H. (2004). Dependence of giant magnetoelectric effect on interfacial bonding for multiferroic laminated composites of rare-earth-iron alloys and lead-zirconate-titanate. *Journal of Applied Physics, 95*, 2660–2664.

Liu, M., Obi, O., Lou, J., Chen, Y., Cai, Z., Stoute, S., Espanol, M., Lew, M., Situ, X., & Ziemer, K. S. (2009). Giant electric field tuning of magnetic properties in multiferroic ferrite/ferroelectric heterostructures. *Advanced Functional Materials, 19*, 1826–1831.

Liu, R., Zhao, Y., Huang, R., Zhao, Y., & Zhou, H. (2010). Multiferroic ferrite/perovskite oxide core/shell nanostructures. *Journal of Materials Chemistry, 20*, 10665–10670.

Lo, C. Y., Choy, S. H., Or, S. W., & Chan, H. L. W. (2010). Magnetoelectric effect in lead-free BNKLBT ceramic/terfenol-D continue fiber composite laminates. *Journal of Applied Physics, 107*(1–4), Article 093907.

Long, Y., Qiu, J., He, X., Chang, Q., Hu, Z., & Liu, H. (2017). Flexible magnetoelectric transducer with high magnetic field sensitivity based on Metglas/poly (vinylidene fluoride) heterostructures. *AIP Advances, 7*.

Lottermoser, T., Lonkai, T., Amann, U., Hohlwein, D., Ihringer, J., & Fiebig, M. (2004). Magnetic phase control by an electric field. *Nature, 430*, 541–544.

Lu, S., Fang, Z., Furman, E., Wang, Y., Zhang, Q., Mudryk, Y., Gschneidner, K., Pecharsky, V., & Nan, C. (2010). Thermally mediated multiferroic composites for the magnetoelectric materials. *Applied Physics Letters, 96*.

Lu, S., Jin, J., Zhou, X., Fang, Z., Wang, Q., & Zhang, Q. (2011). Large magnetoelectric coupling coefficient in poly (vinylidene fluoride-hexafluoropropylene)/Metglas laminates. *Journal of Applied Physics, 110*.

Lu, M.-C., Mei, L., Jeong, D.-Y., Xiang, J., Xie, H., & Zhang, Q. (2015). Enhancing the magnetoelectric response of Terfenol-D/polyvinylidene fluoride/Terfenol-D laminates by exploiting the shear mode effect. *Applied Physics Letters, 106*.

Ma, J., Shi, Z., & Nan, C.-W. (2007). Magnetoelectric properties of composites of single Pb(Zr,Ti)O-3 rods and Terfenol-D/epoxy with a single-period of 1-3-type structure. *Advanced Materials, 19*, 2571–2573.

Martins, P., Gonçalves, R., Lanceros-Mendez, S., Lasheras, A., Gutiérrez, J., & Barandiarán, J. (2014). Effect of filler dispersion and dispersion method on the piezoelectric and magnetoelectric response of CoFe2O4/P (VDF-TrFE) nanocomposites. *Applied Surface Science, 313*, 215–219.

Martins, P., Kolen'ko, Y. V., Rivas, J., & Lanceros-Mendez, S. (2015). Tailored magnetic and magnetoelectric responses of polymer-based composites. *ACS Applied Materials and Interfaces, 7*, 15017–15022.

Martins, P., & Lanceros-Méndez, S. (2013). Polymer-based magnetoelectric materials. *Advanced Functional Materials, 23*, 3371–3385.

Martins, P., Lasheras, A., Gutiérrez, J., Barandiarán, J. M., Orue, I., & Lanceros-Méndez, S. (2011). Optimizing piezoelectric and magnetoelectric responses on CoFe2O4/P (VDF-TrFE) nanocomposites. *Journal of Physics D: Applied Physics, 44*, Article 495303.

Martins, P., Moya, X., Phillips, L., Kar-Narayan, S., Mathur, N., & Lanceros-Mendez, S. (2011). Linear anhysteretic direct magnetoelectric effect in Ni0. 5Zn0. 5Fe2O4/poly (vinylidene fluoride-trifluoroethylene) 0-3 nanocomposites. *Journal of Physics D: Applied Physics, 44*, Article 482001.

Mayeen, A., Kala, M., Jayalakshmy, M., Thomas, S., Philip, J., Rouxel, D., Bhowmik, R., & Kalarikkal, N. (2019). Flexible and self-standing nickel ferrite–PVDF-TrFE cast films: Promising candidates for high-end magnetoelectric applications. *Dalton Transactions, 48*, 16961–16973.

Messing, G. L., Trolier-Mckinstry, S., Sabolsky, E., Duran, C., Kwon, S., Brahmaroutu, B., Park, P., Yilmaz, H., Rehrig, P., & Eitel, K. (2004). Templated grain growth of textured piezoelectric ceramics. *Critical Reviews in Solid State and Materials Sciences, 29*, 45–96.

Mhambi, S., Fisher, D., Tchokonte, M. B. T., & Dube, A. (2021). Permeation challenges of drugs for treatment of neurological tuberculosis and HIV and the application of magneto-electric nanoparticle drug delivery systems. *Pharmaceutics, 13*, 1479.

Miclea, C., Tanasoiu, C., Amarande, L., Miclea, C. F., Plavitu, C., Cioangher, M., Trupina, L., Miclea, C. T., Tanasoiu, T., Susu, M., Voicu, I., Malczanek, V., Ivanov, A., & David, C. (2010). Magnetoelectric properties of multiferroic cobalt ferrite and soft piezoelectric PZT particulate composites. *Journal of Optoelectronics and Advanced Materials, 12*, 272–276.

Mushtaq, F., Chen, X., Torlakcik, H., Steuer, C., Hoop, M., Siringil, E. C., Marti, X., Limburg, G., Stipp, P., & Nelson, B. J. (2019). Magnetoelectrically driven catalytic degradation of organics. *Advanced Materials, 31*, Article 1901378.

Mushtaq, F., Chen, X.-Z., Veciana, A., Hoop, M., Nelson, B. J., & Pané, S. (2022). Magnetoelectric reduction of chromium (VI) to chromium (III). *Applied Materials Today, 26*, Article 101339.

Mushtaq, F., Torlakcik, H., Vallmajo-Martin, Q., Siringil, E. C., Zhang, J., Röhrig, C., Shen, Y., Yu, Y., Chen, X.-Z., & Müller, R. (2019). Magnetoelectric 3D scaffolds for enhanced bone cell proliferation. *Applied Materials Today, 16*, 290–300.

Nan, C.-W. (1994). Magnetoelectric effect in composites of piezoelectric and piezomagnetic phases. *Physical Review B, 50*, 6082.

Nan, C.-W., Bichurin, M. I., Dong, S., Viehland, D., & Srinivasan, G. (2008). Multiferroic magnetoelectric composites: Historical perspective, status, and future directions. *Journal of Applied Physics, 103*(1–35), Article 031101.

Nan, C.-W., Cai, N., Shi, Z., Zhai, J., Liu, G., & Lin, Y. (2005). Large magnetoelectric response in multiferroic polymer-based composites. *Physical Review B, 71*, Article 014102.

Nan, C.-W., Liu, L., Cai, N., Zhai, J., Ye, Y., Lin, Y., Dong, L., & Xiong, C. (2002). A three-phase magnetoelectric composite of piezoelectric ceramics, rare-earth iron alloys, and polymer. *Applied Physics Letters, 81*, 3831–3833.

Nersessian, N., Or, S. W., & Carman, G. P. (2004). Magnetoelectric behavior of Terfenol-D composite and lead zirconate titanate ceramic laminates. *IEEE Transactions on Magnetics, 40*, 2646–2648.

Nguyen, T., Fleming, Y., Bender, P., Grysan, P., Valle, N., El Adib, B., Adjeroud, N., Arl, D., Emo, M., & Ghanbaja, J. (2021). Low-temperature growth of AlN films on magnetostrictive foils for high-magnetoelectric-response thin-film composites. *ACS Applied Materials and Interfaces, 13*, 30874–30884.

Nie, J. W., Xu, G. Y., Yang, Y., & Cheng, C. W. (2009). Strong magnetoelectric coupling in CoFe2O4-BaTiO3 composites prepared by molten-salt synthesis method. *Materials Chemistry and Physics, 115*, 400–403.

O'dell, T. H. (1963). The field invariants in a magnetoelectric medium. *Philadelphia Magazine, 8*, 411–418.

Palneedi, H., Annapureddy, V., Priya, S., & Ryu, J. (2016). Status and perspectives of multiferroic magnetoelectric composite materials and applications. *Actuators*. MDPI, 9.

Palneedi, H., Maurya, D., Geng, L. D., Song, H.-C., Hwang, G.-T., Peddigari, M., Annapureddy, V., Song, K., Oh, Y. S., & Yang, S.-C. (2018). Enhanced self-biased magnetoelectric coupling in laser-annealed Pb (Zr, Ti) O3 thick film deposited on Ni foil. *ACS Applied Materials and Interfaces, 10*, 11018–11025.

Palneedi, H., Maurya, D., Kim, G. Y., Annapureddy, V., Noh, M. S., Kang, C. Y., Kim, J. W., Choi, J. J., Choi, S. Y., & Chung, S. Y. (2017). Unleashing the full potential of magnetoelectric coupling in film heterostructures. *Advanced Materials, 29*, Article 1605688.

Pandey, P., Ghimire, G., Garcia, J., Rubfiaro, A., Wang, X., Tomitaka, A., Nair, M., Kaushik, A., & He, J. (2020). Single-entity approach to investigate surface charge enhancement in magnetoelectric nanoparticles induced by AC magnetic field stimulation. *ACS Sensors, 6*, 340–347.

Park, C. S., Ahn, C. W., Ryu, J., Yoon, W. H., Park, D. S., Kim, H. E., & Priya, S. (2009). Design and characterization of broadband magnetoelectric sensor. *Journal of Applied Physics, 105*(1–6), Article 094111.

Park, C. S., Cho, K. H., Arat, M. A., Evey, J., & Priya, S. (2010). High magnetic field sensitivity in Pb(Zr,Ti)O-3-Pb(Mg1/3Nb2/3)O-3 single crystal/Terfenol-D/Metglas magnetoelectric laminate composites. *Journal of Applied Physics, 107*(1–4), Article 094109.

Park, C.-S., Ryu, J., Choi, J.-J., Park, D.-S., Ahn, C.-W., & Priya, S. (2009). Giant magnetoelectric coefficient in 3–2 nanocomposite thick films. *Japanese Journal of Applied Physics, 48*, Article 080204.

Park, S.-E., & Shrout, T. R. (1997). Ultrahigh strain and piezoelectric behavior in relaxor based ferroelectric single crystals. *Journal of Applied Physics, 82*, 1804–1811.

Pradhan, D. K., Kumari, S., & Rack, P. D. (2020). Magnetoelectric composites: Applications, coupling mechanisms, and future directions. *Nanomaterials, 10*, 2072.

Priya, S., Islam, R., Dong, S., & Viehland, D. (2007). Recent advancements in magnetoelectric particulate and laminate composites. *Journal of Electroceramics, 19*, 149–166.

Priya, S., Ryu, J., Park, C.-S., Oliver, J., Choi, J.-J., & Park, D.-S. (2009). Piezoelectric and magnetoelectric thick films for fabricating power sources in wireless sensor nodes. *Sensors, 9*, 6362–6384.

Priya, S., Yang, S., Maurya, D., & Yan, Y. (2015). *Recent advances in piezoelectric and magnetoelectric materials phenomena*. Woodhead Publishing Series in Electronic and Optical Materials.

Purusothaman, Y., Leng, H., Nanda, A., Levine, I., & Priya, S. (2023). Textured lead-free piezoelectric ceramics for flexible energy Harvesters. *ACS Applied Materials and Interfaces, 15*, 6584–6593.

Pyun, J. (2007). Nanocomposite materials from functional polymers and magnetic colloids. *Polymer Reviews, 47*, 231–263.

Rahaman, M. N. (2017). *Ceramic processing and sintering*. CRC Press.

Rao, B. N., Kaviraj, P., Vaibavi, S., Kumar, A., Bajpai, S. K., & Arockiarajan, A. (2017). Investigation of magnetoelectric properties and biocompatibility of CoFe2O4-BaTiO3 core-shell nanoparticles for biomedical applications. *Journal of Applied Physics, 122*.

Reis, S., Silva, M., Castro, N., Correia, V., Martins, P., Lasheras, A., Gutierrez, J., Barandiarán, J., Rocha, J., & Lanceros-Mendez, S. (2016). Characterization of Metglas/poly (vinylidene fluoride)/Metglas magnetoelectric laminates for AC/DC magnetic sensor applications. *Materials and Design, 92*, 906–910.

Ren, S. Q., Weng, L. Q., Song, S. H., Li, F., Wan, J. G., & Zeng, M. (2005). BaTiO3/CoFe2O4 particulate composites with large high frequency magnetoelectric response. *Journal of Materials Science, 40*, 4375–4378.

Rodzinski, A., Guduru, R., Liang, P., Hadjikhani, A., Stewart, T., Stimphil, E., Runowicz, C., Cote, R., Altman, N., & Datar, R. (2016). Targeted and controlled anticancer drug delivery and release with magnetoelectric nanoparticles. *Scientific Reports, 6*, Article 20867.

Rondinone, A. J., Samia, A. C., & Zhang, Z. J. (1999). Superparamagnetic relaxation and magnetic anisotropy energy distribution in CoFe2O4 spinel ferrite nanocrystallites. *The Journal of Physical Chemistry B, 103*, 6876–6880.

Ruette, B., Zvyagin, S., Pyatakov, A. P., Bush, A., Li, J., Belotelov, V., Zvezdin, A. K., & Viehland, D. (2004). Magnetic-field-induced phase transition in BiFeO 3 observed by high-field electron spin resonance: Cycloidal to homogeneous spin order. *Physical Review B, 69*, Article 064114.

Ryu, J., Carazo, A. V., Uchino, K., & Kim, H. E. (2001a). Magnetoelectric properties in piezoelectric and magnetostrictive laminate composites. *Japanese Journal of Applied Physics, 40*, 4948–4951.

Ryu, J., Carazo, A. V., Uchino, K., & Kim, H. E. (2001b). Piezoelectric and magnetoelectric properties of Lead Zirconate Titanate/Ni-Ferrite particulate composites. *Journal of Electroceramics, 7*, 17–24.

Ryu, H., Murugavel, P., Lee, J., Chae, S., Noh, T., Oh, Y. S., Kim, H. J., Kim, K. H., Jang, J. H., & Kim, M. (2006). Magnetoelectric effects of nanoparticulate Pb (Zr0. 52Ti0. 48) O3–NiFe2O4 composite films. *Applied Physics Letters, 89*.

Ryu, J., Priya, S., Uchino, K., & Kim, H. E. (2002). Magnetoelectric effect in composites of magnetostrictive and piezoelectric materials. *Journal of Electroceramics, 8*, 107–119.

Sahoo, R., Rasna, M., Lisjak, D., Mertelj, A., & Dhara, S. (2015). Magnetodielectric and magnetoviscosity response of a ferromagnetic liquid crystal at low magnetic fields. *Applied Physics Letters, 106*.

Sasmal, A., & Arockiarajan, A. (2023). Recent progress in flexible magnetoelectric composites and devices for Next generation Wearable electronics. *Nano Energy*, Article 108733.

Sasmal, A., Sen, S., Chelvane, J. A., & Arockiarajan, A. (2023). PVDF based flexible magnetoelectric composites for capacitive energy storage, hybrid mechanical energy harvesting and self-powered magnetic field detection. *Polymer, 126141*.

Schofield, D., & Brown, R. F. (1957). Barium titanate compositions for transducer applications. *Canadian Journal of Physics, 35*, 594–607.

Scott, J. F. (2012). Applications of magnetoelectrics. *Journal of Materials Chemistry, 22*, 4567–4574.

Sencadas, V., Lanceros-Mendez, S., & Mano, J. (2004). Behaviour of the ferroelectric phase transition of P (VDF/TrFE)(75/25) with increasing deformation. *Ferroelectrics, 304*, 23–26.
Shi, Z., Ma, J., & Nan, C.-W. (2008). A new magnetoelectric resonance mode in bilayer structure composite of PZT layer and Terfenol-D/epoxy layer. *Journal of Electroceramics, 21*, 390–393.
Shi, Z., Nan, C.-W., Liu, J., Filippov, D., & Bichurin, M. (2004). Influence of mechanical boundary conditions and microstructural features on magnetoelectric behavior in a three-phase multiferroic particulate composite. *Physical Review B, 70*, Article 134417.
Si, S. K., Karan, S. K., Paria, S., Maitra, A., Das, A. K., Bera, R., Bera, A., Halder, L., & Khatua, B. B. (2018). A strategy to develop an efficient piezoelectric nanogenerator through ZTO assisted γ-phase nucleation of PVDF in ZTO/PVDF nanocomposite for harvesting bio-mechanical energy and energy storage application. *Materials Chemistry and Physics, 213*, 525–537.
Singh, R. (1996). *Structure–property correlations in some magnetically ordered Aurivillis phases.* Osmania University Hyderabad.
Song, H., Kim, D.-I., Abbasi, S. A., Gharamaleki, N. L., Kim, E., Jin, C., Kim, S., Hwang, J., Kim, J.-Y., & Chen, X.-Z. (2022). Multi-target cell therapy using a magnetoelectric microscale biorobot for targeted delivery and selective differentiation of SH-SY5Y cells via magnetically driven cell stamping. *Materials Horizons, 9*, 3031–3038.
Song, H., Listyawan, M. A., & Ryu, J. (2022). Core–shell magnetoelectric nanoparticles: Materials, synthesis, magnetoelectricity, and applications. *Actuators*. MDPI, 380.
Srinivas, A., Kim, D.-W., Hong, K. S., & Suryanarayana, S. (2004). Study of magnetic and magnetoelectric measurements in bismuth iron titanate ceramic—Bi8Fe4Ti3O24. *Materials Research Bulletin, 39*, 55–61.
Srinivasan, G., Devreugd, C. P., Flattery, C. S., Laletsin, V. M., & Paddubnaya, N. (2004). Magnetoelectric interactions in hot-pressed nickel zinc ferrite and lead zirconante titanate composites. *Applied Physics Letters, 85*, 2550–2552.
Srinivasan, G., Hayes, R., & Bichurin, M. I. (2003a). Low frequency and microwave magnetoelectric effects in thick film heterostructures of lithium zinc ferrite and lead zirconate titanate. *Solid State Communications, 128*, 261–266.
Srinivasan, G., Laletsin, V. M., Hayes, R., Puddubnaya, N., Rasmussen, E. T., & Fekel, D. J. (2002). Giant magnetoelectric effects in layered composites of nickel zinc ferrite and lead zirconate titanate. *Solid State Communications, 124*, 373–378.
Srinivasan, G., Rasmussen, E. T., Gallegos, J., Srinivasan, R., Bokhan, Y. I., & Laletin, V. M. (2001). Magnetoelectric bilayer and multilayer structures of magnetostrictive and piezoelectric oxides. *Physical Review B: Condensed Matter, 64*(1–6), Article 214408.
Srinivasan, G., Rasmussen, E. T., & Hayes, R. (2003b). Magnetoelectric effects in ferrite-lead zirconate titanate layered composites: The influence of zinc substitution in ferrites. *Physical Review B: Condensed Matter, 67*(1–10), Article 014418.
Stimphil, E., Nagesetti, A., Guduru, R., Stewart, T., Rodzinski, A., Liang, P., & Khizroev, S. (2017). Physics considerations in targeted anticancer drug delivery by magnetoelectric nanoparticles. *Applied Physics Reviews, 4*.
Suryanarayana, S. (1994). Magnetoelectric interaction phenomena in materials. *Bulletin of Materials Science, 17*, 1259–1270.
Tanasă, E., Andronescu, E., Cernea, M., & Oprea, O. C. (2019). Fe 3 O 4/BaTiO 3 composites with core-shell structure. *UPB Scientific Bulletin, Series B: Chemistry and Materials Science, 81*, 171–180.
Trolier-Mckinstry, S., Zhang, S., Bell, A. J., & Tan, X. (2018). High-performance piezoelectric crystals, ceramics, and films. *Annual Review of Materials Research, 48*, 191–217.
van Aken, B. B., Palstra, T. T., Filippetti, A., & Spaldin, N. A. (2004). The origin of ferroelectricity in magnetoelectric YMnO3. *Nature Materials, 3*, 164–170.

van den Boomgaard, J., Terrell, D., Born, R., & Giller, H. (1974). An in situ grown eutectic magnetoelectric composite material: Part I composition and unidirectional solidification. *Journal of Materials Science, 9*, 1705–1709.

Van Suchtelen, J. (1972). Product properties: A new application of composite materials. *Philips Research Reports, 27*, 28–37.

Wan, J. G., Li, Z. Y., Wang, Y., Zeng, M., Wang, G. H., & Liu, J. M. (2005b). Strong flexural resonant magnetoelectric effect in Terfenol-D/epoxy-Pb(Zr,Ti)O-3 bilayer. *Applied Physics Letters, 86*(1–3), Article 202504.

Wan, J. G., Liu, J. M., Chand, H. L. W., Choy, C. L., Wang, G. H., & Nan, C. W. (2003). Giant magnetoelectric effect of a hybrid of magnetostrictive and piezoelectric composites. *Journal of Applied Physics, 93*, 9916–9919.

Wan, H., Shen, R.-F., & Wu, X.-Z. (2005a). *A theoretical study on symmetrical magnetostrictive/piezoelectric laminated composite.*

Wang, Y. J., Leung, C. M., Or, S. W., Zhao, X. Y., & Luo, H. S. (2009a). Resonance converse magnetoelectric effect in a dual-mode bilayered composite of Pb(Mg1/3Nb2/3)O-3-PbTiO3 and Tb0.3Dy0.7Fe1.92. *Journal of Alloys and Compounds, 487*, 450–452.

Wang, Y. J., Leung, C. M., Wang, F. F., Or, S. W., Zhao, X. Y., & Luo, H. S. (2009b). Dual-mode magnetoelectric effect in laminate composite of Terfenol-D alloy and PMN-PT transformer with double output ports. *Journal of Physics D Applied Physics, 42*(1–5), Article 135414.

Wang, J., Neaton, J., Zheng, H., Nagarajan, V., Ogale, S., Liu, B., Viehland, D., Vaithyanathan, V., Schlom, D., & Waghmare, U. (2003). Epitaxial BiFeO3 multiferroic thin film heterostructures. *Science, 299*, 1719–1722.

Wang, Y. J., Or, S. W., Chan, H. L. W., Zhao, X. Y., & Luo, H. S. (2008). Enhanced magnetoelectric effect in longitudinal-transverse mode Terfenol-D/Pb(Mg1/3Nb2/3)O-3-PbTiO3 laminate composites with optimal crystal cut. *Journal of Applied Physics, 103*(1–4), Article 124511.

Wang, R. P., Xie, R. J., Hanada, K., Matsusaki, K., Bando, H., Sekiya, T., & Itoh, M. (2006). Phase diagram of the (Na0.5K0.5)NbO3-ATiO(3) solid solution. *Ferroelectrics, 336*, 39–46.

Wang, Z., Yan, L., Yang, Y., Li, J.-F., Das, J., Geiler, A. L., Yang, A., Chen, Y., Harris, V. G., & Viehland, D. (2011). Magnetoelectric effect in crystallographically textured BaTiO3 films deposited on ferromagnetic metallic glass foils. *Journal of Applied Physics, 109.*

Wang, H., Zhang, Q., Cross, L. E., & Sykes, A. O. (1993). Piezoelectric, dielectric, and elastic properties of poly (vinylidene fluoride/trifluoroethylene). *Journal of Applied Physics, 74*, 3394–3398.

Wang, P., Zhang, E., Toledo, D., Smith, I. T., Navarrete, B., Furman, N., Hernandez, A. F., Telusma, M., Mcdaniel, D., & Liang, P. (2020). Colossal magnetoelectric effect in core–shell magnetoelectric nanoparticles. *Nano Letters, 20*, 5765–5772.

Wu, T., Chung, T. K., Chang, C. M., Keller, S., & Carman, G. (2009). Influence of electric voltage bias on converse magnetoelectric coefficient in piezofiber/Metglas bilayer laminate composites. *Journal of Applied Physics, 106*(1–4), Article 054114.

Wu, J., Xiao, D., & Zhu, J. (2015). Potassium–sodium niobate lead-free piezoelectric materials: Past, present, and future of phase boundaries. *Chemical Reviews, 115*, 2559–2595.

Xing, J., Zheng, T., Wu, J., Xiao, D., & Zhu, J. (2018). Progress on the doping and phase boundary design of potassium–sodium niobate lead-free ceramics. *Journal of Advanced Dielectrics, 8*, Article 1830003.

Yan, Y., Cho, K.-H., Maurya, D., Kumar, A., Kalinin, S., Khachaturyan, A., & Priya, S. (2013). Giant energy density in [001]-textured Pb (Mg1/3Nb2/3) O3-PbZrO3-PbTiO3 piezoelectric ceramics. *Applied Physics Letters, 102.*

Yan, Y., Geng, L. D., Liu, H., Leng, H., Li, X., Wang, Y. U., & Priya, S. (2022). Near-ideal electromechanical coupling in textured piezoelectric ceramics. *Nature Communications, 13*, 3565.

Yan, Y., Zhou, J. E., Maurya, D., Wang, Y. U., & Priya, S. (2016). Giant piezoelectric voltage coefficient in grain-oriented modified PbTiO3 material. *Nature Communications, 7*, Article 13089.

Yan, Y., Zhou, Y., & Priya, S. (2013). Giant self-biased magnetoelectric coupling in co-fired textured layered composites. *Applied Physics Letters, 102.*

Yang, S. C., Ahn, C. W., Cho, K. H., & Priya, S. (2011). Self-bias response of lead-free (1−x) [0.948 K0. 5Na0. 5NbO3−0.052 LiSbO3]−xNi0. 8Zn0. 2Fe2O4−Nickel magnetoelectric laminate composites. *Journal of the American Ceramic Society, 94*, 3889−3899.

Yang, S.-C., Kumar, A., Petkov, V., & Priya, S. (2013). Room-temperature magnetoelectric coupling in single-phase BaTiO3-BiFeO3 system. *Journal of Applied Physics, 113.*

Yang, S., Li, J., Liu, Y., Wang, M., Qiao, L., Gao, X., Chang, Y., Du, H., Xu, Z., & Zhang, S. (2021). Textured ferroelectric ceramics with high electromechanical coupling factors over a broad temperature range. *Nature Communications, 12*, 1414.

Yang, Y., Wang, J., Li, J.-F., Viehland, D., & Nain, A. (2012). *Nanoparticles deposition at specific sites using aligned fiber networks.*

Yu, Z., Chu, Z., Yang, J., Asl, M. J. P., Yuan, X., Yu, Y., Nie, G., Qi, H., & Dong, S. (2020). A magneto-mechano-electric (MME) energy harvester based on rectangular cymbal structure. *Sensors and Actuators A: Physical, 316*, Article 112400.

Yuan, Y., Zhang, S. R., Zhou, X. H., & Liu, J. S. (2006). Phase transition and temperature dependences of electrical properties of [Bi0.5Na1-x-yKxLiy)(0.5)]TiO3 ceramics. *Japanese Journal of Applied Physics, 45*, 831−834.

Yue, K., Guduru, R., Hong, J., Liang, P., Nair, M., & Khizroev, S. (2012). *Magneto-electric nanoparticles for non-invasive brain stimulation.*

Zang, G. Z., Wang, J. F., Chen, H. C., Su, W. B., Wang, C. M., Qi, P., Ming, B. Q., Du, J., & Zheng, L. M. (2006). Perovskite (Na0.5K0.5)(1-x)(LiSb)(x)Nb1-xO3 lead-free piezoceramics. *Applied Physics Letters, 88*(1−3), Article 212908.

Zeng, M., Or, S. W., & Chan, H. L. W. (2010). Large magnetoelectric effect from mechanically mediated magnetic field-induced strain effect in Ni−Mn−Ga single crystal and piezoelectric effect in PVDF polymer. *Journal of Alloys and Compounds, 490*, L5−L8.

Zhai, J. Y., Cai, N., Shi, Z., Lin, Y. H., & Nan, C. W. (2004). Magnetic-dielectric properties of NiFe2O4/PZT particulate composites. *Journal of Physics D Applied Physics, 37*, 823−827.

Zhai, J. Y., Dong, S. X., Xing, Z. P., Li, J. F., & Viehland, D. (2006). Giant magnetoelectric effect in Metglas/polyvinylidene-fluoride laminates. *Applied Physics Letters, 89*(1−3), Article 083507.

Zhai, J., Xing, Z., Dong, S., Li, J., & Viehland, D. (2008). Magnetoelectric laminate composites: An overview. *Journal of the American Ceramic Society, 91*, 351−358.

Zhang, Y., Chen, S., Xiao, Z., Liu, X., Wu, C., Wu, K., Liu, A., Wei, D., Sun, J., & Zhou, L. (2021). Magnetoelectric nanoparticles incorporated biomimetic matrix for wireless electrical stimulation and nerve regeneration. *Advanced Healthcare Materials, 10*, Article 2100695.

Zhang, J., Dai, J., So, L., Sun, C., Lo, C., Or, S. W., & Chan, H. (2009). The effect of magnetic nanoparticles on the morphology, ferroelectric, and magnetoelectric behaviors of CFO/P (VDF-TrFE) 0−3 nanocomposites. *Journal of Applied Physics, 105.*

Zhang, N., Fan, J. F., Rong, X. F., Cao, H. X., & Wei, J. J. (2007). Magnetoelectric effect in laminate composites of Tb1-xDyxFe2-y and Fe-doped BaTiO3. *Journal of Applied Physics, 101*(1−3), Article 063907.

Zhang, N., Ke, W., Schneider, T., & Srinivasan, G. (2006a). Dependence of the magnetoelectric coupling in NZFO-PZT laminate composites on ferrite compactness. *Journal of Physics: Condensed Matter, 18*, 11013−11019.

Zhang, S., Lee, S. M., Kim, D. H., Lee, H. Y., & Shrout, T. R. (2007). Electromechanical properties of PMN−PZT piezoelectric single crystals near morphotropic phase boundary compositions. *Journal of the American Ceramic Society, 90*, 3859−3862.

Zhang, S., Lee, S.-M., Kim, D.-H., Lee, H.-Y., & Shrout, T. R. (2008). Characterization of Mn-modified Pb (Mg1/3Nb2/3) O3–PbZrO3–PbTiO3 single crystals for high power broad bandwidth transducers. *Applied Physics Letters, 93*.

Zhang, S., & Li, F. (2012). High performance ferroelectric relaxor-PbTiO3 single crystals: Status and perspective. *Journal of Applied Physics, 111*, 2.

Zhang, H., Mu, X., Yang, S., Zhang, C., Xu, J., Wang, X., Li, Q., Cao, D., & Li, S. (2020). Evaluation of Metglas/polyvinylidene fluoride magnetoelectric bilayer composites for flexible in-plane resonant magnetic sensors. *Journal of Physics D: Applied Physics, 54*, Article 095003.

Zhang, S. J., Shrout, T. R., Nagata, H., Hiruma, Y., & Takenaka, T. (2007c). Piezoelectric properties in (K0.5Bi0.5)TiO3-(Na0.5Bi0.5)TiO3-BaTiO3 lead-free ceramics. *IEEE Transactions on Ultrasonics, Ferroelectrics, and Frequency Control, 54*, 910–917.

Zhang, S. J., Xia, R., Shrout, T. R., Zang, G. Z., & Wang, J. F. (2006b). Piezoelectric properties in perovskite 0.948(K0.5Na0.5)NbO3-0.052LiSbO(3) lead-free ceramics. *Journal of Applied Physics, 100*(1–6), Article 104108.

Zhang, X., Yao, X., Zhou, J.-P., & Yang, Z. (2018). Comprehensive investigation on direct and converse magnetoelectric effects in longitudinally magnetized and polarized laminate composites by equivalent circuit and experiments. *Journal of Materials Science: Materials in Electronics, 29*, 17706–17713.

Zhao, S., Li, G., Ding, A., Wang, T., & Yin, Q. (2006). Ferroelectric and piezoelectric properties of (Na, K) 0.5 Bi0.5 TiO3 lead free ceramics. *Journal of Physics D Applied Physics, 39*, 2277–2281.

Zhao, C., Wu, H., Li, F., Cai, Y., Zhang, Y., Song, D., Wu, J., Lyu, X., Yin, J., & Xiao, D. (2018). Practical high piezoelectricity in barium titanate ceramics utilizing multiphase convergence with broad structural flexibility. *Journal of the American Chemical Society, 140*, 15252–15260.

Zhou, Y., Zhang, J. C., Li, L., Su, Y. L., Cheng, J. R., & Cao, S. X. (2009). Multiferroic composites in nano-microscale with non-solid solution by Co-ferrite and (K0.5Na0.5)NbO3-based ferroelectric matrix. *Journal of Alloys and Compounds, 484*, 535–539.

Zhu, Q., Xie, Y., Zhang, J., Liu, Y., Zhan, Q., Miao, H., & Xie, S. (2014). Multiferroic CoFe 2 O 4–BiFeO 3 core–shell nanofibers and their nanoscale magnetoelectric coupling. *Journal of Materials Research, 29*, 657–664.

Zubkov, A. (1978a). On the magnetoelectric effect in segnetomagnetic composites based on piezoelectric and magnetostrictive materials. *Elektrichestvo, 10*, 77.

Zubkov, A. S. (1978b). On the magnetoelectric effect in segnetomagnetic composites based on piezoelectric and magnetostrictive materials. *Elektrichestvo, 10*, 77–82.

CHAPTER

6

Multiferroic nanostructures

Peng Zhou[1], *Ying Liu*[1], *Gopalan Srinivasan*[1], *J.F. Li*[2], *and Dwight Viehland*[2]

[1] Oakland University, Rochester, MI, United States; [2] Virginia Tech, Blacksburg, VA, United States

6.1 Introduction

Multiferroic materials with simultaneous ferroelectricity and magnetism provide a pathway to achieving strong magnetoelectric coupling with efficient voltage control of magnetism and compact and power efficient electric field–tunable magnetic devices (Eerenstein et al., 2006; Fetisov & Srinivasana, 2006; Martin et al., 2008; Nan et al., 2008; Sun & Srinivasan, 2012; Vaz et al., 2010). From the viewpoint of structure, nanoscale multiferroic ME composites can essentially be divided into several types: bulk magnetic on ferro/piezoelectric slab, magnetic film on ferro/piezoelectric slab, 2-2, 0-3, and 1-3 type multiferroics nanostructures and one-dimensional multiferroic composites. The bulk magnetic on ferro/piezoelectric slab multiferroic heterostructure will not be discussed in this chapter.

Compared with many bulk ME composites, multiferroic ME nanostructures have some unique advantages. For instance, the ferromagnetic or ferro/piezoelectric phase could be manipulated at the nanoscale. ME coupling mechanisms were explored in a new scale level. Furthermore, in bulk ME composites, the two constituent phases are usually combined by cosintering or adhesive bonding method, which causes loss of the ME coupling strength at the interface. Nevertheless, in ME composite nanostructures, for example, thin films, the different phases are bonded at the atomic level, reducing ME coupling loss at the interface significantly.

Recently, the advances in thin-film growth techniques (Zheng et al., 2004a) and improved theoretical simulations (Duan et al., 2006; Hill, 2000; Lu et al., 2007; Nan et al., 2005; Zhang et al., 2007) accelerated the

https://doi.org/10.1016/B978-0-443-18606-6.00010-1

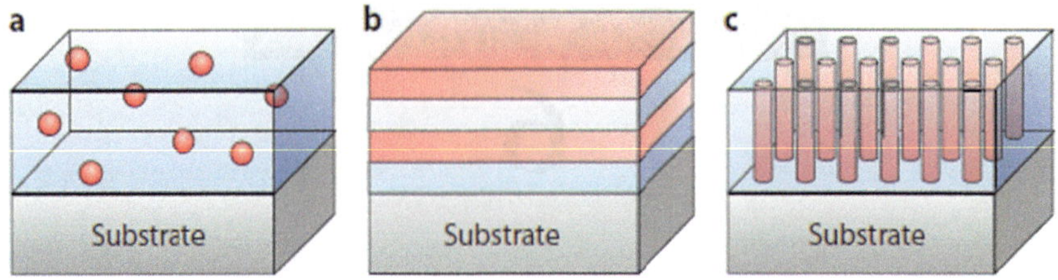

FIGURE 6.1 Schematic illustration of three kinds of ME composite nanostructures with common connectivity schemes: (a) 0–3 particulate nanocomposite films with magnetic particles (0) embedded in a ferroelectric film matrix (3); (b) 2-2 horizontal heterostructure with alternating ferroelectric (2) and magnetic (2) layers, or simply a ferroelectric (or magnetic) thin film grown on a magnetic (or ferroelectric) substrate; and (c) 1-3 vertical heterostructure with one-phase nanopillars (1) embedded in a matrix of another phase (3).

renaissance of multiferroics nanostructures. Novel nanostructures can be achieved by advanced growth techniques and novel design of multiferroic nanostructures were also predicted by theory. Fig. 6.1 shows three basic types of multiferroic nanostructures, 0-3, 2-2, and 1-3. The notation 0-3, 2-2, 1-3, etc. describe the structure arrangement of two-phase composites (Newnham et al., 1978), each number implying the dimension of each phase. For example, a 0-3 composite represents the single-phase particles (represented by 0) embedded in a matrix of another phase (represented by 3), as shown in Fig. 6.1a. In general, 2-2 horizontal heterostructures represent a multilayer thin-film structure, each single phase layer can be denoted by 2, see Fig. 6.1b, and 1-3 vertical heterostructures show the one-dimensional single phase nanopillars, denoted by 1, embedded in the matrix of second phase. These nanostructures, including magnetic thin film/ferroelectric slab structure, are discussed later.

6.1.1 Magnetic/piezoelectric heterostructures

Multiferroic heterostructures with magnetic thin-film deposited on ferro/piezoelectric slab were widely studied [our papers]. Compared with bulk multiferroic composites, interface of ferromagnetic and ferroelectric with atomic level bonding introduced notably less ME coupling loss in thin film/slab heterostructure. Giant ME coupling coefficients were achieved in these multiferroics heterostructure because there exist no sample clamping effect reducing strain/stress mediated ME coupling. Several different deposition methods, such as, spin spray (Liu et al., 2008, 2009a, 2009b) and sputtering (Lou et al., 2009; Zhou et al., 2012; Liu et al., 2012; Nan et al.) were studied to fabricate magnetic thin-film on piezo/ferroelectric slab.

6.1.1.1 *Spin spray deposited ferrite film on piezo/ferroelectric substrate*

The spin-spray deposition process, which was invented in 1983, is a novel technique to directly plate crystalline spinel ferrite films with different compositions from an aqueous solution at a temperature less than 90°C. Conventional ferrite film preparation methods such as sputtering, molecular beam epitaxy, PLD, etc., require high temperatures well above 600°C; however, spin-spray can produce high quality polycrystalline ferrite films at a low temperature between 25 and 90°C. Several chemical reactions are involved in the ferrite spin-spray process, which consist of (a) adsorption of Fe^{2+} and other ions Mn + M = Fe, Co, Ni, Zn, etc. on the substrate surface mediated by OH groups, (b) oxidization process expressed by $Fe^{2+} \rightarrow Fe^{3+} + e$, and (c) ferrite film formation accompanying hydrolytic dissociation. Spin-spray-deposited ferrites provide great opportunities for low-temperature synthesized multiferroic composite materials on ferroelectric substrates, deposited magnetic ferrite thin films on piezo/ferroelectric substrates.

$Ni_{0.23}Fe_{2.77}O_4$(NFO)/Pb(Zr,Ti)O_3 (PZT) multiferroic composites were synthesized by spin-spray deposition of NFO film onto PZT ceramic substrate at 90°C (Liu et al., 2008). Strong interface adhesion between NFO and PZT was observed, which was verified by high-resolution transmission electron microscopy indicating excellent wetting between the NFO and PZT, see Fig. 6.2 and by the strong magnetoelectric coupling in the NFO/PZT multiferroic composite showing an electric field induced remnant magnetization change of 10%. This strong interface adhesion and low-temperature spin-spray synthesis of multiferroic materials provides

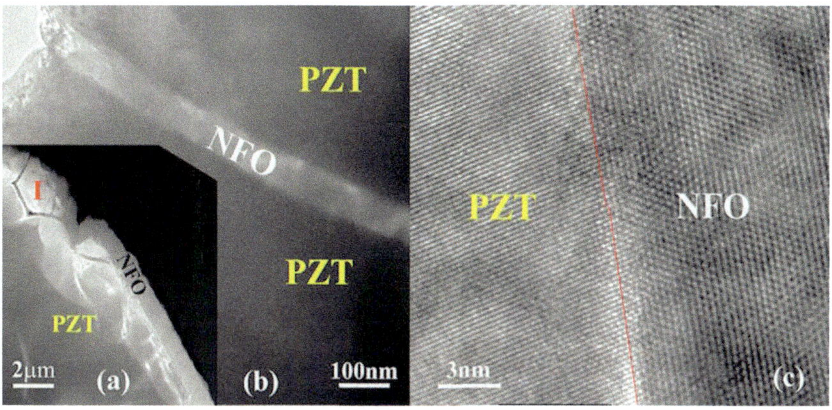

FIGURE 6.2 (Color online) A cross-sectional HAADF image of NFO/PZT (a), HRTEM of Ni ferrite growing in PZT crack (b), and HRTEM ferrite and PZT interface images (c).

an alternative route for novel integrated multiferroic materials and devices.

The interface between NFO and PZT was investigated by high angle annular dark field (HAADF) scanning transmission electron microscopy image (Z-contrast), as shown in Fig. 6.2a. A uniform ferrite film with a thickness of 1 μm can be identified on the PZT phase, no voids were observed at the NFO/PZT interface. It is interesting to note that the NFO ferrite phase was growing into the small surface crack of the PZT phase with a crack width of 100 nm area (I), as shown in Fig. 6.2b, which was proved by the energy dispersive X-ray results. This indicates the excellent wetting between the NFO ferrite phase and the PZT phase during the spin-spray deposition process, which leads to a tight interface and a strong adhesion between the two phases. A high-resolution transmission electron microscopy (HRTEM) image of the NFO/PZT interface was shown in Fig. 6.2c, which shows a clearly tight interface between the ferrite NFO and ferroelectric PZT phases without obvious lattice matching, even though the lattice parameter of the spinel NFO and the perovskite PZT phases are well matched with the NFO lattice parameter being very close to two times of PZT.

The in-plane M-H curves were measured by applying external electric field of 0 and 1.3 MV/m across the thickness direction of PZT substrate, as shown in Fig. 6.3. An obvious upward shift of M-H curve was observed at low external magnetic field and the remnant magnetization was increased by 10% when applying electric field of 1.3 MV/m. In addition, we found both M-H curves reached the same saturation magnetization at high

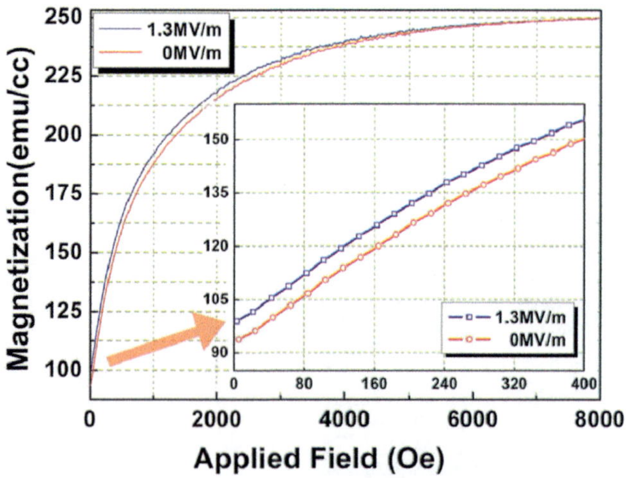

FIGURE 6.3 (Color online) in-plane M-H curves of spin-sprayed Ni ferrite film on PZT when applying 0 and 1.3 MV/m electric fields through the PZT substrate.

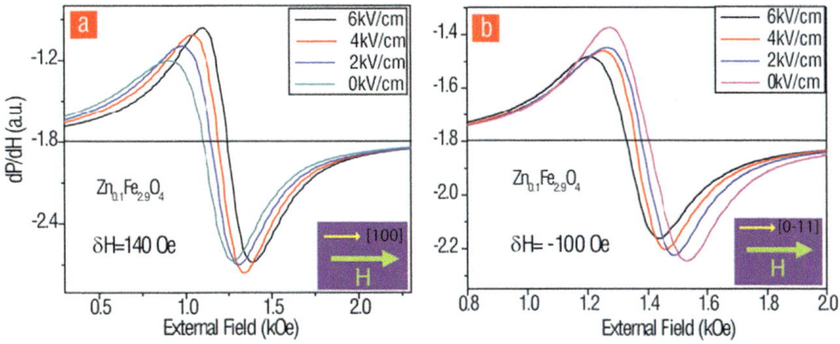

FIGURE 6.4 FMR absorption spectra of multiferroic composites ZFO/PMN−PT under different electric fields while applying an external magnetic field along (a) [1 0 0] and (b) [01−1] directions.

external magnetic field, which indicated ME coupling reduced due to the saturated magnetostriction at high applied magnetic field.

Strong magnetoelectric (ME) interaction was demonstrated at both dc and microwave frequencies in a novel $Zn_{0.1}Fe_{2.9}O_4$/PMN−PT (lead magnesium niobate−lead titanate) multiferroic heterostructure, which was prepared by spin-spray depositing a $Zn_{0.1}Fe_{2.9}O_4$ film on a single-crystal PMN−PT substrate at a low temperature of 90°C (Liu, Obi, et al., 2009). A large electric field−induced ferromagnetic resonance field shift up to 140 Oe was observed, corresponding to an ME coefficient of 23 Oe cm/kV, see Fig. 6.4. In addition, a large electrostatic field tuning of the magnetic hysteresis loops was observed with a large squareness ratio change of 18%. The spin-spray-deposited ferrite/piezoelectric multiferroic heterostructures exhibiting strong ME interactions at both dc and microwave frequencies provide great opportunities for novel electrostatically tunable microwave magnetic devices synthesized at a low temperature.

Multiferroic heterostructures of Fe_3O_4/PZT (lead zirconium titanate), Fe_3O_4/PMN-PT (lead magnesium niobate-lead titanate), and Fe_3O_4/PZN-PT (lead zinc niobate-lead titanate) are prepared by spin-spray depositing Fe_3O_4 ferrite film on ferroelectric PZT, PMN-PT, and PZN-PT substrates at a low temperature of 90°C (Liu et al., 2009b). Strong magnetoelectric coupling (ME) and giant microwave tunability are demonstrated by an electrostatic field induced magnetic anisotropic field change in these heterostructures. Microwave ME interactions and magnetic tunabilities of Fe_3O_4/PMN-PT and Fe_3O_4/PZT composites were demonstrated by electrostatic-field-induced FMR field changes at room temperature, as shown in Fig. 6.5. Here, a microwave cavity operating at TE102 mode at X-band (9.5 GHz) was used to perform FMR measurements of the ferrite/ferroelectric multiferroic composites. The external bias magnetic field was

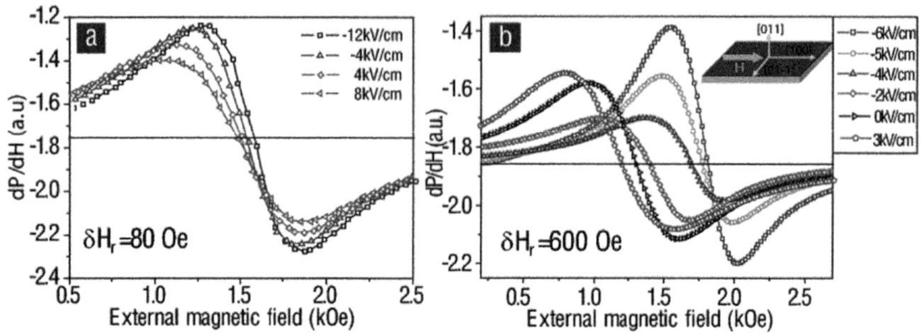

FIGURE 6.5 Ferromagnetic resonance absorption spectra of Fe3O4/PZT (a) and Fe3O4/PMN-PT (b) while applying different electric fields.

applied in the Fe_3O_4 film plane along the PMN-PT [100] or PMN-PT [01−1] directions, with the microwave RF field in-plane and perpendicular to the DC bias field. Clearly, giant microwave ME interaction was observed in Fe_3O_4/PMN-PT, which resulted in a high tunable FMR field range from 1185 Oe to 1786 Oe, or dHr = 600 Oe, when the electric fields across the PMN-PT thickness were changed from 3 kV/cm to 6 kV/cm and when the external magnetic field was applied along the [100] (d_{31}) direction of PMN-PT. This corresponded to a large microwave ME coupling coefficient of 67 Oe cm kV^{-1}. In comparison, the tunable FMR field range in Fe_3O_4/PZT was also measured as 80 Oe, which is much smaller than that of Fe3O4/PMN-PT due to the weak piezoelectric coefficient of PZT compared to PMN-PT.

In order to achieve a larger ratio of tunable FMR field over FMR linewidth, Fe_3O_4 ferrite films were also spin-spray deposited onto a (011)-cut PZN-PT single crystal substrate, which has a higher piezoelectric coefficient and lower loss tangent compared to PMN-PT. The microwave and static ME coupling of the Fe_3O_4/PZN-PT heterostructures were investigated, and the results were shown in Fig. 6.6. A record-high electric field

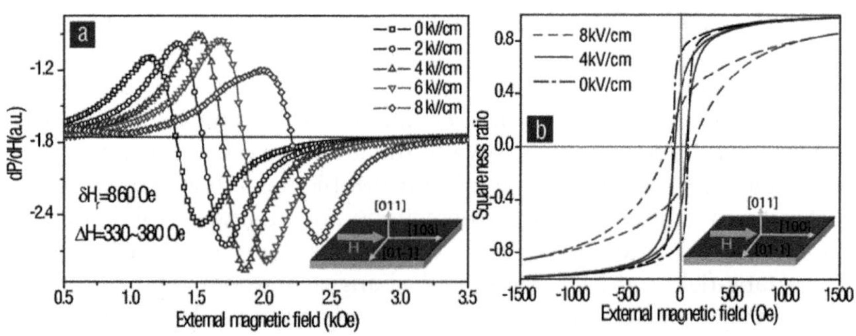

FIGURE 6.6 Ferromagnetic resonance absorption spectra of Fe3O4/PZN-PT (a) and magnetic hysteresis loops of Fe3O4/PZN-PT (b) while applying different electric fields.

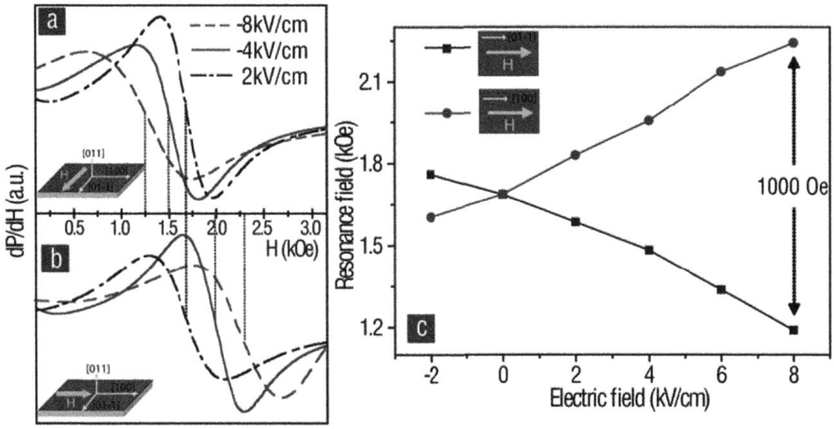

FIGURE 6.7 Ferromagnetic resonance absorption spectra shifts while the external mangnetic field along tensional (a) and compressive (b) directions of PMN-PT. The cooperation of magnetic orientation and electric field generated resonance field change shift is up to 1000 Oe (c).

induced FMR shift up to dHr = 860 Oe was obtained when the external magnetic fields were applied along in-plane [100] direction (d_{31}: −3000 pC/N) and the external electric fields were applied across the thickness of PZN-PT substrate starting from 0 kV/cm to 8 kV/cm, corresponding to an ME coefficient of 108 Oe cm/kV. Clearly, the FMR field tunable range and ME coefficient in Fe_3O_4/PZN-PT are larger than those in Fe_3O_4/PMNPT, which are 600 Oe and 67 Oe cm/kV, respectively. In addition, the FMR linewidth obviously was reduced from $480<\Delta H<620$ Oe in Fe_3O_4/PMN-PT to $330<\Delta H<380$ Oe in Fe_3O_4/PZN-PT, leading to a significantly enhanced ratio of tunable FMR field over FMR linewidth of 2.5. The static ME coupling was shown in Fig. 6.7b, which displayed a significant squareness ratio change of 40% when external magnetic field was along [100] direction, and 8 kv/cm external electric field was applied across the thickness of PZN-PT, which was also higher than that for the Fe_3O_4/PMN-PT.

The Fe_3O_4/PMN-PT can be shifted upward or downward, depending on whether the applied magnetic field is parallel or perpendicular to the effective magnetic field induced by the electric field across the (011)-cut PMN-PT, which is a natural result of the anisotropic piezoelectric coefficient of the (011)-cut PMN-PT. Considering the anisotropic in-plane piezoelectric coefficients of the (011)-cut PMN-PT, which can generate orthogonal tensile and compressive stresses under an electric field, a new concept is introduced that the FMR frequency could be tuned up or down to achieve a larger tunable range by applying the external magnetic field parallel to [100] (d_{31}) or [01-1] (d_{32}) direction of the (011)-cut PMN-PT.

This idea was demonstrated in the Fe_3O_4/PMN-PT multiferroic composite. As shown in Fig. 6.7, an external magnetic field applied along the [01-1] direction led to a reduced FMR field of 1200 Oe (Fig. 6.7a). When the external field was parallel to the [100] direction, the FMR fields were shifted up to 2200 Oe (Fig. 6.7b). Each of them had the same resonance field of 1680 Oe at zero electric field due to the in-plane isotropic magnetic property of the Fe3O4 film. The total resonance field shift was dHr = 1000 Oe when the external magnetic field was applied parallel to [100] and [01-1] directions (Fig. 6.7c). This constitutes a simple but effective approach for achieving twice the tunable FMR frequency range, leading to new opportunities of magnetostatically tunable magnetic device design with large effective magnetic field tunability.

6.1.1.2 Sputtering deposited magnetic alloy film on piezo/ferroelectric substrate

The spin spray method can deposit magnetic thin films at low temperature (<90°C) with good adhesion on a piezo/ferroelectric substrate. Giant ME coupling effects were also observed in these multiferroic heterostructures. Nevertheless, the magnetic material type fabricated by spin spray is limited, only spinel ferrite can be done through this method. In order to discover the magnetoelectric coupling in alloy magnetic thin film/piezoelectric substrate, a sputtering technique is widely used to deposit alloy magnetic thin film onto ferroelectric substrates.

Multiferroic heterostructures of FeGaB/PZN-PT were made by direct deposition of 100 nm FeGaB films using a magnetron sputtering method onto PZN-PT single crystal slabs (Lou et al., 2009). The field-sweep FMR spectra of the FeGaB/PZN-PT multiferroic heterostructure under different electric fields are shown in Fig. 6.8. The external magnetic field was applied parallel to the in-plane [011] direction of the PZN-PT single crystal and was swept between 0 and 1400 Oe. Clearly, the FMR spectra changed dramatically with the external electric field, indicating a giant ME coupling in the FeGaB/PZN-PT heterostructure. The theoretical calculation results were demonstrated in Fig. 6.9. The FMR field (the external magnetic field at which the power absorption spectrum dI/dH crosses zero, i.e., maximum power absorption) shifted from 1172 to 898 Oe on changing the external electric field from (2–5.8 kV/cm). A sudden change of the FMR field from 898 to 425 Oe occurred at electric fields between 5.8 and 6 kV/cm. Such a dramatic change of the FMR field within a very narrow electric field range is closely related to the phase transition from the rhombohedral to the tetragonal phase in the PZN-PT single crystal, which has been reported previously (Zhong et al., 2007). The FMR field barely changed at electric fields higher than 6 kV/cm, which was also consistent with the strain–electric field relation of the PZN-PT single crystal (Zhong et al., 2007). It is important to emphasize

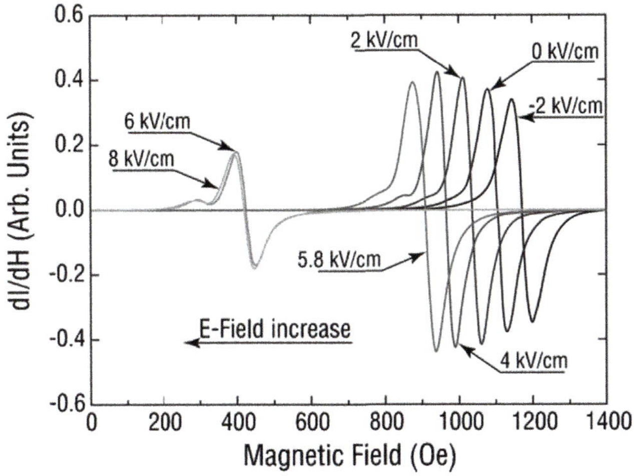

FIGURE 6.8 Electric field dependence of the in-plane field-sweep FMR spectra of the FeGaB/PZN-PT multiferroic heterostructure measured at ca. 9.6 GHz.

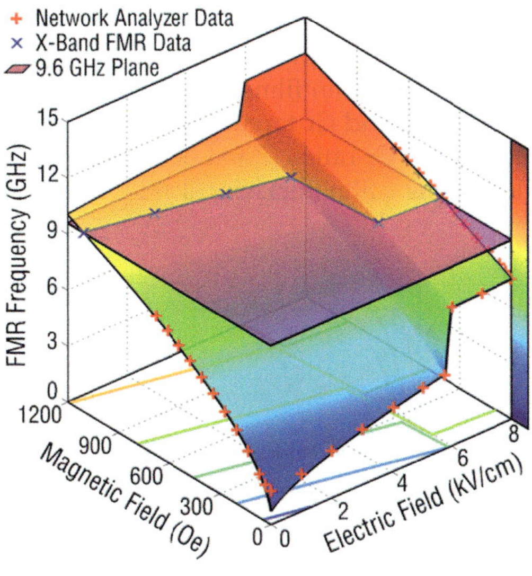

FIGURE 6.9 3D plot showing the FMR frequency of the FeGaB/PZN-PT multiferroic heterostructure as a function of external magnetic and electric bias fields.

that the overall electric field–induced FMR field change (also the total electrostatically induced magnetic anisotropy) is ca. 750 Oe, which is the largest value that has ever been reported, corresponding to a mean ME coupling coefficient 94 Oe cm/kV. Specifically, the ME coupling coefficient amounts to 2365 Oe cm/kV at the phase transition region of the PZN-PT singe crystal. Meanwhile, the FMR linewidth of the FeGaB film, which is a critical parameter for microwave magnetic materials, stays at

50~5 Oe under different electric fields. The ratio between the total tunable magnetic field and the FMR linewidth is as high as ca. 1500%, indicating an excellent figure of merit for tunable microwave devices.

The magnetic hysteresis loops of the FeGaB/PZN-PT multiferroic heterostructure were measured by a vibrating sample magnetometer (VSM) along the [100] direction of the PZN-PT single crystal under different electric fields, which also confirmed the large electric field–induced effective magnetic anisotropy changes, as shown in Fig. 6.10. Clearly, the electric field–induced magnetic anisotropy field is the largest at electric fields of 6 and 8 kV/cm, and lowest at 0 kV/cm. The in-plane magnetic anisotropy fields at different electric fields can be estimated from the hysteresis loops, being ca. 20 Oe at 0 kV/cm and ca. 700 Oe at 6 kV/cm electric field. This leads to an ME coefficient of ca. 113 Oe cm/kV at dc, which is close to the value obtained from the FMR measurements at microwave frequencies.

To achieve a high strain mediated ME coupling effect in magnetic thin film/ferroelectric slab, lots of efforts have been made in developing highly magnetostrictive RF magnetic materials. Recently, in magnetic/ferroelectric heterostructures, a low magnetization approach has been studied to generate large electric field–induced effective magnetic fields. NiCr films having low magnetization and relatively large magnetostriction were discovered. NiCr alloy magnetic thin films were deposited by the DC magnetron cosputtering with Ni and Cr targets at room temperature on Si substrates with different Ni/Cr ratios (Zhou et al., 2012). All films were deposited for 600 s, leading to a film thickness of ~50 nm NiCr

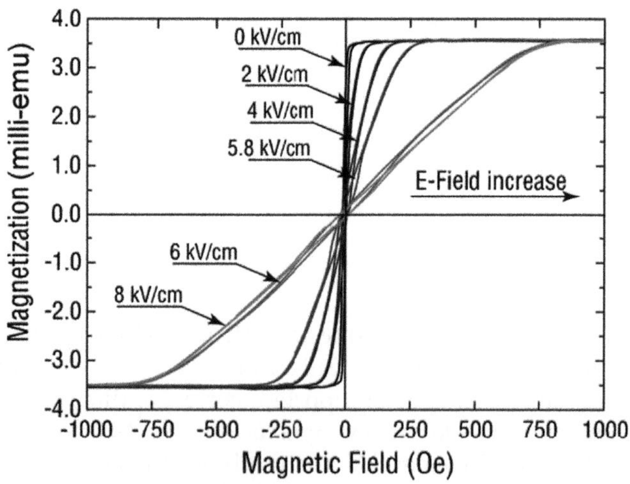

FIGURE 6.10 Magnetic hysteresis loops of the FeGaB/PZN-PT multiferroic heterostructure under different external electric fields measured by VSM.

compositions were measured by X-ray fluorescence (XRF) system. The static NiCr hysteresis loops with different components were measured by vibrating sample magnetometer (VSM). Microwave ME interaction was investigated by a broadband ferromagnetic resonance spectrometer. Static electric field was applied across the NiCr/PZT and NiCr/PZN-PT samples thickness direction.

For achieving electric field tuning of the magnetic properties, strong converse magnetoelectric coupling and large electric field tunable ferromagnetic resonance (FMR) bandwidths are achieved in layered NiCr/lead zirconate titanate (PZT) and NiCr/lead zinc niobate lead titanate (PZN-PT) multiferroic heterostructures. A large electric field induced effective magnetic field of 260 Oe for NiCr/PZT and 756 Oe for NiCr/PZN-PT was observed, corresponding to a giant magnetoelectric coupling coefficient of 13 Oe cm/kV in NiCr/PZT and 75.6 Oe cm/kV in NiCr/PZN-PT multiferroic heterostructures, see Figs. 6.11 and 6.12. A high

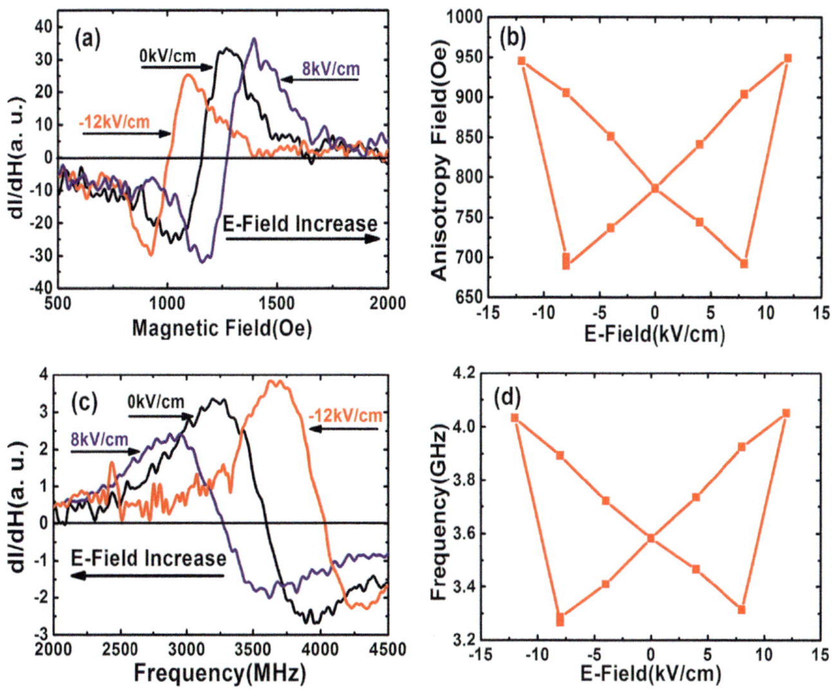

FIGURE 6.11 (a) Electric field dependence of the in-plane magnetic field sweep FMR spectra of NiCr/PZT multiferroic heterostructures measured at 6.85 GHz. (b) Butterfly plot of anisotropy magnetic field as a function of applied electric field from 8 to 12 kV/cm. (c) Electric field dependence of the in-plane frequency sweep FMR spectra of NiCr/PZT multiferroic heterostructure measured at 50 Oe. (d) Butterfly plot of resonance frequency as a function of applied electric field from −8–12 kV/cm.

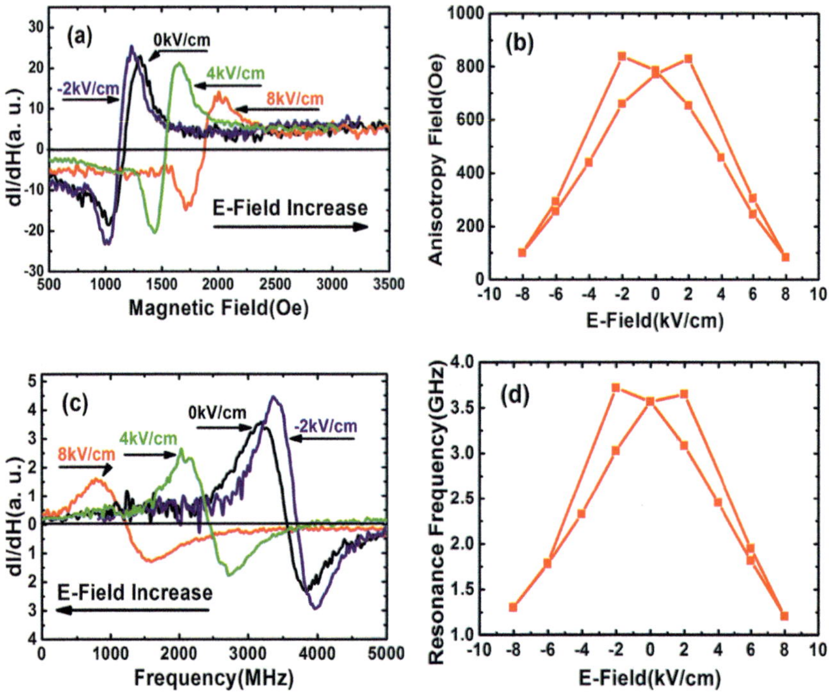

FIGURE 6.12 (a) Electric field dependence of the in-plane magnetic field sweep FMR spectra of NiCr/PZN-PT multiferroic heterostructures measured at 6.85 GHz. (b) Butterfly plot of anisotropy magnetic field as a function of applied electric field from −2–8 kV/cm. (c) Electric field dependence of the in-plane frequency sweep FMR spectra of NiCr/PZN-PT multiferroic heterostructure measured at 50 Oe. (d) Butterfly plot of resonance frequency as a function of applied electric field from −2–8 kV/cm.

voltage tunable FMR frequency range was observed, with f_{max}/f_{min} being 124% and 325% for NiCr/PZT and NiCr/PZN-PT. The strong converse magnetoelectric coupling of NiCr/PZT and NiCr/PZN-PT heterostructures provides great opportunities for electric field tunable RF magnetic devices.

Terfenol-D ($Tb_{0.7}Dy_{0.3}Fe_2$) exhibits the highest room temperature magnetostriction value and has been widely used as actuator and sensor in either bulk or thin film forms (Liu et al., 2012). Based on our previous model, the ME coefficient is proportional to the product of magnetostriction coefficient (k) and piezoelectric coefficient (d), which would enable a giant E-field-induced magnetic anisotropy once incorporating with ferroelectric materials. Terfenol-D will be great candidate of magnetic thin film to realize giant ME coupling coefficient. Amorphous and crystalline Terfenol-D films were grown on single crystal (011) cut PZN-PT substrates that possess giant in-plane piezoelectric coefficients

($d_{31} = -3000$ pC/N [100]; $d_{32} = 1100$ pC/N [01-1]). Amorphous and crystalline Terfenol-D films were deposited on Pt-coated Si substrates and PZN-PT slabs, magnetron sputtering under a variety of temperatures. The base pressure of chamber was 10^{-8} Torr, while the deposition pressure was set at 3 mTorr. Accounting for the relative sputtering yields of two independent targets $Tb_{0.3}Dy_{0.7}$ and Fe, we prepared Terfenol-D films by cosputtering with an estimated composition of $Tb_{0.3}Dy_{0.7}Fe_2$ and thickness of approximately 100 nm.

Strain-engineered E-field tuning of magnetic properties of the Terfenol-D/PZN-PT film-substrate system was examined by E-field-induced changes in magnetization process as shown in Fig. 6.13. For both amorphous (Fig. 6.13a) and crystalline (Fig. 6.13c) or intermediate (Fig. 6.13b) Terfenol-D films, applying an E-field through the thickness direction of PZN-PT, in-plane magnetization process along [100] direction becomes increasingly more difficult, indicating a magnetic anisotropy was produced perpendicular to in-plane [100] orientation. This phenomenon can be interpreted as strong magnetoelastic and piezoelectric effects. A dc field

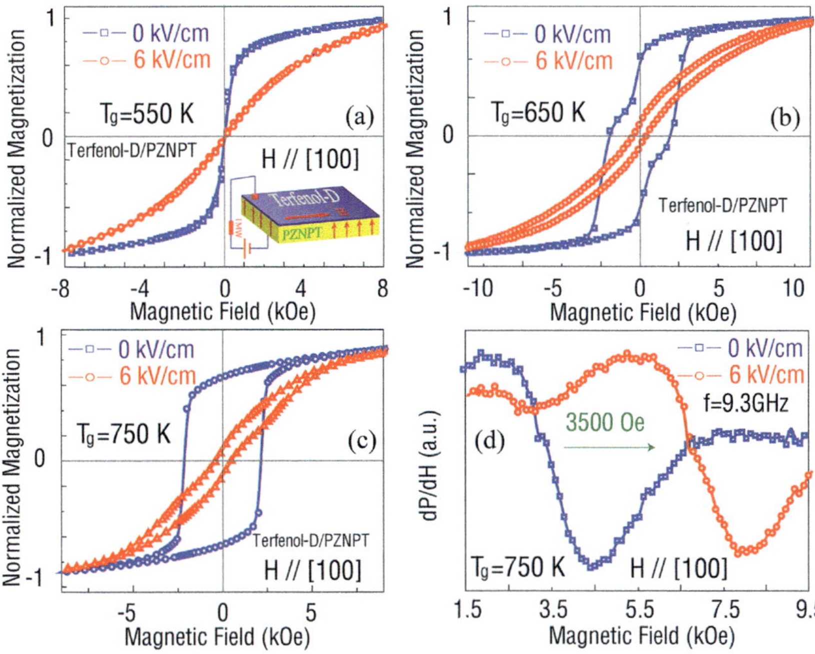

FIGURE 6.13 E-field dependence of in-plane magnetic hysteresis loops of Terfenol-D/PZN-PT deposited at 550 K (a), 650 K(b), and 750 K (c). Inset in (a) is the measurement configuration: E-field is applied through the thickness direction and magnetic field is applied along in-plane [100] direction. (d) FMR spectra of polycrystalline Terfenol-D/PZN-PT at E = 0 kV/cm (blue) and E = 6 kV/cm (red).

applied across the PZN-PT substrate results in a compressive stress along the in-plane [100] direction caused by the inverse piezoelectric effect. Once this mechanical strain coherently transfers to the Terfenol-D film, the magnetic moment undergoes a rotation away from the [100] direction due to the magnetoelastic effect. Given the complicated nature of the magnetization process involving domain wall motion, domain rotation, and nonlinear behaviors, E-field-induced magnetic hysteresis loop change is not sufficient to quantitatively determine this magnetic anisotropy which is essential for fundamental and technologic study. According to the E-field dependence of FMR spectrum plotted in Fig. 6.13d for crystalline Terfenol-D/PZN-PT, the resonance field Hr is shifted upward by 3500 Oe as applying an electric field of 6 kV/m, indicating a record-high E-field-induced magnetic anisotropy of 3500 Oe is achieved, corresponding to a magnetoelectric coefficient of 580 Oe cm/kV.

Besides strain-/stress-mediated ME coupling, Tianxiang et al. revealed that spin polarized interfacial charge effect can also play an important role in magnetic thin film/ferroelectric slab multiferroic heterostructures (Nan et al.). The coexistence of strain and surface charge–mediated magnetoelectric coupling in an ultrathin (1 nm) $Ni_{79}Fe_{21}$ and (011) lead magnesium niobate-lead titanate (PMN-PT) heterostructure was demonstrated. Voltage-controlled ferromagnetic resonance of the $Ni_{79}Fe_{21}$/(011) PMN-PT heterostructures was utilized for quantitatively studying its magnetoelectric behavior. A combination of strain-mediated butterfly curve and surface charge–mediated hysteretic curve was observed with an effective magnetic field change of 375 Oe and a nonvolatile effective magnetic field change at 0 voltage of 104 Oe. For a control sample, a 5 nm Cu layer was inserted into NiFe and PMN-PT, a standard butterfly curve induced by pure strain-mediated ME coupling was observed. By subtracting the change of effective magnetic field induced by strain effect, a pure charge-induced effective magnetic field of 201 Oe was achieved, and the change of effective magnetic field shows a clear response to the polarization of ferroelectric PMN-PT.

As for NiFe/Cu/PMN-PT, the inserted Cu layer played a role as an isolation layer. In that case, no screening charge exists in the NiFe/PMN–PT interface. Thus, pure strain–mediated magnetoelectric coupling was controlled by strain and stress transferred through the Cu thin layer induced by the electric field on PMN-PT due to the piezoelectric effect. As expected, a standard "butter-fly" curve of the FMR effective magnetic field as the function of electric field was observed shown in Fig. 6,14b, having a maximum H_{eff} tunability of 202 Oe. In FMR field sweeping mode, the magnetic field was applied along the in-plane [0-11] direction.

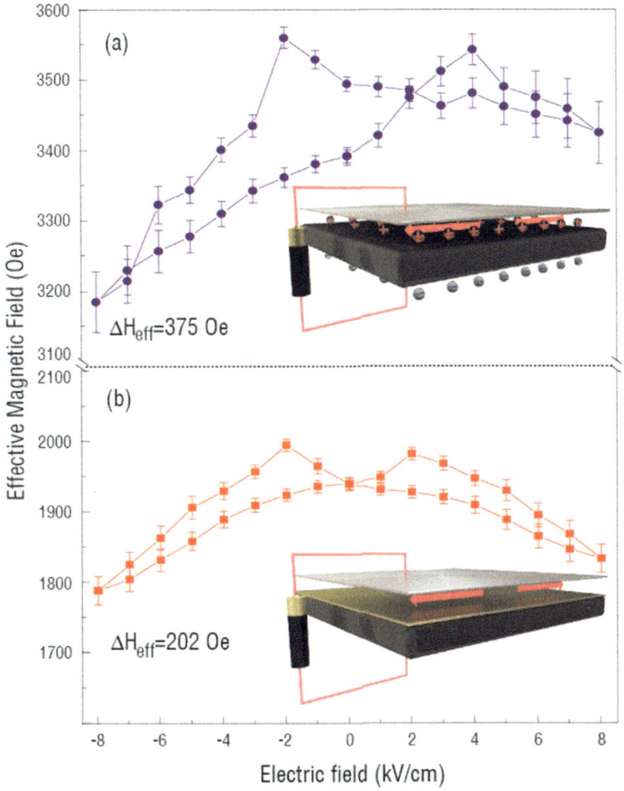

FIGURE 6.14 (a) FMR fields of NiFe/PMN-PT (011) and (b) NiFe/Cu/PMN-PT (011) by applying different electric fields with the bias magnetic field applied along the in-plane [0-11] direction. The blue and red arrow pointed the direction of the applied circling electric field.

Interestingly for comparison, the NiFe/PMN-PT was measured in FMR field sweeping mode under similar experimental conditions including the same circling electric field direction and applied magnetic field direction. By removing the isolating Cu layer, a completely different magnetoelectric coupling phenomenon was observed as shown in Fig. 6.14a. This was possibly resulted from the surface charge induced by the polarization of PMN-PT. The maximum effective magnetic field shift was increased to ~375 Oe. The FMR effective magnetic field was increased due to the different surface anisotropy of NiFe/Cu and NiFe/PMNPT. More importantly, in NiFe/Cu/PMN-PT, the FMR magnetic effective field was unchanged (~1940 Oe) when electric field was 0 due to the linear piezoelectric effect, while in NiFe/PMN-PT, two different magnetic effective fields (~3390 Oe and ~3494 Oe, respectively) were induced depending on the polarization of PMN-PT, specifically, the positive/negative remnant polarization induced by positive/negative electric field. Furthermore, the positive/negative remnant polarization could induce positive/negative charge between NiFe/PMN-PT interfaces thus modifying the magnetic anisotropy via the screening charge effect. While in NiFe/Cu/PMN-PT, because of the conductive Cu layer, the charge was unable to acuminate between the NiFe/Cu interfaces, resulting in nonexistence of charge-mediated magnetoelectric effect. The strain- and charge-mediated magnetoelectric coupling could possibly be involved in NiFe/PMN-PT at the same time. In NiFe/PMN-PT, like magnetic memory, two stable and reversible effective magnetic field states were established at 0 electric field for achieving nonvolatile switching of FMR field by reversing electric field.

In order to further investigate the distinct origin of magnetoelectric coupling effect between NiFe/Cu/PMN-PT and NiFe/PMN-PT, the angular dependence of FMR spectra was acquired by an under varied electric field. In FMR field sweeping mode, magnetic field was first applied along in-plane [0-11] direction defined as 0 degree (360 degrees) and the step of the increment was 15 degrees. Fig. 6.15a and b show the

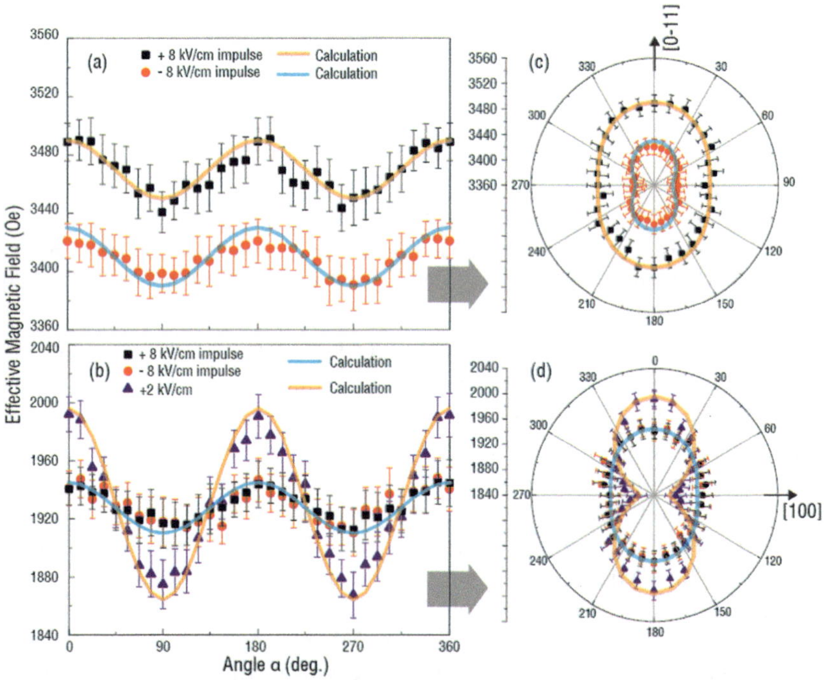

FIGURE 6.15 (a) Angular dependence of FMR effective magnetic fields of NiFe/Cu/PMN-PT and (b) NiFe/PMN-PT under different electric fields and the calculated curves. a is the angle between applied magnetic field and [0-11]direction. (c) and (d) show the polar graph transferred from (a)and (b) respectively.

angular dependence of FMR effective magnetic field of NiFe/PMN-PT and NiFe/Cu/PMN-PT under different electric field; Fig. 6.15c and d show the polar graph transferred form (a) and (b), respectively. In NiFe/Cu/PMN-PT, with the applied electric field impulse of +8 kV/cm and −8 kV/cm electric field impulse, the FMR angular dependence depicts a uniaxial anisotropy with a magnetic easy axis along [100] direction; however, due to the release of piezo-strain at 0 electric field and symmetric behavior of piezoelectric "butterfly" curve (Figure S6.1), there was no magnetic anisotropy change. While applying an electric field of 2 kV/cm, the effective magnetic field increased along [0-11] direction and decreased along [100] direction. The in-plane magnetic anisotropy change is due to the linear piezoelectric effect of (011)-oriented PMN-PT, in particular, anisotropic in-plane piezoelectric coefficients of PMN-PT ($d_{31} = -1800$ pC/N, $d_{32} = 900$ pC/N). The change of effective magnetic field on both directions was ~45 Oe, which is consistent with the FMR field response in Fig. 6.14b. A sharp contrast was observed in NiFe/PMN-PT shown in Fig. 6.15b and d by measuring the angular-dependent FMR spectra after applying +8 kV/cm (black) and −8 kV/cm (red) electric field impulse (from 0 kV/cm through ± 8 kV/cm to 0 kV/cm), respectively. The angular dependence FMR fields both exhibit the uniaxial anisotropy behavior; however, the +8 kV/cm electric field impulse induced a high effective magnetic field, while −8 kV induced a low effective magnetic field. This is direct evidence of the existence of surface charge−induced out-of-plane anisotropy change. The experimental and calculated curve fitted well, which will be discussed later. The change of the effective magnetic field was around 80 Oe, which corresponds with Fig. 6.14a where it exists around 100 Oe discrepancy of FMR field in 0 electric field. Fig. 6.16 shows a voltage-impulse-induced FMR field tuning in NiFe/PMN-PT (011) with a magnetic field along [0-11] direction which demonstrated a robust and repeatable nonvolatile switching of effective magnetic field.

6.1.1.3 0−3 Type particulate nanocomposite films

0−3 Type particulate nanocomposites films refer to the nanostructure, where a magnetic phase is basically embedded in a ferroelectric matrix. Wan et al. (2005) achieved a typical 0−3 particulate ME $CoFe_2O_4$−Pb(Zr, Ti)O_3 composite thin films by a sol-gel process and spin-coating technique. Fig. 6.17a and b shows the X-ray diffraction and scanning electron microscopy images, which imply that there exists local aggregation or phase separation of the $CoFe_2O_4$ and Pb(Zr, Ti)O_3 phases in the films. The ME effect dependence of the magnetic bias and magnetic field frequency (with a high initial ME voltage coefficient for the film, 200−300 mV cm/Oe) of thin films are also demonstrated in Fig. 6.17c.

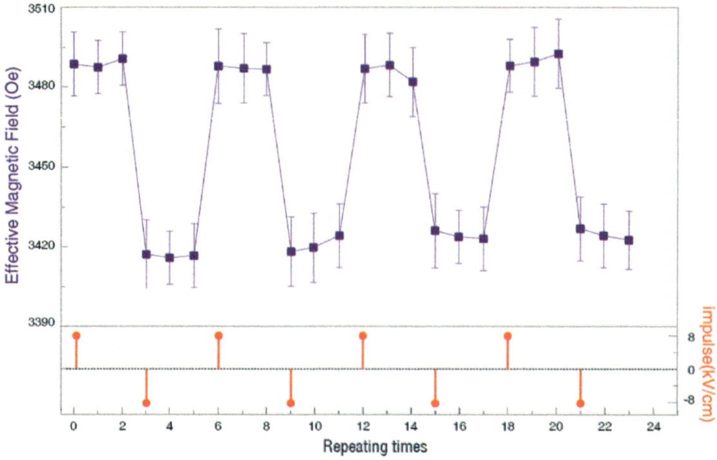

FIGURE 6.16 Electric field impulse induced nonvolatile switching of FMR field of NiFe/PMN-PT with a magnetic field applied along [0-11] direction.

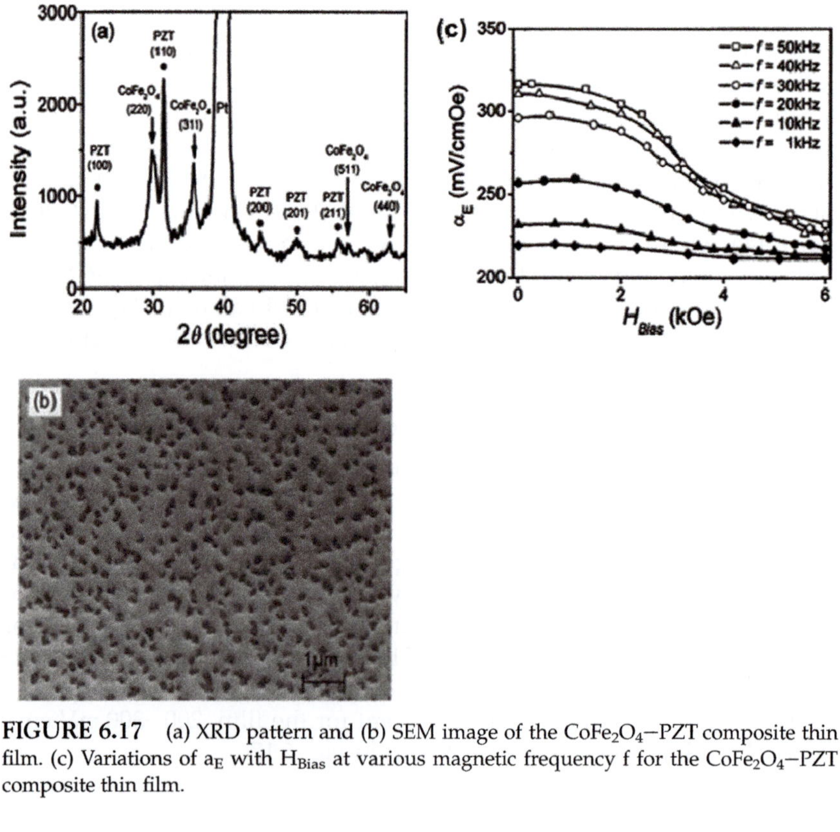

FIGURE 6.17 (a) XRD pattern and (b) SEM image of the $CoFe_2O_4$–PZT composite thin film. (c) Variations of a_E with H_{Bias} at various magnetic frequency f for the $CoFe_2O_4$–PZT composite thin film.

x $Bi_{3.15}Nd_{0.85}Ti_3O_{12}$ (BNTO) - (1-x) $CoFe_2O_4$ (CFO) 0-3 polycrystalline composites films, where BNTO was used as ferroelectric phase, were obtained by Zhong et al. (2007) using a sol-gel process and spin-coating technique as well. The structure of these films was examined by XRD (Fig. 6.18a) and SEM (Fig. 6.18b) images. All the films showed ferroelectric and ferromagnetic behaviors, as shown in Fig. 6.18c and d, well defined ferroelectric and ferromagnetic hysteresis loops were observed in Fig. 6.18c and d. Finally, the ME coefficients were measured in Fig. 6.18e. It is clear that the α_E value increases with decreasing x composite from 0.7 to 0.5 at fixed H_{bias} in the range of 0–6.5 kOe. At zero magnetic bias field, the film has small initial α_E of 5 mV/cm Oe with x = 0.5, which is resulting from the easy-magnetization characteristic with low H_c value, see Fig. 6.18c.

Liu et al. (2007) developed a multiferroic $CoFe_2O_4$–Pb(Zr, Ti)O_3 composite films with a nanoscale mixture of the two phases which was prepared by a modified sol-gel process, in which a mixed precursor solution of both CFO and PZT was used. X-ray diffraction and transmission electron microscopy examinations, see Fig. 6.19a–c revealed the coexistence of perovskite PZT and spinel CFO that were mixed in nanoscale with mean grain sizes of 5–10 nm. Magnetic properties of the CFO-PZT nanocomposite were examined (Fig. 6.19d), which were consistent with their microstructures. The magnetoelectric coupling between CFO and PZT was demonstrated by an external magnetic field induced electric polarization change, as shown in Fig. 6.19e. This modified sol-gel processing provides an alternative for multiferroic composite films, which is simpler and easier to control compared to the conventional layer-layer sol-gel process for multiferroic composite films.

Murugavel et al. (Ryu et al., 2006) fabricated (100)-oriented PZT/NFO composites on (001) $SrTiO_3$ (STO) substrates by PLD method. The NFO nanoparticles were randomly distributed in the PZT matrix in their nanostructure, which are confirmed by structural analysis, see Fig. 6.20. The crystal axes of the NFO are aligned in the ferroelectric matrix with good ferroelectric and magnetic properties. They measured magnetic field-dependent ME coefficient α_E with both transverse and longitudinal components, see Fig. 6.21. The ME effect is resulting from strain/stress coupling between magnetostrictive $NiFe_2O_4$ and piezoelectric $Pb(Zr_{0.52}Ti_{0.48})O_3$ grains. Nevertheless, the highest ME coefficient values for above composites, within the order of 10 mV cm^{-1} Oe^{-1}, are lower than that reported for bulk PZT/NFO composites (Zhai et al., 2004). The reason might be the substrate clamping effects limited the strain/stress induced ME coupling magnitude.

6.1.1.4 1-3 type vertical heterostructures

The 1-3 type vertical heterostructure implies a magnetic nanopillar array embedded in a ferroelectric matrix, as shown in Fig. 6.1c. Zheng

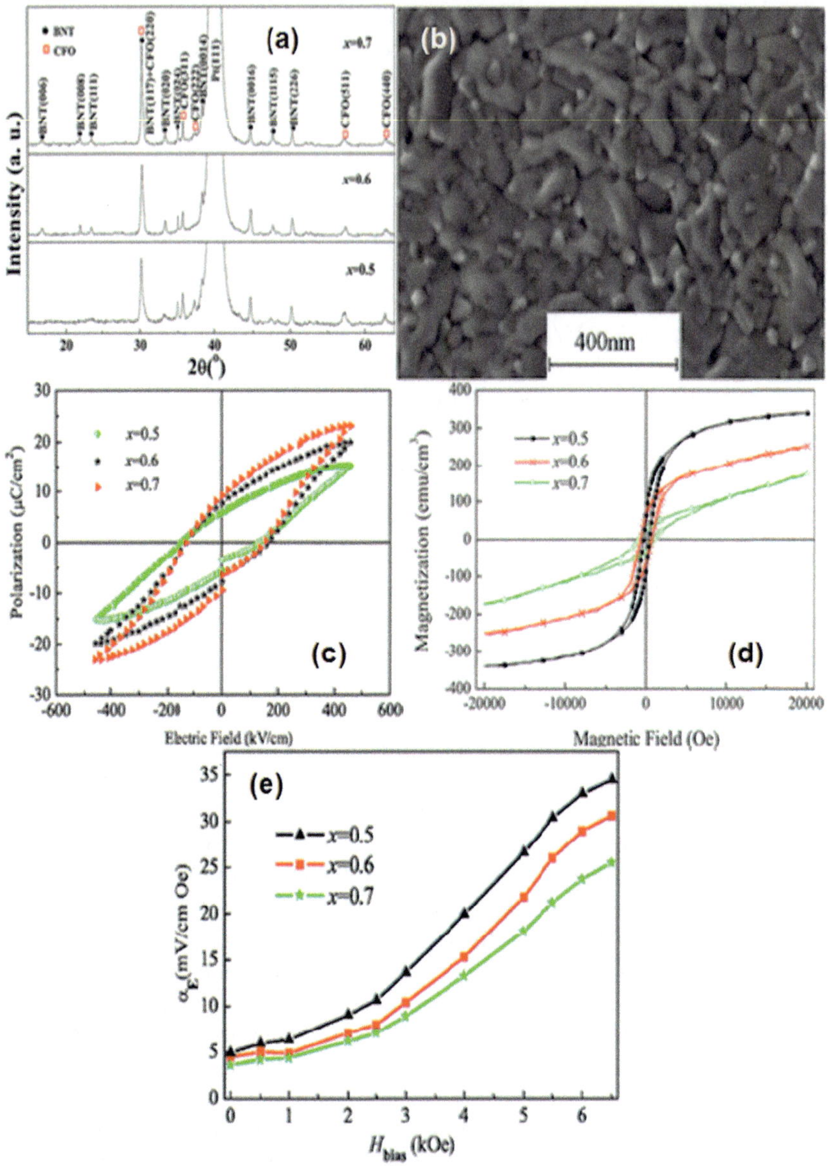

FIGURE 6.18 (a) RD patterns of the xBNT-(1−x) CFO composite thin films with x = 0.5, 0.6, and 0.7 (b) SEM surface morphologies of the xBNT-(1−x) CFO composite thin film with x = 0.6 (c) P-E hysteresis loops of the xBNT-(1−x) CFO composite thin films with x = 0.5, 0.6, and 0.7 (d) magnetic hysteresis loops for the xBNT-(1−x) CFO composite thin films with x = 0.5, 0.6, and 0.7. (e) Variation of a_E with H_{bias} at magnetic frequency f = 20 kHz for the xBNT-(1−x) CFO composite thin films with x = 0.5, 0.6, and 0.7.

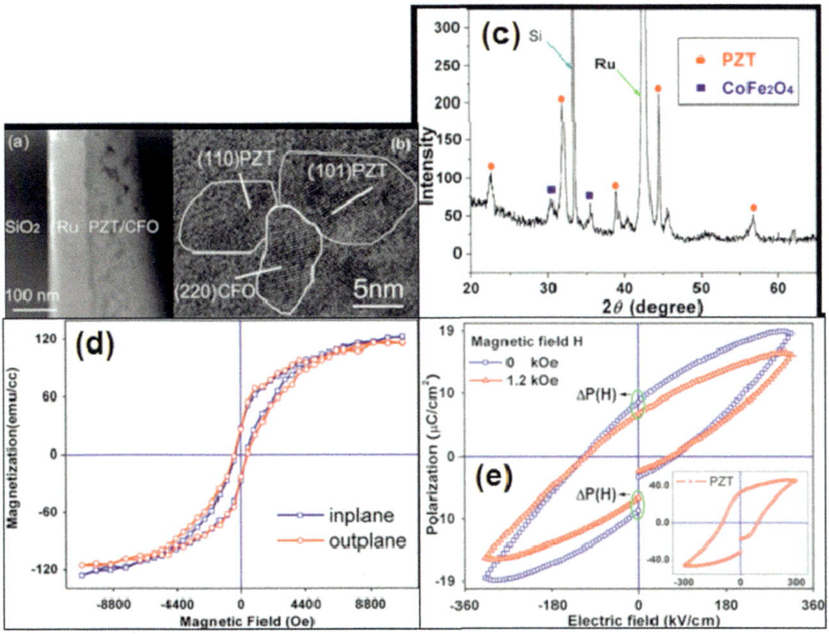

FIGURE 6.19 (a) Cross-section scanning transmission electron microscopy (TEM) image (b) HRTEM image of multiferroic CFO-PZT nanocomposite thin film. (c) X-ray diffraction pattern of multiferroic CFO-PZT nanocomposite thin film. (d) Magnetic hysteresis loops of CFO-PZT nanocomposite thin film. (e) Polarization-electric field hysteresis loops of the multiferroic CFO-PZT nanocomposite film and pure PZT film.

et al., discovered the first 1-3 type of magnetic CFO nanopillars with diameters of 20–30 nm were embedded in a ferroelectric BTO matrix film (Zheng, Wang, et al., 2004). Fig. 6.22 shows the schematic of 1-3 type CFO-BTO multiferroic heterostructure, and Fig. 6.23 shows the magnetic and ferroelectric properties of CFO-BTO multiferroic heterostructure.

Other different 1-3 type multiferroic heterostructure were designed and grown on $SrTiO_3$ single crystal substrates (Levin et al., 2006; Slutsker et al., 2006; Zheng et al., 2006). For example, in (1-x) $PbTiO_3$−x $CoFe_2O_4$ (CFO) heterostructure (Levin et al., 2006; Slutsker et al., 2006), by adjusting the composition of CFO and crystal orientation of STO substrates, the typical 1-3 type multiferroic heterostructure was obtained, see Fig. 6.24. The 1-3 type $BiFeO_3$−CFO multiferroic heterostructure, containing either ferromagnetic or ferroelectric nanopillars, was also achieved by varying the phase fraction, substrate orientation and temperature treatment, see Fig. 6.25. Nevertheless, the design and fabrication of these heterostructures are still a challenge.

A lot of recent work has been done in ME coupling in these 1-3 type heterostructures (BFO-CFO) (Zavaliche et al., 2005, 2007); the localized

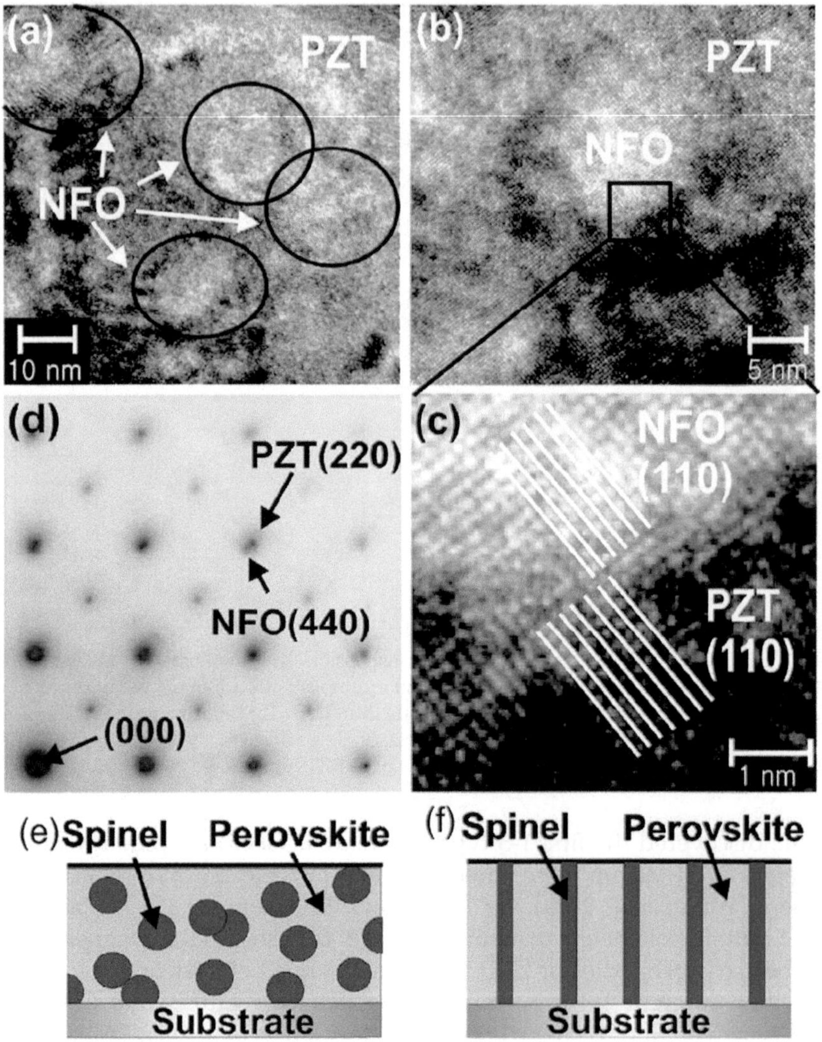

FIGURE 6.20 (a) Low-magnification high-resolution transmission electron microscopy (HRTEM) image of the PZT–NFO film. The marked regions indicate the NFO phase. (b) HRTEM image near the PZT–NFO boundary and (c) an enlarged image of the region marked in (b). (d) Selected-area electron diffraction pattern of the composite film. Schematic representations of a composite film with spinel phase growth involving (e) nanoparticles and (f) self-assembled nanopillars.

magnetization can be switched by reversal of ferroelectric polarization, see Figs. 6.26–6.28. In the 1-3 type vertical architecture, the substrate clamping effect was reduced, and a larger interfacial area was obtained resulting in larger efficient strain coupling. Detailed discussions referred that the

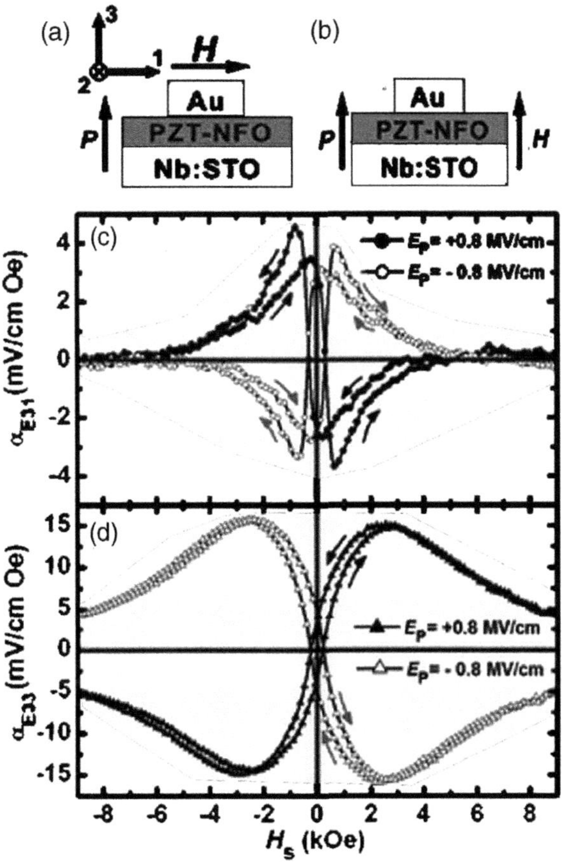

FIGURE 6.21 Schematic diagrams showing the experimental configuration for the (a) transverse a_{E31} and (b) longitudinal a_{E33} components of the magnetoelectric coefficient. (c) a_{E31} and (d)a_{E33} measured on the composite film after poling at +0.8 and −0.8 MV/cm.

magnetization switching is induced by strain/stress coupling between ferromagnetic and ferroelectric phases, leading to time-domain modulation of the magnetic anisotropy in these heterostructures. However, the controllable reversal magnetization switching can only be obtained by adding a weak magnetic field, which breaks the time-reversal symmetry of ME coupling. The MFM and PFM images confirmed the ME coupling directly in 1-3 type vertical heterostructure; however, the ME coefficients are hardly determined due to the leakage problem. The magnetic nanopillars have low resistance penetrating through the magnetic matrix films.

6.1.1.5 2-2 type horizontal heterostructures

2-2 type horizontal nanostructure, see Fig. 6.1b, containing ferroelectric layers and ferromagnetic layers exhibit weak ME effects due to an

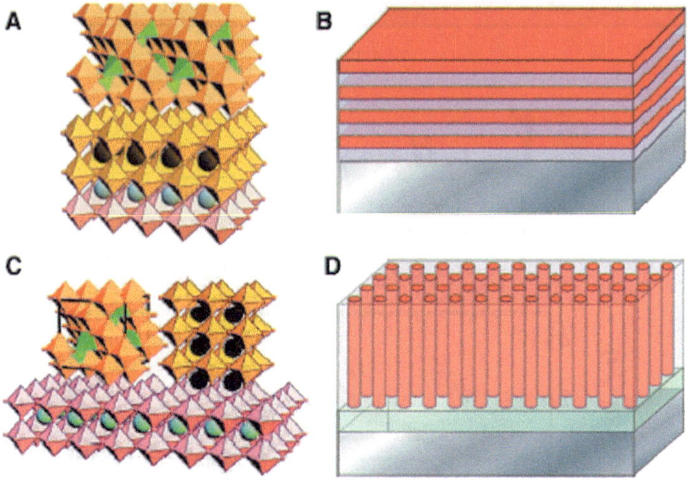

FIGURE 6.22 (a) Superlattice of a spinel (top) and a perovskite (middle) on a perovskite substrate (bottom). (b) Schematic illustration of a multilayer structure on a substrate. (c) Epitaxial alignment of a spinel (top left) and a perovskite (top right) on a perovskite substrate (bottom). (d) Schematic illustration of a self-assembled nanostructured thin film formed on the substrate.

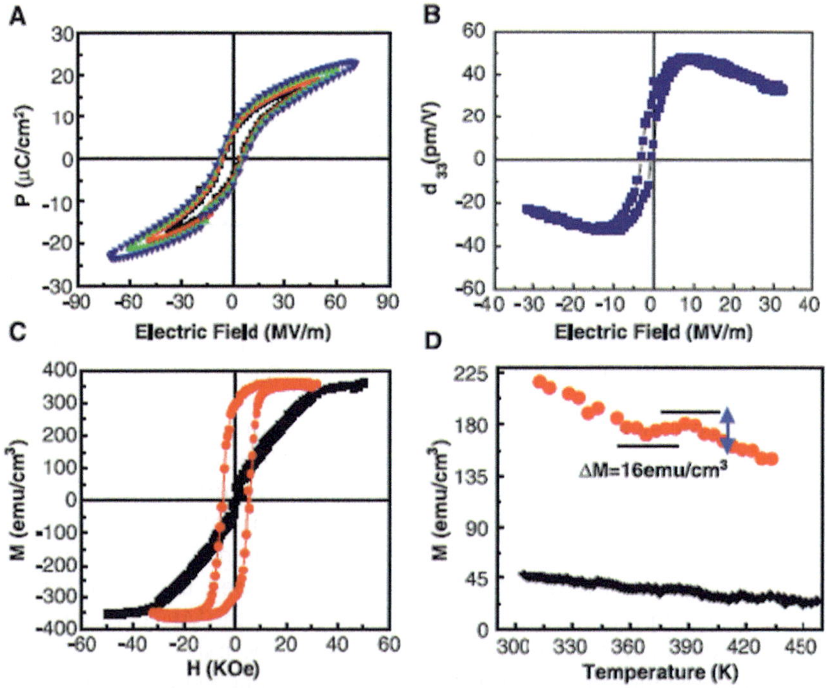

FIGURE 6.23 (a) Polarization–electric field hysteresis loop showing that the film is ferroelectric with a saturation polarization $P = 23\ \mu C/cm^2$. (b) Small-signal piezoelectric d_{33} hysteresis loop for a 50-μm-diameter capacitor. (c) Out-of-plane (red) and in-plane (black) magnetic hysteresis loops depicting the large uniaxial anisotropy. (d) Magnetization versus temperature curve measured at H = 100 Oe, which shows a distinct drop in magnetization at the ferroelectric Curie temperature for the vertically self-assembled nanostructure (red curve); the multilayered nanostructure (black curve) shows negligible change in magnetization.

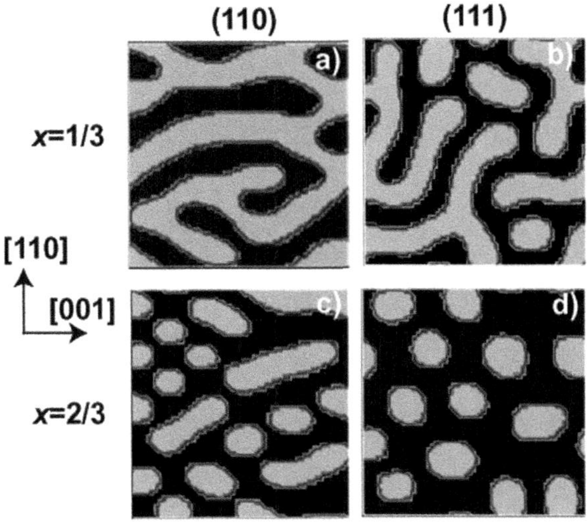

FIGURE 6.24 $xCoFe_2O_4-(1-x)PbTiO_3$ nanostructures simulated using phase-field modeling. a,b) $x = 1/3$; c,d) $x = 2/3$. a,c) Substrate: (110) $SrTiO_3$; b,d) substrate (111) $SrTiO_3$. Bright and dark regions correspond to $CoFe_2O_4$ and $PbTiO_3$, respectively.

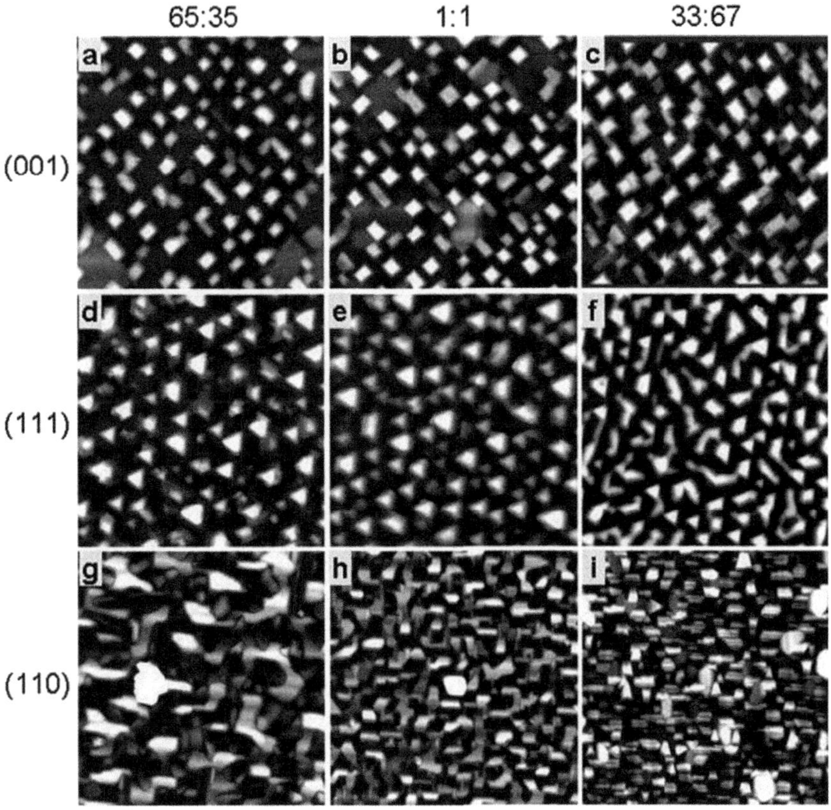

FIGURE 6.25 AFM images of $BiFeO_3-CoFe_2O_4$ nanostructures with volume fractions of 65:35, 1:1, and 33:67 grown on a–c) (001) substrates at 700°C; d–f) (111) substrates at 650°C, and g–i) (110) substrates at 700°C. All images are 3 × 3 μm.

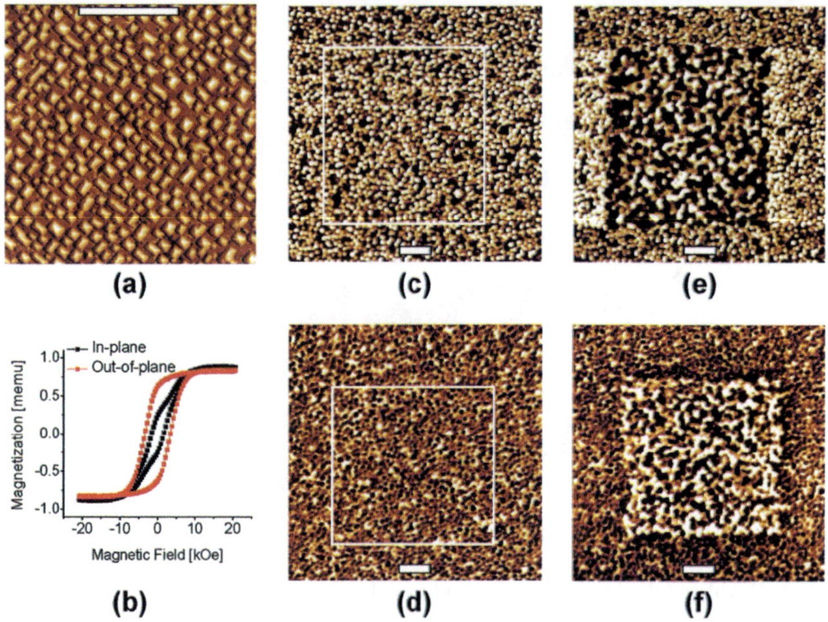

FIGURE 6.26 Morphology and magnetization switching in the $(BiFeO_3)0.65$-$(CoFe_2O_4)$ 0.35 nanostructure. (a) Surface topography as seen by AFM. (b) Out-of-plane (red symbols) and in-plane (black symbols) magnetization loops. (c,d) MFM scans after magnetization in down and up 20 kOe fields, respectively. (e,f) MFM scans taken in the same areas as in (c) and (d), respectively, after electrical poling at −16 V. The bars are 1 μm.

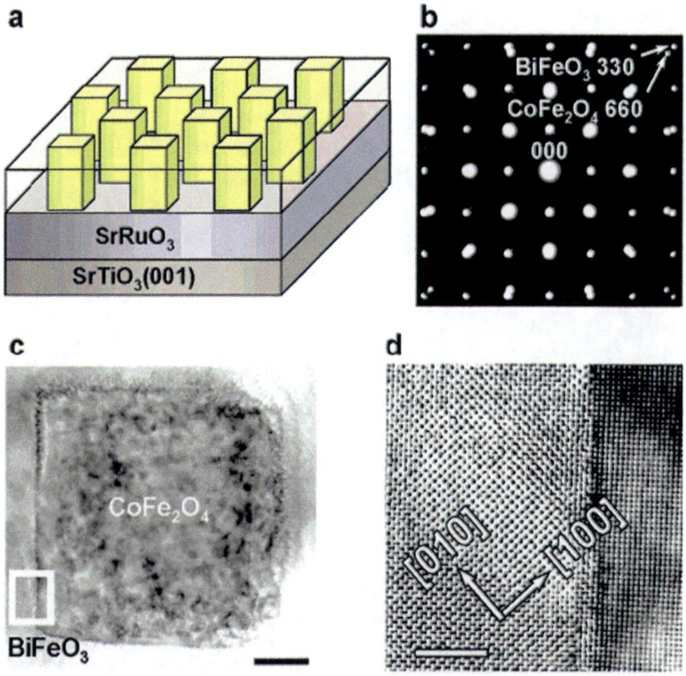

FIGURE 6.27 Structure and morphology of self-assembled $BiFeO_3$–$CoFe_2O_4$ epitaxial nanostructures. (a) A sketch of the heterostructure. (b) Typical selective area electron diffraction pattern. (c) Plan view TEM image of a $CoFe_2O_4$ pillar surrounded by $BiFeO_3$. (d) High resolution plan-view TEM image of the $CoFe_2O_4$–$BiFeO_3$ interface taken in the area marked with a rectangle in c. The bars in c and d are 30 and 3 nm, respectively.

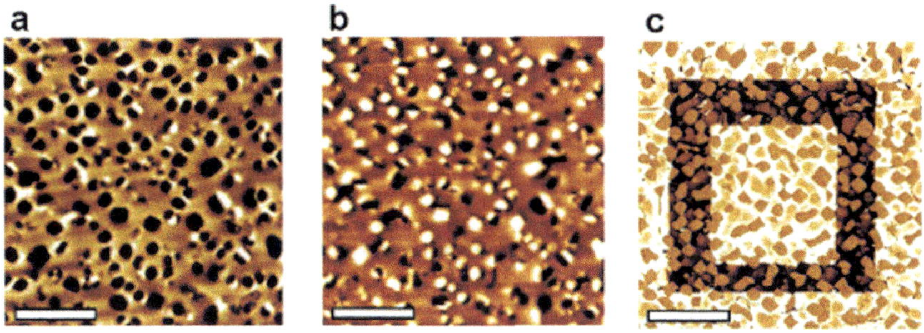

FIGURE 6.28 Magnetic and piezoelectric force microscopy images of a $(BiFeO_3)0.65$-$(CoFe_2O_4)0.35$ film. (a and b) Magnetic force microscopy images of the same area after magnetizing the film perpendicular to the surface in opposite magnetic fields of 20 kOe. (c) Perpendicular piezoelectric force microscopy image taken after poling the film at −8 V (dark frame), and +8 V (white, inside box). The bars are 2 μm.

in-plane substrates constraint, however, which is still comparable with ME effects in the bulk multiferroic heterostructure. The advantage of 2-2 horizontal nanostructures is ease to fabricate and reduced leakage current, due to the highly resistive ferroelectric layers, blocking the current flow compared to 0-3 and 1-3 composites. Therefore, direct ME coupling can be observed in these heterostructures. The 2-2 type heterostructures are easy to process and integrated in a device, a lot of work has been done on these heterostructures (Deng et al., 2008; Eerenstein et al., 2007; Li et al., 2009; Ma et al., 2007; Murugavel et al., 2004; Ryu et al., 2007; Singh et al., 2006; Thiele et al., 2007; Wang et al., 2008; Wu et al., 2006; Zhang et al., 2008, 2009; Ziese et al., 2006). A 2-2 heterostructure of CFO/BTO on a (001) STO substrate prepared by PLD method was reported by Zhang et al. (2008). An enhanced ME voltage coefficient of 66 $mV\ cm^{-1}\ Oe^{-1}$ was obtained, comparable to bulk materials, see Fig. 6.29. Li et al., reported a magnetic field–induced Raman spectra shift in PZT/CFO bilayer 2-2 heterostructure. The reason is because the magnetostriction of CFO layer generated strain/stress change transferred to PZT layer, through magnetic field driving magnetostriction, see Fig. 6.30. The modes change in PZT, giving solid evidence of strain-induced ME coupling in CFO/PZT layer.

2–2 Layered multiferroic heterostructures have been studied by many groups. Ryu et al. (2007) fabricated PZT/$Ni_{0.8}Zn_{0.2}Fe_2O_4$ (NZFO) multilayer structure, Ziese et al. (2006)] studied Fe_3O_4/$BaTiO_3$ (BTO), and Ma et al. (2007) reported PZT/$La_{0.7}Sr_{0.3}MnO_3$ (LSMO), respectively. All these heterostructures are high quality and realized ME coupling within range from 3 to 30 $mV\ cm^{-1}Oe^{-1}$ between ferroelectricity and ferromagnetism, see Figs. 6.31–6.33.

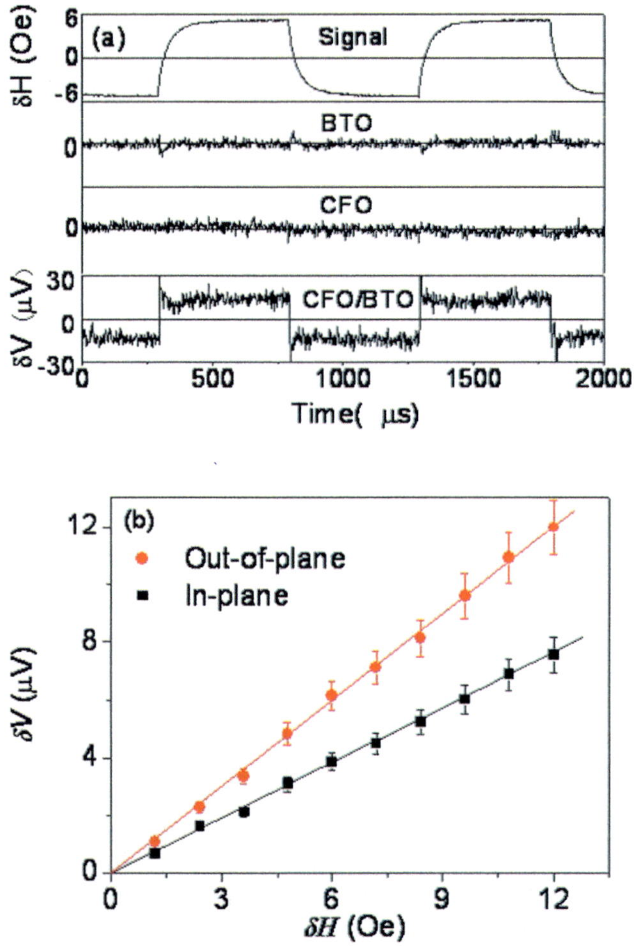

FIGURE 6.29 (a) The ME response of the films when the in-plane magnetic field is applied. (b) Induced ME voltage V as a function of ac magnetic field H at 1 kHz with a bias magnetic field of 100 Oe. The in-plane and out-of-plane in (b) represent the measurement modes with the magnetic fields parallel and perpendicular to the surface of the films, respectively.

Murugavel et al. (2004) and Singh et al. (2006) deposited superlattice–multiferroic heterostructure films. The superlattices of $Pr_{0.85}Ca_{0.15}MnO_3/Ba_{0.6}Sr_{0.4}TiO_3$ and $La_{0.7}Ca_{0.3}MnO_3/BaTiO_3$ on single crystal STO substrate demonstrated good dielectric/ferroelectric and ferromagnetic properties, for instance, enhanced magnetoresistance (M_R) and magnetocapacitance, see Fig. 6.34, the M_R was significantly increased by adding BSTO layer to the PCMO/BSTO multilayer heterostructure. An epitaxial 2-2 horizontal heterostructure has recently been discovered, for

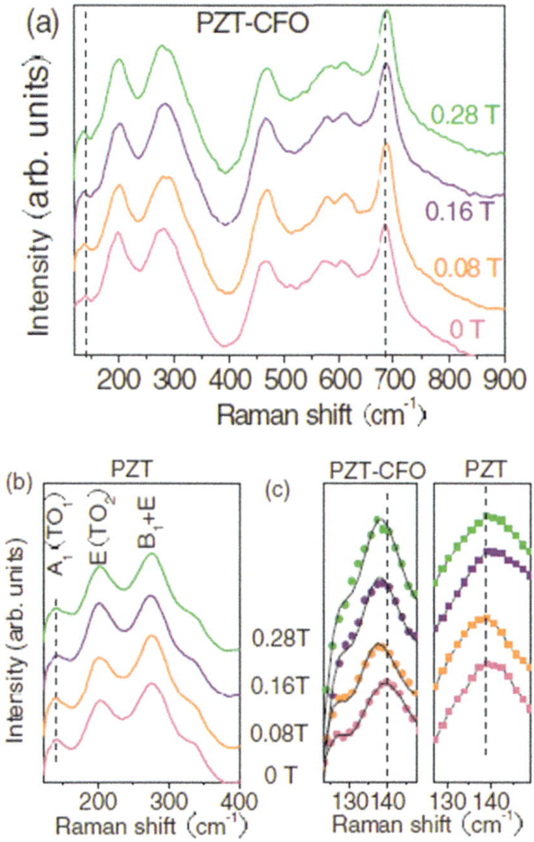

FIGURE 6.30 Raman spectra of (a) the PZT-CFO bilayer and (b) the PZT film in the presence of magnetic field. (c) Comparison of the changes in the soft-mode $A_1(TO_1)$ with magnetic field for the PZT-CFO bilayer and the PZT film. The lines in (c) are the Gaussian fitting.

example, epitaxial PZT film on single crystal LSMO layer (Wu et al., 2006) or epitaxial LSMO film on single crystal BTO (Eerenstein et al., 2007) and $Pb(Mg,Nb)O_3{-}PbTiO_3$ (Thiele et al., 2007). The good interfacial condition improved the efficiency of mechanical ME coupling. In epitaxial PZT/LSMO multilayer, a strong ME coupling effect of 600 mV $cm^{-1}Oe^{-1}$ was obtained at 120 K, see Fig. 6.35. For BTO thin film on CFO ceramics, the ME coefficient was 38 mV $cm^{-1}Oe^{-1}$. Although the CFO ceramics are much cheaper and easier to process than that of single crystal substrates, the ME coefficient, however, is much smaller than single crystal substrates. Therefore, the interfacial strain transfer efficiency is the key point of strain mediated ME coupling in 2-2 type horizontal heterostructure. Besides strain-/stress-mediated ME coupling, other mechanism based on

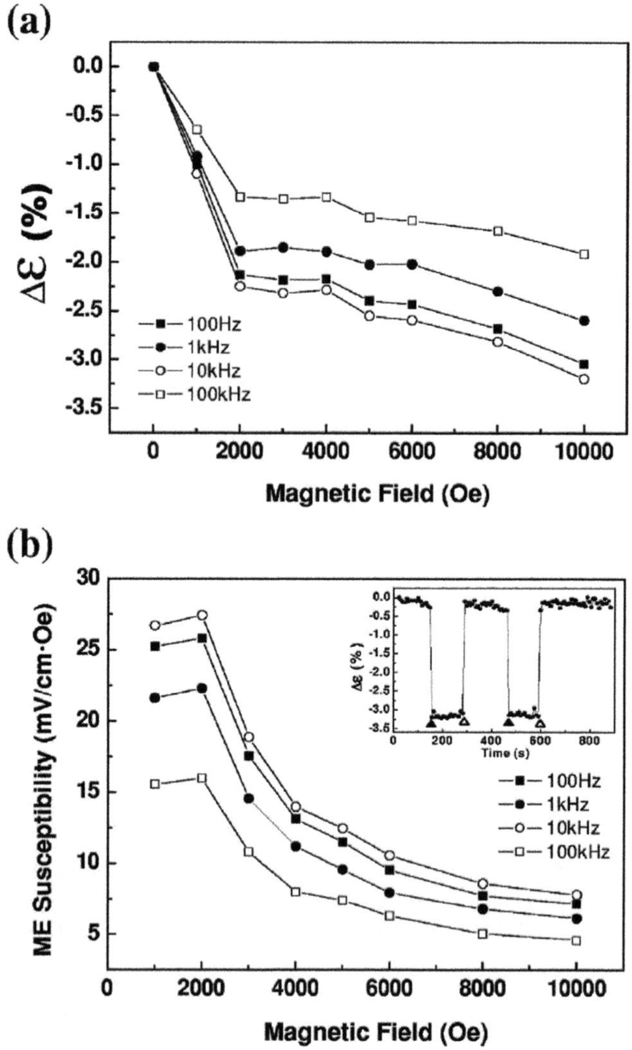

FIGURE 6.31 ME coupling characteristics of the 450-nm-thick PZT/NZFO multilayer consisting of 11 alternating layers. (a) Percent change in the real part of dielectric permittivity at various measuring ac electric-field frequencies. (b) Longitudinal ME susceptibility plotted as a function of ΔH_3.

two to two type heterostructures were also investigated as hot topic. The discussions are listed in chapter 6.

6.1.1.6 Recent advances in thin-film ME composites

Magnetoelectric (ME) materials are of importance for studies on the nature of electromagnetic interactions and for a variety of sensors, energy

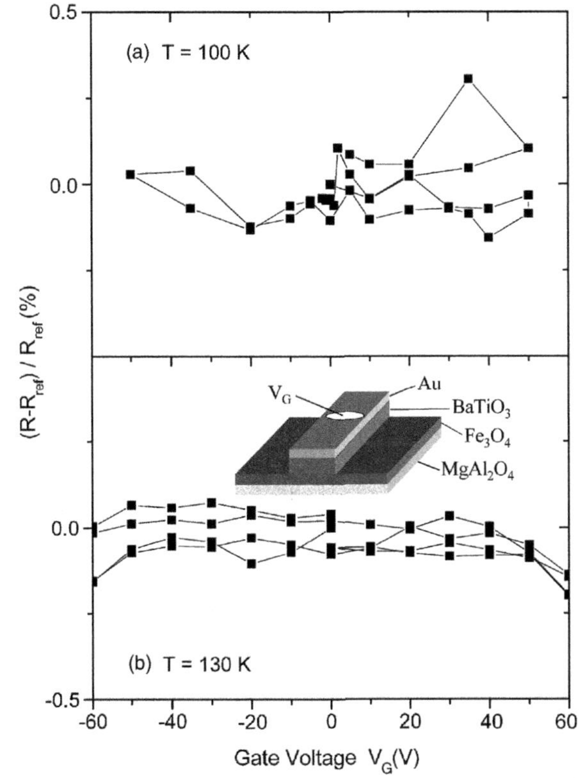

FIGURE 6.32 Dependence of the resistance of the conducting magnetite channel in a direct structure on the gate voltage applied over the BaTiO3 layer. The inset shows a schematic layout of the structure.

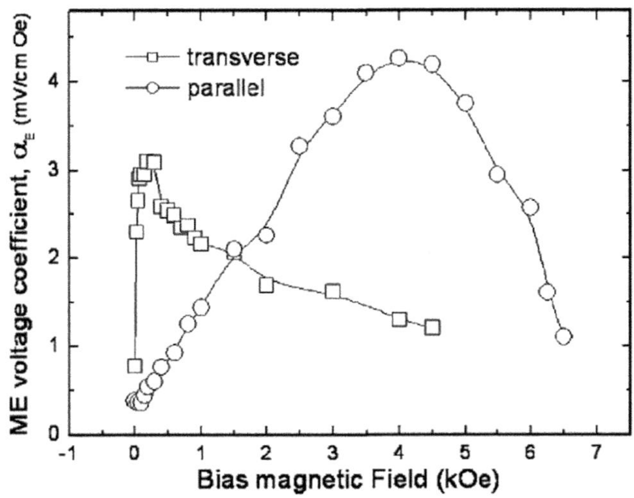

FIGURE 6.33 ME voltage coefficient of PZT/LSMO as a function of the bias magnetic field. The words transverse and parallel, respectively, represent the measurement modes with the magnetic fields $H_{bias}//\delta H$ perpendicular and parallel to the polarization direction of PZT at a frequency of 1 kHz.

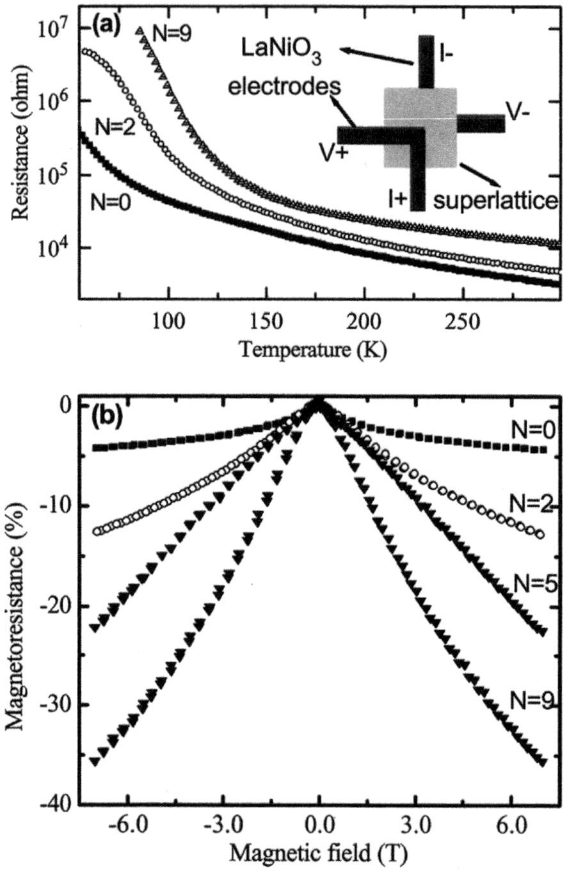

FIGURE 6.34 (a) Temperature dependent resistance for the superlattices $[(PCMO)_{10}/(BST)_N]_{25}$ of N = 0, 2, and 10. The inset shows the junction structure used for the measurement. (b) M_R at high magnetic field measured at 100 K for the superlattices $[(PCMO)_{10}/(BST)_N]_{25}$ of N = 0, 2, 5, and 10. In all measurements, the current is applied perpendicular to the plane of the sample.

harvesters, and gyrators. Recent efforts on thin-film bilayers and multilayers mainly dealt with applications for magnetic sensors for imaging applications and are discussed in Chapters 9, 10 and 14 in this monograph. Here we focus only on recent efforts on ME nanopillars.

Vertically aligned nanopillar composites are appealing due to their unique phase distribution morphologies and enhanced ME response (Gao et al., 2021). However, the interconnected ferromagnetic nanopillars lower the net resistivity of the composites, which gives rise to the deterioration of dielectric and ME properties (Li et al., 2015). Therefore, the key to improving the ME coupling relies on, but not limited to, the enhanced

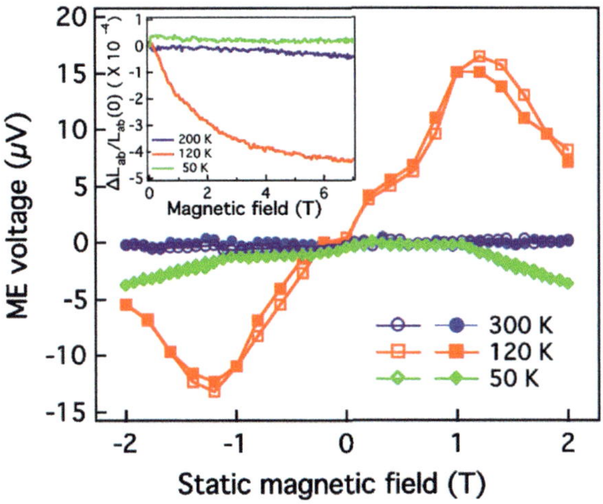

FIGURE 6.35 ME voltage versus static magnetic field measured with a transverse geometry at three different temperatures of 50, 120, and 300 K. The ME voltage is induced by an oscillating magnetic field Hac with a magnitude of 2 Oe and a frequency of 1 kHz. The open (filled) symbols are the data taken with increasing (decreasing) H. The largest peak-to-peak ME voltage swing is ~28 μV, occurring at 120 K.

dielectric and magnetostrictive properties, as well as the interconnection between ferroelectric and ferromagnetic phases. Kim et al. reported the fabrication of self-assembled epitaxial $BiFeO_3-CoFe_2O_4$ (BFO-CFO) nanocomposite thin films and the correlation between strain and magnetic anisotropy (Kim et al., 2018). Tang et al. focused on self-assembled two-phase heterostructures of $BiFeO_3$ and $CuFe_2O_4$ (CuFO−BFO) that had both nanopillars and nanobelts when deposited on $SrRuO_3$-buffered $Pb(Mg_{1/3}Nb_{2/3})O_3-33\%PbTiO_3$ (PMN-PT) single crystals substrates. They compared the magnetic and ME properties of the heterostructures with a single layer of CuFO on PMN-PT. It was found that CuFO/PMN-PT did not show any nanostructure and *M* versus *H* indicated easy plane anisotropy. But with the application of an *E* field *M* versus *H* data did not show any evidence for converse ME coupling. A similar by *M* versus *H* data in the multilayer heterostructures of CuFO−BFO, however, showed out-of-plane magnetic anisotropy likely due to the nanopillar−nanobelt structure in the film. Upon the application of *E*, strong CME with a significant dependence of the remnant magnetization M_r on *E* was measured, i.e., M_r/M_s exhibited typical butterfly-shaped hysteresis with the *E* as shown in Fig. 6.36 (Tang et al., 2019a). These data, therefore, provide clear evidence for ME coupling in the system due to the nanostructure in the film.

Tang et al., also reported on a similar study in self-assembled BFO−CFO films on (100) substrate of PMN-PT (Tang et al., 2018). Data on *M*

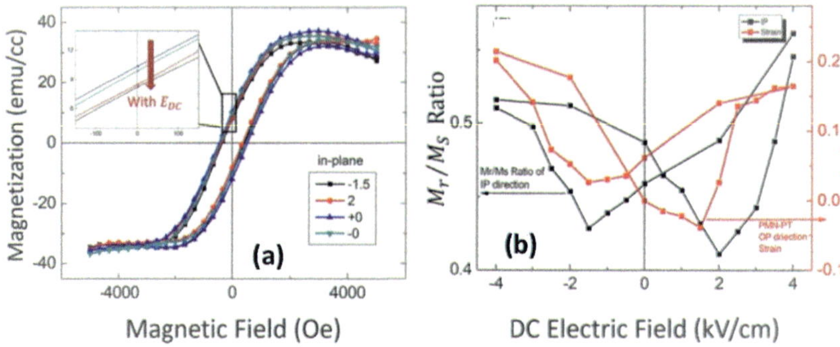

FIGURE 6.36 (a) Data on M versus H as a function of $E = -1.5$ kV/cm, -2 kV/cm and 0 in the heterostructure of BFO–CuFO/SRO/PMN–33PT. The inset shows the change in the remnant magnetization M_r with E. (b) M_r/M_s versus E and strain in PMN-PT versus E for the heterostructure (Tang et al., 2019a).

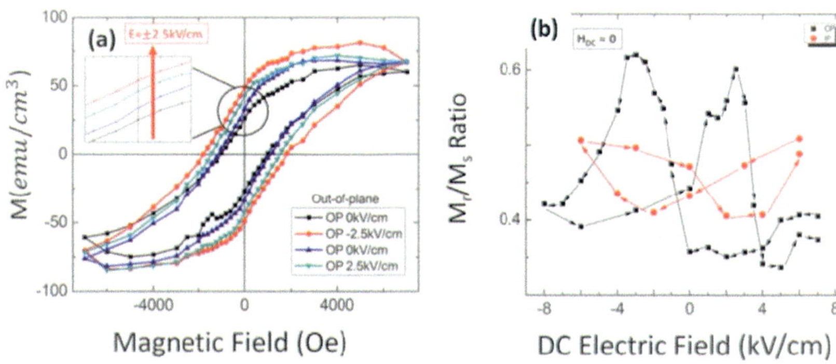

FIGURE 6.37 *M* versus *H* data for heterostructured BFO-CFO for *H* perpendicular to film plane and for a series of electric fields. Inset shows variation in H_r in expanded scale. (b) Ratio M_r/M_s as a function of E for H in-plane or perpendicular to film plane (Tang et al., 2018).

versus *H* shown in Fig. 6.37 for the system showed out-of-plane anisotropy in the system due to shape anisotropy associated with the nanopillars. When a DC electric field, the film showed multiple unique values of a stable magnetization with memory on the removal of electric field. The data in the figure indicated indicate a giant CME effect in the film with a 90% change in the value of M_r/M_s under E (Tang et al., 2018). When vertically integrated two-phase $BiFeO_3$–$CoFe_2O_4$ and $BiFeO_3$–$CuFe_2O_4$ were epitaxially deposited on single crystal substrates of $Pb(Mg_{1/3}Nb_{2/3})_{0.74}Ti_{0.26}O_3$ (PMN-26PT), the nanopillar heterostructures were shown to hinder the rotation of magnetic domains, resulting in nonvolatile electric field tuning of magnetization (Tang et al., 2019b). In another report by the group, two-phase epitaxial $Ni_{0.65}Zn_{0.35}Al_{0.8}Fe_{1.2}O_4$ (NZAFO) and $BiFeO_3$

layers showed nanobelt-like heterostructures. The magnetic anisotropy and coercivity of NZAFO-BFO could be controlled by varying the thickness to alter shape anisotropy of this nanostructure (Tang et al., 2019c).

Most studies of nanopillar based films focused on CFO-BFO on a variety of synthesis procedures and substrates (Jian & Wong, 2017; Kim et al., 2015; Van Lich et al., 2015). Direct measurements of ME exchange between CFO pillars in piezoelectric $BiFeO_3$ matrix using a magnetic cantilever method was reported by Yan et al. (2009). The nanocomposite exhibited a maximum ME coupling coefficient of 20 mV/cm Oe, as well as saturation polarization and magnetization of 60 $\mu C/cm^2$ and 410 emu/cc, respectively. Rafique et al. focused on the ME response in strain-controlled nanocomposite of BFO-CFO grown on a $BaTiO_3$ crystal (Rafique et al., 2017). Temperature-dependent magnetization measurements revealed strong magnetoelastic coupling between the $BaTiO_3$ substrate and the nanocomposite. Moreover, the magnetic anisotropy of the nanocomposite was controlled by the reorientation of ferroelectric domains in $BaTiO_3$ crystal under the applied electric field. Out-of-plane and in-plane ME coupling coefficients calculated from remnant magnetization were 2.6×10^{-7} s m^{-1} and 1.5×10^{-7} s m^{-1}, respectively (Rafique et al., 2017). Wang et al. established a relationship between ME coupling effect and interfacial strain in $BiFeO_3$—$CoFe_2O_4$ composite (Wang, Bian, et al., 2020). Chen et al. measured the transverse and longitudinal ME coupling effect in epitaxial $BaTiO_3/CoFe_2O_4$ nanopillar composites by optical second harmonic generation and piezoresponse force microscopy under magnetic field. They found a correlation between ME coupling strength and vertical interfacial area that was controlled by pillar size (Chen et al., 2019). The ME coupling in the composite was tuned by the competition between the vertical interface coupling effect and the bulk volume conservation effect.

6.1.2 Core—shell nanoparticle composites

Magnetoelectric (ME) nanoparticle and nanofiber composites are of importance due to their potential to exhibit strong strain mediated interactions since they are free of substrate clamping, They are generally encountered in nanobilayers or nanopillars on a substrate. The core—shell nanoparticle or nanofiber composites, with very high surface area to volume ratio compared to thin films, are expected to have efficient strain transfer at the ferroelectric—ferromagnetic interface and strong ME effects. Therefore, they are potential candidates for the next generation of sensors and memory devices (Fiocchi et al., 2022; Henrichs et al., 2020; Smith et al., 2022). This section is devoted to ME effects in core—shell particles. Core—shell fibers are considered in the section that follows.

There have been a few reports dealing with modeling of core—shell particulate composites. Fiocchi et al. (2022) used a nonlinear model based on hysteresis to describe the response of ferromagnetic/piezoelectric core–shell nanoparticles to various magnetic field excitations. Based on the results of the model, one would prefer a large magnetostrictive core diameter and thin piezoelectric shell for strong ME coupling coefficient that can be as high as 15 V/cm•Oe for cobalt ferrite (CFO) and barium titanate (BTO) core–shell nanoparticles.

There have been several reports on multiferroic core–shell nanocomposites, especially with the piezoelectric PZT or BTO and ferrimagnetic CFO or NFO. Kiminami et al. synthesized $Pb(Zr_{0.53}Ti_{0.47})O_3$@$CoFe_2O_4$ two-phase particle composites by a hydrosol gel cosynthesis method (Fernández et al., 2017). They then used the microwave-assisted sintering method to obtain a dense dual-phase powder. The composites showed MEVC of 1.56 mV/cm·Oe (Fernández et al., 2017). Composites with BTO nanoparticles have shown excellent ME coupling because of good connectivity between ferroelectric and magnetic phases. Core–shell particles with $MgFe_2O_4$ and BTO have been reported to have an ME coupling coefficient of 28 mV/cm·Oe (Abraham et al., 2017). Composites of $CuFe_2O_4$ and BTO prepared by sol-gel method had an MEVC of 22.5 mV/cm·Oe (Thankachan et al., 2018). Similarly, $Cu_{0.8}Co_{0.2}Fe_2O_4$@$Ba_{0.8}Sr_{0.2}TiO_3$ core–shell nanoparticles reported to show MEVC as high as 78.14 V/cm·Oe (Wu et al., 2022). The majority of research reports on multiferroic composites focused on the core–shell structure nanoparticles of CFO and BTO showed a maximum ME coefficient of 10 V/cm·Oe (Gao et al., 2020). Khizroev et al., achieved MEVC of more than 5 V/cm·Oe and 2 V/cm·Oe, respectively, in CFO@BTO and $NiFe_2O_4$ (NFO)@BTO core–shell nanoparticles (Wang, Zhang, et al., 2020).

Recently, the magnetoelectric effect in chemically assembled core–shell nanoparticles was reported (Sreenivasulu et al., 2014a, 2014b; Srinivasan et al., 2015). Chemically assembled core–shell nanoparticles offer several advantages over other methods of fabrication. First, they can be produced with a high degree of control over the size, shape, and composition of the nanoparticles. This is important for ensuring that the nanoparticles have the desired properties for a particular application. Second, chemically assembled nanoparticles can be produced in large quantities, which is necessary for commercial applications. Third, chemically assembled nanoparticles can be functionalized with different molecules, which can be used to tune their properties for specific applications. The chemical self-assembly mainly dealt with composites of NFO and BTO. The assembly was carried out in steps: first the NFO and BTO particles were functionalized with complementary coupling groups, such as azide and alkyne groups. Then, the core-shell structure was formed by "click-chemistry" aided by a catalyst (Sreenivasulu, Popov, et al., 2014). Fig. 6.38 shows TEM and MFM images of core-shell particles with 100 nm NFO core and 50 nm BTO shell. The images clearly show the core-shell

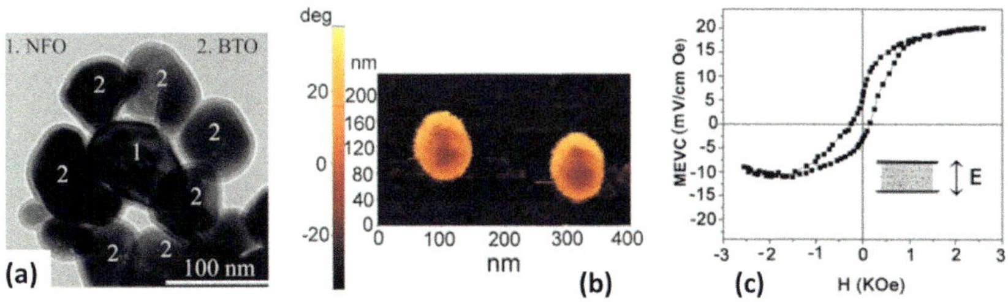

FIGURE 6.38 TEM image showing a chemically assembled core-shell nanoparticle with 100 nm NFO core and 50 nm BTO shell. (b) MFM phase image of the NFO-BTO core shell nanoparticles. (c) ME voltage coefficient measured on a magnetic field assembled thin film of BTO-NFO core shell particles (Sreenivasulu et al., 2014a, 2014b; Srinivasan et al., 2015).

structure for the nanoparticle composites (Sreenivasulu et al., 2014a, 2014b; Srinivasan et al., 2015). Measurements of MEVC indicated a rather weak ME effect in the composites (Sreenivasulu, Popov, et al., 2014). One way to enhance the ME effect is to assemble them into superstructures with the aid of magnetic or electric fields. For example, NFO-BTO core—shell nanoparticles assembled into a 3D film in a magnetic field showed MEVC of ~20 mV/cm Oe as seen in the data of Fig. 6.38. Strong ME coupling in the chemically assembled core—shell particles was also evident from measurements of static magnetic field induced ferroelectric polarization (Popov et al., 2014).

A recent development in studies of ME coupling in the composites is the use of DNA for the assembly of core—shell nanoparticles (Banerjee et al., 2018; Chavez & Gopalan, 2017; Sreenivasulu et al., 2016; Viehland et al., 2018). DNA-assisted assembly of ferrite ferroelectric core—shell nanoparticles is a technique that uses the unique properties of DNA to control the self-assembly of these nanoparticles into desired structures. DNA is a long, double-stranded molecule that is able to bind to itself through complementary base pairing. This property can be used to attach DNA strands to the surface of the nanoparticles, which then allows the nanoparticles to be assembled into specific structures by DNA hybridization. The first step in DNA-assisted assembly is to functionalize the nanoparticles with DNA. This can be done by either covalently attaching DNA to the surface of the nanoparticles or by electrostatically binding DNA to the nanoparticles. Once the nanoparticles are functionalized, they can be assembled into desired structures by hybridization of the DNA strands.

The studies reported involve the use of 200—600 nm BTO for the core and 50—200 nm NFO for the shell. Specifics on the attachment of DNA to the nanoparticles and assembly procedures are described in Refs. 72—75. Figs. 6.39 and 6.40 shows SEM and scanning microwave microscopy

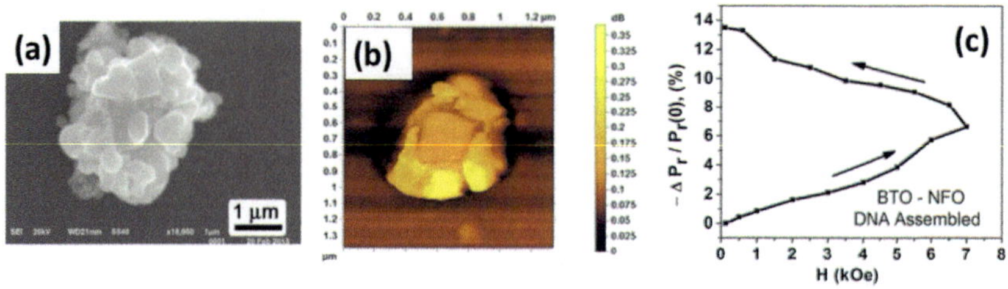

FIGURE 6.39 (a) SEM image showing a DNA-assembled particle with BTO core and NFO shell. (b) Scanning microwave microscopy image at 12 GHz for a 600 nm BTO and 200 nm NFO shell. (c) Fractional change in the remnant polarization due to a static magnetic field applied to a disc of DNA-assembled BTO-NFO core-shell nanoparticle coupling (Banerjee et al., 2018). This may be attributed to the fact that shorter strands would not able to bridge the irregularly shaped nanoparticles into the desired core-shell structure.

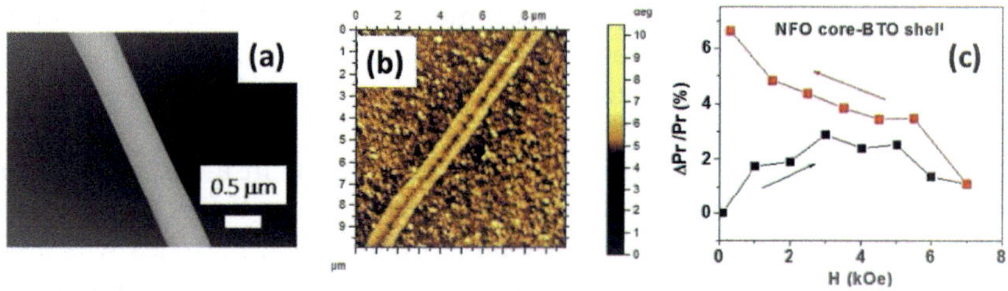

FIGURE 6.40 (a) SEM image of a single core-shell fiber of NFO core and BTO shell. (b) MFM phase image of the fiber. (c) Fractional change in the remnant polarization as a function of static magnetic field H measured in a disk of the fiber (Sreenivasulu et al., 2017).

images of a core-shell particle assembled with the use of DNA (Sreenivasulu et al., 2016). Strength of ME coupling measured by static magnetic field induced polarization on a disc of the core-shell particles (Fig. 6.39) indicate strong coupling in the system (Sreenivasulu et al., 2016). It was reported that long linker region in DNA results in the formation of composites showing strong ME.

6.1.3 Core–shell nanofibers

Core–shell nanowires/nanofibers of ferrites and ferroelectrics are a promising new class of materials with potential applications in a variety of fields, including sensors, and actuators. Core–shell nanofibers have been synthesized using a variety of methods, including electrospinning, sol-gel processing, and chemical vapor deposition. Electrospinning is a versatile method that can be used to produce nanofibers with a wide range of diameters and compositions. The core–shell structure of these materials provides a number of advantages, including enhanced magnetoelectric coupling due to large surface area to volume ratio and improved mechanical properties. The research on these materials is still in its early stages, but the results provided in this section on these materials are very promising. Core–shell nanofibers could be used to create sensors that are sensitive to magnetic fields, electric fields, or temperature. These sensors could be used in a variety of applications, such as medical imaging, security, and environmental monitoring.

Studies reported so far focused mainly on fibers with ferroelectric BTO or PZT and ferrimagnetic spinel or hexagonal ferrites. Petrov et al. modeled the ME coupling in core–shell structures of Y- or W-type hexagonal ferrites (Ni_2Y, Zn_2Y, Zn_2W, Co_2W) and PZT (Petrov et al., 2018). It was predicted that low-frequency MEVC of PZT and Ni_2Y coaxial fibers is expected to be the highest both at low frequency and electromechanical resonance. Caruntu et al. prepared magnetoelectric BTO-$Zn_{1.5}Fe_{1.5}O_4$ core–shell wires by liquid phase deposition method (Yourdkhani et al., 2017). A direct ME coupling coefficient of 1.08 V/cm·Oe was estimated from magnetic field-assisted piezo force microscopy. Chacko et al. (Chacko & Raneesh, 2023), introduced $SrFe_{12}O_{19}$ (SrM) nanoparticles into piezoelectric PVDF-HFP, and successfully prepared SrM-PVDF-HFP multiferroic nanocomposite fibers that had MEVC of 19.9 mV/cm•Oe. A large converse ME coupling coefficient of 1.2×10^{-8} s/m was reported in a single CFO-$PbZr_{0.2}Ti_{0.8}O_3$ core–shell nanofiber through magnetic force microscopy (Liu, Du, et al., 2021). Yadav et al., reported on ME coupling in fibers of $Ba_2Ni_2Fe_{12}O_{22}$-PZT synthesized by electrospinning (Kumar Yadav & Hemalatha, 2022). It was observed that the percentage changes of maximum and remnant polarization under a six kOe field were −8.12% and −12.8%, respectively.

Sreenivasulu and others reported on ME effects in electrospun core–shell fibers of NFO and BTO with the diameter in the range 0.5–1.5 um. The fibers were prepared from sols of the ferrite and BTO and annealed at 600–800 C. SEM image for a single fiber is shown in Fig. 6.39 and the core-shell structure with a well-defined core and shell is seen in the MFM image in the figure. The ME coupling strength measured by fractional change in the remnant polarization was about 2%–7% as shown in Fig. 6.39. They also reported on the nature of converse ME in coaxial nanofiber of NFO and PZT by FMR (Liu, Sreenivasulu, et al., 2020).

Investigations on core–shell fibers of hexagonal ferrites and BTO or PZT also revealed strong ME coupling in the systems (Liu et al., 2020b, 2021b). Fibers with M-type Sr-hexaferrite (SrM) and ferroelectrics were prepared by electrospinning (Liu, Liu, et al., 2020). SEM images, as in Fig. 6.41, showed fibers with uniform diameter and SMM capacitance image showed the core and structure in the fibers (Liu, Liu, et al., 2020) The ME coupling strengths measured by variation of remnant polarization of a disk of the SrM-PZT was −20% (Fig. 6.41) and the MEVC data in

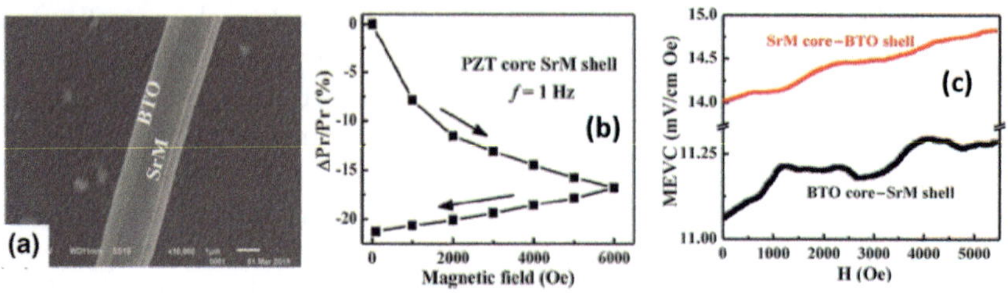

FIGURE 6.41 (a) SEM image of a SrM-BTO core shell fiber. (b) Data on fractional change in the remnant polarization as a function of static magnetic field applied to a disk of fibers of PZT and BTO. (c) Low frequency MEVC versus H for 3D films of fibers of BTO and SrM (Liu, Liu, et al., 2020).

Fig. 6.41 for a 3D film of the fiber was 14–22 mV/cm·Oe. Similarly, studies on fibers with Y-type or W-type hexagonal ferrite and PZT or BTO core–shell indicated strong ME coupling in the composites. The ME coupling strengths reported so far, however, are much lower than predicted by theory. Follow-up efforts are needed to improve the ferroic order parameters of the piezoelectric phase and the magnetostrictive phases in order to improve the ME response.

6.2 Summary

In summary, the advances in sample fabrication, ME response, as well as the coupling mechanisms of nanoscale multiferroic ME composites were discussed in this chapter. The structures of these nanocomposites include magnetic film/piezoelectric substrate heterostructures, 0-3 type particulate nanocomposite films, 1-3 type vertical heterostructures, 2-2 type horizontal heterostructures, ME pillars, core-shell nanoparticle composites, and core-shell nanofibers. Nanoscale ME composites can effectively decrease the size and dissipation, which is the key to the miniaturization of devices. Furthermore, nanoscale ME composites with various structures exhibit properties that may form the basis for a variety of device applications and miniaturization of such devices.

References

Abraham, A. R., Raneesh, B., Woldu, T., et al. (2017). Realization of enhanced magnetoelectric coupling and Raman spectroscopic signatures in 0-0 type hybrid multiferroic core-shell geometric nanostructures. *Journal of Physical Chemistry C, 121*(8), 4352–4362.

Banerjee, A., Zhang, J., Zhou, P., Tuppil, K., Sreenivasulu, G., Qu, H., Zhang, T., Timilsina, R., Chavez, F. A., & Srinivasan, G. (2018). Self-assembly of multiferroic core-shell composites using DNA functionalized nanoparticles. *Journal of Magnetism and Magnetic Materials, 460*, 424–431.

Chacko, S. K., & Raneesh, B. (2023). Dielectric, ferroelectric and magnetoelectric investigations of $SrFe_{12}O_{19}$-embedded PVDF-HFP nanocomposite fiber mats for flexible electronic applications. *Journal of Materials Science, 58*(3), 1158–1170.

Chavez, F. A., & Gopalan, S. (2017). Self-assembly of nanostructures with multiferroic components using nucleic acid linkers. *MRS Communications, 7*(1), 20–26.

Chen, A., Dai, Y., Eshghinejad, A., Liu, Z., Wang, Z., Bowlan, J., Knall, E., Civale, L., MacManus-Driscoll, J. L., Taylor, A. J., Prasankumar, R. P., Lookman, T., Li, J., Yarotski, D., & Jia, Q. (2019). Competing interface and bulk effect–driven magnetoelectric coupling in vertically aligned nanocomposites. *Advanced Science, 6*(19), Article 1901000.

Deng, C., Zhang, Y., Ma, J., Lin, Y., & Nan, C.-W. (2008). *Acta Materialia, 56*, 405.

Duan, C.-G., Jaswal, S. S., & Tsymbal, E. Y. (2006). *Physical Review Letters, 97*, Article 047201.

Eerenstein, W., Mathur, N. D., & Scott, J. F. (2006). *Nature, 442*, 759.

Eerenstein, W., Wiora, M., Prieto, J. L., Scott, J. F., & Mathur, N. D. (2007). *Nature Mater., 6*, 348.

Fernández, C. P., Zabotto, F. L., Garcia, D., et al. (2017). In situ sol-gel co-synthesis at as low hydrolysis rate and microwave sintering of PZT/Fe_2CoO_4 magnetoelectric composite ceramics. *Ceramics International, 43*(8), 5925–5933.

Fetisov, Y. K., & Srinivasana, G. (2006). *Applied Physics Letters, 88*, Article 143503.

Fiocchi, S., Chiaramello, E., Marrella, A., et al. (2022). Modeling of core-shell magneto-electric nanoparticles for biomedical applications: Effect of composition, dimension, and magnetic field features on magnetoelectric response. *PLoS One, 17*(9), 1–17.

Gao, R., Xue, Y. Z., Wang, Z., et al. (2020). Effect of particle size on magnetodielectric and magnetoelectric coupling effect of $CoFe_2O_4$@$BaTiO_3$ composite fluids. *Journal of Materials Science: Materials in Electronics, 31*(11), 9026–9036.

Gao, M., Yang, Y., Rao, W.-F., & Viehland, D. (2021). Magnetoelectricity in vertically aligned nanocomposites: Past, present, and future. *MRS Bulletin, 46*(2), 123–130.

Henrichs, L. F., Mu, X., Scherer, T., et al. (2020). First-time synthesis of a magnetoelectric core-shell composite: Via conventional solid-state reaction. *Nanoscale, 12*(29), 15677–15686.

Hill, N. A. (2000). *J. Phys. Chem. B, 104*, 6694.

Jian, G., & Wong, C. P. (2017). Orientation dependence of magnetoelectric coefficient in BaTiO3/CoFe2O4. *AIP Advances, 7*(7), Article 075307.

Kim, T. C., Ojha, S., Tian, G., Lee, S. H., Jung, H. K., Choi, J. W., Kornblum, L., Walker, F. J., Ahn, C. H., Ross, C. A., & Kim, D. H. (2018). Self-assembled multiferroic epitaxial BiFeO3–CoFe2O4 nanocomposite thin films grown by RF magnetron sputtering. *Journal of Materials Chemistry C, 6*(20), 5552–5561.

Kim, D. H., Sun, X. Y., Aimon, N. M., Kim, J. J., Campion, M. J., Tuller, H. L., Kornblum, L., Walker, F. J., Ahn, C. H., & Ross, C. A. (2015). A three component self-assembled epitaxial nanocomposite thin film. *Advanced Functional Materials, 25*(20), 3091–3100.

Kumar Yadav, S., & Hemalatha, J. (2022). Direct magnetoelectric and magnetodielectric studies of electrospun $Ba_2Ni_2Fe_{12}O_{22}$-$Pb(Zr_{0.52}Ti_{0.48})O_3$ core-shell nanofibers. *Journal of Magnetism and Magnetic Materials, 564*, Article 170174.

Levin, I., Li, J., Slutsker, J., & Roytburd, A. L. (2006). *Adv. Mater., 18*, 2044.

Li, Z., Wang, Y., Lin, Y., & Nan, C.-W. (2009). *Physical Review B: Condensed Matter, 79*, 180406R.

Li, Y., Wang, Z., Yao, J., Yang, T., Wang, Z., Hu, J.-M., Chen, C., Sun, R., Tian, Z., Li, J., Chen, L.-Q., & Viehland, D. (2015). Magnetoelectric quasi-(0-3) nanocomposite heterostructures. *Nature Communications, 6*(1), 6680.

Liu, N., Du, P., Zhou, P., et al. (2021). Magnetoelectric coupling in $CoFe_2O_4$-$Pb(Zr_{0.2}Ti_{0.8})O_3$ coaxial nanofibers. *Journal of American Ceramic Sococitycity, 104*, 948–954.

Liu, M., et al. (2007). *Journal of Applied Physics, 102*, Article 083911.

Liu, M., et al. (2008). *Applied Physics Letters, 92*, Article 152504.

Liu, M., et al. (2009b). *Adv. Funct. Mater, 19*, 1826–1831.

Liu, M., Li, S., Zhou, Z., Beguhn, S., Lou, J., et al. (2012). *Journal of Applied Physics, 112*, Article 063917.

Liu, Y., Liu, Y., Zhou, P., et al. (2020). Strain-mediated magneto-electric interactions in hexagonal ferrite and ferroelectric coaxial nanofibers. *MRS Communications, 10*(2), 230–241.

Liu, M., Obi, O., Lou, J., Stoute, S., Cai, Z., Ziemer, K., & Sun, N. X. (2009). *Journal of Physics D Applied Physics, 42*, Article 045007.

Liu, Y., Sreenivasulu, G., Zhou, P., et al. (2020). Converse magneto-electric effects in a core–shell multiferroic nanofiber by electric field tuning of ferromagnetic resonance. *Scientific Reports, 10*(1), 1–10.

Liu, Y., Zhou, P., Ge, B., et al. (2021). Strain-mediated magneto-electric effects in coaxial nanofibers of y/w-type hexagonal ferrites and ferroelectrics. *Journal of Composites Science, 5*(10).

Lou, J., Liu, M., Reed, D., Ren, Y. H., & Sun, N. X. (2009). *Adv. Mater., 21*, 4711–4715.

Lu, X. Y., Wang, B., Zheng, Y., & Ryba, E. (2007). *Applied Physics Letters, 90*, Article 133124.

Ma, Y. G., Cheng, W. N., Ning, M., & Ong, C. K. (2007). *Applied Physics Letters, 90*, Article 152911.

Martin, L. W., Crane, S. P., Chu, Y. H., Holcomb, M. B., Gajek, M., Huijben, M., Yang, C. H., Balke, N., & Ramesh, R. (2008). *Journal of Physics: Condensed Matter, 20*, Article 434220.

Murugavel, P., Padhan, P., & Prellier, W. (2004). *Applied Physics Letters, 85*, 4992.

Nan, C. W., Bichurin, M. I., Dong, S. X., Viehland, D., & Srinivasan, G. (2008). *Journal of Applied Physics, 103*, 1101.

Nan, C.-W., Liu, G., Lin, Y., & Chen, H. (2005). *Physical Review Letters, 94*, Article 197203.

T. Nan, Z. Zhou, M. Liu, X. Yang, Y. Gao, B. A. Assaf, S. Velu, D. Heiman, H. Luo, Y. H. Lin, C. W. Nan and N. X. Sun, Scientific report, .

Newnham, R. E., Skinner, D. P., & Cross, L. E. (1978). *Materials Research Bulletin, 13*, 525.

Petrov, V. M., Zhang, J., Qu, H., et al. (2018). Theory of magnetoelectric effects in multiferroic core-shell nanofibers of hexagonal ferrites and ferroelectrics. *Journal of Physics D: Applied Physics, 51*(28), Article 284004.

Popov, M., Sreenivasulu, G., Petrov, V. M., Chavez, F. A., & Srinivasan, G. (2014). High frequency magneto-dielectric effects in self-assembled ferrite-ferroelectric core-shell nanoparticles. *AIP Advances, 4*(9), Article 097117.

Rafique, M., Herklotz, A., Dörr, K., & Manzoor, S. (2017). Giant room temperature magnetoelectric response in strain controlled nanocomposites. *Applied Physics Letters, 110*(20), Article 202902.

Ryu, H., et al. (2006). *Applied Physics Letters, 89*, Article 102907.

Ryu, S., Park, J. H., & Jang, H. M. (2007). *Applied Physics Letters, 91*, Article 142910.

Singh, M. P., Prellier, W., Mechin, L., Simon, C., & Raveau, B. (2006). *Journal of Applied Physics, 99*, Article 024105.

Slutsker, J., Levin, I., Li, J. H., Artemev, A., & Royburd, A. L. (2006). *Physical Review B: Condensed Matter, 73*, Article 184127.

Smith, I. T., Zhang, E., Yildirim, Y. A., et al. (2022). Nanomedicine and nanobiotechnology applications of magnetoelectric nanoparticles. *Wiley Interdisciplinary Reviews: Nanomedicine and Nanobiotechnology, 8*, e1849.

Sreenivasulu, G., Lochbiler, T. A., Panda, M., Srinivasan, G., & Chavez, F. A. (2016). Self-assembly of multiferroic core-shell particulate nanocomposites through DNA-DNA hybridization and magnetic field directed assembly of superstructures. *AIP Advances, 6*(4), Article 045202.

Sreenivasulu, G., Petrov, V. M., Chavez, F. A., & Srinivasan, G. (2014). Superstructures of self-assembled multiferroic core-shell nanoparticles and studies on magneto-electric interactions. *Applied Physics Letters, 105*(7), Article 072905.

Sreenivasulu, G., Popov, M., Chavez, F. A., Hamilton, S. L., Lehto, P. R., & Srinivasan, G. (2014). Controlled self-assembly of multiferroic core-shell nanoparticles exhibiting strong magneto-electric effects. *Applied Physics Letters, 104*(5), Article 052901.

Sreenivasulu, G., Zhang, J., Zhang, R., et al. (2017). Multiferroic core-shell nanofibers, assembly in a magnetic field, and studies on magneto-electric interactions. *Materials, 11*(1), 1–15.

Srinivasan, G., Sreenivasulu, G., Benoit, C., Petrov, V. M., & Chavez, F. (2015). Magnetic field directed assembly of superstructures of ferrite-ferroelectric core-shell nanoparticles and studies on magneto-electric interactions. *Journal of Applied Physics, 117*(17), Article 17B904.

Sun, N. X., & Srinivasan, G. (2012). *Spin, 2*(3), Article 1240004.

Tang, X., Gao, M., Leung, C. M., Luo, H., Li, J., & Viehland, D. (2019b). Non-volatility using materials with only volatile properties: Vertically integrated magnetoelectric heterostructures and their potential for multi-level-cell devices. *Applied Physics Letters, 114*(24), Article 242903.

Tang, X., Gao, M., Li, J., & Viehland, D. (2019c). Self-assembled epitaxial BiFeO3-Ni0.65Zn0.35Al0.8Fe1.2O4 nanobelt heterostructures on SrTiO3: Control of magnetic anisotropy, easy axis, and coercivity. *Journal of Applied Physics, 126*(12), Article 123905.

Tang, X., Leung, C. M., Gao, M., Winkler, C., Luo, H., Li, J., & Viehland, D. (2019a). Nanostructure-enhanced magnetoelectric/magnetostrictive properties and reduced losses in self-assembled epitaxial CuFe2O4–BiFeO3 layers on Pb(Mg1/3Nb2/3)O3–33at%PbTiO3 crystals. *Journal of the American Ceramic Society, 102*(9), 5192–5202.

Tang, X., Viswan, R., Gao, M., Leung, C. M., Folger, C., Luo, H., Howe, B., Li, J., & Viehland, D. (2018). Nanopillars with E-field accessible multi-state ($N \geq 4$) magnetization having giant magnetization changes in self-assembled BiFeO3-CoFe2O4/Pb(Mg1/3Nb2/3)-38at%PbTiO3 heterostructures. *Scientific Reports, 8*(1), 1628.

Thankachan, R. M., Raneesh, B., Mayeen, A., et al. (2018). Room temperature magnetoelectric coupling effect in $CuFe_2O_4$-$BaTiO_3$ core-shell and nanocomposites. *Journal of Alloys and Compounds, 731*, 288–296.

Thiele, C., Dörr, K., Bilani, O., Rödel, J., & Schultz, L. (2007). *Physical Review B: Condensed Matter, 75*, Article 054408.

Van Lich, L., Shimada, T., Miyata, K., Nagano, K., Wang, J., & Kitamura, T. (2015). Colossal magnetoelectric effect in 3-1 multiferroic nanocomposites originating from ultrafine nanodomain structures. *Applied Physics Letters, 107*(23), Article 232904.

Vaz, C. A. F., Hoffman, J., Ahn, C. H., & Ramesh, R. (2010). *Adv. Mater., 22*, 2900.

Viehland, D., Fang Li, J., Yang, Y., Costanzo, T., Amin, Y., Gabriel, C., Peng, Z., et al. (2018). Tutorial: Product properties in multiferroic nanocomposites. *Journal of Applied Physics, 124*(6), Article 061101.

Wan, J. G., et al. (2005). *Applied Physics Letters, 86*, Article 122501.

Wang, L., Bian, J., Lu, L., Liang, Z., Zhang, D., Yang, B., Li, L., Lu, G., & Yang, Y. (2020). Interfacial strain driven magnetoelectric coupling in (111)-oriented self-assembled BiFeO3–CoFe2O4 thin films. *Journal of Materials Chemistry C, 8*(10), 3527–3535.

Wang, J., Zhang, Y., Ma, J., Lin, Y., & Nan, C.-W. (2008). *Journal of Applied Physics, 104*, Article 014101.

Wang, P., Zhang, E., Toledo, D., et al. (2020). Colossal magnetoelectric effect in core-shell magnetoelectric nanoparticles. *Nano Letters, 20*(8), 5765–5772.

Wu, T., et al. (2006). *Physical Review B: Condensed Matter, 73*, Article 134416.

Wu, H., Xu, R., Zhou, C., et al. (2022). Effect of core size on the magnetoelectric properties of $Cu_{0.8}Co_{0.2}Fe_2O_4$@$Ba_{0.8}Sr_{0.2}TiO_3$ ceramics. *Journal of Physics and Chemistry of Solids, 160*(12), Article 110314.

Yan, L., Xing, Z., Wang, Z., Wang, T., Lei, G., Li, J., & Viehland, D. (2009). Direct measurement of magnetoelectric exchange in self-assembled epitaxial BiFeO3–CoFe2O4 nanocomposite thin films. *Applied Physics Letters, 94*(19), Article 192902.

Yourdkhani, A., Caruntu, D., Vopson, M., et al. (2017). 1D core-shell magnetoelectric nanocomposites by template-assisted liquid phase deposition. *CrystEngComm, 19*(15), 2079–2088.

Zavaliche, F., Zhao, T., Zheng, H., Straub, F., Cruz, M. P., Yang, P.-L., Hao, D., & Ramesh, R. (2007). *Nano Letters, 7*, 1586.

Zavaliche, F., Zheng, H., Mohaddes-Ardabili, L., Yang, S. Y., Zhan, Q., Shafer, P., Reilly, E., Chopdekar, R., Jia, Y., Wright, P., Schlom, D. G., Suzuki, Y., & Ramesh, R. (2005). *Nano Letters, 5*, 1793.

Zhai, J., Cai, N., Shi, Z., Lin, Y., & Nan, C.-W. (2004). *Journal of Physics D, 37*, 823.

Zhang, Y., Deng, C., Ma, J., Lin, Y., & Nan, C.-W. (2008). *Applied Physics Letters, 92*, Article 062911.

Zhang, J. X., et al. (2007). *Applied Physics Letters, 90*, Article 052909.

Zhang, S.-T., et al. (2009). *Applied Surface Science, 255*, 5092.

Zheng, H., et al. (2004a). *Science, 303*, 661.

Zheng, H., Straub, F., Zhan, Q., Yang, P. L., Hsieh, W. K., Zavaliche, F., Chu, Y. H., Dahmen, U., & Ramesh, R. (2006). *Adv. Mater., 18*, 2747.

Zheng, 79H., Wang, J., Lofland, S. E., Ma, Z., Mohaddes-Ardabili, L., Zhao, T., Salamanca-Riba, L., Shinde, S. R., Ogale, S. B., Bai, F., Viehland, D., Jia, Y., Schlom, D. G., Wuttig, M., Roytburd, A., & Ramesh, R. (2004). *Science, 303*, 661.

Zhong, X. L., et al. (2007). *Applied Physics Letters, 90*, Article 152903.

Zhou, Z., Beguhn, S., Lou, J., Rand, S., Li, M., et al. (2012). *Journal of Applied Physics, 111*, Article 103915.

Ziese, M., Bollero, A., Panagiotopoulos, I., & Moutis, N. (2006). *Applied Physics Letters, 88*, Article 212502.

CHAPTER

7

Magnetoelectric energy harvesting from kHz to GHz

R. Sriramdas[1,2], *Sumanta Kumar Karan*[1], *M. Sanghadasa*[3], *Mirza Bichurin*[4], *O. Sokolov*[4], *Shashank Priya*[1], *Nian Sun*[5], *Zhaoqiang Chu*[6], *Mohammad Javad Pourhosseini Asl*[6], and *Shuxiang Dong*[6]

[1] Department of Materials Science and Engineering, The Pennsylvania State University, University Park, PA, United States; [2] Department of Mechanical Engineering, Amrita School of Engineering, Coimbatore, Amrita Vishwa Vidyapeetham, Coimbatore, Tamil Nadu, India; [3] U.S. Army Combat Capabilities Development Command Aviation & Missile Center, Redstone Arsenal, AL, United States; [4] Politechnic Institute, Novgorod State University, Veliky Novgorod, Russia; [5] Northeastern University, Boston, MA, United States; [6] Peking University, Beijing, China

7.1 Introduction

The need for new type of energy harvesters is growing in demand with the advancement in sensor technologies and electronic devices, particularly in high frequencies. The energy harvesters absorb ambient energy and provide sustainable power to the wireless sensor nodes. Among the most popular sources of ambient energies such as solar, vibration, wind, and electromagnetic waves, magnetoelectric (ME) harvesters absorb the ambient magnetic fields. ME energy harvesting is especially attractive owing to the noncontact type of energy transfer enabling the deployment of IoT systems without the use of external batteries or the associated maintenance. The self-powered IoT systems require compact, lightweight, large bandwidth, and high-power density energy harvesters for their

https://doi.org/10.1016/B978-0-443-18606-6.00004-6

sustained operation. The energy in magnetic fields around any current carrying wire could potentially be harvested by a ME harvester. ME harvesters are composed of piezoelectric and ME material layers forming a resonating structure. The harvesters operating at sub 100 Hz have been extensively studied by several researchers. However, considering the principle of operation of ME harvesters and its performance dependence on the frequency of operation, the present chapter focuses on understanding the performance of the ME harvesters at high frequencies in the range of kHz to GHz.

7.1.1 High-frequency magnetoelectric harvesters

The ME energy harvesters comprising a combination of piezoelectric and ME layers have several key attributes depending on the device geometry and the number of layers of each material. The frequency of operation, size, input AC magnetic field, operating DC bias magnetic field, ME coupling coefficient, mechanical quality factor, and the corresponding power developed are some of the key performance metrics of ME harvesters. A list of ME harvesters operating in kHz frequency range is provided in Table 7.1. It may be noticed that the list is populated in the ascending order of the operating frequency. Furthermore, depending on the materials selected, the ME coupling coefficient, voltage, and the power generated by the ME composite vary from μW to mW. Hence, the principle of operation of the ME composite operation is discussed in the next section.

7.2 Theoretical model for magnetoelectric composites

The ME composite selected to derive the model consists of a piezoelectric layer sandwiched between magnetostrictive (MS) layers. The piezoelectric layer is polarized through thickness, while the MS layers are polarized along the length resulting in a longitudinal-transverse (L-T) combination of the ME composite (Dong et al., 2004a). The MS and piezoelectric materials used for the study are galfenol and PZT-5A.

A pictorial depiction of the galfenol/PZT/galfenol laminate as shown in Fig. 7.1 serves as the ME transducer that generates electrical energy from the magnetic fields. The geometry and the axes convention for the analysis are also shown in the figure. Such a geometry has been analyzed in the literature by constructing a lumped equivalent circuit (Dong et al., 2003; Rupp et al., 2019; Truong et al., 2020). Electromechanical resonance–based modeling of layered composite has been investigated particularly when the sample size is smaller than the electromagnetic

TABLE 7.1 Dependence of output voltage and power on the type of ME composite.

References	Material/type of composite	Frequency (kHz)	ME coupling coefficient (V/cm/Oe)	Device Q	*Bias field* H_{DC} *(Oe)*	*AC field* H_{AC} *(Oe)*	Output voltage (V)/Power (mW)
Li, Wu, et al. (2023)	PZT/Galfenol	1.5	80.2	–	30	2.8	0.11 mW
Onuta et al. (2011)	FeGa/PZT on Si thin film	3.83	–	2000	66.1	6	1 μW
Lima et al. (2019)	All printer P(VDF-TrFE) - CFO	16.2	0.164	–	–	0.75	0.8 V
Dong et al. (2020)	Metglas/PZT/metglas	23.95	2000	–	7.5	56.2μ	0.215 V
Li et al. (2010)	PZT/terfenol-D Bronze substrate	27	–	400	800	1	2V/Oe
Peng et al. (2017)	FeCuNbSiB/terfenol-D/Be-bronze/PZT	35	7.74	–	0	–	0.59 V/Oe
Ma et al. (2017)	Metglas-PZT-SFO	44	4	–	0	3	0.4 V/Oe
Lasheras et al. (2015)	FeCoSiB/PVDF/FeCoSiB	46.8	145.6	–	4.7	0.45	6.4 μW
Li, Zhang, et al. (2023)	Galfenol/PZT/galfenol	62	55	–	150	2.66	1.4 mW
Li, Wen, et al. (2011a)	H-type fork ME composite (Fe–Ni)/PZT	64.6	–	430	–	0.2	61.64 μW

Continued

TABLE 7.1 Dependence of output voltage and power on the type of ME composite.—cont'd

References	Material/type of composite	Frequency (kHz)	ME coupling coefficient (V/cm/Oe)	Device Q	*Bias field* H_{DC} *(Oe)*	*AC field* H_{AC} *(Oe)*	Output voltage (V)/Power (mW)
L'vov et al. (2023)	Ni–Mn-Ga/P(VDF-TrFE)	71.8	18.1	–	485	0.2	Temperature at 323 K
Wang et al. (2010)	Terfenol-D/PMN-PT/terfenol-D	83.6	3.2	–	400	–	1.9 mW/Oe
Annapureddy et al. (2023)	Metglas/Ni/PMN-PZT/Ni/Metglas	88.5	4.2	–	0	1	–
Leung et al. (2022)	Metglas-PZT metglas	95.5	53	–	67.3	–	2.65 V/Oe
Li et al. (2013)	PZT-5/terfenol-D/PZT-8	115	6.6	65/10/1000	178	1	–
Caijiang et al. (2014)	FCNSB–Ni/PZT/Ni–FCNSB	193.3	49.5	–	4	1	1.04 V
Hosur et al. (2023)	Metglas/PZT/metglas	278	–	16.7	125	1	187.3 μW
Swain et al. (2020)	BZT-BCT/Ni laminate	295	22.5	–	145	10	1.5 mW/cm^2
Rajaram Patil et al. (2013)	Ni/PMN-PZT single crystal/Ni	325	190	–	300	–	16 V/cm/Oe over 60 kHz

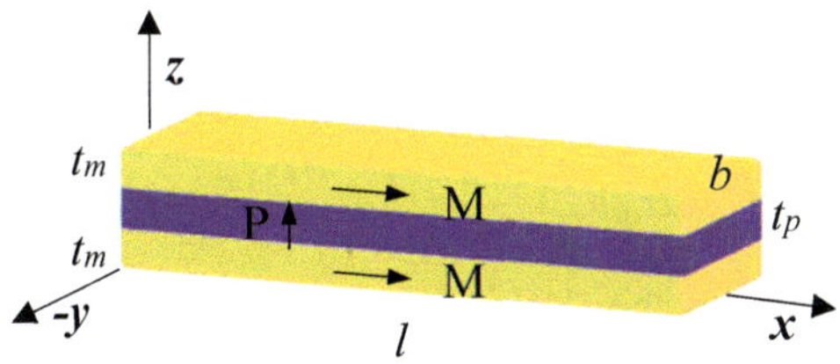

FIGURE 7.1 The ME composite axes convention and geometry.

wavelength (Bichurin et al., 2010). However, the lumped circuit method is inadequate to describe the spatial deformation that occurs along the length of the ME composite. Moreover, the effect of variations in the material properties with DC bias on the power and voltage coefficients (Shi, 2018) is explicitly described in the present analysis. Both temporal and spatial deformations are derived for the ME composite by coupling the MS and piezoelectric effects. The equations of motion are derived from the constitutive relations, force balance, and boundary conditions of the ME composite. Let stress, strain in the MS, and piezoelectric material along x be denoted by T_{11m}, S_{11m}, and T_{11p}, S_{11p}, respectively. The magnetic field along x in the MS material and electric field along $\mathbf{z}$ in the piezoelectric material are represented by H_1 and E_3. It is assumed that H_1 and E_3 are uniform throughout the respective materials. The elastic stiffness at the constant electric field in the piezoelectric material, permittivity at constant strain, and constant magnetic field in MS material are denoted by $c^E{}_{33p}$, $\varepsilon^S{}_{33p}$ and $c^H{}_{33m}$. The effective piezoelectric and piezomagnetic stress constants of interest for the relevant stress components for the piezoelectric and MS materials are e_{31p} and e_{11m}.

7.2.1 Constitutive relations

The constitutive relations among the significant stress components and the electric displacement within the composite are:

$$
\begin{aligned}
T_{11m} &= c^H_{11m}\, S_{11m} - e_{11m}H_1 \\
T_{11p} &= c^E_{11p}\, S_{11p} - e_{31p}E_3 \\
D_3 &= e_{11p}S_{11p} + \varepsilon^S_{33}E_3
\end{aligned}
\tag{7.1}
$$

The length, width of the beam, and thickness of MS and piezoelectric layers are denoted by l, b, t_m, and t_p, respectively. Let the displacement of the infinitesimal element of length dx, the density of MS, and piezoelectric materials be denoted by $u(x,t)$, ρ_m and ρ_p. The equations of motion for the

ME composite using the force balance employing Newton's second law over an infinitesimal element of length d x are derived as:

$$\left(t_p\rho_p + 2t_m\rho_m\right)\frac{\partial^2 u}{\partial t^2} = 2\, t_m\frac{\partial T_{11m}}{\partial x} + t_p\frac{\partial T_{11p}}{\partial x} \tag{7.2}$$

Let thickness ratio, ν, be defined as the ratio of the piezoelectric material thickness t_p and the total thickness $t\ (= t_p + 2\, t_m)$. It is assumed that the deformation is uniform throughout the thickness of the composite resulting in a longitudinal strain component given by $\frac{\partial u}{\partial x}$. The solution to Eq. (7.2) is obtained by using the separation of variables solution method employing Eq. (7.1) with heaviside step function (Erturk & Inman, 2011) for both electric and magnetic fields. Let the solution for displacements be written as $u(x,t) = \sum_{i=0}^{\infty} X_i(t)\psi_i(x)$, where $X_i(t)$ is the temporal displacement and $\psi_i(x)$ is the modal function for the ith mode. The assumed modal solutions for the temporal and spatial displacements (Inman & Singh, 1994) are $X_i(t) = A_i \cos(\omega_i t) + B_i \sin(\omega_i t)$ and $\psi_i(x) = C_i \cos(\omega_i x/a) + D_i \sin(\omega_i x/a)$, where a is given by Eq. (7.3).

7.2.2 Resonance frequency

The nominal resonance frequency in the respective mode is estimated without the coupling terms as:

$$\omega_i = \frac{i\pi a}{l}, \text{ where } \quad a = \sqrt{\frac{c_{11p}\nu + c_{11m}(1-\nu)}{\rho_p\nu + \rho_m(1-\nu)}}, \quad \text{and} \quad i = 1, 2, 3, \ldots \infty \tag{7.3}$$

The natural frequency is determined using the free-free boundary conditions at $x = 0$ and l for the ME composite. In order to describe the equations of motion succinctly in the fundamental mode of vibration ($i = 1$), the integral $\int_0^l \psi^2 \mathrm{d}x$ is denoted by κ. The voltage developed by the ME composite is dependent on the load resistance R_l connected to the ME composite. The current flowing through the resistance is obtained by differentiating the charge estimated by integrating the third equation in (1).

7.2.3 Coupled differential equations

$$M\ddot{X} + C\dot{X} + KX + GV = LH_1 \tag{7.4}$$

$$V + R_l C_p \dot{V} - R_l G\dot{X} = 0 \tag{7.5}$$

The governing equations of motion for the ME composite operating in L-T mode are derived as:

where

$$M = \left(2t_m\rho_m + t_p\rho_p\right)b\kappa, \qquad G = be_{31p}\Psi|_{x=0}^{x=l},$$

$$C = 2\,\zeta\left(2t_m\rho_m + t_p\rho_p\right)b\kappa\omega_n \qquad L = 2\,t_m be_{11m}\Psi|_{x=0}^{x=l}, \text{ and}$$

$$K = (2t_m c_{11m} + t_p c_{11p})b\kappa\omega_n^2/a^2 \qquad C_p = \frac{\varepsilon_{33}^S bl}{t_p}.$$

The type of damping in the composite is assumed to be viscous dissipation with a damping ratio, ζ. The solution to the equations of motion (4) and (5) is obtained by assuming harmonic excitation for the magnetic field as $H_1 = \overline{H}e^{j\omega t}$. The voltage and temporal displacement are also harmonic functions represented by $V = \overline{V}e^{j\omega t}$ and $X = \overline{X}e^{j\omega t}$, respectively. It may be noted that the complex voltage and displacements are given by:

$$\overline{V} = \frac{j\omega GL}{j\omega G^2 + \left(\frac{1}{R_l} + j\omega C_p\right)(K - M\omega^2 + j\omega C)}\overline{H} \tag{7.6}$$

$$\overline{X} = \frac{L}{(K - M\omega^2 + j\omega C) + j\omega G^2/\left(\frac{1}{R_l} + j\omega C_p\right)}\overline{H} \tag{7.7}$$

The complex voltage given by Eq. (7.6) is used to estimate the power developed by the ME composite using $P_{rms} = \overline{V}^2/2R_l$. For brevity in explaining the results, nondimensional parameters—frequency ratio $\Omega = \omega/\omega_n$, piezoelectric coupling factor $k_p^2 = G^2/MC_p\omega_n^2$, quality factor $Q \approx 1/2\zeta$, voltage factor $\Upsilon = \Upsilon[\nu]$, and power factor $\Pi = \Pi[\nu]$ are introduced. It may be noted that the piezoelectric coupling factor $k_p = k_p[\nu]$ is a function of the thickness ratio, ν.

7.2.4 Optimal voltage, geometry, and power output

The open-circuit voltage $\overline{V}_{oc}$ is obtained by setting $R_l \to \infty$ in (6) and rearranging the terms as:

$$\overline{V}_{oc} = \frac{e_{11m}}{e_{31p}}t\,\overline{H}\,\Upsilon, \text{where} \quad \Upsilon = \frac{k_p^2\,(1-\nu)}{1 - \Omega^2 + k_p^2 + j\Omega/Q} \tag{7.8}$$

It may be noted from Eq. (7.8) that the voltage developed by the ME composite is proportional to the total thickness and the ratio of piezomagnetic and piezoelectric stress coupling constants. However, it may be noted that the piezomagnetic stress constant e_{11m} is a function of applied

DC bias magnetic field, that is, $e_{11m} = e_{11m}[H_{DC}]$. Thus, the developed voltage is dependent on both AC and bias magnetic fields. It may also be noted from Eq. (7.8) that the expression excluding U has the units of voltage, while $U[\nu]$ is the dimensionless voltage factor that depends on thickness ratio, driving frequency, quality factor, and coupling factor. As the coupling factor is related to thickness ratio by $k_p^2 \propto \nu / (\rho_m + \nu(\rho_p - \rho_m))$, the higher the thickness ratio, the higher is the coupling factor k_p. With appropriate rearrangement, Eq. (7.6) is indicative of the ME voltage coefficient. However, as the interest is in determining the optimal power from the ME composite, the RMS power is optimized for the load resistance to determine the peak power generated by the ME composite. The P_{rms} is differentiated w.r.t load resistance R_l, and the result is equated to zero to determine the optimal load resistance R_{opt} as:

$$R_{opt} = \frac{1 - \Omega^2 + j\Omega/Q}{j\omega_n C_p \Omega \left(1 - \Omega^2 + k_p^2 + j\Omega/Q\right)} \tag{7.9}$$

The optimal load resistance given by (9) is substituted back into the expression for RMS power P_{rms} to determine the optimal power generated by the composite as:

$$P_{opt} = \omega_n C_p \left(\frac{e_{11m}}{e_{31p}} t\overline{H}\right)^2 \Pi,$$
$$\text{where } \Pi = \frac{k_p^4 \, \Omega (1 - \nu)^2}{8 \left(1 - \Omega^2 + k_p^2 + j\Omega/Q\right) \left(j\left(\Omega^2 - 1\right) + \Omega/Q\right)} \tag{7.10}$$

It may be noticed from Eq. (7.10) that the generated power is proportional to the square of the open-circuit voltage excluding the voltage factor in Eq. (7.8). Thus, the optimal power is proportional to the square of the total thickness, the square of AC magnetic field, natural frequency, device capacitance, and the square of the ratio of piezomagnetic and piezoelectric stress coupling constants, provided the power factor Π is also simultaneously maximized. The effect of the dimensionless parameters k_p, Ω, ν, and Q on the voltage factor U and power factor Π are shown in Fig. 7.2a and b. It may be noted from Fig. 7.2a that the peak voltage occurs at different thickness ratios for different operating frequencies. Nevertheless, the absolute maximum value of the voltage occurs around $\nu = 0.5$ and at a driving frequency slightly above a resonance frequency ($\Omega = 1.02$) as shown in Fig. 7.2a. A similar observation is made by Vopsaroiu et al. (Vopsaroiu et al., 2007). It is worth mentioning that in addition to the power factor, the product $\omega_n C_p$ is also dependent on the thickness ratio. Thus, the optimum thickness ratio for maximizing the power

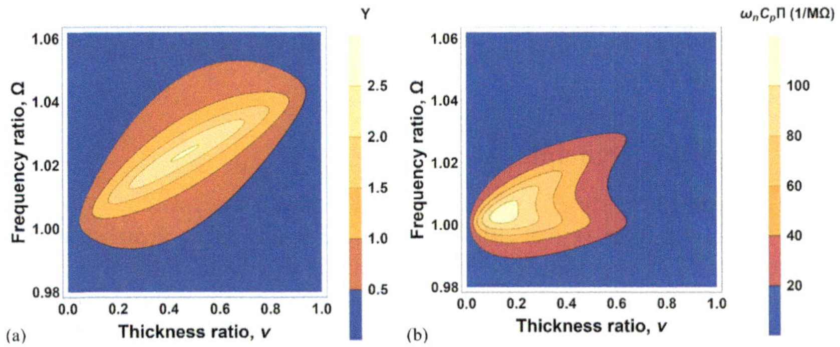

FIGURE 7.2 (a) The voltage factor U as a function of thickness ratio ν, and frequency ratio Ω. The voltage is maximum when the thickness ratio is around 0.5. (b) The power developed by the ME composite for the selected geometry is maximum when the thickness ratio is in the range of 0.12 and 0.19.

developed is governed by the product $\omega_n C_p \Pi$. This product is indicative of the power developed by the ME composite as a function of thickness and frequency ratios. As resonance frequency and the capacitance are needed to determine the distribution, an ME composite of 15 mm length, 1 mm width, and 1 mm total thickness is selected. A quality factor of 100 is assumed for the simulations. It may be noted that, unlike the voltage factor, the power is maximum when the thickness ratio is between 0.12 and 0.19 as shown in Fig. 7.2b. A similar analysis can be adopted to analyze other combinations of ME composites.

7.3 Development of magnetoelectric harvesters

Magnetoelectric (ME) effect is product property of ferromagnetic and piezoelectric materials that is expressed as polarization change in response to an externally applied magnetic field and vice-versa (Fiebig, 2005; Spaldin & Fiebig, 2005). This ferroic coupling provides great opportunity toward developing new applications such as sensors, tunable transformers, filters, memory devices, and energy harvesters (Nan et al., 2020; Priya et al., 2007, 2009). Prior research on ME materials has mainly focused on improving the ME coefficient and sensitivity for their application in sensors. In the past decade, research has started on energy recovery from unused power and developing self-powered sensors and electronics (Beeby et al., 2006; Paradiso & Starner, 2005). ME composites with various connectivity and structures have been developed in order to scavenge the ambient magnetic and mechanical energies (Dong et al., 2008; Lafont et al., 2012; Zhang et al., 2014). These experimental investigations have been complimented by the theoretical modeling, which has led to the understanding of the role of various materials and structure parameters.

7.3.1 Working principle

For energy harvesting applications, by utilizing the product property of the ME composite, one is able to generate the electricity from stray AC magnetic field present in the surrounding and additionally from converting the mechanical vibrations into electricity with higher efficiency. The scavenging mechanism can be described as follows: when the ME composite is placed in an AC magnetic field, the MS layer responds by elongating or contracting, thereby, straining the piezoelectric layer that results in output voltage across the electrical load through a direct piezoelectric effect (Fig. 7.3). Due to the existence of piezoelectric phase in the composite, mechanical oscillation applied on to the composite can directly create electrical voltage. Thus, an ME energy harvester can harness energy from both vibrations and AC magnetic field at the same time. The combination is expected to enhance the power output and conversion efficiency.

7.3.2 Magnetoelectric composite

Magnetoelectric composite can be realized through the interactions between the MS and piezoelectric materials (Zhai et al., 2008). The transfer of magnetic field–induced strain (magnetostriction) from the MS phase to the piezoelectric phase occurs through the elastic coupling. Prior research on ME materials has shown that composites exhibit much higher magnitude of ME coefficient than single-phase materials (Nan et al., 2020; Priya et al., 2007; Zhai et al., 2008). In addition, the design and synthesis of ME composites offer more flexibility resulting in cost-effectiveness and freedom in meeting the shape and size constraints. In the past decade, ME composites with diverse set of materials, connectivity and microstructure have been developed through a variety of processing techniques (Kambale et al., 2012; Martins & Lanceros-Méndez, 2013; Srinivasan, 2010).

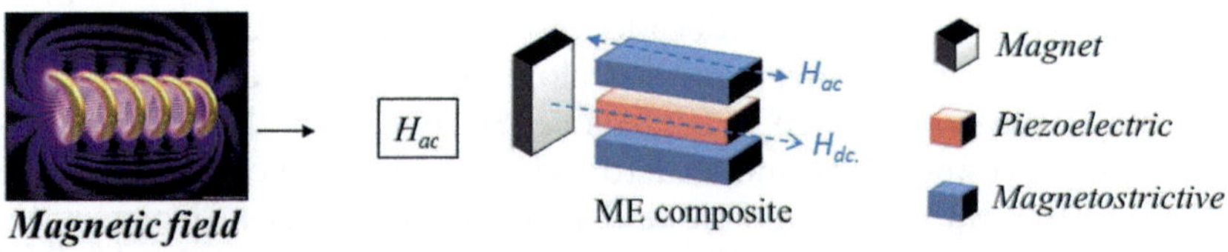

FIGURE 7.3 Schematic diagram depicting magnetic field ME energy harvester.

7.3.2.1 Classification

In general, ME composites can be categorized into different groups based upon phase connectivity, materials selection, operation mode, and synthesis method, as listed in Table 7.2.

Based upon the applied input, the ME effect can be described as: direct ME (DME) effect and converse ME (CME) effect. In DME effect, the coupling coefficient is the ratio of the applied H_{ac} and induced AC electric field (E_{ac}); DME coefficient $= \frac{\text{Magnetic}}{\text{Mechanical}} \times \frac{\text{Mechanical}}{\text{Electric}}$. In CME effect, the coupling coefficient is the ratio between the applied AC electric field and induced AC magnetic field; CME coefficient $= \frac{\text{Electric}}{\text{Mechanical}} \times \frac{\text{Mechanical}}{\text{Magnetic}}$.

Considering the phase connectivity (Newnham et al., 1978), two-phase ME composite system can be typically classified into three common groups: (i) 0-3 type particulate composite, where the first number denotes the magnetic phase (0 dimension particles) embedded in the matrix of piezoelectric phase (3 dimension), (ii) 2-2 type laminate, where the layer-by-layer architecture consists of alternate magnetic and piezoelectric phase, and (iii) 1-3 type cylinder-matrix composite, where nano/micro-rods of magnetic phase are embedded in the piezoelectric matrix. 2-2 type ME laminate structures preserve the physical characteristics of individual phases, minimize the interfacial damping, and exhibit negligible leakage, thereby result in higher ME coupling. A sandwich type configuration with two distinct phases is commonly utilized structure in this category. 1-3 composites have been mainly explored at the micro and nanoscale where they can be easily synthesized through the self-assembly process (Ramesh & Spaldin, 2009). 2-2 type laminates have been widely investigated due to the ease of fabrication and design flexibility. The most common configurations for 2-2 type laminates are MS /piezoelectric (MP) bilayer structure and MS /piezoelectric/ MS (MPM) trilayer structure. Based on different combination of poling and magnetization direction, we can identify four basic operation modes including L-T/T-L, T-T, L-L, where the first letter represents magnetization orientation, the second

TABLE 7.2 Classification of magnetoelectric composites.

Characterization	Classes		
Applied input	DME	CME	
Connectivity	0-3	2-2	1-3
Operation mode	L-T/T-L	T-T	L-L
Material	All ceramics	Ceramic metal	Polymer metal
Synthesis	Bonding	Co-fired	Thin/thick film

letter denotes polarization direction, L stands for longitudinal, and T stands for transverse. All these modes can be easily realized by fabricating structures through epoxy bonding of layers with different magnetization/poling conditions. Both theoretical calculation and experiments have demonstrated that the magnitude of ME voltage coefficient for laminates follows the order (L-L)>(L-T/T-L)>(T-T). Generally, the ME voltage coefficient of L-L mode structure is nearly an order of magnitude higher than those of L-T type laminates (Dong et al., 2006). This is due to the efficient stress transfer ($d_{33} > 2d_{31}$ since Poisson's ratio is smaller than 0.5) when the piezoelectric layer is longitudinally poled. The low value of (T-T) mode is result of the demagnetization effect of the magnetostricitive phase. It is also important to mention here that other more complex modes such as composites in form of bimorph, unimorph, ring, and cylindrical structure have also been developed and provide more choices toward designing suitable harvester structure (Dong et al., 2004b, 2005; Zhai et al., 2006).

7.3.2.2 Materials selection

A suitable selection of piezoelectric and MS materials is an essential step toward realization of high-performance ME composite.

7.3.2.2.1 Piezoelectric materials

Generally, piezoelectric materials exhibiting high magnitude of piezoelectric voltage constant ($g = d/\varepsilon$), low dielectric, and piezoelectric losses are desired. The choice of microstructure is diverse including random ceramics, textured ceramics, single crystals, and polymers, as listed in Table 7.3. In this table, d_{33} is the longitudinal piezoelectric constant, d_{31} is the transverse piezoelectric constant, ε_r is the relative dielectric

TABLE 7.3 List of typical piezoelectric materials (Maurya et al., 2013; Priya et al., 2007; Yan, K-h, et al., 2013; Zhai et al., 2008).

	Materials	d_{33}(pC/N)	d_{31}(pC/N)	ε_r(1 kHz)	k_{33}	T_c(°C)
Random ceramics	$BaTiO_3$	190	−78	1700	0.49	120
	PZT-5	400	−175	1750	0.72	360
Textured ceramics	NBT-BT	322			0.57	90
	PMN-PZT	1100		2310	0.84(k_p)	204
Single crystal	PMN-PZT	1530		4850	0.93	211
Polymer	PVDF		−28	6	0.2(k_t)	170

constant, k_{33} is the longitudinal electromechanical coupling coefficient, and T_c is the Curie temperature.

One of the first ME composites was developed by using $BaTiO_3$ (BTO) as the piezoelectric phase. To further enhance the ME coupling, Pb(Zr,Ti) O_3 (PZT)-based compositions have replaced BTO as the piezoelectric phase due to its larger electromechanical coefficients. With the development of relaxor-based single crystals such as $Pb(Zn_{1/3}Nb_{2/3})O_3-PbTiO_3$ (PZN-PT) and $Pb(Mg_{1/3}Nb_{2/3})O_3-PbTiO_3$ (PMN-PT), the performance of ME composite can be dramatically enhanced based on their ultrahigh electromechanical properties ($d_{33} > 2000$ pC/N and k_{33}>92%). However, the low Curie temperature and high cost hinder their application for ME composites. In addressing this challenge, alternative strategy of texturing ceramics has been proposed. Yan et al. have demonstrated template grain growth (TGG) technique that yields microstructure with <001> preferred grain orientation resulting in high d and g value comparable to that of single crystal. By utilizing this textured ceramics as the piezoelectric layer in the ME composites, a 3.4 times improvement of α_{ME} was realized as compared to that of random ceramics. Lead-free random and textured ceramics have also been considered for fabricating ME composites. Other than these ceramic materials, polymeric piezoelectric compositions such as polyvinylidene difluoride (PVDF) have been utilized for developing highly flexible ME devices (Martins & Lanceros-Méndez, 2013; Pukada, 2000).

7.3.2.2.2 Magnetostrictive materials

Magnetostrictive materials undergo a change in shape due to the variation of the magnetization state under external magnetic field. The magnetic domains in these materials can be reoriented along the direction of applied external magnetic field. During the process of magnetization, the material deformation can be quantified through magnetostriction (λ) coefficient whose saturation value (λs) is used for the comparative material analysis. Table 7.4 provides a list of commonly used MS materials for ME composites.

Among the MS materials listed in Table 7.4, ferrites have larger electrical resistance in comparison to the metal/alloy magnetic materials but smaller magnetostriction. Ferrites are widely used in co-processing with piezoelectric ceramics due to their stability under high temperature. Magnetic alloy terfenol-D exhibits large magnetostriction (1400ppm) with small permeability ($\approx$10). Terfenol-D is suitable for high-performance ME laminate but requires large DC magnetic bias to reach the saturation magnetostriction. Metglas has extremely high relative permeability of >10,000 and high piezomagnetic coefficient (q) but requires low processing temperatures. Metglas-based ME laminates require extremely small H_{bias} (Dong et al., 2006; Wang et al., 2011).

TABLE 7.4 List of typical magnetostrictive materials (Priya et al., 2007; Zhai et al., 2008).

	Materials	λs (ppm)
	Fe_3O_4	40
Ceramic	$NiFe_2O_4$	−26
	$CoFe_2O_4$	−110
	Fe	−9
Metal	Co	−62
	Ni	−33
	Terfenol-D ($Tb_{1-x}Dy_xFe_2$)	1400
Alloy	Galfenol (Fe−Ga)	200
	Metglas ($Fe_{81}Si_{3.5}B_{13.5}C_2$)	40

7.3.2.3 *Synthesis of ME composite*

A variety of synthesis techniques have been used for fabricating ME composites, including epoxy bonding, cofiring, thick-film printing, and thin-film deposition.

7.3.2.3.1 Epoxy bonding

Epoxies bonding is widely used for synthesizing ME laminates due to the ease of fabrication and low cost. Since this process does not require high temperature, most of the material combinations are suitable for the design of ME composites. Epoxy bonding reduces the misfit strain at the interface arising due to the thermal expansion mismatch between the different layers. It also eliminates the atomic interdiffusion and/or chemical reaction between the layers. Thus, this method offers great design flexibility than that of conventional ceramic processing techniques. However, the epoxy layer with low mechanical strength will dampen the generated strain, induce noise floor, and therefore limit the ME response (Wang et al., 2012). Further, manual fabrication process presents another challenge in implementation of industrial manufacturing (Li, Berry, et al., 2011; Wang et al., 2011).

7.3.2.3.2 Cofiring

To optimize the ME interface coupling, direct bonding that completely excludes the epoxy in the composite is necessary. Cofiring technique not only enhances the elastic coupling between phases but also dramatically decreases the cost due to its compatibility with industrial production process that is commonly used for multilayer capacitors. Considering the materials characterization, cofiring of piezoelectric ceramics and ferromagnetic ceramics process is developed (Priya et al., 2007; Srinivasan et al., 2001, 2002). However, it is extremely challenging due to the large difference in shrinkage rates and thermal expansion mismatch among different phases.

7.3.2.3.3 Thin-film deposition

It has been widely accepted that strong interfacial bonding between two phases is required for a large ME response. However, either the low mechanical strength from the epoxy bonding or the possibility of interdiffusion during the high temperature cofiring process may limit the bonding between the two phases. More importantly, these two methods rise another issue that the large sample size may limit its implementation for MEMS-scale functional ME devices. In addressing these issues, thin-film deposition method has been widely employed in the development of ME composites (Jahns et al., 2013; Martin et al., 2008; Prellier et al., 2005). Nanostructure thin-film composites including 0-3 type particular films, 2-2 type layered heterostructures, 1-3 type vertical heterostructures were investigated using either physical vapor deposition (PVD) methods (pulsed laser deposition, sputtering, molecular beam epitaxial) or chemical vapor deposition (CVD) methods (sol-gel spin coating, metal-organic chemical vapor deposition). By optimizing the deposition parameters and materials selection, ME composite thin films with coherent interfaces in various nanostructures were developed by precise control of the lattice mismatch and thickness between the two phases (Vaz et al., 2010).

7.3.2.3.4 Thick-film printing

For practical applications, such as sensors or energy harvesting systems, the weak signal of ME voltage output due to the size effect cannot be neglected, even though large ME coupling can be achieved in the thin-film composites. For a film with 100 nm scale, the real ME voltage detected from the films might be negligibly small at the magnitude of μV level. Therefore, to enhance the voltage output of the ME films, thick films in micron scale are highly demanded. Ryu. et al. demonstrated an efficient method for fabricating thick films using aerosol deposition (AD) (Park et al., 2009). Piezoelectric and MS nano-powders were sprayed on substrate through high pressure gas flow at room temperature. A high print rate with the magnitude of several microns/min can therefore be obtained. Composite films with various materials, thickness, and configuration can also be easily printed on metal/ceramic substrates with high density and good interfacial bonding strength (Kambale et al., 2013; Oh et al., 2012).

7.3.3 ME composite characterization

It can be seen from the above discussion that the possibilities are numerous (materials, connectivity, operation mode, synthesis method and so on) for obtaining the ME composites. To better utilize these composites, investigation of their working behaviors would contribute to understanding the effect of some variables associated with the performance of ME composites. In the following, we will discuss the characteristic behavior of conventional ME composites.

7.3.3.1 *ME versus H_{dc}*

In ME composites, the measured direct ME coupling coefficient is the field conversion ratio between applied H_{ac} and induced E_{ac} under H_{dc}, $\alpha_{ME} = \delta E_{ac}/\delta H_{ac}$. In general, under a constant applied AC magnetic field, the ME coupling coefficient first increases with increasing H_{dc}, reaching a maximum at an optimized DC bias (H_{bias}), and then decreases with further increasing H_{dc}. This peak behavior in response to varying H_{dc} can be further explained by taking the interaction coupling relationship as given below (Cho & Priya, 2011):

$$\alpha_{ME} = \left| \frac{\partial T}{\partial S} \times \frac{\partial D}{\partial T} \times \frac{\partial E}{\partial D} \right| \times \frac{\partial S}{\partial H}, \tag{7.11}$$

where S is the mechanical strain, T is the mechanical stress, D is the electric displacement, E is the electric field, and H is the magnetic field. Since the first part of the above expressions is a nonmagnetic factor, one can rewrite Eq. (7.11) as:

$$\alpha_{ME} \propto \frac{\partial S}{\partial H} = \frac{d\lambda}{dH} = q, \tag{7.12}$$

which indicates that the ME coefficient is directly related to the nature of ferromagnetic phase (λ: magnetostriction, q: piezomagnetic coefficient) and the effectiveness of elastic coupling between the two phases. The dependence of α_{ME} on q indicates the requirement for additional H_{dc}. In summary, typical behavior of the α_{ME}, λ, q and their relationship as a function of H_{dc} can be illustrated as shown in Fig. 7.4.

7.3.3.2 *ME versus f_{ac}*

Another important phenomenon of the ME composite is the ME coupling as a function of applied AC magnetic field (ME vs. f_{ac}). When the AC magnetic field is applied at frequency corresponding to electromechanical resonance (EMR) for the piezoelectric phase or ferromagnetic resonance (FMR) for the magnetic phase in the ME composite, the ME voltage coefficient shows a peak behavior with the magnitude increased by a factor of up to 100×. Bichurin et al. first demonstrate a theory

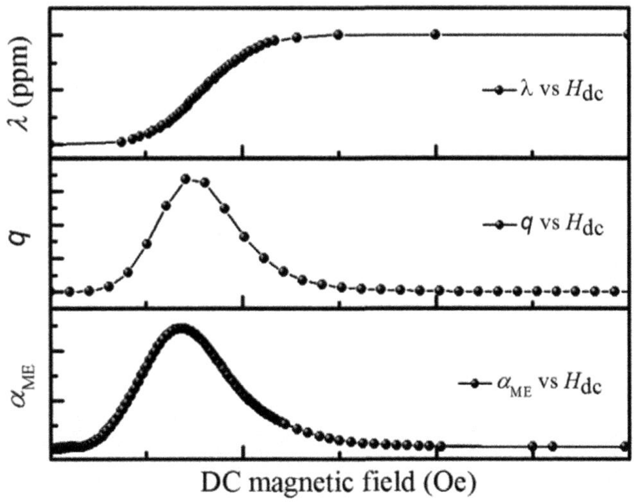

FIGURE 7.4 Typical magnetoelectric behaviors as function of DC magnetic field.

calculation applied to a bilayer NFO-PZT laminates, which produced a 40 times increase in $a_{ME,33}$ at electromechanical resonance compared to that of off-resonance (Bichurin et al., 2003). Similar theories have been developed for the ME effect at ferromagnetic and magnetoacoustic resonance in the composites (Bichurin et al., 2002, 2005). Cho et al. analyzed the direct and converse ME effect in laminate composites and found that DME is maximized at f_a, while the CME is maximized at f_r, as shown in Fig. 7.5 (Cho & Priya, 2011). This phenomenon can be further explained by using piezoelectric constitutive equations and combining it with frequency dependent capacitance of piezoelectric layer.

Generally, for application of ME composite in EMR range, the fundamental resonance frequency can be tuned by the length of the laminates (L) where the first longitudinal resonance $f_L = 1/(2L\sqrt{\rho s_{11}})$ (ρ is the laminate density, s_{11} is the elastic compliance of the laminate). However, for practical use of the composites at low frequency, the corresponding composite dimension would be prohibitively large. Therefore, a model was developed for bilayer ME composite working at bending mode, which shows enhanced coupling at low frequency in an unimorph configuration (Petrov et al., 2009). Alternative approaches which utilize unique laminate configurations have been proposed to achieve low resonance frequency with a limit size of the components. Using these approaches, resonance enhancement of the ME coefficient has been achieved in the frequency range of 100 Hz, using cm-long laminates.

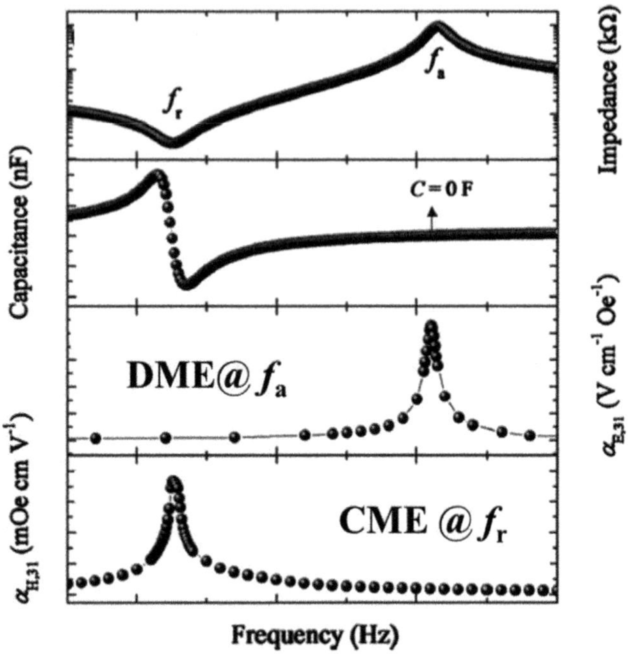

FIGURE 7.5 Impedance, capacitance, and ME behaviors as a function AC frequency.

7.3.3.3 ME versus configuration

It is known that ME response as a function of DC magnetic field is tightly related with the nature of MS materials. From the functional devices point of view, it is important to control the magnitude of ME voltage coefficient and its response. Thus, there are lots of studies on shapes (disk, cylinder, plate, toroid, sphere, and so on) and sizes (numbers of layers, layer thickness, length, and width) of the MS material toward obtaining high ME voltage coefficient. It is widely accepted that ME composite in rectangular shape with large length/width aspect ratio, small thickness and optimized volume ratio will generate high ME voltage coefficient. To better explain the ME versus configuration relationship, the size-induced demagnetization effect was taken into consideration. For ferromagnetic materials, the demagnetization field is directly proportional to demagnetization factor (N_d), $H_d = MN_d$ (M is the magnetization), where N_d is dependent on the dimension and geometry. Thus, the internal effective magnetic field (H_{eff}) in the magnetic phase can be written as $H_{eff} = H_{bias}$-H_d, where H_{bias} is the induced external magnetic field. Therefore, in order to achieve same magnitude of H_{eff}, one needs to apply larger H_{bias} when H_d is larger and vice-versa (Zhou et al., 2012). Furthermore, this can also be corelated with the magnetic flux concentration (B_{eff}) in magnetic phase:

$$B_{eff} = \mu_0\left(H_{eff} + M\right) = \mu_0(H_{bias} + M) - \mu_0 MN_d, \tag{7.13}$$

where a smaller N_d will result in a stronger B_{eff}. It has been shown that high magnetic flux concentration in magnetic phase has positive effect on the ME response of ME laminates (Fang et al., 2009; Gao et al., 2011). Thus, some researchers use the enlarged magnetic layer to enhance the effective magnetic flux density along the piezo/magnetic interface for obtaining larger ME coefficient.

7.4 Self-biased magnetoelectric energy harvester

The dependence of a_{ME} on q indicates the requirement for additional DC magnetic bias. Depending on the optimum magnetic DC bias, ME composites can exhibit a maximum ME voltage coefficient. One needs to use permanent magnet or electromagnet to provide external H_{bias}, which in turn results in problems of large device size and electromagnet interference. Thus, a novel design is desired if one needs to present an ME composite with large tunable α_{ME} in the absence of DC magnetic field, which will pave the way for on-chip devices utilizing ME phenomenon such as energy harvesters, AC magnetic field sensors and high-frequency circuit components. Fig. 7.6 illustrates the schematic of conventional ME composites and self-biased ME composites.

7.4.1 Development of self-biased ME composite

To address this issue, efforts to reduce H_{bias} have been emphasized. One approach to tune the magnitude of optimum DC bias from 6.8 kOe to

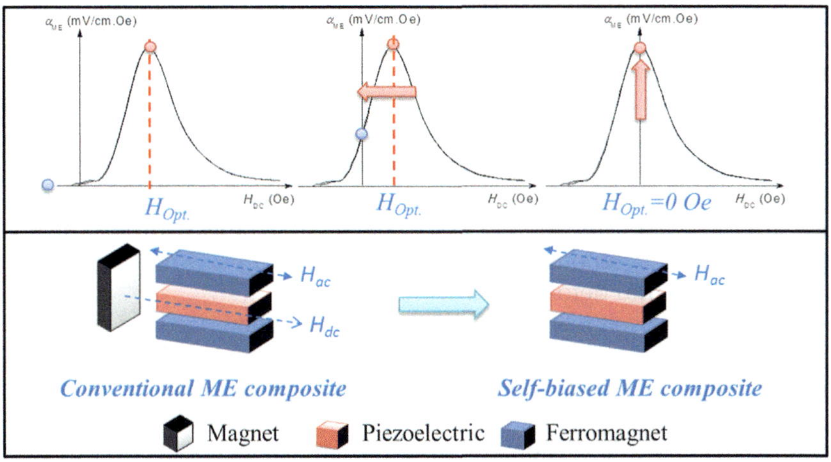

FIGURE 7.6 Schematic of conventional ME effect behavior and self-biased ME behavior.

5Oe is by optimizing the composition and shape of the MS phase. Furthermore, ME composite structures with large tunable a_{ME} in the absence of DC magnetic field, namely "Self-biased ME composites," have been developed in various configuration using different working principles.

The first observation of self-biased ME effect can be originated from G. Srinivasan et al., in 2002. They found hysteresis and remanence of a_{ME} as function of H_{bias} in cofired LSMO-PZT laminates (Srinivasan et al., 2001). Similar behavior has been found in other laminates (Laletin et al., 2005). However, this phenomenon did not attract much attention in the following decade. Many efforts have been focused on optimizing materials, structures, and working mode to enhance the ME voltage coefficient and sensitivity. In 2010, Mandal et al. demonstrated a compositionally graded ferrites/PZT laminate structure, which produces nonzero ME voltage at zero DC magnetic bias (Mandal et al., 2010). They first correlated this nonzero ME behavior with the presence of flexural deformation and grading induced out-of-plane internal magnetic field in the ferrite. Subsequently, Yang et al. presented a methodology that regular ME laminate composites can possess self-bias ME response by changing vibration mode through electrical connections (Yang et al., 2010). Soon after that, Srinivasan et al. and his coworkers systematically investigated the hysteresis and remanence in ME effects in functionally graded MS -piezoelectric layered composites. A theory was also proposed for the self-biased graded composites (Laletin et al., 2012; Mandal et al., 2011; Sreenivasulu et al., 2011). Yang et al. first analytically studied the lead-free based self-biased laminates and proposed an electrically controlled core-free magnetic flux control device by using the self-biased converse ME effect (Lage et al., 2012; Onuta et al., 2011). Alternative approaches have been suggested for the case of thin films that rely on magnetic field dependence of resonance frequency and angular dependence of exchange bias field. This phenomenon is promising for applications such as magnetic field sensor and energy harvester systems. The focus of these studies on bulk structure has been mainly on understanding the nature of the self-biased effect in terms of grading of ferromagnetic composition and bending resonance. However, these structures present another level of practical difficulty for on-chip components as synthesis process has to account for heterogenous graded structure or two-phase magnetic layers along with flexural deformation.

Subsequently, self-biased ME composites with homogenous MS phase have been developed. Yuan et al. first demonstrated a simple Ni/PZT bilayer laminate structure that exhibits tunable self-bias phenomenon with single-phase magnetic material. In this study, fundamental understanding of the role of domain structure and demagnetization field of magnetic materials toward self-biased effect is provided. The self-biased response is related to the nature of magnetic phase and can be controlled by variation in demagnetization state and the resultant differential magnetic flux distribution. Meanwhile, Yan et al. successfully cofired a self-biased composite NCZF/Ag/PMNPT/Ag/NCZF with homogeneous magnetic phase (Fig. 7.7) (Yan, Zhou, & Priya, 2013). Cofired structure provides the opportunity of interfacial engineering to obtain the optimum strain transfer from two phases with a giant self-biased response >1V/cm.Oe. The self-biased response in the cofired laminates is explained by the magnetostriction hysteresis of the magnetic phase and the built-in strain from the internal electrode during cofiring process.

In summary, the working principle of the self-biased ME response can be categorized into three groups: (1) flexural deformation with graded ferromagnetic structure; (2) build-in stress arising from inner electrode; and (3) nature of magnetization hysteresis in the MS phase. The development of self-biased ME composites can be concluded as shown in Fig. 7.7. Self-biased ME composites with functional graded magnetic phase were first reported. With the development of cofired techniques, self-biased ME response can be dramatically enhanced. Finally, with the discovery of tunable self-biased effect in homogenous laminates, the self-biased behavior was well controlled and implemented into an energy harvester device.

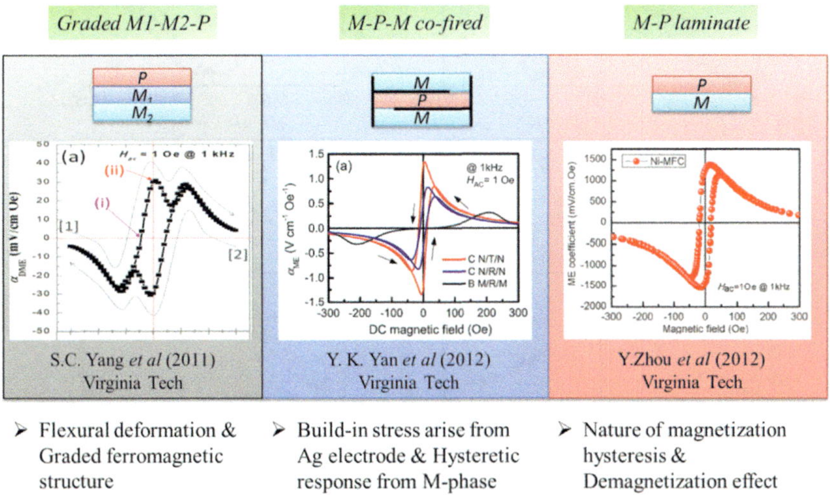

FIGURE 7.7 Development of self-biased magnetoelectric composite (Chul Yang et al., 2011; Yan, Zhou, & Priya, 2013; Zhou et al., 2012).

7.4.2 Working principle of Ni/PZT bilayer self-biased ME composite

Among all these types of self-biased ME composites, the simple Ni/PZT bilayer laminate structure provides the most promising design feasibility and potential toward development of ME energy harvesters and MEMS scalable devices. Thus, we will discuss in detail the working principle of this Ni/PZT self-biased ME composites.

Considering the intrinsic relationship of α_{ME} for composites $= \left|\frac{\partial T}{\partial S} \times \frac{\partial D}{\partial T} \times \frac{\partial E}{\partial D}\right| \times \frac{\partial S}{\partial H}$, where the first part is the nonmagnetic factor, one can derive $\alpha_{ME} \propto \frac{\partial S}{\partial H} = \frac{d\lambda}{dH} = q$. The self-biased hysteretic behavior is related to the nature of the MS materials. To further elucidate the reason for self-biased hysteretic behavior, systematic study was conducted to compare Ni and metglas under applied external magnetic field, as shown in Fig. 7.8 (Chul Yang et al., 2011). The hysteretic behavior of magnetization for Ni

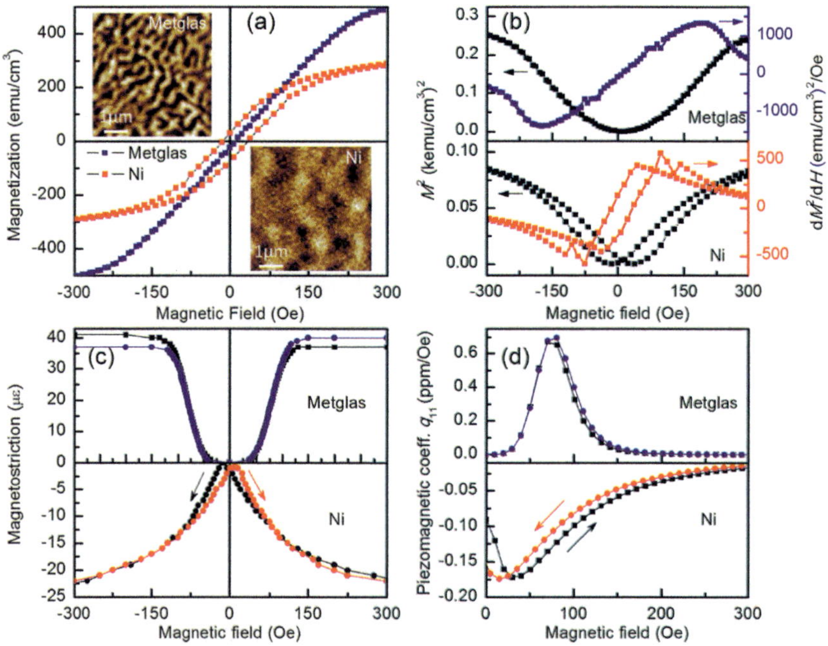

FIGURE 7.8 Magnetization, magnetostriction, and piezomagnetic coefficient performance of metglas and Ni under applied DC magnetic field (Chul Yang et al., 2011). (a) Magnetization-magnetic field (M-H) response. Inset shows magnetic domain images on the surface of Metglas and Ni scanned by MFM; (b) Plot of square magnetization-magnetic field and differential of square magnetization-magnetic field; (c) DC magnetic field dependence of the magnetostriction; (d) DC magnetic field dependence of piezomagnetic coefficients for Ni and Metglas.

can be immediately noticed in comparison with metglas. This phenomenon could be attributed to the differences in the crystallinity and domain structure in both materials, as magnetic force microscopy (MFM) images show in inset of Fig. 7.8a. Nanosized striped domains in metglas can be easily aligned in the direction of applied H-field in comparison to that of macrosized domains in Ni. The absence of grain boundary in amorphous metglas contributes toward the lower coercive field and high reversibility. However, nickel foil with long range ordering required higher field to achieve the random state resulting in larger hysteresis in the magnetization curve. The magnetization behavior can be also correlated with magnetostriction through the relation given as (Ito et al., 1980; Vlasko-Vlasov et al., 2000):

$$\varphi \sim \frac{3\lambda\sigma}{\left(K + 2\pi M^2\right)} \quad or\ \lambda \propto M^2, \tag{7.14}$$

where φ is the angle of magnetic moments, K and σ are the anisotropy constant and stress, respectively. Considering $\alpha_{\mathrm{ME}} \propto q = \mathrm{d}\lambda/\mathrm{d}H$, the relationship $\alpha_{\mathrm{ME}} \propto \mathrm{d}M^2/\mathrm{d}H$ can be used to predict the natural behavior of ME composites with respect to applied magnetic field. It was found that both of the M^2-H curve $\left(\mathrm{d}M^2/\mathrm{d}H\right)$-$H$ behaviors of metglas and Ni matched well with experimental data of λ and q as shown in Fig. 7.8b–d. Therefore, it is believed that difference in the domain and microstructure of metglas and Ni as observed by MFM and resultant different behavior of M^2-H and $\left(\mathrm{d}M^2/\mathrm{d}H\right)$-$H$ behavior is the origin of the self-biased ME response. Thus, a ferromagnetic material like Ni which has obvious hysteretic behavior of magnetostriction, and the nonzero piezomagnetic coefficient is essential for self-biased ME response.

From the point of view of functional ME devices, it is important to control the magnitude of self-biased ME voltage coefficient. As discussed before, ferromagnetic layer geometry plays an important role toward determining the demagnetization factor and the resultant effective internal magnetic field, leading to a shift of H_{bias} and tunability of the ME voltage coefficient. Thus, the self-biased response can be controlled through investigating the MS phase geometry effect, as shown in Fig. 7.9 (Chul Yang et al., 2011).

In all cases, a_{ME} shows hysteretic loop with anticlockwise direction sweep. The values of increasing-field maximum (a_{H1}) and decreasing-field maximum (a_{H2}) with thicker Ni layer were notably higher than that for thinner ones while also requiring larger magnetic biases (increasing-field optimum, *H1* and decreasing-field optimum, *H2*). The magnitude of self-biased point (a_{H0}) slightly decreases as thickness ratio increases as shown in Fig. 7.9c. The maximum self-biased ME response was achieved at $t_{\mathrm{m}/P} = .5$. With fixed thickness ratio at $t_{\mathrm{m}/P} = .5$, as shown

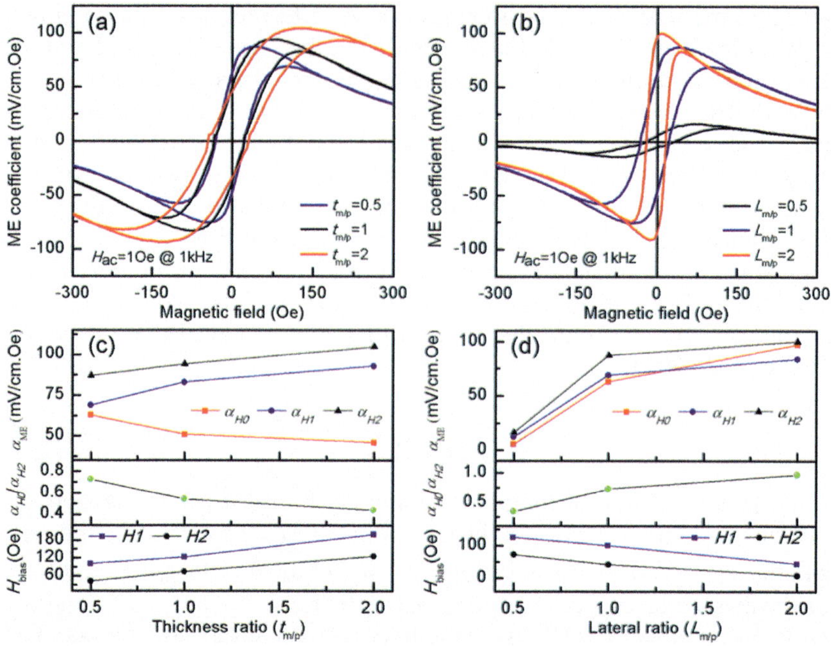

FIGURE 7.9 Magnetostrictive phase geometry effect on the tunability of self-biased response for the Ni/PZT bilayer system (Chul Yang et al., 2011). (a) ME voltage coefficient of Ni-PMT laminates with varied thickness ratio; (b) ME voltage coefficient of Ni-PMT laminates with varied lateral dimension; ME voltage coefficient values of self-biased point, increase-field maximum, decrease-field maximum and self-bias coefficient as function of (c) thickness ratio and (d) lateral dimension.

in Fig. 7.9b, the magnitude of $a_{H0\sim H2}$ increased dramatically as lateral ratio ($L_{m/p}$) increased from 0.5 to 2. Notably, a giant self-bias coefficient (a_{H0}/a_{H2}) of 97% was achieved at $L_{m/P}=2$, compared to 34.4% and 72.5% at $L_{m/P}=.5$ and 1, as shown in Fig. 7.9d. The hysteresis loop was tilted toward y-axis as the optimum magnetic bias (*H1* and *H2*) decreased. Based upon these results, one can obtain and tune the self-bias coefficient by selecting suitable magnetic phase with proper geometry and dimension.

7.4.3 Self-biased magnetoelectric energy harvester

Based on the understanding of Ni/PZT self-biased ME composite, Zhou et al. developed the first ME energy harvester that can scavenge stray AC magnetic field in the absence of DC magnetic field, in terms of self-biased ME energy harvester (Zhou et al., 2013). The device consists of a piezoelectric microfiber composite (MFC) bonded to a Ni plate in a cantilever configuration. Fig. 7.10 illustrate the performance of the Ni-MFC self-biased energy harvester.

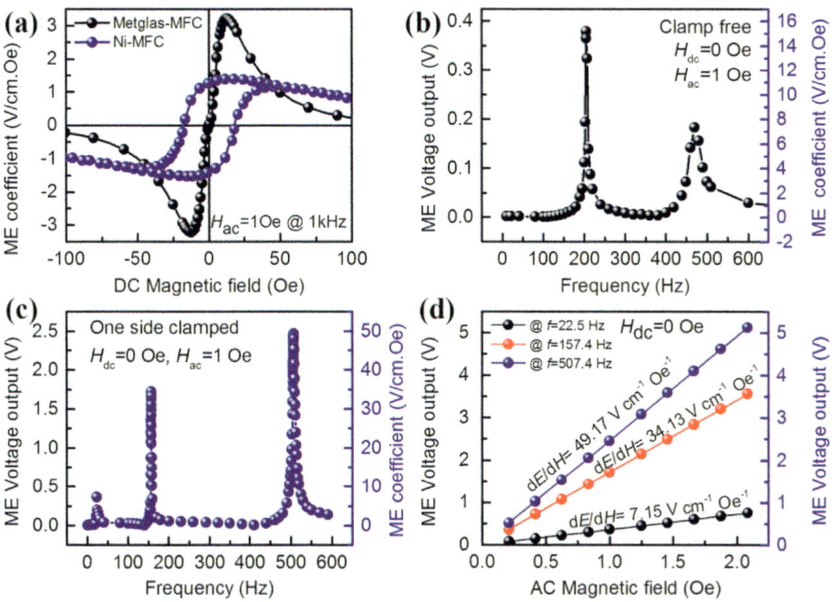

FIGURE 7.10 Performance of the Ni-MFC self-biased energy harvester (Zhou et al., 2013). (a) ME coefficient of Metglas-MFC and Ni- MFC bilayer laminates as function of DC magnetic field; (b) ME voltage output as a function of applied AC magnetic field frequency for the selfbiased ME composite under clamp free condition. (c) ME voltage output as a function of applied AC magnetic field frequency under zerobiased condition; (d) Source free ME voltage output of the dual-phase EH measured at different frequencies under various applied AC magnetic field.

As most of the ferromagnetic materials exhibit zero piezomagnetic coefficient near zero bias, the a_{ME} of a metglas-MFC structure increases from zero to a peak value at the optimum bias and then decreases. In comparison, the a_{ME} of Ni-MFC bilayer shows a hysteretic behavior during H_{dc} sweep (anticlockwise direction) with a large response of $\sim\pm1.25$ V cm^{-1} Oe^{-1} at zero DC magnetic bias, $\sim$98% of the maximum α_{ME} value (1.38 V cm^{-1} Oe^{-1}). This results from the hysteretic behavior of magnetostriction and nonzero piezomagnetic coefficient as discussed in the previous sections.

Under zero-biased condition ($H_{dc} = 0$ Oe), resonance peaks with improved ME voltage output corresponding to bending modes were observed, as shown in Fig. 7.10b. It should be noted that even though there was no applied H_{dc}, self-biased ME response follows the same trend and exhibits enhanced magnitude of voltage coefficient at resonance frequency. The ME laminate was further clamped at one end and attached a Ni tip mass at the free end for enhancing the bending vibration and reducing the resonance frequency, as shown in Fig. 7.10c. The ME voltage output increased with increasing frequency where the maximum ME coefficient with the magnitude of ~50 V cm^{-1} Oe^{-1} was obtained at the resonance frequency of 507.4 Hz. Fig. 7.10d shows that the ME voltage output linearly increased with increasing AC magnetic field at three longitudinal resonance frequencies. The a_{ME} calculated from the slope matches quite well with the measured value in Fig. 7.10c. Large ME voltage with the magnitude of > 5V was achieved at 507.4 Hz, $H_{ac} = 2$ Oe, which can be further increased with increasing H_{ac}.

These results clearly demonstrate that, by using self-biased ME energy harvester, large ME voltage can be generated under zero-biased condition and can be further improved with increasing the AC magnetic field strength at the resonance mode. Although most of the investigations on the ME harvesters revealed that the requirement of DC magnetic field is indispensable, many studies claim zero bias to enhance the power output (Annapureddy et al., 2023; Ma et al., 2017; Peng et al., 2017). In addition, the simple structure of this self-biased ME energy harvester provides great potential toward developing MEMS-scalable high-performance energy harvesting system.

7.5 High-frequency energy harvesting from ME composite

The energy harvesting at GHz frequencies through ME effect is attracting the researchers owing to its applications in implants for neural activity and monitoring that can simultaneously sense and harvest the energy (Das et al., 2020). Considering the frequency of response for the given size of the device, the ME composite has been used as an antenna (Chen et al., 2019). The high Q at high frequencies offers significant advantages in sensitivity as a sensor. The resonant ME composite is also used as a sensor to measure ultralow magnetic fields.

7.5.1 NEMS ME sensors

A NEMS magneto electric sensor was developed by Nan et al. in self-biased configuration. The resonator is based on FeGaB/Al_2O_3 multilayers with the resonance at 215 MHz.

The change in the natural frequency as shown in Fig. 7.11 as a function of the bias magnetic field enormously improves the sensitivity of the magnetic field sensor. It is observed that both admittance amplitude and the resonance frequency follow the same trend with the ambient bias field (Nan et al., 2013). The sensor is found to have excellent sensitivity and linearity at very low bias fields particularly in the range of 50 pT to 4 nT.

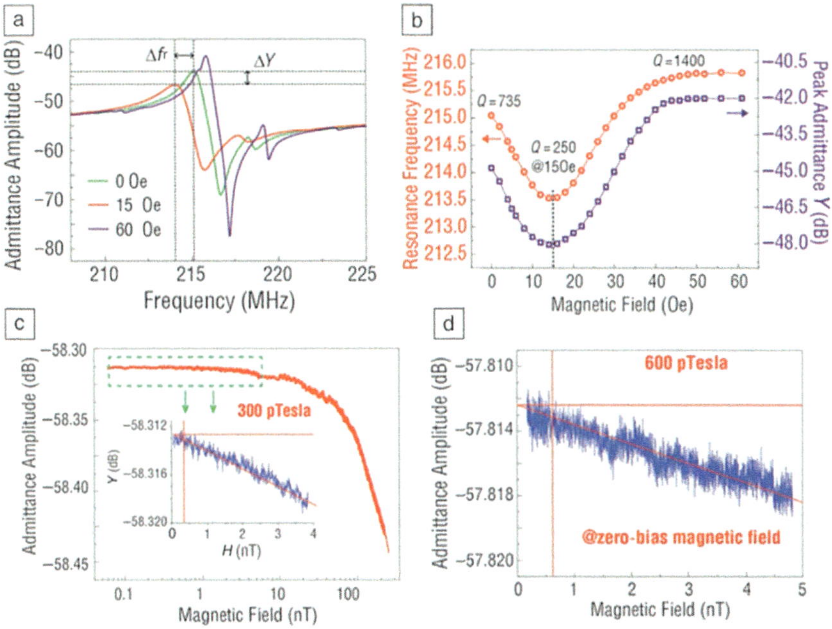

FIGURE 7.11 Admittance profile as a function of frequency and the bias magnetic field (Nan et al., 2013). (a) Admittance curve and Butterworth–van Dyke (BVD) model fitting of the fabricated NEMS magnetic field sensor. (b) Resonance frequency and admittance amplitude at the resonance frequency as a function of DC magnetic field. The sensitivity and linearity of the magnetic field sensor, (c) showing the admittance amplitude as a function of a minute varied DC bias magnetic field superimposing a fixed DC field of 5 Oe. (d) at zero bias magnetic field in a magnetically unshielded environment.

However, the detecting range for the sensor is observed to be above 300 pT. When a zero bias is enforced, the detection range has moved down to 600 pT.

7.5.2 ME antenna with acoustic actuation

The miniaturization of conventional antennas is a key challenge than can be achieved using ME composites. At high frequencies of 30–300 MHz and ultrahigh frequencies of 0.3–3 GHz, the conventional approach entails in generating oscillating electric fields. Considering the coupling between electric field and magnetic fields, electromagnetic radiation can also be achieved using changing magnetic fields. The strong ME coupled devices can be designed to radiate electromagnetic waves.

These devices act as acoustically actuated ME antenna with customized transceiver. The receiver is designed to have RF magnetic field transformed to appropriate voltage or charge output through the ME effect. Furthermore, the associated difference in the wavelength between acoustic and RF wave frequencies enables shrinking the size of the antenna significantly (Nan et al., 2017). Fig. 7.12 shows the effect of width (w) of NPRs and thickness (t) of FeGaB/AlN thin-film FBARs on the resonance frequency. It is observed that when both transmitter and receiver are given the same power input, ME antenna offers higher gain over coil antenna. Thus, ME coupling enhances the performance of the antenna at high frequencies. The compact implantable medical devices using ME antenna offer compliant specific absorption rate compared with miniaturized coils. A wireless power transfer efficiency that is up to two orders higher than the microcoils using ME antenna spanning $250 \times 174\ \mu m^2$ has been demonstrated by Zaeimbashi et al. (Zaeimbashi et al., 2021). They have demonstrated the simultaneous sensing and harvesting of magnetic field using the developed ME antenna. Moreover, brain–computer interfacing has been realized with a highest figure of merit of 2701 using an ME antenna with an on-chip energy harvesting circuit operating at a frequency of 2.6 GHz over a bandwidth spanning 3.37 MHz (Nasrollahpour et al., 2021a). An array of ME antennas was

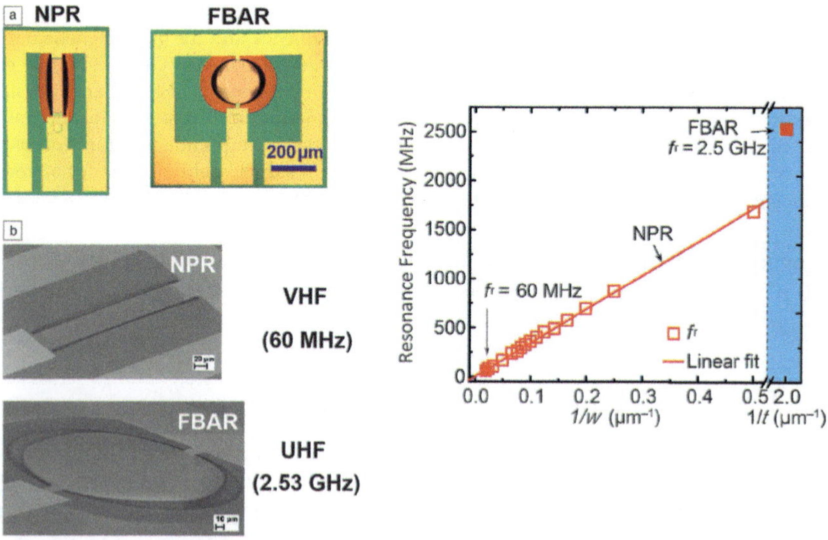

FIGURE 7.12 Configuration and wide bandwidth of nanoplate resonators (NPRs) and FBARs and in the range of very-high-frequency and ultra-high-frequency range (Nan et al., 2017). (a) The optical images showing the device layout of the (ME) NPR and FBAR, (b) Scanning electron microscopy (SEM) images of the fabricated the ME NPR and ME FBAR showing the suspended plate and AlN anchors. The plot shows the resonance frequency as the function of one over device width (1/W) for NPR devices and one over AlN thickness for FBAR devices.

proposed for harvesting electromagnetic energy to power an implantable device called NanoNeuroRFID that can detect RF magnetic fields as low as 40 pT operating in the range of 10–100 MHz (Zaeimbashi et al., 2019). Additionally, circuit level simulations for estimating and characterizing the energy harvested by a 10-stage ME energy harvester spanning 400 × 75 μm were verified (Das et al., 2023). It has been demonstrated that acoustically coupled nano ME antennas provide highest antenna gain with high sensitivity to magnetic field (Lin et al., 2018). An ME harvester has been used in dosimetry applications to sense a sound noise of almost 73 dB with area as small as 0.0435 mm^2 (Nasrollahpour et al., 2021b). The ME antennas not only provide low profile conformality with good impedance matching but also miniaturize the antenna size.

In summary, the ME coupling in ME composites enables their application to wireless power transfer in addition to magnetic field sensing at high frequencies. Self-biased ME composites offer the advantage of operating with large ME coupling at no additional DC magnetic field. The usage of bilayer ME layers is a potential approach for achieving self-biasing in ME composites. A Ni-MFC ME composite has been shown to achieve a remarkable self-biasing with 49.1 V/cm/Oe operating at 0.5 kHz. A theoretical model of an M-P-M ME composite is presented to estimate the optimum impedance and power generated by the composite, which can be extended to any other combinations of MS and piezoelectric layers. The compact ME antennas acoustically coupling the transmitting and receiving units offer up to two orders of size benefit. Furthermore, ME antenna resonators at MEMS scale operating in MHz to GHz range can be connected together for developing broadband reconfigurable antennas. The ME antennas have been demonstrated to sense magnetic field and simultaneously harvest electromagnetic energy. The ME antennas have been designed to achieve a figure of merit as high as 2701 for wireless power transfer in GHz frequencies extending their use to implantable devices. The structural designs coupled with ME antenna extend the wireless sensing network and energy harvesting applications to 5G frequency ranges. With ultralow size combined with the large figure of merit, ME composites stand as one of the best candidates while simultaneously sensing the magnetic field and harvesting the electromagnetic energy.

Acknowledgments

The authors acknowledge S. Abrishami, N. Lobandi, and M. Nasrollahpour for their contribution to enhancing the quality of the book chapter.

References

Annapureddy, V., Park, S. H., Song, H., & Ryu, J. (2023). Tunable self-biased magnetoelectric effect in magnetization-graded magnetoelectric composites. *Journal of Alloys and Compounds, 957*, Article 170121. https://doi.org/10.1016/j.jallcom.2023.170121

Beeby, S. P., Tudor, M. J., & White, N. M. (2006). Energy harvesting vibration sources for microsystems applications. *Measurement Science and Technology, 17*(12), R175–R195. https://doi.org/10.1088/0957-0233/17/12/R01

Bichurin, I., Filippov, A., Petrov, M., Laletsin, M., Paddubnaya, N., & Srinivasan, G. (2003). Resonance magnetoelectric effects in layered magnetostrictive-piezoelectric composites. *Physical Review B: Condensed Matter and Materials Physics, 68*(13). https://doi.org/10.1103/PhysRevB.68.132408

Bichurin, M. I., Petrov, V. M., Averkin, S. V., & Filippov, A. V. (Oct. 2010). Electromechanical resonance in magnetoelectric layered structures. *Physics of the Solid State, 52*(10), 2116–2122. https://doi.org/10.1134/S1063783410100161/METRICS

Bichurin, M. I., Petrov, V. M., Kiliba, Y. V., & Srinivasan, G. (2002). Magnetic and magnetoelectric susceptibilities of a ferroelectric/ferromagnetic composite at microwave frequencies. *Physical Review B: Condensed Matter and Materials Physics, 66*(13), 1–10. https://doi.org/10.1103/PhysRevB.66.134404

Bichurin, M. I., Petrov, V. M., Ryabkov, O. V., Averkin, S. V., & Srinivasan, G. (2005). Theory of magnetoelectric effects at magnetoacoustic resonance in single-crystal ferromagnetic-ferroelectric heterostructures. *Physical Review B: Condensed Matter and Materials Physics, 72*(6). https://doi.org/10.1103/PhysRevB.72.060408

Caijiang, L., et al. (2014). *Zero-biased magnetoelectric composite Fe73.5Cu1Nb3Si13.5B9-Ni-Pb(Zr1−x,Tix)O3 for current sensing.pdf* (pp. 498–501).

Chen, W. S., Cheng, C. M., & Zhang, G. R. (2019). Multi-input multi-output magnetoelectric dipole antennas comprising dual monopole feed antennas for energy harvesting. *Sensors and Materials, 31*(2), 509–518. https://doi.org/10.18494/SAM.2019.2107

Cho, K. H., & Priya, S. (2011). Direct and converse effect in magnetoelectric laminate composites. *Applied Physics Letters, 98*(23), 4. https://doi.org/10.1063/1.3584863

Chul Yang, S., Cho, K. H., Park, C. S., & Priya, S. (2011). Self-biased converse magnetoelectric effect. *Applied Physics Letters, 99*(20). https://doi.org/10.1063/1.3662420

Das, D., et al. (2020). A radio frequency magnetoelectric antenna prototyping platform for neural activity monitoring devices with sensing and energy harvesting capabilities. *Electron, 9*(12), 1–11. https://doi.org/10.3390/electronics9122123

Das, D., et al. (2023). Circuit-level modeling and simulation of wireless sensing and energy harvesting with hybrid magnetoelectric antennas for implantable neural devices. *IEEE Open Journal of Circuits and Systems, 4*(March), 139–155. https://doi.org/10.1109/ojcas.2023.3259233

Dong, S., et al. (2005). Circumferential-mode, quasi-ring-type, magnetoelectric laminate composite-a highly sensitive electric current andor vortex magnetic field sensor. *Applied Physics Letters, 86*(18), 1–3. https://doi.org/10.1063/1.1923184

Dong, C., et al. (2020). A portable very low frequency (VLF) communication system based on acoustically actuated magnetoelectric antennas. *IEEE Antennas and Wireless Propagation Letters, 19*(3), 398–402. https://doi.org/10.1109/LAWP.2020.2968604

Dong, S., Li, J. F., & Viehland, D. (2004a). Characterization of magnetoelectric laminate composites operated in longitudinal-transverse and transverse–transverse modes. *Journal of Applied Physics, 95*(5), 2625–2630. https://doi.org/10.1063/1.1644027

Dong, S., Li, J. F., & Viehland, D. (2004b). Voltage gain effect in a ring-type magnetoelectric laminate. *Applied Physics Letters, 84*(21), 4188–4190. https://doi.org/10.1063/1.1756676

Dong, S., Li, J. F., & Viehland, D. (Oct. 2003). Longitudinal and transverse magnetoelectric voltage coefficients of magnetostrictive/piezoelectric laminate composite: Theory. *IEEE*

Transactions on Ultrasonics, Ferroelectrics, and Frequency Control, 50(10), 1253–1261. https://doi.org/10.1109/TUFFC.2003.1244741

Dong, S., Zhai, J., Li, J., & Viehland, D. (2006). Near-ideal magnetoelectricity in high-permeability magnetostrictive/piezofiber laminates with a (2-1) connectivity. *Applied Physics Letters, 89*(25), 4. https://doi.org/10.1063/1.2420772

Dong, S., Zhai, J., Li, J. F., Viehland, D., & Priya, S. (2008). Multimodal system for harvesting magnetic and mechanical energy. *Applied Physics Letters, 93*(10). https://doi.org/10.1063/1.2982099

Erturk, A., & Inman, D. J. (Mar. 2011). Piezoelectric energy harvesting. *Piezoelectric Energy Harvest*, 54–63. https://doi.org/10.1002/9781119991151

Fang, Z., Lu, S. G., Li, F., Datta, S., Zhang, Q. M., & El Tahchi, M. (2009). Enhancing the magnetoelectric response of Metglas/polyvinylidene fluoride laminates by exploiting the flux concentration effect. *Applied Physics Letters, 95*(11). https://doi.org/10.1063/1.3231614

Fiebig, M. (2005). Revival of the magnetoelectric effect. *Journal of Physics D Applied Physics, 38*(8), R123–R152. https://doi.org/10.1088/0022-3727/38/8/R01

Gao, J., Gray, D., Shen, Y., Li, J., & Viehland, D. (2011). Enhanced dc magnetic field sensitivity by improved flux concentration in magnetoelectric laminates. *Applied Physics Letters, 99*(15). https://doi.org/10.1063/1.3650713

Hosur, S., Karan, S. K., Priya, S., & Kiani, M. (2023). Short-range communication for small biomedical implants using magnetoelectric effect. In *2023 IEEE int. Symp. Circuits syst* (pp. 11–15). https://doi.org/10.1109/ISCAS46773.2023.10181393

Inman, D. J., & Singh, R. C. (1994). *Engineering vibration* (Vol 3). Englewood Cliffs, NJ: Prentice Hall.

Ito, S., Aso, K., Makino, Y., & Uedaira, S. (1980). Magnetostriction and magnetization of iron-based amorphous alloys. *Applied Physics Letters, 37*(7), 665–666. https://doi.org/10.1063/1.92029

Jahns, R., et al. (2013). Giant magnetoelectric effect in thin-film composites. *Journal of the American Ceramic Society, 96*(6), 1673–1681. https://doi.org/10.1111/jace.12400

Kambale, R. C., et al. (2013). Colossal magnetoelectric response of PZT thick films on Ni substrates with a conductive LaNiO3 electrode. *Journal of Physics D Applied Physics, 46*(9). https://doi.org/10.1088/0022-3727/46/9/092002

Kambale, R. C., Jeong, D. Y., & Ryu, J. (2012). Current status of magnetoelectric composite thin/thick films. *Advances in Condensed Matter Physics, 2012.* https://doi.org/10.1155/2012/824643

Lafont, T., et al. (2012). Magnetostrictive-piezoelectric composite structures for energy harvesting. *Journal of Micromechanics and Microengineering, 22*(9). https://doi.org/10.1088/0960-1317/22/9/094009

Lage, E., et al. (2012). Biasing of magnetoelectric composites. *Nature Materials, 11*(6), 523–529. https://doi.org/10.1038/nmat3306

Laletin, V. M., et al. (2005). Frequency and field dependence of magnetoelectric interactions in layered ferromagnetic transition metal-piezoelectric lead zirconate titanate. *Applied Physics Letters, 87*(22), 1–3. https://doi.org/10.1063/1.2137450

Laletin, U., et al. (2012). Hysteresis and remanence in magnetoelectric effects in functionally graded magnetostrictive-piezoelectric layered composites. *Physical Review B: Condensed Matter and Materials Physics, 85*(10). https://doi.org/10.1103/PhysRevB.85.104404

Lasheras, A., et al. (2015). Energy harvesting device based on a metallic glass/PVDF magnetoelectric laminated composite. *Smart Materials and Structures, 24*(6). https://doi.org/10.1088/0964-1726/24/6/065024

Leung, C. M., Liu, Y., Li, Q., & Wang, F. (Jul. 2022). A giant dual-resonant in magnetoelectric composite based on electrical and electromechanical coupling towards enhanced working bandwidth and frequency tuning. *Physica Status Solidi - Rapid Research Letters, 16*(7), Article 2200046. https://doi.org/10.1002/PSSR.202200046

Li, M., Berry, D., Das, J., Gray, D., Li, J., & Viehland, D. (2011). Enhanced sensitivity and reduced noise floor in magnetoelectric laminate sensors by an improved lamination process. *Journal of the American Ceramic Society, 94*(11), 3738–3741. https://doi.org/10.1111/j.1551-2916.2011.04659.x

Li, P., et al. (2013). Wide-bandwidth high-sensitivity magnetoelectric effect of magnetostrictive/piezoelectric composites under adjustable bias voltage. *Sensors and Actuators A: Physical, 201*, 164–171. https://doi.org/10.1016/j.sna.2013.07.005

Li, P., Wen, Y., Jia, C., & Li, X. (2011a). A magnetoelectric composite energy harvester and power management circuit. *IEEE Transactions on Industrial Electronics, 58*(7), 2944–2951. https://doi.org/10.1109/TIE.2010.2076308

Li, P., Wen, Y., Liu, P., Li, X., & Jia, C. (2010). A magnetoelectric energy harvester and management circuit for wireless sensor network. *Sensors and Actuators A: Physical, 157*(1), 100–106. https://doi.org/10.1016/j.sna.2009.11.007

Li, F., Wu, T., Ma, R., Li, J., Gao, X., & Zhu, J. (2023a). Multi-frequency response magnetoelectric cantilever beam devices for magnetic field energy harvesting and detection. *IEEE Sensors Journal, 23*(14), 15489–15494. https://doi.org/10.1109/JSEN.2023.3280203

Li, F., Zhang, X., Wu, T., Li, J., Gao, X., & Zhu, J. (2023b). Enhancement of magnetoelectric coupling and anisotropy by Galfenol/PZT/Galfenol magnetoelectric sandwich device. *Sensors and Actuators A: Physical, 349*(2022), Article 114020. https://doi.org/10.1016/j.sna.2022.114020

Lima, A. C., et al. (2019). All-printed multilayer materials with improved magnetoelectric response. *Journal of Materials Chemistry C: Materials for Optical and Electronic Devices, 7*(18), 5394–5400. https://doi.org/10.1039/c9tc01428d

Lin, H., et al. (2018). Future antenna miniaturization mechanism: Magnetoelectric antennas. In *, 2018. IEEE MTT-S int. Microw. Symp. Dig* (pp. 220–223). https://doi.org/10.1109/MWSYM.2018.8439678

L'vov, V. A., et al. (2023). Giant magnetoelectric effect of Ni–Mn-Ga/piezopolymer composites tailored by a martensitic transformation. *Composites Science and Technology, 241*, Article 110101. https://doi.org/10.1016/j.compscitech.2023.110101

Ma, J. N., Xin, C. Z., Ma, J., Lin, Y. H., & Nan, C. W. (2017). Design and analysis of a self-biased broadband magnetoelectric cantilever operated at multi-frequency windows. *AIP Advances, 7*(3). https://doi.org/10.1063/1.4978872

Mandal, S. K., Sreenivasulu, G., Petrov, V. M., & Srinivasan, G. (2010). Flexural deformation in a compositionally stepped ferrite and magnetoelectric effects in a composite with piezoelectrics. *Applied Physics Letters, 96*(19). https://doi.org/10.1063/1.3428774

Mandal, S. K., Sreenivasulu, G., Petrov, V. M., & Srinivasan, G. (2011). Magnetization-graded multiferroic composite and magnetoelectric effects at zero bias. *Physical Review B: Condensed Matter and Materials Physics, 84*(1). https://doi.org/10.1103/PhysRevB.84.014432

Martin, L. W., et al. (2008). Multiferroics and magnetoelectrics: Thin films and nanostructures. *Journal of Physics: Condensed Matter, 20*(43). https://doi.org/10.1088/0953-8984/20/43/434220

Martins, P., & Lanceros-Méndez, S. (2013). Polymer-based magnetoelectric materials. *Advances in Functional Materials, 23*(27), 3371–3385. https://doi.org/10.1002/adfm.201202780

Maurya, D., Zhou, Y., Yan, Y., & Priya, S. (2013). Synthesis mechanism of grain-oriented lead-free piezoelectric Na 0.5Bi0.5TiO3-BaTiO3 ceramics with giant piezoelectric response. *Journal of Materials Chemistry C: Materials for Optical and Electronic Devices, 1*(11), 2102–2111. https://doi.org/10.1039/c3tc00619k

Nan, C. W., Bichurin, M. I., Dong, S., Viehland, D., & Srinivasan, G. (2020). Multiferroic magnetoelectric composites: Historical perspective, status, and future directions. *Progress in Advanced Dielectrics, 103*, 191–293. https://doi.org/10.1142/9789811210433_0005

Nan, T., et al. (2017). Acoustically actuated ultra-compact NEMS magnetoelectric antennas. *Nature Communications, 8*(1), 1–7. https://doi.org/10.1038/s41467-017-00343-8

Nan, T., Hui, Y., Rinaldi, M., & Sun, N. X. (2013). Self-biased 215MHz magnetoelectric NEMS resonator for ultra-sensitive DC magnetic field detection. *Scientific Reports, 3*, 1–6. https://doi.org/10.1038/srep01985

Nasrollahpour, M., et al. (2021a). Magnetoelectric (ME) antenna for on-chip implantable energy harvesting. In *Proc. Annu. Int. Conf. IEEE eng. Med. Biol. Soc. EMBS* (pp. 6167–6170). https://doi.org/10.1109/EMBC46164.2021.9629823

Nasrollahpour, M., et al. (2021b). Magnetoelectric antenna for miniaturized acoustic noise dosimetry applications. *IEEE Sensors Letters, 5*(10), 1–4. https://doi.org/10.1109/LSENS.2021.3109529

Newnham, R. E., Skinner, D. P., & Le, C. (1978). Connectivity and piezoelectric-pyroelectric composites. *Materials Research Bulletin, 13*, 525–536.

Oh, N. K., et al. (2012). 2-2 structured magnetoelectric composites by aerosol deposition. *Journal of the American Ceramic Society, 95*(3). https://doi.org/10.1111/j.1551-2916.2011.05050.x

Onuta, T. D., Wang, Y., Long, C. J., & Takeuchi, I. (2011). Energy harvesting properties of all-thin-film multiferroic cantilevers. *Applied Physics Letters, 99*(20), 10–13. https://doi.org/10.1063/1.3662037

Paradiso, J. A., & Starner, T. (2005). Energy scavenging for mobile and wireless electronics. *IEEE Pervasive Computer, 4*(1), 18–27. https://doi.org/10.1109/MPRV.2005.9

Park, C. S., Ryu, J., Choi, J. J., Park, D. S., Ahn, C. W., & Priya, S. (2009). Giant magnetoelectric coefficient in 3-2 nanocomposite thick films. *Japanese Journal of Applied Physics, 48*(8 Part 1), 0802041–0802043. https://doi.org/10.1143/JJAP.48.080204

Peng, J., Zhou, H., & Lu, C. (2017). Colossal multi-resonant magnetoelectric response in Fe73.5Cu1Nb3Si13.5B9/Terfenol-D/Be-bronze/Pb(Zr,Ti)O3 structure at zero bias magnetic field. *Journal of Materials Science: Materials in Electronics, 28*(21), 16227–16232. https://doi.org/10.1007/s10854-017-7525-y

Petrov, V. M., Srinivasan, G., Bichurin, M. I., & Galkina, T. A. (2009). Theory of magnetoelectric effect for bending modes in magnetostrictive- piezoelectric bilayers. *Journal of Applied Physics, 105*(6), Article 63911. https://doi.org/10.1063/1.3087766

Prellier, W., Singh, M. P., & Murugavel, P. (2005). The single-phase multiferroic oxides: From bulk to thin film. *Journal of Physics: Condensed Matter, 17*(30), R803–R832. https://doi.org/10.1088/0953-8984/17/30/R01

Priya, S., Islam, R., Dong, S., & Viehland, D. (2007). Recent advancements in magnetoelectric particulate and laminate composites. *Journal of Electroceramics, 19*(1), 147–164. https://doi.org/10.1007/s10832-007-9042-5

Priya, S., Ryu, J., Park, C. S., Oliver, J., Choi, J. J., & Park, D. S. (2009). Piezoelectric and magnetoelectric thick films for fabricating power sources in wireless sensor nodes. *Sensors, 9*(8), 6362–6384. https://doi.org/10.3390/s90806362

Pukada, E. (2000). History and recent progress in piezoelectric polymers. *IEEE Transactions on Ultrasonics, Ferroelectrics, and Frequency Control, 47*(6), 1277–1290. https://doi.org/10.1109/58.883516

Rajaram Patil, D., et al. (2013). Multiple broadband magnetoelectric response in thickness-controlled Ni/[011] Pb(Mg1/3Nb2/3)O3-Pb(Zr,Ti)O 3 single crystal/Ni laminates. *Applied Physics Letters, 103*(5). https://doi.org/10.1063/1.4817383

Ramesh, R., & Spaldin, N. A. (2009). Multiferroics: Progress and prospects in thin films. *Nanoscience and Technology, 6*, 20–28. https://doi.org/10.1142/9789814287005_0003

Rupp, T., Truong, B. D., Williams, S., & Roundy, S. (Feb. 2019). Magnetoelectric transducer designs for use as wireless power receivers in wearable and implantable applications. *Materials, 12*(3). https://doi.org/10.3390/MA12030512

Shi, Y. (Feb. 2018). Modeling of nonlinear magnetoelectric coupling in layered magnetoelectric nanocomposites with surface effect. *Composite Structures, 185*, 474–482. https://doi.org/10.1016/J.COMPSTRUCT.2017.11.019

Spaldin, N. A., & Fiebig, M. (2005). The renaissance of magnetoelectric multiferroics. *Science (80), 309*(5733), 391–392. https://doi.org/10.1126/science.1113357

Sreenivasulu, G., Mandal, S. K., Bandekar, S., Petrov, V. M., & Srinivasan, G. (2011). Low-frequency and resonance magnetoelectric effects in piezoelectric and functionally stepped ferromagnetic layered composites. *Physical Review B: Condensed Matter and Materials Physics, 84*(14). https://doi.org/10.1103/PhysRevB.84.144426

Srinivasan, G. (2010). Magnetoelectric composites. *Annual Review of Materials Research, 40*, 153–178. https://doi.org/10.1146/annurev-matsci-070909-104459

Srinivasan, G., Rasmussen, E. T., Gallegos, J., Srinivasan, R., Bokhan, Y. I., & Laletin, V. M. (2001). Magnetoelectric bilayer and multilayer structures of magnetostrictive and piezoelectric oxides. *Physical Review B: Condensed Matter and Materials Physics, 64*(21). https://doi.org/10.1103/PhysRevB.64.214408

Srinivasan, G., Rasmussen, E. T., Levin, B. J., & Hayes, R. (2002). Magnetoelectric effects in bilayers and multilayers of magnetostrictive and piezoelectric perovskite oxides. *Physical Review B: Condensed Matter and Materials Physics, 65*(13), 1–7. https://doi.org/10.1103/PhysRevB.65.134402

Swain, A. B., Dinesh Kumar, S., Subramanian, V., & Murugavel, P. (2020). Engineering resonance modes for enhanced magnetoelectric coupling in bilayer laminate composites for energy harvesting applications. *Physical Review Applied, 13*(2), 1–11. https://doi.org/10.1103/PhysRevApplied.13.024026

Truong, B. D., Andersen, E., Casados, C., & Roundy, S. (2020). Magnetoelectric wireless power transfer for biomedical implants: Effects of non-uniform magnetic field, alignment and orientation. *Sensors and Actuators A: Physical, 316*, Article 112269. https://doi.org/10.1016/j.sna.2020.112269

Vaz, C. A. F., Hoffman, J., Ahn, C. H., & Ramesh, R. (2010). Magnetoelectric coupling effects in multiferroic complex oxide composite structures. *Advances in Materials, 22*(26–27), 2900–2918. https://doi.org/10.1002/adma.200904326

Vlasko-Vlasov, V. K., Lin, Y. K., Miller, D. J., Welp, U., Crabtree, G. W., & Nikitenko, V. I. (2000). Direct magneto-optical observation of a structural phase transition in thin films of manganites. *Physical Review Letters, 84*(10), 2239–2242. https://doi.org/10.1103/PhysRevLett.84.2239

Vopsaroiu, M., Blackburn, J., & Cain, M. G. (Aug. 2007). A new magnetic recording read head technology based on the magneto-electric effect. *Journal of Physics D Applied Physics, 40*(17), 5027. https://doi.org/10.1088/0022-3727/40/17/003

Wang, Y., et al. (2010). Electrical resistance load effect on magnetoelectric coupling of magnetostrictive/piezoelectric laminated composite. *Journal of Alloys and Compounds, 500*(2), 224–226. https://doi.org/10.1016/j.jallcom.2010.04.009

Wang, Y., et al. (2011). An extremely low equivalent magnetic noise magnetoelectric sensor. *Advances in Materials, 23*(35), 4111–4114. https://doi.org/10.1002/adma.201100773

Wang, Y., et al. (2012). Influence of interfacial bonding condition on magnetoelectric properties in piezofiber/Metglas heterostructures. *Journal of Alloys and Compounds, 513*, 242–244. https://doi.org/10.1016/j.jallcom.2011.10.029

Yan, Y., K-h, C., Maurya, D., Kumar, A., Kalinin, S., & others. (2013). Giant energy density in [001]-textured Pb(Mg1/3Nb2/3)O3-PbZrO3-PbTiO3 piezoelectric ceramics. *Applied Physics Letters, 1*(102), Article 042903.

Yan, Y., Zhou, Y., & Priya, S. (2013). Giant self-biased magnetoelectric coupling in co-fired textured layered composites. *Applied Physics Letters, 102*(5). https://doi.org/10.1063/1.4791685

Yang, S. C., Park, C. S., Cho, K. H., & Priya, S. (2010). Self-biased magnetoelectric response in three-phase laminates. *Journal of Applied Physics, 108.*

Zaeimbashi, M., et al. (2019). NanoNeuroRFID: A wireless implantable device based on magnetoelectric antennas. *IEEE Journal of Electromagnetics, RF and Microwaves in Medicine and Biology, 3*(3), 206–215. https://doi.org/10.1109/JERM.2019.2903930

Zaeimbashi, M., et al. (2021). Ultra-compact dual-band smart NEMS magnetoelectric antennas for simultaneous wireless energy harvesting and magnetic field sensing. *Nature Communications, 12*(1). https://doi.org/10.1038/s41467-021-23256-z

Zhai, J., Xing, Z., Dong, S., Li, J., & Viehland, D. (2006). Detection of pico-Tesla magnetic fields using magneto-electric sensors at room temperature. *Applied Physics Letters, 88*(6), Article 62510. https://doi.org/10.1063/1.2172706

Zhai, J., Xing, Z., Dong, S., Li, J., & Viehland, D. (2008). Magnetoelectric laminate composites: An overview. *Journal of the American Ceramic Society, 91*(2), 351–358. https://doi.org/10.1111/j.1551-2916.2008.02259.x

Zhang, J., Li, P., Wen, Y., He, W., Yang, A., & Lu, C. (2014). Enhanced magnetoelectric effects in composite of piezoelectric ceramics, rare-earth iron alloys, and shape-optimized nanocrystalline alloys. *Review of Scientific Instruments, 85*(3). https://doi.org/10.1063/1.4867670

Zhou, Y., Apo, D. J., & Priya, S. (2013). Dual-phase self-biased magnetoelectric energy harvester. *Applied Physics Letters, 103*(19). https://doi.org/10.1063/1.4829151

Zhou, Y., Chul Yang, S., Apo, D. J., Maurya, D., & Priya, S. (2012). Tunable self-biased magnetoelectric response in homogenous laminates. *Applied Physics Letters, 101*(23). https://doi.org/10.1063/1.4769365

CHAPTER

8

Magnetoelectric current/magnetic field sensor

Y. Zhou[1], Shashank Priya[1], Yisi Liu[2], Xianfeng Liang[2], Cunzheng Dong[2], and Nian Sun[2]

[1] Virginia Tech, Blacksburg, VA, United States; [2] Northeastern University, Boston, MA, United States

8.1 Introduction

8.1.1 State-of-the-art current sensors

A current sensor is a system or device that can measure the magnitude of an electrical current in a cable/wire under either AC or DC conditions. Usually, the current sensors encompass many aspects of physics and electronics. Based on their working mechanism, there are mainly four types of electrical current sensing techniques, including the Hall effect, induction transformer, resistive shunt, and Rogowski coil sensors, as shown in Fig. 8.1.

1. Hall effect sensor (Fig. 8.1a): It is a transducer that varies its output voltage in response to a magnetic field via the Hall effect, given as

$$V_H = -\frac{IB}{nte} \tag{8.1}$$

where V_H is the Hall voltage, I is the current across the plate length, B is the magnetic field, t is the thickness of the plate, e is the elementary charge, and n is the charge carrier density of the carrier electrons. Generally, the Hall effect-based current sensor is mounted in a core at a right angle to the concentrated magnetic field along with a signal conditioning circuitry. When exposed to a magnetic

Magnetoelectric Composites
https://doi.org/10.1016/B978-0-443-18606-6.00002-2

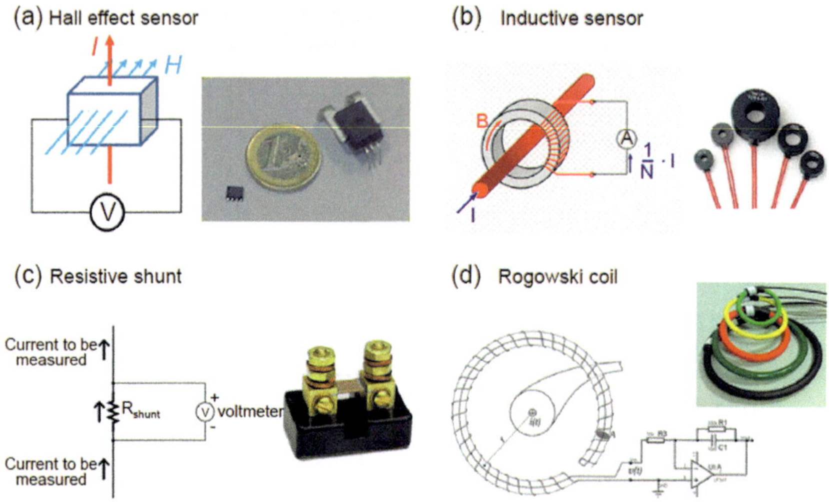

FIGURE 8.1 Schematic of the types of current sensors, including (a) Hall effect sensor, photograph of ± 20A and ±200 A current sensors; (b) Inductive sensor, photograph of wire lead current transformers/sensors; (c) Rogowski coil, examples of various Rogowski coils; (d) Resistive shunt, photograph of a DC ammeter shunt. *(a) Allegro MicroSystems, LLC; (b) Thai Lin; (c) Yueqing Tengtai Electric Co., Ltd. (b) (d) Deltec Shunts, LLC.*

field generated from a cable with current flow, the energized Hall sensor produces a potential difference that can be measured to identify the magnitude of the magnetic field or current. However, the use of iron cores dramatically enhances the electromagnetic interference, and the high sensitivity is obtained at the expense of high-quality signal conditioning electronics to monitor the inherently weak Hall voltage (Ramsden, 2006).

2. Inductive sensor (Fig. 8.1b): It monitors the electrical energy changes between two or more circuits through electromagnetic induction, also named a current transformer. Generally, a current transformer consists of a wire-wound core and a signal conditioner. A varying current signal passing through the core magnifies the magnetic field; in return, the magnetic field induces a current in the winding. Thus, this type of current sensing possesses power supply-free measurement nature. However, it can only deal with an AC current with a limited operating frequency range (40–60 Hz) due to the increased magnetic losses at elevated frequencies with a low detection sensitivity.
3. Resistive shunt (Fig. 8.1c): It is a calibrated resistor placed in a current path that produces a voltage drop proportional to the current flow. Since its resistance is known, the measured voltage across the shunt can be scaled to reflect the current value. However, the shunt resistor will generate heat as current is passed through it.

Therefore, this kind of sensing technique is generally used in applications where the emphasis is on accuracy and repeatability under low-current conditions. In addition, for practical application, although the shunt resistor is small in size and inexpensive, the need for cutting and reterminating the current carrier can be expensive to operate.

4. Rogowski coil (Fig. 8.1d): It is a toroid of wire for measuring the AC current or high-speed current pulses. This type of coil has an air core with a helical coil of wire with the lead from one end returning through the center of the coil to the other end.

The changes in magnetic field induce an electromotive force (EMF) within a wire loop, which is a voltage signal that is proportional to the changes in the magnitude field inside the loop: $EMF \propto \frac{dB(t)}{dt}$. Since $B(t) \propto i(t)$, the voltage output of the loop is therefore proportional to the time-differentiation of the current. The voltage produced by a Rogowski coil is

$$V = -\frac{A\,N\mu_0}{l}\frac{dI}{dt} \tag{8.2}$$

where $A = \pi a^2$ is the area of one of the small loops, N is the number of turns, $l = 2\pi a$ is the length of the winding, and dI/dt is the rate of change of the current. Since no iron core is needed, it is highly linear even when subjected to a large current. However, this type of coil can only be used with the support of an integrator circuit to obtain the current waveform, which needs external power support.

In summary, a comparison of these different current sensing technologies is listed in Table 8.1. The technology for sensing current has evolved, driven by the need for improved sensitivity, smaller size, and compatibility with electronic systems. However, none of the above mentioned current sensors satisfy all the needs related to self-powered, miniature size, high performance, low detection limits, reliability, and cost-

TABLE 8.1 Comparison between different current sensing technologies.

Sensing type	Hall sensor	Inductive sensor	Resistance shunt	Rogowski coil
Cost	High	Medium	Very low	Low
Linearity	Poor	Fair	Very good	Very good
High current capability	Good	Good	Very poor	Very good
Power consumption	Medium	Low	High	Low
Output variation with temperature	High	Low	Medium	Very low

effectiveness. The development of a compact, low-noise, and low-cost passive current sensor is technologically important and much needed for variety of electronic systems.

8.1.2 Magnetoelectric sensors

Magnetoelectric (ME) composites can directly generate an electric signal in response to the external magnetic field, leading to a passive nature without the need of external power sources. In addition, by optimizing the composite materials and configurations, it can produce significantly large voltage and sensitivity over a wide range of frequency. Therefore, the development of ME composites and devices toward current/magnetic field sensing applications is technologically important. In this chapter, we will systematically describe the development of ME current sensors with respect to following aspects:

1. Sensing mechanisms, sensor designs, and challenges of typical ME-sensing elements.
2. Development of novel self-biased ME composites that can be used as ME passive current sensors in the absence of a DC magnetic field.
3. Development of ME transformers that can be used over a wide frequency range of magnetic signals based on their tunable features.
4. Classification of magnetic noises and corresponding noise-reduction methods.

8.2 Material in bulk sensor

8.2.1 Bulk ME material

The birth of ME effect was marked by two independent events: (i) the magnetization or polarization of a moving dielectric in an electric field (Röntgen, 1888) or a magnetic field (Wilson, 1904); (ii) Curie proposed the possibility of the intrinsic ME effect in 1894 (Curie, 1894). More than 60 years later, it was proved on the basis of crystal symmetry by Landau et al. (Landau et al., 2013). Since the experimental demonstration of ME effect in single-phase material Cr2O3 (Astrov, 1960, 1961; Dzyaloshinskii I, 1960), many efforts have been made to explore the possibility of achieving strong ME coupling in single-phase materials.

Several families of the single-phase compounds were found with ME behavior: Bi compounds (Caicedo et al., 2008; Chu et al., 2007; Eerenstein et al., 2006; Wang et al., 2003), rare earth (RE) manganites (Kajimoto et al., 2004; Kimura et al., 2003), Pb family perovskite oxides (Martinez et al., 2010; Palkar & Malik, 2005; Sanchez et al., 2013; Schmid, 1994), $REMn_2O_5$

family (Cheong & Mostovoy, 2007; Hur et al., 2004; Uga et al., 1998), fluoride family (Ederer & Spaldin, 2006; Fox & Scott, 1977), etc. However, they are reported in small numbers due to the contradiction between the formation of magnetic moments with partially filled d-orbitals and the requirements of empty d-orbitals in ferroelectric (FE) oxides (Hill, 2000; Prellier et al., 2005). In addition, these single-phase multiferroic materials are also suffering from the weak ME coupling effects and the low Curie temperature (Catalan & Scott, 2009).

The magnetodielectric (MD) effect in (ferro)magnetic dielectrics has been investigated extensively to control the dielectric properties with a magnetic field for applications such as field-tunable capacitive resonators (Huang et al., 1997; Ikonen et al., 2006). A detailed magnetic field dependence of the dielectric constant (ε) and electric polarization in $DyMn_2O_5$ single crystal was investigated by Hur et al. (Hur et al., 2004). Due to the high sensitivity of the incommensurate state to external perturbation, an unprecedented large change in the dielectric constant with a magnetic field was discovered for more than 100% in a wide temperature range below 20 K. This "colossal magnetodielectric" (CMD) effect near a unique commensurate–incommensurate magnetic transition clearly shows an exciting interaction between spin and lattice degrees of freedom. An extraordinary change in the dielectric constant with the increasing temperature (T) under different magnetic fields (H) was discovered. The zero-field curve exhibits clear dielectric anomalies occurring at magnetic transition states, which suggests a significant spin/lattice coupling. A maximum change for ε of ~109% at 3 K with *H* of 7 T was achieved. When H is applied along the b axis, the magnetic easy axis, the MD effect is large as well, but not as significant as that for the a axis. The change in $\Delta\varepsilon/\varepsilon_0$ as a function of H at 3 and 20 K for magnetic field cooling (FC) and zero-magnetic field cooling (ZFC) was also measured. Compared to the largest change of 109% for FC at 3 K, the change for ZFC is slightly less, ~90%. The maximum value at 20 K is reduced to ~40%. In the measurements of the change in the electric polarization (P) with increasing H at 3 K, a hysteresis loop and an abrupt change at ~1.8 T were observed, which are closely correlated with the magnetization loop and $\Delta\varepsilon/\varepsilon_0$ curve at 3 K.

Kimura et al. reported the discovery of ME effect in perovskite manganite $TbMnO_3$ and realized the magnetic control ferroelectric polarization (Kimura et al., 2003). Via the magnetoelastically induced lattice modulation, the magnetic structure was modulated with the emergence of spontaneous polarization. The schematic of the crystal structure of $TbMnO_3$, which is the orthorhombically distorted perovskite structure, at room temperature and spatial variation along the b axis of Mn magnetic moment is shown in Fig. 8.2a. The magnetic field dependence of the change in electric polarization along the a axis at various temperatures

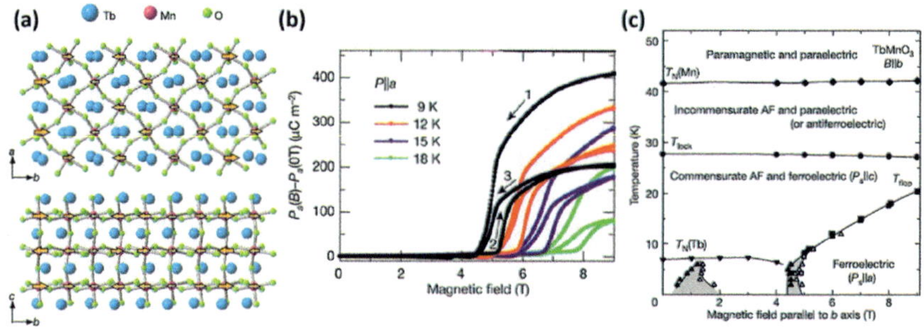

FIGURE 8.2 (a) Rough sketches of the $TbMnO_3$ crystal structure at room temperature and spatial variation along the b axis of Mn magnetic moment. *Mn* magnetic moment below T_N is denoted by using orange arrows. (b) Magnetic field-induced change in electric polarization along the a axis in $TbMnO_3$. The data were collected after the magnetic field cooling. (c) Temperature versus magnetic field phase diagram for $TbMnO_3$ with a magnetic field applied along the b axis. *Reprinted with permission from Kimura et al., Nature 426(6962), 55 (2003).*

is displayed in Fig. 8.2b. A remarkable change in the polarization ($\sim 6 \times 10^{-4}$ C m^{-2}) is caused by the magnetic field–induced polarization with the nature of first-order phase transition. Fig. 8.2c shows the phase diagram of $TbMnO_3$ with magnetic phase boundaries denoted by triangles for the magnetic field applied along the b axis. An extraordinary change in the dielectric constant with increasing temperature (T) under different magnetic fields (H) was observed. The zero-field curve exhibits clear dielectric anomalies at magnetic transition states, indicating significant spin/lattice coupling. A maximum change in ε of approximately 109% at 3 K with an H of 7 T was achieved. When H is applied along the b axis, the magnetic easy axis, the MD effect remains large, though not as significant as along the a axis. The change in $\Delta\varepsilon/\varepsilon_0$ as a function of H at 3 and 20 K for both magnetic field cooling (FC) and zero-magnetic field cooling (ZFC) was also measured. Compared to the largest change of 109% for FC at 3 K, the change for ZFC is slightly less, around 90%. The maximum value at 20 K is reduced to approximately 40%. The measurements of the change in electric polarization (P) with increasing H at 3 K showed a hysteresis loop and an abrupt change at around 1.8 T, closely correlating with the magnetization loop and $\Delta\varepsilon/\varepsilon_0$ curve at 3 K.

8.2.2 Bulk ME composites

In this section, we first introduce the basic idea about ME composites and their common structures. Then, a brief history of the development of bulk ME composites is given. After that, the pros and cons of bulk ME composites based on different materials are discussed. Finally, a specific example of an ME laminate composite with PZT and Terfenol-D disks is presented for further understanding. In order to compensate for the deficiencies of the single-phase materials, a giant ME coupling at room temperature was realized by tailoring the properties of artificial multiferroic composites, which are consisted of ferroelectric (FE) and ferromagnetic (FM) compounds (Islam et al., 2008; Praveen et al., 2018; Rajaram Patil et al., 2013). The purpose of this method is to generate cross coupling between the parent compounds via elastic interaction. This novel structure owns both properties of their parent materials, which is known as a product-property of the composite (Nan, 1994; Van Suchtelen, 1972). These multiferroic composites were found to have a large ME coefficient that is one to two orders higher than that of the single-phase materials. The practical applications of the ME effect become true as a result of the appearance of the ME composites with a much larger ME coefficient above room temperature. There are different structures for multiferroic composites that create strong ME coupling for multifunctional applications(Nan et al., 2008): (i) 0–3-type mixed phase of particulate nano-composites, (ii) 1–3-type vertically mediated architecture with three-dimensional ferromagnetic nano-pillars embedded in a ferroelectric phase, and (iii) 2–2-type classic ME multilayers that are widely used for devices due to the advances in thin-film growth techniques. By properly choosing the properties such as connectivity, microstructure, volume fraction, of individual phases, the composite structures open up new paths for tailoring the ME coupling. This technique was initially investigated in the 0–3-type ceramic composites (Van den Boomgaard & Born, 1978; Van Run et al., 1974; Van Suchtelen, 1972) and later in the 2–2-type laminated multilayers (Lopatin et al., 1994; Mori & Wuttig, 2002; Ryu et al., 2002; Srinivasan et al., 2001). More recently, Dong et al. proposed ME composites structures with 2–1 and 1–1 connectivity, which exhibited an enhanced ME coupling coefficient (Chu et al., 2017; Dong et al., 2006). They reported a high off-resonance ME coupling coefficient of 22 $Vcm^{-1}\ Oe^{-1}$ in a 2–1 connectivity composite consisting of an amorphous Metglas alloy and multiple piezoelectric fibers (Dong et al., 2006). An enhanced resonant ME coupling coefficient of ≈7000 Vcm−1 Oe−1 was later found in a 1–1 connectivity L–T (longitudinally magnetized and transversely poled) ME composite consisting of a PMN-PT single-crystal fiber and laser-treated amorphous Metglas alloy (Chu et al., 2017).

The first work on the ME composites was reported at Philips laboratory about sintered $BaTiO_3$ (BTO)/CoFe2O4 (CFO), which were fabricated by unidirectional solidification (Van den Boomgaard & Born, 1978; Van Run et al., 1974; Van Suchtelen, 1972). The ME coefficient of BTO/CFO is much larger than that of single-phase multiferroic materials such as Cr_2O_3, which is 130 $mV\ cm^{-1}\ Oe^{-1}$. These research studies did not attract too much attention at first, but they did motivate a significant theoretical effort on ME ceramic composites before 2000. In the early 2000s, the experiments on the ME effect of the giant magnetostrictive material $Tb1-xDyxFe_2$ (Terfenol-D) became a milestone in the development of bulk ME composites. After that, various ME bulk composites consisting of the ferroelectric phase, such as BTO, Pb(Zr, Ti)O3 (PZT), $Pb(Mg, Nb)O_3-PbTiO_3$ (PMN-PT), and polyvinylidene fluoride (PVDF) and the ferromagnetic phase, such as CFO, Terfenol-D, and $LaMnO_3$ (LMO), have been investigated over the last few years.

In the last decade, various bulk ME composites such as ceramic composites, magnetic alloy-based composites, and polymer-based composites have been extensively investigated. To achieve bulk ceramic composites, cosintering at high temperatures is generally used to get different combinations of ferroelectric oxides and magnetic oxides. The low ME coupling so far observed in ceramic composites is mainly due to the preparation problems such as interdiffusion, reaction, and thermal expansion mismatch between two ME phases at high temperature. One of the main problems for the particulate ceramic composites is their leakage problem that results from the conductive ferrites. The core–shell structured particles with a high concentration of the ferrite-core well into the piezoelectric-shell, which can prevent the ferrite-core from direct contact during the sintering process, are an effective method to avoid this problem (Curecheriu et al., 2010; Duong et al., 2006). However, it is very difficult to prepare good core–shell structured particles with ferrite core and piezoelectric shell. More work was also done on other fabrication techniques to improve the properties of particulate ceramic composites (Agrawal et al., 2009; Jiang et al., 2007). High ME effect can be achieved in the 2–2-type laminate composite ceramics due to the elimination of the leakage problem. However, the ME signal strength induced from the piezoelectric layer is reduced because the ferrite layers are not conductive enough to serve as electrodes. To solve this problem, internal electrodes between piezoelectric and magnetostrictive layers are introduced to reduce the loss. Among bulk ME composites, magnetic alloy–based composites such as the laminate structures of Terfenol-D and different piezoelectric layers have shown the strongest ME effect. With further optimization for such magnetic alloy-based composites, a larger ME voltage coefficient can be obtained up to $\sim 57\ Vcm^{-1}\ Oe^{-1}$ (Jia et al., 2008). Furthermore, by optimizing structural designs for considering the mechanical behavior, a larger ME response was shown in some novel structures of the bulk ME composites such as an ultrasonic horn with more mechanical energy converted into electric energy (Li et al., 2007). However, due to the high saturation field and low permeability, Terfenol-D-based composites are not appropriate for low field applications. Instead, soft magnetic alloys, such as Metglas (Zhai et al., 2006) and

Permendur (Sreenivasulu et al., 2012), are preferred. The polymer binders, which are used to bond the piezoelectric layer and magnetic alloys together, can drastically influence the ME response of the magnetic alloy-based composites. In order to eliminate the adhesive fatigue and aging effect of the interfacial binders, magnetic alloys can be directly deposited on top of the piezoelectric materials via various deposition techniques such as magnetron sputtering (Liu et al., 2013; Lou et al., 2009). Due to the direct and tight bonding effect between the piezoelectric and magnetic phases, where the interfacial binders are not desired anymore in such composites, the ME response can be greatly enhanced. Polymer-based composites are also a great choice for realizing large ME coupling because they are easily fabricated with improved mechanical properties. Good stability during the fatigue measurement and the aging test has been demonstrated in a polymer-based composite with a PZT rod embedded in the Terfenol-D/epoxy matrix (Ma et al., 2007). Compared to the ceramic and magnetic alloy-based ME composites, the 0–3-type particulate polymer-based composites with magnetic particles embedded in the piezoelectric polymer phase attract wide interest due to the easy fabrication process.

Based on the product property, an ME material that was composed of piezoelectric oxide $BaTiO_3$ and magnetostrictive ferrites such as $CoFe_2O_4$ was developed by Philips in 1972 (Van Suchtelen, 1972). By stacking and bonding PZT and Terfenol-D disks, Ryu et al. investigated an ME laminate composite with no chemical reaction involved in the fabrication process (Ryu et al., 2001). The schematic structure and photograph of the device are shown in Fig. 8.3a and b. Fig. 8.3c displays the measured ME voltage coefficients of PZT and Terfenol-D laminate composites as a function of the DC magnetic bias field. It was found that ME coefficient increases with the decreasing thickness of the PZT layer, which can be explained by the increasing compressive stress in the thinner PZT layer. A maximum value for the ME coefficient at 1 kHz of 4 .68 $V/cm.Oe$ at room temperature was obtained. The theoretical ME coefficient equation for the laminate composite can be derived as follows $\frac{dE}{dH} = \frac{V_{out}}{H_{ac} \times t_p} = \frac{2 \times g_{31} \times \sigma_{31p}^{E}}{H_{ac}}$ (V/cm Oe) where g_{31} is the piezoelectric voltage constant and σ_{31p}^{E} is the compressive stress in the PZT layer. Therefore, a larger ME voltage coefficient can be acquired when the compressive stress in the PZT layer is higher. The theoretical calculation for the output voltage (dV/dH) and the ME coefficient (dE/dH) as a function of the thickness ratio (t_t/t_p) between Terfenol-D and PZT is presented in Fig. 8.3d. The output voltage decreases, and the ME voltage coefficient increases with the increasing thickness ratio. Both values saturate when the thickness ratio is larger than 10.

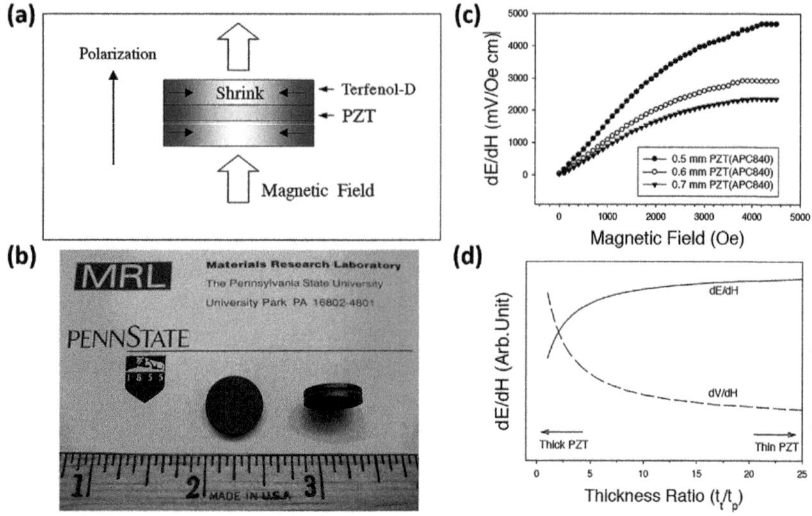

FIGURE 8.3 Magnetoelectric (ME) laminate composite with PZT and Terfenol-D disks. (a) Schematic structure. (b) Photograph of the composite. (c) ME voltage coefficient at 1 KHz versus DC magnetic field with various thicknesses of the PZT layer. (d) Thickness ratio ($t_t/\ t_p$) dependence of theoretically calculated ME voltage coefficients for PZT and the Terfenol-D laminate composite. *Reproduced with permission from Ryu et al., Jpn. J. Appl. Phys., Part 1 40(8), 4948 (2001).*

8.3 Development of magnetoelectric sensors

ME composites as magnetic field detection sensors have attracted lots of attention due to their simple implementation mechanism. Operation of conventional ME composites requires two input components: a magnetic DC bias and a magnetic AC field. Either of these two components can be detected by providing one of the components, leading to a magnetic sensor for detecting AC or DC fields. Since the strength of the magnetic field depends on the current *I* in the wire, the electric current can be monitored by detecting the corresponding magnetic flux.

8.3.1 Sensing mechanism

For the case of a long, straight wire carrying a current I, the magnetic field generated is wrapped around the wire, as shown in Fig. 8.4. Depending upon the strength of the AC or DC current, the magnitude of the AC or DC magnetic field varies as a function of the distance from the wire, given as

$$H = \frac{I}{2\pi r} \tag{8.3}$$

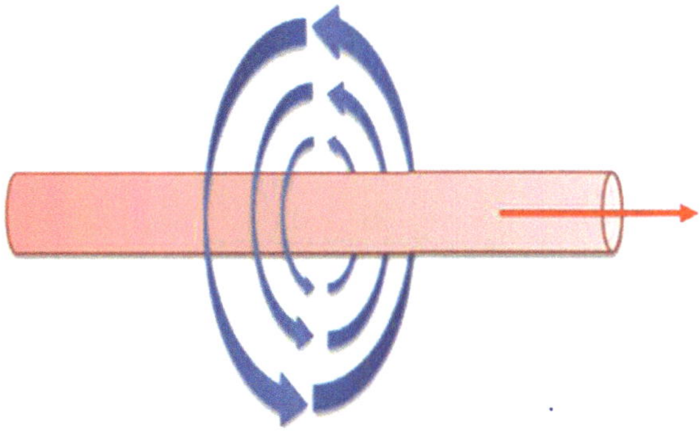

FIGURE 8.4 Schematic of current passes through a conductor induces a magnetic field.

where r is the radius of the vortex magnetic field. Accordingly, by using the ME composites, one can detect the change in magnetic field by monitoring the variation of the ME voltage and use that measurement to derive the status of the corresponding I. Therefore, the current sensing mechanism is similar to the magnetic field detection principle based on the ME coupling effect. The ME coupling can be explained as the following: on application of a magnetic field to the composite, a strain will be induced and transferred to the piezoelectric phase, and a proportional charge in the piezoelectric layer will be generated. The input magnetic field on the ME composite consists of an AC magnetic field along with a DC magnetic bias. One of these two variables can be detected by providing the other component. Thus, the ME composite can be utilized as an AC/DC magnetic field sensor (Dong et al., 2006; Wang et al., 2011). By selecting the high-performance piezoelectric and magnetostrictive materials and combining them into a suitable structure, one can detect a small magnetic field as low as $\sim 1\ pT$ at a frequency of only several Hz (Wang et al., 2011).

8.3.1.1 DC magnetic field–dependent response

For ME composites, the measured direct ME coupling coefficient is the field conversion ratio between applied H_{ac} and induced E_{ac} under applied H_{dc}, given as $\alpha_{\mathrm{ME}} = \delta \mathrm{E}_{\mathrm{ac}}/\delta \mathrm{H}_{\mathrm{ac}}$. In general, under a constant applied AC magnetic field, the ME coupling coefficient first increases with increasing H_{dc}, reaching a maximum at an optimized DC bias (H_{bias}), and then decreases with further increasing H_{dc}. This peak behavior in response to varying H_{dc} can be further explained by taking into account the relationship given below (Cho & Priya, 2011):

$$\alpha_{\mathrm{ME}} = \left| \frac{\partial \mathrm{T}}{\partial \mathrm{S}} \times \frac{\partial \mathrm{D}}{\partial \mathrm{T}} \times \frac{\partial \mathrm{E}}{\partial \mathrm{D}} \right| \times \frac{\partial \mathrm{S}}{\partial \mathrm{H}} \tag{8.4}$$

where S is the mechanical strain, T is the mechanical stress, D is the electric displacement, E is the electric field, and H is the magnetic field. Since the first part of the above expression is a nonmagnetic factor, one can rewrite Eq. (8.4) as

$$\alpha_{ME} \propto \frac{\partial S}{\partial H} = \frac{d\lambda}{dH} = q \tag{8.5}$$

which indicates that the ME coefficient is directly related to the nature of the ferromagnetic phase (λ: magnetostriction, q: piezomagnetic coefficient) and the effectiveness of elastic coupling between the two phases. The dependence of α_{ME} on q indicates the requirement for additional H_{dc}. In summary, typical behavior of the α_{ME}, λ, q, and their relationship as a function of H_{dc} can be illustrated as shown in Fig. 8.5.

From the above discussion, we can clearly notice that the ME coupling coefficient essentially tracks the strength of q, which is proportional to the rate of change in λ, and then vanishes when magnetostriction attains saturation. Depending on the magnitude of optimum magnetic DC bias, ME composites can exhibit maximum ME coefficient. The composition and geometry of the ferromagnetic phase can be tuned to tailor the magnitude of optimum DC bias ranging from five to 6.8 kOe (Dong et al., 2006; Nan et al., 2008). Depending upon the materials utilized, ME composites will have different ME behavior due to the difference in coercive field, permeability, and magnetization. The volume fraction of the

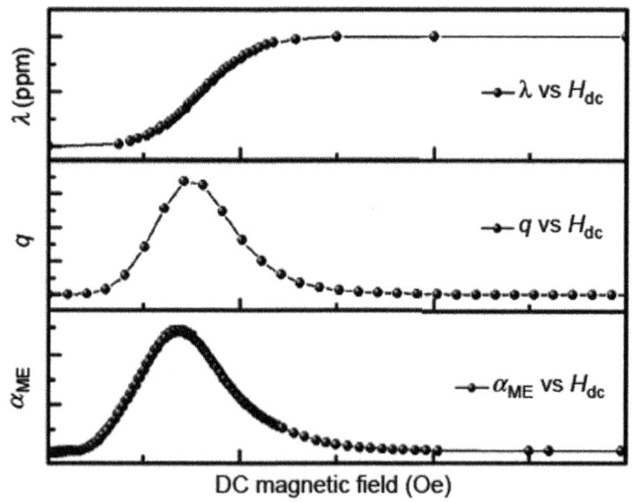

FIGURE 8.5 Typical magnetoelectric behaviors as a function of a DC magnetic field for a longitudinal magnetized and transversely poled ME composite.

magnetostrictive phase affects the ME behavior and optimum H_{bias} and could also be used as a control parameter.

8.3.1.2 AC magnetic field–dependent response

Another important phenomenon of the ME composites is the change in ME coupling as a function of the applied AC magnetic field (ME vs. f_{ac}). When the AC magnetic field is applied at a frequency corresponding to electromechanical resonance (EMR) for the piezoelectric phase or ferromagnetic resonance for the magnetic phase in the ME composite, the ME voltage coefficient shows a peak behavior with the magnitude increased by a factor of up to 100 (Zhai et al., 2008a). Bichurin et al. presented a theoretical analysis for the bilayer of NFO-PZT laminates and predicted a 40-times increase in $\alpha_{ME,33}$ at EMR compared to that of off-resonance conditions. Similar theoretical calculations have been developed for the ME effect at ferromagnetic and magnetoacoustic resonance. Cho et al. analyzed the direct and converse ME effect in laminate composites and showed that DME is maximized at f_a, while the CME is maximized at f_r, as shown in Fig. 8.6 (Cho & Priya, 2011). This phenomenon can be further

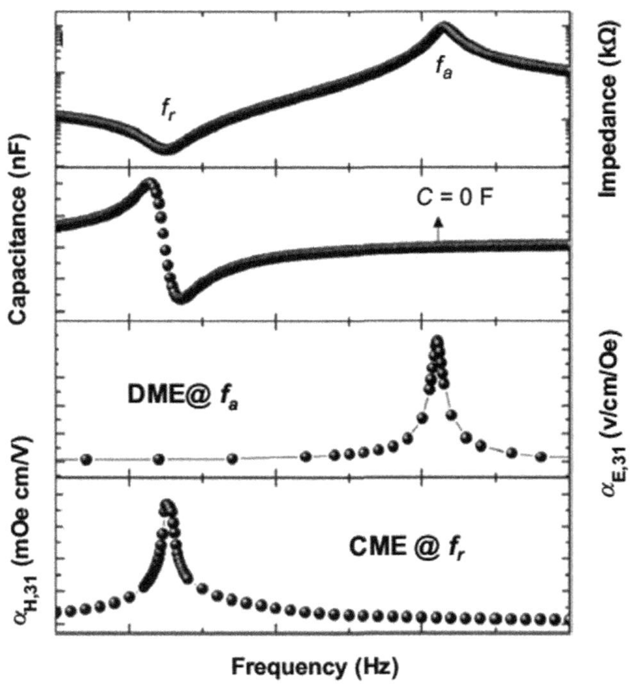

FIGURE 8.6 Impedance, capacitance, and ME voltage coefficient as a function of AC frequency. *Reprinted with permission from Cho and Priya (2011), Copyright 2011, AIP Publishing LLC.*

explained by using piezoelectric constitutive equations and combining them with frequency-dependent capacitance of the piezoelectric layer. The fundamental resonance frequency can be tuned by the length of the laminates (*l*), where the first longitudinal resonance is expressed as $f_L = \frac{1}{(2l\sqrt{\rho s_{11}})}$ (ρ is the laminate density and s_{11} is the elastic compliance of the laminate) (Zhai et al., 2008b). However, for practical use of the composites at low frequency, the corresponding composite dimension would be prohibitively large. In order to overcome this problem, an unimorph configuration was proposed for bilayer ME composites working at the bending mode that showed enhancement in the coupling at low frequencies. Alternative approaches that utilize unique laminate configurations have been proposed to achieve low resonance frequency with a limited size of the components. Using these approaches, resonance enhancement of the ME coefficient has been achieved in the frequency range of 100 Hz, using cm-long laminates.

8.3.1.3 Delta-E effect

Another sensing theory for the DC or low-frequency magnetic field is based on the **delta-E effect as the third type of the ME sensor** (Gojdka et al., 2011; Jahns et al., 2014; Li et al., 2017; Nan et al., 2013; Zabel et al., 2016). The delta-E effect represents Young's modulus change in the magnetostrictive materials with the magnetization under an external magnetic field. Hence, the DC signal is mixed up to the resonance frequency, and the low-frequency 1/f noise and acoustic disturbance are excluded from the measurement. An extremely low magnetic noise bulk ME sensor was demonstrated by Wang et al. in 2011 (Wang et al., 2011), with the ME composite consisted of Metglas and the PMN-PT fiber. The resonance frequency was set to 1 kHz with a maximum ME coefficient of 52 $V/(Oe\ cm)$. The noise sources of this ME sensor are dielectric loss noise and DC leakage resistance noise. The measured equivalent magnetic noise is 5.1 pT/Hz at 1 Hz. Fig. 8.7 shows the equivalent magnetic noise floor of the Metglas/PZT/Metglas laminate. At $f = 1$ Hz, the equivalent magnetic noise was 13.3 $pT/Hz^{0.5}$. To further decrease the equivalent magnetic noise floors, PMN−PT fibers with higher piezoelectric constants and lower dielectric loss factors could be used (Wang et al., 2011a).

8.3.2 Theoretical model

A high field/current sensitivity can be obtained from an ME composite that exhibits high ME voltage coefficient and low signal-to-noise ratio. The sensitivity can be further increased by a factor of 10 or more at the mechanical resonance frequency. By optimizing material combination and composite geometry, one is able to obtain detection sensitivity on the

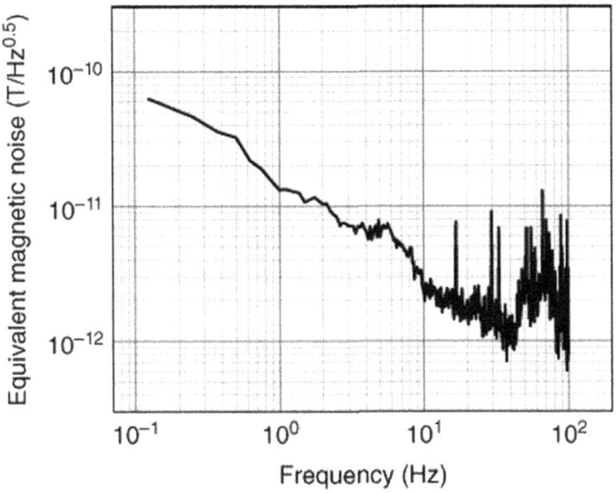

FIGURE 8.7 Equivalent magnetic noise floor of ME sensor unit with Metglas/PZT/Metglas laminates.

order of $pT/Hz^{1/2}$. The ME coefficients for the laminates can be determined by taking into account the constitutive piezoelectric and piezomagnetic behavior. A figure of merit for a conventional 2-2 type laminate ME composite can be derived as the following (Srinivasan et al., 2001):

$$\text{L}-\text{T mode: } \alpha_{E31} = \frac{\delta E_3}{\delta H_1} = \frac{-2d_{31}^{p} q_{11}^{m} v^{m}}{\left(s_{11}^{m} + s_{12}^{m}\right)\varepsilon_{33}^{T,P} v^{p} + \left(s_{11}^{p} + s_{12}^{p}\right)\varepsilon_{33}^{T,P} v^{m} - 2\left(d_{31}^{p}\right)^{2} v_{m}} \tag{8.6}$$

$$\text{L}-\text{L mode: } \alpha_{E33} = \frac{\delta E_3}{\delta H_3} = \frac{-2d_{31}^{p} q_{13}^{m} v^{m}}{\left(s_{11}^{m} + s_{12}^{m}\right)\varepsilon_{33}^{T,P} v^{p} + \left(s_{11}^{p} + s_{12}^{p}\right)\varepsilon_{33}^{T,P} v^{m} - 2\left(d_{31}^{p}\right)^{2} v_{m}} \tag{8.7}$$

where E_k and H_k are vector components of the electric and magnetic field, d_{31}^{p} is the piezoelectric coefficient, v^m and v^p are volumes of magnetostrictive and piezoelectric phases, s_{12}^{p}, s_{12}^{p} are elastic compliances for the piezoelectric phase, s_{11}^{m}, s_{12}^{m} are elastic compliances for the magnetostrictive phase, q_{11} is piezomagnetic coefficient of the magnetostrictive phase, and $\varepsilon_{33}^{T;P}$ is the permittivity of the piezoelectric phase.

The superscripts "*m*" and "*p*" represent the magnetostrictive and piezoelectric phase, respectively. This relationship clearly reflects the important role of material parameters of the piezoelectric phase (g_{31}, S_{11}^{P}) and the magnetostrictive phase (q_{11}, S_{11}^{m}) toward achieving a large ME response.

8.3.3 Design concepts and challenges

There are several fundamental challenges in the realization of high-performance ME current sensors. The performance of ME composites is tightly related with the materials selection, structure, processing method, operating mode, dimensionality, and other dynamic parameters. One of the biggest challenges in the implementation of ME composites has been the magnitude of ME voltage coefficient α_{ME}, which represents the magnetic field output voltage conversion. The characteristic ME behavior can be tuned by improving the constitutive piezoelectric and magnetostrictive behavior. Thus, the selection of appropriate piezoelectric and magnetostrictive materials is the primary step toward achieving a high-performance ME transducer. Next, the effect of synthesis techniques on the ME response should be considered. The epoxy bonding method provides an easy and simple fabrication process and provides more feasibility for the device configuration. However, the low mechanical strength of epoxy hampers the effective coupling between the magnetostrictive and piezoelectric phases. In addressing this issue, direct bonding between the consecutive phases through a low temperature cofiring technique can be used to enhance the effective coupling.

Magnetic noise in the environment (extrinsic noise: electromagnetic interference and thermal noise) and in the detection circuits (intrinsic noise: Johnson noise and $1/f$ noise) play an important role toward the detection capabilities of the ME-laminated sensors (Wang et al., 2014). In addressing the extrinsic noise, appropriate rejection techniques can be used to reduce or eliminate the electromagnetic interference or thermal noise. By using a metallic shield placed between the ME component and external environment, it is possible to control the propagation of electromagnetic fields. A good thermal insulation helps in extending the sensor stability in a wide range of working temperatures. On the other hand, both the measurement circuit and sensor configuration need to be considered for sensor performance with regard to the inevitable intrinsic noise.

In the following sections, we will discuss the development of ME current sensors, covering conventional ME composite-based current sensors, self-biased ME composite-based current sensors, ME transformer-based current sensors, and noise-reduction methods.

8.4 Conventional ME composite–based sensor

In considering the effective ME coupling, so far only bulk-sized ME laminate composites that exhibit a large ME effect above room temperature have the possibility to be implemented as current sensors for practical usage. There are three main promising device configurations that can be used for current sensing, including rectangular shaped, ring-type shaped, and polymer-based flexible structures.

8.4.1 Rectangular-shaped sensor

It is well-known that the ME response as a function of a DC magnetic field is tightly related with the nature of magnetostrictive materials. From the functional device point of view, it is important to control the magnitude of the ME voltage coefficient and its response. Thus, there are several studies related to shapes (disk, cylinder, plate, toroid, sphere, etc.) and sizes (numbers of layers, layer thickness, length, and width) of the magnetostrictive material toward obtaining high ME voltage coefficient (Nan et al., 2008). It is widely accepted that ME composites with a rectangular shape, large length/width aspect ratio, and small thickness with optimized volume ratio will generate high ME voltage coefficient (Zhai et al., 2008a). To better explain the ME versus configuration relationship, the size-induced demagnetization effect has been taken into consideration. For ferromagnetic materials, the demagnetization field is directly proportional to the demagnetization factor (N_d), $H_d = MN_d$ (M is the magnetization), where N_d is dependent on the dimension and geometry. Thus, the internal effective magnetic field (H_{eff}) in the magnetic phase can be written as $H_{eff} = H_{bias} - H_d$, where H_{bias} is the induced external magnetic field. Therefore, in order to achieve the same magnitude of H_{eff}, one needs to apply larger H_{bias} when H_d is larger and vice versa (Zhou, Yang, et al., 2012). Furthermore, this can also be correlated with the magnetic flux concentration (B_{eff}) in the magnetic phase:

$$B_{eff} = \mu_0\left(H_{eff} + M\right) = \mu_0(H_{bias} + M) - \mu_0 MN_d \tag{8.8d}$$

where a smaller N_d will result in a stronger B_{eff}. It has been shown that high magnetic flux concentration in the magnetic phase has a positive effect on the ME response of ME laminates. Thus, some researchers use the enlarged magnetic layer to enhance the effective magnetic flux density along the piezo/magnetic interface for obtaining larger ME coefficient.

Fig. 8.8 illustrates a case study that describes the influence of Metglas length on the magnetic field concentration in ME laminate sensors. As expected, it was found that the lengthening of the magnetostrictive layer increases the magnetic flux density along the center of the Metglas layer (Fig. 8.8a) and therefore leads to an enhanced ME voltage coefficient (Fig. 8.8b). As a result, the corresponding DC magnetic field detection sensitivity of Metglas/PZT laminates improved from 15 nT (Fig. 8.8c) to 6 nT (Fig. 8.8d) with respect to the Metglas length. Such high sensitivities

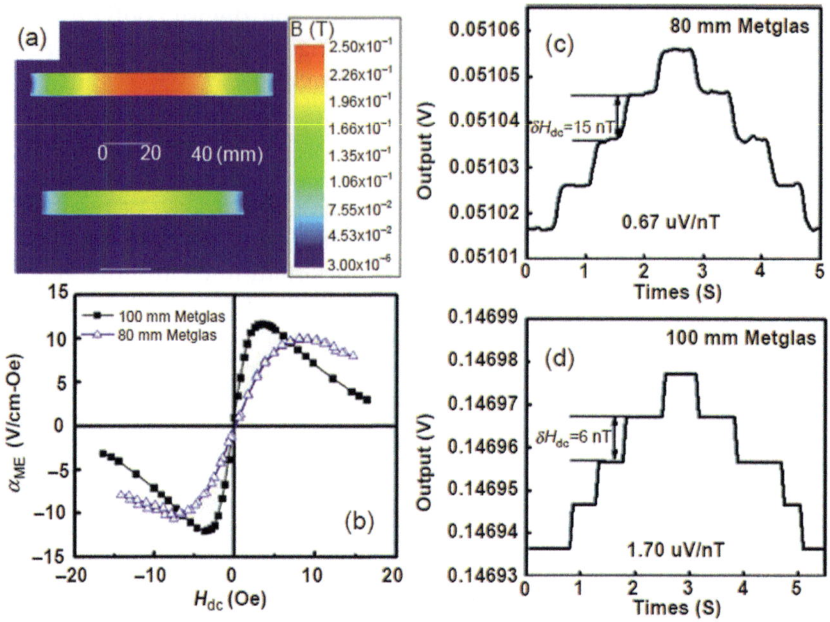

FIGURE 8.8 (a) In-plane magnetic field strength along the center-plane of Metglas foil in response to a low DC bias field, as simulated by Maxwell 3D; (b) the corresponding magnetoelectric voltage coefficient of Metglas/PZT laminate sensor with different Metglas length as a function of DC bias in response to $H_{ac} = 1\ Oe$ at 1 kHz; (c) and (d) comparison of the DC field sensitivity for Metglas/PZT laminate sensor with different size Metglas under AC drive conditions at $f = 1\ kH$ and $H_{ac} = 0.1\ Oe$. *Reprinted with permission from Gao, Gray et al. (2011), Copyright 2011, AIP Publishing LLC.*

can be further affected based on the magnitude and orientation of the local geomagnetic field.

For the implementation of the rectangular-shaped ME composites, Fig. 8.9 shows a conceptual design of the ME current sensor. In this design, the rectangular-shaped ME laminate is mounted on the surface of an electrical cable. A pair of permanent magnets was placed at the two ends of the ME laminate to provide a DC bias H_{dc}.

By optimizing the H_{dc} with the selected magnet, an enhanced ME voltage coefficient and the resultant high sensitivity was obtained. This design configuration provides a simple and flexible way for sensor installation and detection. However, the need of the permanent magnet hinders the way toward device miniaturization. More importantly, the sensor placement, with respect to the orientation of the cable, plays an important role toward maximum sensitivity and may produce another level of difficulty for sensor implementation.

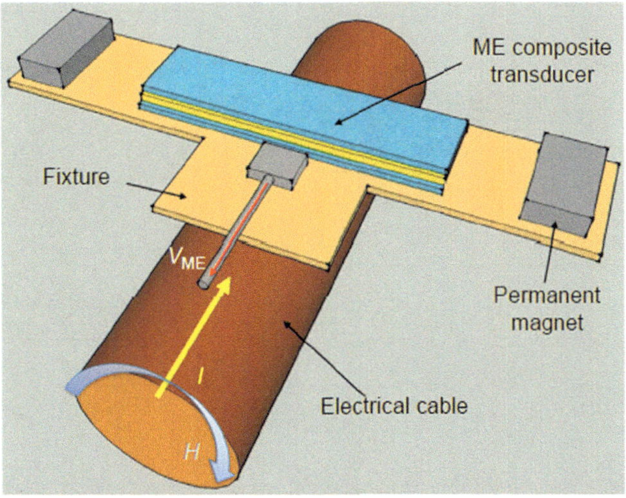

FIGURE 8.9 Conceptual design of a rectangular-shaped ME current sensor mounted on the surface of an electrical cable.

8.4.2 Ring-type sensor

Previously, investigations on ME composites have focused on the simple construction of square or rectangular shapes based on the consideration of an effective demagnetization field and the leading magnetic flux density across the magnetostrictive phases. As-fabricated ME composites can be operated in either a transverse-transverse (T-T) mode, L-T mode, or L-L mode, which exhibits large ME effects for a specific direction of the applied magnetic field. Thus, the best sensitivity can only be achieved for the detection of an electric current (magnetic field) flowing in a certain direction. However, in most cases, it is important to utilize a stable and high-sensitivity sensor for detecting a current in electrical cables surrounded with AC vortex magnetic fields for a power integrated circuit. Accordingly, ring-type ME laminates were developed as ideal configurations for a vortex magnetic field or current detection. Dong et al. demonstrated a ring-type ME laminate consisting of two Terfenol-D rings magnetized in the circumferential direction and one PZT ring consisting of segments poled in the circumference direction, as shown in Fig. 8.10a (Dong, Li, & Viehland, 2004). In this design, the cable can be placed in the center of the ring-type sensor. An intensified strain will be generated and transferred to the piezoelectric layer when a vortex magnetic field is generated. Thus, by monitoring the induced ME voltage, one would be able to identify the electrical current. Fig. 8.10b illustrates the induced ME voltage response to a square-wave current through the ring-type current sensor. In comparison with a toroidal-type reluctance

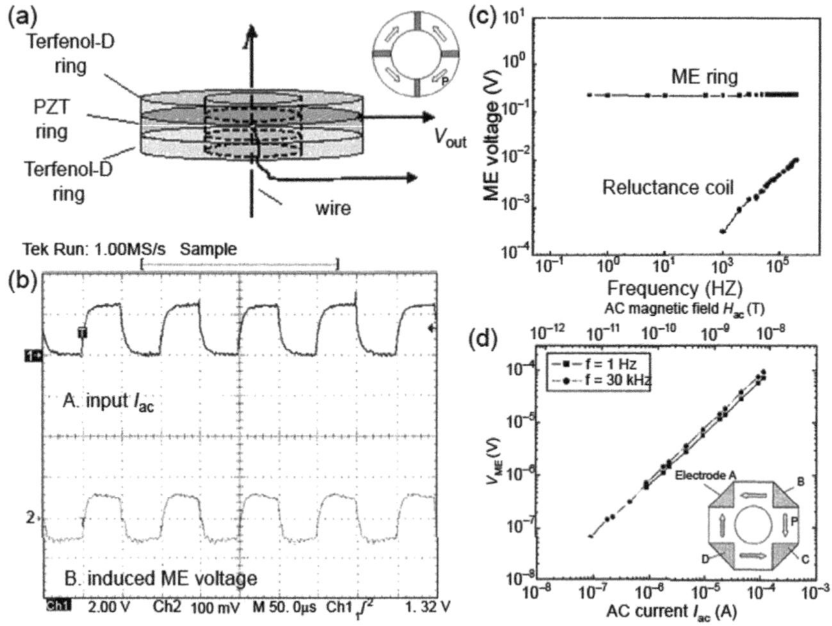

FIGURE 8.10 (a) Configuration of the ring-type Terfenol-D/PZT laminate composite, inset, illustrates the piezoelectric polarization direction in the circumferential direction; (b) induced ME response to switching currents; (c) ME response of a ring-type ME sensor and a reluctance coil as a function of driving frequency; (d) ME response as a function of AC magnetic field and current. *Reprinted with permission from Dong et al. (2005), Dong, Li, & Viehland (2004), Copyright 2004, AIP Publishing LLC.*

coil with $N = 100$ turns, this ring-type sensor shows a near-flat ME response with higher voltage output as a function of frequency (Fig. 8.10c), indicating its promise and advantage as a highly sensitive and stable current sensor. Furthermore, by selecting single-crystal materials with higher piezoelectric properties and optimizing the configuration of a ring-type structure, Dong et al. proposed a quasi-ring-type ME sensor, which is able to detect AC current as small as 10^{-7} A, and/or a vortex magnetic field as small as 6 pT, as shown in Fig. 8.10d (Dong et al., 2005).

However, the requirement of a DC magnetic field source is still a concern. In addressing this challenge, Chung et al. proposed a modified ring-type ME current sensor, as shown in Fig. 8.11a. In this design, instead of using the whole Terfenol-D ring, an epoxy-bonded Terfenol short-fiber ring with an embedded NdFeB magnetic plate was used as the magnetostrictive phase. The permanent magnetic plates not only provide an initial inner magnetic bias for a circumferential magnetization but also eliminate the need of an external DC bias source for ME coupling optimization. Fig. 8.11b plots the comparison of the electric current sensitivity

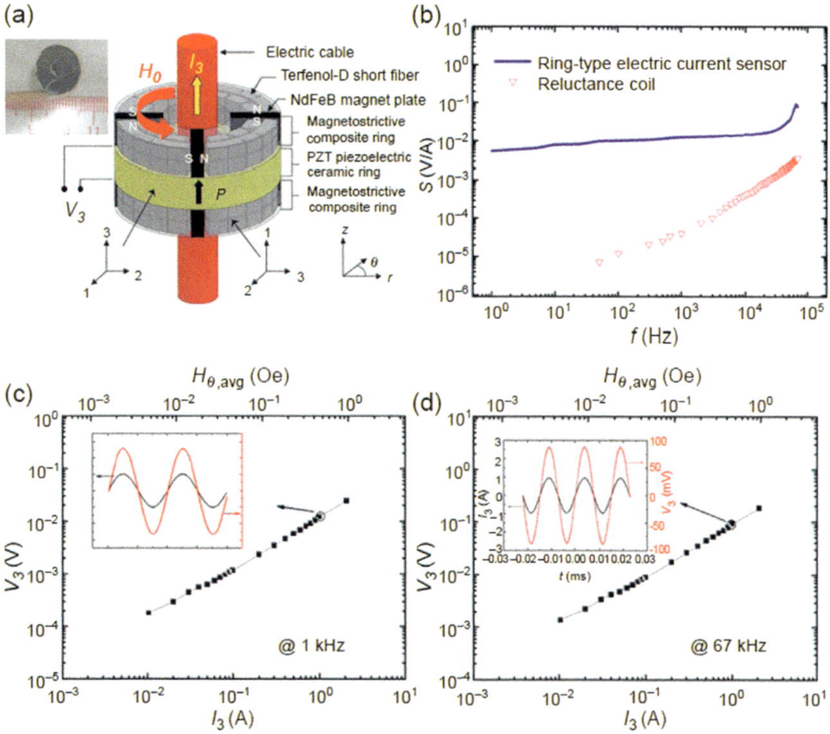

FIGURE 8.11 Design, ME response, and detection sensitivity of the modified ring-type ME current sensor. *Reprinted with permission from Leung et al. (2010), Copyright 2010, AIP Publishing LLC.*

(S) between the modified ring-type sensor and a reluctance coil with 100 turns. It can be seen that the sensitivity of the ring-type sensor possesses a stable response in a wide range of AC frequency. More details of the ring-type sensor current detection response can be found in Fig. 8.11c.

8.4.3 Polymer-based flexible sensor

Despite the large ME coefficient and high sensitivity achieved in bulk ceramic-based ME composites, there is concern with the brittleness of these materials. In particular, the piezoelectric material is usually rigid, fragile, and costly, leading to a high cost and difficulty in implementation. From a practical point of view, the need for some tolerance to external stresses is important. Accordingly, ME polymer–based composites with high flexibility and nonbrittle features were developed as promising

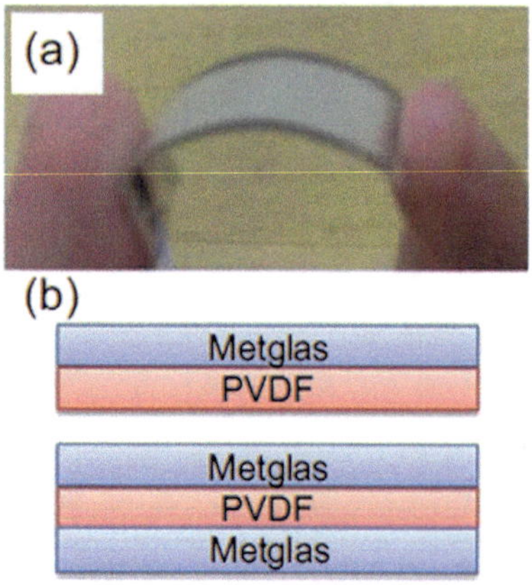

FIGURE 8.12 (a) Image and schematic of flexible Metglas/PVDF laminate composites; (b) DC sensitivity of a bimorph Metglas/PVDF sensor measured under $H_{ac} = 0.1\ Oe$ at $f = 50\ kHz$. *Adapted from Zhai et al. (2006), Copyright 2006, AIP Publishing LLC.*

alternatives for the ceramic-based sensors. Zhai et al. developed thin ($< 100\ \mu m$) and flexible Metglas/polyvinylidene-fluoride (PVDF) ME laminates in both unimorph and bimorph configurations (Fig. 8.12) (Zhai et al., 2006). Besides the flexible nature of this sensor, the composite was also able to provide a giant ME voltage coefficient of 7.2 V/cm.Oe at low frequency, and up to 310 $V/cm.Oe$ under resonance frequency, leading to a high sensitivity (Zhai et al., 2006). Alternatively, Le et al. developed a flexible current sensor based on a bilayer structure consisting of the Cytop polymer and a magnetic tape filled with magnetically soft particles. The Cytop polymer is an electret (or ferroelectret) material, which exhibits a high charge density and a large piezoelectric coefficient ($25 - 700\ pC/N$). It was found that the laminate of a transversely charging electret, along with bias magnetic tape, could attain a large ME coefficient of 4.58 $V/cm\ Oe$ at 1 kHz. In addition, two other types of polymer-based particulate nanocomposites and polymers as binder composites have been developed.

8.5 Self-biased ME composites-based current sensors

Although the above discussed current sensors based on conventional ME composites provide great potential for practical sensing application, most of them require an external DC magnetic bias field to obtain an optimum ME response and high current sensitivity. Considering the composition variation and geometry change, one can tune the optimum DC bias as low as several Oe. However, it is still a requirement to use a permanent magnet or electromagnet to provide DC magnetic bias, which, in turn, is a disadvantage for sensor implementation and miniaturization. Therefore, it is necessary to eliminate these limitations from DC bias. In particular, a novel design is desired if one needs to present an ME composite with large, tunable α_{ME} in the absence of a DC magnetic field. In this section, self-biased ME composites that can provide a giant ME coefficient at zero bias are introduced, and their application as current sensors have been demonstrated.

8.5.1 Self-biased ME composites

The self-biased ME effect is defined as ME composites that exhibit large nonzero α_{ME} in the absence of a DC magnetic field. Fig. 8.13 illustrates the schematic of self-biased ME composites compared to that of conventional ME composites. Recently, self-biased ME composites have been developed in various configurations using different strategies.

The highlights in the development of self-biased ME composites can be summarized in Fig. 8.14. Observation of self-biased ME effect was first reported by G. Srinivasan et al., in 2002, as they found hysteresis and remanence of α_{ME} as a function of H_{bias} in cofired LSMO-PZT laminates. Similar behavior was also found in other laminates. However, this phenomenon did not attract much attention in the following decade, and most of the efforts have been focused on optimizing materials, structures and working modes to enhance the ME voltage coefficient and sensitivity (Nan et al., 2008; Zhai et al., 2008a).

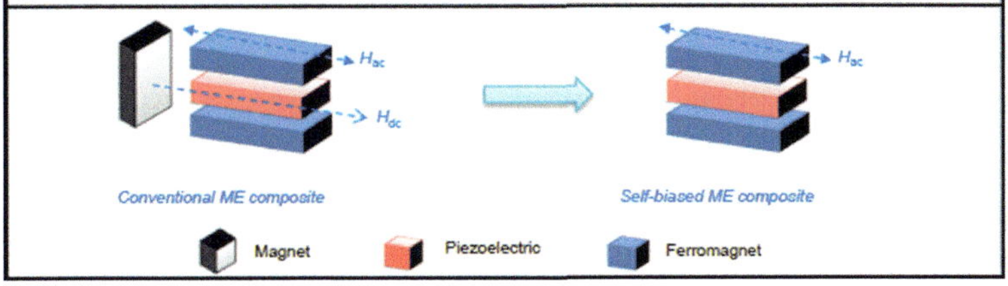

FIGURE 8.13 Schematic of conventional ME effect behavior and self-biased ME behavior.

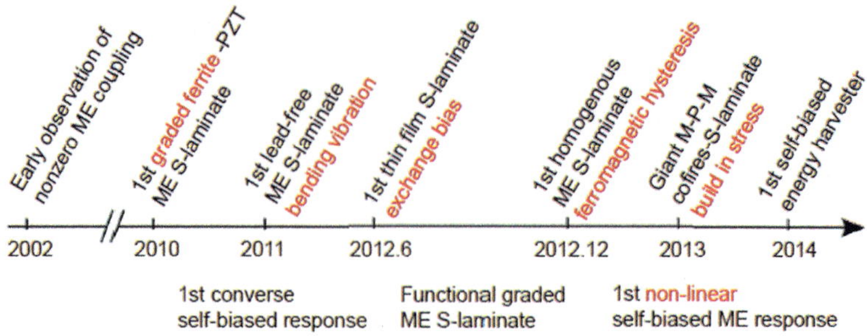

FIGURE 8.14 Milestones for the development of self-biased magnetoelectric composites.

In 2010, Mandal et al. demonstrated a compositionally graded ferrites/PZT laminate structure that produced nonzero ME voltage at zero DC magnetic bias. They first correlated this nonzero ME behavior with the presence of flexural deformation and grading induced out-of-plane internal magnetic field in the ferrite. Subsequently, Yang et al. presented a methodology that regular ME laminate composites can possess a self-biased ME response by changing vibration mode through electrical connections, as shown in Fig. 8.15a (Yang et al., 2010). Soon after that, Srinivasan et al. systematically investigated the hysteresis and remanence in ME effects in functionally graded magnetostrictive-piezoelectric layered composites; they further proposed a theoretical model for the

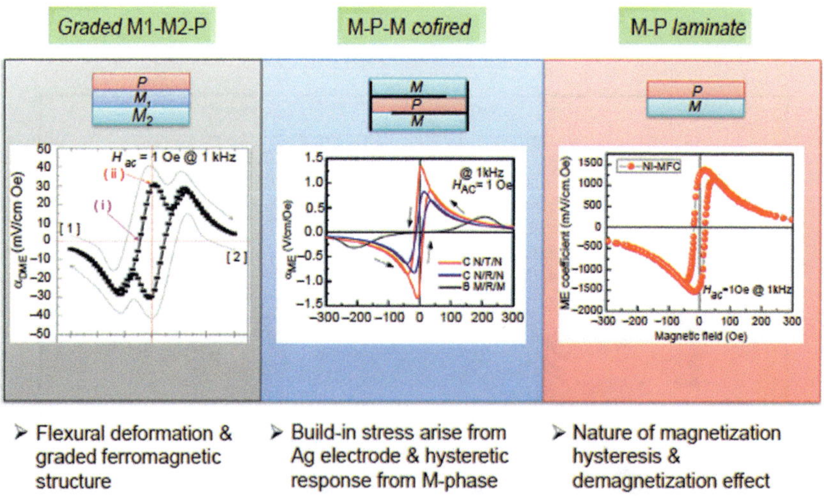

FIGURE 8.15 Development of self-biased magnetoelectric composite. *Reprinted with permission from Yang, Cho et al. (2011); Yan et al. (2013); Zhou, Yang, et al. (2012), Copyright 2012, AIP Publishing LLC.*

self-biased graded composites. Yang et al. first analytically studied the lead-free based self-biased laminates and proposed electrically controlled core-free magnetic flux control devices by using the self-biased converse ME effect (Yang, Ahn, et al., 2011; Yang, Cho, et al., 2011; Yang et al., 2010). Alternative approaches have been suggested for the case of thin films that rely on the angular dependence of exchange bias field from a multicomponent magnetic system to provide a nonzero piezomagnetic coefficient at zero bias (Lage et al., 2012; Onuta et al., 2011). The focus of these studies was mainly on understanding the nature of self-biased effect from multicomponent magnetic systems.

Subsequently, self-biased ME composites with a homogenous magnetostrictive phase have been developed. Yuan et al. first demonstrated a simple Ni/PZT bilayer laminate structure that exhibits tunable self-bias phenomenon with single-phase magnetic material (Fig. 8.15c) (Zhou, Yang, et al., 2012). In this study, fundamental understanding of the role of domain structure and the demagnetization field of magnetic materials toward a self-biased effect was provided. The self-biased response was found to be related with the nature of the magnetic phase and hence could be controlled by variation in the demagnetization state and the resultant differential magnetic flux distribution. Meanwhile, Yan et al. successfully cofired a self-biased composite NCZF/Ag/PMNPT/Ag/NCZF with a homogeneous magnetic phase (Yan et al., 2013). Cofired structure provides the opportunity of interfacial engineering to obtain the optimum strain transfer from two phases with a giant self-biased response $> 1\ V/cm/Oe$, as shown in Figure (b). The self-bias response in the cofired laminates was explained by the low field magnetostriction hysteresis and the build-in stress from the internal electrode during the cofiring process. Further, by considering the nonlinear magnetic properties of the ferromagnetic phase, the self-biased phenomenon could also be observed in conventional ME composites that measured under the nonlinear region.

In summary, the working principle of the self-biased ME response can be categorized into three main groups: (1) flexural deformation with graded ferromagnetic structure; (2) build-in stress arising from an inner electrode; and (3) the nature of magnetization hysteresis in the magnetostrictive phase. The development of self-biased ME composites can be concluded, as shown in Fig. 8.15. These discoveries of self-biased ME effect are technically important for applications of functional ME devices, especially for magnetic field/current sensors, core-free magnetic flux control devices, and electrically controlled magnetic memory devices (Yang, Cho et al., 2011). This phenomenon not only provides great potential toward devices miniaturization but also reduced the electromagnetic interference that might be induced from the DC bias sources.

8.5.2 Application as current sensor

The early demonstration of using self-biased ME composites as an electric current sensor can be originated from the work done by Lu et al., as shown in Fig. 8.16. In this design, a PZT plate is bonded with graded ferromagnetic layers (*Ni* and *FeCuNbSiB* foils) in a cantilever configuration with one end clamped (Fig. 8.16a). The self-biased effect is evoked due to the build-in magnetic bias formed in the graded ferromagnetic layers with different magnetic properties. A proof mass was added on the free end of the cantilever, which can be used to modify the natural resonance frequency of the ME cantilever to enhance the ME coefficient. By locating the current-carrying cable near the laminate sensor, the generated vortex magnetic field will be captured via the functional graded magnetic layers, thereby creating an ME voltage though the piezoelectric plate for current sensing in the absence of DC magnetic bias.

In order to detect the current of conventional power-line cables (50–60 Hz), the cantilever ME laminate was configured with adjusted geometry and tip mass (M_q). Moreover, the number of *FeCuNbSiB* layers

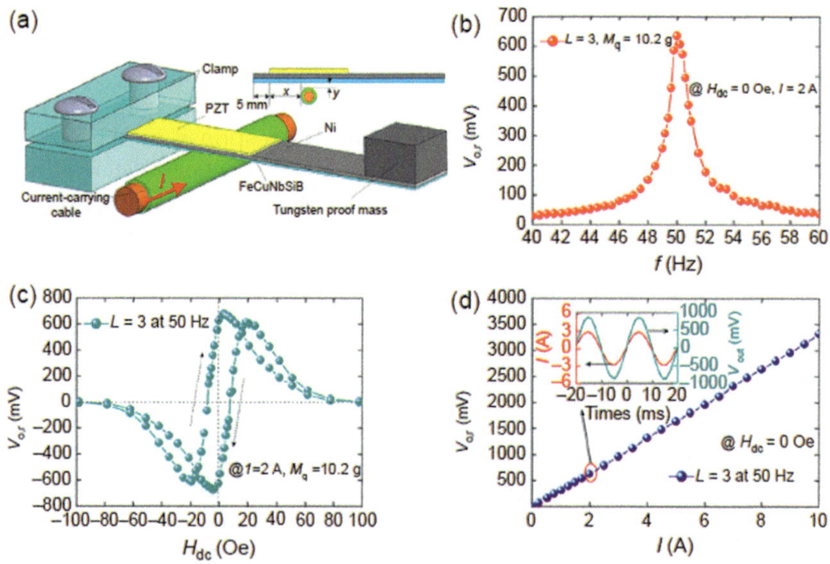

FIGURE 8.16 Demonstration of self-biased ME composites-based current sensor. (a) Schematic diagram of the cantilever-type ME current sensor design; (b) zero-biased ME voltage output as a function of frequency; (c) self-biased ME response as a function of DC bias; and (d) self-biased ME response as a function of detected current. *Reprinted with permission from Lu et al. (2014), Copyright 2014, Elsevier.*

(L) was optimized to achieve the largest self-biased ME coefficient. As a consequence, as shown in Fig. 8.16b, the self-biased ME voltage shows a peak behavior at 50 Hz in response to a power-line cable carried with an AC current of 50 Hz, indicating a good match of the optimized current sensor. Fig. 8.16c plots the DC bias dependence of the ME voltage measured with $I = 2A$ at 50 Hz. It can be clearly noticed that the maximum ME voltage output was obtained at zero-biased condition. Fig. 8.16d illustrates the zero-biased ME voltage output under varied AC currents, and the inset shows the corresponding time-domain plot of I and V_{out} dependence. From these results, the self-biased ME sensor exhibits an excellent linearity and high sensitivity (330 mV/A) for monitoring the electric current in power-line cable.

Compared with the conventional ME current sensor, the presented self-biased-based sensor has high sensitive and DC-bias-free features. These features can be further expanded toward other types of self-biased ME composites. The development of self-biased-based ME composites is promising toward high sensitivity and miniaturized current sensor applications.

8.6 ME transformer–based current sensors

ME transformers as voltage gain devices have been investigated in literature, and it has been shown that a giant voltage gain can be obtained at the resonance frequency (Dong, Li, Viehland, et al., 2004; Dong et al., 2009). With the application of a small DC magnetic bias, the ME transformer can provide a tunable voltage gain and/or a shift of resonance frequency, which is promising as a magnetic sensor. By configuring the characterization of ME composites and piezoelectric transformers, a variety of ME transformers have been developed (Dong, Li, Viehland, et al., 2004; Dong et al., 2009; Jia et al., 2008). In the design proposed by Dong, Li, Viehland, et al. (2004), Dong et al. (2009) as shown in Fig. 8.17a, an input AC voltage was applied to the coil that generated an oscillating AC magnetic field (H_{ac}). The generated field interacted with the ME laminate and amplified the voltage output. In this design, the use of winding coil around the laminated ME composite poses challenges in its implementation. Alternatively, Zhang et al. (2013) demonstrated an ME transformer consisting of a ring-type Terfenol-D/PZT laminate composite directly connected with a Rosen-type PZT ceramic transformer, as shown in Fig. 8.17b (Zhang et al., 2013). The ME transformer was characterized as a current sensor. The current sensitivity (S_I) in two concurrent operational

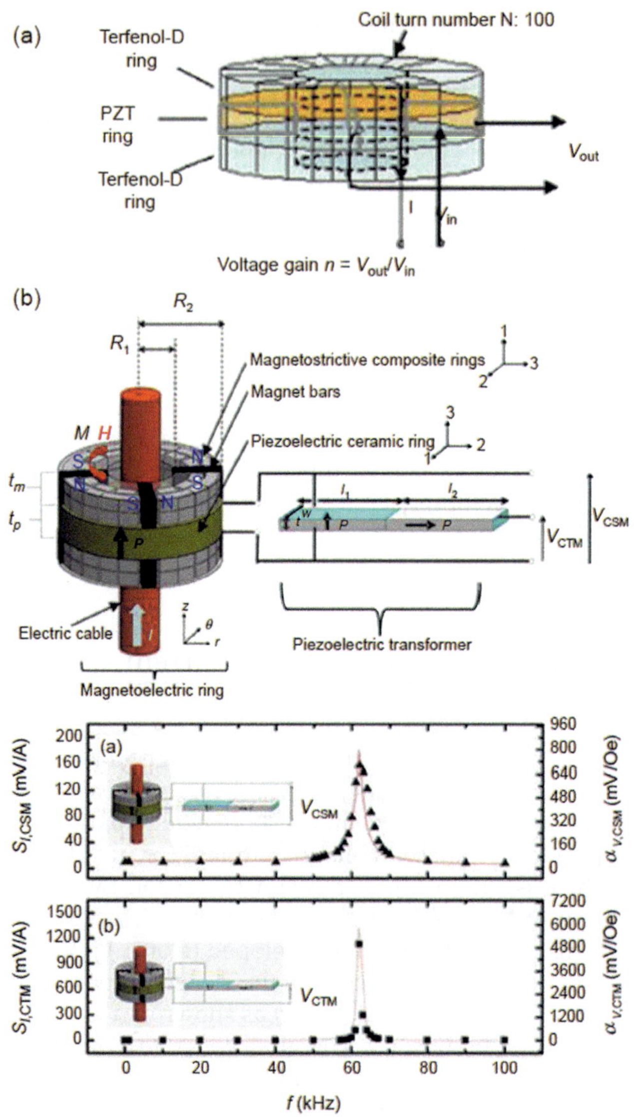

FIGURE 8.17 (a) Coil-winded-based ME transformer. (b) Ring-type ME composites connected with Rosen-type PZT ceramic transformer. *(a) Reprinted with permission from Dong, Li, Viehland, et al., 2004, Copyright 2004: AIP Publishing LLC. (b) Reprinted with permission from Zhang et al. (2013), Copyright 2013: AIP Publishing LLC.*

modes was evaluated both theoretically and experimentally. A large enhancement in the resonance $S_I \sim 1000mV/A$ at 62 kHz (Fig. 8.17b) was obtained as a result of the amplified vortex ME effect caused by the resonance voltage step-up effect in the piezoelectric transformer.

However, the separate components of this ME transformer bring in another level of difficulty toward sensor implementation. Thus, even though these structures are intriguing, their implementation for realization of miniature transformer and magnetic sensors will be limited. In this chapter, we will discuss two different types of ME tunable transformers in addressing these challenges.

8.6.1 Epoxy-bonded ME transformer

Considering the high ME coupling of ME laminate and the easy-fabricated unipoled piezoelectric transformer, epoxy-bonded ME transformers with different configurations were investigated (Jia et al., 2008; Zhou, Wang, et al., 2012). A 2-2-laminated ME transformer was designed and fabricated by directly bonding magnetostrictive layers onto the output port of a cofired multilayer piezoelectric transformer, as shown in Fig. 8.18. The multilayered piezoelectric transformer was synthesized via a low temperature cofiring process. Piezoelectric composition $0.2\,Pb\left(Zn_{1/3}Nb_{2/3}\right. -0.8\,Pb(Zr_{0.5}Ti_{0.5})O_3\ (PZNT)$ was chosen and synthesized by a conventional mixed oxide method due to its good piezoelectricity and low sintering temperature. Fig. 8.18a illustrates

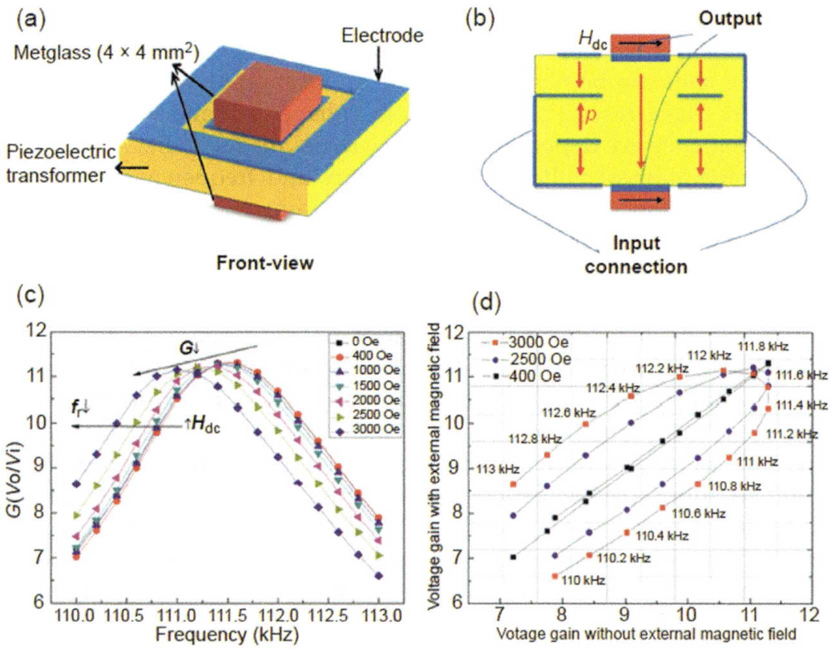

FIGURE 8.18 (a) and (b) Schematic diagram of the ME transformer, consisting of a multilayer piezoelectric ceramic transformer and two layers of Metglas. (c) and (d) ME transformer voltage gain variation as a function of frequency under varied DC magnetic bias.

the schematics of the ME transformer. In this design, the input and output electrodes are on the same side of the square plate and are isolated from each other by a fixed gap of 2 mm. The electrode pattern is a square/square structure with a floating connection among each layer. Metglas, with a dimension of $4 \times 4\ mm^2$, are attached on both side of the output section. Fig. 8.18b shows the electrical connection of the transformer in the step-up mode. In this case, on applying the electrical excitation to the external rectangle frame (input), the contour extensional vibration is generated and transferred to the center square (output) through the gap. At the output, the mechanical strain is converted to electrical voltage through a direct piezoelectric effect. The multilayer section is equivalent to connecting n units in parallel, where the input motional capacitance ($C_{a,input}$) is increased by n times. Simplistically, the gain of the transformer is given by: $Gain = \sqrt{\frac{C_{a,input}}{C_{a,output}}}$. Thus, the multilayered structure can provide higher voltage gain. By locating the longitudinally poled transformer under a transverse DC magnetic field (H_{dc}), the peak value of the voltage gain was found to change with respect to different resonant frequency, as shown in Figure 8.15(c). Figure 8.15(d) plots the voltage gain with and without an external magnetic field, indicating the effectivity of voltage gain tunability under an induced magnetic field. There is a significant change in the magnitude of the voltage gain with an applied external magnetic field, which increases as the frequency approaches the resonance point. Furthermore, for the same magnitude of voltage gain, a large frequency shift can be noticed with the application of an external magnetic field. These results prove that both voltage gain and working resonance frequency are dependent on the external magnetic field.

The generation of voltage gain tunability and frequency shift under induced H_{dc} allows us to tailor the ME transformer with desired performance, offering the potential for the design of a magnetic field–tunable sensor. The working mechanism can be simply explained as follows: when an external magnetic field is applied to the composite, the Metglas induces strain in terms of shape change on the center square section through magnetostriction. This strain is transferred to the output section of the transformer through elastic coupling. Next, a mechanical stress is generated in the center square, resulting in an electric polarization by direct piezoelectric effect. Meanwhile, the electric field–induced strain from the input piezoelectric phase is interfered with this magnetic field–induced strain from the magnetostrictive phase. The resulting differential strain is converted into voltage through the direct piezoelectric effect and will induce change in the magnitude of output voltage and working resonance frequency. In this case, external magnetic field affects the transformer performance through the direct ME effect. Thus, it is expected that the magnitude of the voltage gain and working resonance frequency of the transformer will be tuned and shifted under an external magnetic field through the ME effect. In return, one can detect the ambient magnetic field/current by monitoring the performance of the ME transformer. This sensing mechanism can be further extended to other configurations by selecting high-performance magnetostrictive materials and piezoelectric transformers.

8.6.2 Cofired ME transformer

To achieve the best ME tunable performance, giant ME coupling and more efficient stress transfers are desired. However, the problems with epoxy-bonded laminate ME composites could not be neglected. (1) Device miniaturization is a concern since the processing requires manual operations like dicing, cutting, bonding, and wiring. (2) Low mechanical strength of the epoxy may deteriorate the ME coupling and sensitivity. The ideal solution to these problems is discovering an all-in-one ME composite, which possesses both giant ME coupling coefficient and transformer function.

In addressing these concerns, cofired ME transformers were developed. Islam et al. demonstrated a sintered ME transformer consisting of a particulate $Pb(Zr_{0.52}Ti_{0.48})O_3 - Ni_{0.8}Zn_{0.2}Fe_2O_4$ $(PZT - NZF)$ composite and utilizing the electrode pattern of a conventional unipoled transformer, as shown in Fig. 8.19a. The unipoled ME transformer possesses a ring-dot type electrode pattern with an electrical connection in the step-up mode. On applying an electrical drive to the external ring (input), the radial extensional vibration is generated through converse piezoelectric effect and further transferred to the center dot (output). At the output, the mechanical strains are converted into electrical voltage through direct piezoelectric effect. Since NZF magnetostrictive particles were spread all over the transformer disk through the mixed oxide sintering route, the transformer performance is sensitive with respect to an external magnetic field. Fig. 8.19b demonstrates a significant shift in the magnitude of resonance frequency with respect to the corresponding maximum voltage output under applied DC magnetic bias. In addition, the electric field–induced strain will also interact with the magnetic particles, thereby generating magnetic flux through the ME transformer. This magnetic flux can interact with the external magnetic field, producing a flux gradient that can be detected through the tunable feature of the transformer. On the other hand, the change in voltage gain can be modulated by optimizing input voltage, as shown in Fig. 8.19c. The variation in voltage gain before and after applied magnetic bias was clearly demonstrated in Fig. 8.19d.

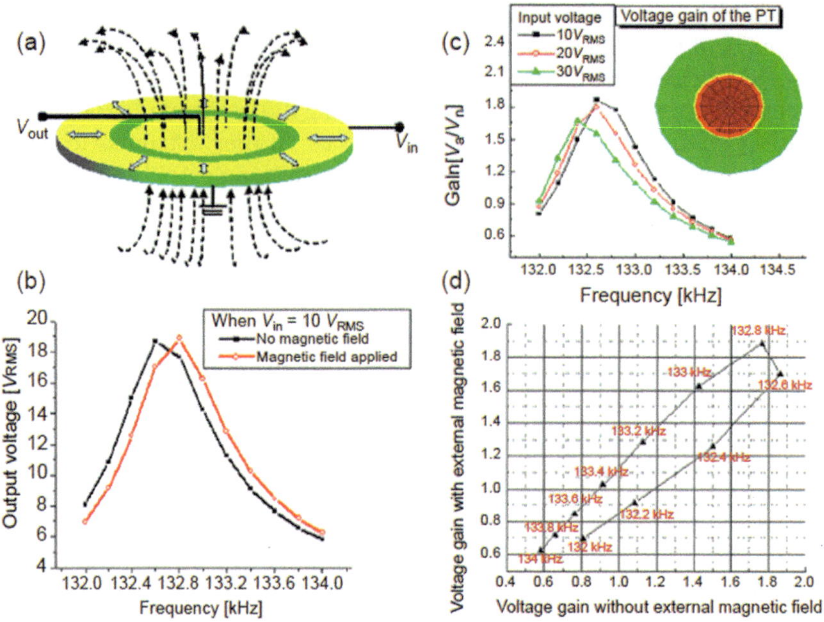

FIGURE 8.19 (a) schematic diagram of the cofired particulate magnetoelectric transformer and its performance (b) with and without an applied DC magnetic field. (c) Voltage gain variation as a function of frequency under varied input voltage. (d) Change in voltage gain with and without the external magnetic field. *(a) and (b) Reprinted with permission from Islam et al. (2006); Copyright: 2006 AIP Publishing LLC; (c) and (d) Reprinted with permission from Kim et al. (2007), Copyright: 2007 AIP Publishing LLC.*

These results indicate that the cofired ME transformer not only possesses a simple structure for fabrication and implementation but also can be easily manipulated by changing the input driving voltage and ring-dot pattern. However, the small ME coupling of the particulate composites hinders its tunable feature under an applied DC magnetic field.

In order to further enhance the ME response, Yuan et al. designed and demonstrated a cofired ME transformer consisting of two unipoled piezoelectric transformers and a magnetostrictive layer in a laminate configuration (Zhou et al., 2014). The ME transformer was designed by combining magnetostrictive material and a piezo-electric transformer in a piezoelectric/magnetostrictive/piezoelectric (PMP) laminate structure with an Ag inner electrode (Fig. 8.20a). Composition corresponding to $0.2\,Pb\left(Zn_{1/3}Nb_{2/3}\right) - 0.8\,Pb(Zr_{0.5}Ti_{0.5})O_3$ $(PZNT)$ and $Ni_{0.6}Cu_{0.2}Zn_{0.2}Fe_2O_4$ $(NCZF)$ were selected as piezoelectric and magnetostrictive materials, respectively. In this unipoled transformer design, the input electrode "square ring" and the center output elec-trode

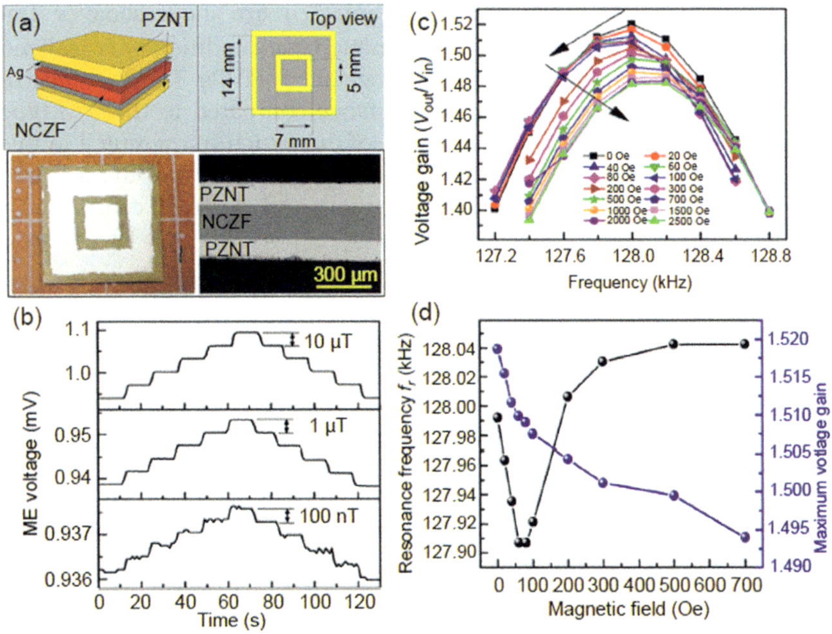

FIGURE 8.20 (a) schematic and images of cofired laminate ME transformer; (b) DC sensitivity of the cofired ME laminate; (c) voltage gain variation as a function of frequency under varies DC magnetic bias; and (d) variation of resonance frequency and maximum voltage gain as a function of external DC magnetic field. *Reprinted with permission from Zhou et al. (2014), Copyright 2014, AIP Publishing LLC.*

"square" are on the same side of the piezoelectric plate and are isolated from each other by a fixed gap of 1 mm. Two unipoled transformers were electrically connected in a series. Fig. 8.20b illustrates the DC magnetic field sensitivity of the cofired transformer; magnetic field variation as small as $H_{dc} = 100\,nT$ could be detected, indicating an enhanced ME coupling and sensitivity in the cofired 2-2 type laminate.

Fig. 8.20c shows the measured voltage gain G as a function of frequency with H_{dc} varying from 0 to 2500 Oe. It is clearly noticed that the magnitude of maximum voltage gain decreased and the peak position (resonance frequency) shifted as a function of H_{dc}. The magnitude of maximum voltage gain and corresponding resonance frequency shift are summarized as a function of H_{dc} in Fig. 8.20d. It is obvious that the resonance frequency decreases first with increasing H_{dc}, showing a minimum magnitude of 127.9 kHz in the range of 60–80 Oe, and then increases with increasing H_{dc}. These results demonstrate that by optimizing the ME transformer configuration, large resonance frequency tunability

$(f_{r.o} - f_{r.H})/H = 1.4Hz/Oe$, where $f_{r.o}$ and $f_{r.H}$ are the resonance frequencies under zero and nonzero magnetic field condition was realized with a small DC magnetic field in the range of 60–80 Oe. The 2-2-type ME laminates presented a superior performance compared to the 0–3 type particulate composite. The performance of the structure presented in this study can be further improved by optimizing the input/output area ratio through multilayering, material composition, and electrical connectivity.

8.7 Note on magnetic sensor detection limits and terminology (Liang et al., 2021)

Given the purpose of such devices, one of their defining properties is how small a magnetic field signal such a device can reasonably detect. However, on this point, some notable inconsistencies exist in the body of work of the communities studying various types of magnetometers. Thus, a word of caution is in order about the methods and terminology related to the detection of magnetic fields.

There are actually two issues: first, how to quantify the smallest magnetic field a sensor or antenna can detect, and second, what term(s) to use for such a measure. Mainly, two ways have been used to report the detection limit: in dimensions of [field] or in dimensions of [field]/$\sqrt{\ }$Hz. The former approach adopts a fixed set of parameters relevant to the sensor (such as the voltage sustaining its operation) and then measures the sensor output at decreasing magnetic field strengths until the signal flat lines, thereby establishing the minimum field detected. The latter method is superior in terms of quantifying the sensor's intrinsic properties that are independent of experimental parameters but is more complex, requiring a measurement of the power spectral density (PSD) of the intrinsic noise floor. The detection limit under this approach is obtained by dividing the noise $\sqrt{PSD}$ by the sensor's sensitivity (Fermon & Pannetier-Lecoeur, 2013; Guo et al., 2015; Stutzke et al., 2005) (change in output signal per change in a magnetic field), and the dimension of $Hz^{-\frac{1}{2}}$ in the resulting quantity arises from the noise spectrum in this ratio. This characterization is more suitable for making comparisons across sensors that have different structures and detection mechanisms and may be tested under different parameters. In particular, under the first method, the lowest detectable magnetic field signal can be arbitrarily improved by methods like increasing acquisition time or averaging. However, the PSD is already normalized by acquisition time (Fermon & Pannetier-Lecoeur, 2013), thus removing this source of arbitrariness.

The second issue is the lack of agreement on terminology. Although it is common to find magnetic sensor performance reported in terms of [field]/√Hz, this result has been variously referred to as "field equivalent noise" (Fermon & Pannetier-Lecoeur, 2013), "equivalent magnetic noise" (Wang et al., 2014), "magnetic noise" (Dong et al., 2020), "detectivity" (Stutzke et al., 2005), "limit of detection" (Yarar et al., 2016), "minimum detectable signal" (Guo et al., 2015), "sensitivity"(Wolf et al., 2015), and perhaps by other names. It would benefit the magnetic sensor community to adopt a shared standard of terms. To this end, as well as to unify the discussion of results covered in this review, we propose the following. First, we suggest that "sensitivity" be reserved for quantifying the change in the sensor's output signal per unit of the magnetic field (Vig & Walls, 2000), and that "magnetic noise" be used to designate the noise from magnetization fluctuations and other magnetic effects. Second, we propose a new term: "minimum detected field" (MDF) to designate the first kind of result discussed above, given in dimensions of the magnetic field only. Previously, "limit of detection" has been used in this case (Nan et al., 2013), but this may create confusion when the same term is used for both types of measurements. For the other quantity, [field]/√ Hz, we like "detectivity" for its concision and the implication that it is not an absolute detection limit but rather an estimate or metric (Guo et al., 2015; Stutzke et al., 2005) used for standardized comparison to other sensors. Thus, these are the definitions employed in the text to follow except where the authors were unable to alter the labels within the figures that have been reproduced here.

8.8 Magnetic noise and elimination

The detection capability of ME sensors not only depends on the ME voltage signals in response to an incident magnetic field but is also affected by the noise that exists in the detection component, both internally and externally. In the case of ME sensors, the dominate noise sources can be classified as (1) extrinsic noises (vibration noise, thermal noise, and electromagnetic interference) and (2) intrinsic noises (Johnson noise and $1/f$ noise). Therefore, it is of great importance to identify the noise sources and eliminate their influence toward better detection sensitivity of the sensors.

8.8.1 Extrinsic noise

Extrinsic noise is referred to as any unexpected or unwanted disturbances that are originated from the environment. These disturbances can be mechanical vibration, magnetic field, temperature fluctuation, and so on. The main noise sources that can disturb the ME composites are thermal and vibration noises. In addressing these issues, Xing et al. systematically investigated the ability of various fundamental modes to eliminate environmental noise from thermal and vibration noise. The idea is based on designing the sensor signal and the noise signal in different fundamental modes. In a symmetrical structure, the neutral line sits in the center of the bending piezoelectric, where the vibration-induced charge can be canceled. In an unsymmetrical structure, both bending and longitudinal modes will be excited due to thermal expansion difference, where the thermal strain can be partially canceled. Fig. 8.21 summarized various ME laminate configurations and their capability to reject thermal or vibration noises. Accordingly, SS-UN (symmetrical signal, unsymmetrical noise) modes have the ability to cancel vibration noise, whereas US-SN (unsymmetrical signal, symmetrical noise) modes have the capability to reject thermal noise.

Thermal fluctuation noise affects the detection unit via the pyroelectric effect. This phenomenon can be originated from the nature of pyroelectric materials with lower crystal symmetry that can generate electricity in response to temperature changes. When considering the crystal structure,

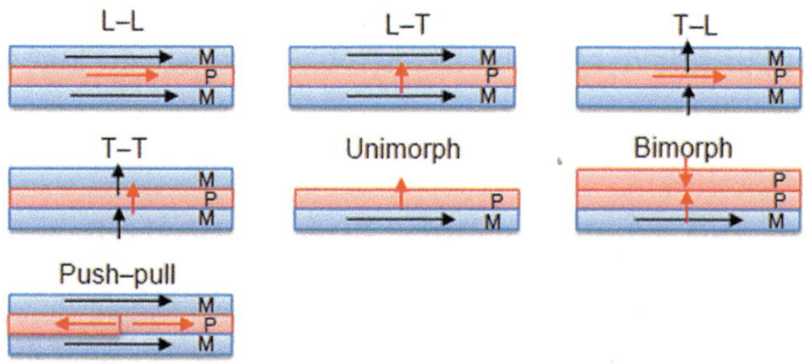

Model	Vibration cancellation	Thermal cancellation
Unimorph	No	No
Bimorph	Partially	Partially
L–T, T–T, L–L, T–L	Yes	No
Push–pull	Yes	No
SS–UN[a]	Yes	No
US–SN[b]	No	Yes

[a]SS–UN:symmetrical-signal, unsymmetrical noise
[b]US–SN:unsymmetrical-signal, symmetrical noise

FIGURE 8.21 ME laminate composite structures and their effect in noise cancellation.

10 of these 20 piezoelectric crystal classes are pyroelectric. In particular, the widely used PZT, 4 mm point group, has a pyroelectric effect. Thus, the ME composite sensors would be affected by the thermal noise in the piezoelectric (PZT-based) layer.

As discussed above, US-SN modes have the capability to reject thermal noise; one of the candidate configurations is ME laminate in a bimorph structure (Fig. 8.21). Based on this concept, Zhai et al. demonstrate a design of Terfenol-D/PZT bimorph ME laminate that operates in a bending mode under an unsymmetrical magnetic bias (Zhai et al., 2008b), Zhai et al., 2008bs shown in Fig. 8.14. (a). Unlike conventional bimorph, the magnetization directions are opposite in the two Terfenol-D layers by using two U-shaped magnetic biases. Thus, under an applied AC magnetic field, the bimorph laminate operates in bending modes with top and bottom PZT layers generating charges of opposite signs, where the differential output of the ME voltage will be doubled. However, when the temperature is changed, each layer will change shape in a longitudinal mode with induced thermal charges of the same sign, where the differential output of the thermal noise will be null. The comparative results in Fig. 8.22a clearly demonstrate the capability of thermal noise cancellation by using this bimorph structure. Alternatively, Israel et al. proposed a different thermal-noise rejection ME sensor by connecting two ME multilayer $BaTiO_3 : Ni$ capacitors, as shown in Fig. 8.22b. In this design, the two ME elements were electrically connected in parallel with oppositely oriented polarization and possess stable temperature dependence performance compared to that of a single ME element.

Vibration noise affects the ME sensor via the piezoelectric effect, which refers to spontaneous polarization coupled to pressure and/or stress changes. One can use a vibration isolation table to minimize the effect of vibration noise in a lab condition; however, this is not suitable for practical applications. According to Xing's investigation, it conceptually demonstrated that vibrational noise can be eliminated by using an ME laminate with a mechanically symmetrical structure. Gao et al. demonstrated a differential-mode Metglas/PZT ME laminate, where two layers of PZT with the same polarization were parallelly connected to create a differential symmetric structure. Under an applied longitudinal magnetic field, the PZT layer contracts or expands longitudinally with identical signs of a generated charge, leading to a doubling of the signal. Conversely, an applied vibration noise will tend to cause an asymmetric bending deformation; the corresponding opposite signs of charges generated from top and bottom PZT layers will result in an attenuation of the output signal. Thus, the differential structure has the ability to reject external vibrational noise by separating the signal and noise with a different vibration mode.

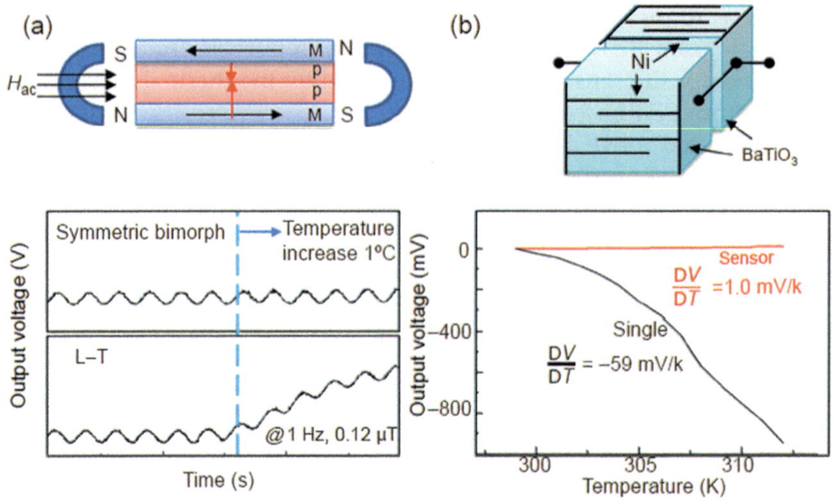

FIGURE 8.22 (a) Schematic illustration of symmetric ME bimorph design and its output signals under thermal fluctuations in time domain. (b) Schematic illustration of a temperature compensated device based on two ME multilayer capacitors (MLCs) and its temperature dependence performance. *Adapted from Zhai et al. (2008b).*

8.8.2 Intrinsic noise

Intrinsic noise refers to disturbances that are originated from the detection unit and/or detection circuits inherently. In detection circuits, the main noise sources are (1) Johnson noise and (2) Flicker ($1/f$) noise.

Johnson noise is electronic noise generated by thermal agitation of electrons in resistors, which is random in nature. It can be represented as either a voltage noise source or a current noise source in series/parallel with a noiseless resistor, as shown in Fig. 8.23. For a given bandwidth, the root mean square (RMS) voltage noise (V_n) and current noise (I_n) can be expressed as

$$V_n = \sqrt{4k_B T \Delta f R} \tag{8.9}$$

$$I_n = \sqrt{4k_B T \Delta f / R} \tag{8.10}$$

where $k_B = 1.38 \times 10^{23}\ J/K$ is Boltzmann's constant, T is absolute temperature in kelvins, D_f is the bandwidth in Hz over which the noise is measured, and R is resistance value in ohms. From these equations, we can see that both temperature and resistance play an important role toward the Johnson noise density at a fixed bandwidth.

Flicker noise ($1/f$ noise), or contact noise, is present in all devices consisting of imperfect contact between two conductors. The various

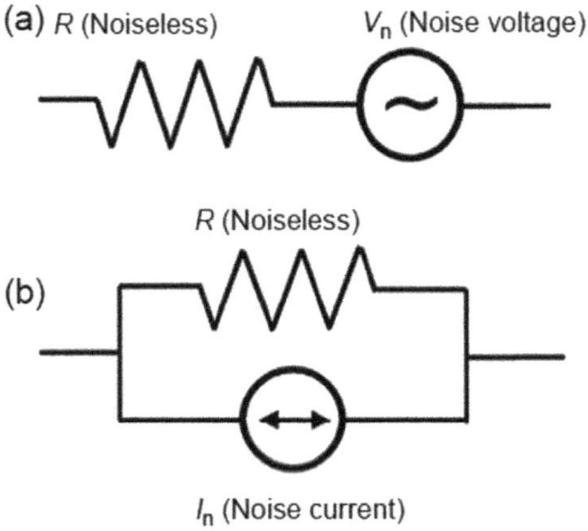

FIGURE 8.23 Equivalent circuit models of (a) noise voltage and (b) noise current.

kinds of mechanisms for Flicker noise depends on the device type, where the random generation and recombination of electron-hole pairs in semiconductors are the most commonly found. It can be modeled by a noise current source in parallel with the device. For a given bandwidth Δf, the RMS flicker-noise current (I_f) can be expressed as

$$I_f = \sqrt{\frac{K_f I^m \Delta f}{f}} \tag{8.11}$$

where K_f is the flicker-noise constant, I is the DC current, and m is the flicker-noise exponent. It can be noticed that the flicker noise is sensitive with respect to frequency, which is specifically high at a low frequency range.

Fig. 8.24a shows an equivalent noise circuit model for an ME-based sensor (Wang et al., 2014). In this noise model, five noise sources were considered as main contributions to the total noise density. The ME composite was modeled as a magnetic field-induced charge source in parallel with a capacitor C and resistor R. There are two noise sources for the ME detection unit: (1) the dielectric loss noise (N_{loss}) source and (2) the leakage resistance noise (N_R) source. The corresponding charge noise density is given as

$$N_{loss} = \sqrt{\frac{4k_B TC \tan \delta}{2\pi f}} \tag{8.12}$$

$$N_R = \frac{1}{2\pi f}\sqrt{\frac{4k_B T}{R}} \tag{8.13}$$

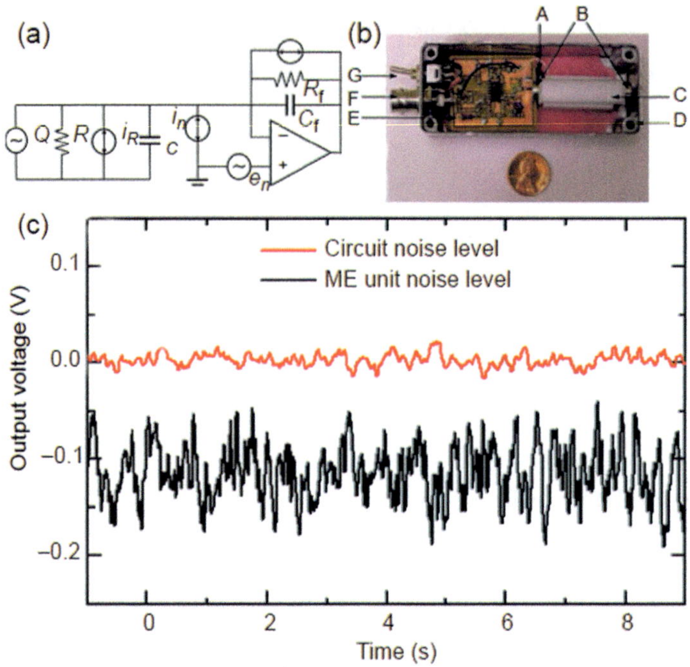

FIGURE 8.24 (a) Equivalent noise model for a magnetic sensor. (b) Photograph of a prototype ME detection system, consisting of an ME sensor and low-noise charge amplifier. (c) Output noise level of circuit and sensor unit. *Adapted from Wang et al. (2014) (a); Reprinted with permission from Xing et al. (2007), Copyright 2007, AIP Publishing LLC.*

where T is the temperature in kelvin, tan d is the dielectric loss, R is the DC resistance of the ME sensor, and f is the frequency in Hz. For the detection circuit, there are three noise sources, including (1) current (N_i) and (2) voltage (N_v) noise from low-noise amplifiers and (3) thermal noise (N_{R_f}) from feedback resistors. The noise charge density can be expressed as

$$N_i = \frac{1}{2\pi f} i_n \tag{8.14}$$

$$N_v = \frac{e_n \left| 1 + \frac{Z}{Z_f} \right|}{|H(s)|} \tag{8.15}$$

$$N_{R_f} = \frac{1}{2\pi f} \sqrt{\frac{4k_B T}{R_f}} \tag{8.16}$$

where in is the current noise density, e_n is the voltage noise density of the low-noise amplifier, and Z and Z_f represent the ME laminate and the feedback impedance. Thus, the total noise charge density (N^T) can be given as

$$N^T = \sqrt{N_{loss}^2 + N_R^2 + N_i^2 + N_v^2 + N_{R_f}^2} \tag{8.17}$$

Thus, one can use Eqs. (8.12)-(8.17) to estimate and analyze the contributions from various noise sources.

Fig. 8.24b illustrates a previous prototype ME detection system consisting of an ME composite (component C) and low-noise detection circuit (component E), as discussed above. In this prototype, the measured RMS value of the circuit is dramatically smaller than that of the ME detection unit (circuit + sensor), as shown in Fig. 8.24c. Accordingly, Wang et al. systematically reviewed the development of equivalent magnetic noise reduction for ME sensors (Wang et al., 2014). There are mainly three strategies toward noise reduction, including (1) reducing the detection circuit noise via a low-noise amplifier, (2) reducing the ME laminate noise via reducing dielectric loss and increasing resistance of piezoelectric materials, and (3) increasing the ME charge coefficient without introducing addition noise via interfacial bonding/poling condition/geometry design optimization.

Acknowledgments

The authors gratefully acknowledge the financial support from the Office of Basic Energy Science, the Department of Energy (DOE), and the Air Force Office of Scientific Research (AFOSR).

References

Agrawal, S., Cheng, J., Guo, R., Bhalla, A. S., Islam, R. A., & Priya, S. (2009). Magnetoelectric properties of microwave sintered particulate composites. *Materials Letters, 63*(26), 2198–2200.

Astrov, D. (1960). The magnetoelectric effect in antiferromagnetics. *Soviet Physics Journal of Experimental and Theoretical Physics, 11*(3), 708–709.

Astrov, D. (1961). Magnetoelectric effect in chromium oxide. *Soviet Physics Journal of Experimental and Theoretical Physics, 13*(4), 729–733.

Caicedo, J., Zapata, J., Gómez, M., & Prieto, P. (2008). Magnetoelectric coefficient in BiFeO3 compounds. *Journal of Applied Physics, 103*(7).

Catalan, G., & Scott, J. F. (2009). Physics and applications of bismuth ferrite. *Advanced Materials, 21*(24), 2463–2485.

Cheong, S.-W., & Mostovoy, M. (2007). Multiferroics: A magnetic twist for ferroelectricity. *Nature Materials, 6*(1), 13–20.

Cho, K. H., & Priya, S. (2011). *Applied Physics Letters, 98*(23), 232904.

Chu, Y.-H., Martin, L. W., Holcomb, M. B., & Ramesh, R. (2007). Controlling magnetism with multiferroics. *Materials Today, 10*(10), 16–23.

Chu, Z., Shi, H., Shi, W., Liu, G., Wu, J., Yang, J., & Dong, S. (2017). Enhanced resonance magnetoelectric coupling in (1-1) connectivity composites. *Advanced Materials, 29*(19), Article 1606022.

Curecheriu, L. P., Buscaglia, M. T., Buscaglia, V., Mitoseriu, L., Postolache, P., Ianculescu, A., & Nanni, P. (2010). Functional properties of BaTiO3–Ni0. 5Zn0. 5Fe2O4 magnetoelectric ceramics prepared from powders with core-shell structure. *Journal of Applied Physics, 107*(10).

Curie, P. (1894). Sur la symétrie dans les phénomènes physiques, symétrie d'un champ électrique et d'un champ magnétique. *J. Phys. Theor. Appl., 3*(1), 393–415. https://doi.org/10.1051/jphystap:018940030039300

Dong, S. X., Bai, J. G., Zhai, J. Y., Li, J. F., Lu, G. Q., Viehland, D., et al. (2005). *Applied Physics Letters, 86*(18).

Dong, C., He, Y., Li, M., Tu, C., Chu, Z., Liang, X., Chen, H., Wei, Y., Zaeimbashi, M., Wang, X., Lin, H., Gao, Y., & Sun, N. X. (2020). A portable very low frequency (VLF) communication system based on acoustically actuated magnetoelectric antennas. *IEEE Antennas and Wireless Propagation Letters, 19*(3), 398–402. https://doi.org/10.1109/LAWP.2020.2968604

Dong, S. X., Li, J. F., & Viehland, D. (2004). *Journal of Applied Physics, 96*(6), 3382.

Dong, S. X., Li, J. F., Viehland, D., Cheng, J., & Cross, L. E. (2004). *Applied Physics Letters, 85*(16), 3534.

Dong, S., Zhai, J., Li, J., & Viehland, D. (2006). Near-ideal magnetoelectricity in high-permeability magnetostrictive/piezofiber laminates with a (2-1) connectivity. *Applied Physics Letters, 89*(25).

Dong, S. X., Zhai, J. Y., Priya, S., Li, J. F., & Viehland, D. (2009). *IEEE Transactions on Ultrasonics Ferroelectrics and Frequency Control, 56*(6), 1124.

Duong, G. V., Groessinger, R., & Turtelli, R. S. (2006). Magnetoelectric properties of CoFe $ _2 $ O $ _4 $-BaTiO $ _3 $ core-shell structure composite. *IEEE Transactions on Magnetics, 42*(10), 3611–3613.

Dzyaloshinskii, I.,E. (1960). On the magneto-electrical effects in antiferromagnets. *Soviet Physics - JETP, 10*, 628–629. https://cir.nii.ac.jp/crid/1571980074033746432.

Ederer, C., & Spaldin, N. A. (2006). Origin of ferroelectricity in the multiferroic barium fluorides Ba M F 4: A first principles study. *Physical Review B, 74*(2), Article 024102.

Eerenstein, W., Mathur, N., & Scott, J. F. (2006). Multiferroic and magnetoelectric materials. *Nature, 442*(7104), 759–765.

Fermon, C., & Pannetier-Lecoeur, M. (2013). Noise in GMR and TMR sensors. In C. Reig, S. Cardoso, & S. C. Mukhopadhyay (Eds.), *Giant magnetoresistance (GMR) sensors: From basis to state-of-the-art applications* (pp. 47–70). Springer Berlin Heidelberg. https://doi.org/10.1007/978-3-642-37172-1_3

Fox, D. L., & Scott, J. (1977). Ferroelectrically induced ferromagnetism. *Journal of Physics C: Solid State Physics, 10*(11), Article L329.

Gojdka, B., Jahns, R., Meurisch, K., Greve, H., Adelung, R., Quandt, E., Knöchel, R., & Faupel, F. (2011). Fully integrable magnetic field sensor based on delta-E effect. *Applied Physics Letters, 99*(22). https://doi.org/10.1063/1.3664135

Guo, Y., Wang, J., White, R. M., & Wang, S. X. (2015). Reduction of magnetic 1/f noise in miniature anisotropic magnetoresistive sensors. *Applied Physics Letters, 106*(21). https://doi.org/10.1063/1.4921764

Hill, N. A. (2000). *Why are there so few magnetic ferroelectrics?* (Vol 104, pp. 6694–6709). ACS Publications. In.

Huang, Z., Cao, Y., Sun, Y., Xue, Y., & Chu, C. (1997). Coupling between the ferroelectric and antiferromagnetic orders in YMnO 3. *Physical Review B, 56*(5), 2623.

Hur, N., Park, S., Sharma, P. A., Guha, S., & Cheong, S.-W. (2004). Colossal magnetodielectric effects in D y M n 2 O 5. *Physical Review Letters, 93*(10), Article 107207.

Ikonen, P. M., Rozanov, K. N., Osipov, A. V., Alitalo, P., & Tretyakov, S. A. (2006). Magneto-dielectric substrates in antenna miniaturization: Potential and limitations. *IEEE Transactions on Antennas and Propagation, 54*(11), 3391–3399.

Islam, R. A., Rong, C.-b., Liu, J., & Priya, S. (2008). Effect of gradient composite structure in cofired bilayer composites of Pb (Zr 0.56 Ti 0.44) O 3–Ni 0.6 Zn 0.2 Cu 0.2 Fe 2 O 4 system on magnetoelectric coefficient. *Journal of Materials Science, 43*, 6337–6343.

Jahns, R., Zabel, S., Marauska, S., Gojdka, B., Wagner, B., Knöchel, R., Adelung, R., & Faupel, F. (2014). Microelectromechanical magnetic field sensor based on ΔE effect. *Applied Physics Letters, 105*(5). https://doi.org/10.1063/1.4891540

Jia, Y., Luo, H., Zhao, X., & Wang, F. (2008). Giant magnetoelectric response from a piezoelectric/magnetostrictive laminated composite combined with a piezoelectric transformer. *Advanced Materials, 20*(24), 4776–4779.

Jiang, Q., Shen, Z., Zhou, J., Shi, Z., & Nan, C.-W. (2007). Magnetoelectric composites of nickel ferrite and lead zirconnate titanate prepared by spark plasma sintering. *Journal of the European Ceramic Society, 27*(1), 279–284.

Kajimoto, R., Yoshizawa, H., Shintani, H., Kimura, T., & Tokura, Y. (2004). Magnetic structure of TbMnO 3 by neutron diffraction. *Physical Review B, 70*(1), Article 012401.

Kimura, T., Goto, T., Shintani, H., Ishizaka, K., Arima, T.-h., & Tokura, Y. (2003). Magnetic control of ferroelectric polarization. *Nature, 426*(6962), 55–58.

Lage, E., Kirchhof, C., Hrkac, V., Kienle, L., Jahns, R., Knochel, R., et al. (2012). *Nature Materials, 11*(6), 523.

Landau, L. D., Bell, J. S., Kearsley, M. J., Pitaevskii, L. P., Lifshitz, E. M., & Sykes, J. B. (2013). *Electrodynamics of continuous media*. Elsevier Science. https://books.google.com/books?id=jedbAwAAQBAJ.

Li, M., Matyushov, A., Dong, C., Chen, H., Lin, H., Nan, T., Qian, Z., Rinaldi, M., Lin, Y., & Sun, N. X. (2017). Ultra-sensitive NEMS magnetoelectric sensor for picotesla DC magnetic field detection. *Applied Physics Letters, 110*(14). https://doi.org/10.1063/1.4979694

Li, P., Wen, Y., & Bian, L. (2007). Enhanced magnetoelectric effects in composite of piezoelectric ceramics, rare-earth iron alloys, and ultrasonic horn. *Applied Physics Letters, 90*(2).

Liang, X., Matyushov, A., Hayes, P., Schell, V., Dong, C., Chen, H., He, Y., Will-Cole, A., Quandt, E., Martins, P., McCord, J., Medarde, M., Lanceros-Mendez, S., van Dijken, S., Sun, N. X., & Sort, J. (2021). Roadmap on magnetoelectric materials and devices. *IEEE Transactions on Magnetics, 57*(8), 1–57. https://doi.org/10.1109/TMAG.2021.3086635

Liu, M., Zhou, Z., Nan, T., Howe, B. M., Brown, G. J., & Sun, N. X. (2013). Voltage tuning of ferromagnetic resonance with bistable magnetization switching in energy-efficient magnetoelectric composites. *Advanced Materials, 25*(10), 1435–1439.

Lopatin, S., Lopatina, I., & Lisnevskaya, I. (1994). Magnetoelectric PZT/ferrite composite material. *Ferroelectrics, 162*(1), 63–68.

Lou, J., Liu, M., Reed, D., Ren, Y., & Sun, N. X. (2009). Giant electric field tuning of magnetism in novel multiferroic FeGaB/lead zinc niobate–lead titanate (PZN-PT) heterostructures. *Advanced Materials, 21*(46), 4711–4715.

Ma, J., Shi, Z., & Nan, C.-W. (2007). Magnetoelectric properties of composites of single Pb(Zr,Ti)O3 rods and terfenol-D/epoxy with a single-period of 1-3-type structure. *Advanced Materials, 19*(18), 2571–2573.

Martinez, R., Palai, R., Huhtinen, H., Liu, J., Scott, J., & Katiyar, R. (2010). Nanoscale ordering and multiferroic behavior in Pb (Fe 1/2 Ta 1/2) O 3. *Physical Review B, 82*(13), Article 134104.

Mori, K., & Wuttig, M. (2002). Magnetoelectric coupling in Terfenol-D/polyvinylidenedifluoride composites. *Applied Physics Letters, 81*(1), 100–101.

Nan, C.-W. (1994). Magnetoelectric effect in composites of piezoelectric and piezomagnetic phases. *Physical Review B, 50*(9), 6082.

Nan, C., Bichurin, M., Dong, S., Viehland, D., & Srinivasan, G. (2008). Magnetoelectric multiferroic composites. *Journal of Applied Physics, 103*, Article 031101.

Nan, T., Hui, Y., Rinaldi, M., & Sun, N. X. (2013). Self-biased 215MHz magnetoelectric NEMS resonator for ultra-sensitive DC magnetic field detection. *Scientific Reports, 3*(1), 1985. https://doi.org/10.1038/srep01985

Onuta, T. D., Wang, Y., Long, C. J., & Takeuchi, I. (2011). *Applied Physics Letters, 99*(20), 203506.

Palkar, V., & Malik, S. (2005). Observation of magnetoelectric behavior at room temperature in Pb (FexTi1−x) O3. *Solid State Communications, 134*(11), 783−786.

Praveen, J. P., Monaji, V. R., Chandrakala, E., Indla, S., Subramanian, V., & Das, D. (2018). Enhanced magnetoelectric coupling in Ti and Ce substituted lead free CFO-BCZT laminate composites. *Journal of Alloys and Compounds, 750*, 392−400.

Prellier, W., Singh, M., & Murugavel, P. (2005). The single-phase multiferroic oxides: From bulk to thin film. *Journal of Physics: Condensed Matter, 17*(30), Article R803.

Rajaram Patil, D., Chai, Y., Kambale, R. C., Jeon, B.-G., Yoo, K., Ryu, J., Yoon, W.-H., Park, D.-S., Jeong, D.-Y., & Lee, S.-G. (2013). Enhancement of resonant and non-resonant magnetoelectric coupling in multiferroic laminates with anisotropic piezoelectric properties. *Applied Physics Letters, 102*(6).

Ramsden, E. (2006). *Hall-effect sensor: Theory and applications*. Elsevier.

Röntgen, W. C. (1888). Ueber die durch Bewegung eines im homogenen electrischen Felde befindlichen Dielectricums hervorgerufene electrodynamische Kraft. *Annalen der Physik, 271*(10), 264−270.

Ryu, J., Carazo, A. V., Uchino, K., & Kim, H.-E. (2001). Magnetoelectric properties in piezoelectric and magnetostrictive laminate composites. *Japanese Journal of Applied Physics, 40*(8R), 4948.

Ryu, J., Priya, S., Uchino, K., & Kim, H.-E. (2002). Magnetoelectric effect in composites of magnetostrictive and piezoelectric materials. *Journal of Electroceramics, 8*, 107−119.

Sanchez, D. A., Ortega, N., Kumar, A., Sreenivasulu, G., Katiyar, R. S., Scott, J., Evans, D. M., Arredondo-Arechavala, M., Schilling, A., & Gregg, J. (2013). Room-temperature single phase multiferroic magnetoelectrics: Pb (Fe, M) x (Zr, Ti)(1−x) O3 [M= Ta, Nb]. *Journal of Applied Physics, 113*(7).

Schmid, H. (1994). Multi-ferroic magnetoelectrics. *Ferroelectrics, 162*(1), 317−338.

Sreenivasulu, G., Laletin, U., Petrov, V. M., Petrov, V. V., & Srinivasan, G. (2012). A permendur-piezoelectric multiferroic composite for low-noise ultrasensitive magnetic field sensors. *Applied Physics Letters, 100*(17). https://doi.org/10.1063/1.4705305

Srinivasan, G., Rasmussen, E., Gallegos, J., Srinivasan, R., Bokhan, Y. I., & Laletin, V. M. (2001). Magnetoelectric bilayer and multilayer structures of magnetostrictive and piezoelectric oxides. *Physical Review B, 64*(21), Article 214408.

Stutzke, N. A., Russek, S. E., Pappas, D. P., & Tondra, M. (2005). Low-frequency noise measurements on commercial magnetoresistive magnetic field sensors. *Journal of Applied Physics, 97*(10). https://doi.org/10.1063/1.1861375

Uga, M., Iwata, N., & Kohn, K. (1998). Magnetoelectric effect of TmMn2O5. *Ferroelectrics, 219*(1), 55−62.

Van den Boomgaard, J., & Born, R. (1978). A sintered magnetoelectric composite material BaTiO 3-Ni (Co, Mn) Fe 2 O 4. *Journal of Materials Science, 13*, 1538−1548.

Van Run, A., Terrell, D., & Scholing, J. (1974). An in situ grown eutectic magnetoelectric composite material: Part 2 physical properties. *Journal of Materials Science, 9*, 1710−1714.

Van Suchtelen, J. (1972). Product properties: A new application of composite materials. *Philips Research Reports, 27*, 28−37.

Vig, J. R., & Walls, F. L. (2000, 9-9 June 2000). A review of sensor sensitivity and stability. In *Proceedings of the 2000 IEEE/EIA international frequency control symposium and exhibition (cat. No.00CH37052)*.

Wang, Y., Gray, D., Berry, D., Gao, J., Li, M., Li, J., & Viehland, D. (2011). An extremely low equivalent magnetic noise magnetoelectric sensor. *Advanced Materials, 23*(35), 4111.

Wang, Y., Li, J., & Viehland, D. (2014). Magnetoelectrics for magnetic sensor applications: Status, challenges and perspectives. *Materials Today, 17*(6), 269–275. https://doi.org/10.1016/j.mattod.2014.05.004

Wang, J., Neaton, J., Zheng, H., Nagarajan, V., Ogale, S., Liu, B., Viehland, D., Vaithyanathan, V., Schlom, D., & Waghmare, U. (2003). Epitaxial BiFeO3 multiferroic thin film heterostructures. *Science, 299*(5613), 1719–1722.

Wilson, H. A. (1904). On the electric effect of rotating a dielectric in a magnetic field. *Proceedings of the Royal Society of London, 73*(488), 490–492. https://doi.org/10.1098/rspl.1904.0066

Wolf, T., Neumann, P., Nakamura, K., Sumiya, H., Ohshima, T., Isoya, J., & Wrachtrup, J. (2015). Subpicotesla diamond magnetometry. *Physical Review X, 5*(4), Article 041001. https://doi.org/10.1103/PhysRevX.5.041001

Yan, Y. K., Zhou, Y., & Priya, S. (2013). *Applied Physics Letters, 102*(5).

Yang, S. C., Ahn, C. W., Cho, K. H., & Priya, S. (2011a). *Journal of the American Ceramic Society, 94*(11), 3889.

Yang, S. C., Cho, K. H., Park, C. S., & Priya, S. (2011b). *Applied Physics Letters, 99*(20).

Yang, S. C., Park, C. S., Cho, K. H., & Priya, S. (2010). *Journal of Applied Physics, 108*(9).

Yarar, E., Salzer, S., Hrkac, V., Piorra, A., Höft, M., Knöchel, R., Kienle, L., & Quandt, E. (2016). Inverse bilayer magnetoelectric thin film sensor. *Applied Physics Letters, 109*(2). https://doi.org/10.1063/1.4958728

Zabel, S., Reermann, J., Fichtner, S., Kirchhof, C., Quandt, E., Wagner, B., Schmidt, G., & Faupel, F. (2016). Multimode delta-E effect magnetic field sensors with adapted electrodes. *Applied Physics Letters, 108*(22).

Zhai, J., Dong, S., Xing, Z., Li, J., & Viehland, D. (2006). Giant magnetoelectric effect in Metglas/polyvinylidene-fluoride laminates. *Applied Physics Letters, 89*(8), Article 083507. https://doi.org/10.1063/1.2337996

Zhai, J. Y., Xing, Z. P., Dong, S. X., Li, J. F., & Viehland, D. (2008a). *Journal of the American Ceramic Society, 91*(2), 351.

Zhai, J. Y., Xing, Z. P., Dong, S. X., Li, J. F., & Viehland, D. (2008b). *Applied Physics Letters, 93*(7).

Zhang, S. Y., Leung, C. M., Kuang, W., Or, S. W., & Ho, S. L. (2013). *Journal of Applied Physics, 113*(17).

Zhou, J. P., Wang, P., Yang, J., Liu, P., & Zhang, H. W. (2012). *Journal of Applied Physics, 111*(3).

Zhou, Y., Yan, Y., & Priya, S. (2014). *Applied Physics Letters, 104*(23).

Zhou, Y., Yang, S. C., Apo, D. J., Maurya, D., & Priya, S. (2012). *Applied Physics Letters, 101*(23).

CHAPTER

9

Magnetoelectric sensors based on magnetoelectric MEMS

Dirk Meyners, Jeffrey McCord, and Eckhard Quandt
Kiel University, Kiel, Germany

9.1 The MEMS approach to magnetoelectric cantilever sensors

Micro-electro-mechanical systems (MEMS) is the discipline that builds on the processes of semiconductor technology to enable the miniaturization of components that combine electrical functions with other functions, such as sensing environmental variables. MEMS-based fabrication benefits from the fact that large quantities can be produced and material and energy consumption is minimized. The MEMS approach is particularly promising for sensor fabrication because miniaturization enables densely packed arrays of sensor elements, smaller sensors minimize the influence of the measurement on the quantity to be measured, and surface effects are becoming more important than volume effects, which can be used to introduce new sensing mechanisms and to enhance important sensor properties such as sensitivity. Therefore, it makes sense to aim to profitably exploit the advantages of MEMS-based production for magnetoelectric (ME) magnetic field sensors. Silicon micromechanics in particular, as a subset of MEMS technology, has become important in realizing the resonator structures that are essential for many ME concepts (Marauska et al., 2013; Nan et al., 2013; Su, Niekiel, Fichtner, Thormaehlen, et al., 2020; Zhao et al., 2009). Following this approach, the relevance of layer stress engineering was quickly recognized (Onuta et al., 2011). This is especially necessary to avoid stress-induced effects on the magnetic phase due to inverse magnetostriction. With this in mind and taking into account the findings on the losses in micromechanical cantilevers (Yasumura et al., 2000), silicon-based ME structures with a high

https://doi.org/10.1016/B978-0-443-18606-6.00011-3

mechanical quality factor Q can be fabricated. For resonantly operated ME sensors, the Q factor has a linear effect on the effective magnetoelectric coefficient and the sensitivity for direct detection of resonant magnetic fields (Table 9.1). It is therefore attractive to apply certain measures to further increase Q. For example, vacuum encapsulation, a common step in MEMS, reduces energy exchange losses with the surrounding atmosphere, thereby increasing the sensitivity of encapsulated resonant ME sensors. Another approach is to apply dedicated excitation schemes. Wang et al. report an increase in Q from 3000 to 35000 and an ME coefficient of 2MV/cmOe when the oscillation of the ME microcantilever is parametrically amplified at twice the resonant frequency (Wang et al., 2015).

The limit of detection (LOD) of resonant ME magnetic field sensor improves, at best, in the absence of excitation-dependent disturbing mechanisms such as magnetic losses, with the square root of the quality factor (Durdaut et al., 2017). However, due to the bandwidth decreasing with $\Delta f = f_{res}/Q$ (f_{res} : resonance frequency), the applicability of high Q ME resonators remains limited to applications where a low bandwidth can be used advantageously, for example, for sensor noise suppression. A medically relevant example is magnetic motion sensing, where artificial sources of fixed frequency magnetic fields are localized and tracked for patient motion analysis in unshielded environments (Hoffmann et al., 2021).

Moreover, the MEMS approach allows the integration of piezoelectric actuators to tune the resonant frequency. Fig. 9.1 illustrates two examples of such tunable ME magnetic field sensors (Su, Niekiel, Fichtner, Kirchhof, et al., 2020). In addition to the inner ME beam, the MEMS structures have two outer beams coated with AlN as the piezoelectric actuator material. The individual beams are mechanically coupled via thin poly-Si bars at their tips. Applying a DC voltage induces a mechanical stress in the actuators, which changes the effective stiffness of the resonator system. The desired effect on the resonant frequency is seen in Fig. 9.1b showing frequency sweeps of ME voltage recordings for various actuator voltages. Depending on the sign of the electrical voltage applied across the AlN, the mechanical resonant frequency is reduced or increased, and the maximum of the ME response is shifted accordingly. In the selected voltage range of ± 100 V, the resonance shift is nearly linear and achieves a sensitivity of 0.41% per 100 V for the type II design (Fig. 9.1c). The bandwidth of the sensors is not increased, but the MEMS concept demonstrates the capability to precisely adjust the resonant frequency and compensate for unavoidable manufacturing tolerances.

TABLE 9.1 Properties of ME MEMS cantilevers.

References	Marauska et al. (2013)	Zabel et al. (2015)	Kirchhof et al. (2013)	Kirchhof et al. (2013)	Wang et al. (2015)
f_{res}	7.1 kHz	7.6 kHz	152 Hz	4.7 kHz	3.6 kHz
Q	160	905	1700 at $P < 10^{-3}$ mbar	5800 at $P < 10^{-3}$ mbar	35000
Sensitivity	3.8 V/mT	76 Hz/mT	24 V/mT	40 V/mT	1000 V/mT
Mode of operation	Direct detection	ΔE effect	Direct detection	Direct detection	Parametric amplification

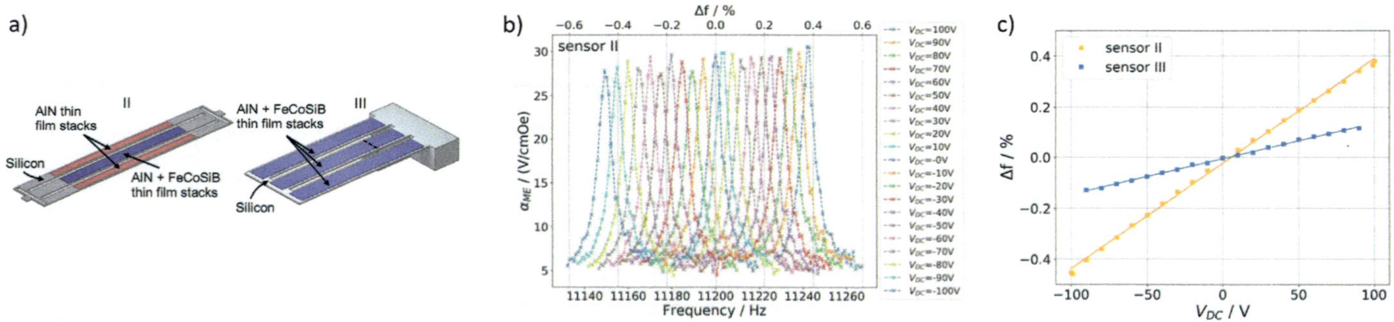

FIGURE 9.1 (a) Schematic illustration of two types of ME sensors with tunable resonance frequency. Type II: Double-side clamped coupled resonator. The inner beam which serves as the magnetic sensor element, consists of a 12 μm thick poly-Si substrate and is coated with 1 μm AlN and 2 μm FeCoSiB. It has 2234 μm × 105 μm lateral dimension. Type III: Three parallel clamped-free cantilevers with functional layers as in the middle beam of type II. The desired cantilever sizes were 1500 μm × 200 μm (more details in (Su, Niekiel, Fichtner, Kirchhof, et al., 2020)). (b) Measured ME response of sensor type II for different tuning voltages V_{DC} applied to the outer beams. (c) Change of resonance frequency versus V_{DC}. *From Su, Niekiel, Fichtner, Kirchhof, et al. (2020), with part (a) of the figure cropped to the images of the sensors of type II and III.*

The bandwidth limitation can be partly overcome by using several oscillation modes (Huang et al., 2015) or by the simultaneous use of several ME cantilevers of different dimensions, as has been proposed for energy harvesters (Xue et al., 2008; Zhang & Chen, 2010), the latter at the expense of miniaturization. An elegant and promising approach to extend the sensitivity range of an ME sensor to low frequency signals is the use of concepts that take advantage of the ΔE effect. The ΔE effect describes the change in Young's modulus due to a magnetostrictive contribution that depends on the magnetization and thus on the applied magnetic field (Lee, 1955). ME composite resonators therefore exhibit field-dependent resonant frequencies that can be monitored by frequency, phase, or amplitude modulation techniques.

The feasibility of ΔE-effect based DC magnetic field sensor was successfully demonstrated in 2005 for bulk resonators composed of a quartz crystal laminated with amorphous FeCoB (Yoshizawa et al., 2005). A first demonstration with an MEMS-based system followed a few years later (Gojdka et al., 2011; Zabel et al., 2015). Then, Zabel et al. fabricated 3 mm $\times$ 1 mm $\times$ 0.05 mm polysilicon cantilevers by bulk micromachining and integrated an AlN layer and a FeCoSiB/Cr multilayer to form the MEMS ME sensors (Zabel et al., 2015). The ME cantilevers were piezoelectrically excited in the first bending mode via the integrated AlN layer and read out using a simultaneous current measurement. By sweeping the frequency of the piezoelectric drive, the resonance curve was recorded to identify the excitation frequency at the maximum slope. Using this optimum operating point with the highest sensitivity and amplitude modulation, a detection limit of 140 $pT/Hz^{1/2}$ for 20 Hz magnetic signals was achieved. Exploiting the second bending mode at 46 kHz and an adapted electrode design, the detection limit and bandwidth were further improved to 100 $pT/Hz^{1/2}$ with a frequency span from 20 Hz to 100 Hz (Zabel et al., 2016).

In a subsequent development, the MEMS ME cantilevers were equipped with an exchange bias multilayer based on magnetostrictive FeCoSiB (Spetzler et al., 2021). A total layer thickness of 4 μm FeCoSiB was achieved by 20 iterations of the Ta 5/Cu 3/MnIr 8/FeCoSiB 200 [nm] stack unit. By annealing the stack in a magnetic field, the unidirectional anisotropy was imprinted tilted to the long bending axis to allow the sensor to operate in zero field and to define the directionality of the sensor response (see Chapter 2, Section 2.4 *Magnetostrictive multilayers for magnetoelectrics—Exchange bias for zero field operation*). Moreover, patterning several electrodes on the AlN enabled the acquisition of the resonances of the first bending mode at 7.73 kHz and the first torsional mode at 47.96 kHz. The goal of the sensor design was to simultaneously detect magnetic fields and determine the position and orientation of the ME cantilevers in space. Such simultaneous field detection and sensor localization is of major relevance, because the exact knowledge of the sensor position is required for successful reconstruction of field sources from measured field distributions (Dong et al., 2018). This knowledge is even more significant for the measurement of moving magnetic field sources for the purpose of the aforementioned medical application of patient motion analysis. In the experiment, for simultaneous test field measurement and sensor localization, the first bending mode was measured in direct detection, while a ΔE-effect readout method had been chosen for the torsion mode. In this way and without magnetic field bias, a $LOD = 350\ \mathrm{pT/Hz^{1/2}}$ was achieved for fields with frequencies of $f = 25$ Hz. In the first bending mode, the detection limit for resonant fields was about $30\ \mathrm{pT/Hz^{1/2}}$. As a proof of concept, the ME sensor was placed on a 56 cm × 56 cm test field and moved along three different paths (Fig. 9.2a). The measurements were conducted twice with different orientations of the sensor either along the x-axis or the y-axis. To localize the sensor, four coils were mounted at the outer corners of the field, sending orthogonal signals at the frequency of the first bending mode. Determining the sensor position on the basis of the detected coil signals is equivalent to solving an inverse problem, which was accomplished here by a minimum least square algorithm (Wang et al., 1992). At the same time, a further coil was moved along with the sensor to generate a low frequency test signal that contained frequency components similar to those of a human heart signal. The transmitted power was adjusted so that the field amplitude at the location of the sensor for the r-wave was approximately 250 nT.

The result of the sensor localization is summarized in Fig. 9.2b and c for the path drawn in blue (Fig. 9.2a). Here, μ_e denotes the mean localization error, and m_e is the median of the localization error. d_e specifies the percentage of correct sensor orientation estimates. When solving the inverse problem, best results are obtained if a tilt angle $\gamma = 20^\circ \pm 10^\circ$ is assumed for the sensitivity axis of the ME sensor (dashed line in Fig. 9.2b and c). Both the mean localization error and the median then amount to about 1 cm.

With a mean localization error $\mu_e = 5$ cm and a median error $m_e =$ 2.5 cm for all position-orientation pairs in Fig. 9.2a, the achieved quality of localization in the centimeter range is comparable to the quality of other magnetic localization approaches (de Angelis et al., 2015). However, in the work presented here, an artificial heart signal could also be reconstructed successfully and simultaneously to the localization (Fig. 9.2d).

According to Fig. 2.7 in Chapter 2, Section 2.4 *Magnetostrictive multilayers for magnetoelectrics—Exchange bias for zero field operation*, a tilt of the sensitivity axis is to be expected due to the directionality of the exchange-

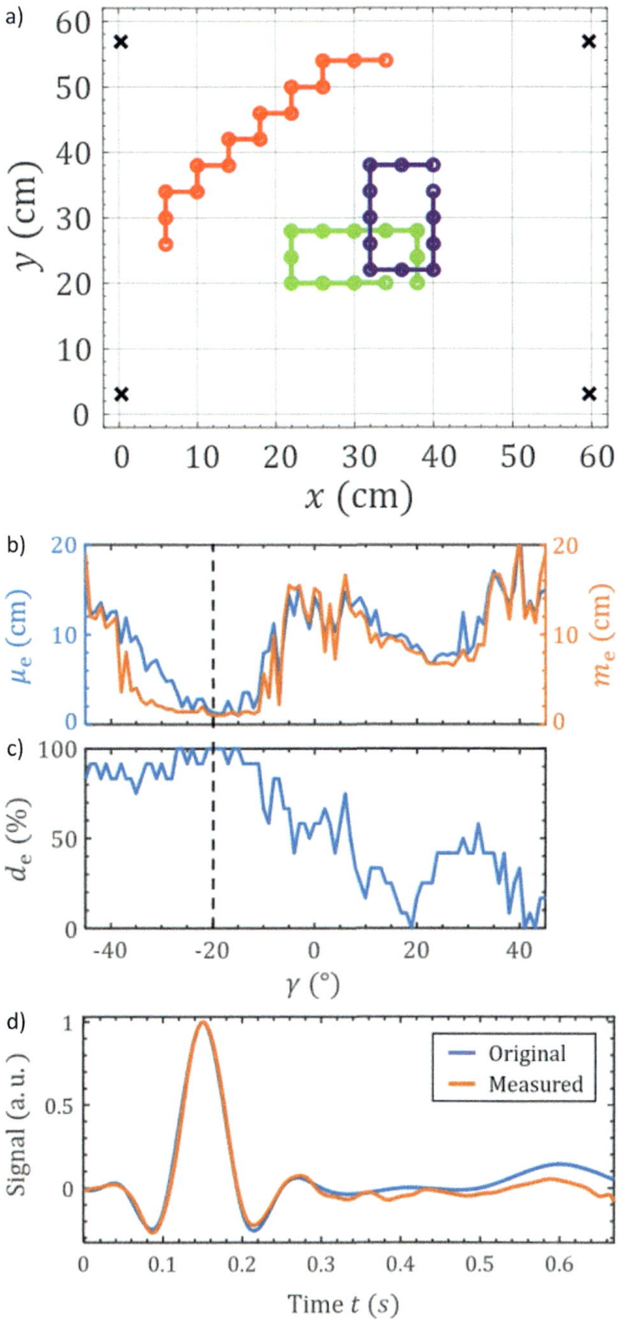

FIGURE 9.2 (a) In a 56 cm × 56 cm sized test field with a grid size of 1 cm, the ME sensor was moved along three different paths. The sensor positions are indicated by red, green, and blue circles. Black crosses at the edges mark the positions of the transmitting coils. (b) Mean localization error μ_e and median error m_e as a function of the assumed sensitivity axis tilt γ. (c) Percentage d_e of correctly estimated sensor orientations. (d) Reconstructed (demodulated, detrended, and lowpass-filtered) signal recorded simultaneously during sensor localization. *From Spetzler et al. (2021).*

biased ME sensor. Furthermore, the sensitivity axis should be perpendicular to the unidirectional anisotropy and form an angle $\gamma = 45°$ with the long cantilever axis. The deviation of γ with optimum localization is an indication of the limitation of the layer concept (Lage et al., 2014).

MEMS-based fabrication enables micromechanical resonators with high mechanical quality factors, which can be used profitably for resonant ME sensors in direct detection mode in particular. In addition, the MEMS approach allows the integration of piezoelectric actuators and complex magnetic multilayers, which can be designed to tune the desired sensor properties such as frequency, directionality, and sensitivity at zero magnetic field bias. To achieve high bandwidths, however, other ME sensor concepts are required, such as the converse ME sensors and surface acoustic wave sensors discussed in the following two Sections. An extremely attractive application of miniaturized ME structures, namely, the use as receivers and transmitters of electromagnetic waves, is discussed in Chapter 11 *Magnetoelectric Antennas for VLF* and Chapter 12 *Magnetoelectric Antennas for RF.*

9.2 Converse magnetoelectric sensors

An obvious way to overcome the bandwidth limitation of resonant ME cantilever sensors is to use high-frequency vibration modes in combination with a modulation scheme to increase the sensitivity to low-frequency signals. This is feasible even without changing the design of the sensor elements. Due to the weaknesses of high-frequency sensor excitation by applied magnetic fields in terms of energy consumption, possible crosstalk in arrays and miniaturization, electrical excitation via the piezoelectric in the ME composite is preferable. The electrical excitation and modulation of bulk laminated ME composite sensors was investigated in detail experimentally and theoretically in a collaboration between the groups of Christophe Dolabdjian, Peter Finkel, and Dwight Viehland in 2015 and 2016 (Zhuang, Sing, et al., 2015, 2016). During the same period, the experimental proof of concept was successfully demonstrated for thin-film composites (Hayes et al., 2016). The sensor mechanisms involved are similar to those of previously introduced ΔE-effect sensors (Gojdka et al., 2011; Yoshizawa et al., 2005). Hayes et al. used 2 μm unpolarized PZT to excite a 25 mm × 2.3 mm × 0.3 mm Si cantilever near its first bending resonance at $f_{res} = 689$ Hz. The oscillation was read out via a second piezoelectric AlN layer. A multilayer of 10 x Ta 5/Cu 3/MnIr 8/FeCoSiB 200 [nm] was used to measure magnetic fields without applying magnetic field bias. For this purpose, the magnetic anisotropy was induced by 45° to the long cantilever axis. In a first experiment, the ME transducer was excited exactly at resonance with a sinusoidal peak-to-peak voltage of 0.3 V_{PP}. In the presence of a low-frequency magnetic measurement field, the magnetostrictive modulation of the FeCoSiB multilayer leads to two sidebands that appear in the output spectrum of the sensor next to the dominant carrier signal (Fig. 9.3a). These sidebands have a frequency spacing from the driving carrier signal equal to the signal frequency f_{sig}. To detect signals whose frequency is greater than the bandwidth of the resonator, the carrier frequency is adjusted to satisfy the modulation condition $f_{mod} = f_{res} \pm f_{sig}$ so that a side band amplified by the mechanical resonance is generated in the output spectrum (see Fig. 9.3b).

The measurements demonstrate the electrically induced frequency conversion of a low-frequency magnetic signal into the resonance of a thin film ME cantilever type sensor. However, in order to achieve a breakthrough in terms of overcoming the bandwidth limitation, the use of high-frequency oscillation modes is essential. High frequencies also enable inductive readout via a coil surrounding the ME resonator, which results in a sensor concept presented by Mermelstein as early as 1986 (Mermelstein, 1986) and has been further investigated by other research groups in the following years (Fetisov et al., 2007; Hayes et al., 2018; Zhuang, Lam Chok Sing, et al., 2015). The stress-induced magnetization change caused by the piezoelectric excitation leads to a periodic flux change, which in turn is modulated in the presence of the magnetic field to be measured. The modulated flux change is then detected inductively. To distinguish them from conventional, resonant ME sensors operated in direct detection, these magnetic field sensors are also called inverse or converse ME sensors due to the inversion of the piezoelectric and magnetostrictive effects used.

A specific so-called U-mode oscillation was used by Hayes et al. (2019). It can be excited piezoelectrically with a simple plate capacitor design for the electrodes and occurs in 25 mm × 2.45 mm × 0.3 mm single-sided clamped cantilevers at approximately 515 kHz. The existence of the U-mode was demonstrated both by finite element analysis and experimentally using high-speed vibrometry (see Fig. 9.4a). A characteristic feature of this mode is that the bending occurs symmetrically along the short cantilever axis, with the displacement increasing continuously with the distance from the clamping. When excited through a 2 μm thick AlN layer with a voltage of amplitude $V_{ext} = 100$ mV, the maximum displacement at the free end of the beam is about 30 nm. Due to this small displacement amplitude, the attenuation due to air damping remains low and the quality factor is relatively high at approximately 1000. At the same time, because of the bending along the short axis, the radius of curvature is small and the mechanical stress is therefore large. During an oscillation

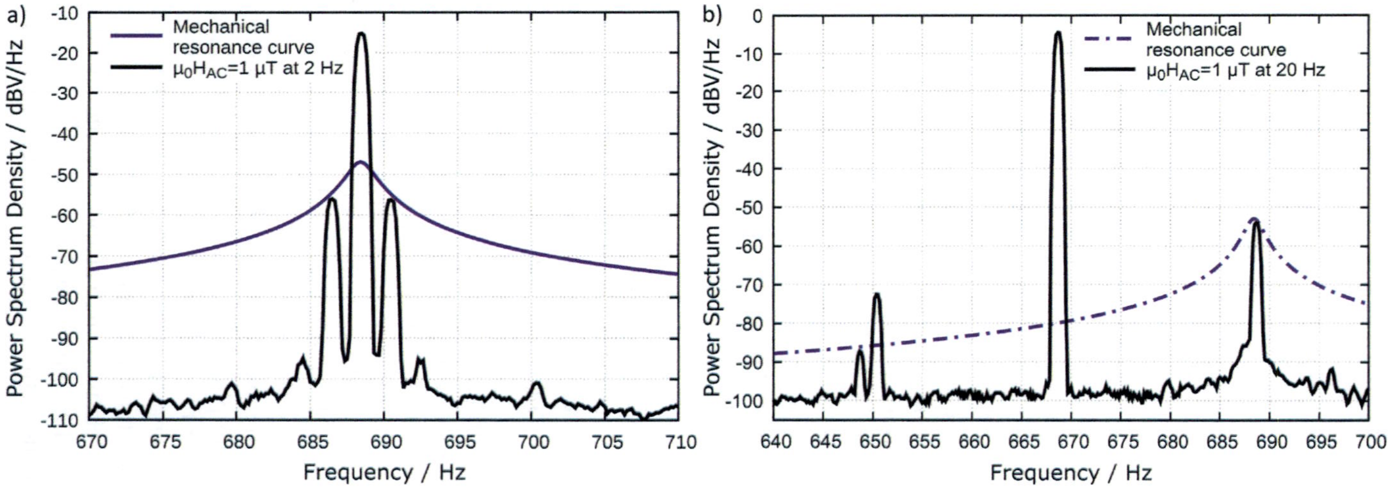

FIGURE 9.3 (a) Spectrum picked up at the AlN layer of a piezoelectrically excited ME thin-film composite in resonance. In addition to the carrier signal, two sidebands appear which originate from a magnetic 2 Hz test signal with an amplitude of 1 μT. The blue curve shows the resonance behavior qualitatively. (b) Spectrum for test signal with frequency $f_{\mathrm{sig}} = 20$ Hz The sensor is driven at a frequency $f_{\mathrm{mod}} = 669$ Hz so that the sideband of the frequency $f_{\mathrm{mod}} + f_{\mathrm{sig}} = f_{\mathrm{res}}$ is amplified. *From Hayes et al. (2016).*

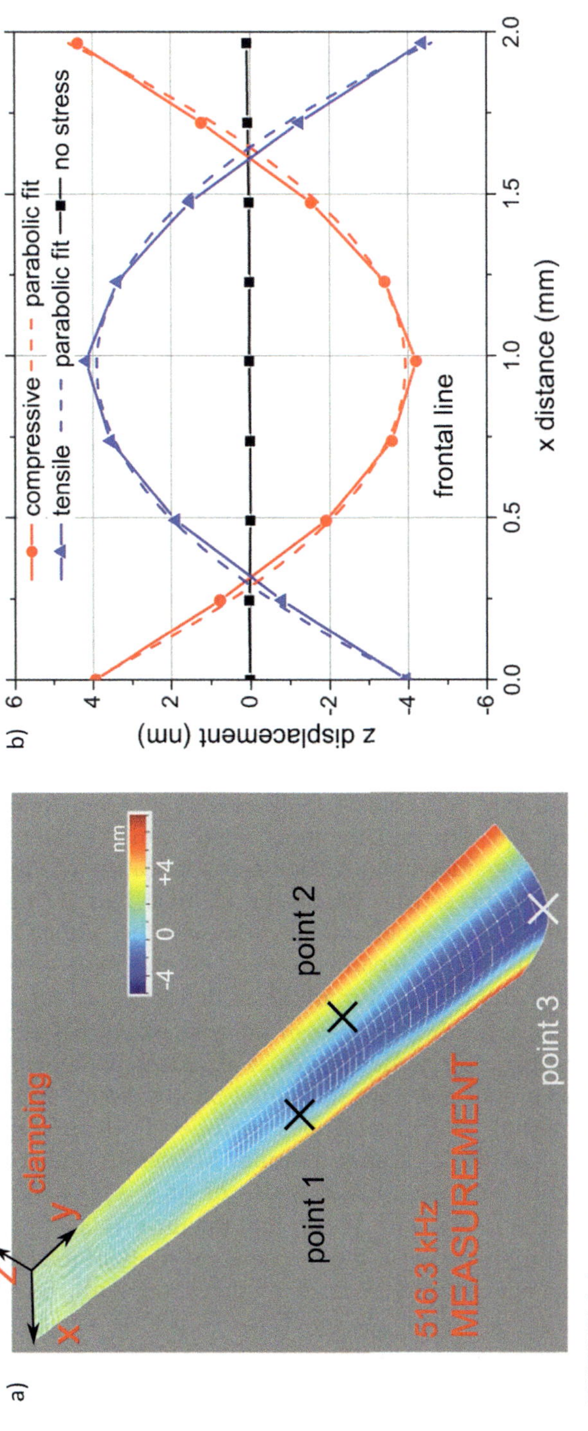

FIGURE 9.4 (a) Vibrometry of a piezoelectrically excited ME composite showing the so-called U-mode characterized by bending along the short cantilever axis. The color code indicates the measured displacement. (b) Displacement along the short axis at the tip of the cantilever, showing the extremes of one cycle of motion, leading to alternatingly compressive and tensile stress in the magnetostrictive film. *From Hayes et al. (2019).*

period, the mechanical stress changes its sign and by means of inverse magnetostriction, the effective anisotropy of a magnetic film placed on the oscillator is modulated in time accordingly (see Fig. 9.4b). In the experiment, Hayes et al. used a 2 μm thick single layer of FeCoSiB with magnetic anisotropy set in an annealing step along the short cantilever axis. Assuming a saturation magnetostriction of 30 ppm of the FeCoSiB layer, they estimated the stress-induced anisotropy at the peak of the vibration to be $K_\sigma = 165\,\mathrm{J/m^3}$, which corresponds to about a quarter of the uniaxial magnetic anisotropy of the film $K_u = 620\,\mathrm{J/m^3}$.

Obviously, this modulation of magnetic anisotropy is sufficient to induce a periodic variation of the magnetization. Fig. 9.5 shows the evolution of the coil voltage under piezoelectric excitation and as a function of a slowly varying magnetic field. Starting from large fields and magnetic saturation, the coil voltage passes through a maximum before the magnetic field changes sign. At a magnetic field of ± 0.05 Oe, the induced voltage becomes minimal and a phase change occurs due to a change in the sign of the excited net magnetization. Due to the hysteretic properties and inhomogeneous magnetization of the FeCoSiB layer, the coil voltage does not disappear at zero field (Fig. 9.5b). At this point, the sensor system reveals an enormous sensitivity to small DC magnetic fields with a slope of the characteristic of about 40 kV/T. Examining the sensor response to static fields in the nanotesla range, Hayes et al. determine the detection limit for DC fields to be $210\,\mathrm{pT/Hz^{1/2}}$.

The signal spectrum of the same converse sensor system in the presence of low-frequency AC fields (10 Hz or 23 Hz signal frequency) is illustrated in Fig. 9.6. The test magnetic fields can be determined by evaluating the arising sidebands. A characteristic feature of the modulation concept is that the sensitivity can be at most half as high as the sensitivity to constant magnetic fields because the signal energy is distributed over the two sidebands. Owing to the high resonant frequency, the bandwidth of the mechanical oscillator $\Delta f_{\mathrm{mech}} = f_{\mathrm{res}}/(2Q) = 260$ Hz is about 50 times higher than in the case of piezoelectric modulation in the first bending mode (Hayes et al., 2016). Moreover, the mechanical resonance does not limit the signal bandwidth of converse ME sensors (Spetzler et al., 2025), as can be concluded from an investigation by Elzenheimer et al. (2023). Using a similar ME sensor, they reported a significantly higher 3 dB signal bandwidth of $\Delta f_{\mathrm{sig}} = 6.2$ kHz. A closer look at the spectrum in Fig. 9.6a reveals that the uncorrelated component in the sensor output increases in the frequency vicinity of the carrier. The result is a frequency-dependent limit of detection with, for example, $\mathrm{LOD} = 70\,\mathrm{pT/Hz^{1/2}}$ at 10 Hz and $52\,\mathrm{pT/Hz^{1/2}}$ at 53 Hz (Fig. 9.6b). This may be the consequence of an up-conversion of low-frequency noise and also of carrier-induced magnetization processes. During sensor operation,

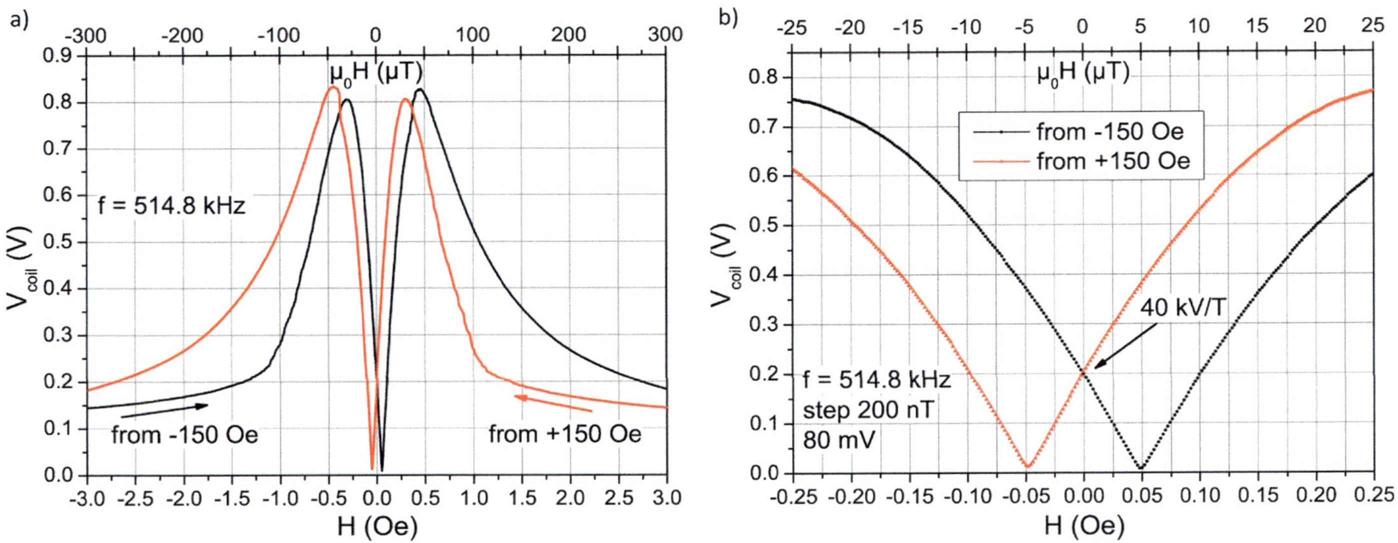

FIGURE 9.5 Characteristic of a converse ME sensor with AlN and FeCoSiB functional layers, each 2 μm thick. The 25 mm × 2.45 mm × 0.3 mm, single-sided clamped cantilever is driven in the U-mode (cf. Fig. 9.4). The excitation voltage on the AlN structure designed as a plate capacitor is $V_{ext} = 80$ mV. (a) Coil voltage after buffer amplifier (unity gain) as a function of a slowly varying external magnetic field. (b) Close-up of the low field region. *From Hayes et al. (2019).*

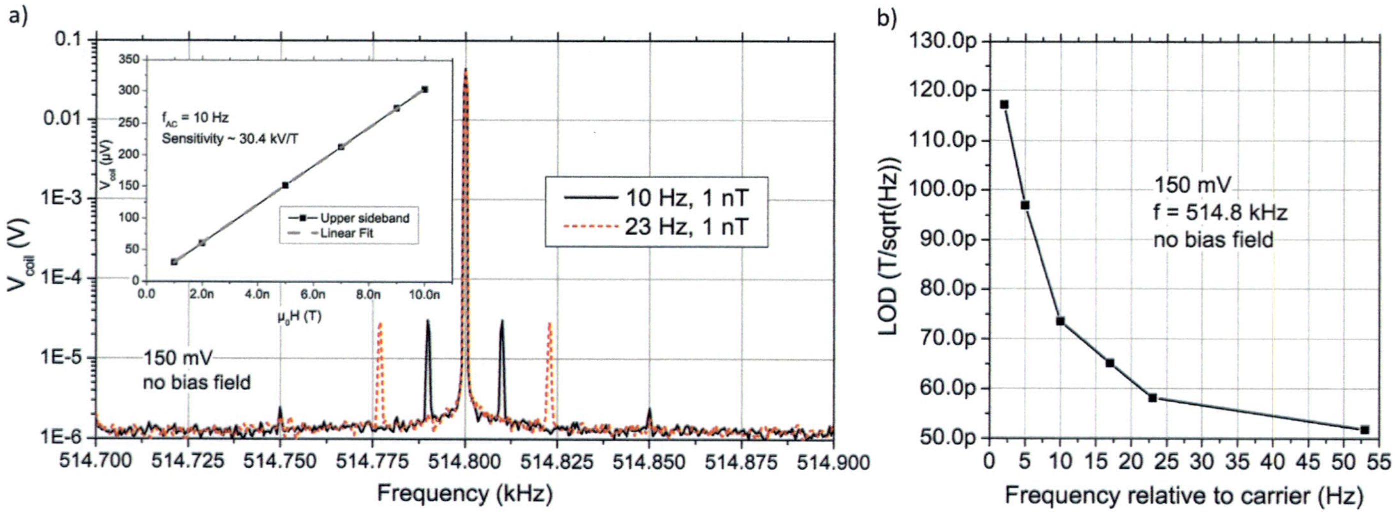

FIGURE 9.6 (a) Signal spectra of a thin film converse ME sensor for 10 Hz and 23 Hz test signals with an amplitude of 1 nT. In a 10 nT field range, the sensitivity is constant (inset). (b) Limits of detection achieved for optimum excitation and low-frequency signals. *From Hayes et al. (2019).*

the FeCoSiB single layer is in a multidomain state typical for soft magnetic materials. In fact, it is observed that the noise level near the carrier increases sharply when a threshold value (V_{ext} approximately equal to 200 mV at 2 μm AlN) is exceeded, while the sensitivity increases approximately linearly with the excitation amplitude. Thus, there is an optimal excitation characterized by the best signal-to-noise ratio (SNR) that maximizes sensitivity while avoiding excessive stress-induced remagnetization accompanied by domain wall movements. Consequently, further improvement of the detection requires magnetic domain engineering to regulate the excited magnetization processes.

In a further development, an attempt was made to improve the sensor properties by introducing a magnetic phase with the possibility to form flux closure structures without decaying into multi-domain states (Thormählen et al., 2024). Based on the layer solution for magnetic frequency conversion (MFC, see Section 2.4.1 *The antagonism of sensitivity versus noise—Magnetostrictive multilayers for magnetic domain control* (Chapter 2)), the underlying concept was a magnetostrictive multilayer whose magnetic FeCoSiB layers were separated by nonmagnetic intermediate layers. In addition, the magnetization was stabilized by exchange-bias coupling to thin MnIr layers. In order to achieve high magnetic responsivity, the thickness of the individual magnetic layers was chosen to be 500 nm, so that the overall system consisted of an 8-fold repetition of the Ta 10/Cu 3/MnIr 8/FeCoSiB 500 [nm] layer stack [thicknesses in nanometers]. To initialize the magnetic anisotropies, a simplified, 30-minute temperature treatment at $T = 270\,°C$ was carried out in an alternating magnetic field with an amplitude of 20 mT and a frequency of 5 Hz. Although this did not result in layer-wise single domain magnetization, it did result in a magnetic domain state with significantly reduced domain wall density and reduced domain wall mobility because the walls were stabilized by the exchange bias coupling. Furthermore, the stray field coupling between the individual layers prevented typical closure domains from forming. With the modified magnetic layer system, total sensitivities of the sensor system of $S_{DC} = 175$ kV/T for constant magnetic fields and $S_{AC} = 85$ kV/T for low-frequency alternating fields were achieved. However, when comparing to the results in reference (Hayes et al., 2019), it should be noted that the electronic amplifier gain was increased by a factor of 10 and the total layer thickness of the FeCoSiB was increased by a factor of 2. In addition, a higher driving voltage $V_{ext} = 700$ mV was chosen for optimal piezoelectric excitation. The responsivity of the multilayer system normalized to the FeCoSiB thickness is thus reduced to a value 30 times lower than the responsivity of the single layer. In the experiment, this is partially compensated by the stronger excitation and the layer stack scalable in thickness. The reduced responsivity of stray field-coupled layer systems has already been observed in the antiparallel layer systems developed for the MFC (Jovičević Klug et al., 2019). On the other hand, the amplifier gain-adjusted noise level of the converse sensor system is also reduced by a factor of 33, despite the 4.6-fold piezoelectric excitation. Overall, the use of the exchange-biased magnetic multilayer system with flux closure structures improved the detection limits by about one order of magnitude to $LOD_{DC} = 18\,pT/Hz^{1/2}$ for constant magnetic fields and $LOD = 8\,pT/Hz^{1/2}$ at 10 Hz. In addition, a highly linear response for AC fields was proven for amplitudes ranging from 50 pT up to 100 nT. Because of these promising properties, converse ME sensors were selected for a field study to test the feasibility of cardiological measurements (Elzenheimer et al., 2022). A further LOD improvement by about one order of magnitude is required to achieve QRS-complex recovery in heart-signal recordings of test persons without averaging successive measurements (Elzenheimer et al., 2023).

9.3 Magnetoelastic surface acoustic wave sensors

In addition to high-frequency ME resonators such as the converse ME sensors, magnetic field sensors based on surface acoustic waves (SAWs) represent a sensor class that is attractive due to its combination of low detection limits and wide signal bandwidth. SAWs can be excited in piezoelectric materials when finger-like electrodes (interdigital transducers, IDTs) are used to generate high-frequency alternating electric fields that lead to mechanical oscillation of the substrate via the inverse piezoelectric effect (White & Voltmer, 1965). If a second pair of electrodes is included, the propagating wave is detected using the direct piezoelectric effect. The two IDT electrode pairs form a delay line that reacts extremely sensitively to physical changes on its surface with variations in propagation time. If the delay line is functionalized with an additional layer, various types of sensors such as gas, bio, and chemical sensors can be implemented (Mandal & Banerjee, 2022). The idea of inserting magnetic layers into the delay line led to the development of magnetically tunable RF components (Ganguly et al., 1975; Webb et al., 1979; Yamamoto et al., 1979). The measurement of magnetic fields has also been successfully demonstrated experimentally (Elhosni et al., 2016; Hanna, 1987; Kadota et al., 2011; Kittmann et al., 2018; Liu et al., 2018; Polewczyk et al., 2017) and modeled (Mazzamurro et al., 2020; Zhou et al., 2014) by several research groups. The principle of magnetic field measurement with an SAW delay line is based on the magnetoelastic coupling between magnetization and stiffness of the ferromagnetic layer located between the IDTs. If the layer is magnetized by an external field, its mechanical waveguiding properties change, and therefore also the wave attenuation and phase velocity. Both the phase difference and the change in amplitude of the signals detected at the IDTs can be used as a measure of the external field. If the SAW sensor is excited with a carrier signal s_c of frequency f_c, in the presence of an external magnetic field $H_{ex}(t)$ the signal taken at the output IDTs becomes

$$s(t) \propto \cos\left(2\pi f_c t + SH_{ex}(t) + \phi_c(t) + \phi_{SAW}(t)\right), \tag{9.1}$$

with the phase sensitivity S and the phase noise ϕ_{SAW} of the sensor and the phase fluctuation of the carrier ϕ_c. Any scattering that affects the propagation of the wave contributes to the phase noise. Possible causes are imperfections in the piezoelectric substrate and the IDTs, reflections at the device ends, and impurities on the surface (Jungerman et al., 1985; Parker, 1979). Magnetoelastically induced losses (Berry & Pritchet, 1976) are also to be expected and are discussed in more detail below.

The implementation of Love waves, named after Edward Hough Love, in magnetic SAW sensors appears particularly promising. Love waves are guided, purely horizontal shear waves (Jakoby & Vellekoop, 1997). The time-varying shear stress induces an alternating magnetic anisotropy, which evokes a deflection of the magnetization in the direction of $\pm 45°$ to the wave propagation. Accordingly, the phase velocity for this wave type is determined by the shear modulus of the magnetic layer, which can lead to particularly large magnetic field induced effects (Sárközi et al., 2000). Furthermore, the strong confinement of the wave energy at the surface of the device enhances the sensor sensitivity.

The design of a magnetoelastic Love SAW sensor, which has been intensively investigated in the recent past (Durdaut, 2020) is shown schematically in Fig. 9.7. The sensor is based on an alpha-quartz crystal cut in such a way that its surface normal forms an angle of 42.75° with the crystallographic Z-axis. This frequently used ST-cut has the advantage of a vanishingly small temperature dependence of the delay time (Schulz et al., 1970). In addition, the IDTs are aligned in such a way that the wave propagation is orthogonal to the crystallographic X-axis, resulting in the desired pure shear wave excitation (Jakoby & Vellekoop, 1997). The IDTs consist of a lithographically structured Au layer, which has been provided with thin Cr layers to improve adhesion. The IDTs are embedded in a 4.5 μm thick amorphous SiO_2 layer grown by plasma enhanced chemical vapor deposition (PECVD). Due to the lower propagation velocity, the SiO_2 layer acts as a waveguide and concentrates the wave energy at the surface of the sensor. The length of the delay line L_D is defined by the center-to-center distance of the electrode pairs and in this example is 4.62 mm or 165 times the wavelength. The magnetoelastically active magnetic layer is

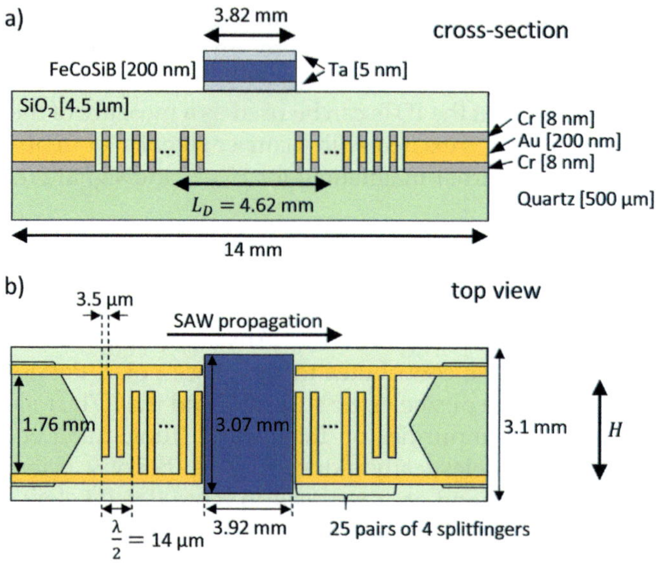

FIGURE 9.7 Layout of a magnetoelastic Love SAW sensor based on ST-cut quartz. The wave propagation is perpendicular to the crystallographic *X*-axis. In a typical configuration, the field *H* to be measured is applied in-plane with the device and perpendicular to the direction of wave propagation. Further details are given in the text. (a) Cross-section (b) top view.

prepared on top of the guiding layer. An overlap with the electrode structures is avoided in order to keep the capacitive coupling of the input and output IDTs small. Details of the electrode design are sketched in the schematic top view (Fig. 9.7b). The electrodes are of split-finger type to avoid mechanical Bragg reflections at the SAW resonance frequency (Hashimoto, 2000). The period of the finger electrodes determines the wavelength of the SAW and is 28 μm. To avoid the recoupling of reflected waves, the guiding layer is angled at its edges.

When discussing the sensor properties, it is advantageous to factorize the phase sensitivity S, that is, the change in phase φ per unit of the field H to be measured. With the shear modulus G and the phase velocity v, the following equation is obtained (Kittmann et al., 2018):

$$S = \frac{\partial \varphi}{\partial H} = \frac{\partial G}{\partial H} \cdot \frac{\partial v}{\partial G} \cdot \frac{\partial \varphi}{\partial v} \tag{9.2}$$

In the equation on the right, the first factor is a measure of the intensity of the magnetoelastic effect. It essentially depends on the choice of the magnetic material and its properties, in particular the orientation of the magnetization with respect to the wave propagation direction. The second factor describes the change of the effective phase velocity on the delay line in response to the magnetic field–induced change of the shear modulus. It reflects how effectively the change in the magnetoelastic material affects the waveguiding properties of the overall structure. This parameter, also referred to as structural sensitivity, therefore depends on the choice and thickness of the guiding layer material, the thickness of the magnetic film, and the elastic coupling between the two layers. The last term is determined by the layout of the SAW sensor and results as the geometric sensitivity S_{lay} from the frequency f and the path length l of the magnetoacoustic wave through the part of the delay line influenced by the magnetic material:

$$S_{\mathrm{lay}} = -\frac{2\pi l f}{v^2}$$

Considering the best possible detection limit, the sensor layout can be used to achieve an optimum compromise between high sensitivity S_{lay} and high damping of the wave for a given material combination. The sensitivity of the magnetic material $S_{\mathrm{mag}} = \partial G/\partial H$ (first factor in Eq. 9.2) can be expressed as the product of the reduced magnetic susceptibility χ and the shear modulus change per magnetization change:

$$S_{\mathrm{mag}} = \frac{\partial M}{\partial H}\frac{\partial G}{\partial M} = M_{\mathrm{S}}\,\chi\frac{\partial G}{\partial M}$$

Here, M_{S} denotes the material's saturation magnetization. Early in the development of magnetic SAWs (magneto-surface acoustic wave, MSAW), the good suitability of amorphous materials such as $Fe_{100\text{-}x}B_x$ with $15 < x < 20$ and $Fe_{78}Si_{10}B_{12}$ and magnetic multilayers such as TbFe/CoFe was pointed out (Inoue et al., 1990; Webb et al., 1979; Zhou et al., 2014). As with the classical ME composites discussed above (see Section 2.3 *Magnetostrictive thin film materials for magnetoelectrics* in Chapter 2), the objective is to achieve a high level of magnetic susceptibility. However, for the SAW-based devices, the challenge is to combine soft magnetic properties with a high magnetostrictively induced change of the shear modulus. In this context, both Webb et al. (1979) and Yokokawa et al. (1992) pointed out the importance of preparation-related layer stresses in the magnetoelastic material, which can easily occur especially when using ST-cut quartz substrates due to their anisotropic thermal expansion. Taking into account the trigonal quartz structure and the ST-cut, the thermal expansion coefficient α'_{11} perpendicular to the crystallographic X-axis, that is, in the direction of Love wave propagation, is (Schell et al., 2020):

$$\alpha'_{11} = \alpha_{11}\sin^2\theta + \alpha_{33}\cos^2\theta,$$

with α_{11} und α_{33} denoting thermal expansion coefficients along the X- and Z-axis and θ labeling the angle formed by the surface normal of the SAW sensor and the Z-axis. Under the assumption that the magnetic layer is deposited on a heated substrate, a stress-induced magnetic anisotropy will occur after the magnetic layer cools down to room temperature by ΔT:

$$K_{\sigma} = \frac{3}{2}\lambda_{S} E_{f} \Delta\alpha \Delta T, \tag{9.3}$$

with the difference of the thermal expansion coefficients $\Delta\alpha = \alpha_{11} - \alpha'_{11}$ and the Young's modulus E_f of the magnetic layer. For a positive magnetostrictive amorphous metal with typically isotropic and smaller thermal expansion than quartz, this anisotropy is aligned along the delay line.

The strong dependence of the performance of magnetic SAW sensors on the preparation conditions of sputtered FeCoSiB layers was investigated in reference Schell et al., 2020, which emphasizes the necessity of taking into account the magnetic anisotropy contributions introduced by the fabrication processes. The design of the quartz-based sensors corresponded to the scheme in Fig. 9.7. A total of three cases were considered: sputter deposition without an additional magnetic field, deposition under the same conditions but with subsequent thermal treatment in a magnetic field, and deposition under the influence of an additional magnetic field to align the magnetic anisotropy during the growth process. In all cases, the deposition was carried out at a low deposition rate with cooling pauses to keep the heating of the quartz substrates to temperatures below 40 °C. The preparation conditions and the most relevant sensor properties are summarized in Table 9.2.

For all three cases, there is a preference in the magnetic domain orientation. In the case of deposition without magnetic field, however, the domain orientation appears tilted toward the delay line (see magneto-optical Kerr effect (MOKE) image of the demagnetized state for case 1 in Table 9.2). Due to the deposition at only moderate self-heating, the proportion of the stress-induced anisotropy is small compared to the effective total anisotropy of the FeCoSiB layer. Apparently, a further treatment is required for a controlled magnetic anisotropy alignment. However, the temperature posttreatment in a magnetic field with the typical annealing temperature of 270 °C for FeCoSiB is not an option, as this leads to an enormously increased magnitude of uniaxial anisotropy, which, according to Eq. (9.3) is 90% stress induced (see Table 9.2). The magnetic susceptibility in the hard direction decreases accordingly, and the SAW

TABLE 9.2 Comparison of magnetoelastic Love SAW sensors with Ta 5/FeCoSiB 200/Ta 5 [nm] sensing layer prepared under various conditions. In all cases the magnetic layer deposition was carried out in 30 *s* lasting magnetron sputtering cycles (residual magnetic field at the substrate position not exceeding 25 *Oe*) with a low deposition rate of 0.16 nm/*s* and interrupted by 300 *s* long cooling breaks. In case 2, the coated SAW substrate was annealed at 270 °C for 30 min and subsequently cooled down to room temperature in a saturating magnetic field of $H_{anneal} = 1300\ Oe$. K_σ is the magnetic anisotropy calculated according to Eq. (9.3). For ΔT, either the substrate heating determined for the deposition or the annealing temperature was assumed. K_{eff} denotes the effective magnetic anisotropy derived from integral magnetization curves. For maximal phase sensitivity S_{max} and optimal LOD a magnetic field bias H_{bias} was optionally applied perpendicular to the delay line.

Case	(1) As-prepared	(2) Annealed	(3) Deposited with magnetic field bias
Magnetic field bias during layer growth	No	No	$H_{dep} = 180$ Oe along delay line
Postdeposition annealing	No	270°C for 30 min at $H_{anneal} = 1300$ Oe along delay line	No
Magnetization after demagnetization along to the delay line (left-to-right in the shown images), white and black arrows indicate prime magnetization directions	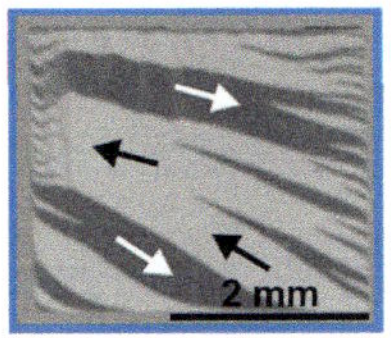	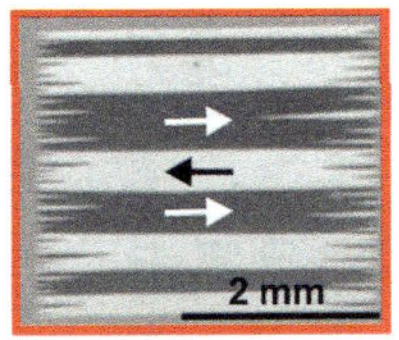	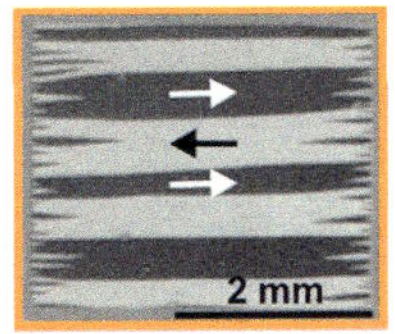
K_σ	$270\ J/m^3$	$4995\ J/m^3$	$378\ J/m^3$
K_{eff}	$1230\ J/m^3$	$5600\ J/m^3$	$840\ J/m^3$
S_{max}	340 °/mT at $H_{bias} = 0$	17 °/mT at $H_{bias} = 100$ Oe	2000 °/mT at $H_{bias} = 1$ Oe
LOD at 10 Hz	$275\ pT/Hz^{1/2}$	$7.500\ pT/Hz^{1/2}$	$70\ pT/Hz^{1/2}$

Figures and data from Schell et al. (2020).

sensor shows a dramatically reduced sensitivity and deteriorated detection limit. The best performance is achieved when the magnetic anisotropy is imprinted during film growth. For magnetic fields applied in the film plane and perpendicular to the delay line, the SAW sensor achieves a maximal phase sensitivity of $S_{\max} = 2000\ {}^{\circ}/\text{mT}$ and $LOD = 70\ \text{pT}\big/\text{Hz}^{1/2}$ for 10 Hz signals. Interestingly, all SAW sensors show the same $1/f$ phase noise characteristic in the low frequency range from about 1 to 1000 Hz, and the differences in noise level of 2 to 3 dB are also within the range of typical device-to-device variations (see Fig. 9.8a). One explanation may be found in the similarity of the magnetic domain configurations (see Table 9.2). The fact that the sensor phase noise in the low frequency range is dominated by magnetic or magnetoelastic interactions is evident from the reduction of the phase noise in the state of magnetic saturation. Due to the high bandwidth and the frequency-independent sensitivity up to the kHz range, the $1/f$ characteristic of the phase noise is directly transferred to the frequency-dependent LOD (Fig. 9.8b).

A systematic investigation of the phase noise of SAW sensors with the design shown in Fig. 9.7 can be found in the articles by Durdaut et al. (2019, 2021). To ensure that the phase noise is generated solely by the SAW delay line, the SAW sensors are operated with heterodyne electronics. To generate the SAW excitation $c(t)$, the signal output of a low-frequency (50 kHz) and phase-stable oscillator is raised into the bandpass around 145 MHz of the delay line by mixing it with a high-frequency local oscillator. The signal modulated by the delay line is then mixed down

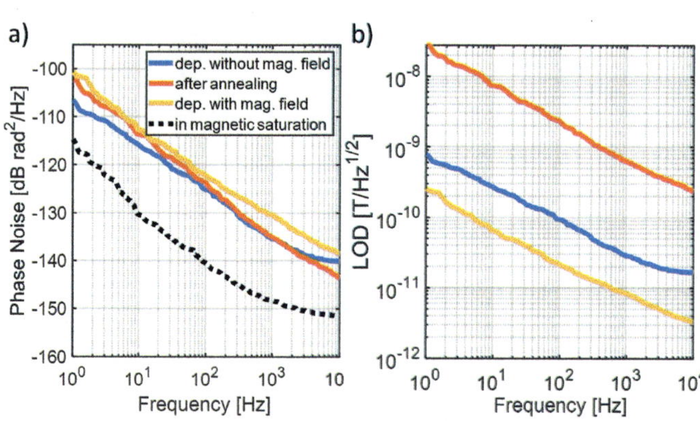

FIGURE 9.8 (a) Power spectral densities of random phase noise fluctuations and (b) calculated detection limits for SAW magnetic field sensors with FeCoSiB layers prepared under various process conditions (details see text). *From Schell et al. (2020) with adjustment of the ordinates and change of colors. All original data points are displayed.*

again with the same local oscillator in order to perform demodulation and obtain the measurement signal. The phase noise of the high-frequency oscillator can thus be avoided, and ϕ_c in Eq. (9.1) can be neglected. Durdaut et al. measured the scattering parameter S_{12} and the insertion losses (IL) as a function of an externally applied magnetic field bias (see Fig. 9.9a). They found that hysteretic effects and asymmetries in the remagnetization of the FeCoSiB layer caused by a slightly tilted magnetic anisotropy axis are reflected in IL. Field regions of large insertion losses coincide with the occurrence of a multidomain state and maximum losses occur only on one side of the field after the zero crossing of the field. Conversely, domain observations show that the domain switching fields depend on the SAW excitation energy and that there is an energy transfer channel from the electric realm to the magnetic realm mediated by the acoustic wave (see Fig. 9.9b).

Following the common approach, Durdaut et al. represent the generated phase noise as a polynomial in frequency:

$$\phi(t) = b_0 + b_{-1}f^{-1} + b_{-2}f^{-2} + \ldots,$$

where the coefficients b_0, b_{-1}, and b_{-2} describe the white noise, the flicker noise, and the random walk of phase. They observe a significant dependence on the external magnetic field for both the white noise and the flicker noise. The white noise is large in the magnetic field regions in which the insertion losses are also large and associated with the occurrence of the multidomain states. Field regions of highest flicker noise in turn show the same dependence on the SAW excitation as the insertion losses themselves. This parallelism can be explained by the fact that both quantities are related to the magnetic loss factor $\tan\delta = \mu_r''/\mu_r'$ via the complex magnetic permeability $\mu_r = \mu_r' + i\mu_r''$. The flicker noise can be up to 46 dB (factor 40000!) greater than in magnetic saturation and limits the sensor performance for low frequencies up to the 1 kHz range. Due to the common dependencies shown by the insertion losses and the flicker noise, it follows that the optimum magnetic field bias for sensor operation depends on both the magnetic field history and the SAW excitation power.

At very strong SAW excitations above 4 dBm, Durdaut et al. found an increase in f^{-2} noise that is correlated with hopping of domain walls. Excessive SAW excitation therefore degrades sensor performance and should be avoided. The authors rule out eddy current losses as negligible due to the comparatively low SAW frequency of 145 MHz and the thin metal coating of around 200 nm. At higher frequencies or metallizations in the 1 micrometer thickness range, however, these can become

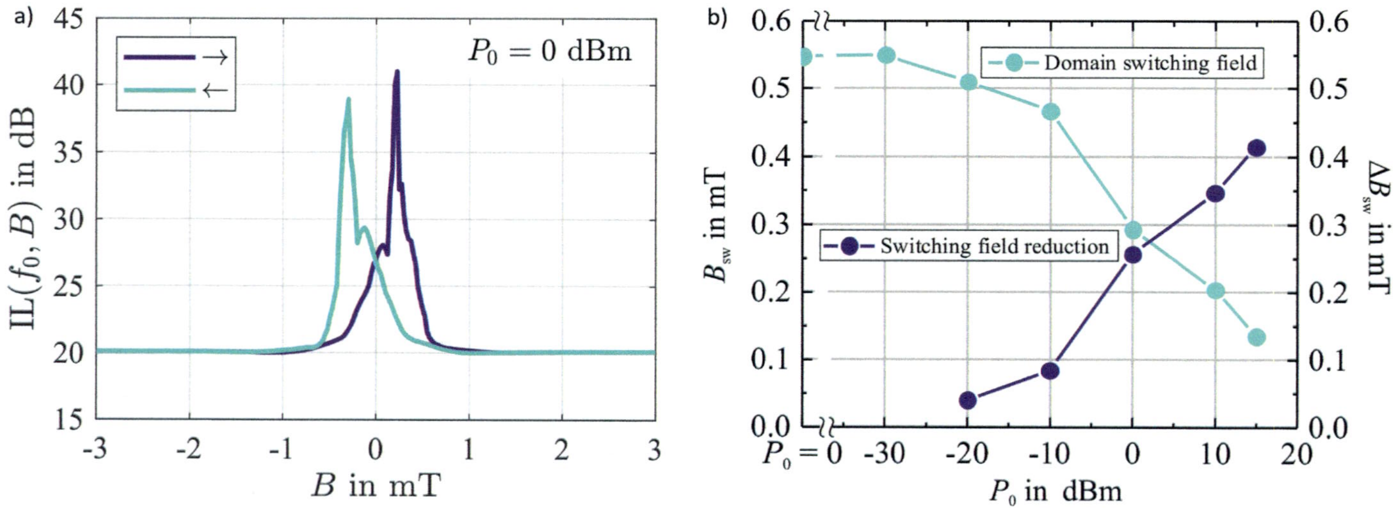

FIGURE 9.9 (a) Insertion loss as a function of an external static magnetic field for a typical SAW excitation of 1 mW. While the losses in magnetic saturation are 20 dB, they can rise to over 40 dB depending on the magnetic field. (b) Domain switching fields and change as a function of the SAW excitation power P_0. *From Durdaut et al. (2021).*

significant. Following the argumentation of Inoue et al., thin insulating intermediate layers in the magnetostrictive phase should then be able to effectively suppress the eddy currents, since the Love wave induces currents perpendicular to the layer plane due to its property as a pure horizontal shear wave (Inoue et al., 1990).

In order to be able to trace the strong interdependencies between magnetically induced insertion losses, phase noise, excitation power, and magnetic field bias back to specific magnetoelastic interactions, a stroboscopic method has been developed that allows the observation of the magnetoacoustic part of the SAW wave during sensor operation (Müller et al., 2022). Based on the magnetooptical Kerr effect, this microscopy method combines a temporal resolution in the picosecond range with a spatial resolution of better than 300 nm. The use of wide-field optics also enables imaging of the entire magnetically coated delay line. Müller et al. investigated an SAW sensor with the design shown in Fig. 9.7 in magnetic saturation caused by a small external magnetic field. The imaged wavelength is significantly reduced with $\lambda_{FeCoSiB} = 27.2\ \mu m$ compared to the periodicity of the IDTs with $\lambda = 28\ \mu m$. This underlines the influence of the SiO_2 guiding layer and the magnetic film with an effectively lower phase velocity of around 4000 m/s. Even more remarkable is the fact that even in the magnetically saturated state, an attenuation of the wave amplitude with an attenuation factor $d = -0.30$ dB/mm is measured. This wave attenuation is a combined effect of causes, which lie in the quartz substrate, guiding layer, and magnetic material. The wave reflection at structure edges, which was already a concern in the development of earlier SAW components (Jungerman et al., 1985), is also observed by Müller et al. at the edges of the FeCoSiB structure with a significant reflection coefficient of $r = 0.12$. This effect reduces the transmission in addition to the wave attenuation mentioned above and leads to a reduction in the signal $s(t)$ from Eq. (9.1) but not necessarily to a significant increase in the phase noise.

In the operating point of the SAW sensor, that is, in small magnetic fields around zero, the situation changes dramatically with the appearance of spike domains and domain walls in the single layer FeCoSiB (Fig. 9.10a). Using the example of a 180° wall that extends over the entire magnetically coated part of the delay line, stroboscopic microscopy reveals that the interaction with the time-varying stress field leads to magnetoelastic excitation of the domain wall structure itself. This interaction manifests in a spatially and temporarily varying wall orientation. In addition, there is a reduction in the magnetic contrast accompanying the wave in the vicinity of the domain wall. Müller et al. interpret this as

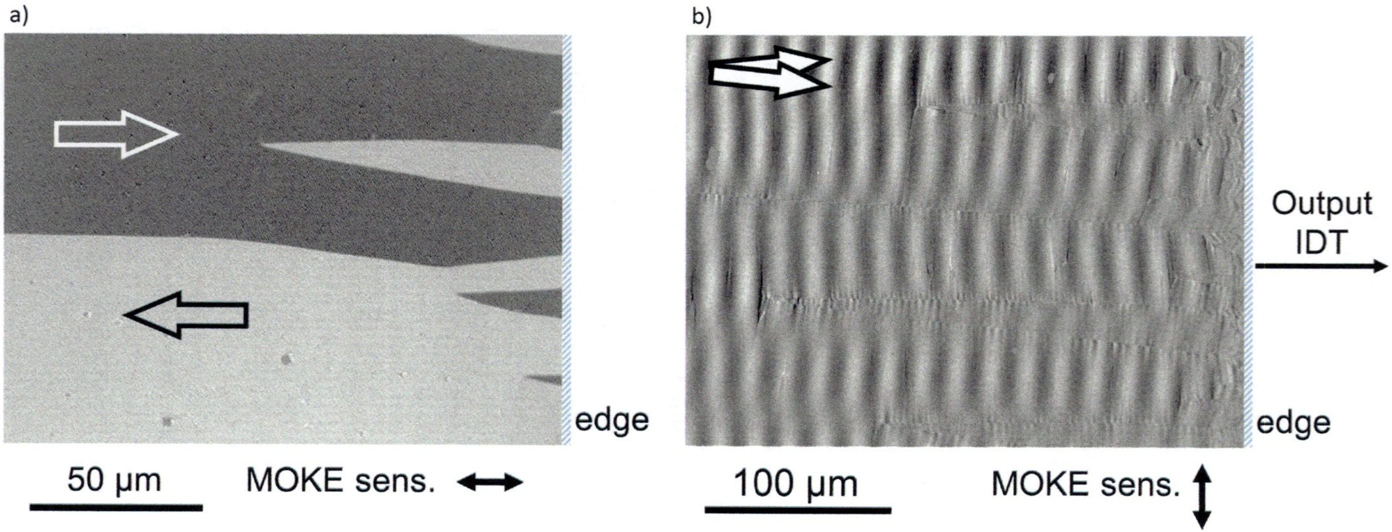

FIGURE 9.10 (a) Remanent magnetization featuring 180° walls and spike domains typically forming at the structural FeCoSiB edge. (b) Snapshot recorded with rotated MOKE sensitivity. The magnetoelastic wave is visible through the periodic magnetic contrast. The stroboscopic imaging is insensitive to irreproducible and nonsynchronous effects resulting in image blur. Magnetization and MOKE microscope sensitivity (MOKE sens.) directions are indicated by arrows. *(b) From Müller et al. (2022) with modified annotations.*

an indication of the presence of a magnetic noise-inducing loss mechanism caused by nonreproducible domain wall responses.

The perturbation of wave propagation at inhomogeneous magnetization structures is further exemplified in Fig. 9.10b. It shows a snapshot of the magnetoelastic wave coming from the left and reaching a 180° wall and three spike domains protruding into the FeCoSiB structure from the right edge. The diverse perturbating effects develop, for example, in a clear reduction of the wave amplitude within the spike domains as well as in a shift and tilt of the wave fronts. In the immediate vicinity of the domain walls, waves with a much smaller wavelength become visible, which result from the above-mentioned wall excitation. The reduction of the magnetic wave contrast in front of the spike domains is a consequence of the superposition with waves reflected at the domains (see also videos of magnetoelastic wave propagation in the supplement of (Müller et al., 2022)). Clearly, the interaction with the magnetization structures leads to signal reduction and uncorrelated phase variation. By stroboscopic imaging, Müller et al. reveal the cause of phase noise of magnetic origin and the link to micromagnetic structures and SAW propagation.

An obvious approach to improve the SAW sensor performance is to avoid the formation of domains by utilizing tailored magnetic multilayers. Schell et al. investigated the approach of antiparallel exchange biased FeCoSiB layers (Schell et al., 2023), which has already shown promising results for magnetic frequency conversion and converse ME sensors (see Section 2.4.1 *The antagonism of sensitivity versus noise—Magnetostrictive multilayers for magnetic domain control* (Chapter 2) and Section 9.2 *Converse Magnetoelectric Sensors*). Since a thermal treatment on the ST-cut quartz is not applicable, the exchange bias has to be initialized during the layer growth. Therefore, the layer sequence is reversed, and the ferromagnetic layers are deposited before the antiferromagnetic MnIr layers (see Fig. 9.11a). The integrated NiFe seed layers are known to be used to induce a {111} texture along the growth direction (Tsunoda et al., 2000). By reversing a magnetic field $H_{\mathrm{dep}} = \pm 600$ Oe applied in the sample's plane during deposition, the arising unidirectional anisotropies in the two adjacent FeCoSiB layers can be made nearly antiparallel with exchange bias fields of opposite sign (see Fig. 9.11b). The marginally different coercive fields and loop displacements in the easy axis magnetization curve indicate that the two layers differ slightly in other details. However, magneto-optical microscopy reveals that the remanent magnetization state of at least the optically accessible top layer is single domain due to the magnetic flux-closing layer structure. Furthermore, the magnetization reversal in a hard axis loop occurs via a coherent

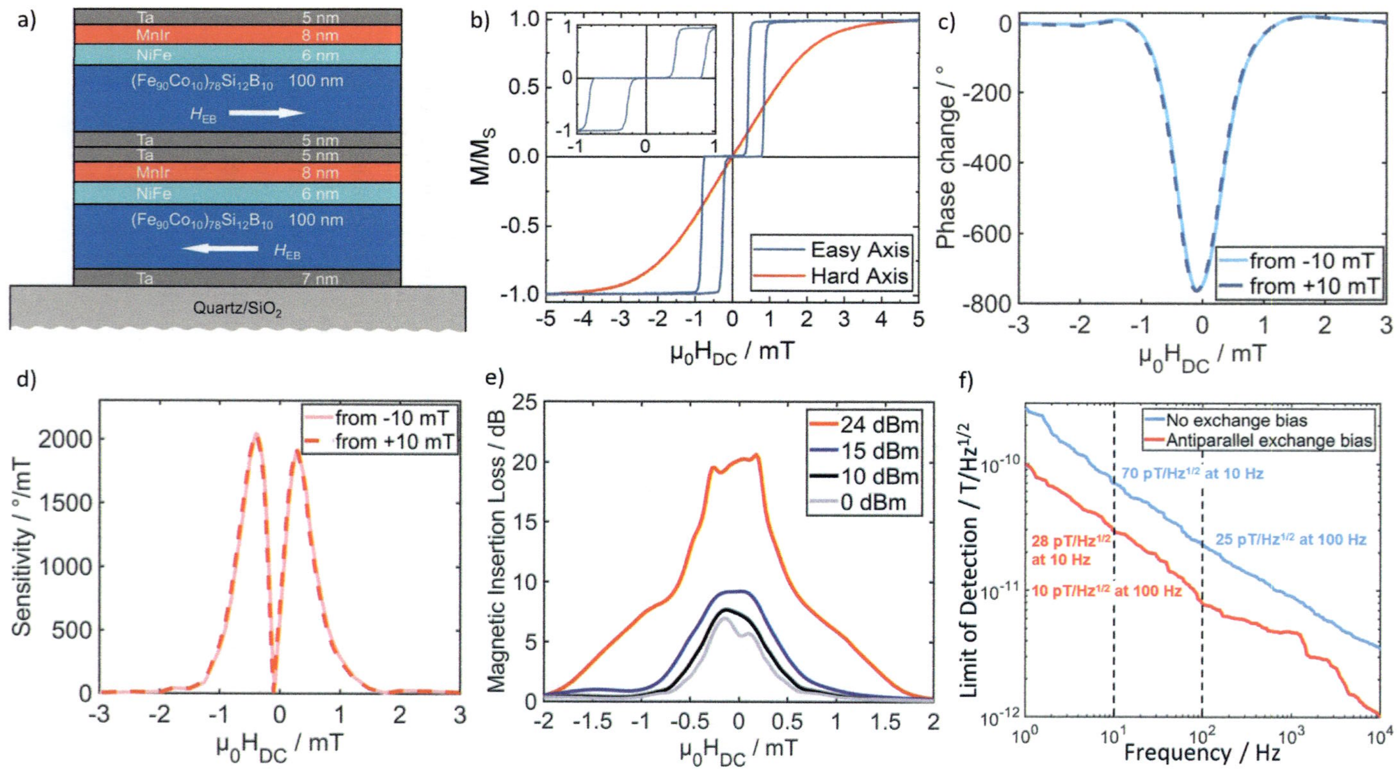

FIGURE 9.11 (a) Magnetoelastic layer stack with magnetic flux closure for the application in SAW magnetic field sensors. The antiparallel magnetization (indicated by arrows) is to be achieved during layer growth by deposition in magnetic fields with opposite directions. (b) Magnetization curves of the stack shown in (a) recorded along the easy (*blue*) and hard (*red*) axis of magnetization. The inset displays the easy axis magnetization in a narrow field range. (c) Change of phase φ versus applied static magnetic field at SAW excitation power $P_0 = 10$ dBm. (d) Phase sensitivity S measured in response to an AC magnetic field of 0.01 Oe amplitude and 10 Hz frequency ($P_0 = 10$ dBm). (e) Magnetic insertion losses normalized to the insertion losses at magnetic saturation for various P_0. (f) LOD at optimum working point of SAW sensors with the layer stack shown in (a) and a single 200 nm thick FeCoSiB layer. *Figures (a)–(e) from Schell et al. (2023).*

magnetization rotation without substantial domain wall nucleations. The phase and phase sensitivity curves therefore exhibit only a small hysteresis and a field dependence (Fig. 9.11c and d), which is expected for an SAW sensor with dominant magnetic anisotropy along the delay line (Mazzamurro et al., 2020). The field regions with maximum slope correspond to the operating points with maximum sensitivity $S_{max} = 2040\ ^\circ/mT$, which is virtually the same as the maximum sensitivity for SAW sensors of the same design but with a single 200 nm thick FeCoSiB layer (Schell et al., 2020).

The insertion losses relative to the losses in magnetic saturation are shown in Fig. 9.11e against the external magnetic field. These additional losses become large for small magnetic fields around zero but show a symmetrical field curve due to the changed magnetization behavior in contrast to the sensor with a single layer (compare Fig. 9.9a). At very high excitation power P_0, the sensitivity function of the sensor changes and the magnetic losses increase strongly ($P_0 = 24$ dBm $= 0.25$ W in Fig. 9.11e). In the absence of a multidomain magnetic state, the phase noise at the optimum operating point ($P_0 = 5$ dBm…8 dBm) is 8 dB lower and the detection limit (LOD $= 28\ pT/Hz^{1/2}$ at 10 Hz) is improved by a factor of about 2.5 compared with SAW sensors exhibiting a single magnetic sense layer (Fig. 9.11f). The specific causes of the phase noise and its dependence on the SAW excitation must be further investigated for the single-domain SAW sensor.

An alternative could be SAW sensors based on piezoelectric thin films. Such thin-film SAWs have attracted research interest in the past (Lee et al., 2001; Okano et al., 1994; Yokokawa et al., 1992; Teshigahara et al., 2012), in particular because the temperature stability of the devices can be increased (Tsubouchi et al., 1982) and the fabrication on silicon wafers should facilitate integration in MEMS and CMOS processes (Moench et al., 2022). The development of thin-film SAW magnetic field sensors should also benefit from the extended possibilities for tailoring magnetic layers. However, research on SAW magnetic field sensors on silicon substrates is still at an early stage (Meyer et al., 2021, 2024). Table 9.3 summarizes ME sensor performance data demonstrated for the MEMS sensors discussed in the previous Sections 9.1 *The MEMS approach to magnetoelectric cantilever sensors*, 9.2 *Converse Magnetoelectric Sensors*, and 9.3 *Magnetoelastic Surface Acoustic Wave Sensors*. The consistent evolution of the approach to further miniaturized ME devices is discussed in the following Chapter 10 *Magnetoelectric sensors based on magnetoelectric NEMS*.

TABLE 9.3 MEMS ME magnetic field sensors and their performance data.

Sensor concept	MEMS cantilever	Converse ME	Surface acoustic wave
Magnetostrictive layer, thicknesses in nm	20 iterations of Ta 5/Cu 3/MnIr 8/FeCoSiB 200	8 iterations of Ta 10/Cu 3/MnIr 8/FeCoSiB 500	2 iterations of Ta 5/FeCoSiB 100/NiFe 6/MnIr 8
LOD at low frequency	350 pT/Hz$^{1/2}$ *at* 25 Hz	8 pT/Hz$^{1/2}$ *at* 10 Hz	28 pT/Hz$^{1/2}$ *at* 10 Hz
Bandwidth	20 Hz	6.2 kHz	10 kHz
Magnetic field bias for optimal LOD	0	0.031 Oe	$<$ 5 Oe
References	Spetzler et al. (2021)	Elzenheimer et al. (2023), Thormählen et al. (2024)	Schell et al. (2023)

References

Berry, B. S., & Pritchet, W. C. (1976). Magnetoelasticity and internal friction of an amorphous ferromagnetic alloy. *Journal of Applied Physics, 47*, 3295–3301. https://doi.org/10.1063/1.323128

de Angelis, G., Pasku, V., de Angelis, A., Dionigi, M., Mongiardo, M., Moschitta, A., & Carbone, P. (2015). An Indoor AC magnetic positioning system. *IEEE Transactions on Instrumentation and Measurement, 64*, 1267–1275. https://doi.org/10.1109/TIM.2014.2381353

Dong, J., Zhang, G., Geng, Y., & Wang, J. (2018). Influence of magnetic measurement modeling on the solution of magnetostatic inverse problems applied to current distribution reconstruction in switching air arcs. *IEEE Transactions on Magnetics, 54*, 1–4. https://doi.org/10.1109/TMAG.2017.2773265. Article 8000704.

Durdaut, P. (2020). *Ausleseverfahren und Rauschmodellierung für magnetoelektrische und magnetoelastische Sensorsysteme*. Dissertation. Düren: Shaker Verlag.

Durdaut, P., Kittmann, A., Rubiola, E., Friedt, J.-M., Quandt, E., Knöchel, R., & Höft, M. (2019). Noise analysis and Comparison of phase- and frequency-detecting readout systems: Application to SAW delay line magnetic field sensor. *IEEE Sensors Journal, 19*, 8000–8008. https://doi.org/10.1109/JSEN.2019.2914965

Durdaut, P., Müller, C., Kittmann, A., Schell, V., Bahr, A., Quandt, E., Knöchel, R., Höft, M., & McCord, J. (2021). Phase noise of SAW delay line magnetic field sensors. *Sensors, 21*(16), 5631. https://doi.org/10.3390/s21165631

Durdaut, P., Salzer, S., Reermann, J., Röbisch, V., Hayes, P., Piorra, A., Meyners, D., Quandt, E., Schmidt, G., Knöchel, R., & Höft, M. (2017). Thermal-mechanical noise in resonant thin-film magnetoelectric sensors. *IEEE Sensors Journal, 17*, 2338–2348. https://doi.org/10.1109/JSEN.2017.2671442

Elhosni, M., Elmazria, O., Petit-Watelot, S., Bouvot, L., Zhgoon, S., Talbi, A., Hehn, M., Aissa, K. A., Hage-Ali, S., Lacour, D., Sarry, F., & Boumatar, O. (2016). Magnetic field SAW sensors based on magnetostrictive-piezoelectric layered structures: FEM modeling and experimental validation. *Sensors and Actuators A: Physical, 240*, 41–49. https://doi.org/10.1016/j.sna.2015.10.031

Elzenheimer, E., Bald, C., Engelhardt, E., Hoffmann, J., Hayes, P., Arbustini, J., Bahr, A., Quandt, E., Höft, M., & Schmidt, G. (2022). Quantitative evaluation for magnetoelectric sensor systems in biomagnetic diagnostics. *Sensors, 22*(3), 1018. https://doi.org/10.3390/s22031018

Elzenheimer, E., Hayes, P., Thormählen, L., Engelhardt, E., Zaman, A., Quandt, E., Frey, N., Höft, M., & Schmidt, G. (2023). Investigation of converse magnetoelectric thin-film sensors for magnetocardiography. *IEEE Sensors Journal, 23*, 5660–5669. https://doi.org/10.1109/JSEN.2023.3237910

Fetisov, Y. K., Petrov, V. M., & Srinivasan, G. (2007). Inverse magnetoelectric effects in a ferromagnetic–piezoelectric layered structure. *Journal of Materials Research, 22*, 2074–2080. https://doi.org/10.1557/jmr.2007.0262

Ganguly, A. K., Davis, K. L., Webb, D. C., Vittoria, C., & Forester, D. W. (1975). Magnetically tuned surface-acoustic-wave phase shifter. *Electronics Letters, 11*, 610–611. https://doi.org/10.1049/el:19750465

Gojdka, B., Jahns, R., Meurisch, K., Greve, H., Adelung, R., Quandt, E., Knöchel, R., & Faupel, F. (2011). Fully integrable magnetic field sensor based on delta-E effect. *Applied Physics Letters, 99*, Article 223502. https://doi.org/10.1063/1.3664135

Hanna, S. M. (1987). Magnetic field sensors based on SAW propagation in magnetic films. *IEEE Transactions on Ultrasonics, Ferroelectrics, and Frequency Control, 34*(2), 191–194. https://doi.org/10.1109/T-UFFC.1987.26931

Hashimoto, K. (2000). *Surface acoustic wave devices in telecommunications* (Vol. 116). Berlin: Springer-Verlag.

Hayes, P., Jovičević Klug, M., Toxværd, S., Durdaut, P., Schell, V., Teplyuk, A., Burdin, D., Winkler, A., Weser, R., Fetisov, Y., Höft, M., Knöchel, R., McCord, J., & Quandt, E. (2019). Converse magnetoelectric composite resonator for sensing small magnetic fields. *Scientific Reports, 9*, Article 16355. https://doi.org/10.1038/s41598-019-52657-w

Hayes, P., Salzer, S., Reermann, J., Yarar, E., Röbisch, V., Piorra, A., Meyners, D., Höft, M., Knöchel, R., Schmidt, G., & Quandt, E. (2016). Electrically modulated magnetoelectric sensors. *Applied Physics Letters, 108*, Article 182902. https://doi.org/10.1063/1.4948470

Hayes, P., Schell, V., Salzer, S., Burdin, D., Yarar, E., Piorra, A., Knöchel, R., Fetisov, Y. K., & Quandt, E. (2018). Electrically modulated magnetoelectric AlN/FeCoSiB film composites for DC magnetic field sensing. *Journal of Physics D Applied Physics, 51*, Article 354002. https://doi.org/10.1088/1361-6463/aad456

Hoffmann, J., Elzenheimer, E., Bald, C., Hansen, C., Maetzler, W., & Schmidt, G. (2021). Active magnetoelectric motion sensing: Examining performance metrics with an experimental setup. *Sensors, 21*(23), 8000. https://doi.org/10.3390/s21238000

Huang, D., Lu, C., & Bing, H. (2015). Multipeak self-biased magnetoelectric coupling characteristics in four-phase Metglas/Terfenol-D/Be-bronze/PMN-PT structure. *AIP Advances, 5*, Article 047140. https://doi.org/10.1063/1.4919248

Inoue, M., Tsuboi, Y., Yokokawa, N., & Fujii, T. (1990). Love-type magneto-surface-acoustic-wave in multilayered highly magnetostrictive films separated by insulating layers. *IEEE Transactions on Magnetics, 26*, 1465–1467. https://doi.org/10.1109/20.104413

Jakoby, B., & Vellekoop, M. J. (1997). Properties of Love waves: Applications in sensors. *Smart Materials and Structures, 6*, 668–679. https://doi.org/10.1088/0964-1726/6/6/003

Jovičević Klug, M., Thormählen, L., Röbisch, V., Toxværd, S. D., Höft, M., Knöchel, R., Quandt, E., Meyners, D., & McCord, J. (2019). Antiparallel exchange biased multilayers for low magnetic noise magnetic field sensors. *Applied Physics Letters, 114*, Article 192410. https://doi.org/10.1063/1.5092942

Jungerman, R. L., Baer, R. L., & Bray, R. C. (1985). Delay dependence of phase noise in SAW filters. In *IEEE 1985 ultrasonics symposium, San Francisco, CA, USA* (pp. 258–261).

Kadota, M., Ito, S., Ito, Y., Hada, T., & Okaguchi, K. (2011). Magnetic sensor based on surface acoustic wave resonators. *Japanese Journal of Applied Physics, 50*(7S), 07HD07. https://doi.org/10.1143/JJAP.50.07HD07

Kirchhof, C., Krantz, M., Teliban, I., Jahns, R., Marauska, S., Wagner, B., Knöchel, R., Gerken, M., Meyners, D., & Quandt, E. (2013). Giant magnetoelectric effect in vacuum. *Applied Physics Letters, 102*, Article 232905. https://doi.org/10.1063/1.4810750

Kittmann, A., Durdaut, P., Zabel, S., Reermann, J., Schmalz, J., Spetzler, B., Meyners, D., Sun, N. X., McCord, J., Gerken, M., Schmidt, G., Höft, M., Knöchel, R., Faupel, F., & Quandt, E. (2018). Wide band low noise love wave magnetic field sensor system. *Scientific Reports, 8*, 278. https://doi.org/10.1038/s41598-017-18441-4

Lage, E., Urs, N. O., Röbisch, V., Teliban, I., Knöchel, R., Meyners, D., McCord, J., & Quandt, E. (2014). Magnetic domain control and voltage response of exchange biased magnetoelectric composites. *Applied Physics Letters, 104*, Article 132405. https://doi.org/10.1063/1.4870511

Lee, E. W. (1955). Magnetostriction and magnetomechanical effects. *Reports on Progress in Physics, 18*, 184–229. https://doi.org/10.1088/0034-4885/18/1/305

Lee, S.-H., Jeong, H.-H., Bae, S.-B., Choi, H.-C., Lee, J.-H., & Lee, Y.-H. (2001). Epitaxially grown GaN thin-film SAW filter with high velocity and low insertion loss. *IEEE Transactions on Electron Devices, 48*, 524–529. https://doi.org/10.1109/16.906446

Liu, X., Tong, B., Ou-Yang, J., Yang, X., Chen, S., Zhang, Y., & Zhu, B. (2018). Self-biased vector magnetic sensor based on a Love-type surface acoustic wave resonator. *Applied Physics Letters, 113*, Article 082402. https://doi.org/10.1063/1.5044478

Müller, C., Durdaut, P., Holländer, R. B., Kittmann, A., Schell, V., Meyners, D., Höft, M., Quandt, E., & McCord, J. (2022). Imaging of Love waves and their interaction with magnetic domain walls in magnetoelectric magnetic field sensors. *Advanced Electronic Materials, 8*, Article 2200033. https://doi.org/10.1002/aelm.202200033

Mandal, D., & Banerjee, S. (2022). Surface acoustic wave (SAW) sensors: Physics, materials, and applications. *Sensors, 22*(3), 820. https://doi.org/10.3390/s22030820

Marauska, S., Jahns, R., Kirchhof, C., Claus, M., Quandt, E., Knöchel, R., & Wagner, B. (2013). Highly sensitive wafer-level packaged MEMS magnetic field sensor based on magnetoelectric composites. *Sensors and Actuators A: Physical, 189*, 321–327. https://doi.org/10.1016/j.sna.2012.10.015

Mazzamurro, A., Dusch, Y., Pernod, P., Bou Matar, O., Addad, A., Talbi, A., & Tiercelin, N. (2020). Giant magnetoelastic coupling in a Love acoustic waveguide based on TbCo2/FeCo nanostructured film on ST-cut quartz. *Physical Review Applied, 13*(4), 044001. https://doi.org/10.1103/PhysRevApplied.13.044001

Mermelstein, M. D. (1986). Magnetoelastic amorphous metal fluxgate magnetometer. *Electronics Letters, 22*, 525–526. https://doi.org/10.1049/el:19860358

Meyer, J. M., Ghori, M. Z., Rickers, M., Fichtner, S., Bodduluri, M. T., Blohm, L., Yarar, E., Giese, T., & Lofink, F. (2024). *MikroSystemTechnik Kongress 2023: Mikroelektronik/Mikrosystemtechnik und ihre Anwendungen - Nachhaltigkeit und Technologiesouveränität 23. - 25. Oktober 2023, Dresden, Neuerscheinung*. Berlin: VDE VERLAG.

Meyer, J. M., Schell, V., Su, J., Fichtner, S., Yarar, E., Niekiel, F., Giese, T., Kittmann, A., Thormählen, L., Lebedev, V., Moench, S., Žukauskaitė, A., Quandt, E., & Lofink, F. (2021). Thin-film-based SAW magnetic field sensors. *Sensors, 21*(24), 8166. https://doi.org/10.3390/s21248166

Moench, S., Meyer, J. M., Zukauskaite, A., Lebedev, V., Fichtner, S., Su, J., Niekiel, F., Giese, T., Thormählen, L., Quandt, E., & Lofink, F. (2022). AlScN-based SAW magnetic field sensor for isolated closed-loop hysteretic current control of switched-mode power converters. *IEEE Sensors Letters, 6*(10), 1–4. https://doi.org/10.1109/LSENS.2022.3205853. Article 2500904.

Nan, T., Hui, Y., Rinaldi, M., & Sun, N. X. (2013). Self-biased 215 MHz magnetoelectric NEMS resonator for ultra-sensitive DC magnetic field detection. *Scientific Reports, 3*(1), 1985. https://doi.org/10.1038/srep01985

Okano, H., Tanaka, N., Takahashi, Y., Tanaka, T., Shibata, K., & Nakano, S. (1994). Preparation of aluminum nitride thin films by reactive sputtering and their applications to GHz-band surface acoustic wave devices. *Applied Physics Letters, 64*, 166–168. https://doi.org/10.1063/1.111553

Onuta, T.-D., Wang, Y., Long, C. J., & Takeuchi, I. (2011). Energy harvesting properties of all-thin-film multiferroic cantilevers. *Applied Physics Letters, 99*, Article 203506. https://doi.org/10.1063/1.3662037

Parker, T. E. (1979). 1/f phase noise in quartz s.a.w. devices. *Electronics Letters, 15*(10), 296. https://doi.org/10.1049/el:19790211

Polewczyk, V., Dumesnil, K., Lacour, D., Moutaouekkil, M., Mjahed, H., Tiercelin, N., Petit Watelot, S., Mishra, H., Dusch, Y., Hage-Ali, S., Elmazria, O., Montaigne, F., Talbi, A., Bou Matar, O., & Hehn, M. (2017). Unipolar and bipolar high-magnetic-field sensors based on surface acoustic wave resonators. *Physical Review Applied, 8*(2), 024001. https://doi.org/10.1103/PhysRevApplied.8.024001

Sárközi, Z., Mackay, K., & Peuzin, J. C. (2000). Elastic properties of magnetostrictive thin films using bending and torsion resonances of a bimorph. *Journal of Applied Physics, 88*, 5827–5832. https://doi.org/10.1063/1.1318388

Schell, V., Müller, C., Durdaut, P., Kittmann, A., Thormählen, L., Lofink, F., Meyners, D., Höft, M., McCord, J., & Quandt, E. (2020). Magnetic anisotropy controlled FeCoSiB thin films for surface acoustic wave magnetic field sensors. *Applied Physics Letters, 116*, Article 073503. https://doi.org/10.1063/1.5140562

Schell, V., Spetzler, E., Wolff, N., Bumke, L., Kienle, L., McCord, J., Quandt, E., & Meyners, D. (2023). Exchange biased surface acoustic wave magnetic field sensors. *Scientific Reports, 13*, 8446. https://doi.org/10.1038/s41598-023-35525-6

Schulz, M. B., Matsinger, B. J., & Holland, M. G. (1970). Temperature dependence of surface acoustic wave velocity on α quartz. *Journal of Applied Physics, 41*, 2755–2765. https://doi.org/10.1063/1.1659311

Spetzler, B., Bald, C., Durdaut, P., Reermann, J., Kirchhof, C., Teplyuk, A., Meyners, D., Quandt, E., Höft, M., Schmidt, G., & Faupel, F. (2021). Exchange biased delta-E effect enables the detection of low frequency pT magnetic fields with simultaneous localization. *Scientific Reports, 11*, 5269. https://doi.org/10.1038/s41598-021-84415-2

Spetzler, E., Spetzler, B., Seidler, D., Arbustini, J., Thormählen, L., Elzenheimer, E., Höft, M., Bahr, A., Meyners, D., & McCord, J. (2025). On the origin of signal and bandwidth of converse magnetoelectric magnetic field sensors. *Advanced Sensor Research, 4*(1), 2400109. https://doi.org/10.1002/adsr.202400109

Su, J., Niekiel, F., Fichtner, S., Kirchhof, C., Meyners, D., Quandt, E., Wagner, B., & Lofink, F. (2020). Frequency tunable resonant magnetoelectric sensors for the detection of weak magnetic field. *Journal of Micromechanics and Microengineering, 30*, Article 75009. https://doi.org/10.1088/1361-6439/ab8dd0

Su, J., Niekiel, F., Fichtner, S., Thormaehlen, L., Kirchhof, C., Meyners, D., Quandt, E., Wagner, B., & Lofink, F. (2020). AlScN-based MEMS magnetoelectric sensor. *Applied Physics Letters, 117*, Article 132903. https://doi.org/10.1063/5.0022636

Teshigahara, A., Hashimoto, K., & Akiyama, M. (2012). Scandium aluminum nitride: Highly piezoelectric thin film for RF SAW devices in multi GHz range 1–5. *IEEE International Ultrasonics Symposium*. https://doi.org/10.1109/ULTSYM.2012.0481

Thormählen, L., Hayes, P., Elzenheimer, E., Spetzler, E., Schmidt, G., Höft, M., McCord, J., Meyners, D., & Quandt, E. (2024). Low-noise inverse magnetoelectric magnetic field sensor. *Applied Physics Letters, 124*, Article 172402. https://doi.org/10.1063/5.0196637

Tsubouchi, K., Sugai, K., & Mikoshiba, N. (1982). Zero temperature coefficient surface-acoustic-wave devices using Epitaxial AlN films. *Ultrasonics Symposium*, 340–345. https://doi.org/10.1109/ULTSYM.1982.197842

Tsunoda, M., Tsuchiya, Y., Hashimoto, T., & Takahashi, M. (2000). Magnetic anisotropy and rotational hysteresis loss in exchange coupled Ni–Fe/Mn–Ir films. *Journal of Applied Physics, 87*, 4375–4388. https://doi.org/10.1063/1.373081

Wang, Y., Onuta, T.-D., Long, C. J., Geng, Y., & Takeuchi, I. (2015). Colossal magnetoelectric effect induced by parametric amplification. *Applied Physics Letters, 107*, Article 192902. https://doi.org/10.1063/1.4935332

Wang, J. Z., Williamson, S. J., & Kaufman, L. (1992). Magnetic source images determined by a lead-field analysis: The unique minimum-norm least-squares estimation. *IEEE Transactions on Biomedical Engineering, 39*, 665–675. https://doi.org/10.1109/10.142641

Webb, D., Forester, D., Ganguly, A., & Vittoria, C. (1979). Applications of amorphous magnetic-layers in surface-acoustic-wave devices. *IEEE Transactions on Magnetics, 15*, 1410–1415. https://doi.org/10.1109/TMAG.1979.1060442

White, R. M., & Voltmer, F. W. (1965). Direct piezoelectric coupling to surface elastic waves. *Applied Physics Letters, 7*, 314–316. https://doi.org/10.1063/1.1754276

Xue, H., Hu, Y., & Wang, Q.-M. (2008). Broadband piezoelectric energy harvesting devices using multiple bimorphs with different operating frequencies. *IEEE Transactions on Ultrasonics, Ferroelectrics, and Frequency Control, 55*, 2104–2108. https://doi.org/10.1109/TUFFC.903

Yamamoto, H., Naoe, M., Yamanaka, S., Yamaguchi, M., & Kogo, H. (1979). Variable surface acoustic wave delay line consisting of a magnetic thin film on a LiNbO 3 substrate. *Japanese Journal of Applied Physics, 18*, 209–213. https://doi.org/10.7567/JJAPS.18S1.209

Yasumura, K. Y., Stowe, T. D., Chow, E. M., Pfafman, T., Kenny, T. W., Stipe, B. C., & Rugar, D. (2000). Quality factors in micron- and submicron-thick cantilevers. *Journal of Microelectromechanical Systems, 9*, 117–125. https://doi.org/10.1109/84.825786

Yokokawa, N., Tanaka, S., Fujii, T., & Inoue, M. (1992). Love-type surface-acoustic waves propagating in amorphous iron-boron films with multilayer structure. *Journal of Applied Physics, 72*, 360–366. https://doi.org/10.1063/1.351863

Yoshizawa, N., Yamamoto, I., & Shimada, Y. (2005). Magnetic field sensing by an electrostrictive/magnetostrictive composite resonator. *IEEE Transactions on Magnetics, 41*, 4359–4361. https://doi.org/10.1109/TMAG.2005.855168

Zabel, S., Kirchhof, C., Yarar, E., Meyners, D., Quandt, E., & Faupel, F. (2015). Phase modulated magnetoelectric delta-E effect sensor for sub-nano tesla magnetic fields. *Applied Physics Letters, 107*, Article 152402. https://doi.org/10.1063/1.4932575

Zabel, S., Reermann, J., Fichtner, S., Kirchhof, C., Quandt, E., Wagner, B., Schmidt, G., & Faupel, F. (2016). Multimode delta-E effect magnetic field sensors with adapted electrodes. *Applied Physics Letters, 108*, Article 222401. https://doi.org/10.1063/1.4952735

Zhang, C. L., & Chen, W. Q. (2010). A wideband magnetic energy harvester. *Applied Physics Letters, 96*, Article 123507. https://doi.org/10.1063/1.3360218

Zhao, P., Zhao, Z., Hunter, D., Suchoski, R., Gao, C., Mathews, S., Wuttig, M., & Takeuchi, I. (2009). Fabrication and characterization of all-thin-film magnetoelectric sensors. *Applied Physics Letters, 94*, Article 243507. https://doi.org/10.1063/1.3157281

Zhou, H., Talbi, A., Tiercelin, N., & Bou Matar, O. (2014). Multilayer magnetostrictive structure based surface acoustic wave devices. *Applied Physics Letters, 104*, Article 114101. https://doi.org/10.1063/1.4868530

Zhuang, X., Lam Chok Sing, M., Dolabdjian, C., Wang, Y., Finkel, P., Li, J., & Viehland, D. (2015). Dynamic sensitivity and noise floor of a bonded magneto(Elasto)Electric laminate for low frequency magnetic field sensing under strain modulations. *KEM, 644*, 236–239. https://doi.org/10.4028/www.scientific.net/KEM.644.236

Zhuang, X., Sing, M. L. C., Dolabdjian, C., Wang, Y., Finkel, P., Li, J., & Viehland, D. (2015). Sensitivity and noise evaluation of a bonded magneto(elasto) electric laminated sensor based on in-plane Magnetocapacitance effect for Quasi-static magnetic field sensing. *IEEE Transactions on Magnetics, 51*(1), 1–4. https://doi.org/10.1109/TMAG.2014.2356852. Article 2500204.

Zhuang, X., Yang, M.-T., Lam Chok Sing, M., Dolabdjian, C., Finkel, P., Li, J., & Viehland, D. (2016). Tunable magnetoelectric bending resonance for sensing static magnetic fields. *IEEE Sensors Journal, 16*, 662–669. https://doi.org/10.1109/JSEN.2015.2490581

CHAPTER

10

Magnetoelectric NEMS sensors

Bin Luo, Cunzheng Dong, and Nian Sun

Northeastern University, Boston, MA, United States

10.1 Magnetoelectric sensor overview

Magnetometers are widely used in biomagnetic sensing, navigation, magnetic anomaly detection, communications, etc. Biomedical magnetometry is an application space that requires detection of magnetic field signals in frequency ranges of 1–1000 Hz and amplitudes of ~20 fT—tens of pT to monitor brain, nerve, skeletal muscle, and heart functions (Cohen & Givler, 2003; Elzenheimer et al., 2018; Uchikawa & Kotani, 1992; Zuo et al., 2020a). Superconducting quantum interference device (SQUID) magnetometers are the current state of the art in terms of biomedical magnetometry. An SQUID can detect magnetic signals in the single-digit fT range; however, the cryogenic cooling requirement drastically limits the overall form factor, power efficiency, and applications (Granata et al., 2011; Jovičević Klug et al., 2019). Magnetoelectric magnetometers can operate at room temperatures with demonstrated detection of single-digit fT magnetic fields at kHz (Dong et al., 2025), therefore are well suited to lower the cost, power consumption, and form factor for biomedical magnetometry systems.

The principal mechanism of ME sensors is based on the direct ME effect (DME), that is, a voltage generation in the piezoelectric phase due to the strain deformation induced from the magnetostrictive phase by external magnetic fields (Carman & Sun, 2018; Viehland et al., 2018). Based on the magnetization reorientation direction in the magnetostrictive phase and the polarization oscillation direction in the piezoelectric phase, ME sensors can be configured in one of various operational modes, including L–L, L–T, T–T, and T–L (where L is longitudinal, T is transverse). Among these modes, due to the shape anisotropy of ME laminate composites, the L–L and L–T modes provide the highest ME voltage coefficient (α_{ME}), which primarily depends upon the materials'

https://doi.org/10.1016/B978-0-443-18606-6.00003-4

properties, interfacial bonding, magnetic bias, and operational mode (Dong et al., 2006; Zhai et al., 2008). In 2003, Dong et al. proved that α_{ME} can be dramatically enhanced by 100× at the electromechanical resonance frequency (EMR) (Dong, Cheng, et al., 2003). Since then, researchers have been devoted to increase α_{ME} by designing new coupling structures, improving lamination processes, and looking for better materials (Dong et al., 2003b, Dong et al., 2005a, 2005b; Li et al., 2011; Mori & Wuttig, 2002; Wang et al., 2011; Zhai et al., 2006). To date, the highest reported value of α_{ME} in laminated composites is 52 V cm^{-1}•Oe at off-resonance, and 7000 V cm^{-1}•Oe at EMR (Chu et al., 2017; Wang et al., 2011). This resonance enhancement of ME coefficient results in the possibility of making ME laminated composites as sensitive magnetic sensors.

Since 2008, Wang et al. have been devoted to developing L−T and L−L multi push-pull mode bulk ME sensors, and have pushed the detection limit down to 5.1 pT $Hz^{-1/2}$ at 1 Hz (Viehland et al., 2018; Wang et al., 2012, 2014a, 2014b). However, for bulk ME sensors, the dimensions are in the order of ~10 cm, and the usage of hard magnets are required on both ends of ME sensors to provide bias magnetic field. These factors greatly reduced the spatial resolution for magnetic sensing and increased the risk of mutual interference between sensor arrays. Compared to bulk ME sensors, microfabricated thin film ME sensors have perfect interfacial bonding between piezoelectric and magnetostrictive layers at atomic level. Furthermore, thin film ME sensors demonstrate reduction in size down to the order of mm or even μm range, which is essential for the integration of sensor arrays and the improvement of spatial resolution. In 2010, Greve et al. developed the first AlN/FeCoSiB based thin film ME heterostructure with giant ME coefficient of 737 V cm^{-1}•Oe at EMR (Greve et al., 2010). Later on, this cantilever type thin film ME heterostructure was investigated for a potential application for magnetic sensors (Jahns et al., 2011). In 2013, Nan et al. reported a 215 MHz piezoelectric contour mode nano plate resonator (NPR) coated with magnetostrictive FeGaB thin films as a DC magnetic field sensor (Nan et al., 2013). With their high sensitivity and miniature size, these thin film ME heterostructures have paved a practical way toward the application of ME sensors for magnetic sensing.

For thin film ME sensors, numerous strategies have been investigated in order to improve the LOD. By improving the deposition process flow, magnetostrictive layers can be deposited on the smooth Si surface before the growth of AlN layers, which preserves the amorphous nature of magnetostrictive thin films and thereby enhances the piezomagnetic coefficient (Yarar et al., 2016). By the adoption of the interdigital electrodes scheme, an increase of signal-to-noise ratio (SNR) has been observed by enabling d_{33} mode on piezoelectric layers (Piorra et al., 2013). To reduce the cross sensitivities to acoustic and vibration noise, a tuning fork structure with differential output has been developed (Salzer et al., 2016a). Furthermore, the exchange-biased ferromagnetic/antiferromagnetic multilayer structure enables an optimal ME response without the need of an external bias magnetic field, which effectively reduces the magnetic domain activity in the magnetostrictive phase (Hrkac et al., 2014; Lage et al., 2012; Röbisch et al., 2015). Using these approaches, a LOD of 400 fT $Hz^{-1/2}$ has been achieved so far at the EMR frequency.

10.2 Magnetoelectric sensor operation scheme for magnetic sensing

There are three main schemes of magnetoelectric magnetometry: direct detection, frequency modulation, and those based on the ΔE-effect (i.e., magnetic field induced change in Young's modulus) (Dong et al., 2022; Luo et al., 2024). Fig. 10.1 highlights the corresponding operating principle and noise spectral density (NSD, or magnetic noise, in T/$\sqrt{Hz}$) for each of the schemes (Hayes et al., 2016; Kittmann et al., 2018; Nan et al., 2013; Salzer et al., 2018; Shen et al., 2013; Yarar et al., 2016). Direct detection relies on detecting AC magnetic fields via the direct ME effect—an induced alternating magnetostriction translates to a stress variation in the piezoelectric layer and thus a change in electrical polarization, which can be monitored via an electrical voltage readout. The primary metric that indicates the promise of a direct detection ME magnetometer is the dependence of the ME voltage coefficient on the applied magnetic field (a schematic diagram is shown in Fig. 10.1).

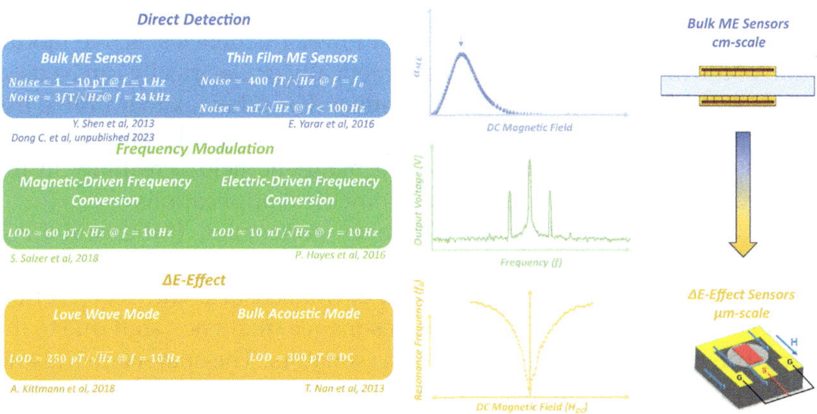

FIGURE 10.1 Magnetoelectric magnetometer operating schemes and their corresponding noise spectral density (NSD). *Adapted from Luo et al. (2024).*

Typically, these ME magnetometers require a DC magnetic bias field to push sensor operation to the peak operating regime of the ME voltage coefficient response to applied magnetic field. Significantly reduced magnetic noise can be achieved through direct detection of mechanical resonance because the ME sensor output voltage is improved for magnetic signals that coincide with the mechanical resonance frequency, with a demonstrated magnetic noise down to 400 $fT/\sqrt{Hz}$ (Yarar et al., 2016). Additionally, operating the ME sensor at mechanical resonance and in vacuum (e.g., pressure below 10^{-5} mbar), rather than in air, can lead to a tenfold reduced noise due to reduced air damping, and therefore, an increased quality factor (Kirchhof et al., 2013).

Thin film ME sensors have demonstrated equivalent magnetic noise as low as 400 fT $Hz^{-1/2}$ at their electromechanical resonance (Yarar et al., 2016). The EMR frequencies of thin film ME sensors are in the range of kHz to MHz, depending on the resonators' structure and dimension (Viehland et al., 2018). However, most of the biomagnetic signals are quasi-static magnetic fields ranging from 0.01 to 1000 Hz. At such a low frequency range, the noise floor increases rapidly, and a significant reduction of the ME coefficient happens when the devices are working in off-resonance mode. While these direct detection, on-resonant ME sensors have a limit of detection in the single-digit fT—pT regime, the resonance enhancement comes at the cost of reduced bandwidth, and therefore detecting low-frequency magnetic fields can be challenging (e.g., below resonance and in the low frequency regime the sensors suffer from $1/f$ noise) (Viehland et al., 2018; Yarar et al., 2016). Therefore, direct detection of biomagnetic signals would not be a practical option for ME sensors. To overcome the limitations at low frequency range of ME sensors, three indirect detection schemes have been proposed and investigated extensively. To circumvent operation at mechanical resonance, while still maintaining the necessary sensitivity at lower frequencies, active modulation of the magnetic phase either magnetically or electrically can be used (Jahns et al., 2012), as well as the ΔE-effect. For schemes that utilize frequency modulation, the resonant magnetic field signals can be down-converted to low frequencies, enabling magnetic field sensitivity in the low-frequency regimes required for biomedical applications.

The magnetic and electrical frequency conversion (MFC and EFC) methods allow for the low-frequency magnetic signals being up-converted to the side band of the EMR of ME sensors. The delta-E effect (ΔE) method enables the detection of resonance frequency shift due to the change of the elastic properties of ME resonators. The characteristics and the figure of merits of thin film ME sensors with different operation mechanisms are summarized and compared with other state-of-the-art magnetometers in Table 10.1. Details of these detection strategies are reviewed and discussed in this section.

TABLE 10.1 Comparison of the figure of merits of ME sensors with different operation schemes.

	Magnetic noise [T Hz$^{-1/2}$]					
Magnetometers	**1 Hz**	**10 Hz**	**100 Hz**	**Typical element size [cm^3]**	**Special requirements**	**Ref.**
ME MFC	1×10^{-9}	6×10^{-11}	–	8.7×10^{-3}	Modulation coil	(Jahns et al., 2012; Jovičević Klug et al., 2019)
ME EFC	1.2×10^{-10}	7×10^{-11}	–	1.25×10^{-4}	Pick up coil	(Hayes et al., 2019)
ME ΔE	3×10^{-10}	1×10^{-10}	7×10^{-11}	1.7×10^{-4}	Acoustic shielding	(Zabel et al., 2016)
ME SAW	–	2.5×10^{-10}	8×10^{-11}	3.8×10^{-2}	Phase detector	(Kittmann et al., 2018)

Reproduced with permission from Dong et al. (2022).

10.3 Magnetoelectric sensor utilizing frequency modulation scheme

For the frequency modulation approach, there are two operation schemes including magnetic and electric frequency conversion. Magnetic frequency conversion exploits the nonlinearity of the magnetic bias response of magnetostrictive materials. An advantage of magnetic modulation frequency conversion is that the sum of the modulation and signal frequencies can be selected to be equal to the mechanical resonance of the sensor, thus resonant operation can be exploited for any arbitrary signal frequency (Viehland et al., 2018). The magnetic noise is therefore reduced because the resonance enhancement can be exploited (Dong et al., 2003). This has been experimentally demonstrated for an ME sensor based on FeCoSiB/AlN heterostructures with a magnetic noise down to 60 pT/$\sqrt{\text{Hz}}$ at 10 Hz (Salzer et al., 2018). The other approach is electric frequency conversion which relies on a piezoelectrically induced AC actuation of the sensor, which through the strain mediated interface results in an additional magnetic oscillation in the magnetostrictive component. This also leads to side band generation (Viehland et al., 2018). This has been experimentally demonstrated for an ME sensor based on an exchange biased FeCoSiB stack with a magnetic noise in the low nT/$\sqrt{\text{Hz}}$ range (Hayes et al., 2016). The electric frequency conversion approach has a smaller form factor and lower power consumption than the magnetic modulation approach, because it does not require an external driving coil. For both approaches, the up conversion of the measurement signal frequency and thus an increased sensor operation frequency make the ME sensors based on frequency modulation less susceptible to the $1/f$ noise that dominates at lower frequencies. As a result, they have improved magnetic noise compared to passive operation of the sensors (Petrie et al., 2012).

10.3.1 Magnetoelectric sensor utilizing magnetic frequency modulation scheme

The principle of the MFC technique is based on the magnetostriction effect, which has a quadratic response to the applied magnetic field of small amplitude. If an alternating modulation magnetic field is applied, the superposition of the modulation field and the signal field contains a product term, which expresses a frequency conversion (Gillette et al., 2011; Li et al., 2013; Shen et al., 2011).

This technique offers the possibility to achieve resonance enhanced sensitivities at low frequencies. The MFC method also exhibits advantages of higher slew rate, lower mechanical noise, and simplicity for sensor arrays equipment.

In 2012, Jahns et al. demonstrated the first MFC method for up-converting low frequency magnetic signals with the carrier frequency on thin film ME sensors (Jahns et al., 2012). Although the measurement results show that at 1 Hz the sensitivity can be enhanced by a factor of 1000 compared to the direct detection method, however, MFC method introduces additional magnetic noise from the modulation field (Zhuang et al., 2011). During the excitation of the magnetostrictive film, the magnetic domain (MD) walls rotate and move periodically, which results in magnetic domain induced Barkhausen noise and impairs the sensor performance (Durdaut et al., 2017; Salzer et al., 2016b; Salzer et al., 2018). For ME sensors, MD can be reduced by exchange biased antiferromagnetic/ferromagnetic multilayer stacks (Hrkac et al., 2014; Lage et al., 2012, 2013; Röbisch et al., 2015). However, due to demagnetization effect, MDs still form with the increment of magnetic layer thicknesses, which increases sensor noise and limits the sensor sensitivity.

In 2019, Klug et al. demonstrated an approach to overcome these limitations by introducing an antiparallel exchange biased (APEB) multilayer scheme (Jovičević Klug et al., 2019), which enabled single domain films without the restriction on magnetic layer thickness. The APEB multilayer stack consists of 20 repetitions of [Ta/Cu/Mn_3Ir/$(Fe_{90}Co_{10})_{78}Si_{12}B_{10}$/Ta/Cu/$Mn_3Ir$/$(Fe_{90}Co_{10})_{78}Si_{12}B_{10}$]. A single domain in the top layer of the whole structure has been observed by Magneto-optical Kerr effect (MOKE) imaging (McCord, 2015). In contrast, a parallel exchange biased (PEB) ME sensor displayed domain development on sensor edge. Upon applying magnetic fields along the sensing direction, high-resolution micrographs revealed a distinct higher stability of the magnetization configuration for the APEB compared to the PEB sensor. As coherent magnetization rotation happened, no signs of domain wall activity were found in the APEB sensor, whereas the PEB sensor displayed an immediate domain wall activity.

With the same modulation field applied, a distinct magnetic noise suppression has been observed in the APEB scheme. As the modulation field amplitude increased, the noise level of the PEB sensor continued to increase, while the APEB sensor remained constant. For the PEB sensor, compared to the direct measurement, an increase in the noise level with the MFC measurement has been observed. In contrast, for the APEB sensor, no difference has been found in the noise level when comparing the direct and MFC measurement. Therefore, it has been proved that by applying the APEB scheme, more than one order of magnitude noise reduction can be achieved. This APEB scheme is also applicable to operation mechanisms based on the ΔE effect and the electrical frequency conversion method.

10.3.2 Magnetoelectric sensor utilizing electric frequency modulation scheme

Although magnetic frequency conversion enables low frequency measurement with resonance-enhanced sensitivities, it still suffers from drawbacks such as size, current consumption, crosstalk between modulation fields in sensor arrays, domain-activity-induced noise by large modulation fields. To overcome those issues, the electrical frequency conversion method has been proposed and investigated. The principle of the EFC is based on the converse ME effect, that is, applying modulation electric field to the piezoelectric layer as the carrier signal, whereby piezomagnetic effect results in periodic change of the magnetization in the magnetostrictive layer (Chu et al., 2019a, 2019b). Consequently, this leads to amplitude modulation by superposition with the magnetic field signals.

In 2016, Hayes et al. demonstrated an electrical frequency conversion method by up-converting low frequency magnetic signals to the EMR of an ME sensor (Hayes et al., 2016). In the study, two different piezoelectric layers are independently prepared on both sides of the Si cantilever. The PZT layer was used for modulation field excitation, while the AlN layer was used for output voltage readout. For the characteristics of piezoelectric displacement, AlN behaved linearly, while PZT showed a quadratic and hysteretic displacement with respect to the applied electric field. Therefore, PZT is used for excitation due to its nonlinear inverse piezoelectric behavior, which is essential for electrical frequency conversion. In the study, an asymmetric modulation technique was used, where the upper sideband peak was set to match the EMR frequency at 689 Hz. For a 20 Hz magnetic field signal, the modulation frequency was set to 669 Hz. In this mode, the 20 Hz 1 mT signal shows a 40 dB SNR, which greatly benefits from the enhancement of the mechanical resonance approximately 30 dB over the lower sideband, leading to an LOD of 10 nT $Hz^{-1/2}$. The results demonstrate the feasibility of the EFC method, while remedying the drawbacks associated with MFC method. Since the excitation as well as the readout was conducted by separate piezoelectric layers, there is no need of a pickup coil and the system size can be effectively reduced. Moreover, without the need of driving current for the modulation coil, the power consumption is much lower, and no stray fields interfere with each other in array configuration. Furthermore, the EFC technique has the advantage of high frequency operation, which would be impossible for the MFC method due to the required large amplitude modulation field.

In 2019, Hayes et al. demonstrated a high-frequency EFC method with a U mode mechanical resonance (Hayes et al., 2019). The high resonance frequency excitation has the advantage of wide bandwidth and low sensitivity to acoustic vibrations. The ME sensor was placed in a pick-up coil, which was parallelly connected to a trimmer capacitor so that its resonance frequency can be tuned to match the excited mechanical resonance. When the sensor oscillates under the excitation at 514.8 kHz with an amplitude of 100 mV, the so-called "U mode" bends symmetrically along the short axis, which induces a high resonance frequency of 514.8 kHz compared to the low-frequency flexural modes. For cantilever type ME sensors oscillating at low frequency flexural modes, the dominant mechanical loss lies in the air damping, thus rarely exceeding a mechanical quality factor Q_m of few hundred. For ME sensors oscillating in U mode, the interaction with air molecules is much less, leading to a lower damping and a higher value of Q_m.

When excited moderately, the U mode ME sensor proves to be very sensitive to weak magnetic fields. Under zero bias magnetic field, the sensitivity shows a linear relationship with the carrier voltage amplitude. However, this linear relation is not valid for the noise floor, particularly for the frequency regime of most interests below 20 Hz. The noise increase around the carrier frequency is much higher than the outer region frequency band. The strong noise in this regime stems from the up-conversion of low-frequency noise and the periodic magnetization processes in the magnetic phase. Furthermore, this noise near carrier demonstrates distinct regimes with regard to the carrier amplitude. For carrier amplitude below about 200 mV, the noise near carrier remains relatively flat between 2 and 4 μV $Hz^{-1/2}$. However, as the modulation voltage amplitude further increases, the noise reaches nearly 26 μV $Hz^{-1/2}$ at 260 mV. Therefore, to avoid strong noise contributions, it is essential to keep the modulation voltage amplitude below 200 mV, where the modulation of the magnetic anisotropy stays effective, while no periodic magnetization sweeping is induced. The output spectra of the carrier amplitude at 514.8 kHz and the symmetric sidebands of the modulated 1 nT AC magnetic field signals at 10 and 23 Hz show a linear sensitivity of 30.4 kV T^{-1} at 10 Hz with respect to different AC magnetic field amplitudes. The LOD gradually decreases from 210 pT $Hz^{-1/2}$ at 2 Hz to 52 pT $Hz^{-1/2}$ at 53 Hz.

10.4 Magnetoelectric sensor utilizing ΔE-effect scheme

The ΔE effect was first used as magnetic sensor by Osiander et al., in 1996 (Osiander et al., 1996). ΔE-effect is the change in the Young's modulus E of a magnetic material with an applied magnetic field which adds additional magnetic field-induced magnetostrictive strain to the elastic Hookean strain (Spetzler et al., 2021). Upon the application of an external magnetic field, the Young's modulus of a ME resonator will be changed due to the ΔE-effect, which leads to a changed electromechanical resonance frequency and impedance. A ΔE effect of up to 50 and 153 GPa for magnetic field-annealed FeCoSiB and FeGaB thin films was reported by Quandt and Sun's group, respectively (Dong et al., 2018; Ludwig & Quandt, 2002). MEMS have received great interest in sensor applications recently as they are promising for integrated magnetic sensors in different applications. Based on the MEMS technology and ΔE effect, a variety of structures such as resonators (Li et al., 2017; Nan et al., 2013), cantilevers (Gojdka et al., 2011; Zabel et al., 2015; Zabel et al., 2016), and delay lines (Kittmann et al., 2018; Schell et al., 2020) have been utilized to explore the performance of integrated ME magnetic sensors.

Self-biased NEMS magnetoelectric sensors based on the ΔE-effect have been demonstrated in an FeGaB/AlN nano-plate resonator with a DC magnetic field detection limit down to 300 pT (Nan et al., 2013). The self-biased 215 MHz Magnetoelectric MEMS magnetic sensor was based on a FeGaB/Al_2O_3 nano-plate resonator with well-designed interdigital electrodes (Nan et al., 2013). By utilizing the ME MEMS resonator with a high-quality factor of 735, whose admittance is very sensitive to DC magnetic fields at the EMR, a LOD for DC magnetic fields of ~300 pT was achieved in the ultrasensitive RF ME sensor. While a DC magnetic field is applied to the ME heterostructure, a magnetostrictive strain is generated in the FeGaB film, which results in the change of FeGaB film's Young's modulus due to ΔE effect. Therefore, the EMR frequency and admittance value are varied under a DC magnetic field because of the altered equivalent Young's modulus of the MEMS ME resonator. The microscale MEMS magnetic sensor based on magnetostrictive/piezoelectric heterostructure is energy-efficient, self-biased, and CMOS fabrication compatible, which makes it promising for future applications of DC and low frequency AC magnetic sensors.

As shown in Fig. 10.2a, the DC magnetic fields generated by a pair of Helmholtz coils are applied along the length direction of the ME MEMS magnetic sensor. Fig. 10.2b shows the details of the layered structure of the magnetic sensor, where the piezoelectric AlN layer is sandwiched between the magnetostrictive FeGaB/Al_2O_3 multilayers and the Pt interdigital electrodes (IDEs). The sensing part with area of 100 μm (W) × 200 μm (L) is fully covered by IDEs and magnetostrictive material as displayed by the SEM image in Fig. 10.2c. An extensional vibration mode was excited in AlN film and the EMR frequency is expressed by

$$f_0 = \frac{1}{2W_0}\sqrt{\frac{E_{eq}}{\rho_{eq}}}$$

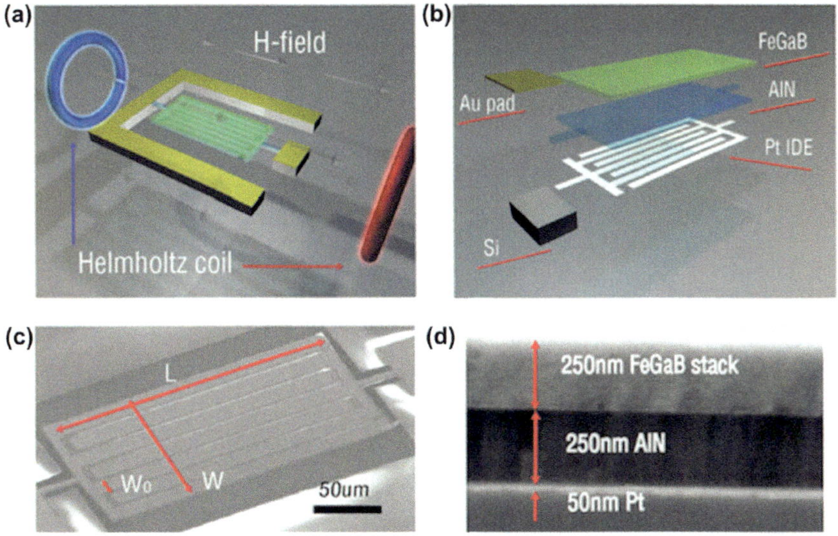

FIGURE 10.2 Self-biased NEMS magnetoelectric sensors based on the ΔE-effect. (a) Schematic and (b) the layered structure of the ME sensor. Scanning electron microscopy images of the (c) fabricated MEMS ME sensor and the (d) cross-section. *Reproduced with permission from (Nan et al., 2013).*

where, W_0 represents the pitch width between IDEs, E_{eq}, and r_{eq} are the equivalent Young's modulus and density of the device, which can be calculated as $E_{eq} = \sum E_i v_i$ and $\rho_{eq} = \sum \rho_i v_i$, where E_i, r_i, and v_i are the Young's modulus, density, and volume ratio for each layer in the ME sensor, respectively. A very high EMR frequency of 215 MHz of the MEMS magnetic field was induced by setting the pitch width W_0 of the Pt IDEs. By completely removing the Si substrate underneath the MEMS resonator, the substrate clamping effect is diminished, therefore, a strong ME coupling and high sensitivity can also be accomplished.

The experimental data and Butterworth–van Dyke (BVD) model fitting for admittance curve of the miniaturized ME magnetic sensor is depicted in Fig. 10.3a, which shows an EMR frequency of 215 MHz. The Q_m and electromechanical coupling coefficient (k_t^2) at zero bias magnetic field in air are extracted from the BVD model fitting as 735% and 1.54%, respectively. The value of k_t^2 is comparable to what was reported in conventional AlN NPR with the same electrodes configuration (Rinaldi et al., 2011), while the Q_m is much larger than what is typically obtained in ME magnetic sensors that operate at low frequencies (Marauska et al., 2012). Fig. 10.3b shows the equivalent BVD electrical circuit of the AlN/FeGaB based MEMS magnetic sensor, where, R_s is the resistance of metal electrodes, C_0 is the resonator capacitance of the ME heterostructure, R_{op}

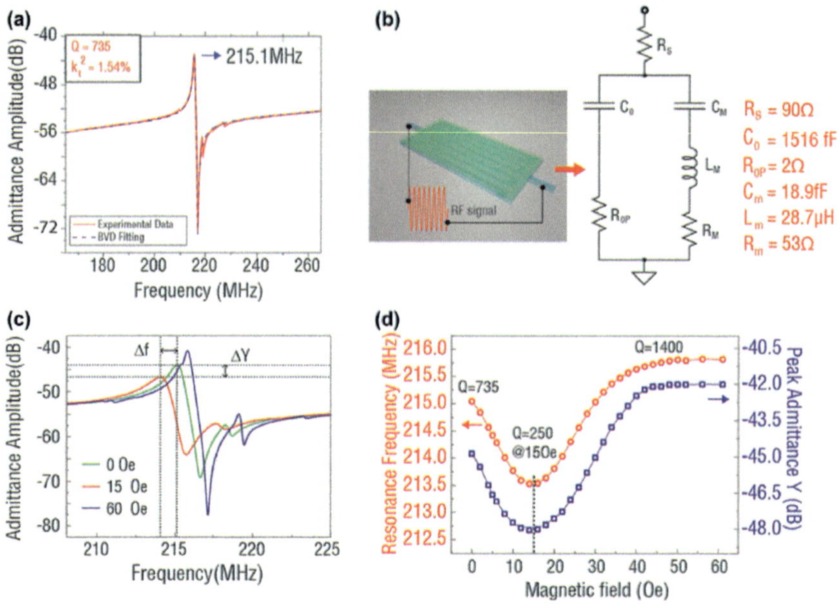

FIGURE 10.3 Electromechanical and magnetic sensing performance of self-biased NEMS magnetoelectric sensors based on the ΔE-effect. (a) Experimental data and BVD model fitting for admittance curve of the ME sensor. (b) The equivalent BVD circuit of the MEMS resonator. (c) Admittance curves of the MEMS resonator under different DC bias magnetic fields. (d) EMR frequency and peak admittance amplitude at the resonance frequency versus DC magnetic fields. *Reproduced with permission from (Nan et al., 2013).*

represents the parasitic resistance from substrate, C_M, L_M, and R_M denote the motional capacitance, inductance, and resistance, respectively. A strong ME coupling was demonstrated in the admittance curves at various DC bias magnetic fields of the magnetic sensor as shown in Fig. 10.3c, where both EMR frequency and admittance amplitude are changed due to ΔE effect. The values of EMR frequency and peak admittance at the resonance frequency were measured and exhibited in Fig. 10.3d, in which both curves first decreased and then increased with increasing DC bias field. The Q_m extracted from BVD model fitting also followed this trend that the minimum value of 250 was achieved at the transition field of 15 Oe, which can be attributed to the magnetic loss of domain wall activities.

Despite the well-known phenomenon as the ΔE effect in applications of ME magnetic sensor, changes in other elastic moduli such as the shear modulus G are normally more prominent (Sárközi et al., 2000). Depending on the waves excited in the piezoelectric layer, different kinds of deformation and velocity involving various elastic moduli are generated. Rayleigh waves, which propagate at the surface with out-of-plane shear and longitudinal deformation, are magnetically sensitive to the changes of Young's modulus E. The propagation of horizontal shear waves that are only sensitive to the alterations of shear modulus G is parallel to the magnetostrictive film. However, Love waves, which are concentrated at the surface with horizontal shear deformation by a phase velocity gradient toward the surface, incorporate the advantages of both types of waves (Jakoby & Vellekoop, 1997). Kittmann et al. designed a novel Love wave-type surface acoustic wave (SAW) magnetic field sensor with a dedicated guiding layer and finite element modeling simulations (Kittmann et al., 2018). The Love wave SAW sensor, which operates at around 150 MHz in a delay line configuration, measures the time delay and associated phase shift and translates them to the sensed magnetic field. Love wave sensors based on FeCoSiB have demonstrated a magnetic noise of $\approx 100\ \mathrm{pT}/\sqrt{\mathrm{Hz}}$ over a 50 kHz bandwidth and a dynamic range of 120 dB (Kittmann et al., 2018). A magnetic noise of $28\ \mathrm{pT}/\sqrt{\mathrm{Hz}}$ at 10 Hz was demonstrated for a ΔE-effect Love wave surface acoustic wave sensor based on an exchange biased stack of FeCoSiB and $Ni_{81}Fe_{19}$ coupled to an antiferromagnetic (AFM) $Mn_{80}Ir_{20}$ layer. The exchange biased stack with FeCoSiB coupled to an AFM layer was designed to eliminate domain walls and therefore the magnetic $1/f$ noise was reduced by 8 dB compared to a comparable control device without an exchange bias material stack (Schell et al., 2023).

The proposed Love wave SAW sensor is designed to use ST-cut quartz as the substrate. A guiding layer of SiO_2 is utilized to convert the acoustic wave energy within the magnetostrictive layer. Two IDEs are used to excite the shear acoustic waves and form a delay line structure. The magnetostrictive thin film FeCoSiB is deposited between the two delay lines to sense magnetic fields. The total sensitivity of the SAW sensor is composed of three major contributions

$$S = \frac{\partial \varphi}{\partial H} = S_{mag} \cdot S_{str} \cdot S_{geo} = \frac{\partial G}{\partial H} \cdot \frac{\partial v}{\partial G} \cdot \frac{\partial \varphi}{\partial v}$$

where, magnetic sensitivity S_{mag} is introduced to express the change of shear modulus G under an external magnetic field H; structural sensitivity S_{str} represents the influence of G on the surface wave velocity v; geometric sensitivity S_{geo} shows the phase shift ϕ influenced by the wave velocity. The noise level at 10 and 100 Hz are 250 and $80\ \mathrm{pT\ Hz}^{-1/2}$, respectively. It is obvious that the noise floor is degraded by $1/f$ noise in the low-frequency range.

Due to the large influence and difficulty of the deposition for soft magnetic thin films with high magnetostriction, Schell et al. investigated the effect of magnetic anisotropy controlled magnetostrictive FeCoSiB thin films on SAW magnetic field sensors (Schell et al., 2020). The uniaxial stresses resulting from thermally induced anisotropic expansion during the deposition or magnetic annealing processes make it hard to precisely control the magnetic anisotropy of FeCoSiB thin films. By analyzing the impact of the deposition process and heat treatment, Schell et al. implemented a low-temperature deposition under an in-situ magnetic field and improved the LOD down to 70 and 25 pT $Hz^{-1/2}$ at 10 and 100 Hz, respectively (Schell et al., 2020).

10.5 Outlook of magnetoelectric sensor

In the past 10 years, thin film ME sensors have attracted great interest as a promising candidate for biomagnetic sensing. In order to improve their sensitivity, extensive efforts have been made to develop new materials, structures, fabrication processes, detection schemes, and to build up theories on noise analysis. With the employment of the frequency conversion method and the ΔE effect, thin film ME sensors have achieved a detection limit to tens of pT $Hz^{-1/2}$ in the low frequency regime of interest.

However, there is still a gap of sensing capability between thin film ME sensors and state-of-the-art magnetometers like SQUID and optically pumped magnetometers (OPM) (Elzenheimer et al., 2020). To further improve the performance of ME sensors, soft magnetostrictive thin films with higher piezomagnetic coefficient are essential to be developed. Moreover, new structures/vibration modes for ME resonators working at higher frequencies with high Q_m can effectively increase the SNR. To suppress the intrinsic noise of ME sensors, the magnetic domain activity should be carefully controlled through the engineering of magnetic layers and the tuning of excitation fields.

In terms of the applications for biomagnetic measurement, thin film ME sensors provide a low profile, high spatial resolution, contactless solution over the electrode-based measurements, but they are currently not ready for real life clinical diagnosis. The feasibility of magnetocardiography (MCG) (Reermann et al., 2018) and magnetic particle mapping (MPM) (Friedrich et al., 2019) using ME sensors have been successfully demonstrated and the results are promising, while magnetoneurography (MNG) (Hughes et al., 1981; Mackert, 2004; Nan et al., 2017; Zaeimbashi et al., 2019; Zaeimbashi et al., 2021) and magnetomyography (MMG) (Zuo et al., 2020b) have proved to be feasible but still in the stage of a practical plan. To apply ME sensors for in vivo and in vitro test on chip, magnetic shielding and active magnetic field cancellation are needed to replace the bulky and expensive shielding room. Furthermore, the cross sensitivity to acoustic noise can be reduced by the application of gradiometry configurations such as the tuning fork structure (Salzer et al., 2016a). From the signal processing perspective, advanced data analysis and algorithms are highly demanded for rejecting non-stationary noise burst. In the future, the integration and employment of ME sensor arrays can greatly reduce the screening time and increase reliability.

Overall, thin film ME sensors offer high sensitivity, passive detection scheme, low cost, and room temperature operation. While much work remains to be accomplished toward clinical applications, it is anticipated that thin film ME sensors with improved performance would be promising for next generation superior magnetic sensing for biomedical applications. ME magnetometers enable the detection of magnetic fields at levels of approximately 5 $fT/\sqrt{Hz}$ at 24 kHz (Dong et al., 2025); the goal, however, is to extend this sensitivity to a broader frequency range, spanning from 0.01 Hz to 1 kHz, while achieving detection thresholds in the order of femtotesla. To achieve this, novel ME sensor designs and new schemes for noise cancellation need to be explored.

References

Carman, G. P., & Sun, N. (2018). Strain-mediated magnetoelectrics: Turning science fiction into reality. *MRS Bulletin, 43*, 822–828.

Chu, Z., Dong, C., Tu, C., He, Y., Liang, X., Wang, J., Wei, Y., Chen, H., Gao, X., Lu, C., & Zhu, Z. (2019a). Voltage-driven nonlinearity in magnetoelectric heterostructures. *Physical Review Applied, 12*, 044001.

Chu, Z., Dong, C., Tu, C., Liang, X., Chen, H., Sun, C., Yu, Z., Dong, S., & Sun, N. X. (2019b). A low-power and high-sensitivity magnetic field sensor based on converse magnetoelectric effect. *Applied Physics Letters, 115*.

Chu, Z., Shi, H., Shi, W., Liu, G., Wu, J., Yang, J., & Dong, S. (2017). Enhanced resonance magnetoelectric coupling in (1-1) connectivity composites. *Advanced Materials, 29*, 1606022.

Cohen, D., & Givler, E. (2003). Magnetomyography: Magnetic fields around the human body produced by skeletal muscles. *Applied Physics Letters, 21*, 114–116.

Dong, S. X., Cheng, J. R., Li, J. F., & Viehland, D. (2003). Enhanced magnetoelectric effects in laminate composites of Terfenol-D/$Pb(Zr,Ti)O_3$ under resonant drive. *Applied Physics Letters, 83*, 4812–4814.

Dong, C., Liang, X., Gao, J., Chen, H., He, Y., Wei, Y., Zaeimbashi, M., Matyushov, A., Sun, C., & Sun, N. X. (2022). Thin film magnetoelectric sensors toward biomagnetism: Materials, devices, and applications. *Advanced Electronic Materials, 8*, 2200013.

Dong, C., Li, M., Liang, X., Chen, H., Zhou, H., Wang, X., Gao, Y., McConney, M. E., Jones, J. G., Brown, G. J., Howe, B. M., & Sun, N. X. (2018). Characterization of magnetomechanical properties in FeGaB thin films. *Applied Physics Letters, 113*, 262401.

Dong, S., Li, J.-F., & Viehland, D. (2003). Ultrahigh magnetic field sensitivity in laminates of TERFENOL-D and Pb $(Mg_{1/3}\ Nb_{2/3})\ O_3$–$PbTiO_3$ crystals. *Applied Physics Letters, 83*, 2265–2267.

Dong, C., Sun, C., Chen, L., He, Y., Liu, Y., Luo, B., & Sun, N. X. (2025). Ultra-compact magnetoelectric sensor for Femto-Tesla VLF signal reception. *Materials Today Electronics*, 100135 SSRN 5019077.

Dong, S., Zhai, J., Bai, F., Li, J. F., & Viehland, D. (2005a). Push-pull mode magnetostrictive/piezoelectric laminate composite with an enhanced magnetoelectric voltage coefficient. *Applied Physics Letters, 87*.

Dong, S., Zhai, J., Wang, N., Bai, F., Li, J., Viehland, D., & Lograsso, T. A. (2005b). Fe–Ga/Pb ($Mg_{1/3}Nb_{2/3}$) O_3–$PbTiO_3$ magnetoelectric laminate composites. *Applied Physics Letters, 87*.

Dong, S. X., Zhai, J. Y., Li, J. F., & Viehland, D. (2006). Near-ideal magnetoelectricity in high-permeability magnetostrictive/piezofiber laminates with a (2-1) connectivity. *Applied Physics Letters, 89*(25).

Durdaut, P., Salzer, S., Reermann, J., Röbisch, V., McCord, J., Meyners, D., Quandt, E., Schmidt, G., Knochel, R., & Höft, M. (2017). Improved magnetic frequency conversion approach for magnetoelectric sensors. *IEEE Sensors Letters, 1*(3), 1–4.

Elzenheimer, E., Laufs, H., Sander-Thömmes, T., & Schmidt, G. (2018). Magnetoneurograhy of an electrically stimulated arm nerve. *Usability of Magnetoelectric (ME) Sensors for Magnetic Measurements of Peripheral Arm Nerves, 4*, 363–366.

Elzenheimer, E., Laufs, H., Schulte-Mattler, W., & Schmidt, G. (2020). Magnetic measurement of electrically evoked muscle responses with optically pumped magnetometers. *IEEE Transactions on Neural Systems and Rehabilitation Engineering, 28*, 756–765.

Friedrich, R.-M., Zabel, S., Galka, A., Lukat, N., Wagner, J.-M., Kirchhof, C., Quandt, E., McCord, J., Selhuber-Unkel, C., Siniatchkin, M., & Faupel, F. (2019). Magnetic particle mapping using magnetoelectric sensors as an imaging modality. *Scientific Reports, 9*, 2086.

Gillette, S. M., Geiler, A. L., Gray, D., Viehland, D., Vittoria, C., & Harris, V. G. (2011). Improved sensitivity and noise in magneto-electric magnetic field sensors by use of modulated AC magnetostriction. *IEEE Magnetics Letters, 2*, 2500104.

Gojdka, B., Jahns, R., Meurisch, K., Greve, H., Adelung, R., Quandt, E., Knöchel, R., & Faupel, F. (2011). Fully integrable magnetic field sensor based on delta-E effect. *Applied Physics Letters, 99*(22).

Granata, C., Vettoliere, A., & Russo, M. (2011). An ultralow noise current amplifier based on superconducting quantum interference device for high sensitivity applications. *Review of Scientific Instruments, 82*.

Greve, H., Woltermann, E., Quenzer, H.-J., Wagner, B., & Quandt, E. (2010). Giant magneto-electric coefficients in (Fe90Co10) 78Si12B10-AlN thin film composites. *Applied Physics Letters, 96*.

Hayes, P., Jovičević Klug, M., Toxværd, S., Durdaut, P., Schell, V., Teplyuk, A., Burdin, D., Winkler, A., Weser, R., Fetisov, Y., Höft, M., Knöchel, R., McCord, J., & Quandt, E. (2019). Converse magnetoelectric composite resonator for sensing small magnetic fields. *Scientific Reports, 9*, 16355.

Hayes, P., Salzer, S., Reermann, J., Yarar, E., Röbisch, V., Piorra, A., Meyners, D., Höft, M., Knöchel, R., Schmidt, G., & Quandt, E. (2016). Electrically modulated magnetoelectric sensors. *Applied Physics Letters, 108*(18).

Hrkac, V., Lage, E., Köppel, G., Strobel, J., McCord, J., Quandt, E., Meyners, D., & Kienle, L. (2014). Amorphous FeCoSiB for exchange bias coupled and decoupled magnetoelectric multilayer systems: Real-structure and magnetic properties. *Journal of Applied Physics, 116*(13).

Hughes, G. B., Josey, A. F., Glasscock III, M. E., Jackson, C. G., Ray, W. A., & Sismanis, A. (1981). Clinical electroneurography: Statistical analysis of controlled measurements in twenty-two normal subjects. *The Laryngoscope, 91*, 1834–1846.

Jahns, R., Greve, H., Woltermann, E., Quandt, E., & Knochel, R. H. (2011). Noise performance of magnetometers with resonant thin-film magnetoelectric sensors. *IEEE Transactions on Instrumentation and Measurement, 60*, 2995–3001.

Jahns, R., Greve, H., Woltermann, E., Quandt, E., & Knochel, R. (2012). Sensitivity enhancement of magnetoelectric sensors through frequency-conversion. *Sensors and Actuators A: Physical, 183*, 16–21.

Jakoby, B., & Vellekoop, M. J. (1997). Properties of Love waves: Applications in sensors. *Smart Materials and Structures, 6*, 668.

Jovičević Klug, M., Thormählen, L., Röbisch, V., Toxværd, S. D., Höft, M., Knöchel, R., Quandt, E., Meyners, D., & McCord, J. (2019). Antiparallel exchange biased multilayers for low magnetic noise magnetic field sensors. *Applied Physics Letters, 114*(19).

Kirchhof, C., Krantz, M., Teliban, I., Jahns, R., Marauska, S., Wagner, B., Knöchel, R., Gerken, M., Meyners, D., & Quandt, E. (2013). Giant magnetoelectric effect in vacuum. *Applied Physics Letters, 102*(23).

Kittmann, A., Durdaut, P., Zabel, S., Reermann, J., Schmalz, J., Spetzler, B., Meyners, D., Sun, N. X., McCord, J., Gerken, M., Schmidt, G., Höft, M., Knöchel, R., Faupel, F., & Quandt, E. (2018). Wide band low noise Love wave magnetic field sensor system. *Scientific Reports, 8.*

Lage, E., Kirchhof, C., Hrkac, V., Kienle, L., Jahns, R., Knöchel, R., Quandt, E., & Quandt, D. (2012). Exchange biasing of magnetoelectric composites. *Nature Materials, 11*, 523–529.

Lage, E., Woltering, F., Quandt, E., & Meyners, D. (2013). Exchange biased magnetoelectric composites for vector field magnetometers. *Journal of Applied Physics, 113.*

Li, M., Wang, Y., Shen, Y., Gao, J., Li, J., & Viehland, D. (2013). Structural dependence of nonlinear magnetoelectric effect for magnetic field detection by frequency modulation. *Journal of Applied Physics, 114.*

Li, M., Matyushov, A., Dong, C., Chen, H., Lin, H., Nan, T., Qian, Z., Rinaldi, M., Lin, Y., & Sun, N. X. (2017). Ultra-sensitive NEMS magnetoelectric sensor for picotesla DC magnetic field detection. *Applied Physics Letters, 110.* https://doi.org/10.1063/1.4979694

Li, M., Berry, D., Das, J., Gray, D., Li, J., & Viehland, D. (2011). Enhanced sensitivity and reduced noise floor in magnetoelectric laminate sensors by an improved lamination process. *Journal of the American Ceramic Society, 94*, 3738–3741.

Ludwig, A., & Quandt, E. (2002). Optimization of the/spl Delta/E effect in thin films and multilayers by magnetic field annealing. *IEEE Transactions on Magnetics, 38*, 2829–2831.

Luo, B., Will-Cole, A. R., Dong, C., He, Y., Liu, X., Lin, H., Huang, R., Shi, X., McConney, M., Page, M., Sanghadasa, M., Ramesh, R., & Sun, N. X. (2024). Magnetoelectric microelectromechanical and nanoelectromechanical systems for the IoT. *Nature Reviews Electrical Engineering, 1*, 317–334.

Mackert, B.-M. (2004). Magnetoneurography: Theory and application to peripheral nerve disorders. *Clinical Neurophysiology, 115*, 2667–2676.

Marauska, S., Jahns, R., Greve, H., Quandt, E., Knöchel, R., & Wagner, B. (2012). MEMS magnetic field sensor based on magnetoelectric composites. *Journal of Micromechanics and Microengineering, 22*, 065024.

McCord, J. (2015). Progress in magnetic domain observation by advanced magneto-optical microscopy. *Journal of Physics D: Applied Physics, 48*, 333001.

Mori, K., & Wuttig, M. (2002). Magnetoelectric coupling in Terfenol-D/polyvinylidenedifluoride composites. *Applied Physics Letters, 81*, 100–101.

Nan, T. X., Hui, Y., Rinaldi, M., & Sun, N. X. (2013). Self-biased 215MHz magnetoelectric NEMS resonator for ultra-sensitive DC magnetic field detection. *Scientific Reports, 3*, 1985. https://doi.org/10.1038/srep01985

Nan, T. X., Lin, H., Gao, Y., Matyushov, A., Yu, G., Chen, H., Sun, N., Wei, S., Wang, Z., Li, M., Wang, X., Belkessam, A., Guo, R., Chen, B., Zhou, J., Qian, Z., Hui, Y., Rinaldi, M., McConney, M. E., Howe, B. M., Hu, Z., Jones, J. G., & Sun, N. X. (2017). Acoustically actuated ultra-compact NEMS magnetoelectric antennas. *Nature Communications, 8*, Article 296.

Osiander, R., Ecelberger, S. A., Givens, R. B., Wickenden, D. K., Murphy, J. C., & Kistenmacher, T. J. (1996). A microelectromechanical-based magnetostrictive magneto meter. *Applied Physics Letters, 69*(19), 2930–2931.

Petrie, J., Gray, D., Viehland, D., Sreenivasulu, G., Srinivasan, G., Mandal, S., & Edelstein, A. S. (2012). Shifting the operating frequency of magnetoelectric sensors. *Journal of Applied Physics, 111*(7).

Piorra, A., Jahns, R., Teliban, I., Gugat, J. L., Gerken, M., Knöchel, R., & Quandt, E. (2013). Magnetoelectric thin film composites with interdigital electrodes. *Applied Physics Letters, 103*(3).

Reermann, J., Durdaut, P., Salzer, S., Demming, T., Piorra, A., Quandt, E., Frey, N., Höft, M., & Schmidt, G. (2018). Evaluation of magnetoelectric sensor systems for cardiological applications. *Measurement, 116*, 230–238.

Rinaldi, M., Zuo, C., Van der Spiegel, J., & Piazza, G. (2011). Reconfigurable CMOS oscillator based on multifrequency AlN contour-mode MEMS resonators. *IEEE Transactions on Electron Devices, 58*, 1281–1286.

Röbisch, V., Yarar, E., Urs, N. O., Teliban, I., Knöchel, R., McCord, J., Quandt, E., & Meyners, D. (2015). Exchange biased magnetoelectric composites for magnetic field sensor application by frequency conversion. *Journal of Applied Physics, 117*(17).

Salzer, S., Jahns, R., Piorra, A., Teliban, I., Reermann, J., Höft, M., Quandt, E., & Knöchel, R. (2016a). Tuning fork for noise suppression in magnetoelectric sensors. *Sensors and Actuators A: Physical, 237*, 91–95.

Salzer, S., Durdaut, P., Röbisch, V., Meyners, D., Quandt, E., Höft, M., & Knöchel, R. (2016b). Generalized magnetic frequency conversion for thin-film laminate magnetoelectric sensors. *IEEE Sensors Journal, 17*, 1373–1383.

Salzer, S., Röbisch, V., Klug, M., Durdaut, P., McCord, J., Meyners, D., Reermann, J., Höft, M., & Knöchel, R. (2018). Noise limits in thin-film magnetoelectric sensors with magnetic frequency conversion. *IEEE Sensors Journal, 18*, 596–604.

Sárközi, Z., Mackay, K., & Peuzin, J. (2000). Elastic properties of magnetostrictive thin films using bending and torsion resonances of a bimorph. *Journal of Applied Physics, 88*, 5827–5832.

Schell, V., Müller, C., Durdaut, P., Kittmann, A., Thormählen, L., Lofink, F., Meyners, D., Höft, M., McCord, J., & Quandt, E. (2020). Magnetic anisotropy controlled FeCoSiB thin films for surface acoustic wave magnetic field sensors. *Applied Physics Letters, 116*(7).

Schell, V., Spetzler, E., Wolff, N., Bumke, L., Kienle, L., McCord, J., Quandt, E., & Meyners, D. (2023). Exchange biased surface acoustic wave magnetic field sensors. *Scientific Reports, 13*, 8446.

Shen, L. G., Li, M., Gao, J., Shen, Y., Li, J. F., Viehland, D., Zhuang, X., Lam Chok Sing, M., Cordier, C., Saez, S., & Dolabdjian, C. (2011). Magnetoelectric nonlinearity in magnetoelectric laminate sensors. *Journal of Applied Physics, 110*(11).

Shen, Y., McLaughlin, K. L., Gao, J., Li, M., Li, J., & Viehland, D. (2013). Effective optimization of magnetic noise for a Metglas/Pb(Zr,Ti)O-3 magnetoelectric sensor array in an open environment. *Materials Letters, 91*, 307–310.

Spetzler, B., Golubeva, E. V., Friedrich, R. M., Zabel, S., Kirchhof, C., Meyners, D., McCord, J., & Faupel, F. (2021). Magnetoelastic coupling and delta-E effect in magnetoelectric torsion mode resonators. *Sensors, 21*, 2022.

Uchikawa, Y., & Kotani, M. (1992). Measurement of magnetic field produced from the human body. *IEEE Translation Journal on Magnetics in Japan, 7*, 600–607.

Viehland, D., Wuttig, M., Mccord, J., & Quandt, E. (2018). Magnetoelectric magnetic field sensors. *MRS Bulletin, 43*, 834–840.

Wang, Y. J., Gao, J. Q., Li, M. H., Shen, Y., Hasanyan, D., Li, J. F., & Viehland, D. (2014a). A review on equivalent magnetic noise of magnetoelectric laminate sensors. *Philosophical Transactions of the Royal Society A: Mathematical, Physical and Engineering Sciences, 372*.

Wang, Y., Li, J., & Viehland, D. (2014b). Magnetoelectrics for magnetic sensor applications: Status, challenges and perspectives. *Materials Today, 17*, 269–275.

Wang, Y., Gao, J., Li, M., Hasanyan, D., Shen, Y., Li, J., Viehland, D., & Luo, H. (2012). Ultra-low equivalent magnetic noise in a magnetoelectric Metglas/Mn-doped Pb (Mg1/3Nb2/3) O3-PbTiO3 heterostructure. *Applied Physics Letters, 101*(2).

Wang, Y., Gray, D., Berry, D., Gao, J., Li, M., Li, J., & Vieland, D. (2011). An extremely low equivalent magnetic noise magnetoelectric sensor. *Advanced Materials, 23*, 4111.

Yarar, E., Salzer, S., Hrkac, V., Piorra, A., Höft, M., Knöchel, R., Kienle, L., & Quandt, E. (2016). Inverse bilayer magnetoelectric thin film sensor. *Applied Physics Letters, 109*(2).

Zabel, S., Kirchhof, C., Yarar, E., Meyners, D., Quandt, E., & Faupel, F. (2015). Phase modulated magnetoelectric delta-E effect sensor for sub-nano tesla magnetic fields. *Applied Physics Letters, 107*(15).

Zabel, S., Reermann, J., Fichtner, S., Kirchhof, C., Quandt, E., Wagner, B., Schmidt, G., & Faupel, F. (2016). Multimode delta-E effect magnetic field sensors with adapted electrodes. *Applied Physics Letters, 108*(22).

Zaeimbashi, M., Lin, H., Dong, C., Liang, X., Nasrollahpour, M., Chen, H., Sun, N., Matyushov, A., He, Y., Wang, X., Tu, C., Wei, Y., Zhang, Y., Cash, S. S., Onabajo, M., Shrivastava, A., & Sun, N. (2019). NanoNeuroRFID: A wireless implantable device based on magnetoelectric antennas. *IEEE Journal of Electromagnetics, RF and Microwaves in Medicine and Biology, 3*, 206–215.

Zaeimbashi, M., Nasrollahpour, M., Khalifa, A., Romano, A., Liang, X., Chen, H., Sun, N., Matyushov, A., Lin, H., Dong, C., Xu, Z., Mittal, A., Martos-Repath, I., Jha, G., Mirchandani, N., Das, D., Onabajo, M., Shrivastava, A., Cash, S., & Sun, N. X. (2021). Ultra-compact dual-band smart NEMS magnetoelectric antennas for simultaneous wireless energy harvesting and magnetic field sensing. *Nature Communications, 12*, Article 3141.

Zhai, J., Dong, S., Xing, Z., Li, J., & Viehland, D. (2006). Giant magnetoelectric effect in Metglas/polyvinylidene-fluoride laminates. *Applied Physics Letters, 89*.

Zhai, J., Xing, Z., Dong, S., Li, J., & Viehland, D. (2008). Magnetoelectric laminate composites: An overview. *Journal of the American Ceramic Society, 91*, 351–358.

Zhuang, X., Sing, M. L. C., Cordier, C., Saez, S., Dolabdjian, C., Shen, L., Li, J. F., Li, M., & Viehland, D. (2011). Evaluation of applied axial field modulation technique on ME sensor input equivalent magnetic noise rejection. *IEEE Sensors Journal, 11*, 2266–2272.

Zuo, S., Heidari, H., Farina, D., & Nazarpour, K. (2020a). Miniaturized magnetic sensors for implantable magnetomyography. *Advanced Materials Technologies, 5*, 2000185.

Zuo, S., Schmalz, J., Özden, M.Ö., Gerken, M., Su, J., Niekiel, F., Lofink, F., Nazarpour, K., & Heidari, H. (2020b). Ultrasensitive magnetoelectric sensing system for pico-tesla magnetomyography. *IEEE Transactions on Biomedical Circuits and Systems, 14*, 971–984.

CHAPTER

11

Magnetoelectric antennas for VLF

Yisi Liu, Cunzheng Dong, Yifan He, and Nian Sun

Northeastern University, Boston, MA, United States

11.1 Introduction

As a significant component of wireless communication systems, antennas are ubiquitous in our daily life. However, a critical challenge of the conventional antennas driven by electrical current or voltage is miniaturization. For conventional antennas, the dimension should be at least 1/10th of the free space wavelength of electromagnetic (EM) waves, to maintain good radiation efficiency and enhance the directivity (Chen et al., 2020; Volakis et al., 2009). This makes compact antenna designs difficult, as they require small antenna volumes and high efficiency simultaneously. The magnetoelectric (ME) antenna, utilizing mechanical strain as an intermediary (Viehland et al., 2018), offers a new approach to address this issue through impressive miniaturization, low power consumption, and high theoretical radiation efficiency (Domann & Carman, 2017; Yao et al., 2015). The practical ME antenna, based on the suspended acoustic resonator, was first designed by Nan et al. for very-high-frequency (VHF, 30–300 MHz) band and ultrahigh frequency (UHF, 0.3–3 GHz) band in 2017 (Nan et al., 2017). In addition, the ME antenna based on the bulk surface acoustic wave (SAW) resonator was demonstrated by Dong et al. for very-low-frequency (VLF, 3–30 kHz) band in 2020 (Dong et al., 2020). These ME antennas utilize the acoustic resonance inside the resonator, the wavelength of which five orders of magnitude smaller than the wavelength of the leading to a significantly smaller volume compared to conventional antennas (Dong et al., 2020)

Magnetoelectric Composites
https://doi.org/10.1016/B978-0-443-18606-6.00001-0

11.2 Background of very low frequency antenna

In these Radio wave bands, since the free-space wavelength of VLF (very low frequency, 3–30 kHz) ranges from 10 to 100 km, physically huge antennas are required to achieve acceptable radiation efficiency. VLF communication is widely used in ionosphere guided communication, submarine, and underground communication due to its low signal attenuation versus distance and the capability of obstruction penetration brought by the large wavelength. Conventional electromagnet-driven antennas utilize the oscillating charge to generate electric and magnetic field radiation (Miron, 2006). Their sizes are typically comparable to the wavelength corresponding to the operating frequency of the antennas. Since the wavelength of VLF is in the range of 10–100 km, the conventional VLF antenna would be enormous and unsuitable for portable device, additionally requiring large amounts of feeding power to function properly. To tackle these issues, new mechanism must be proposed to achieve antenna miniaturization.

Mechanical antennas, a newly emerging category of compact low frequency antenna, can be divided into motor-driven antenna and acoustically driven antenna. Motor-driven antennas utilize motors to drive permanent magnets or electret materials pieces (Fawole & Tabib-Azar, 2017; Liu et al. 2020; Wang et al., 2021). The magnetic dipole or electric charge contained in the materials is physically driven to motion to generate EM radiation in space. However, the performance of motor-driven antennas is highly constrained by the motor's performance in terms of rotation speed and the ability to switch fast between two operating states (Barani et al., 2021). Research efforts have focused on modulation schemes to overcome this constraint.

Acoustically driven antennas typically are piezoelectric/electrostrictive material-based mechanical resonator. The coupling between the elastic wave and the electromagnetic wave induces volume polarization charge and surface polarization charge, which can be approximated as an electric dipole for radiation analysis (Yang et al., 2018). Unlike conventional antennas, acoustically driven antennas are actuated at their electromechanical resonance (EMR) rather than the EM wave resonance. Since the speed of sound is much slower than the EM wave, at same frequency, the wavelength of acoustic wave is much smaller than the EM wave, which makes the acoustically driven antennas one–two order smaller than their conventional counterpart.

ME antenna, a category of acoustically driven antenna, contains a piezoelectric/electrostrictive material phase and a high permeability magnetostrictive material phase. The ME antenna can work in both receiving and transmitting mode taking advantage of direct ME effect

(DME) and converse ME effect (CME) respectively. In receiving mode, the magnetostrictive material generates strain due to the external magnetic field signal. This strain is coupled to the piezoelectric/electrostrictive material phase and converted to electric voltage output through piezoelectric/electrostrictive effect. In transmitting mode, externally applied electric excitation is converted to the strain in piezoelectric/electrostrictive material phase. This strain is coupled to the magnetostrictive material phase, inducing the magnetization oscillation due to the free energy minimization, thereby generating a magnetic field signal in space.

To quantify the strength of ME effect, the direct and converse ME coefficient are defined as (Liang et al., 2020):

$$\alpha_{\text{Direct}} = \frac{\partial P}{\partial H} \tag{11.1}$$

$$\alpha_{\text{converse}} = \frac{\partial M}{\partial E} \tag{11.2}$$

where, P is the electric polarization, H is the applied magnetic field, M is the magnetization and E is the applied electric field. To better describe the ME coupling condition in designing practical applications or analyzing the experimental results, the magnetically induced voltage ME coefficient is defined and widely used (Vopson et al., 2017):

$$\alpha_{\text{ME}} = \frac{\partial E}{\partial H} \tag{11.3}$$

where, H is the applied magnetic field and E is the induced electric field.

Due to the advantages of ME antenna in respect of compact size, low power consumption, impedance matching (Kemp et al., 2019) and cost, they have been intensively studied during the most recent years since the early work in 2020 by Dong et al. (2020). Research on ME antenna has primarily focused on antenna modeling (Luong & Wang, 2022; Xu et al., 2021), performance enhancement though structure or fabrication optimization (Deng et al., 2021; Lasheras et al., 2021), broadband operation (Yun et al., 2022; Xiao et al., 2023; Xu et al., 2023), and specific applications such as underwater communication (Dong et al., 2020; Du et al., 2023) and bio-implant (Chen et al., 2022; Hayes et al., 2019; Zaeimbashi et al., 2021).

11.3 Acoustically actuated VLF ME antennas

VLF antennas have unique advantages in underwater and underground communications due to the large electromagnetic wavelength and good penetration distance in dense media. Traditional VLF antennas, however, suffer from relatively large size and power consumption.

Recently, mechanical resonators based on ME heterostructures have been utilized for compact and power-efficient antennas. In these devices, the external applied voltage across the electrostrictive/piezoelectric material induces a deformation in this layer, transferring the strain to the magnetostrictive material through ME coupling. Energy minimization taking place in the magnetic material layer drives the magnetization in a periodic manner and generates a near-field electromagnetic signal for communication along with the piezoelectric phase itself.

In a study by Dong et al. (2020), a VLF communication system was demonstrated using a pair of ME heterostructures excited at their electromagnetic resonance (EMR) frequency. The structure and the image of devices are shown in Fig. 11.1a and b, respectively. Three layers of highly-magnetostrictive Metglas laminates were bonded to a piece of PZT (piezoelectric material) fibers with interdigital electrodes (IDE) on it. The bonding and compacting process was conducted in a vacuum bag, and the volume of a single device was 0.33 cm^3. The ME transmitter and receiver were tuned by an external DC bias magnetic field of 7.5 Oe to resonate at the same frequencies of 23.95 kHz. As shown in Fig. 11.1c, the ME receiver was able to deliver an output voltage of 215 μV with an AC

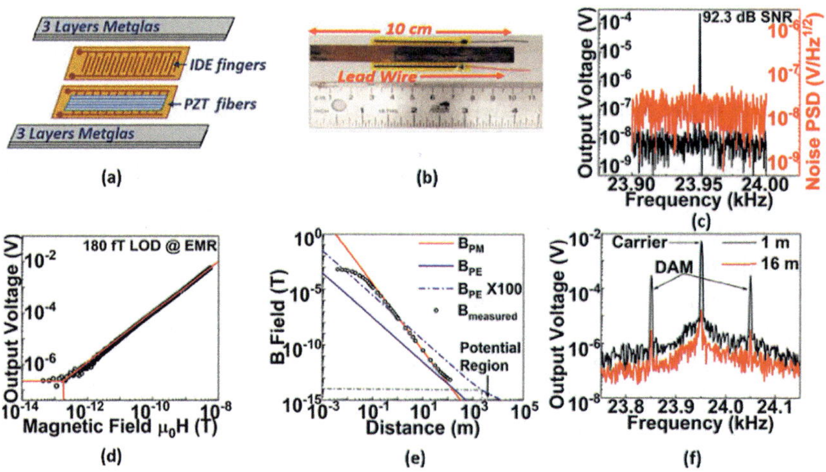

FIGURE 11.1 (a). Schematic of the ME antenna. (b) Top view of a fabricated ME antenna prototype. (c). Output voltage spectrum of the ME Rx at EMR. (d) MDF measurement of the ME Rx at EMR. (e). Predicted and measured B field distribution along with distance. (f). ME Rx output voltage spectrum shows the received 100 Hz DAM signal with carrier signal at EMR. *Reproduced with permission from Dong, C., He, Y., Li, M., Tu, C., Chu, Z., Liang, X., Chen, H., Wei, Y., Zaeimbashi, M., Wang, X., Lin, H., Gao, Y., & Sun, N. X., (2020). A portable very low frequency (VLF) communication system based on acoustically actuated magnetoelectric antennas.* IEEE Antennas and Wireless Propagation Letters, 19*(3), 398–402. https://doi.org/10.1109/LAWP.2020.2968604. Copyright [2020], IEEE.*

magnetic field of 5.62 nT. After measuring the magnetic noise floor at the receiver output by a spectrum analyzer, a 92.3 dB signal-to-noise ratio was obtained; hence the receiver was calculated to have a magnetic signal (limit of detection (LOD)) of 136 fT at EMR. Fig. 11.1d shows the magnetic signal detection measurement results given by an SR830 lock-in amplifier, and the measured LOD is 180 fT at EMR. The communication test was conducted with a driving voltage of 80 V_{rms} at 23.95 KHz. As shown in Fig. 11.1e, the magnetic field generated by the ME transmitter was predicted by a combination of near-field electric dipole and magnetic dipole model, corresponding to the piezoelectric phase and magnetostrictive phase, respectively. It was shown that with a detectivity of 260 $fT/Hz^{1/2}$, a communication distance of 120 m was achieved, and it was also predicted that a potential communication distance of 2.5–10 km could be achieved with a 100-elements antenna array, using the current receiver setup. The authors also showed a unique modulation scheme (direct antenna modulation, DAM) enabled by the nonlinear response of the devices. Fig. 11.1f indicates that, with a driven carrier signal at EMR frequency and a magnetic modulation signal at baseband frequency of 100 Hz, the antenna itself generated a frequency mixing tone in the spectrum, and a modulation signal to noise ratio (SNR) of 29 dB was achieved at 16 m communication distance.

11.3.1 Materials and structures

The magnetoelectric effect is defined as the magnetic field manipulation of electric polarization (direct ME effect), or the electric field controlling of magnetization (converse ME effect). Since the discovery of two-phase ME composite, strong ME coupling in magnetostrictive/piezoelectric heterostructures has been realized for potential applications in memory, sensors, and antennas. As the basic theory of ME transmitters, the converse ME (CME) coupling explains the mechanism of energy transfer from electricity to magnetism through strain deformation. Recently, ME antennas which combine the acoustic resonance with strong magnetoelectric coupling have been demonstrated as a new approach to potentially break the limitations in efficiency and dimensions of conventional ESAs (Chen et al., 2020). Unlike the conventional antennas which use accelerating charges to induce EM waves, the ME antennas use magnetic dipole moment oscillation which are acoustically actuated at their electromechanical resonance (EMR) rather than the EM wave resonance. The limitation of the antenna size to the wavelength of EM waves can thereby be decoupled. Since at the same operating frequency, acoustic waves present orders of magnitude smaller wavelength than EM waves, this leads to a one to two orders of miniaturization in antenna size (Nan et al., 2017).

The idea of the VLF mechanical antenna based on ME coupling was initially brought up and patented by Sun and Li in 2016, and a practical device was reported by Sun and Dong in 2020 (Dong et al., 2020). The ME antenna is a combination of three layers sandwich structure, which are Metglas ribbons (piezomagnetic material), PZT fibers (the piezoelectric material), and interdigital electrode (IDE) fingers. With the applied in-plane electric field, a strain deformation is generated in the piezoelectric phase, which is transferred to the piezomagnetic phase and results in a magnetization oscillation at the mechanical resonance, and thereby, a magnetic current for radiation. In this mechanism, the excitation electric field direction and the strain deformation direction are both in-plane, which is based on the d_{33} type ME coupling. The magnetization oscillation performs analogously to magnetic dipoles, and its main radiation direction is also in-plane.

Based upon the preceding works in ME antennas, a modified type of ME antenna based on a high-Q factor mechanical resonator with new material and structure. As shown in Fig. 11.2, the mechanical resonator is still based on the sandwich structure with piezomagnetic layers (Metglas) on both sides and piezoelectric layer (PZT-8) in between them. However, the IDE fingers are removed, and the piezomagnetic and piezoelectric layers are directly coupled through the bonding of epoxy resin. The excitation electric field direction is out-of-plane, and the strain deformation direction is in-plane, which makes the mechanical resonator working on d_{31} type ME coupling. Although for same type of PZT material, its d_{33}

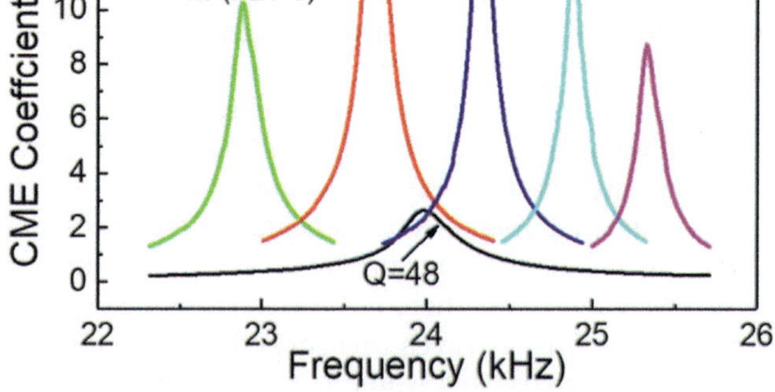

FIGURE 11.2 Comparison of the CME coefficient of PZT-8/Metglas and PZT-5A/Metglas ME antennas with different layers of Metglas.

piezoelectric constant is always higher than d_{31} piezoelectric constant, however, since the new structure has full coupling area throughout the surface of both piezoelectric and piezomagnetic phases, the effective transfer of strain deformation is higher than the previous reported d_{33} type ME antenna. Furthermore, PZT-8 has a much higher mechanical quality factor of 1200 than previously used PZT-5A, which is only 100, and the dissipation factor is six times less. This means at mechanical resonance, PZT-8 based ME antenna has much stronger amplification of strain deformation and less energy loss during the transfer from electricity to mechanics.

To test and optimize the performance of the proposed new ME antennas, Dong applied AC voltage on the resonator at its electromechanical resonance (EMR), the CME coupling induced magnetization oscillation in the piezomagnetic layers, resulting in an induction of AC magnetic field at the EMR frequency. As shown in Fig. 11.2, the PZT-8 based ME antenna has orders of magnitude with higher CME coefficient than the previous reported PZT-5A based ME antenna. The optimized 12 layers of Metglas on PZT-8 based ME antenna achieved an overall Q factor 120, compared to the PZT-5A based ME antenna of 48. These results indicate that by using same level of driving voltage, the new ME antenna can convert much more energy from electric field to magnetic field radiation, with lower level of energy loss. In other words, at the same level of magnetic field radiation, the new ME antenna can save more power and enhance the efficiency.

11.3.2 Field strength and efficiency

As the main radiation fields from ME antennas have proved to be equivalent to magnetic dipoles, therefore, the far field radiation and antenna efficiency can be estimated by the magnetic dipole model. The magnetic and electric fields radiated by a small magnetic dipole, as given by (11.5)–(11.7), are valid everywhere except at the origin.

$$B_r = -\frac{\mu_0 m_0 k^3}{4\pi}\left[\frac{2j}{(kr)^2} - \frac{2}{(kr)^3}\right] \cos\theta e^{jkr} \tag{11.5}$$

$$B_\theta = -\frac{\mu_0 m_0 k^3}{4\pi}\left[\frac{1}{kr} + \frac{j}{(kr)^2} - \frac{1}{(kr)^3}\right] \sin\theta e^{jkr} \tag{11.6}$$

$$E_\varphi = \frac{\mu_0 \omega m_0 k^2}{4\pi}\left[\frac{1}{kr} + \frac{j}{(kr)^2}\right] \sin\theta e^{jkr} \tag{11.7}$$

The power in the near field region ($k_r \ll 1$) very close to the antenna is predominantly reactive and in the far field region ($k_r \gg 1$) is predominantly real. In contrast to the near field, the dominant term in (11.5)–(11.7) for far field is the first term within the parentheses. Since for $k_r \gg 1$ the B_r component will be inversely proportional to r_2 whereas B_θ will be inversely proportional to r. Therefore, for far field radiation at large distances ($k_r \gg 1$), the B_r component will be smaller compared to B_θ, which can be assumed equal to zero. For the infinitesimal dipole, the E and H field components of the magnetic dipole in the far-field region are perpendicular to each other and transverse to the direction of propagation.

To estimate the antenna efficiency, we need first to find the total radiated power. To illustrate this for the magnetic dipole, the complex power density is formed by the time average Poynting vector:

$$W = \frac{1}{2}(E \times H^*) = \frac{1}{2}\left(- E_\varphi H_\theta^* + E_\varphi H_r^*\right) \tag{11.8}$$

The real power Prad radiated by the ME antenna can therefore be found by integrating (11.8) over a closed sphere of radius r at any far field distance, whose real part is given by:

$$P_{\text{rad}} = \frac{\mu_0 \omega^4 |m_0|^2}{12\pi c^3} \tag{11.9}$$

Here, the only unknown factor is the equivalent magnetic moment m_0 of the ME antenna. The value of this parameter from the measurement results of the near field magnetic strength. As shown in Fig. 11.3, with 30

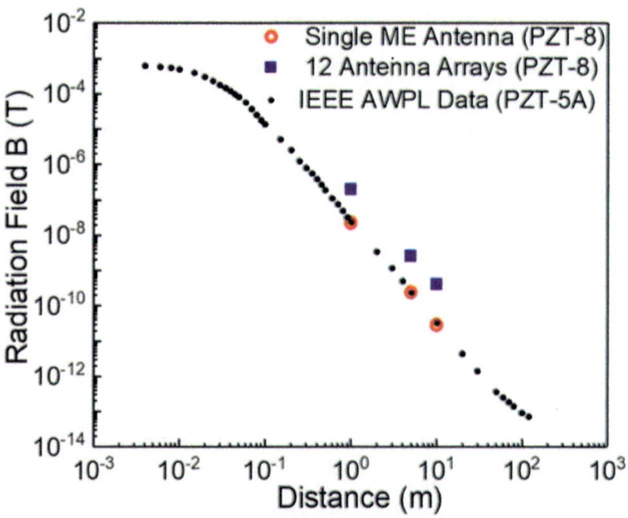

FIGURE 11.3 Comparison of the radiation field over distance from 12 ME antenna arrays with a single ME antenna unit.

Vrms driving voltage, a single PZT-8 based ME antenna can generate the same level of magnetic field as the PZT-5A based ME antenna at 80 Vrms. From this curve, we determined the equivalent magnetic moment as 0.16 A m^2. Using this value, we can calculate the radiated power from a single ME antenna unit from Eq. (11.9) as 1.08×10^{-14} W. The input power can be determined by monitoring the feed current and voltage on the ME antenna. The efficiency of a single ME antenna element was finally estimated as 5.4×10^{-14}.

11.4 ME antenna impedance analysis

The ME antenna can be modeled by the modified Butterworth-Van Dyke (MBVD) model as shown in Fig. 11.4. The MBVD model is a typical model to describe the resonance performance of acoustic resonators (Larson et al. 2000; Uzunov et al. 2017). The top RCL arm consists of the motional resistance Rm, capacitance C_m, and inductance L_m, representing the acoustic properties and resonance of the resonator. C_0 is the static electrical capacitance. For the FBAR antenna, it is the capacitance between top and bottom electrodes, while for the SAW antenna, it is the parallel capacitance between interdigital electrodes. R_s is the electrical

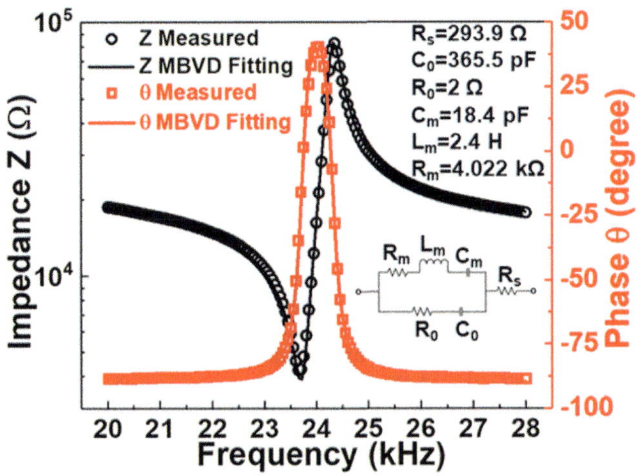

FIGURE 11.4 MBVD model of the VLF ME antenna with measured and fitted input impedance. *Reproduced with permission from Dong, C., He, Y., Li, M., Tu, C., Chu, Z., Liang, X., Chen, H., Wei, Y., Zaeimbashi, M., Wang, X., Lin, H., Gao, Y., & Sun, N. X., (2020). A portable very low frequency (VLF) communication system based on acoustically actuated magnetoelectric antennas.* IEEE Antennas and Wireless Propagation Letters, *19(3), 398–402. https://doi.org/10.1109/LAWP.2020.2968604. Copyright 2020 IEEE.*

resistance from the electrodes. These five components make up a Butterworth-Van Dyke (BVD) model. Additionally, resistance R_0 is added in series with C_0 to define the material loss. With this R_0, the equivalent circuit is called the MBVD model.

The resonance frequencies are defined by:

$$f_s = \frac{1}{2\pi\sqrt{L_m C_m}} \tag{11.10}$$

$$f_p = f_s\sqrt{\frac{C_m + C_0}{C_0}} \tag{11.11}$$

Here, f_s is the mechanical resonance frequency, while f_p is the anti-resonance frequency. Hence, Q_s (Q-factor of series resonance), Q_{s0} (Q-factor of series resistance with R0 and Rs loss), and Q_p (Q-factor of parallel resonance with loss) can be expressed as (Larson et al., 2000)

$$Q_s = \frac{1}{2\pi f_s R_m C_m} \tag{11.12}$$

$$Q_{s0} = \frac{Q_s}{1 + \frac{Rs}{R_m}} \tag{11.13}$$

$$Q_p = Q_s \frac{f_s}{f_p} \frac{1}{1 + \frac{R_0}{R_m}} \tag{11.14}$$

and the input impedance is described by

$$Z(f) = \frac{X_p}{j\frac{f}{f_p}} \bullet \frac{\left[1 - \left(\frac{f}{f_s}\right)^2 + j\frac{f}{f_s} \bullet \frac{1}{Q_{s0}}\right]}{\left[1 - \left(\frac{f}{f_p}\right)^2 + j\frac{f}{f_p} \bullet \frac{1}{Q_{s0}}\right]} \tag{11.15}$$

Where, $X_p = \frac{1}{2\pi f_p C_0}$.

Unlike conventional electrical antennas with a comparable volume and operating frequency, the ME antenna has a much smaller reactive impedance, which is mainly introduced by the electrical capacitor. This electrical capacitor can be controlled by the geometric design of the resonator area (for FBAR) or pitch shape (for the SAW), and so ME antennas can be easily matched to any electric circuit with any output impedance, leading to a low return loss at resonance. Also, with the minimized reactive part, more energy is radiated to the environment, instead of being stored in the near field. For these reasons, the ME antenna has much higher radiation efficiency than conventional antennas. In addition, since the mechanical resonance is affected by the permittivity and permeability of the surroundings, the impedance matching of the ME antenna will be stable in any environment, such as in the human body or close to magnetic materials.

11.5 Bandwidth and nonlinear antenna modulation

For traditional EM antenna with an impedance matching network, a high antenna quality factor and limited impedance bandwidth can cause serious impedance mismatch in the frequency out of the operating bandwidth. In this frequency range, minimal real power is consumed by the antenna for heat dissipation and radiation, hence, the far field radiation magnetic and electric field is weak. The frequency dependent antenna impedance matching condition would induce transmitting waveform distortion in time domain and subsequently diminishes the communication capability of the system.

In 1948, Chu (1948) derived the antenna quality factor as a function of antenna size under maximum gain condition for different order spherical waves. Harrington (1960) later considered the antenna loss for quality factor calculation. Sievenpiper et al. (2012) summarized the relations between the antenna electric size and the lower limit of antenna quality factor proposed earlier in (Collin & Rothschild, 1964; McLean, 1996) for linearly and circularly polarized linear time-invariant (LTI) antennas shown in Eqs. (11.16) and (11.17), respectively.

$$Q = \frac{1}{ka} + \frac{1}{(ka)^3} \tag{11.16}$$

$$Q = \frac{1}{ka} + \frac{1}{2(ka)^3} \tag{11.17}$$

Where, k is the wave number corresponding to the antenna operating frequency, and a is the minimum radius of the sphere that can contain the antenna.

The high antenna quality factor indicates that most of the energy fed into the antenna is confined in the near field region without radiation. Also, since for antennas with rather small fractional bandwidth defined by a certain voltage standing wave ratio (VSWR), the operating bandwidth is inversely proportional to Q, the trade-off between the antenna size and communication capability raises a dilemma for the researchers especially for the design of electrically small antenna with the largest structure size less than $\lambda/10$. Numerous methodologies have been proposed to overcome the transmitting waveform distortion imposed by the small bandwidth, mostly by violating one or more assumptions (LTI) of the limit proposed by Chu (1948), Wheeler (1947) and elaborated in (Sievenpiper et al., 2012). Currently most of these works focus on implementing the switching mechanism in the antenna system to violate the aforementioned time-invariant assumption.

The undesirable transmitting waveform distortion caused by the high antenna quality factor can be overcome by properly configuring the switching sequence and changing the resonance condition of the antenna in real time, specifically, the long "ring down" and "ramping up" time of transmitting wave amplitude limited by the energy discharging and charging time of the antenna has been shortened to the switching time of the switch. The technique was usually referred to as direct antenna modulation (DAM) and has been intensively studied in the recent 20 years (Azad & Wang, 2014; Keller et al., 2010; Weijun & Yuanxun, 2004).

For piezoelectric material based acoustically driven mechanical antennas, a similar frequency dependent response is induced by the intrinsic mechanical resonance behavior. The most prevalent methodology to bypass the unfavorable impact of high-quality factor is so-called capacitive loading shift keying (CLSK). With a capacitor series connected to the mechanical antenna, the resonance frequency of the antenna would change, and the binary data can be modulated into the switching signal of the capacitor to achieve a frequency shift keying (FSK) communication with carrier frequency equal to the resonance frequency at different switching states. Kemp et al. (2019) proposed this binary FSK and implemented the modulation scheme with a high-Q piezoelectric rod transmitter, Yu et al. (2022) applied this modulation scheme in an ME antenna based wireless bio-implant. However, CLSK suffers from the increasing system complexity and large on-chip device area caused by the extra switching components, capacitor as well as the synchronization circuit for carrier and switching signal.

11.6 VLF ME antenna array

Since the wavelength of VLF antenna is in the range of tens of kilometers, which is much larger than the size of ME antennas, therefore, it is possible to implement ME antenna arrays within a small area to enhance the total radiation field strength without inducing any phase difference caused by the path difference between each antenna element. However, it is also crucial to ensure that the driving voltage on each antenna unit has the same frequency and phase. A power amplifiers used to drive the ME antennas, but also have limitations on either maximum voltage or current or both. It is essential to arrange the way of connection and the number of antenna arrays within the power handling capability for each power amplifier. Dong has reported an ME antenna array (Dong et al. 2022), a 30 Vrms driving voltage for a single antenna unit can achieve the same level of radiation field strength as the previously demonstrated PZT-5A based ME antenna at 80 Vrms driving voltage.

The system connection diagram is shown in the inset of Fig. 11.5, a signal generator sent out a VLF signal at 23.7 kHz through custom designed power amplifiers. Each power amplifier was driving a single ME antenna unit in parallel. All of the 12 ME antenna units were biased with optimal DC bias field and tuned with same EMR frequency at 23.7 kHz. To test and quantify the radiation enhancement by using ME antenna arrays, Dong measured the radiated magnetic field at 1 m distance for different number of antenna elements. The results are shown in Fig. 11.5, starting from single ME antenna unit to 12 elements, the magnetic field strength displayed a notably linear increment form 20–210 nT. This result implies that for VLF ME antennas, a proper arrangement of antenna array can yield linear enhancement of radiation field strength.

Though the value of efficiency of a single ME antenna seems extremely low, however, considering its size, which is also extremely small compared to the tens of kilometers long wavelength, the ME antennas still have potential to greatly enhance the radiation efficiency by implementing large number of antenna arrays. As pointed out in Eq. (11.5), the radiated power follows a square law with the total magnetic moment,

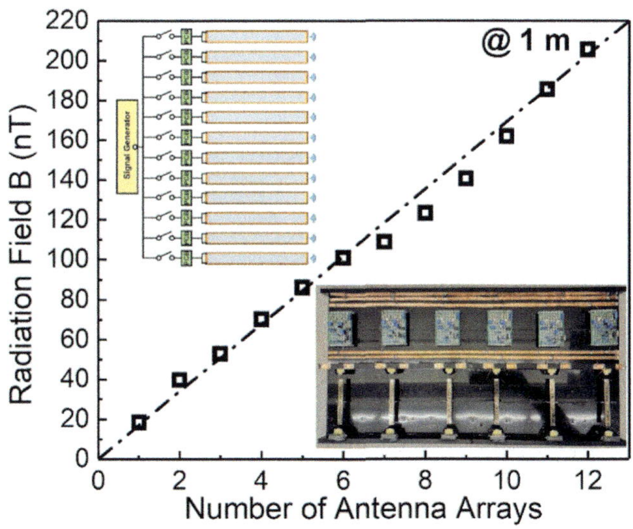

FIGURE 11.5 Linear increment of magnetic field strength with number of ME antenna arrays, the insets show the schematic and real image of the implemented ME antenna arrays.

which is linearly proportional to the number of antenna arrays. In other words, by deploying about three million of these ME antenna arrays, the efficiency of the system can be increased up to 50%, which is close to the conventional electric VLF antennas, while the power consumption will be much lower.

As the near-field strength of ME antennas has proved to be equal to magnetic dipoles, and the equivalent magnetic dipole moment can be determined by the measurement results at certain distance in open space. Therefore, even though there is no simulator that can directly predict the radiation from ME heterostructure so far, the transmission behavior of magnetic field in underwater environment can still be simulated using an equivalent magnetic dipole. In this section, a finite element method (FEM) simulator is used to simulate and analyze the proposed ME antennas.

The FEM simulation software COMSOL multiphysics is used to simulate the magnetic field distribution of the proposed ME antennas. As shown in the inset of Fig. 11.6, a cylindrical coil is located at the coordinate origin, which generate an equivalent magnetic dipole moment of 0.16 A m^2 as a single ME antenna unit. It has a length of 10 cm and a diameter of 10 cm and is enclosed in an air bubble with a radius of 10 cm. Out of the air bubble, the medium is set as seawater with conductivity of 5 S/m, relative permittivity of 81. The outer boundary is set as the perfectly matched layer.

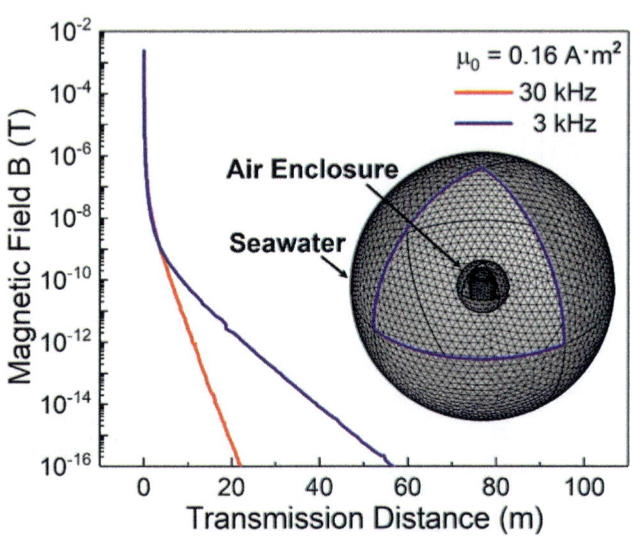

FIGURE 11.6 Simulation of the equivalent magnetic dipole in seawater from 3 to 30 kHz, the inset shows the 3D FEM structure.

The strength of the magnetic field versus transmission distance at different frequencies in VLF band are shown in Fig. 11.6. The equivalent dipole m_0 of a single ME antenna unit can transmit magnetic field in seawater ranging from 20 to 50 m, given the receiver sensitivity of ~1 fT level. As shown in Fig. 11.7, when the operating frequency of m_0 is set as 24 kHz. It can be found that there is almost no difference of the transmission behavior between the air and seawater within the range of 5 m, i.e., there is reflection on the boundary of seawater and air bubble. Furthermore, as the equivalent dipole m_0 increased by 1000 times, the transmission distance can be achieved to 30 m at 24 kHz. Therefore, with the implementation of ME antennas arrays and adjusting the mechanical resonance frequency toward the lower bond of VLF band, it is expected to achieve a wireless underwater communication range of 50 m or beyond.

11.7 Challenges

11.7.1 Challenges of the mechanical antenna

Even though a large amount of effort has been spent on mechanical antennas, there still exist many challenges in the antenna design, numerical simulations, fabrication, and measurements, and yet the opportunities for performance improvement and resulting applications for compact, high date communication systems are many too.

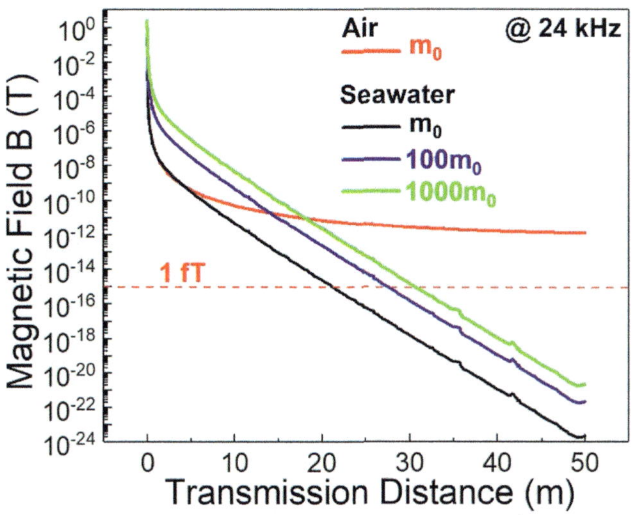

FIGURE 11.7 Comparison of the transmission of magnetic field from different magnetic dipole moments at 24 kHz in air and seawater.

11.7.2 Challenges on antenna measurement

After device fabrication, the antenna measurement setup poses several challenges as well. For the VLF ME antenna, at present, the field measurement is restricted between 1 and 1000 m distance, to avoid the "size effect" for a short distance and large noise for a long distance. In the future, an extremely sensitive search coil magnetometer with an ultralow noise level can be used to decrease the magnetic noise for long distance measurement for a more accurate magnetic field measurement.

References

Azad, U., & Wang, Y. E. (2014). Direct antenna modulation (DAM) for enhanced capacity performance of near-field communication (NFC) link. *IEEE Transactions on Circuits and Systems. I, Regular Papers, 61*(3), 902–910. https://doi.org/10.1109/TCSI.2013.2283995

Barani, N., Kashanianfard, M., & Sarabandi, K. (2021). A mechanical antenna with frequency multiplication and phase modulation capability. *IEEE Transactions on Antennas and Propagation, 69*(7), 3726–3739. https://doi.org/10.1109/TAP.2020.3044385

Chen, H., Liang, X., Dong, C., He, Y., Sun, N., Zaeimbashi, M., He, Y., Gao, Y., Parimi, P. V., Lin, H., & Sun, N.-X. (2020). Ultra-compact mechanical antennas. *Applied Physics Letter, 117*(17), Article 170501. https://doi.org/10.1063/5.0025362

Chen, J. C., Kan, P., Yu, Z., Alrashdan, F., Garcia, R., Singer, A., Lai, C. S. E., Avants, B., Crosby, S., Li, Z., Wang, B., Felicella, M. M., Robledo, A., Peterchev, A. V., Goetz, S. M., Hartgerink, J. D., Sheth, S. A., Yang, K., & Robinson, J. T. (2022). A wireless millimetric magnetoelectric implant for the endovascular stimulation of peripheral nerves. *Nature Biomedical Engineering, 6*(6), 706–716. https://doi.org/10.1038/s41551-022-00873-7

Chu, L. J. (1948). Physical limitations of omni-directional antennas. *Journal of Applied Physics, 19*(12), 1163–1175. https://doi.org/10.1063/1.1715038

Collin, R., & Rothschild, S. (1964). Evaluation of antenna Q. *IEEE Transactions on Antennas and Propagation, 12*(1), 23–27. https://doi.org/10.1109/TAP.1964.1138151

Deng, T., Chen, Z., Di, W., Chen, R., Wang, Y., Lu, L., Luo, H., Han, T., Jiao, J., & Fang, B. (2021). Significant improving magnetoelectric sensors performance based on optimized magnetoelectric composites via heat treatment. *Smart Materials and Structures, 30*(8), Article 85005. https://doi.org/10.1088/1361-665X/ac0858

Domann, J. P., & Carman, G. P. (2017). Strain powered antennas. *Journal of Applied Physics, 121*(4), Article 044905. https://doi.org/10.1063/1.4975030

Dong, C., He, Y., Li, M., Tu, C., Chu, Z., Liang, X., Chen, H., Wei, Y., Zaeimbashi, M., Wang, X., Lin, H., Gao, Y., & Sun, N. X. (2020). A portable very low frequency (VLF) communication system based on acoustically actuated magnetoelectric antennas. *IEEE Antennas and Wireless Propagation Letters, 19*(3), 398–402. https://doi.org/10.1109/LAWP.2020.2968604

Dong, C., Xu, Y., & Wu, J. (2022). *VLF mechanical antenna arrays for underwater wireless communications* (New York, NY, USA).

Du, Y., Xu, Y., Wu, J., Qiao, J., Wang, Z., Hu, Z., Jiang, Z., & Liu, M. (2023). Very-low frequency magnetoelectric antennas for portable underwater communication: Theory and experiment. *IEEE Transactions on Antennas and Propagation, 71*(3). https://doi.org/10.1109/TAP.2022.3233665

Fawole, O. C., & Tabib-Azar, M. (2017). An electromechanically modulated permanent magnet antenna for wireless communication in harsh electromagnetic environments. *IEEE Transactions on Antennas and Propagation, 65*(12), 6927–6936. https://doi.org/10.1109/TAP.2017.2761555

Harrington, R. F. (1960). Effect of antenna size on gain, bandwidth, and efficiency. *Journal of Research of the National Bureau of Standards, 64D*(1). Paper 64Dl- 31.

Hayes, P., Jovičević Klug, M., Toxværd, S., Durdaut, P., Schell, V., Teplyuk, A., Burdin, D., Winkler, A., Weser, R., Fetisov, Y., Höft, M., Knöchel, R., McCord, J., & Quandt, E. (2019). Converse magnetoelectric composite resonator for sensing small magnetic fields. *Scientific Reports, 9*(1). https://doi.org/10.1038/s41598-019-52657-w

Keller, S. D., Palmer, W. D., & Joines, W. T. (2010). Switched antenna circuit with increased information bandwidth. *IEEE Antennas and Wireless Propagation Letters, 9*, 1045–1048. https://doi.org/10.1109/LAWP.2010.2089775

Kemp, M. A., Franzi, M., Haase, A., Jongewaard, E., Whittaker, M. T., Kirkpatrick, M., & Sparr, R. (2019). A high Q piezoelectric resonator as a portable VLF transmitter. *Nature Communications, 10*(1). https://doi.org/10.1038/s41467-019-09680-2

Larson, J. D., Bradley, P. D., Wartenberg, S., & Ruby, R. C. (2000). *Modified Butterworth-Van Dyke circuit for FBAR resonators and automated measurement system.*

Lasheras, A., Saiz, P. G., Porro, J. M., Quintana, I., Polak, C., & Lopes, A. C. (2021). Enhanced performance of magnetoelectric laminated composites by geometry engineering for high frequency applications. *Journal of Alloys and Compounds, 884*, Article 161065. https://doi.org/10.1016/j.jallcom.2021.161065

Liang, X., Dong, C., Chen, H., Wang, J., Wei, Y., Zaeimbashi, M., He, Y., Matyushov, A., Sun, C., & Sun, N. (2020). A review of thin-film magnetoelastic materials for magnetoelectric applications. *Sensors, 20*(5), 1532. https://doi.org/10.3390/s20051532

Liu, Y., Hou, M., & Gong, S. (2020). *A rotating permanent magnet transmitter for magnetic induction communication in RF-impenetrable environment.*

Luong, K. Q. T., & Wang, Y. E. (2022). Analysis of dynamic magnetoelastic coupling in mechanically driven magnetoelectric antennas. *Sensors, 22*(2), 455. https://doi.org/10.3390/s22020455

McLean, J. S. (1996). A re-examination of the fundamental limits on the radiation Q of electrically small antennas. *IEEE Transactions on Antennas and Propagation, 44*(5), 672. https://doi.org/10.1109/8.496253

Miron, D. B. (2006). *Small antenna design.* Newnes/Elsevier.

Nan, T., Lin, H., Gao, Y., Matyushov, A., Yu, G., Chen, H., Sun, N., Wei, S., Wang, Z., Li, M., Wang, X., Belkessam, A., Guo, R., Chen, B., Zhou, J., Qian, Z., Hui, Y., Rinaldi, M., McConney, M. E., … Sun, N. X. (2017). Acoustically actuated ultra-compact NEMS magnetoelectric antennas. *Nature Communications, 8*(1), 296. https://doi.org/10.1038/s41467-017-00343-8

Sievenpiper, D. F., Dawson, D. C., Jacob, M. M., Kanar, T., Kim, S., Long, J., & Quarfoth, R. G. (2012). Experimental validation of performance limits and design guidelines for small antennas. *IEEE Transactions on Antennas and Propagation, 60*(1), 8–19. https://doi.org/10.1109/TAP.2011.2167938

Uzunov, I. S., Terzieva, M. D., Nikolova, B. M., & Gaydazhiev, D. G. (2017). *Extraction of modified butterworth - Van Dyke model of FBAR based on FEM analysis.*

Viehland, D., Wuttig, M., Mccord, J., & Quandt, E. (2018). Magnetoelectric magnetic field sensors. *MRS Bulletin, 43*(11), 834–840. https://doi.org/10.1557/mrs.2018.261

Volakis, J., Chen, C. C., & Fujimoto, K. (2009). *Small antennas:miniaturization techniques & applications.* McGraw Hill LLC. https://doi.org/https://books.google.com/books?id=HOYUqJbi3fwC

Vopson, M. M., Fetisov, Y. K., Caruntu, G., & Srinivasan, G. (2017). Measurement techniques of the magneto-electric coupling in multiferroics. *Materials, 10*(8), 963. https://doi.org/10.3390/ma10080963

Wang, C., Cui, Y., Song, X., Zhong, J., Wei, M., Wu, M., & Yuan, H. (2021). Model, design, and testing of an electret-based portable transmitter for low-frequency applications. *IEEE*

Transactions on Antennas and Propagation, 69(9), 5305–5314. https://doi.org/10.1109/TAP.2021.3061007

Weijun, Y., & Yuanxun, W. (2004). *Direct antenna modulation - a promise for ultra-wideband (UWB) transmitting* (Piscataway NJ).

Wheeler, H. A. (1947). Fundamental limitations of small antennas. *Proceedings of the IRE, 35*(12), 1479–1484. https://doi.org/10.1109/JRPROC.1947.226199

Xiao, N., Wang, Y., Chen, L., Wang, G., Wen, Y., & Li, P. (2023). Low-frequency dual-driven magnetoelectric antennas with enhanced transmission efficiency and broad bandwidth. *IEEE Antennas and Wireless Propagation Letters, 22*(1), 34–38. https://doi.org/10.1109/LAWP.2022.3201070

Xu, G., Xiao, S., Li, Y., & Wang, B.-Z. (2021). Modeling of electromagnetic radiation-induced from a magnetostrictive/piezoelectric laminated composite. *Physics letters. A, 385*, Article 126959. https://doi.org/10.1016/j.physleta.2020.126959

Xu, J., Li, Z., Pan, X., Wen, X., Cao, J., Gong, W., Yang, S., Lei, M., Yao, F., & Bi, K. (2023). Ultra-wideband electrostrictive mechanical antenna. *Advanced Functional Materials, 33*(8). https://doi.org/10.1002/adfm.202210868

Yang, G., Du, J., Wang, J., & Yang, J. (2018). Frequency dependence of electromagnetic radiation from a finite vibrating piezoelectric body. *Mechanics Research Communications, 93*, 163–168. https://doi.org/10.1016/j.mechrescom.2017.04.001

Yao, Z., Wang, Y. E., Keller, S., & Carman, G. P. (2015). Bulk acoustic wave-mediated multiferroic antennas: Architecture and performance bound. *IEEE Transactions on Antennas and Propagation, 63*(8), 3335–3344. https://doi.org/10.1109/TAP.2015.2431723

Yu, Z., Alrashdan, F. T., Wang, W., Parker, M., Chen, X., Chen, F. Y., Woods, J., Chen, Z., Robinson, J. T., & Yang, K. (2022). *Magnetoelectric backscatter communication for millimeter-sized wireless biomedical implants* (New York, NY, USA).

Yun, X., Lin, W., Hu, R., Liu, Y., Wang, X., Yu, G., Zeng, Z., Zhang, X., & Zhang, B. (2022). Bandwidth-enhanced magnetoelectric antenna based on composite bulk acoustic resonators. *Applied Physics Letters, 121*(3). https://doi.org/10.1063/5.0098323

Zaeimbashi, M., Nasrollahpour, M., Khalifa, A., Romano, A., Liang, X., Chen, H., Sun, N., Matyushov, A., Lin, H., Dong, C., Xu, Z., Mittal, A., Martos-Repath, I., Jha, G., Mirchandani, N., Das, D., Onabajo, M., Shrivastava, A., Cash, S., & Sun, N. X. (2021). Ultra-compact dual-band smart NEMS magnetoelectric antennas for simultaneous wireless energy harvesting and magnetic field sensing. *Nature Communications, 12*(1). https://doi.org/10.1038/s41467-021-23256-z

CHAPTER

12

Magnetoelectric antennas for RF applications

Bin Luo, and Nian Sun

Northeastern University, Boston, MA, United States

12.1 Magnetoelectric antenna mechanism and highlight

Antennas have been one of the largest components in an electronic system, which have so far been resistant to size miniaturization, thus leading to electronics with a large cost, size, weight, and power (CSWaP). Conventional compact antennas operate at their electromagnetic resonance, leading to antenna sizes comparable to their electromagnetic wavelength. In addition, conventional antennas are based on electric currents that suffer from platform effects with a high profile. Magnetoelectric (ME) antennas are mechanical antennas based on layered ME piezoelectric/magnetostrictive heterostructures operating at their electromechanical resonance, leading to ultracompact ME antenna sizes comparable to their mechanical or acoustic wavelength, which is ~5 orders of magnitude smaller than electromagnetic wavelength. At the same time, ME antennas are based on the magnetic currents, instead of electric currents; thus, ME antennas are immune from platform effect and show enhanced radiation on a ground plane (Liang et al., 2020a, 2022, 2023; Zaeimbashi et al., 2021).

The characteristics and roadmap of ME antenna (Luo et al., 2024a) are summarized in Fig. 12.1. By applying an AC voltage on the piezoelectric layer, oscillating mechanical strain is induced and transferred to the magnetostrictive film. This electric-mechanical-magnetic energy transduction at the electromechanical resonance triggers AC magnetization oscillation in the magnetic layer, leading to magnetic dipole–induced electromagnetic radiation (Fig. 12.1a). Additionally, the ME antenna can transmit and receive electromagnetic waves with the same ME heterostructure. Operating at their electromechanical resonance, ultracompact ME antennas have dimensions comparable to their mechanical or acoustic

Magnetoelectric Composites
https://doi.org/10.1016/B978-0-443-18606-6.00013-7

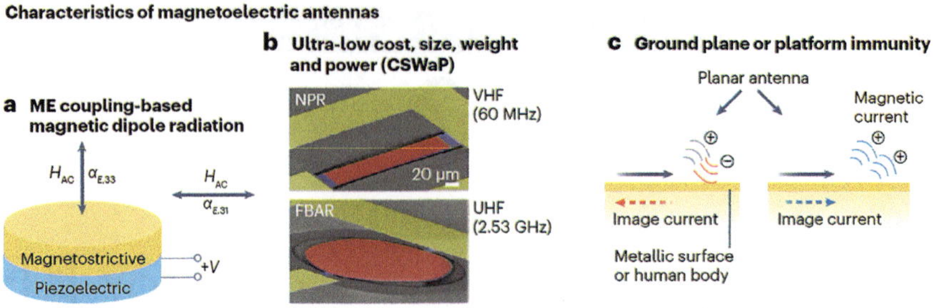

FIGURE 12.1 Smart ME antenna characteristics. (a) Magnetoelectric (ME) antenna piezoelectric–magnetostrictive composites. In the transmitting mode, an alternating current (AC) voltage is applied to the piezoelectric layer, generating a strain wave via piezoelectric effect. The strain wave induces the magnetostrictive film deformation and triggers the magnetization dynamics for AC magnetic field radiation (H_{AC}). In the receiving mode, the magnetization in response to the radiofrequency (RF) magnetic fields from electromagnetic waves leads to magnetic film deformation owing to magnetostriction effect. The strain in magnetic film gives rise to the deformation of the piezoelectric layer and generates RF voltage output via inverse piezoelectric effect. The strain-mediated transverse ($\alpha_{E,31}$) and longitudinal ($\alpha_{E,33}$) ME coupling leads to in-plane and out-of-plane radiation magnetic field H_{AC}. (b) Scanning electron microscopy (SEM) image of ME antenna at very-high-frequency (VHF) and ultrahigh-frequency (UHF) bands, showing ultralow cost, size, weight, and power (CSWaP) feature of ME antennas. (c) Illustration of ground plane or platform immunity of ME antennas. The out-of-phase imaging electric current of a conventional electrical antenna mounted on a metallic surface, ground or human body cancels the antenna radiation (left side), whereas a ME antenna utilizes magnetic current for radiation and its in-phase imaging current contributes to 3 dB gain enhancement of the ME antenna radiation (right side). *Reproduced from Luo et al. (2024a).*

wavelength, which is five orders of magnitude smaller than the electromagnetic wavelength (Fig. 12.1b).

Mounting a conventional dipole antenna on a ground plane, such as a metallic surface or human body, can lead to a detuned antenna impedance and detrimental effects on the performance, as the ground plane can reflect, absorb, and scatter electromagnetic waves, leading to destructive interference. ME antennas, however, are based on magnetic currents, instead of electric currents, which use magnetic dipole for radiation. ME antennas have been demonstrated to show ground plane immunity as the image dipole of the ME antenna on a ground plane is in-phase with the magnetic dipole itself (Fig. 12.1c). Therefore, ME antennas are immune from the platform effect and exhibit enhanced radiation on a ground plane (Liang et al., 2020a, 2022, 2023; Nan et al., 2017; Zaeimbashi et al., 2021). With the demonstrated ground plane immunity, ME antennas are ideally suited as conformal antennas for platform-independent deployments, such as biomedical health monitoring (Chen et al., 2022), etc.

12.2 Single radiofrequency magnetoelectric antenna

After the first simulation prediction of efficient bulk acoustic wave (BAW)-mediated multiferroic antenna by finite-difference time-domain

(FDTD) technique (Yao et al., 2015), Nan et al. pioneered the first batch of laterally vibrating nanoplate resonator (NPR) and vertically vibrating thin-film bulk acoustic resonator (FBAR) ME antenna structures experimentally, which is the first experimental demonstration and marks a significant milestone (Nan et al., 2017). This first demonstrated integrated FBAR ME antenna operates at 2.53 GHz and shows an antenna gain of −18 dBi with a FeGaB/AlN ME resonating disk with a thickness of ~1 μm and diameter of 200 μm released but anchored on a 0.5 mm thick Si substrate (Nan et al., 2017), as shown in Fig. 12.2. The NPR ME antenna consisted of a rectangular resonating plate with a single-finger bottom Pt electrode and a thin-film FeGaB/AlN active resonant heterostructure, as shown in Fig. 12.2a. As shown in Fig. 12.2c, a high ME coupling coefficient of $\alpha_{ME} = 6$ kVOe^{-1} cm^{-1} can be obtained at EMR frequency $f_{r,\text{NPR}}$ of 60.68 MHz without a DC bias magnetic field, indicating a strong self-biased ME coupling. The strong ME coupling leads to a high electromechanical transduction efficiency, low loss, and effective interaction between acoustic resonance and RF magnetic field component of EM waves, which contributes to a high quality factor Q of 930 and an electromechanical coupling coefficient k_t^2 of 1.35% (Zaeimbashi et al., 2021), as shown in Fig. 12.2b. The ME NPR antenna achieves 1−2 orders of magnitude miniaturization over state-of-the-art compact antennas without performance degradation.

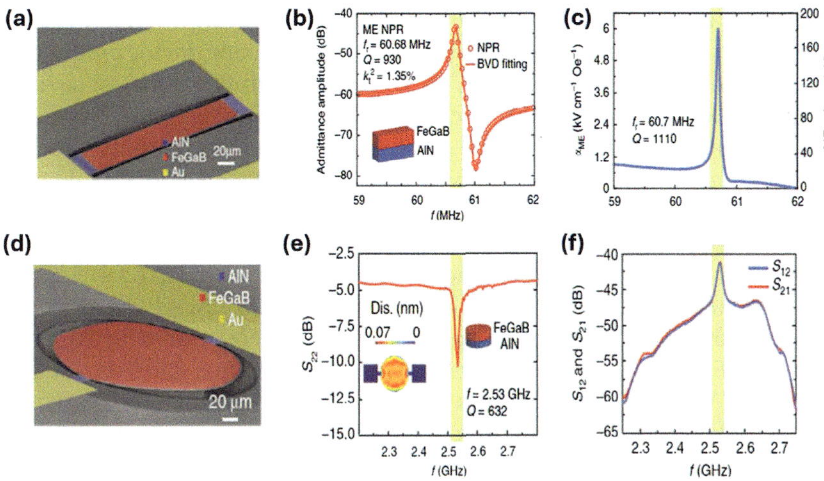

FIGURE 12.2 Magnetoelectric (ME) nanoplate resonator (NPR) and thin-film bulk acoustic resonator (FBAR) antennas with gigantic ME coupling. ME NPR antenna: (a) Scanning electron microscopy (SEM) images of the fabricated ME NPR antenna. (b) Admittance curve and Butterworth−van Dyke model fitting of the ME NPR. The inset shows the schematic of the cross-section of the ME heterostructure. (c) ME coupling coefficient (left axis) and the induced ME voltage (right axis) versus the frequency of RF magnetic field excitation H_{RF}. FBAR antenna: (d) SEM photo. (e) Return loss S_{22} and (f) radiating characteristic (S_{12}) and receiving characteristic (S_{21}) at the resonance of the FBAR device. *Reproduced from Nan et al. (2017).*

The ME FBAR antenna composed of a suspended FeGaB/AlN ME circular disk is shown in Fig. 12.2d. Different from ME NPR antennas, the ME FBAR antenna operated at GHz in thickness extensional vibration mode, as shown in Fig. 12.2e. The ME FBAR antennas exhibited a peak return loss of 10.26 dB, a *Q*-factor of 632, and a mechanical resonance frequency of 2.53 GHz. The operating frequency (same as the mechanical resonance frequency) $f_{0,FBAR}$ can be changed by adjusting the thickness T of the ME heterostructure due to the relation:

$$f_{0,FBAR} \propto \frac{1}{T}\sqrt{\frac{E_{eq}}{\rho_{eq}}}$$

where T the thickness of the FBAR resonator; E_{eq} is the equivalent Young's modulus; and ρ_{eq} is the equivalent density of the ME composite. The FBAR antenna achieved an antenna gain of −18 dBi at the resonance frequency of 2.53 GHz based on the measured radiated signal (S_{12}) and received signal (S_{21}), as shown in Fig. 12.2f. Compared with the state-of-art conventional compact antennas operating at the same frequency, the dimension of the ME antenna is significantly smaller due to the guided acoustic wave with much lower velocity compared to the EM wave.

The designed and fabricated NPR and FBAR ME antennas could realize different modes of vibration and radiation at both very high frequency (VHF, 60 MHz) and ultrahigh-frequency (UHF, 2.525 GHz) operation frequency bands. Moreover, the similar microfabrication process of both NPR and FBAR-based antennas on the same silicon wafer allows the integration of broadband ME antenna arrays working in the frequency range from tens of MHz (NPR with large lateral dimensions) to several GHz (FBAR with thinner AlN thickness) on one chip. These ultracompact ME antennas serve as potential candidates for future communication systems, the Internet of things, wearable antennas, bio-implantable and bio-injectable antennas, smartphones, and wireless communication systems.

In order to realize the multifunctional ME antennas, a dual-band smart ME antenna for wireless energy harvesting and magnetic sensing (Zaeimbashi et al., 2021) was experimentally demonstrated in 2020, as shown in Fig. 12.3. The demonstrated smart ME antenna consists of three FeGaB/AlN ME strip resonators for impedance matching, as shown in Fig. 12.3a,b. The three resonators exhibit two modes: out-of-plane thickness vibration mode at ~2.5 GHz and in-plane width at 63.6 MHz, as shown in Fig. 12.3c,d. The two modes can be used for wireless power transfer energy harvesting and magnetic sensing, respectively. The coil to ME antenna power transfer efficiency as a function of separation distance was measured, as shown in Fig. 12.3e. Compared to the state-of-the-art coil-to-coil energy harvester, the dual-band smart ME antenna shows an

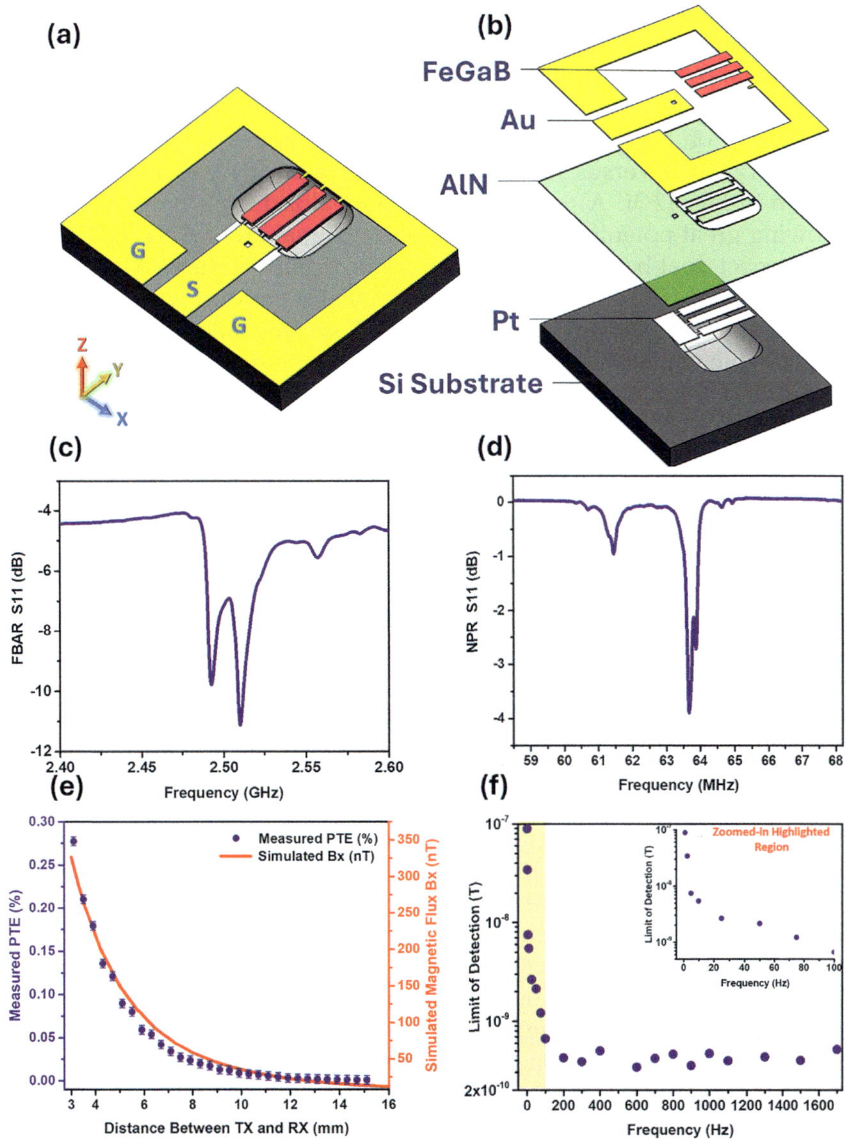

FIGURE 12.3 Dual-band smart magnetoelectric (ME) antenna for simultaneous wireless energy harvesting and magnetic sensing. (a) 3D schematic of the fabricated dual-band smart ME antenna with three nanoplate resonators (NPRs). (b) Exploded structure of the ME antenna showing all the layers. (c) S_{11} of the thickness mode of the smart ME antenna showing a resonance frequency at 2.51 GHz (d) S_{11} of the width mode of the smart ME antenna showing a resonance frequency at 63.6 MHz. (e) The left-axis shows the measured power transfer efficiency (PTE) versus the distance between ME antenna and Tx coil, and the right axis shows the simulated magnetic flux density along the x-axis (B_x) generated by the Tx coil versus distance from Tx coil. (f) Limit of Detection (LoD) of ME antenna versus frequency of external alternating magnetic field. *Reproduced from Zaeimbashi et al. (2021).*

overwhelming Figure of Merit (FOM) of 3721, which holds great promise for ultracompact efficient wireless energy harvester for wireless power transfer and communication systems. In terms of magnetic sensing capability characterization, a 1 kHz, 9.4 μT external magnetic field was modulated on the carrier frequency of the ME antenna, and the limit of detection (LoD) versus modulated signal frequency was measured, as shown in Fig. 12.3f. A ~400 pT LoD was realized from 200−1600 MHz, showing great potential for sensing neural signal around 1 kHz. The ultracompact dual-band smart nanoelectromechanical systems ME antenna with a size of 250 × 174 μm^2 that can efficiently perform wireless energy harvesting and ultrasmall magnetic field sensing allows the wireless implantable medical devices to be compliant with the specific absorption rate (SAR) limit. Furthermore, the antenna's limit of detection within 300−500 pT enables the implantable medical devices to record neural magnetic fields for health monitoring and holds great promise for NeutroRFID ME sensor node (Zaeimbashi, 2019).

A challenge with the demonstrated ultrahigh-frequency (UHF) FBAR ME antennas is their physical fragility because they are released/suspended devices. A solution to this challenge has been recently experimentally demonstrated, namely, a solidly mounted resonator (SMR) ME antenna, which has comparable performance and a much more robust structure with 24 dB enhanced power handling compared to FBAR ME antennas (Liang et al., 2020a). The SMR ME antennas consist of a FeGaB/ZnO ME resonator deposited on a distributed acoustic Bragger reflector (Chen et al., 2020; Liang et al., 2020a, 2023), as shown in Fig. 12.4a and b. The Bragg reflector is an acoustic wave reflector consisting of several periods of thin films with either low acoustic impedance or high acoustic impedance (Dubois et al., 1998; Lakin et al., 1995; Lanz & Muralt, 2005). The high reflection coefficient makes most acoustic energy be reflected instead of being dissipated into the substrate. Strong displacement occurs at the piezoelectric and magnetostrictive layers, indicating acoustic energy confinement in the ME stack, as shown in Fig. 12.4c and d. The low dissipation of acoustic energy contributes to the enhancement of radiation efficiency and antenna gain. The piezoelectric layer is 600 nm ZnO, and the magnetostrictive layer is (FeGaB 85 nm/SiO_2 5 nm)$_{10}$. The Bragg reflector shows a high reflection from 1 to 3 GHz and leads to the S_{11} peak of an electromechanical resonance frequency at 1.75 GHz, as shown in Fig. 12.4e. The SMR antenna shows a high antenna gain of −18.8 dBi, 1.0% electromechanical coupling coefficient, and $Q = 95$. The SMR ME antenna also has a high P1dB (1 dB compression point) of 31 dBm, compared to 7 dBm of the FBAR, showing 24 dB enhanced power handling capability of the SMR ME antennas, as shown in Fig. 12.4f. This is because acoustic Bragg reflector stacks in the SMR antenna have a much larger thermal conductivity than the air below the FBAR resonator, indicating the SMR

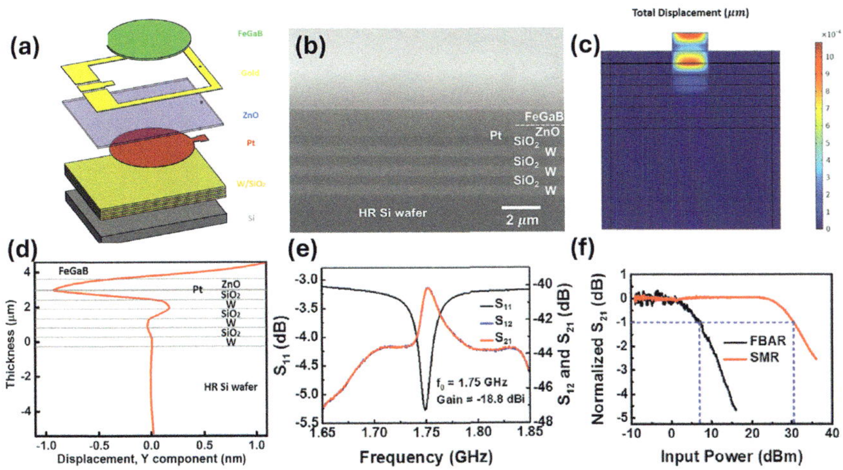

FIGURE 12.4 High gain solidly mounted resonator (SMR) ME antenna with high power handling capability. (a) 3D schematic drawing of the SMR ME antenna. (b) Cross-sectional scanning electron microscopy (SEM) image of the SMR ME antenna. (c) Displacement profile at the resonant frequency of distributed acoustic Bragg reflectors. (d) Standing wave amplitude as a function of depth. (e) Return loss curve (S_{11}), receiving (S_{12}) and transmitting behavior (S_{21}) of the SMR antenna with a resonant frequency of 1.75 GHz and an antenna gain of −18.8 dBi. (f) Power sweep of the FBAR and SMR antennas. The FBAR antenna has a 1 dB compression point (P1dB) of 7.1 dBm, while the P1dB of the SMR antenna is 30.4 dBm, which shows the better power handling capability of the SMR antenna. *Reproduced from Liang et al. (2023).*

ME antenna can more effectively release joule heating than the FBAR ME antenna. It was noted that the resonator of the SMR antenna was fixed to the Bragger reflector with no inherent curvature, thereby avoiding the disorder and multi-direction of magnetic domains due to a tensor stress.

12.3 Radiofrequency magnetoelectric antenna array

The performance of the ME antenna can be further improved through an array topology. The simulated induced output voltage of the mechanically driven ME antenna can be tripled through use of 3 ME resonators in series compared to a single resonator due to nonlinear coupling (Liang et al., 2018; Zaeimbashi et al., 2021), as shown in Fig. 12.5a. Additionally, as shown in Fig. 12.5b, the bandwidth of the ME antenna can be larger through the construction of an array with resonators with varying resonance frequencies (either through varying the thickness or the width of the resonator depending on the acoustic mode) (Liang et al., 2018). As shown in Fig. 12.5c, the FBAR ME antenna array has been recently demonstrated (Luo et al., 2024b), realizing a 10-dBi gain

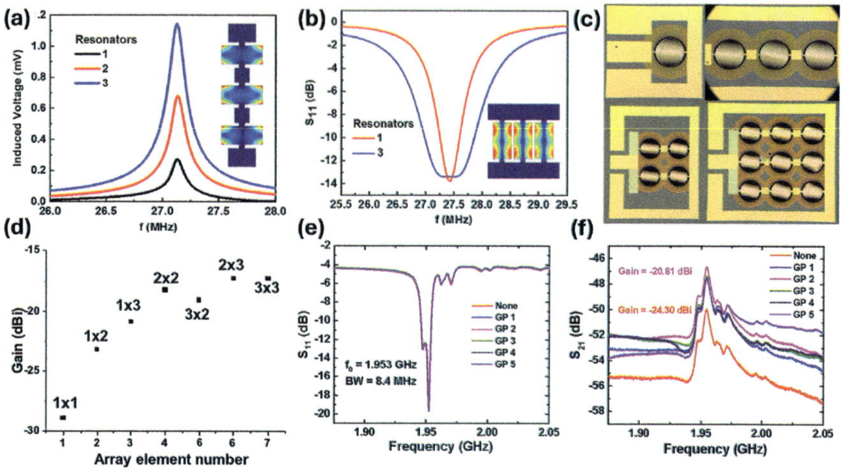

FIGURE 12.5 FBAR ME antenna array. (a) Increased ME voltage output as a function of ME resonator in series. (b) Increased bandwidth as a function of ME resonator in parallel. (c) Optic images of FBAR ME antenna arrays with various array topology (d) Antenna gain versus array element number, showing a 10-dBi gain enhancement for 3 × 3 ME antenna array. (e) Return loss S_{11} of FBAR ME antenna mounted on various sizes of ground planes. and (f) receiving characteristic (S_{21}) of FBAR ME antenna mounted on various sizes of ground planes, showing a consistent 3-dBi gain enhancement and ground plane immunity. *Fig. 12.5a–b reproduced from Liang et al. (2018); Fig. 12.5c–f reproduced from Luo et al. (2024b).*

enhancement over a single FBAR ME antenna and showing consistent ground plane immunity and 3-dBi gain enhancement when mounted on various sizes of ground planes, as shown in Fig. 12.5d–f. Unlike conventional electrical antenna arrays, the ME antenna array increases the antenna gain in a way that radiation effciency is enhanced, but the directivity is not quite affected, owing to its ultracompact dimension much less than one-quarter of electromagnetic wavelength. Through careful selection of the number of series elements, parallel resonators with varying anchor widths, and ensuring an impedance of 50 Ω at the resonant frequency, the compact and energy-efficient FBAR ME antenna array can potentially exhibit exceptionally high gain and radiation efficiency as well as wide frequency bandwidth, opening up exciting possibilities in multiband or wideband UHF wire-less communication, wireless power transfer, and portable electronic and biomedical devices. Furthermore, the consistent 3 dBi gain enhancement when mounted on different shapes of ground planes position the FBAR ME antenna array as an outstanding platform-independent conformal antenna candidate for portable implantable biomedical devices and various applications requiring compatibility with diverse ground plane configurations.

12.4 Outlook of radiofrequency magnetoelectric antenna

Pushing the electromechanical resonance in an ME resonator to a magnetoacoustic resonance (Bichurin et al., 2005) and realize strong magnon-phonon coupling is crucial to boosting antenna gain and radiation efficiency for power-efficient wireless communication, wireless power transfer, and magnetic sensing. Magnons and phonons are integral quasiparticles in the realm of solid-state physics. Magnons, the result of collective electron spin wave excitation, hold immense potential for applications in advanced information processing, communication, and sensing. Their attributes, including tunable frequencies through magnetic fields, rapid operation, and extended transduction distances, stem from their high spin density and magnetic dipolar coupling (Awschalom et al., 2021). Phonons, which denote collections of acoustic vibrations, excel as carriers of coherent and quantum information due to their long lifetimes within solid-state platforms. The small wavelengths associated with acoustic waves make phonons particularly suitable for on-chip hybrid quantum systems (Li et al., 2020), with their frequencies conveniently adjustable through geometric manipulation.

Coherent magnon–phonon coupling is a critical mechanism for seamless information transfer between magnon and phonon excitations (Huang & Tian, 2021). These fundamental quasiparticles play a pivotal role in bridging various quantum information carriers. The coupling not only holds the promise for revolutionizing information technologies but also sparks interdisciplinary investigations into hybrid multiphysical cavities such as cavity optomechanics, optomagnonics, and magnomechanics (Clerk et al., 2020; Zhang et al., 2016). A central aspect of strong magnon–phonon coupling, identified by anticrossings in magnon and phonon dispersion relations, gives rise to hybrid magnetoacoustic modes termed magnon polarons. The magnon–phonon coupling is quantified by the cooperativity parameter.

$$C = \frac{g^2}{\kappa_M \kappa_P} = \frac{(\Delta f)^2}{4\kappa_M \kappa_P}$$

where g is the coupling strength of magnons and phonons, Δf is the mode-splitting frequency gap. κ_M and κ_P are the loss rates of the magnon and phonon modes, respectively (Li et al., 2020).

Recent research has focused on robust magnon–phonon coupling with improved quality factors (Q) of acoustic and spin resonances, coupling strength (g), and cooperativity (C). This shift is driven by its potential contributions to magnon spintronics, spin caloritronics, and emerging phenomena such as the spin Seebeck effect, spin pumping, magnon–phonon conversion, and thermal Hall effect (Bauer et al., 2012; Bozhko et al., 2020). To enhance *Q*, *g*, and *C*, researchers are adopting innovative design methodologies. Strategies include reducing energy dissipation, localizing energy in specific coupling regions, and enhancing magnetoelastic coupling. Notable examples include fabricating Ni nanomagnets on Si substrates to achieve strong coupling using ultrafast pump-probe techniques (Berk et al., 2019) and leveraging perpendicular standing acoustic waves mediated by GGG substrates to enhance coupling between magnons in top and bottom low-loss YIG films (An et al., 2020). In 2020, a Galfenol nanograting with a high magnetoelastic coupling constant was introduced, leading to enhanced cooperativity (An et al., 2020; Berk & Schmidt, 2020). Techniques such as Gaussian focusing IDTs in surface acoustic wave devices and Bragg reflectors for magnetoacoustic systems further enhance coupling (Bolgar et al., 2018; Satzinger et al., 2018; Whiteley et al., 2019). A strong magnon−phonon coupling in bulk YIG/AlN ME resonators, simulated in 2022, revealed insights into radiation efficiency enhancement with two split magnetoelastic radiation modes (Ji et al., 2022). In the strong coupling region, the radiation efficiency can be enhanced by over two orders of magnitudes with a maximum value of around 3.5% (Ji et al., 2022). This coupling can also significantly boost radiation efficiency in ME antennas, offering the potential for highly efficient IoT applications.

Key design parameters for robust magnon−phonon coupling are piezoelectric materials with high electromechanical coupling efficiency and magnetostrictive materials with low loss and high magnetostriction. The potential of ion slicing to create high-quality released YIG films for FBAR and SMR antennas, enhancing radiation efficiency, is also evident (Levy et al., 1997, 1998; Rachford et al., 2000). Additionally, the replacement of AlN with AlScN having six times enhanced piezoelectric constant (Fichtner et al., 2019) or use of materials with higher magnetostriction constants such as boron- or carbon-doped FeGa and FeCo alloys (Liang et al., 2019, 2020b; Schoen et al., 2016), spinel NiZnAl-ferrite (Emori et al., 2017), and magnesium aluminum ferrite oxide $MgAl_{0.5}Fe_{1.5}O_4$ (Li et al., 2022) hold promise for further enhanced coupling. These advancements can enable highly efficient ultra-compact ME antennas with high radiation efficiency and magnon−phonon quantum transduction devices.

In summary, UHF ME antennas are self-biased and passive devices that have the highest gains at the same frequency among nano-scale passive antennas, making them ideal for a variety of applications, including portable and wearable IoT devices, biomedical and implantable devices, etc. Future iterations seek to incorporate low magnetic loss materials such as yttrium iron garnet into the ME antennas, aiming to boost antenna gain and radiation efficiency. Besides, 2D and 3D micromagnetic and acoustic simulation platforms need to be developed for the fundamental understanding of the complex interactions of spin waves and acoustic waves and optimization design of UHF ME antenna units and array structures.

12.5 Challenges of radiofrequency magnetoelectric antenna

12.5.1 Challenges on antenna simulation

At present, there are no numerical tools or softwares to simulate mechanical vibration, antenna radiation, and efficiency at the same time. The mechanical vibration can be simulated by COMSOL, which employs the finite element method (FEM) for electromechanical resonance simulation. Hence, the antenna return loss and vibration mode can be obtained by COMSOL simulation. However, the lack of a multiphysics model for ME coupling and electromagnetic radiation makes it difficult to simulate the conversion of magnetic domain oscillations into antenna radiation. On the other hand, a high-frequency structural simulator (HFSS) is widely used for return loss, gain, and radiation pattern analysis of conventional antennas, but there is an absence of mechanical vibration and ME coupling models. Similarly, the numerical calculation of efficiency, radiation pattern, and gain can also be done with an FDTD method (Yao et al., 2015).

In conclusion, for a complete simulation method on ME antennas, a reliable ME coupling model, a piezoelectricity model, and a radiation calculation model are necessary. Unfortunately, there is no single simulation tool with all of these components.

12.5.2 Challenges on device fabrication

During the antenna fabrication, magnetic domain control is the main issue. The magnetic film with a single domain has the strongest magnetostriction and lowest magnetic loss, which can improve radiation efficiency. The magnetic stacks of ME devices are still multidomain (Liang et al., 2020a, 2023; Nan et al., 2017) due to the deposition process and film stress. The thin-film stress is related to plasma energy during the film deposition process (Chason et al., 2002; Müller, 1987). Large film stress results in a huge anisotropy field and thereby disarranged domain directions. This adversely affects the magnetic domain rotation, leading to a lowered antenna gain and radiation efficiency. Hence, the deposition conditions still need to be optimized for a lower stress. Besides, exchange biased stacks (Hrkac et al., 2014; Jovičević Klug et al., 2019; Lage et al., 2012, 2013, 2014; Röbisch et al., 2015; Schell et al., 2023; Thormählen et al., 2024) show great potential for realizing single magnetic domain for improving antenna radiation performance.

12.5.3 Challenges on improving antenna performance

Currently, the antenna gain and radiation efficiency are low for practical applications. For example, the gain of the FBAR antenna is −18 dBi (Nan et al., 2017). The tightly packed large array of the ME antenna elements is another scalable approach, which leads to enhanced radiation efficiency. This is contrary to conventional antenna arrays, for which the spacing between the elements is half wavelength or comparable. Using the FBAR antenna as an example, an array structure with 3 single resonators was introduced (Lin et al., 2018), in which the simulation results show a more than 3 times enhancement in induced output voltage from the ME antenna array, which is due to more equivalent magnetic dipoles in the whole antenna structure.

Another challenge is the narrow operating bandwidth due to high quality factor Q, since the 3-dB bandwidth equals to $1/(\sqrt{2}\ Q)$. For a given Q of 632, the bandwidth is 0.228% (Nan et al., 2017). Since the overall Q-factor of the antenna includes the device loss (Q_{loss}) and radiation (Q_{rad}), decreasing the Q_{loss} of the resonator is a reliable choice for a wider bandwidth, and AlScN or ZnO is a typical alternative of AlN for a lower Q. By using AlScN or ZnO, more mechanical loss is introduced to lower down the overall Q. However, the antenna efficiency may be deteriorated by this approach if no other measure is taken to reduce Q_{rad}, since efficiency $= \frac{Q_t}{Q_{rad}}$. Moreover, an antenna array with different resonator thicknesses for each element is another way to improve the bandwidth.

The last problem is the fragile mechanical anchors between the antenna resonator and substrate. For the FBAR antenna, an encapsulation structure is needed to protect the device. The most common packaging method is wafer bonding. A better solution has been the SMR design, where the whole resonator is fixed to the substrate with no floating components. Therefore, the SMR device has a higher structural integrity and can withstand higher power inputs and extreme physical shocks.

References

An, K., et al. (2020). Coherent long-range transfer of angular momentum between magnon Kittel modes by phonons. *Physical Review B: Condensed Matter, 101*, Article 060407.

Awschalom, D. D., et al. (2021). Quantum engineering with hybrid magnonic systems and materials (invited paper). *IEEE Transactions on Quantum Engineering, 2*, 1−36.

Bauer, G. E. W., Saitoh, E., & van Wees, B. J. (2012). Spin caloritronics. *Nature Materials, 11*, 391−399.

Berk, C., et al. (2019). Strongly coupled magnon-phonon dynamics in a single nanomagnet. *Nature Communications, 10*.

Berk, C. R., & Schmidt, H. (2020). A quantum tango between magnons and phonons. *Physics, 13*, 167.

Bichurin, M., Petrov, V., Ryabkov, O., Averkin, S., & Srinivasan, G. (2005). Theory of magnetoelectric effects at magnetoacoustic resonance in single-crystal ferromagnetic-ferroelectric heterostructures. *Physical Review B: Condensed Matter, 72*, Article 060408.

Bolgar, A. N., et al. (2018). Quantum regime of a two-dimensional phonon cavity. *Physical Review Letters, 120*, Article 223603.

Bozhko, D. A., Vasyuchka, V. I., Chumak, A. V., & Serga, A. A. (2020). Magnon-phonon interactions in magnon spintronics (Review article). *Low Temperature Physics, 46*, 383–399.

Chason, E., Sheldon, B. W., Freund, L. B., Floro, J. A., & Hearne, S. J. (2002). Origin of compressive residual stress in polycrystalline thin films. *Physical Review Letter, 88*(15), 156103.

Chen, H. H., et al. (2020). Ultra-compact mechanical antennas. *Applied Physics Letters, 117*.

Chen, J. C., et al. (2022). A wireless millimetric magnetoelectric implant for the endovascular stimulation of peripheral nerves. *Nature Biomedical Engineering, 6*, 706–716.

Clerk, A. A., Lehnert, K. W., Bertet, P., Petta, J. R., & Nakamura, Y. (2020). Hybrid quantum systems with circuit quantum electrodynamics. *Nature Physics, 16*, 257–267.

Dubois, M.-A., Muralt, P., Matsumoto, H., & Plessky, V. (1998). *IEEE ultrasonics symposium. Proceedings (cat. No. 98CH36102)* (pp. 909–912) (IEEE).

Emori, S., et al. (2017). Coexistence of low damping and strong magnetoelastic coupling in epitaxial spinel ferrite thin films. *Advanced Materials, 29*, Article 1701130.

Fichtner, S., Wolff, N., Lofink, F., Kienle, L., & Wagner, B. (2019). AlScN: A III-V semiconductor based ferroelectric. *Journal of Applied Physics, 125*.

Hrkac, V., et al. (2014). Amorphous FeCoSiB for exchange bias coupled and decoupled magnetoelectric multilayer systems: Real-structure and magnetic properties. *Journal of Applied Physics, 116*, 134302.

Huang, H., & Tian, Z. T. (2021). Topological phonon-magnon hybrid excitations in a two-dimensional honeycomb ferromagnet. *Physical Review B: Condensed Matter, 104*.

Ji, Y., Zhang, C., & Nan, T. (2022). Magnon-phonon-interaction-induced electromagnetic wave radiation in the strong-coupling region. *Physical Review Applied, 18*, Article 064050.

Jovičević Klug, M., et al. (2019). Antiparallel exchange biased multilayers for low magnetic noise magnetic field sensors. *Applied Physics Letters, 114*, 192410.

Lage, E., et al. (2012). Exchange biasing of magnetoelectric composites. *Nature Materials, 11*, 523–529.

Lage, E., et al. (2013). Exchange biased magnetoelectric composites for vector field magnetometers. *Journal of Applied Physics, 113*(17).

Lage, E., et al. (2014). Magnetic domain control and voltage response of exchange biased magnetoelectric composites. *Applied Physics Letters, 104*, 132405.

Lakin, K., McCarron, K., & Rose, R. (1995). *IEEE ultrasonics symposium. Proceedings. An international symposium* (pp. 905–908) (IEEE).

Lanz, R., & Muralt, P. (2005). Bandpass filters for 8 GHz using solidly mounted bulk acoustic wave resonators. *IEEE Transactions on Ultrasonics, Ferroelectrics, and Frequency Control, 52*, 938–948.

Levy, M., Osgood, R. M., Kumar, A., & Bakhru, H. (1997). Epitaxial liftoff of thin oxide layers: Yttrium iron garnets onto GaAs. *Applied Physics Letters, 71*, 2617–2619.

Levy, M., Osgood, R. M., Kumar, A., & Bakhru, H. (1998). Crystal ion slicing of single-crystal magnetic garnet films. *Journal of Applied Physics, 83*, 6759–6761.

Li, Y., et al. (2020). Hybrid magnonics: Physics, circuits, and applications for coherent information processing. *Journal of Applied Physics, 128*.

Li, R. F., et al. (2022). Anisotropic magnon spin transport in ultrathin spinel ferrite thin films-evidence for anisotropy in exchange stiffness. *Nano Letters, 22*, 1167–1173.

Liang, X. F., Chen, H. H., Sun, N., Lin, H., & Sun, N. X. (2018). Novel acoustically actuated magnetoelectric antennas. In *2018 IEEE antennas and propagation society international symposium on antennas and propagation & USNC/URSI National Radio Science Meeting* (pp. 2189–2190).

Liang, X. F., et al. (2019). Soft magnetism, magnetostriction, and microwave properties of Fe-Ga-C alloy films. *Ieee Magnetics Letters, 10*.

Liang, X. F., et al. (2020a). Mechanically driven SMR-based MEMS magnetoelectric antennas. In *2020 IEEE international symposium on antennas and propagation and north American radio science meeting* (pp. 661–662).

Liang, X. F., et al. (2020b). A review of thin-film magnetoelastic materials for magnetoelectric applications. *Sensors, 20*.

Liang, X. F., et al. (2022). Experimental demonstration of ground plane immunity for magnetoelectric antennas. In *2022 IEEE Mtt-S international microwave workshop series on advanced materials and processes for Rf and Thz applications, IMWS-AMP*.

Liang, X., et al. (2023). Mechanically driven solidly mounted resonator-based nanoelectromechanical systems magnetoelectric antennas. *Advanced Engineering Materials, 25*(21), 2300425.

Lin, H., et al. (2018). NEMS magnetoelectric antennas for biomedical application. *IEEE International Microwave Biomedical Conference (IMBioC)*.

Luo, B., et al. (2024a). Magnetoelectric microelectromechanical and nanoelectromechanical systems for the IoT. *Nature Reviews Electrical Engineering, 1–18*.

Luo, B., et al. (2024b). Gain enhancement and ground plane immunity of mechanically driven thin-film bulk acoustic resonator magnetoelectric antenna arrays. *Advanced Functional Materials, 2403244*.

Müller, K. H. (1987). Stress and microstructure of sputter-deposited thin films: Molecular dynamics investigations. *Journal of Applied Physics, 62*(5), 1796–1799.

Nan, T., et al. (2017). Acoustically actuated ultra-compact NEMS magnetoelectric antennas. *Nature Communications, 8*.

Rachford, F. J., Levy, M., Osgood, R. M., Kumar, A., & Bakhru, H. (2000). Magnetization and FMR studies of crystal-ion-sliced narrow linewidth gallium-doped yttrium iron garnet. *Journal of Applied Physics, 87*, 6253–6255.

Röbisch, V., et al. (2015). Exchange biased magnetoelectric composites for magnetic field sensor application by frequency conversion. *Journal of Applied Physics, 117*(17).

Satzinger, K. J., et al. (2018). Quantum control of surface acoustic-wave phonons. *Nature, 563*, 661–665.

Schell, V., et al. (2023). Exchange biased surface acoustic wave magnetic field sensors. *Scientific Reports, 13*, 8446.

Schoen, M. A. W., et al. (2016). Ultra-low magnetic damping of a metallic ferromagnet. *Nature Physics, 12*, 839–842.

Thormählen, L., et al. (2024). Low-noise inverse magnetoelectric magnetic field sensor. *Applied Physics Letters, 124*, 172402.

Whiteley, S. J., et al. (2019). Spin-phonon interactions in silicon carbide addressed by Gaussian acoustics. *Nature Physics, 15*, 490–+.

Yao, Z., Wang, Y., Keller, S., & Carman, G. (2015). Bulk acoustic wave-mediated multiferroic antennas: Architecture and performance bound. *IEEE Transactions on Antennas and Propagation, 63*, 3335–3344.

Zaeimbashi, M., et al. (2019). NanoNeuroRFID: A wireless implantable device based on magnetoelectric antennas. *IEEE Journal of Electromagnetics, RF and Microwaves in Medicine and Biology, 3*(3), 206–215.

Zaeimbashi, M., et al. (2021). Ultra-compact dual-band smart NEMS magnetoelectric antennas for simultaneous wireless energy harvesting and magnetic field sensing. *Nature Communications, 12*.

Zhang, X., Zou, C.-L., Jiang, L., & Tang, H. X. (2016). Cavity magnomechanics. *Science Advances, 2*, Article e1501286.

CHAPTER

13

Power, RF, microwave, and millimeter-wave magnetoelectric devices (tunable inductors, filters, and phase shifters)

Gopalan Srinivasan[1], *Bin Luo*[2], *Yifan He*[2], *Xiaxin Liu*[2], *Yisi Liu*[2], *and Nian Sun*[2]

[1] **Oakland University, Rochester, MI, United States;** [2] **Northeastern University, Boston, MA, United States**

13.1 Introduction

Ferrites are preferred materials due to low losses at high frequencies for use as inductors and microwave and millimeter wave devices including resonators, filters, phase shifters, and RF sensors. The key challenge in miniaturization of these devices is the need for a tunable bias magnetic field. The recent efforts on electric field tuning of such devices facilitated by the magneto-electric effects in a ferrite-ferroelectric is a significant development in this regard. A wide variety of new applications, especially for high-frequency integrated electronics and magnetic storage which include waveguides, magnetoresistive random-access memory (MRAM), spin-based transistors, oscillators, and sensors, etc. have emerged in recent years. This chapter provides key details on these device efforts in recent years (Che et al., 2004; Gao, Zardareh, et al., 2014; Gupta et al., 2021; Haldar et al., 2016; He et al., 2018, 2021; Kaka et al., 2005; Nikonov & Bourianoff, 2005; Zand et al., 2017).

Magnetoelectrics (ME) refers to the coupling between magnetism and electricity and is one of the aspects of multiferroics, which indicates materials present two or more ferroic (ferroelectricity, ferroelasticity, and

https://doi.org/10.1016/B978-0-443-18606-6.00007-1

ferromagnetism) properties. Compared to their single-phase counterparts, two-phase ME heterostructure that couples through the mechanical strain has much larger ME coupling (Spaldin & Fiebig, 2005). The ME coupling effect provides a new methodology to mediate and monitor the magnetism in the magnetic material and hence brings a number of new potential features to the devices mentioned above. Thanks to the emergence and prosperity of complementary metal-oxide-semiconductor (CMOS) and thin-film technology, ME materials have been integrated into batch fabrication and incorporated in a number of compact, power-efficient on-chip devices. The invention of integrated electric field tunable magnetic devices including inductors (Chen et al., 2020), filters (Chen et al., 2020), etc. benefit from the converse ME effect (CME), which means electrical field control of magnetization.

The nature of strain-mediated, high-frequency ME coupling and hybrid spin-electromagnetic modes in ferromagnetic/ferroelectric composites led to their use for a new family of dual E- and H-tunable microwave and millimeter-wave signal processing devices (Bichurin et al., 2007; Lawes & Srinivasan, 2011; Srinivasan, 2010; Sun & Srinivasan, 2012). The ferromagnetic phase in the composites for such high-frequency devices, in general, are low-loss ferrites, such as single-crystals or liquid phase epitaxy (LPE) grown films of yttrium iron garnet (YIG) and single-crystal nickel ferrite (NFO), lithium ferrite (LFO), Y-type hexagonal ferrites, and M-type hexagonal barium-aluminum and strontium-aluminum ferrites. Ferroelectric phases of interest are lead zirconate titanate (PZT), lead magnesium niobate-lead titanate (PMN-PT), and lead zinc niobate-lead titanate (PZN-PT) (Sun & Srinivasan, 2012).

A majority of devices discussed here are based on ferromagnetic resonance (FMR) in a layered ferrite/ferroelectric composite. FMR is a powerful tool for studies on the converse-ME (CME) effects in multiferroic composites. An electric field E applied to a ferrite/ferroelectric composite produces a mechanical deformation in the ferroelectric phase, which in turn is coupled to the ferrite, resulting in a shift in the resonance field/frequency. Information on the nature of CME coupling could therefore be obtained from data on the resonance field shift δH_E or frequency shift δf_E versus E. The first such study was on bulk samples of 90% YIG/10% PZT, and the ME coupling was quite weak because of low PZT concentration (Bichurin et al., 1997). But in samples with any higher PZT, FMR line broadening masked the electric field effect on the resonance field. Such line broadening, however, could easily be eliminated in layered samples. Theoretical models for CME in ferrite/ferroelectric bilayers predict a strong interaction (Bichurin et al., 2001, 2002). The effect also opens up the possibility of novel voltage-tunable ferrite signal processing devices (Sun & Srinivasan, 2012).

The CME effect discussed above takes place in ferrite/ferroelectric layered structures when the phases are *tightly bound*, that is, when the mechanical stress created in one is transferred to the second. There are, however, other ME phenomena that do not require bonding between the layers and take place simply due to the proximity of two materials having different dielectric and magnetic properties. An example of such a phenomenon is the formation of hybrid spin-electromagnetic waves in the layered structures (Demidov et al., 2002). Theoretical models for wave propagation in ferromagnetic/ferroelectric slabs predict intensive hybridization between the electromagnetic waves propagating in the ferroelectric film and the spin waves propagating in the ferromagnetic film. Also, the theory shows that the dispersion characteristics of hybrid waves can be tuned with an electric field or a magnetic field (Ustinov et al., 2006).

The CME at FMR and the hybrid waves in multiferroic composites are of importance for dual E- and H-tunable microwave and millimeter-wave devices. Tunable microwave devices are widely used in radar, telecommunication, and RF devices. One class of such devices is based on ferrites (Adam et al., 1991, 2002). In particular, resonators fabricated from YIG films are attractive due to their planar geometry, small size (area on the order of several square millimeters), and high Q-factor (up to 4000). Their resonance frequency could be tuned over a wide frequency range (0.5–12 GHz) through changes in the magnetic permeability by varying the bias magnetic field created by an electromagnet (Adam et al., 2002). However, such "magnetic" tuning is relatively slow and is associated with large power consumption. Another class of tunable microwave devices is based on ferroelectric materials (Cole et al., 2003). The tuning of such devices is realized by applying an electric field that reduces the dielectric permittivity of the ferroelectric material. Advantages of this "electrical" tuning are fast resonant frequency adjustment and low power consumption because ferroelectric elements, essentially, draw zero static current. But ferroelectric devices in general have low Q and high insertion loss. The multiferroic composites provide a new promising technology for tunable microwave and millimeter-wave devices, which combines the advantages of ferrite and ferroelectric devices (Sun & Srinivasan, 2012). The devices are based on CME or the excitations of hybrid spin-electromagnetic waves in ferrite/ferroelectric layered structures. Such devices are dual-tunable, with electrical tuning realized through the application of an electric field to the ferroelectric layer, while the magnetic tuning is realized through the application of a bias magnetic field.

This chapter is organized as follows. First, we discuss converse ME effects at FMR and the nature of hybrid waves in the composites. Composite materials for high-frequency devices, including tunable inductors, resonators, filters, phase shifters, and beam-forming antennas, are considered next. Finally, the details on multiferroic microwave and millimeter-wave devices and miniature antennas are provided. Integrated magnetic and ME devices such as high-performance transformers, inductors with laminated magnetic core, voltage tunable inductors, E- and H- field tunable bandpass filters are included in this chapter. These applications are good representatives for power components, basic circuit components, RF and microwave signal processing blocks, and RF and microwave signal transmitting/receiving blocks, respectively (He et al., 2021).

13.2 CME effect at ferromagnetic resonance

Ferrites are the materials of choice for microwave signal processing devices due to low losses, and the ferrite/piezoelectric composite is a promising candidate for high-frequency devices based on CME interactions. Theoretical models of the effects are provided in Bichurin et al. (2001, 2002). In general, the composite susceptibility under an applied magnetic field H and electric field E and the order of parameters are related by the following equations:

$$\begin{aligned} \boldsymbol{P} &= \chi^{E}\mathbf{E} + \chi^{EM}\mathbf{H} \\ \boldsymbol{M} &= \chi^{ME}\mathbf{E} + \chi^{M}\mathbf{H} \end{aligned} \tag{13.1}$$

Here P is the electric polarization, M is the magnetization, χ^{E} and χ^{M} are the electric and magnetic susceptibilities, and χ^{EM} and χ^{ME} are the ME susceptibilities. One has to consider two important interactions: (1) coupling between microwave magnetic field and DC electric field and (2) coupling between microwave magnetic and electric fields. The former is expressed in terms of electric field effects on magnetic susceptibility, whereas the latter is described by ME susceptibility. Here we are primarily concerned with the influence of the electric field that produces a piezoelectric strain in the composites that manifests as an induced magnetic anisotropy in the ferromagnetic phase, leading to a shift in the FMR profile.

A general theory, applicable to either bulk or layered composites subjected to DC and AC magnetic fields and a static electric field E, was considered in Bichurin et al. (2001, 2002) and applied to layered lithium ferrite (LFO)-PZT, NFO-PZT, and YIG-PZT. The theoretical FMR profiles in terms of the real part of the magnetic susceptibility χ' versus H for $E = 0$ and 300 kV/cm are shown in Fig. 13.1. The results are for a bilayer with H and E perpendicular to the sample plane and for a frequency of 9.3 GHz. The static field range is chosen to include FMR in the ferrite. For $E = 0$, one observes the expected resonance in the profiles. With the application of $E = 300$ kV/cm, a downshift in the resonance field occurs, since E essentially gives rise to an internal magnetic field. The condition for FMR, the resonance frequency ω_0 for H perpendicular to the plane of a ferrite-ferroelectric disk, is given by

$$\omega_0 = \gamma(\mathrm{H} + \mathrm{H_A} - 4\pi M) \tag{13.2}$$

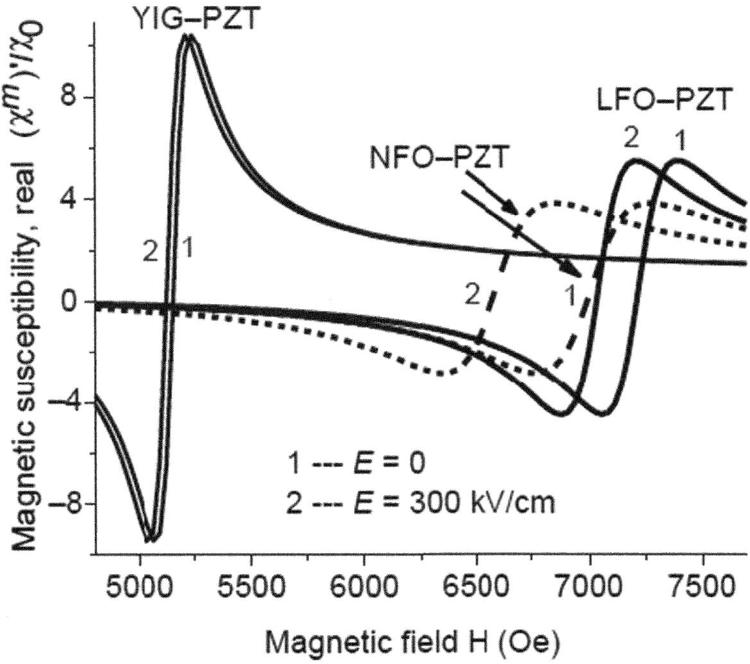

FIGURE 13.1 Susceptibility versus H profiles for bilayer ferrite-PZT at 9.3 GHz with (1) E = 0 and (2) E = 300 kV/cm. Fields E and H are perpendicular to the plane of disk-shaped samples of nickel ferrite (NFO)-PZT, lithium ferrite (LFO)-PZT, and yttrium iron garnet (YIG)-PZT (Bichurin et al., 2001).

where γ is the gyromagnetic ratio, H_A is the anisotropy field, and $4\pi M$ is the saturation induction. For an electric field E applied to a ferrite/piezoelectric bilayer, the piezoelectric deformation is equivalent to the effect of magnetostriction on FMR, leading to an internal magnetic field $\delta H_E = AE$, where A is the ME coupling constant. The ME effects at FMR can be taken into account by

$$\omega_0 = \gamma(H + H_A - 4\pi M + AE) \tag{13.3}$$

The E-induced shift δH_E in the resonance field is determined by the magnetostriction λ and piezoelectric coupling d. When E and H are directed along a symmetry axis, <100> for example, perpendicular to the plane (1, 2) of a bilayer, one obtains (Bichurin et al., 2001)

$$\delta H = \frac{3\lambda_{100} d_{31} E_3}{M_0 \left[\left(p_{s_{11}} + p_{s_{12}} \right)(1 - v) + \left(m_{s_{11}} + m_{s_{12}} \right) v \right]} \tag{13.4}$$

where s is the compliance for magnetostrictive (m) and piezoelectric (p) phases and v is the volume (or thickness) fraction for the piezoelectric phase. The field shift δH_E in Fig. 13.1 varies from a maximum of 330 Oe for NFO-PZT to a minimum of 22 Oe for YIG-PZT and is due to $\lambda \sim 2$ ppm for YIG versus −30 ppm for NFO (Bichurin et al., 2001).

The profiles in Fig. 13.1 also reveal the possibility of voltage-controlled ferrite devices. In particular, one can envision E-tunable FMR-based devices such as ferrite resonators, filters, phase shifters, and attenuators. Such devices are discussed in Section 13.5.

13.3 Hybrid spin-electromagnetic waves in composites

Recently, a new promising technology for tunable microwave devices, which combines the advantages of ferrite and ferroelectric devices, has been developed (Sun & Srinivasan, 2012; Ustinov et al., 2006). The technique involves the excitations of hybrid spin-electromagnetic waves in ferrite−ferroelectric layered structures. Such excitations are dual E- and H-tunable, with electrical tuning realized through the application of an electric field to the ferroelectric layer, while the magnetic tuning is realized through the application of a bias magnetic field.

A theory of the hybrid waves in ferrite−ferroelectric bilayers based on a coupled-mode approach was developed in Ustinov et al. (2006). To find the resonance frequency of the hybrid modes in a rectangular bilayer, for example, one has to solve the Maxwell equations for the microwave electric and magnetic fields and the linearized Landau−Lifshits equation for the microwave component of the magnetization. For simplicity, one could ignore the exchange interactions in the ferrite, as well as any crystallographic magnetic anisotropy. The equations were solved together with the appropriate boundary conditions. The microwave magnetization can be expanded in a series using the system of magnetostatic eigenmodes of a rectangular ferrite resonator. Similarly, the electromagnetic field of the ferroelectric resonator can also be developed in a series of Eigen functions. A system of equations thus developed allows one to determine resonance frequencies of the composite resonator using only the information related to the uncoupled ferrite and ferroelectric subsystems, namely, their resonance frequencies and the distributions of microwave electromagnetic fields of the corresponding eigenmodes (Demidov et al., 2002; Ustinov et al., 2006).

In the general case, the system of equations for an infinite system describing coupling of *all* the ferromagnetic eigenmodes with *all* the ferroelectric eigenmodes could be solved only numerically. The frequency separation between the electromagnetic modes of the ferroelectric resonator is usually rather large (of the order of the frequency of the fundamental mode of the ferroelectric resonator) and, therefore, in most cases, one can ignore the possible excitation of the higher ferroelectric modes. In contrast, the frequency separation between the modes of the ferromagnetic resonator is much smaller, and a large number of these modes could be excited simultaneously. It should be noted that the simultaneous excitation of the several modes of a ferromagnetic resonator leads to a rather complicated shape and inhomogeneous broadening of the resonance curve of the hybrid ferrite-ferroelectric resonator, and, if possible, should be avoided (Ustinov et al., 2006).

While the reduction of the spurious influence of the higher ferromagnetic modes can be achieved by the use of small ferromagnetic resonators, having a larger frequency separation between the eigenmodes by using a highly symmetrical resonator geometry is also preferred. Namely, if the structures of the microwave electromagnetic fields of the ferromagnetic and ferroelectric eigenmodes are very similar, then each ferroelectric mode will be, for the most part, coupled to only one ferromagnetic mode. In particular, such a situation is approximately realized in the rectangular resonator geometry with equal in-plane sizes of the ferromagnetic and ferroelectric slabs. In this case, the coupling of the lowest ferroelectric mode to higher ferromagnetic modes is caused only by edge effects and the finite thickness of the ferromagnetic film, and, in most cases, can be neglected.

In the case when one (λ-th) ferroelectric mode of frequency Ω_λ interacts with only one (v-th) ferromagnetic mode of frequency ω_v, one obtains the following expression for the resonance frequencies of the hybrid ferrite–ferroelectric resonator (Ustinov et al., 2006):

$$\omega_{1,2}^2 = \frac{1}{2}\left(\omega_\nu^2 + \Omega_\lambda^2 + \xi\omega_M\omega_\nu \pm \sqrt{\left(\omega_\nu^2 + \Omega_\lambda^2 + \xi\omega_M\omega_\nu\right)^2 - 4\omega_\nu^2\Omega_\lambda^2}\right) \tag{13.5}$$

Here ξ is a dimensionless coupling parameter for the ferroelectric and ferromagnetic eigenmodes and $\omega_M = \gamma 4\pi M$. For typical experimental conditions, ξ is on the order of 0.001–0.01. The above equation allows one to also determine the magnetic tuning $\partial\omega/\partial \mathrm{H}$ and electric tuning $\partial\omega/\partial \mathrm{E}$ of the resonance frequencies of the coupled hybrid resonator and to estimate the resonance linewidth $\Delta\omega \approx |\frac{\partial\omega}{\partial H_0}|\Delta H + \varepsilon|\frac{\partial\omega}{\partial\varepsilon}|\tan\delta$, where $\Delta \mathrm{H}$ is the linewidth of the FMR in the ferrite layer and δ is the angle characterizing the dielectric losses in the ferroelectric layer.

The theory predicts superior tuning characteristics when one optimizes the resonator geometry, that is, equal in-plane dimensions of the ferrite and ferroelectric layers of this hybrid resonator. This geometry maximizes coupling between the ferromagnetic and dielectric resonant modes because the electromagnetic fields of the modes overlap efficiently. In addition, due to the strong dipolar pinning of magnetization at the lateral edges of a thin ferromagnetic layer (Cole et al., 2003) and the high dielectric permittivity of the ferroelectric medium, resonance modes of both the ferromagnetic and ferroelectric components have approximately the same spatial structure of electromagnetic fields. As a result, the main dielectric mode is coupled mostly to the main ferromagnetic mode, that is, the proposed geometry substantially reduces spurious influence of the higher-order resonance modes of the layered structure (Ustinov et al., 2006). Resonators and phase shifters based on hybrid waves are discussed later in this chapter.

13.4 Composites for high-frequency devices

Ferrites are the materials of choice for high-frequency devices due to low losses at microwave and millimeter-wave frequencies (Adam et al., 1991, 2002). Composites with spinel ferrites, YIG, and Y-, Z-, and M-type hexagonal ferrites are appropriate for use in E− and H-tunable devices over 1−110 GHz. The choice of ferrite for a specific frequency range is determined by (1) the desired anisotropy field for operation of the device without or with nominal bias magnetic field, (2) the magnetostriction λ that determines the strength of ME coupling, (3) saturation magnetization $4\pi M$, (4) FMR line-width ΔH, and (v) high-frequency dielectric and magnetic losses. The parameters of importance for the ferrites are listed in Table 13.1 (Hellwege & Springer, 1970). Thin films grown by liquid phase epitaxy and single-crystals of YIG have ΔH on the order of 0.5-1 Oe at frequencies up to 12 GHz and low magnetic and dielectric losses. For frequencies above X-band, however, the need of large external magnetic fields makes the use of YIG less attractive for ferromagnetic resonance-based devices. Two types of ferrites (1) nickel and lithium ferrites, $Ni\ Fe_2\ O_4$ and $Li_{0.5}\ Fe_{2.5}\ O_4$, are appropriate for devices up to 15 GHz and (2) hexaferrites for 15−110 GHz devices. Spinel ferrites, including nickel ferrite and lithium ferrite, have ΔH = 5−50 Oe. A large uniaxial or planar anisotropy field in the hexaferrites facilitates FMR over 15−110 GHz for moderate bias magnetic fields. The Y- and Z-type hexagonal ferrites with easy-plane anisotropy, $Ba_2Zn_2Fe_{12}O_{22}$ (Zn_2Y) and $Ba_3Co_2Fe_{24}O_{41}$ (Co_2Z), are preferred for use at 10−40 GHz. M-type hexaferrites, $Me\ Fe_{12\text{-}x}\ Al_x\ O_{19}$ (Me = Ba, Sr), have uniaxial anisotropy, and $BaFe_{12}O_{19}$ (BaM) or SrM are ideal for the frequency range 40−75 GHz. Aluminum-substituted hexaferrites, including (Ba, Al) M and (Sr, Al) M, have a uniaxial anisotropy field HA on the order of 17−40 kOe and are ideal for 50−110 GHz devices (Hellwege & Springer, 1970) .

TABLE 13.1 Key material parameters for ferromagnetic and ferroelectric phases for use in high-frequency multiferroic devices.

Ferromagnetic/ ferroelectric phase	**$4\pi M$ (G)**	ΔH (Oe)	H_A (kOe)	λ (ppm)	d (pm/V)
YIG	1750		0.05	−2	
Nickel ferrite				−30	
Lithium ferrite				−18	
Zn_2Y					
Ba−Al-M		6-50	17-37	−6 to −15	
Sr−Al-M		10-50	19-40	−6 to −10	
PZT					400-600
PMN-PT					1500-2500
PZN-PT					1600-3200

The saturation magnetization $4\pi M$, FMR line-width ΔH, anisotropy field H_A, magnetostriction λ parallel to the applied magnetic field direction, and the piezoelectric coupling coefficient d are listed (Hellwege & Springer, 1970) 13.5 Multiferroic high-frequency devices.

Similarly, low-loss ferroelectrics with high piezoelectric coupling (PZT, PMN-PT, PZN-PT) and dielectrics that show rapid changes in ε with E are important for use as a ferroelectric/dielectric phase (Cross, 1993). The ferroelectric nature of PZT, $Pb(Zr_{1-x}Tix)O_3$, is critically dependent on the Zr/Ti ratio, with PbZrO3 being an antiferroelectric and PbTiO3 ferroelectric at room temperature. Much interest is focused on compositions close to the morphotropic phase boundary (MPB) for Ti ratio of ~47% because of their large piezoelectric constants (Cross, 1993). PMN-PT, (1 -x) $Pb(Mg_{1/3}Nb_{2/3})O_3$-x $PbTiO_3$, is a relaxor ferroelectric with a diffused phase transition, frequency-dependent dielectric maxima, nonlinear dielectric response for bias voltage and shows an MPB around 33% Ti. Single-crystal lead magnesium-lead titanates (0:68 PMN-0.32 PT) have desirable characteristics ($\varepsilon = 5000-7000$; $d_{33} = 2500-3200$ pC/N, $\tan\delta = 0.005$) (Cross, 1993). PZTs have high d_{33} (400−600 pC/N) for use in composites but have high losses at microwave frequencies. Similarly, pure barium titanate (BTO) experiences losses due to its domain structure, which can be reduced by forming a solid solution with strontium titanate (Cross, 1993). Bismuth strontium titanate (BST) compositions (0.5BTO-0.5STO) with a high dielectric constant and E-dependence are used in devices based on hybrid modes (Cross, 1993).

13.5 Multiferroic high-frequency devices

This section is devoted to discussions on multiferroic high-frequency devices. The devices could be classified into two broad categories, as in Table 13.2: E-tuning due to mechanical strain and voltage tuning due to variation in dielectric constant or permeability. First, we discuss devices that utilize strain-mediated ME coupling. This includes resonators, band-pass and bandstop filters, phase shifters, and attenuators. Hybrid-wave resonators, phase shifters, and delay lines that utilize variation in ε with E for tuning are considered next. Multiferroic tunable inductors that take advantage of the variation in permeability with E are also discussed. Finally, we focus on antenna miniaturization with the use of high permeability-high permittivity multiferroic composites.

TABLE 13.2 Multiferroic high-frequency devices.

Ferromagnetic/ ferroelectric composite	Frequency range	Devices	Reference
Strain-mediated E-tuning			
YIG/PZT/PMN-PT/ PZN-PT	1–12 GHz	Resonator	
Nickel ferrite/PZT/ PMN-PT/PZN-PT	3–15 GHz	Filter	
Lithium ferrite/PZT/ PMN-PT/PZN-PT	3–15 GHz	Phase shifter	
Zn2Y/PZT/PMN-PT/ PZN-PT	12–24 GHz	Attenuator	
Ba–Al-M/PZT/PMN-PT/PZN-PT	47–110 GHz		
Sr–Al-M/PZT/PMN-PT/PZN-PT	47–110 GHz		
Hybrid waves			
YIG/PZT/PMN-PT	1–12 GHz	Delay lines	
YIG/BST	1–12 GHz	Resonator phase shifter	
E-induced permeability			
FeB/PZT	1–4 GHz	Inductor	
High ε and μ -composites			
Ferrite/BST	10–100 MHz	Antenna	

13.5.1 Strain-mediated E-tuning

Ferrite-ferroelectric resonators, filters, phase shifters, and attenuators that employ voltage tuning of FMR are considered (Fetisov & Srinivasan, 2006, 2008; Li et al., 2011; Shastry et al., 2004; Tatarenko & Srinivasan, 2011; Tatarenko et al., 2006a, 2006b, 2007, 2010, 2012; Yang et al., 2012, 2013).

13.5.1.1 Resonators

Resonators are front-end elements in phased array radars and are also used in filters and phase shifters. One has to choose the appropriate ferrite for the specific frequency range. The fabrication and characterization of an electric field–tunable 1–10 GHz YIG-PZT microwave planar resonator is discussed in Fetisov & Srinivasan (2006). Fig. 13.2 shows the schematics of a YIG-PZT microstripline resonator that is based on voltage tuning of FMR by CME. An LPE YIG film of thickness of 15 μm and lateral dimensions of $1 \times 2.2\ mm^2$ on 2-mm-thick gallium gadolinium garnet (GGG) substrate of (111) orientation was used. The film had an FMR linewidth of ~0.6 Oe, measured at 5 GHz. A PZT plate with dimensions of $4 \times 4 \times 0.5\ mm^3$ and with 5-mm-thick silver electrodes on both sides was bonded to the YIG film with a fast-dry epoxy. The bilayer was placed on a microstrip transducer and was subjected to a bias magnetic field H parallel or perpendicular to the bilayer plane. A (DC) electrical field E = 0–10 kV/cm was generated across PZT by applying a voltage.

Microwave measurements were carried out using a vector network analyzer. A CW input signal with f = 2–10 GHz and power $P_{in} = 0.1$ mW was applied to the microstrip transducer. Low input power was chosen to prevent heating of the sample due to power absorption at FMR. Profiles of reflected power $P_{ref}(f)$ versus f were recorded for a series of H and E. Fig. 13.3 shows representative scattering parameter $S_{11}(f) = 10 \log [(P_{ref}(f)/P_{in}(f)]$ versus f for an out-of-plane H = 3.01 kOe. For E = 0, the spectra contained a well-defined FMR absorption peak at ~3.57 GHz with a maximum insertion loss of −45 dB and a 3-dB linewidth $\Delta f = 3.4$ MHz. The off-resonance loss could be partially due to ferroelectric losses in PZT. With the application of E = 10 kV/cm, the FMR peak shifted to a lower frequency by $\delta f_1 = 40$ MHz. When the polarity of E was reversed by reversing the voltage applied to PZT, the peak upshifted by $\delta f_1 = 38$ MHz. The CME coupling constant $A = \delta f/E = \sim 4$ MHz cm/kV and the voltage tuning for $E = \pm 10$ kV/cm is around 2% of the central frequency.

The E-tuning discussed above could be nonlinear for large electric fields due to ferroelectric hysteresis, as discussed next. Fig. 13.4 shows

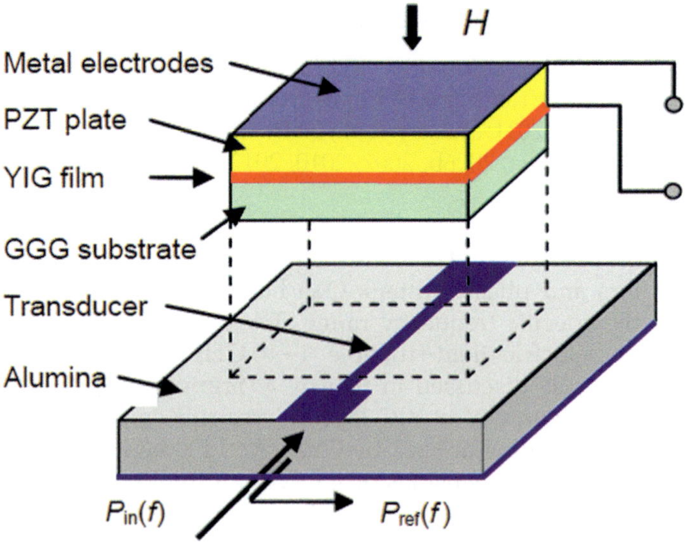

FIGURE 13.2 YIG-PZT resonator (Fetisov & Srinivasan, 2006).

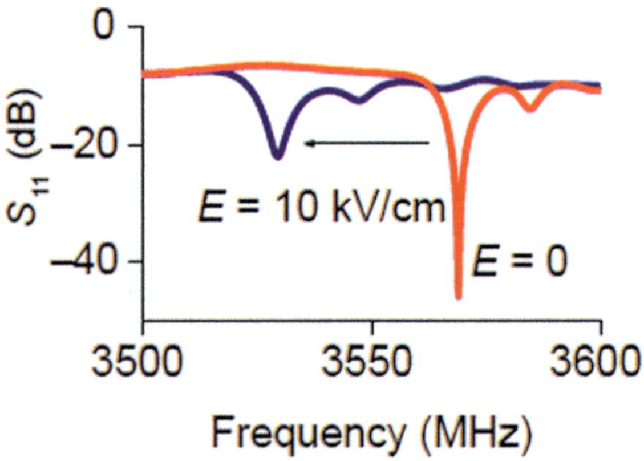

FIGURE 13.3 Electric-field-tunable YIG-PMN-PT resonator with perpendicular magnetization (H = 3010 Oe) (Fetisov & Srinivasan, 2006).

measured dependence of the frequency shift df as a function of E for small, intermediate, and large amplitudes (Fetisov & Srinivasan, 2008). For $|E| < 2$ kV/cm, it is shown in Fig. 13.4a that δf first increases nearly linear with a slope A = 1 MHz/(kV/cm) as E increases from 0 to 2 kV/cm in the positive and negative directions.

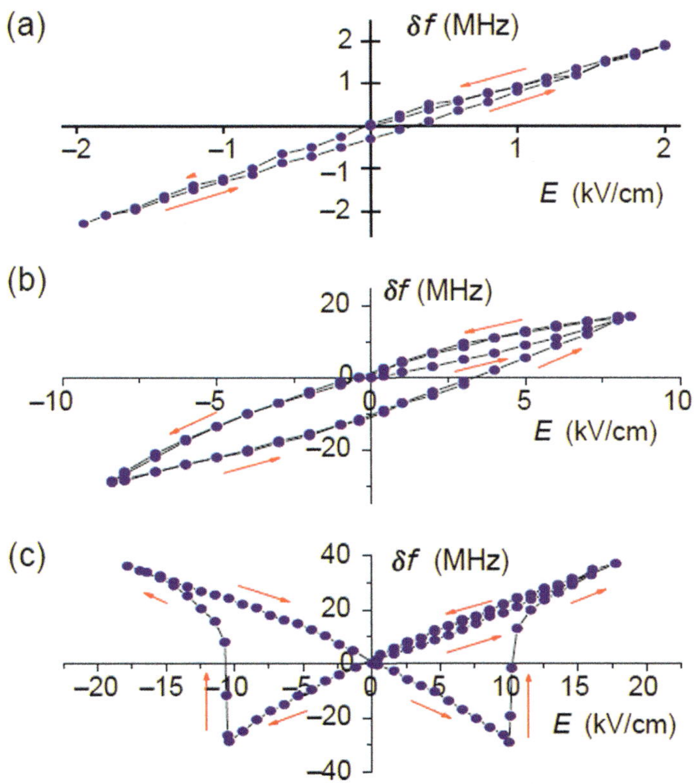

FIGURE 13.4 Shift in resonator frequency of a YIG-PZT resonator versus E for small, intermediate, and large electric-field values that indicate hysteresis and a flip in the frequency shift due to reversal of polarization (Fetisov & Srinivasan, 2008).

Fig. 13.4b shows similar δf versus E for intermediate values of E, for $|E| < 8$ kV/cm. One observes a nonlinear increase in δf as E increases from 0 to 8 kV/cm in the positive direction. After subsequent decrease of E to 0, δf decreases to zero as well but along a different path. Then, as E reverses sign and increases in the negative direction, δf changes its sign and increases nonlinearly. As a result, after the variation in E = -8 kV/cm to 8 kV/cm, the tuning characteristics of hysteretic type is formed, and the loop is also shifted down on the vertical axis. For this nonlinear case, there is no single valued tuning coefficient A.

Fig. 13.4c shows measured dependence of df on E for large amplitudes of E, for $|E| < 18$ kV/cm. As E increases up to the maximum value in the positive direction and then decreases to zero, the shift δf similar to the intermediate case is measured. After reversal of E, δf changed its sign and at a threshold field $E = -10$ kV/cm, δf switched abruptly from -0 MHz to $+10$ MHz. The time required for this transition at E_1 was $= 10-100$ s, and it decreased exponentially when E exceeded the threshold value. With further increase in the magnitude of E, δf increased and reached its maximum value of 40 MHz. Then, as E was decreased to zero and subsequently increased in the positive direction, δf was negative and increased with E until an abrupt switch from −30 MHz to +10 MHz occurred at $E_2 = +10$ kV/cm. As a result, a "butterfly-like" tuning characteristic, shown in Fig. 13.4c, was formed.

To understand the results in Fig. 13.4, one has to consider the piezoelectric strain S versus E in PZT. The strain S in PZT to a linear approximation is given by $S = d_{13}E$, where $d_{13} \sim d_{33}/2$ is the in-plane piezoelectric coefficient of the PZT. The effective anisotropy field in YIG is proportional to the strain $Ha = 3(\lambda_{11}Y/M)S$, where λ11 is the magnetostriction, Y is Young's modulus, and M is the magnetization of YIG. Therefore, the theory (Bichurin et al., 2001, 2002) predicts a linear δf versus E since $\delta f = A \cdot \lambda_{11} \cdot d_{13}\, E$, where A is a coefficient that depends on material and bilayer parameters. An estimation based on the above formula and E = 2 kV/cm gives δf = 1.3 MHz, which is in good agreement with the data of Fig. 13.4a.

In order to account for the nonlinearity in the frequency shift in Fig. 13.4, E dependence of the piezoelectric deformation of PZT was measured using a strain gauge (Fetisov & Srinivasan, 2008). Fig. 13.5 shows S(E) versus E data for intermediate and high amplitudes of 9 kV/cm and 18 kV/cm, respectively. The arrows show direction of the field change. It is seen in Fig. 13.5a that for intermediate amplitudes of E, the variation of S with E is linear only at the initial part of the curve, for $0 < E < 9$ kV/cm. After reversal of the field direction and cyclic variation in E, we observe hysteresis in S(E) that resembles the dependence δf(E) on E in Fig. 13.4b. Data on S for high amplitudes of E in Fig. 13.5b are a "butterfly-like" type, symmetrical with respect to the vertical axis. An abrupt change in the sign and value of S is measured at the same threshold field $|E| = 10$ kV/cm, as for the frequency shift data in Fig. 13.4c. This transition corresponds to repolarization of an initially electrically poled PZT plate. It is obvious that deformation data in Fig. 13.5b have an identical shape as the shift δf(E) in Fig. 13.4c. Thus, the E-tuning of the resonator could be linear or nonlinear, depending on the strength of E (Fetisov & Srinivasan, 2008).

Similar X-band resonator studies were reported for 1.5–2.0-μm-thick nickel ferrite ($NiFe_2O_4$) films grown heteroepitaxially on PZN-PT or PMN-PT substrates using direct liquid injection chemical vapor deposition (Li et al., 2011). A strong ME coupling resulting in large shifts in the FMR profile was observed and is due to electrostatic field induced magnetic anisotropic field changes in these heterostructures. The data indicated an ME coefficient $A = \delta H/E \sim 20$ Oe cm/kV (~60 MHz cm/kV) at 9 and 11 GHz, which is an order of magnitude higher than for YIG-PZT. But

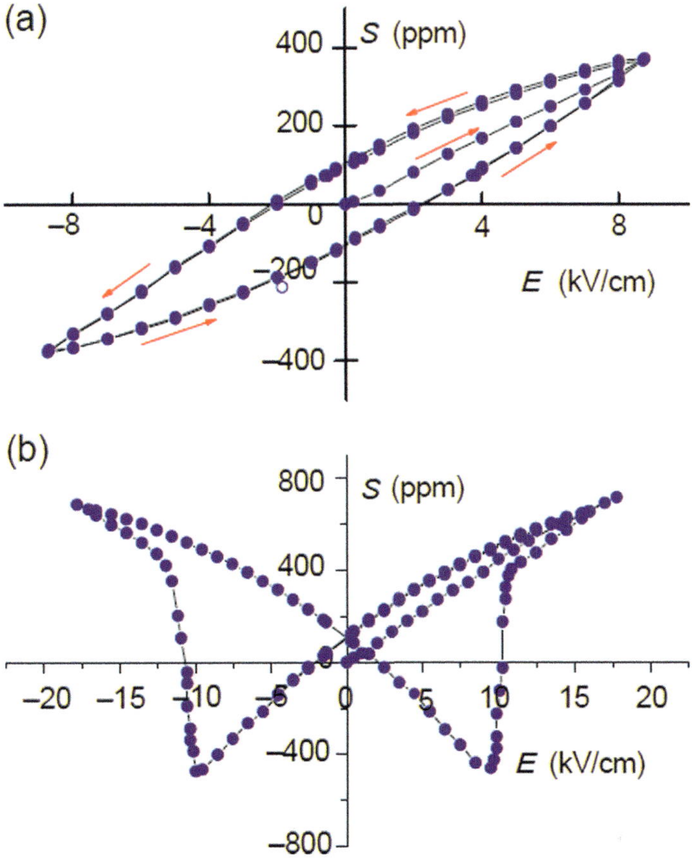

FIGURE 13.5 Piezoelectric strain *S* versus *E* for PZT (Fetisov & Srinivasan, 2008).

the main drawback with the use of NFO is the large FMR linewidth of around 300 Oe for the chemical vapor deposited (CVD) films, which is somewhat comparable to E-tuning with a field of 10 kV/cm (Li et al., 2011).

For resonators over the frequency range 12–110 GHz, hexagonal ferrites are used. Converse ME effects over 8–25 GHz were reported in bilayers of single-crystal Y-type hexagonal ferrite $Ba_2Zn_2Fe_{12}O_{22}$ (Zn_2Y) and polycrystalline PZT or single-crystal PMN-PT (Tatarenko et al., 2012). The bilayers were made by epoxy bonding or eutectic bonding the ferrite and piezoelectric. The strength of ME interactions A was measured from data, as in Fig. 13.6, on electric field E tuning of FMR in the ferrite-ferroelectric in a stripline transducer. The resonator could be tuned by 120 MHz with E = 12 kV/cm, corresponding to ME coupling strength A = 10 MHz cm/kOe (Tatarenko et al., 2012).

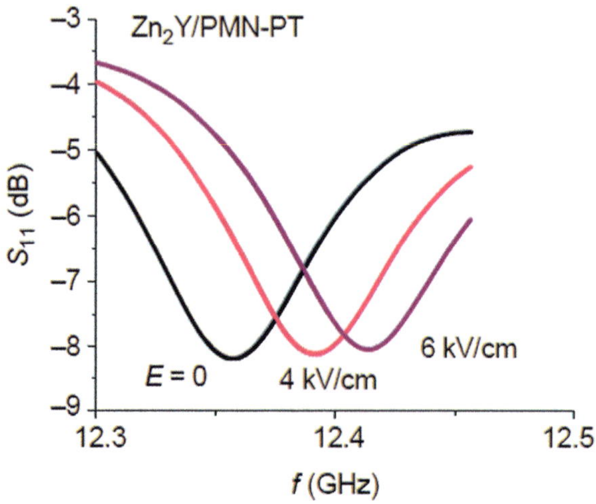

FIGURE 13.6 S_{11} versus f profiles for a series of E showing the shift in FMR in the ferrite-PMN-PT bilayer (Tatarenko et al., 2012).

Millimeter-wave resonators of barium or strontium hexaferrite and PZT were reported (Srinivasan et al., 2006; Ustinov & Srinivasan, 2008). Since BaM has the desired high uniaxial anisotropy fields H_A, FMR and magnetic excitations could be observed at nominal external field H. For BaM with $H_A = 17$ kOe, FMR is expected in the frequency range 47–60 GHz for external fields H_0 = five to nine kOe. Studies were performed on bilayers of single-crystal BaM prepared by floating-zone method and polycrystalline PZT. The measurement cell consisted of a coaxial-to-(WR-17) waveguide adapter. The bilayer was placed inside the waveguide section of the adapter. Studies were performed with E and H perpendicular to the bilayer plane and parallel to the magnetic easy axis of BaM. A vector network was used for data on scattering matrix element S_{11} versus f for a series of H_0 and for E = 0–10 kV/cm. Fig. 13.7 shows representative data on S_{11} versus f. One observes a series of absorption peaks for Ho = 5980 Oe and E = 0 that are identified as electromagnetic modes in BaM (Srinivasan et al., 2006). Due to the specific H orientation, that is, perpendicular to the sample plane, magnetostatic forward volume waves are hybridized with dielectric modes (Ustinov et al., 2006). The mode in Fig. 13.7 shows a 3-dB linewidth of 1 MHz that corresponds to a quality factor $Q = 5 \times 10^4$.

The electric field effects on absorption profiles for the mode of interest are shown in Fig. 13.7. One notices an upshift in the mode frequency with the application of E. Two significant changes are observed as E is increased from 0 to 8 kV/cm: a frequency shift of 4 MHz and an increase in absorption by 10–20 dB. The E-induced shift δf is much higher than the

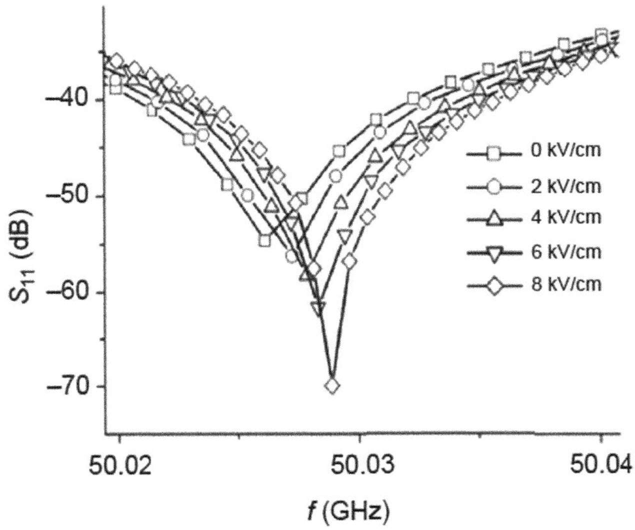

FIGURE 13.7 Electric-field effects on magnetic modes in barium hexaferrite PZT bilayer (Srinivasan et al., 2006).

width of excitation. Data on the E-dependence of δf showed hysteresis and remanence, and the variation in df with E tracks the dependence of the piezoelectric deformation on E in PZT. A maximum shift of 8 MHz is measured for $E = \pm 10\ kV/cm$. Thus, the ME coupling constant A = δf/E = 0.8 MHz cm/kV must be compared with A = 3–15 MHz cm/kV for single-crystal YIG-PMN-PT and Zn_2Y-PMN-PT bilayers.

A frequency-agile hexaferrite–piezoelectric composite resonator for potential device applications at subterahertz frequencies was reported (Ustinov & Srinivasan, 2008). The bilayer was composed of aluminum-substituted barium hexagonal ferrite ($BaAl_2Fe_{10}O_{19}$) and lead zirconate titanate (PZT). The bilayer showed FMR at 108 GHz and demonstrated a ME interaction coefficient of about 1 MHz cm/kV (Ustinov & Srinivasan, 2008).

To summarize, strain-mediated E-tuning of ferrite resonators has been demonstrated in composites with PZT, PMN-PT, or PZN-PT. The resonators could be stripline planar structures for frequencies up to 40 GHz. For higher frequencies, one has to utilize waveguide structures.

13.5.1.2 Filters

Filter is essential in RF/microwave signal processing circuits and blocks. There are several types of filters where the integrated magnetic and ME materials function as a core part. In a typical filter based on lumped element (Fang et al., 2017; Jiang et al., 2020), the magnetic materials serve as the core of inductors in the filter topology, and the function is similar as forementioned inductor components. FMR absorption and magnetostatic wave manipulation properties of magnetic materials are also widely utilized for achieving band-stop and band-pass characteristics. Band-stop filters utilizing FMR absorption typically consist of a transmission line that couples with the magnetic materials. When the frequency of electromagnetic wave propagating in the transmission line coincides with FMR frequency of the material, a maximum attenuation and hence a stop band is generated. Multiple materials including transition metal (Wei et al., 2000) ferrite (Harward et al., 2014; Song et al., 2009; Tsai et al., 2005) and soft magnetic composite (He et al., 2018) are adopted for these band-stop filters. Magnetostatic wave manipulation property in ferrite is utilized to achieve nonreciprocal band-pass characteristic (Lin et al., 2015), which will be elaborated late. The third type of integrated filter utilizes the mechanical resonance to realize frequency selectivity (Lin et al., 2016), where the input excitation signal is transformed into acoustic wave at the input resonator and propagate to the output resonator thorough the released material layer.

Reconfigurability is a crucial feature in filter design considering the need for wide band operation, the size of the signal processing block and the yielding cost. MEMS switches (Chih-Chieh & Rebeiz, 2011; Entesari & Rebeiz, 2005; Sekar et al., 2011) are adopted to realize the reconfigurability. However, this mechanism suffers from the tuning discontinuity which limits its flexibility. The integrated filters that incorporate the magnetic materials mentioned above are usually magnetically tunable. He presented an integrated tunable band-stop filter that achieved a tunability of 70% under a low magnetic field of 400 Oe working in C-band (He et al., 2018). However, for some higher frequency filter applications, the relatively lossy ferrite has to be adopted and a large magnetic field is required for tenability (Harward et al., 2014). Meanwhile, the instrument for applying a magnetic field is usually bulky and power consuming, which makes the magnetically tunable filters an inferior choice. Integrated ME materials offer a possibility to achieve on-chip electric field tunable filter devices. The advantages of ME tuning mechanism are continuous tuning, low power consumption, and easily defined tuning effect range (determined by the size of ME materials). Current efforts are still committed to mitigate the magnetic material performance degradation induced by the deposition or bonding of the piezoelectric phase in ME materials.

In 2015, the first integrated dual E- and H-field tunable band-pass filter was demonstrated in an inverted-L-shape microstrip structure with a multiferroic heterostructure formed by a spin-sprayed NiZn ferrite film and a PMN-PT piezoelectric slab, as shown in Fig. 13.8a (Lin et al., 2015). The signal from the input port is transmitted as magnetostatic surface wave (MSSW). The forward and backward wave transmissions happen

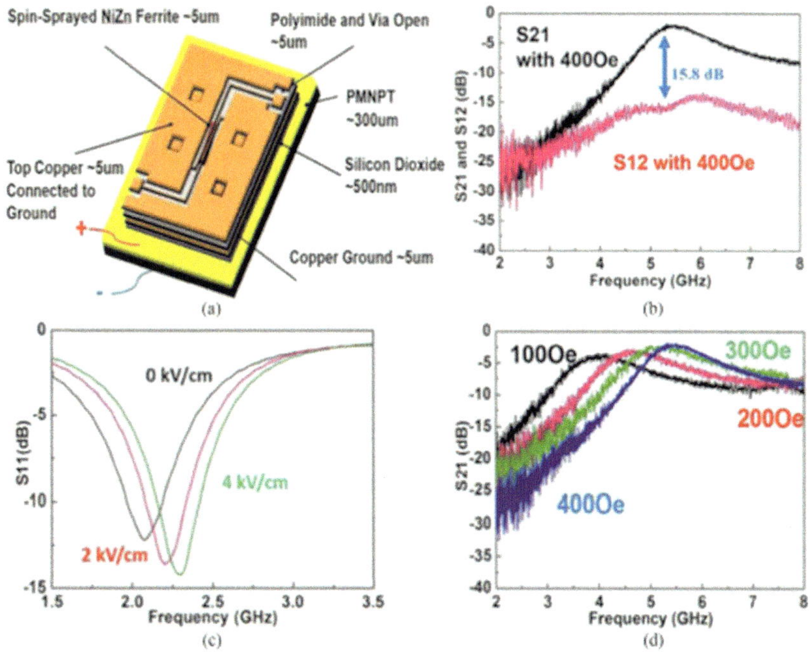

FIGURE 13.8 Integrated NiZn/PMN-PT dual E− and H-field tunable ME band-pass filter. (a) Schematics of integrated tunable inductor. (b) S_{12} and S_{21} under a bias magnetic field of 400 Oe. (c) S_{11} under different bias electric fields. (d) S_{21} under different bias magnetic fields. *Reproduced from Lin et al. (2015).*

on bottom and top surface of the ferrite film. The passband frequency of the filter is determined by the dispersion relation of the MSSW. Due to the discrepancy of radiation resistance on the top and bottom surfaces of the ferrite film, a nonreciprocal transmission characteristic can be observed (Chu et al., 2018; Weng et al., 2017). By ideally rotating the NiZn ferrite film by 45°, the standing wave resonance and the multipassband are diminished since the reflected wave at the edge of ferrite film is converted to magnetostatic back volume wave (MSBVW), which will decay fast at the filter operating frequency. The detail of the filter operation principle is presented in (Zhang et al., 2020). As shown in Fig. 13.8b, the 15.8 dB difference of S12 and S21 in band shows the nonreciprocal transmission. The central frequency could be electrically tuned from 2.075 to 2.295 GHz when the out-of-plane electric field rises from 0 kV/cm to 4 kV/cm, as shown in Fig. 13.8c, namely, an E-field frequency tunability of 55 MHz/(kV/cm). In addition, a tunable central frequency from 3.78 to 5.27 GHz could be realized by increasing the DC magnetic field from 100 to 400 Oe, corresponding to an H-field frequency tunability of about 5 MHz/Oe, as shown in Fig. 13.8d.

In 2016, an integrated RF tunable bandpass filter based on two coupled elliptic-shape nano-mechanical resonators with FeGaB/Al2O3/AlN multiferroic heterostructure on a bottom Pt electrode was reported (Lin et al., 2016) and schematically shown in Fig. 13.9a. The filter relies on the acoustic wave propagation to achieve signal transmission between the two ports, the two resonators work in the contour mode whose resonance frequency is determined by the width of the coupled structure. The filter exhibited a return loss of −11.2 dB, an insertion loss of 3.4 dB and a quality factor of 252 at the central frequency of 93.165 MHz, as shown in Fig. 13.9b. The tunability test showed an H-field frequency tunability of 5 kHz/Oe and an E-field frequency tunability of 2.3 kHz/V, presented in Fig. 13.9c and d, respectively. The tunability is realized based on the Δ E effect, where the Young's modulus of FeGaB changes with applied fields.

An electric field–tunable microwave band-pass filter based on FMR in a bilayer of YIG-PZT was designed and characterized. The device showed a 125-MHz tuning range for E = 0–3 kV/cm and an insertion loss of 5 dB at 6.5 GHz (Tatarenko et al., 2006a). The single-cavity ME filter, shown in Fig. 13.10, consisted of a 1-mm-thick dielectric ground plane (permittivity

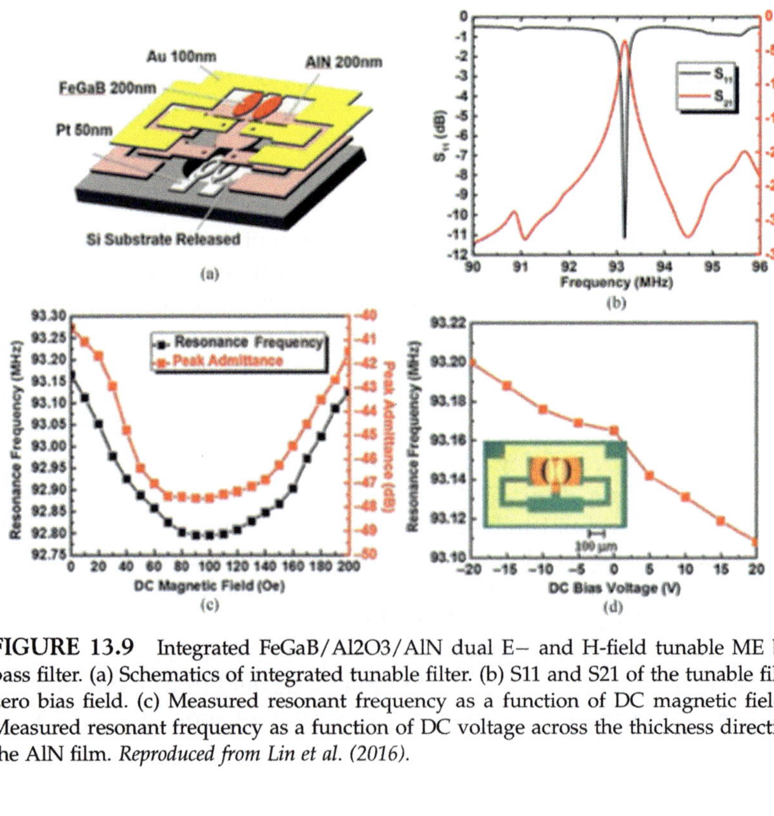

FIGURE 13.9 Integrated FeGaB/Al2O3/AlN dual E– and H-field tunable ME band-pass filter. (a) Schematics of integrated tunable filter. (b) S11 and S21 of the tunable filter at zero bias field. (c) Measured resonant frequency as a function of DC magnetic field. (d) Measured resonant frequency as a function of DC voltage across the thickness direction of the AlN film. *Reproduced from Lin et al. (2016).*

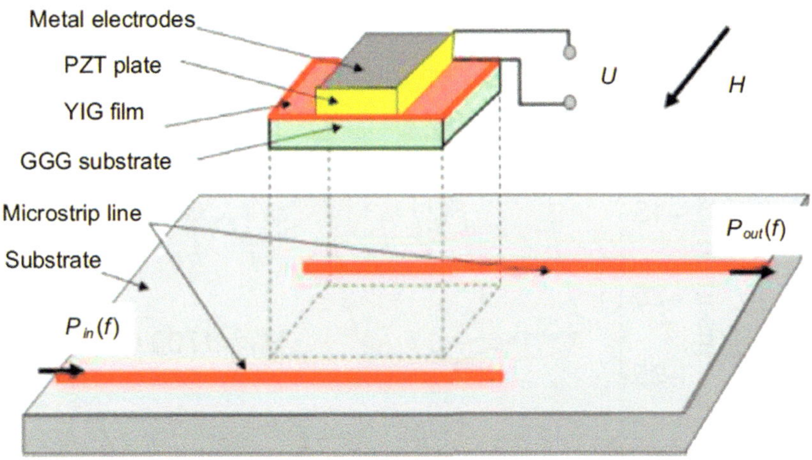

FIGURE 13.10 A magnetoelectric (ME) band-pass filter. The ME resonator consisted of a 110 μm thick (111) YIG on GGG bonded to PZT (Tatarenko et al., 2006a).

of 10), input and output microstrips of nonresonance lengths, and an ME element. The input–output decoupling is determined by the gap between the microstrips. Power is coupled from input to output under FMR in the ME element.

The ME element consisted of an epitaxial YIG film bonded to PZT. A 110-μm-thick YIG film grown by LPE on a (111) GGG substrate was used. The film had dimensions of 5.5×1.5 mm^2, saturation induction $4\pi M$ of 1750 G, and FMR linewidth of 1 Oe. A PZT plate with the dimensions $4 \times 1 \times 0.5$ mm^3 was initially poled by heating up to 150°C and cooling back to room temperature in an electric field of 10 kV/cm perpendicular to the sample plane. The layered structure was made by bonding the YIG film surface to PZT with a 0.08-mm-thick layer of ethyl cyanoacrylate, a fast-dry epoxy. The sample was placed between the transducers, as in Fig. 13.11, and was subjected to a field H parallel to the sample plane and perpendicular to the microstrips.

The frequency f dependence of the insertion loss L, that is, the transmitted power through the ME element, was measured at 4–10 GHz as a function of H and E. Representative results on L versus f are shown in Fig. 13.11. Consider first the profile for $E = 0$. The maximum input–output coupling is observed at $f_r = 6.77$ GHz that corresponds to FMR in YIG for an in-plane $H = 1.7$ kOe, $4\pi M = 1.75$ kG, and gyromagnetic ratio $\gamma = 2.8$ GHz/kOe. The loss increases sharply $f < f_r$ and the off-resonance isolation is 20–25 dB. For $f > f_r$, secondary maxima due to magnetostatic modes are seen in Fig. 13.11. A significant modification of L versus f profile is observed in Fig. 13.11 when $E = 1$ kV/cm is applied across PZT; f_r is upshifted by $\delta f = 28$ MHz. The shift arises due to strain at the YIG-

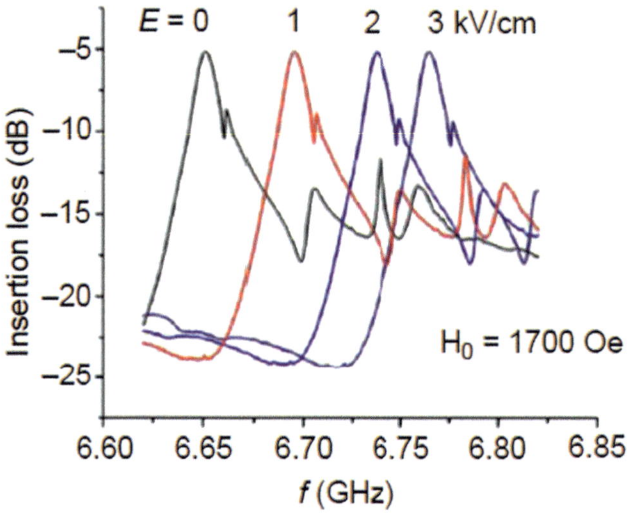

FIGURE 13.11 Loss versus f characteristics for a series of E for the YIG-PZT filter (Tatarenko et al., 2006a).

PZT interface caused by the piezoelectric deformation in PZT. A further increase in E results in an increase in the magnitude of the up shift, as shown in Fig. 13.11. We observed a down shift in f_r when the direction of E was reversed by reversing the polarity of applied voltage. This is attributed to a switch from compressive to tensile strain in YIG. The filter can be tuned by 2% of the central frequency with a nominal electric field of 3 kV/cm (Tatarenko et al., 2006a).

A majority of resonators and filters utilizing CME in ferrite–ferroelectric composites so far, however, were carried out for FMR in magnetically saturated ferrites. But similar studies under resonance in unsaturated ferrites are of interest for high-frequency device applications. Recent efforts in this regard include resonances in partially magnetized YIG and in YIG-ferroelectric composites for use as band-pass filters (Yang et al., 2012, 2013). Yang et al. studied H-tuning of FMR over 2.7–3.7 GHz and 5.9–6.4 GHz in YIG and showed that the resonance frequency f_r scales as $1/\sqrt{\mu_r}$, where μ_r is the real part of the permeability. They also demonstrated H-tunable filters based on the modes (Yang et al., 2012, 2013). Their work was extended to E-tuning of the FMR and 10% tuning of the pass-band through CME in YIG/PZN-PT (Yang et al., 2013). Such filters under low-bias field resonances in ferrites are of particular interest for microwave device applications since providing a high bias magnetic field is usually the biggest challenge for ferrite devices. One also could obtain a large E-tuning under such conditions.

13.5.1.3 *Phase shifters*

Microwave phase shifters are important elements for use in oscillators and phased array antenna systems and beamforming networks. Traditional ferrite phase shifters are based on the Faraday rotation for electromagnetic waves in axially magnetized ferrite rods in waveguides. The devices require an electromagnet or solenoidal magnetic field for phase tuning and involve large power dissipation (Adam et al., 2002). A desirable alternative is the latching ferrite phase shifter that operates at the remanent magnetization for the ferrite element and requires current pulses for switching the magnetization state (Adam et al., 1991). Phase shifts could also be realized through magnetostatic wave (MSW) propagation or FMR in planar ferrites in microstripline structures. The MSW wavelength and group velocity are 2–3 order in magnitude smaller than that for electromagnetic waves of the same frequency. This makes possible a phase shift of decades of π for propagation distances of several millimeters. In the case of phase shifters operating close to FMR, a rapid change in the permeability and phase shift can be achieved by tuning the bias magnetic field (Tatarenko et al., 2006b). But all of the above ferrite phase shifters, in general, require high electrical power for operation and cannot be miniature in size or compatible with the integrated circuit technology. Ferroelectric phase shifters, on the other hand, can be tuned with an electric field (Cross, 1993). But such phase shifters are very lossy at frequencies above 1–5 GHz. With a ferrite–ferroelectric layered structure, it is possible to achieve both magnetic- and electric field–tunability of the phase shift. Giant voltage-controlled FMR has been demonstrated in strain-mediated magnetic/piezoelectric heterostructures providing routes for applications in voltage-tunable RF and microwave devices (phase shifters) (Gupta & Kotnala, 2022). Among ferrite materials, yttrium iron garnet (YIG) possesses the advantage of low absorption, narrow FMR line width, and high Q factor at microwave frequencies. YIG/piezoelectric heterostructures have been also frequently utilized for synthesizing microwave phase shifters (Ustinov et al., 2014). An electric field–tunable YIG-PZT phase shifter based on FMR was designed and characterized. The electric field control of the phase shift $\delta\varphi$ arises through ME interactions (Tatarenko et al., 2006b). The piezoelectric deformation in PZT in an electric field E leads to a shift in the FMR frequency in YIG and a phase shift. For E = 5–8 kV/cm applied across PZT, $\delta\varphi = 90–180$ degrees and an insertion loss of 1.5–4 dB was obtained. It was possible to realize linear or nonlinear $\delta\varphi$ versus E characteristics with proper choice for bias magnetic field or frequency. The insertion loss is on the order of 1.5–4 dB, somewhat close to the desirable 0.5 dB for practical applications. Theoretical estimates of $\delta\varphi$ versus E are in very good agreement with the data (Tatarenko et al., 2006b).

The design for the ME microwave phase shifter is shown in Fig. 13.12. It consists of a 1-mm-wide microstrip on a 1-mm-thick aluminum oxide ground plane (relative permittivity of 10), a YIG-PZT resonator, and microstrip loops of length $\lambda/8$ and $3\lambda/8$ that produce a circularly polarized microwave magnetic field in the resonator. The ME resonator consisted of a 0.5-mm-thick, 5 × 5-mm plate of PZT and a 4 × 2.5 mm epitaxial (100) YIG film of thickness 124 μ m on a 0.5-mm-thick gadolinium gallium garnet (GGG) substrate. Silver electrodes were deposited on PZT that was poled by heating up to 150°C and cooling back to room temperature in an electric field of 20 kV/cm perpendicular to the sample plane. The YIG film with a saturation magnetization of 1750 Oe and an FMR linewidth of 1 Oe was bonded to PZT with a 0.08-mm-thick layer of ethyl cyanoacrylate, a fast-dry epoxy.

The device characteristics were studied with a vector network analyzer. The operating frequency was chosen by applying appropriate bias magnetic field H parallel to the sample plane. An electromagnet was used for the magnetic field generation. For a given frequency f, FMR is expected in YIG for $H = H_r$. One has to choose $H > H_r$, to obtain low-insertion loss and a linear or nonlinear variation of the real part of the permeability μ' with H. An input cw signal $P_{in}(f)$ of frequency $f = 1\text{–}10$ GHz and power $P_{in} = -10$ dBm was applied to the input transducer. The frequency dependences of the insertion loss $L(f) = 10 \log_{10}[P_{out}(f)/P_{in}(f)]$ and phase shift $\pi(f)$ of the output signal were measured as a function of H and E. Representative data on the electric-field-tunable YIG-PZT phase shifters are shown in Fig. 13.13. The bias field $H = 2700$ Oe corresponds to the region in which the variation of μ'

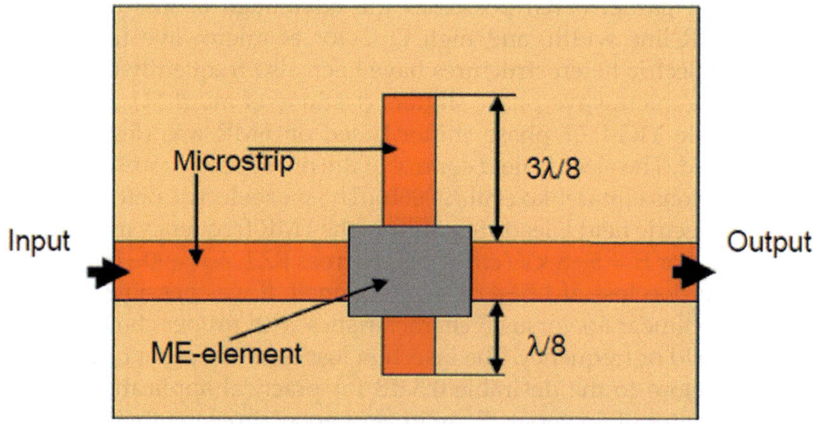

FIGURE 13.12 Diagram showing the schematics of a YIG/PZT phase shifter (Tatarenko et al., 2006b).

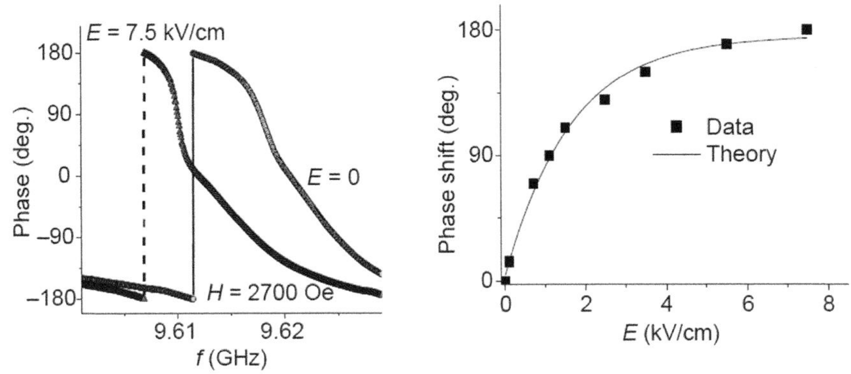

FIGURE 13.13 The phase angle ∅ versus frequency *f* at 9.6 GHz for *E* = 0 and 7.5 kV/cm and differential phase shift versus E at 9.615 GHz (Tatarenko et al., 2006b).

versus H is nonlinear. For E = 0, one observes a smooth variation in f versus f and an expected discontinuity at f = 9.6 GHz. Application of E = 7.5 kV/cm to PZT resulted in a downshift in frequency δf = 6 MHz and a corresponding differential phase shift of $\delta\phi = 180$ degrees. By increasing the bias field to 2720 Oe, it was demonstrated that $\delta\phi$ showed a linear dependence on E (Tatarenko et al., 2006b).

Similar phase shifters with hexaferrite/ferroelectric elements in waveguides could be realized. Bilayers of Co_2Z (20–40 GHz), Zn_2Y (15–40 GHz) (Tatarenko & Srinivasan, 2011), or BaM (45–67 GHz) and PZT could be used. For a stripline Zn_2Y/PMN-PT phase shifter, data on differential phase shift versus E indicated a differential phase shift of 50 degrees for E = 12 kV/cm, a linear variation in the phase shift with E, and an insertion loss of 4–8 dB. Device optimization will have to include studies on (1) ferroelectric leakage current, which will affect the electrical tuning, (2) ferroelectric fatigue, and (3) ferroelectric and loss parameters of PZT.

Other devices on strain mediated E-tuning include an attenuator. An electric field–tunable YIG/PMN-PT attenuator operating at 7.2 GHz was reported (Tatarenko et al., 2007). The device is based on FMR in YIG. Electric field (E) tuning of the attenuation was accomplished through ME interactions in the bilayer. Theoretical estimates of attenuation versus E were in very good agreement with data.

13.5.1.4 Inductors

The inductor is a type of device whose operations rely on magnetic flux variation and energy conversion. With the rapid development of integrated RF, microwave, and power circuits, the high Q compact inductor with sufficiently large inductance is highly demanded. In the next two sub-sections, we will focus on introducing two different types of inductors (on-chip inductor with integrated magnetic core and voltage tunable inductor with ME composite. The integrated magnetic materials, which serve as the magnetic core by concentrating the magnetic flux generated during the operation, can boost the inductance. High-permeability materials including permalloy (NiFe) (Yun et al., 2004), NiFeW (Mundotiya et al., 2017), NiFeZn (Mariappan et al., 2016), NiZn ferrite (Wang et al., 2017), FeCoSiB (Kurata et al., 1993), FeCoCu (Flynn et al., 2007), FeGaB (Gao, Zardareh et al., 2014; Gao, Zare et al., 2014), etc. are mostly used magnetic materials for the core. However, there are two limitations brought by the integrated magnetic core: (1). The relatively thick magnetic core in on-chip devices induces large eddy current loss during high frequency operation, which limits the quality factor; (2) the FMR of the magnetic core causes energy absorption in the corresponding frequency range, which diminishes the power efficiency and working frequency range.

To mitigate the high eddy current loss issue, the integrated magnetic core was engineered as a laminated structure (El-Ghazaly et al., 2017; Gao, Zare et al., 2014) with thinner magnetic material layers and isolation spacer in between. Thinner magnetic material has higher equivalent resistance and hence reduces the eddy current induced by the time-varying magnetic field. As for the loss caused by FMR, the Kittel (Kittel, 1948) equation suggests that by determining proper magnetic film dimensions, the demagnetizing factor can be manipulated such that the FMR frequency is higher, which subsequently expands the operational frequency range (El-Ghazaly et al., 2017). The most commonly adopted inductor structures are solenoid and spiral/planner structures, where the solenoid structure (El-Ghazaly et al., 2017; Gao, Zare et al., 2014; Wang et al., 2017; Yun et al., 2004) incorporates the magnetic core between the wires at the upper and lower layers, whereas the spiral/planer (Gardner et al., 2009; Salvia et al., 2005; Yamaguchi et al., 2000, 2005) structure of the magnetic core layer is deposited beneath or between the planer wires to form the magnetic circuit. In 2014, Gao (Gao, Zardareh et al., 2014) reported a series of integrated inductors using FeGaB (100 nm)/Al_2O_3 (5 nm)/FeGaB (100 nm) multilayer sandwich structure as the magnetic core. Solenoid structure was adopted for the inductor design. The magnetic core boosted the inductance by nearly 100% in the frequency range of up to 3 GHz in comparison to the air core inductor with the same dimensions. The inductance showed a flat characteristic with a peak inductance of over 3 nH, which indicated that the laminated magnetic core was of help to eliminate the eddy current loss in high frequency range and the energy loss contributed by FMR. The quality factor in the frequency range of up to 3 GHz, where the peak quality factor of the inductor with magnetic core was over 10 and showed a better performance compared to the air core inductor in the relatively low frequency.

Tunable RF inductors provide another possible solution for tackling the trade-off between inductance, quality factor and operating frequency range, since the external electric signal can lead to the control and improvement of these 3 parameters at the same time. There have been two main branches of control strategies for tunable inductors in microelectromechanical systems (MEMS). One mechanism is to adjust the turns or numbers of inductor coils. Based on this, Park changed the turns or length of inductor coils by MEMS switch and achieved an operation frequency of 2.4 GHz, a Q-factor of 3, and an extremely large tunability of 187.5%. The main drawback of this MEMS switch control strategy is the discrete tunability of inductance limited by the number of switches (Piljae et al., 2004; Shirane et al., 2012). In order to overcome this shortcoming, the inductor coupling control method with a secondary coil was put forward and applied to the fabrication of related devices, with an operating frequency ranging from 4.35 to 7 GHz, a quality factor of ~10 and an inductance tunability from ~30% to ~90% (Jeong-Il & Peroulis, 2009; Zine-El-Abidine et al., 2005). However, the related work was greatly limited by the complicated structure and fabrication procedure. Another perspective is to change the properties of the magnetic core or the magnetic field distribution in it. Following this direction, researchers tried to control the magnetic flux by thermal actuation (Bhattacharya et al., 2015) or a metal shield (Fang et al., 2010). DC current was also widely used to modify the properties of the magnetic core. For example, Salvia realizes a 40% increase in inductance, an inductance tunability of 15% and a Q-factor from 5 to 11 in the tunable on-chip inductors operating up to a frequency of 5 GHz using patterned permalloy laminations (Salvia et al.) Unfortunately, both methods exhibit limited inductance tunability, high power consumption, significant thermal effect, and excessive noise (Salvia et al.; Tengxing et al., 2017; Vroubel et al., 2004).

With the development of ME materials, researchers tended to use ME coupling to adjust the permeability of magnetic core and tune the inductance, aimed at improving energy efficiency, reducing joule heating, and increasing inductance tunability. In order to ensure higher inductance density and efficient utilization of magnetic film in the electrostatically tunable inductor, the most popular spiral, and solenoid structures have been widely used.

A solenoid type ME inductor fabricated with a layered multiferroic composite core was proposed by Lou, which consisted of two layers of Metglas magnetic ribbons and one PZT piezoelectric slab (Lou et al., 2009). The inductance showed a strong electric field tunability, with maximum inductance change of 450%, 250%, and 50% for operation frequencies of 1 kHz, 100 kHz, and 5 MHz, respectively. The estimated energy consumption to achieve maximum tuning 450% is only about 1.3 mJ, indicating that the fabricated tunable ME inductors were essentially passive and energy efficient and achieved a giant tunability. The quality factor is improved from about three at zero electric field to about 8.5 at an external field of 12 kV/cm due to the decreased permeability, increased skin depth, and reduced core eddy current loss. The giant tunability achieved in such ME inductors with low power consumption makes them ideal candidates for novel power efficient and compact electronic devices.

In order to extend the operation frequency range and enhance the quality factor by reducing the eddy current loss, Lin fabricated PZT/Metglas/PZT voltage tunable inductors with single layer Metglas film and multiple piezoelectric slabs (Hwaider et al., 2015). An 80% increase in Q-factor (10 times of extension in operational frequency and 3 times of enhanced quality factor) was observed. In addition, a 150% improvement in quality factor and 100 times extension in operational frequency could be found in ME inductors with a single PZT slab, whereas a 235% improvement in quality factor and 200 times of extension in operational frequency were realized in its double PZT slabs counterpart. Though the PZT/Metglas/PZT ME inductors show a high performance in quality factor, inductance density and a giant tunability of 450% in the kilohertz frequency band, the Metglas/PZT multiferroic heterostructures are not applicable for the gigahertz applications due to their low FMR frequency. These inductors fail to realize compatible integration on CMOS circuits, which prevents their applications in RF/microwave electronics.

To realize gigahertz range frequency operation, Chen has recently demonstrated a voltage tunable GHz integrated magnetic inductor by using the FeGaB/Al2O3/PMN-PT ME composites, as shown in Fig. 13.14a (Chen et al., 2020). Compared with materials like NiZn, the high permeability further improved by annealing, FMR frequency of 1.85 GHz with a narrow FMR bandwidth of 16~20 Oe, large magnetostriction of FeGaB films lead to high inductance density, operation band in GHz range, relatively low magnetic loss and high frequency tunability, respectively. The inductor showed a relatively constant inductance and quality factor value in the frequency range of 0.5–2 GHz. The alumina spacer reduced the thickness of single-layer magnetic material, thereby increasing the sheet resistance and reducing the eddy current loss. The E-field tunability test results shown in Fig. 13.14b and c indicate that the inductance rises from 1.2 to 3.5 nH and the quality factor rises from 2.7 to 5.2 with an applied electric field up to 10 kV/cm in this frequency range. The a maximum inductance continuous tunability of 191% at 1.5 GHz was achieved, which was significantly larger than other integrated inductors tuned by magnetic flux, DC bias current, and coil coupling.

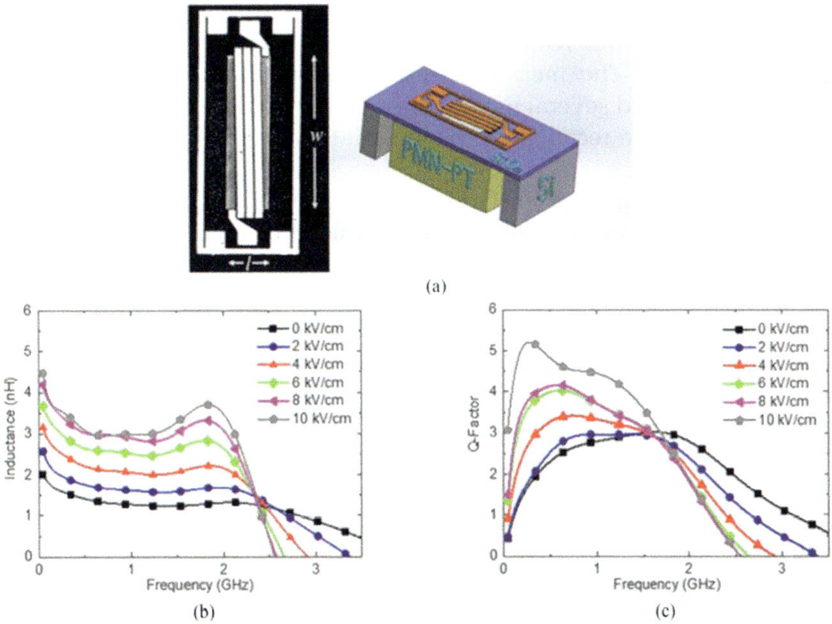

FIGURE 13.14 Integrated gigahertz FeGaB/Al2O3/PMN-PT voltage tunable ME inductor. (a) Optical image and structure model of integrated voltage tunable inductor with 3.5 turn and a magnetic film of 340 × 800 μm. (b) Measured inductance and (c) Q-factor of integrated FeGaB/PMN-PT tunable inductor with applied electric field from 0 to 10 kV/cm. *Reproduced from Chen et al. (2020).*

13.5.1.5 Hybrid spin-electromagnetic wave devices

Here we discuss resonators, phase shifters, and delay lines of multiferroic composites, in which hybrid waves are excited and E-tuning is achieved by changes in the dielectric permittivity (Fetisov & Srinivasan, 2005a, 2005b; Kim et al., 2000; Pettiford et al., 2007; Semenov et al., 2006; Ustinov & Srinivasan, 2010; Ustinov et al., 2007, 2008).

An electric field–tunable, microwave resonator with a YIG-BST-layered structure was investigated (Semenov et al., 2006). In this case, the ferrite and the ferroelectric were placed on top of each other since bonding for strain transfer is not a requirement in this tuning. The measured characteristics at 3.6 GHz, corresponding to the proximity of magnetic and dielectric resonance modes, showed a broad-band electrical tunability. A tuning range of 1.5% of the central resonance frequency was obtained for nominal electric fields.

The device was constructed in two steps. First, a straight-edge YIG film resonator was fabricated and characterized. The resonator was made from a 7-μ m-thick (111) YIG film on a 500-μ m-thick GGG substrate. Its in-plane dimensions were chosen to be 0.5 mm by 1.5 mm in order to excite hybridized wave modes with the wave number that would allow efficient electrical tuning. The measured resonance spectrum of the YIG film resonator exhibited several well-defined peaks with all of them identified as eigenmodes of a rectangular magnetostatic wave (MSW) straight-edge resonator.

Second, a rectangular BST resonator was fabricated and characterized. Through an electrodynamical analysis of dispersion in relation to TE waves in a dielectric layer, the dimensions of the rectangular BST resonator were calculated for a frequency range of around 3650 MHz. Then, the amplitude–frequency characteristic of the fabricated BST resonator was measured, and the resonant modes were identified. For the specific design, which is described below, the main working ferroelectric mode was labeled as (032). This mode demonstrated a maximum quality factor.

A schematic diagram of the coupled resonator construction is given in Fig. 13.15. To excite the ferrite–ferroelectric resonator, a microstrip transmission line patterned on 3-mm-thick copper film deposited on alumina substrate was used. The YIG film straight-edge resonator was placed on the microstrip line, which was directed along its longer edge. The resonator was magnetized in its plane with a bias magnetic field H

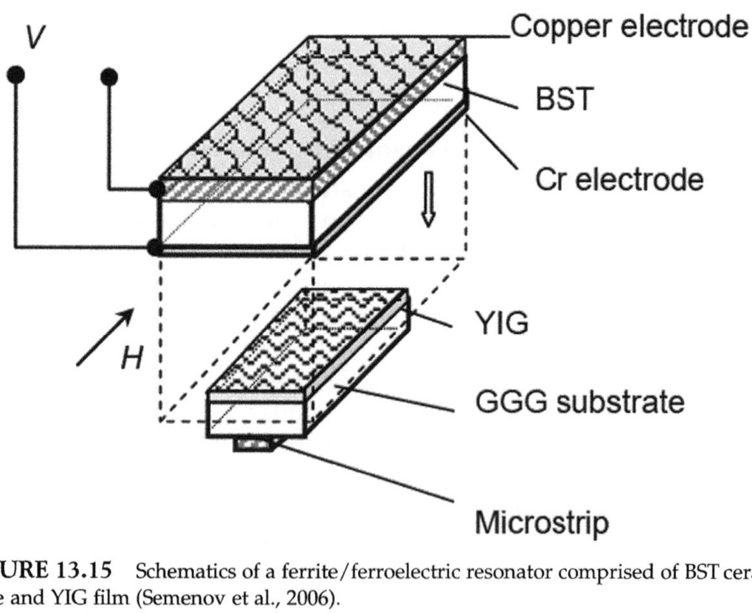

FIGURE 13.15 Schematics of a ferrite/ferroelectric resonator comprised of BST ceramic plate and YIG film (Semenov et al., 2006).

parallel to the microstrip line main axis. The rectangular ferroelectric resonator with in-plane dimensions $3 \times 5\ mm^2$ was cut from the $0.5BaTiO_3$-$0.5SrTiO_3$ ceramic layer of thickness 500 m, prepared from the carbonate and oxide precursors, and annealed at 1400 °C. Metal electrodes were deposited on both sides of the ferroelectric resonator: the upper 3-mm-thick Cu by magnetron sputtering and the lower 50-nm-thick Cr by vacuum evaporation. The Cr thickness was smaller than the skin depth, so there was no attenuation of the microwave fields at the YIG film. The Cu and Cr electrodes were used to apply a voltage to tune the permittivity of BST.

In the bilayer resonator, electromagnetic coupling among the MSW modes and the ferroelectric mode occurred, and the resonance branches repulsed each other and became hybridized in the region of the frequency degeneration of the MSW eigenmodes and the ferroelectric mode. The degree of the repulsion depended on the value of the bias magnetic field. Fig. 13.16 presents the reflected power versus frequency profiles showing the resonance for voltages 0–900 V across BST. These profiles were recorded for the main, (011)-mode of the YIG film. One observes a tuning range of 54 MHz, corresponding to 1.5% of the central resonant frequency of the YIG/BST resonator.

Electrical tunability could be substantially improved and losses could be reduced by choosing the ferroelectric layer to have resonant properties (Semenov et al., 2006; Ustinov et al., 2006). To improve the tuning characteristics, resonators with equal in-plane dimensions of the ferrite and

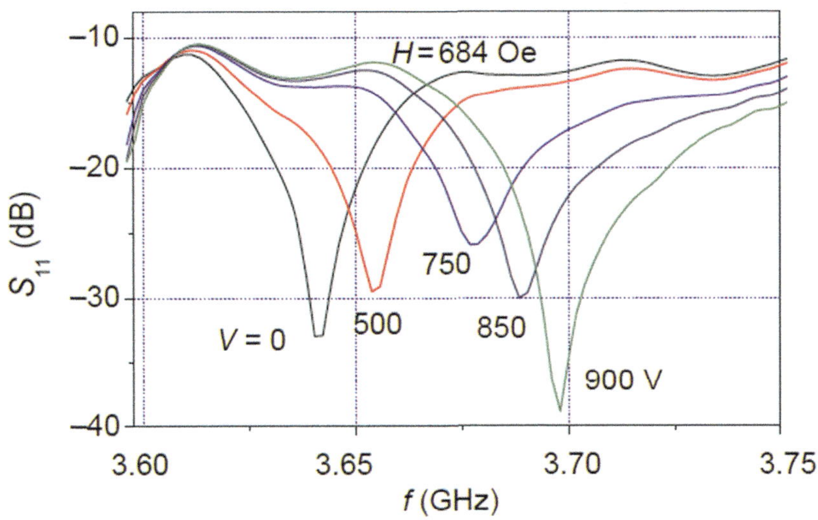

FIGURE 13.16 Electrical tuning characteristics of the YIG/BST composite resonator (Semenov et al., 2006).

ferroelectric layers were proposed (Ustinov et al., 2006). This geometry maximizes coupling between the ferromagnetic and dielectric resonant modes because the electromagnetic fields of the modes overlap efficiently. A tuning range of 100 MHz for the resonator frequency was realized at 5 GHz through the variation of magnetic permeability and dielectric permittivity of the YIG-BST structure.

A microwave planar phase shifter based on a ferrite–ferroelectric bilayer as a waveguide for hybrid spin-electromagnetic waves was studied (Fetisov & Srinivasan, 2005a, 2005b; Ustinov et al., 2007). The ferroelectric layer was a ceramic BST slab of the composition $Ba_{0.6}Sr_{0.4}TiO_3$ of dimensions 5 × 10 × 0.5 mm. Chromium electrodes of thickness 50 nm were deposited on both surfaces of BST. The dependence of the permittivity versus applied voltage measured for BST showed a 20% variation for E = 20 kV/cm. A 5.7-μm-thick (111) YIG film on GGG was placed on BST in a microstrip line phase shifter, as shown in Fig. 13.17. Representative data on the return loss, insertion loss, and differential phase shift

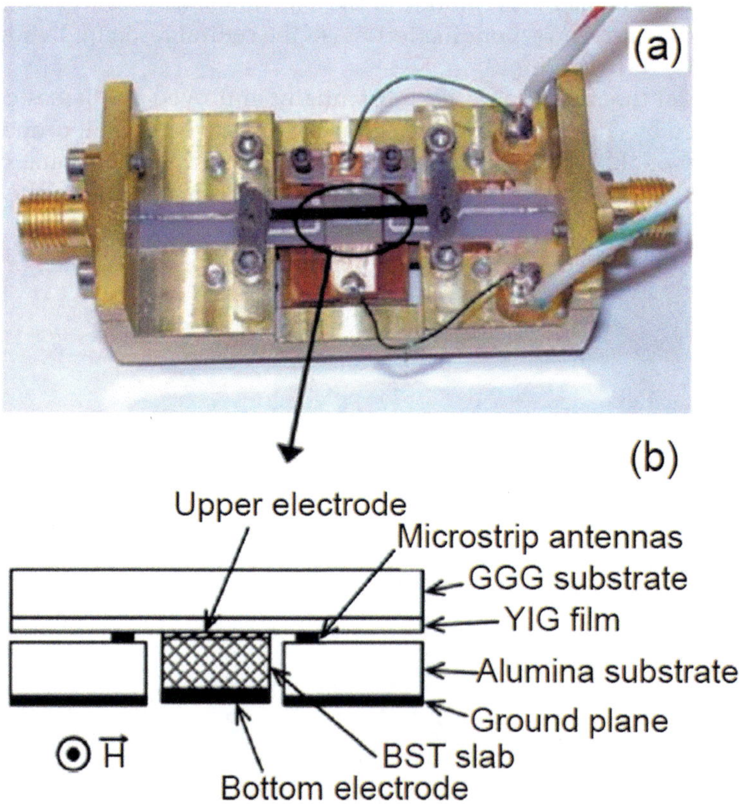

FIGURE 13.17 YIG/BST hybrid wave phase shifter (Ustinov et al., 2007).

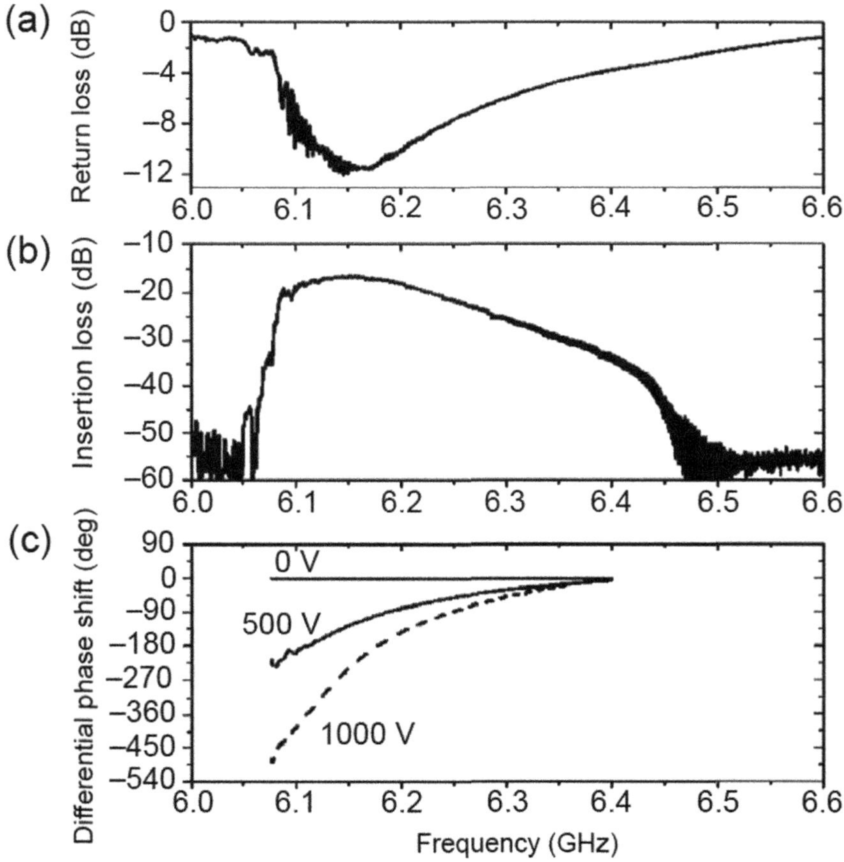

FIGURE 13.18 Frequency dependences of the return loss (a), insertion loss (b), and electrically induced differential phase shift.

are shown in Fig. 13.18. These data were obtained for H = 1413 Oe. As is clear from the figure, the return loss of −10 dB and the insertion loss of −15 dB were observed at 6.15 GHz. The differential phase shift $\Delta\varphi$ in Fig. 13.18 was calculated from the measured phase-frequency characteristics for 0, 500, and 1000 V. The electric field tunability data indicate a continuous variable $\Delta\varphi$ with increasing voltage. The values of $\Delta\varphi$ depended on the signal frequency. In other words, they were a function of the spin wave number. Note that the value of the differential phase shift more than 180 degrees was obtained for the frequencies around 6.15 GHz. A substantial reduction in the insertion loss is required for practical use of the device.

Finally, the *electric field* control of delay time was observed in a ferrite–ferroelectric microwave delay line. A microstrip delay line with a bilayer of (111) YIG film and (001) PMN-PT was studied (Fetisov & Srinivasan, 2005a, 2005b). A 10%–25% variation in delay time was measured when the electric field applied to the PMN-PT layer was increased from 0 to 8 kV/cm. The tunability was attributed to variations in the permittivity for PMN-PT in an electric field and its effect on the dispersion characteristics of hybrid spin-electromagnetic waves that were excited in the bilayer.

13.5.2 Other multiferroic microwave device applications

Several other devices that utilize unique properties of multiferroic composites were reported in recent years (Petrov et al., 2008, 2013; Yang et al., 2009, 2010, 2012). Notable among them include tunable inductors, ferrite–ferroelectric substrates for antenna miniaturization, and ferrite–ferroelectric phase shifters in slot antenna for beam forming.

13.6 Summary

Multiferroic composite materials having mechanically coupled magnetic/ferroelectric phases can lead to strong ME coupling, which allows for electric field control of magnetic properties (converse ME coupling), as well as magnetic field tuning of electrical polarization (direct ME coupling). Of particular interest is the converse ME coupling, which leads to voltage-tunable FMR or voltage-tunable permeability with minimum power consumption, thereby enabling voltage-tunable magnetic devices. Voltage control of magnetism in multiferroic composites has led to a variety of novel devices, including microwave resonators, band-pass and band-stop filters, and phase shifters.

Similarly, a microwave ME coupling of importance is there is excitation of hybridized spin-electromagnetic waves in multiferroic composites. Such excitations arise due to overlap of magnetic modes in the ferromagnetic and dielectric resonance in the ferroelectric, and the two layers need not be bonded. Microwave devices, including resonators, phase shifters, and delay lines, have been demonstrated based on the hybrid modes. There are several challenges, however, for these voltage-tunable RF multiferroic devices. Primary limitations for practical use include (1) a limited frequency tuning range of 1–10%, although the theory predicts a strong ME coupling, and (2) very high insertion loss for hybrid wave devices. Thus, the full potential of voltage tunable RF multiferroic devices is yet to be realized.

References

Adam, J. D., Daniel, M. R., Emtage, P. R., & Talisa, S. H. (1991). Magnetostatic waves. In , *15. Physics of Thin Films* (pp. 1–141). Elsevier.

Adam, J. D., Davis, L. E., Dionne, G. F., Schloemann, E. F., & Stitzer, S. N. (2002). Ferrite devices and materials. *IEEE Transactions on Microwave Theory and Techniques, 50*(3), 721–737.

Bhattacharya, A., Mandal, D., & Bhattacharyya, T. K. (2015). A 1.3-2.4-GHz 3.1-mW VCO using electro-thermo- mechanically tunable self-assembled MEMS inductor on HR substrate. *IEEE Transactions on Microwave Theory and Techniques, 63*(2), 459–469. https://doi.org/10.1109/TMTT.2014.2380357

Bichurin, M. I., Kornev, I. A., Petrov, V. M., Tatarenko, A. S., Kiliba, Y. V., & Srinivasan, G. (2001). *Physical Review B, 64*, Article 094409.

Bichurin, M. I., Petrov, R. V., & Kiliba, Y. V. (1997). *Ferroelectrics, 204*, Article 311e319.

Bichurin, M. I., Petrov, V. M., Kiliba, Y. V., & Srinivasan, G. (2002). *Physical Review B, 66*, Article 134404.

Bichurin, M. I., Viehland, D., & Srinivasan, G. (2007). *Journal of Electroceramics, 19*, 243.

Che, W., Yung, E. K.-N., Wu, K., & Nie, X. (2004). Design investigation on millimeter-wave ferrite phase shifter in substrate integrated waveguide. *Electromagnetic Waves, 45*, 263–275. https://doi.org/10.2528/PIER03082801

Chen, H., Wang, X., Gao, Y., Shi, X., Wang, Z., Sun, N., Zaeimbashi, M., Liang, X., He, Y., Dong, C., Wei, Y., Jones, J., McConney, M., Page, M., Howe, B., Brown, G., & Sun, N. (2020). Integrated tunable magnetoelectric RF inductors. *IEEE Transactions on Microwave Theory and Techniques, 68*(3), 951–963. https://doi.org/10.1109/TMTT.2019.2957472

Chih-Chieh, C., & Rebeiz, G. M. (2011). High- Q 4-6-GHz suspended stripline RF MEMS tunable filter with bandwidth control. *IEEE Transactions on Microwave Theory and Techniques, 59*(10), 2469–2476. https://doi.org/10.1109/TMTT.2011.2163166

Chu, Z., PourhosseiniAsl, M., & Dong, S. (2018). Review of multi-layered magnetoelectric composite materials and devices applications. *Journal of physics. D, Applied Physics, 51*(24), Article 243001. https://doi.org/10.1088/1361-6463/aac29b

Cole, M. W., Nothwang, W. D., Hubbard, C., Ngo, E., & Ervin, M. (2003). Low dielectric loss and enhanced tunability of $Ba_{0.6}Sr_{0.4}TiO_3$ based thin films via material compositional design and optimized film processing methods. *Journal of Applied Physics, 93*(11), 9218–9225.

Cross. (1993). Ferroelectric ceramics: Tailoring properties for specific applications. In N. Setter (Ed.), *Ferroelectric Ceramics*. Basel: Birkhäuser.

Demidov, V. E., Kalinikos, B. A., & Edenhofer, P. (2002). Dipole-exchange theory of hybrid electromagnetic-spin waves in layered film structures. *Journal of Applied Physics, 91*(12), 10007–10016.

El-Ghazaly, A., White, R. M., & Wang, S. X. (2017). Gigahertz-band integrated magnetic inductors. *IEEE Transactions on Microwave Theory and Techniques, 65*(12), 4893–4900. https://doi.org/10.1109/Tmtt.2017.2728040

Entesari, K., & Rebeiz, G. M. (2005). A differential 4-bit 6.5-10-GHz RF MEMS tunable filter. *IEEE Transactions on Microwave Theory and Techniques, 53*(3), 1103–1110. https://doi.org/10.1109/TMTT.2005.843501

Fang, J., Li, H., & Tang, Y. (2017). A magnetic integrated LLCL filter for grid-connected voltage-source converters. *IEEE Transactions on Power Electronics, 32*(3), 1725–1730. https://doi.org/10.1109/TPEL.2016.2613578

Fang, D.-M., Yuan, Q., Li, X.-H., & Zhang, H.-X. (2010). Electrostatically driven tunable radio frequency inductor. *Microsystem Technologies: Sensors, Actuators, Systems Integration, 16*(12), 2119–2122. https://doi.org/10.1007/s00542-010-1138-1

Fetisov, Y. K., & Srinivasan, G. (2005a). Ferrite/piezoelectric microwave phase shifter: studies on electric field tunability. *Electronics Letters, 41*(19), 1.

Fetisov, Y. K., & Srinivasan, G. (2005b). Electrically tunable ferrite-ferroelectric microwave delay lines. *Applied Physics Letters, 87*(10), 103502.

Fetisov, Y. K., & Srinivasan, G. (2006). Electric field tuning characteristics of a ferrite-piezoelectric microwave resonator. *Applied Physics Letters, 88*(14), 143503.

Fetisov, Y. K., & Srinivasan, G. (2008). Nonlinear electric field tuning characteristics of yttrium iron garnet–lead zirconate titanate microwave resonators. *Applied Physics Letters, 93*(3), Article 033508.

Flynn, D., Toon, A., Allen, L., Dhariwal, R., & Desmulliez, M. P. Y. (2007). Characterization of core materials for microscale magnetic components operating in the megahertz frequency range. *IEEE Transactions on Magnetics, 43*(7), 3171–3180. https://doi.org/10.1109/Tmag.2007.895700

Gao, Y., Zardareh, S. Z., Yang, X., Nan, T. X., Zhou, Z. Y., Onabajo, M., Liu, M., Aronow, A., Mahalingam, K., Howe, B. M., Brown, G. J., & Sun, N. X. (2014). Significantly enhanced inductance and quality factor of GHz integrated magnetic solenoid inductors with FeGaB/Al2O3 multilayer films. *IEEE Transactions on Electron Devices, 61*(5), 1470–1476. https://doi.org/10.1109/Ted.2014.2313095

Gao, Y., Zare, S., Onabajo, M., Li, M., Zhou, Z. Y., Nan, T. X., Yang, X., Liu, M., Mahalingam, K., Howe, B. M., Jones, J. G., Brown, G. J., & Sun, N. X. (2014). Power-efficient voltage tunable RF integrated magnetoelectric inductors with FeGaB/Al_2O_3 multilayer films. In *2014 ieee mtt-S international microwave symposium (ims)*.

Gardner, D. S., Schrom, G., Paillet, F., Jamieson, B., Karnik, T., & Borkar, S. (2009). Review of on-chip inductor structures with magnetic films. *IEEE Transactions on Magnetics, 45*(10), 4760–4766. https://doi.org/10.1109/TMAG.2009.2030590

Gupta, S., Bhatter, P., & Kakkar, V. (2021). Point-of-care detection of tuberculosis using magnetoresistive biosensing chip. *Tuberculosis, 127*. https://doi.org/10.1016/j.tube.2021.102055

Gupta, R., & Kotnala, R. K. (2022). A review on current status and mechanisms of room-temperature magnetoelectric coupling in multiferroics for device applications. *Journal of Materials Science, 57*(27), 12710–12737. https://doi.org/10.1007/s10853-022-07377-4

Haldar, A., Kumar, D., & Adeyeye, A. O. (2016). A reconfigurable waveguide for energy-efficient transmission and local manipulation of information in a nanomagnetic device. *Nature Nanotechnology, 11*(5), 437–443. https://doi.org/10.1038/nnano.2015.332

Harward, I., Camley, R. E., & Celinski, Z. (2014). On-wafer magnetically tunable millimeter wave notch filter using M -phase Ba hexagonal ferrite/Pt thin films on Si. *Applied Physics Letters, 105*(17). https://doi.org/10.1063/1.4900519

He, Y. F., Gao, Y., Chen, H. H., Lin, H., Wei, Y. Y., Yang, X., & Sun, N. X. (2018). Integrated tunable bandstop filter using self-biased FeGaB/Al2O3 multilayer thin film. *IEEE Transactions on Magnetics, 54*(9).

He, Y., Luo, B., & Sun, N.-X. (2021). Integrated magnetics and magnetoelectrics for sensing, power, RF, and microwave electronics. *IEEE Journal of Microwaves, 1*(4), 908–929.

Hellwege, K.H., & Springer, A.M. (Eds.) Landolt-Bornstein; *Numerical data and functional relationships in science and technology, Group III, Crystal and Solid State Physics,* vol 4(b), *Magnetic and Other Properties of Oxides*, Springer-Verlag, New York (1970).

Hwaider, L., Jing, L., Yuan, G., Hasegawa, R., Ming, L., Howe, B., Jones, J., Brown, G., & Sun, N. X. (2015). Voltage tunable magnetoelectric inductors with improved operational frequency and quality factor for power electronics. *IEEE Transactions on Magnetics, 51*(1), 1–5. https://doi.org/10.1109/TMAG.2014.2326619

Jeong-Il, K., & Peroulis, D. (2009). Tunable MEMS spiral inductors with optimized RF performance and integrated large-displacement electrothermal actuators. *IEEE Transactions on Microwave Theory and Techniques, 57*(9), 2276–2283. https://doi.org/10.1109/TMTT.2009.2027153

Jiang, S., Liu, Y., Mei, Z., Peng, J., & Lai, C.-M. (2020). A magnetic integrated LCL-EMI filter for a single-phase SiC-mosfet grid-connected inverter. *IEEE Journal of Emerging and Selected Topics in Power Electronics, 8*(1), 601–617. https://doi.org/10.1109/JESTPE.2019.2937816

Kaka, S., Pufall, M. R., Silva, T. J., Katine, J. A., Rippard, W. H., & Russek, S. E. (2005). Mutual phase-locking of microwave spin torque nano-oscillators. *Nature, 437*(7057), 389–392. https://doi.org/10.1038/nature04035

Kim, W. J., Chang, W., Qadri, S. B., Wu, H. D., Pond, J. M., Kirchoefer, et al. (2000). Electrically and magnetically tunable microwave device using (Ba, Sr) $TiO_3/Y_3Fe_5O_{12}$ multilayer. *Applied Physics A, 71*, 7–10.

Kittel, C. (1948). On the theory of ferromagnetic resonance absorption. *Physical Review, 73*(2), 155–161. https://doi.org/10.1103/PhysRev.73.155

Kurata, H., Shirakawa, K., Nakazima, O., & Murakami, K. (1993). Study of thin film micro transformer with high operating frequency and coupling coefficient. *IEEE Transactions on Magnetics, 29*(6), 3204–3206. https://doi.org/10.1109/20.281137

Lawes, G., & Srinivasan, G. (2011). Introduction to magnetoelectric coupling and multiferroic films. *Journal of Physics D: Applied Physics, 44*(24), 243001.

Li, N., et al. (2011). Electrostatic tuning of ferromagnetic resonance and magnetoelectric interactions in ferrite-piezoelectric heterostructures grown by chemical vapor deposition. *Applied Physics Letters, 99*(19), 192502.

Lin, H., Nan, T. X., Qian, Z. Y., Gao, Y., Hui, Y., Wang, X. J., Guo, R. D., Belkessam, A., Shi, W., Rinaldi, M., & Sun, N. X. (2016). Tunable RF band-pass filters based on NEMS magnetoelectric resonators. In *2016 ieee mtt-S international microwave symposium (ims)*. Go to ISI>://WOS:000390313200394.

Lin, H., Wu, J., Yang, X., Hu, Z. Q., Nan, T. X., Emori, S., Gao, Y., Guo, R. D., Wang, X. J., & Sun, N. X. (2015). Integrated non-reciprocal dual H- and E-field tunable bandpass filter with ultra-wideband isolation. In *2015 ieee mtt-S international microwave symposium (ims)*. Go to ISI>://WOS:000370722900338.

Lou, J., Reed, D., Liu, M., & Sun, N. X. (2009). Electrostatically tunable magnetoelectric inductors with large inductance tunability. *Applied Physics Letters, 94*(11). https://doi.org/10.1063/1.3103273

Mariappan, S. G., Moazenzadeh, A., & Wallrabe, U. (2016). Polymer magnetic composite core based microcoils and microtransformers for very high frequency power applications. micromachines. *Micromachines, 7*(6). https://doi.org/10.3390/mi7060100

Mundotiya, B. M., Dinulovic, D., Rissing, L., & Wurz, M. C. (2017). Fabrication and characterization of a Ni-Fe-W core microtransformer for high-Frequency power applications. *Sensors and Actuators. A. Physical, 267*, 42–47. https://doi.org/10.1016/j.sna.2017.09.032

Nikonov, D. E., & Bourianoff, G. I. (2005). Spin gain transistor in ferromagnetic semiconductors-the semiconductor Bloch-equations approach. *IEEE Transactions on Nanotechnology, 4*(2), 206–214. https://doi.org/10.1109/TNANO.2004.837847

Petrov, R. V., Tatarenko, A. S., Srinivasan, G., & Mantese, J. V. (2008). Antenna miniaturization with ferrite ferroelectric composites. *Microwave and Optical Technology Letters, 50*(12), 3154–3157.

Petrov, R. V., Murthy, D. V., Sreenivasulu, G., & Srinivasan, G. (2013). A slot antenna with magnetoelectric elements. *Microwave and Optical Technology Letters, 55*(3), 533–535.

Pettiford, C., Dasgupta, S., Lou, J., Yoon, S. D., & Sun, N. X. (2007). Bias field effects on microwave frequency behavior of PZT/YIG magnetoelectric bilayer. *IEEE Transactions on Magnetics, 43*(7), 3343–3345.

Piljae, P., Cheon Soo, K., Mun Yang, P., Sung Do, K., & Hyun Kyu, Y. (2004). Variable inductance multilayer inductor with MOSFET switch control. *IEEE Electron Device Letters, 25*(3), 144–146. https://doi.org/10.1109/LED.2003.822670

Salvia, J., Bain, J. A., & Patrick Yue, C. (2005). *Tunable on-chip inductors up to 5 GHz using patterned permalloy laminations.*

Sekar, V., Armendariz, M., & Entesari, K. (2011). A 1.2–1.6-GHz substrate-integrated-waveguide RF MEMS tunable filter. *IEEE Transactions on Microwave Theory and Techniques, 59*(4), 866–876. https://doi.org/10.1109/TMTT.2011.2109006

Semenov, A. A., Karmanenko, S. F., Kalinikos, B. A., Srinivasan, G., Slavin, A. N., & Mantese, J. V. (2006). Dual-tunable hybrid wave ferrite-ferroelectric microwave resonator. *Electronics Letters, 42*(11), 1.

Shastry, S., Srinivasan, G., Bichurin, M. I., Petrov, V. M., & Tatarenko, A. S. (2004). Microwave magnetoelectric effects in single crystal bilayers of yttrium iron garnet and lead magnesium niobate-lead titanate. *Physical Review B—Condensed Matter and Materials Physics, 70*(6), Article 064416.

Shirane, A., Ito, H., Ishihara, N., & Masu, K. (2012). Planar solenoidal inductor in radio frequency micro-electro-mechanical systems technology for variable inductor with wide tunable range and high quality factor. *Japanese Journal of Applied Physics, 51*(5). https://doi.org/10.1143/JJAP.51.05EE02

Song, Y.-Y., Ordóñez-Romero, C. L., & Wu, M. (2009). Millimeter wave notch filters based on ferromagnetic resonance in hexagonal barium ferrites. *Applied Physics Letters, 95*(14). https://doi.org/10.1063/1.3246170

Spaldin, N. A., & Fiebig, M. (2005). The renaissance of magnetoelectric multiferroics. *Science (American Association for the Advancement of Science), 309*(5733), 391–392. https://doi.org/10.1126/science.1113357

Srinivasan, G. (2010). Magnetoelectric composites. *Annual Review of Materials Research, 40*(1), 153–178.

Srinivasan, G., Zavislyak, I. V., & Tatarenko, A. S. (2006). Millimeter-wave magnetoelectric effects in bilayers of barium hexaferrite and lead zirconate titanate. *Applied Physics Letters, 89*(15), 152508.

Sun, N. X., & Srinivasan, G. (2012, September). Voltage control of magnetism in multiferroic heterostructures and devices. In, *Vol. 2, No. 03. Spin* (p. 1240004). World Scientific Publishing Company.

Tatarenko, A. S., Gheevarughese, V., & Srinivasan, G. (2006a). Magnetoelectric microwave bandpass filter. *Electronics Letters, 42*(9), 540–541.

Tatarenko, A. S., Srinivasan, G., & Bichurin, M. I. (2006b). Magnetoelectric microwave phase shifter. *Applied Physics Letters, 88*(18), 183507.

Tatarenko, A. S., Murthy, D. V. B., & Srinivasan, G. (2012). Hexagonal ferrite-piezoelectric composites for dual magnetic and electric field tunable 8–25 GHz microstripline resonators and phase shifters. *Microwave and Optical Technology Letters, 54*(5), 1215–1218.

Tatarenko, A. S., & Srinivasan, G. (2011). A strain engineered voltage tunable millimeter-wave ferrite phase shifter. *Microwave and Optical Technology Letters, 53*(2), 261–264.

Tatarenko, A. S., Srinivasan, G., & Filippov, D. A. (2007). Magnetoelectric microwave attenuator. *Electronics Letters, 43*, 674–675.

Tatarenko, A. S., Ustinov, A. B., Srinivasan, G., Petrov, V. M., & Bichurin, M. I. (2010). Microwave magnetoelectric effects in bilayers of piezoelectrics and ferrites with cubic magnetocrystalline anisotropy. *Journal of Applied Physics, 108*(6), Article 063908.

Tengxing, W., Wei, J., Divan, R., Rosenmann, D., Ocola, L. E., Yujia, P., & Guoan, W. (2017). Novel electrically tunable microwave solenoid inductor and compact phase shifter utilizing permaloy and PZT thin films. *IEEE Transactions on Microwave Theory and Techniques, 65*(10), 3569–3577. https://doi.org/10.1109/TMTT.2017.2731765

Tsai, C. S., Qiu, G., Gao, H., Yang, L. W., Li, G. P., Nikitov, S. A., & Gulyaev, Y. (2005). Tunable wideband microwave band-stop and band-pass filters using YIG/GGG-GaAs layer structures. *IEEE Transactions on Magnetics, 41*(10), 3568–3570. https://doi.org/10.1109/TMAG.2005.855191

Ustinov, A. B., & Srinivasan, G. (2008). Subterahertz excitations and magnetoelectric effects in hexaferrite-piezoelectric bilayers. *Applied Physics Letters, 93*(14), 142503.

Ustinov, A. B., & Srinivasan, G. (2010). Speed of a ferrite—ferroelectric microwave planar resonator. *Technical Physics, 55*, 900—903.

Ustinov, A. B., et al. (2006). Electric field tunable ferrite-ferroelectric hybrid wave microwave resonators: Experiment and theory. *Journal of Applied Physics, 100*(9), Article 093905.

Ustinov, A. B., Srinivasan, G., & Kalinikos, B. A. (2007). Ferrite-ferroelectric hybrid wave phase shifters. *Applied Physics Letters, 90*(3), Article 031913.

Ustinov, A. B., Kalinikos, B. A., Tiberkevich, V. S., Slavin, A. N., & Srinivasan, G. (2008). Q factor of dual-tunable microwave resonators based on yttrium iron garnet and barium strontium titanate layered structures. *Journal of Applied Physics, 103*(6), Article 063908.

Ustinov, A. B., Kalinikos, B. A., & Srinivasan, G. (2014). Nonlinear microwave phase shifter on electromagnetic-spin waves. *Technical Physics, 59*(9), 1406—1408. https://doi.org/10.1134/S1063784214090266

Vroubel, M., Yan, Z., Rejaei, B., & Burghartz, J. N. (2004). Integrated tunable magnetic RF inductor. *IEEE Electron Device Letters, 25*(12), 787—789. https://doi.org/10.1109/LED.2004.839208

Wang, X., Chen, H., Shi, X., Gao, Y., Lin, H., Jones, J. G., Howe, B. M., Brown, G. J., & Sun, N. X. (2017). A novel NiZn ferrite integrated magnetic solenoid inductor with a high quality factor at 0.7—6 GHz. *AIP Advances, 7*(5). https://doi.org/10.1063/1.4973283

Yamaguchi, M., Baba, M., Suezawa, K., Moizumi, T., Arai, K. I., Shimada, Y., Haga, A., Tanabe, S., & Ito, K. (2000, 11-16 June 2000). Magnetic RF integrated thin-film inductors. In *2000 IEEE MTT-S international microwave symposium digest (cat. No.00CH37017).*

Yamaguchi, M., Bae, S., Kim, K. H., Tan, K., Kusumi, T., & Yamakawa, K. (2005). Ferromagnetic RF integrated inductor with closed magnetic circuit structure. *IEEE MTT-S international microwave symposium digest.*

Yang, G. M., Xing, X., Daigle, A., Liu, M., Obi, O., Stoute, S., … Sun, N. X. (2009). Tunable miniaturized patch antennas with self-biased multilayer magnetic films. *IEEE Transactions on Antennas and Propagation, 57*(7), 2190—2193.

Yang, G. M., Xing, X., Obi, O., Daigle, A., Liu, M., Stoute, S., … Sun, N. X. (2010). Loading effects of self-biased magnetic films on patch antennas with substrate/superstrate sandwich structure. *IET Microwaves, Antennas & Propagation, 4*(9), 1172—1181.

Yang, G. M., Obi, O., & Sun, N. X. (2012). Enhancing ground plane immunity of dipole antennas with spin—spray-deposited lossy ferrite films. *Microwave and Optical Technology Letters, 54*(1), 230—233.

Yang, X., Wu, J., Beguhn, S., Nan, T., Gao, Y., Zhou, Z., & Sun, N. X. (2013). Tunable bandpass filter using partially magnetized ferrites with high power handling capability. *IEEE Microwave and Wireless Components Letters, 23*(4), 184—186.

Yun, E. J., Jung, M., Cheon, C. I., & Nam, H. G. (2004). Microfabrication and characteristics of low-power high-performance magnetic thin-film transformers. *IEEE Transactions on Magnetics, 40*(1), 65—70. https://doi.org/10.1109/Tmag.2003.821119

Wei, W., Tsai, C. S., Lee, C. C., Yoo, H. J., Hopster, H., & Mills, D. L. (2000). *Ferromagnetic Fe/Ag-GaAs waveguide structures for wideband microwave integrated notch filter devices.*

Weng, L., Cao, X., Xv, H., Huang, W., Wang, B., & Yang, Z. (2017). *Frequency-dependent complex magnetic permeability and magnetic losses of Fe-Ga alloy.*

Zand, R., Roohi, A., & DeMara, R. F. (2017). Energy-efficient and process-variation-resilient write circuit schemes for spin Hall effect MRAM device. *IEEE Transactions on Very Large Scale Integration Systems, 25*(9), 2394—2401. https://doi.org/10.1109/TVLSI.2017.2699579

Zhang, Z., Wu, Y., Sang, L., Wu, H., Huang, J., Wang, L., Takahashi, Y., Li, R., Koizumi, S., Toda, M., Akita, I. M., Koide, Y., & Liao, M. (2020). Coupling of magneto-strictive FeGa film with single-crystal diamond MEMS resonator for high-reliability magnetic sensing at high temperatures. *Materials Research Letters, 8*(5), 180–186. https://doi.org/10.1080/21663831.2020.1734680

Zine-El-Abidine, I., Okoniewski, M., & McRory, J. G. (2005). *RF MEMs tunable inductor using bimorph microactuators*.

CHAPTER

14

Magnetoelectric composites for biomedical applications

Abbey Knoepfel[1], *Sumanta Kumar Karan*[1], *S. Sharma*[1], *S. Hosur*[2], *M. Kiani*[2], *and Shashank Priya*[1,3]

[1] Department of Materials Science and Engineering, The Pennsylvania State University, University Park, PA, United States; [2] School of Electrical Engineering and Computer Science, The Pennsylvania State University, University Park, PA, United States; [3] Chemical Engineering and Materials Science, University of Minnesota, Minneapolis, MN, United States

14.1 Introduction

The magnetoelectric (ME) phenomenon has been extensively researched over the past few years to tackle complex challenges and meet current and emerging needs in the biomedical field. The ME materials and devices have been used for various biomedical applications such as on-demand release of attached drug, specific targeting, stability, and biocompatibility (Fig. 14.1) (Kolishetti et al., 2022). This chapter describes the promising applications of ME materials and composites for targeted drug delivery, noninvasive brain imaging and artificial nerve activity stimulation, wireless capsule endoscopies (WCEs), and devices for wireless energy harvesting and data transferring (Hannachi & Slimani, 2022).

There has been a recent surge in application of ME nanoparticles and micro-to-nanoscale devices for biomedical applications. Early-generation ME materials were found to exhibit a strong ME coupling effect at temperatures well below room temperature. However, in the past couple of decades, several ME materials have been

https://doi.org/10.1016/B978-0-443-18606-6.00006-X

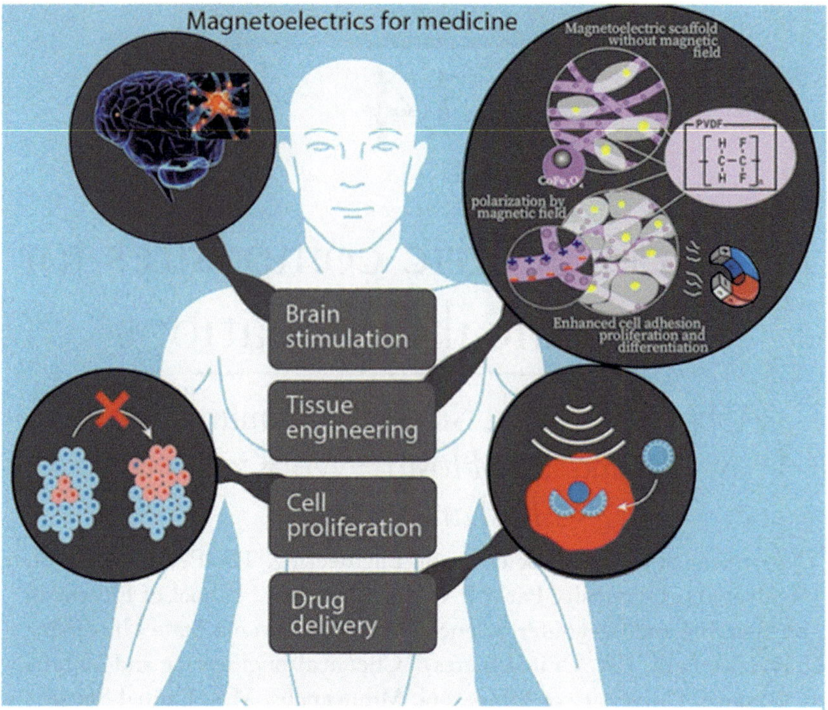

FIGURE 14.1 Summary of biomedical applications of magnetoelectric (ME) composites. *Reprinted from Kopyl et al. (2021).*

synthesized with a strong ME coupling effect at room temperature (Kopyl et al., 2021). In a human body, there are several types of electric fields. The ME material's ability to generate an electric field via the application of a magnetic field, which can be tuned for different types of cells, opens potential for application in bio-stimulants, targeted therapeutics, biosensors, nerve stimulation, and more (Kopyl et al., 2021). Traditional devices require bulky power sources and invasive surgical procedures. ME devices would avoid the need for such bulky implants and thereby open opportunity for minimally invasive diagnostic and therapeutic applications. With the development of room-temperature biocompatible ME nanoparticles, in vivo studies have successfully demonstrated the delivery of ME nanoparticles across the blood−brain barrier for wireless deep brain stimulation, and in vitro studies have shown success in tissue regeneration and high-specificity cancer treatments with ME nanoparticles (Hannachi & Slimani, 2022).

In broader terms, the ME composites can comprise of single-phase or multiphase materials. Single-phase ME materials have a single component which exhibits both magnetostrictive and piezoelectric properties simultaneously, examples of which include multiferroics, magnetoelectrically coupled oxides, relaxor ferroelectrics, and organic—inorganic hybrid materials (Hannachi & Slimani, 2022; Liang, Chen et al., 2021; Smith et al., 2023). Single-phase multiferroic ME materials are not ideal choices for biomedical applications as they typically exhibit weak ME coupling that is only present at low temperatures (Liang, Matyushov et al., 2021; Mitoseriu, 2005). The ME coupling coefficient, α, of single-phase ME materials is typically $<<0.1\ V\ cm^{-1}\ Oe^{-1}$ near room temperature (Hannachi & Slimani, 2022). On the other hand, multiphase ME materials are composed of distinct magnetostrictive and piezoelectric layers. These multiphase ME composites offer many advantages over single-phase ME materials. The ME response of multiphase ME systems can be more easily tailored and enhanced based on optimizing the properties of the constituent material layers. There have been reported values of the ME coefficient of multiphase ME composites, which are two orders of magnitude higher than those of the single-phase materials (Smith et al., 2023). There are many ways to combine the piezoelectric and magnetostrictive phases to form ME composites as shown in Fig. 14.2 (Smith et al., 2023). The properties of the constituent materials, including magnetostriction constant, resistivity, permeability, permittivity, and piezoelectric strain, along with the structure, will dictate the ME properties of the resultant composite (Priya & Inman, 2009).

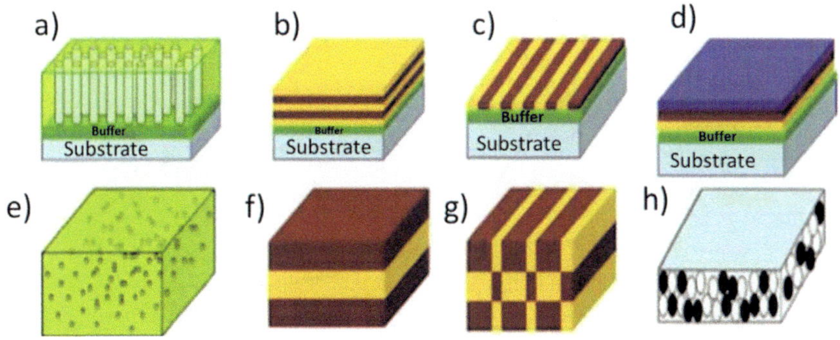

FIGURE 14.2 Schematic of possible multi-phase magnetoelectric (ME) composite structures: (a) Self-assembled nanopillar, (b) layer-by-layer deposition, (c) layer deposition with masking and patterning, (d) three-phase deposition for gradient materials, (e) particles dispersed in polymer matrix, (f) lamination, (g) checkerboard arrangement, and (h) sintered particulates. *Adapted from Kopyl et al. (2021).*

One study by Hosur et al. examined the effect of different ME transducer designs on the wireless power transfer performance (Priya & Inman, 2009). The effect of the transducer dimensions, d.c. bias magnetic field, loading and operation frequency on resonance frequency, quality factor, and received power were determined via modeling and testing (Hosur et al., 2021). As shown in Fig. 14.3a, an ME device harvested over 20 times as much power as compared to a classic coil for an applied load between 1 and 100 kΩ, demonstrating that for a given device size, an ME transducer can harvest much more power at a lower frequency than a traditional coil (Hosur et al., 2021). They also demonstrated the change in received power upon changing the angle of rotation, θ, and for small θ (<45 degrees), ME transducer outperformed the traditional coil of similar volume (Fig. 14.3b) (Hosur et al., 2021).

Thermal mismatching between the phases may arise in multiphase ME composites, which can result in thermal stresses causing a decrease in the composite's performance. Another challenge is the complexity in

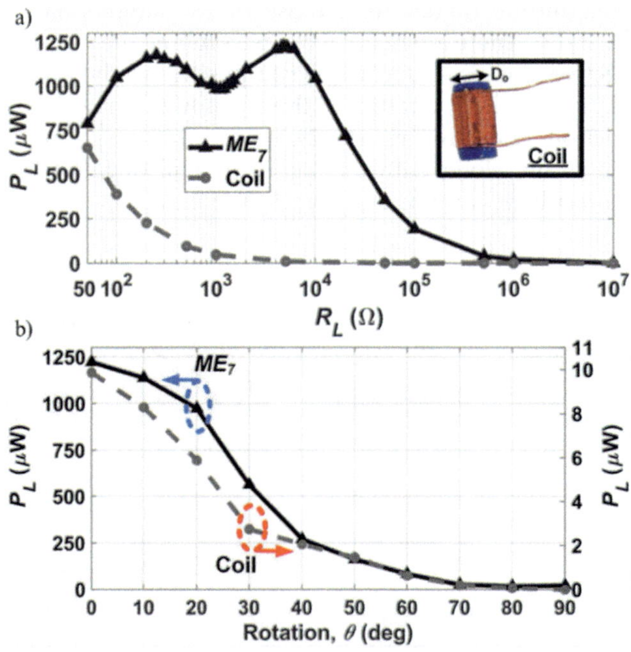

FIGURE 14.3 (a) Measured received power, P_L, versus loading, R_L of ME_7 and optimized solenoid coil of comparable volume ($V = 150\ mm^3$); (b) measured P_L versus angle of rotation, θ, in the xy plane at $R_L = 4.8\ k\Omega$. Peak applied a.c. magnetic field, $H_{AC,peak}$, = 1.41 Oe. *Reprinted with permission from Hosur et al. (2021).*

fabrication, as the interfacing between individual constituents can require extensive optimization to achieve enhanced ME coupling. Multiphase multiferroic materials have been studied in various biomedical applications, including drug delivery and drug release processes, magnetic hyperthermia (MH), magnetic resonance imaging (MRI), magnetic particle imaging (MPI), and in-vitro cell separation and diagnostics.

A strong ME effect at room temperature in multiphase ME composites can only be achieved by the combination of high-performance piezoelectric and magnetostrictive phases. For strain-mediated ME composites, the ME coefficient can be expressed as: $\alpha_{ME} = \frac{k_c d_{ij,m} d_{ki,p}}{\varepsilon_r \varepsilon_0}$ (Amirov, 2022). The ME coefficient is directly proportional to the interface coupling coefficient (k_c) between the magnetostrictive and piezoelectric phases, the magnetostrictive coefficient ($d_{ij,m} = \frac{\partial S}{\partial H}$) and the piezoelectric coefficient ($d_{ki,p} = \frac{\partial P}{\partial S}$), and inversely proportional to the relative permittivity (ε_r) (Chu et al., 2018). The highest piezoelectric response has been seen in lead-based piezoelectric PMN-PT and PZN-PT single crystals; however, there are some concerns regarding toxicity and biocompatibility due to the presence of lead element. Therefore, lead-free piezoelectric alternatives are good candidates for biomedical applications, including $K_{0.5}Na_{0.5}NbO_3$ (KNN), $BaTiO_3$ (BTO), $Ba_{0.85}Ca_{0.15}Zr_{0.1}Ti_{0.9}O_3$ (BCZT), piezo-polymer composites (i.e., PVDF), and nonferroelectric piezoelectric crystals (i.e., AlN, ZnO) (Chu et al., 2018). The harmful effects of Pb-containing materials on the environment and human health have been well-studied. Humans can experience negative effects on the central nervous, renal, reproductive, and cardiovascular systems following exposure to high Pb concentrations (Pullarao et al., 2019).

ME composites for biomedical applications must consider this toxicity and emphasize Pb-free piezoelectric materials. Recent research in this area has been focused on the improvement of the piezoelectric coefficients in Pb-free materials (Assi et al., 2016). Not only does the material system need to be nontoxic, but it also needs to consider any immune response mounted due to the presence of the foreign object in contact with the body. Internal devices should be developed to avoid any biofouling response resulting in the formation of a vascular tissue that encapsulates the device and decreases the functionality and longevity of the device (Pullarao et al., 2019).

Core-shell ME nanoparticles (MENs) are an attractive option for biomedical applications due to their ability to produce a strong ME effect at room temperature (Morais et al., 2010). The ME effect is achieved via the elastic coupling of magnetostrictive and piezoelectric phases under an applied magnetic field (Fiocchi et al., 2022). The core typically consists of a ferromagnetic material with a spinel structure (AB_2O_4) that possesses a high magnetic moment and magnetostrictive properties. The shell is typically a ferroelectric material with a perovskite structure (ABO_3) that possesses piezoelectric properties and a spontaneous electric polarization (Song et al., 2022). Current research efforts for biomedical applications of MENs have mainly utilized a magnetostrictive $CoFe_2O_4$ (CFO) core and a piezoelectric $BaTiO_3$ (BTO) shell due to the composite's high ME effect and well-documented synthesis procedures. As discussed in Chapter 2, the dielectric asymmetry in MENs results in an electric field that spreads to the local area when an a.c. magnetic field is applied. Two common synthesis routes are the sol-gel method that is helpful for obtaining desired particle size, and the hydrothermal method that offers increased solubility and faster reaction of precursors at elevated temperature and pressure. Additional core–shell MEN synthesis routes have been reported, including solid-state reaction, microemulsion-based process, and solvent evaporation methods (Song et al., 2022).

Magnetic fields are the only wireless communication method able to access deep tissues with minimal attenuation due to the low magnetic susceptibility of biological material, Fig. 14.4 (Song et al., 2022). Optical stimulation offers very high spatial and temporal resolution but is limited to penetration depths of less than 1 mm due to scattering and absorbance by surrounding tissue. Acoustic waves can penetrate one to two orders of magnitude deeper depending upon the frequency. Ultrasound signals with lower frequencies offer deeper penetration depths without significant attenuation; however, the resolution is proportional to the wavelength resulting in modulation volumes greater than 1 mm^3 (Chen et al., 2017). ME devices penetrate the human body with little absorption when operating at low-frequency magnetic fields, on the order of kHz. The safety limits for d.c. and a.c. magnetic field strengths are 550 μT and 1670 Oe, respectively, as set by the Institute of Electrical and Electronics Engineers (IEEE) standards (Chen et al., 2017). Studies on toxicity of ME

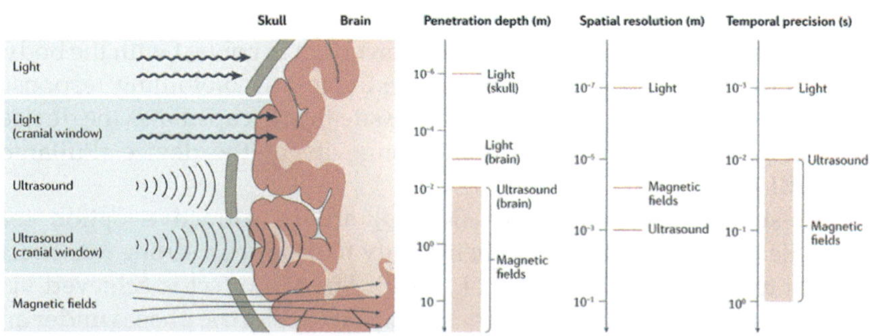

FIGURE 14.4 Schematic diagram depicting penetration of light, ultrasound, and magnetic field waves. Associated penetration depth, spatial resolution, and temporal precision of the respective waves. *Reprinted with permission from Chen et al. (2017).*

core–shell nanoparticles have been conducted. For example, the peripheral blood toxicity and organ-specific toxicity was analyzed in adult mice for $BaTiO_3$–CoF_2O_4 core–shell nanoparticles, and no toxicity was observed (Hosur et al., 2021).

14.2 Application of ME nanoparticles for targeted drug delivery

The utilization of ME nanoparticles (MENs) has become increasingly popular as noninvasive drug delivery systems due to their unique properties, which address notable challenges in modern treatments of various maladies (i.e., cancers, viral infections, bacterial infections) (Guduru et al., 2013; Hannachi & Slimani, 2022; Stimphil et al., 2017).

Each year, the American Cancer Society releases an estimate of the number of new cancer diagnoses and deaths based on the most recent data on cancer occurrences collected by national cancer registries and the National Center for Health Statistics (Mhambi et al., 2021). Cancer is the second leading cause of death in the United States, with an estimated 1.95 million cancer diagnoses and 609,820 deaths due to cancer this year (Siegel et al., 2023). The mortality rate has been decreasing, with an overall reduction of 33% since 1991, which reflects the advancements in early diagnosis and treatments (Siegel et al., 2023). However, due to the coronavirus disease in 2019 (COVID-19), delays in diagnosis and treatment could result in an increase of advanced-stage diagnoses and mortality in the coming years, but that will not be directly quantified for at least 2–3 years due to the lag in population-based cancer data (Siegel et al., 2023).

A main challenge in the treatment of various cancers is the inability to destroy the cancerous cells while preserving the surrounding healthy cells, which leads to the severe side effects many patients experience during treatments (Siegel et al., 2023; Smith et al., 2023). The low specificity also results in a low probability that every cancerous cell has been eradicated, which may lead to future recurrence of the cancer (Schmidt et al., 2020). MENs offer a targeting mechanism driven by a physical force, as opposed to antigen/antibody matching, as well as a high-specificity delivery mechanism, which enhances cellular uptake of the attached drug across the targeted cell membranes and spares normal cells (Smith et al., 2023). MENs also offer an externally controlled mechanism for on-demand drug delivery, release, and capability to integrate image-guided delivery via the application of d.c. and a.c. magnetic fields (Abdelazim, 2017; Smith et al., 2023; Stimphil et al., 2017). MENs can intrinsically generate electric fields, enabling the on-demand release of drugs. They can also induce electroporation when the electric field is able to overcome the membrane potential of the cell. This results in the opening of the cell membrane pores to enhance the uptake of a drug and guide it inside the cell or tissue, enhancing the treatment efficacy (Rodzinski et al., 2016). The separate control of drug delivery and release is due to the nonzero ME effect of the MENs. The nonzero ME effect is a result of the strongly coupled magnetostrictive and piezoelectric layers, while traditional magnetic nanoparticles, such as iron oxides, lack this coupling effect (Kaushik et al., 2017). The electrical properties of cancer cells differ from those of healthy cells (Rodzinski et al., 2016). Ovarian cancer cells have a membrane potential −3 mV, compared to a membrane potential of −50 mV for healthy cells. Therefore, the electric field reaction would differ between MENs and cancer cells versus healthy cells (Rodzinski et al., 2016). The working mechanism for drug delivery can be explained as follows, assuming initially there is a symmetric ionic bond prior to applying a magnetic field and an electric dipole moment forms (Rodzinski et al., 2016). Once the magnetic field is applied, the induced electric dipole breaks the symmetry which breaks the already weak bond, while the bond at the other end of the dipole becomes stronger. The dipole moment direction changes under the ideal excitation of the a.c. magnetic field and uniformly breaks all bonds and releases the drug while simultaneously inciting electroporation (Song et al., 2022).

Rodzinski et al. studied the efficacy of a $CoFe_2O_4$−$BaTiO_3$ core−shell MENs-based drug delivery for metastatic (stage III) ovarian cancer (Song et al., 2022). The current gold-standard treatment protocol is cytoreductive surgery followed by chemotherapy, consisting of paclitaxel and carboplatin (Rodzinski et al., 2016). Improved survival rates for stage III ovarian cancer have been observed in patients that received intraperitoneal chemotherapy via a surgically implanted catheter, but this approach has not been widely adopted due to surgical complications, infections, and toxicity (Rodzinski et al., 2016). Paclitaxel is a common chemotherapeutic agent in the treatment of multiple types of cancer so a successful nanotechnology-based drug delivery system could drastically improve treatment options and mortality rates for many cancer diagnoses (Rodzinski et al., 2016). Under an applied magnetic field, the paclitaxel-loaded MENs formed a negative surface charge, and the conductivity of the cancer cell membranes temporarily increased, facilitating the movement of MENs to the cancer cells (Rodzinski et al., 2016). Once the MENs approach the cancer cells, they are able to be transferred into the cell due to their high-specificity electroporation effect (Fig. 14.5c). At this point, an a.c. field is applied, which releases the paclitaxel from the MENs inside of the cancer cells and the healthy cells remain intact as their electroporation threshold is significantly higher than the cancer cells. The time dependency of tumor volume was closely monitored for the IV treatment of paclitaxel-loaded MENs and three control nanocarriers: magnetic

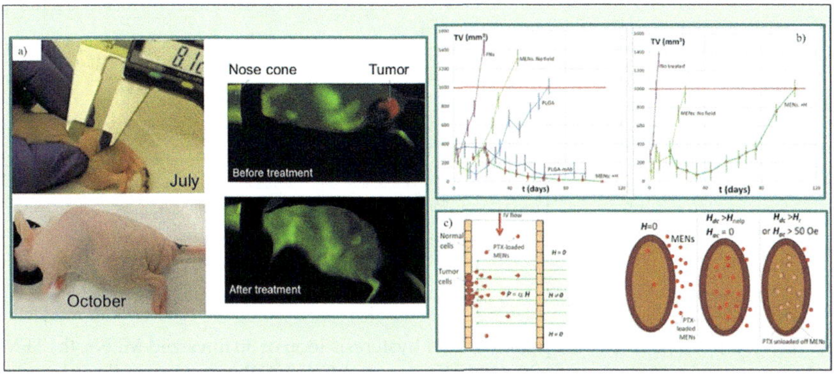

FIGURE 14.5 (a) Photographs of tumor in July at its peak volume (268 mm^3) (top left) and in October after treatment (bottom left), and IR images taken with fluorescent agent Her2Sense 645 before starting (top right) and after (bottom right) completion of MEN treatment; (b) comparison of tumor volume (TV) change after systemic IV administration of paclitaxel via MENs and control nanocarriers with and without magnetic field (left) or localized subcutaneous administration of MENs with and without local magnetic field (right); (c) schematic of magnetic field controlled paclitaxel delivery via MENs through a capillary injected into the circulatory system (Chen et al., 2017). *Reprinted with permission from Rodzinski et al. (2016).*

nanoparticles (FN), poly(D, L-lactide-co-glycolide) (PLGA), and PLGA conjugated with HER2-neu-monoclonal antibody (PLGA-mAb) (Rodzinski et al., 2016). The time dependency on tumor volume was also monitored for the subcutaneous injection of paclitaxel-loaded MENs with and without an applied magnetic field (Fig. 14.5b). The intravenous (IV) administration was more effective than the subcutaneous delivery, due to limited mobility which decreased the amount of drug able to transfer into the cancer cells. The only mice that were cured were those that were subjected to 3 months of weekly IV injections of paclitaxel-loaded MENs followed by localized magnetic field treatments (Fig. 14.5a) (Rodzinski et al., 2016). The results were in agreement with the proposed hypothesis that the underlying mechanism for targeting cancer cells is due to the magnetically controlled electricity from the ME effect rather than from pure magnetism. The study further confirmed that the high-specificity targeted delivery is possible due to application of a local d.c. magnetic field. The local d.c. magnetic field is responsible for the localization of loaded MENs near the tumor site and the formation of a negatively charged surface from the induced electric dipole (Rodzinski et al., 2016).

Nagesetti et al. used MENs with nuclear magnetic resonance (NMR) to identify different cancer cells in vitro (Rodzinski et al., 2016). $CoFe_2O_4$–$BaTiO_3$ core–shell MENs were prepared via a solution method and characterized using transmission electron microscopy (TEM) and

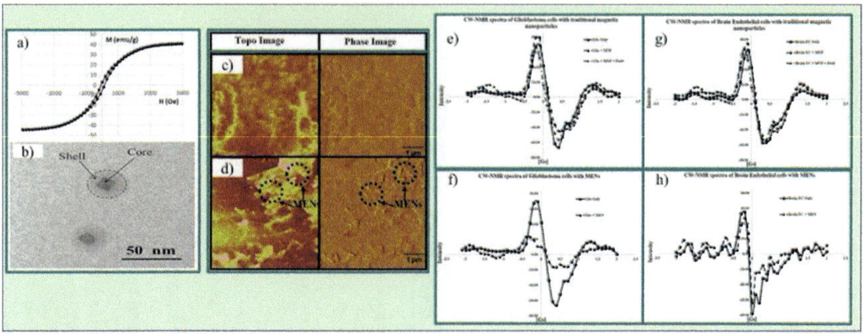

FIGURE 14.6 (a) Room temperature M-H hysteresis loop of 30 nm sized MENs; (b) TEM image of core-shell structure of 3 nm MENs; (c–e, g) CW-NMR spectra of cells only, cells incubated with traditional magnetic nanoparticles (MNPs), and cells incubated with MNPs under an applied magnetic field of 100 Oe for (e) glioblastoma cells U87-MG, (g) normal brain endothelial cells; (f, h) CW-NMR spectra of cells only, cells incubated with MENs, and cells incubated with MENs under an applied magnetic field of 100 Oe for (f) glioblastoma cells U87-MG, (h) normal brain endothelial cells (Rodzinski et al., 2016). *Reprinted with permission from Nagesetti et al. (2017).*

atomic force microscopy (AFM) (Fig. 14.6b–d) (Nagesetti et al., 2017). The NMR spectra in the presence of ovarian carcinoma cells (SKOV-3), glioblastoma cells (U87-MG), and breast adenocarcinoma cells (MDF-7) as well as their noncancerous counterparts (brain endothelial cells) all differed when in the presence of the MENs (Nagesetti et al., 2017). The effect of traditional magnetic nanoparticles was also measured but had a much smaller observable effect on the spectra measured, as shown in Fig. 14.6e and g. Due to the MENs ability to enhance the electric field configuration on each cell membrane, which differ depending on the type and stage of cancer cell present, there was a significant change in the NMR spectra observed (Fig. 14.6f and h) (Nagesetti et al., 2017). This can be attributed to the cell membrane properties of each type of cell resulting in distinct MENs organization and particle binding to the cells which minimizes electric energy (Nagesetti et al., 2017). The core–shell MENs have shown promising results and have the potential for researching as drug delivery vessels in addition to improving cancer diagnostics via their ability to produce distinct spectra in the presence of different types of cancer cells. The type of high specificity provided by MENs opens the door for the development of cancer treatments and cures (Nagesetti et al., 2017).

MENs have been studied as carriers for targeted delivery across the blood–brain barrier (BBB) via the application of a direct current (d.c.) magnetic gradient on the order of 1000 Oe/cm, which pulled the MENs into the brain from the blood stream after intravenous (IV) injection of the MENs (Kaushik et al., 2016, 2017; Kaushik, Rodriguez et al., 2019;

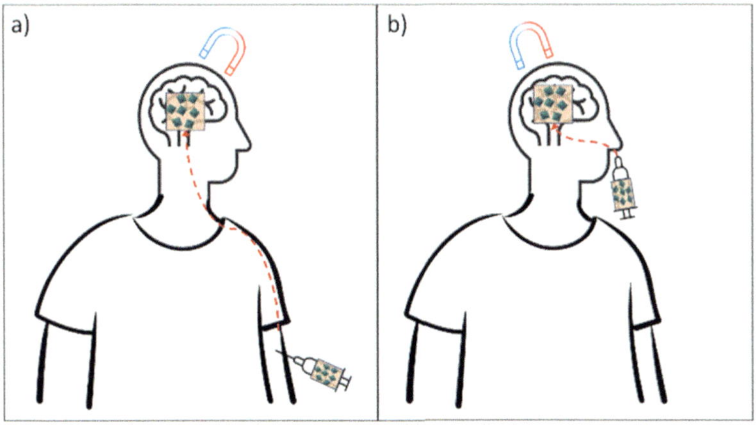

FIGURE 14.7 Schematic of MENs delivery via (a) IV injection, (b) IN inhalation under a d.c. magnetic gradient. *Adapted from Nagesetti et al. (2017).*

Kaushik, Yndart et al., 2019; Smith et al., 2023). The MENs have been shown to be nontoxic to the brain, hepatic, kidney, and respiratory systems (Nguyen et al., 2021). Pardo et al. demonstrated an alternative delivery pathway into the central nervous system (CNS) via intranasal (IN) inhalation under a d.c. magnetic gradient of the same order of magnitude as the IV delivery (Fig. 14.7), which was shown to be over four times more effective under equivalent conditions (Kaushik et al., 2016; Pardo & Khizroev, 2022).

14.3 Noninvasive brain imaging and neural stimulation

Noninvasive brain imaging and neural stimulation offer promising diagnostic and treatment options for applications ranging from the treatment of neurodegenerative diseases to brain–machine interfacing with robotic prosthetics. The utilization of MENs as well as nano- and microsized devices have both shown to be potential candidates for improvements in brain imaging and neural stimulation. Brain mapping is of great interest in the medical community, as we still have much to learn about the mechanisms involved in decision-making. Furthering our understanding of its complex mechanisms will improve the ways we treat various maladies. In this section, we provide brief description of the components of the brain before designing ME-based neural stimulation and brain mapping devices (Fig. 14.8) (Smith et al., 2023). It is important to note that the surface of both the cerebrum and cerebellum are coated in a layer of tissue called the cerebral cortex, which is where most of the

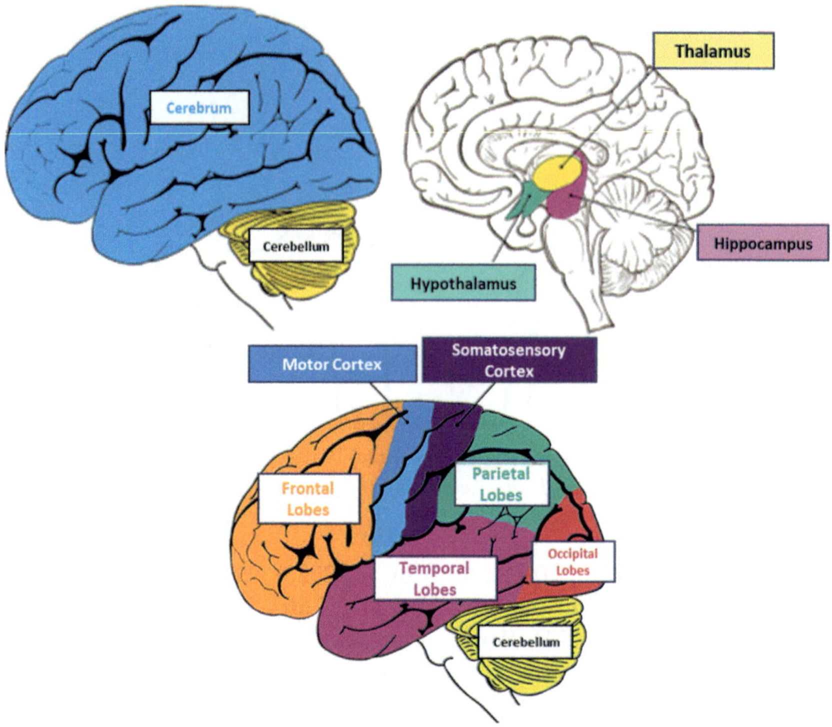

FIGURE 14.8 Schematics of the regions of the human brain. *Adapted with permission courtesy of the Smith et al. (2023).*

information processing in the brain occurs (NIH, 2023). This coating is also referred to as gray matter since most of the other regions in the brain appear white when imaged. The three main parts of the brain are the *cerebrum*, *cerebellum*, and *brainstem* (Maldonado & Alsayouri, 2023).

Cerebrum: The cerebrum is the largest and most highly developed portion of the brain. It is in the uppermost anterior region of the brain and possesses a left and right hemisphere. The two hemispheres are nearly mirror images of each other and are connected via the corpus callosum, a deep fissure of nerve fibers. The corpus callosum is responsible for communication between the two hemispheres, as well as with various organs in the body (NIH, 2023).

Within the cerebrum, the brain can be further separated into lobes that specialize in performing different functions.

Frontal lobe: The left and right frontal lobes are directly behind the forehead, and they are responsible for cognition, motor function, and language. It is known that the left frontal lobe is responsible for motor function of the right side of the body, and the right frontal lobe is responsible for motor function of the left side of the body (NIH, 2023).

Parietal lobe: The parietal lobe is located posteriorly to the frontal lobes and is responsible for the interpretation of vision, hearing, motor, sensory, and memory formation functions. The somatosensory cortex is located in this lobe and receives movement information via the brain stem from the spinal cord and peripheral nerves. It is also where the sense of touch is processed, including pressure, temperature, vibrations, and texture (NIH, 2023).

Temporal lobe: The temporal lobe is located inferiorly to the frontal lobe and processes sensory information to save memories, languages, and emotions. It has a significant role in visual, hearing, and spatial perception. The hippocampus is located in this lobe, which processes memory formation and controls thirst, hunger, and mood. In the deep brain, within the temporal lobe, are the thalamus and hypothalamus, and these regions contain glands which are responsible for the secretion and coordination of hormone levels in the body (NIH, 2023). The pineal gland can be found in a groove slightly above the thalamus and controls the body's circadian rhythm as it is responsible for the production of melatonin (Maldonado & Alsayouri, 2023). The hypothalamus is attached to the pituitary gland, which controls most of the endocrine glands in the body, like the adrenal glands located on top of the kidneys in the abdomen, to produce more or less of specific hormones, such as cortisol or adrenaline (Hilton, 2002).

Occipital lobe: The occipital lobe is in the posterior-most region of the cerebrum and contains the visual cortex. This region is responsible for the processing, controlling, and interpretation of visual information (NIH, 2023).

Cerebellum: The cerebellum is located inferior to the cerebrum and superior to the brainstem. It coordinates voluntary movement and controls the precision and accuracy of motor activity via the sensory information received and processed from the brain and spinal cord. The cerebellum also controls balance and the learning of new motor tasks. It aids in multiple cognitive functions, including language and attention (NIH, 2023).

Brain stem: The brainstem connects the cerebrum and cerebellum to the spinal cord and is the primary center for involuntary functions, such as breathing, regulation of body temperature, respiration, digestion, and swallowing. It can be further separated into three distinct parts: the midbrain, pons, and medulla. There is both gray and white matter in the brain stem, with the white matter composed of neuronal cell axons that travel down from the cerebral cortex down the spinal cord and peripheral nerves. This white matter is responsible for voluntary motor function and allows for somatosensory information to travel to higher regions of the brain (NIH, 2023).

The potential for brain–computer interfacing (BCI) is that it would directly use a person's brain to send and receive signals for various electronic devices (Rosenfeld & Wong, 2017). Applications for BCI technology include neural prosthetics for hearing, vision, and motor control. There are many challenges to address for BCI to become a relevant technology. They require electrodes to be placed on or within specific regions of the brain, typically in the cerebral cortex or within the subcortical nuclei that necessitates a plug connection via the skull and scalp with cables attached, which is not a viable permanent solution (Hiller-Sturmhöfel & Bartke, 1998). Microelectrodes are the most invasive form, but offer the highest resolution, while electrocorticography (ECoG) use electrodes placed in the subdural space and electroencephalography (EEG) uses scalp electrodes to detect cortical waveforms which are processed via advanced computer algorithms (Rosenfeld & Wong, 2017). The further the electrodes are from the brain, the safer the implantation procedure is but at the cost of a higher signal-to-noise ratio (Rosenfeld & Wong, 2017). BCI needs to be able to decode movement in three-dimensional space in addition to the desired grasp state of the robotic system (Rosenfeld & Wong, 2017). Hochberg et al. demonstrated the ability of a paralyzed individual to use BCI for controlling a robotic arm (Vilela & Hochberg, 2020). Since then, multiple studies have been completed demonstrating improvements in various BCI control of more advanced prosthetics (Hochberg et al., 2006).

14.3.1 Noninvasive brain imaging

Brain imaging is critical for the diagnosis and treatment of various diseases and disorders. Current techniques utilized for brain imaging include MRI, EEG, positron emission tomography, magnetoencephalography (MEG), and MRI (Vilela & Hochberg, 2020). There are different techniques for MRI depending on the type of imaging, such as functional MRI (fMRI), diffusion tensor imaging (DTI), and combined diffusion tensor-functional MRI (DfMRI) (Guduru et al., 2018). There are many disadvantages to current imaging techniques. The duration of an MRI typically ranges between 15 and 45 min but may last 60 min or longer, depending on the region of the body, the number of images to be captured, and gadolinium-based contrast agents being administered as resolution enhancement (Ibrahim et al., 2024). MRI also heavily relies on a patient's ability to remain immobile, as small movements can result in distortions or artifacts on the images leading to inaccurate anatomical images and misdiagnoses. MRI provides structural and anatomical details of deep tissue, but its resolution is limited even with contrast agents (Guduru et al., 2018). The ability to map the brain wirelessly via electric fields would greatly reduce the burden on the patient and allow for higher resolution imaging, potentially also reducing the time required for the procedure (Zhang et al., 2022). Fig. 14.9 depicts the relative size of MRI

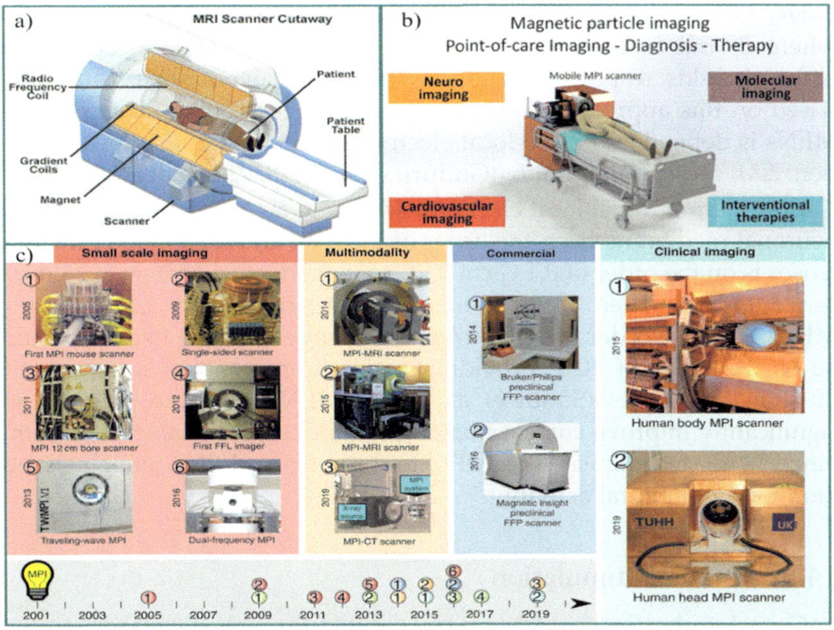

FIGURE 14.9 (a) Diagram of MRI scanner; (b) schematic of portable MPI for brain imaging; c) history of MPI, from small-scale imaging for animal models to commercial animal imaging models to models suitable for clinical testing. *(a) Reprinted with permission from Grant and Coyne (2012). (b) Reprinted with permission from Ludewig et al. (2022). (c) Reprinted with permission from Ludewig et al. (2022).*

and MPI scanners as well as the history of MPI scanner development over the past 20 years. MRI has temporal resolution on the order of one second, which is not fast enough for real-time imaging, since much of the neural activity occurs on the order of a millisecond, or faster. Magnetic nanoparticle imaging (MPI) has recently emerged as a tool for wirelessly mapping the electric field, and although the set up resembles MRI, the physics of the operating principle vary drastically (Guduru et al., 2018). MRI detects the change in water proton nuclear magnetization while MPI detects the change in iron electronic magnetization, which is a much more sensitive method with lower background signal and minimal tissue attenuation (Zhang et al., 2022). MPI lacks anatomical reference information; therefore, it requires a complimentary technique to collect meaningful information. The utilization of MENs rather than traditional magnetic nanoparticles for contrast enhancement would generate an electric field map in addition to the structural image (Zhang et al., 2022). MENs integrated with MPI can provide significantly faster detection rates compared to MRI. The induced magnetization change of MENs is dependent on the local electric field, which can be described by Landau's theory of multiferroics via the following equation: $G(E,H) = -\alpha_{ij}E_iH_j$, where E_i and H_j are the i-th and j-th components of local electric or magnetic fields, respectively, and α_{ij} is the ME tensor (Guduru et al., 2018). Based on this approximation, the induced magnetization change of the MENs is dependent on the local electric field following the linear equation: $\Delta M_i = -\partial G/\partial H_i = \alpha_{ij}E_i$ (Guduru et al., 2018; Zhang et al., 2022).

Considering that a local electric field due to an action potential at the neuronal membrane is on the order of 1 V/cm, and the value of the ME tensor is on the order of 0.1 G cm V^{-1}, a magnetization change of 1 emu/cc would be induced. If we assume the saturation magnetization of the MENs is on the order of 10 emu/cc, the change in magnetization would be on the order of 10%, providing a significant contrast for magnetic imaging techniques (Smith et al., 2023). MENs have the potential to significantly improve current imaging techniques, especially when combined with techniques like MPI, to provide real-time imaging and monitoring of neural activity.

14.3.2 Neural stimulation

Complex electrical signals are transmitted through synapses between the hundreds of billions of neurons in the human body (Guduru et al., 2018; Hoque Apu et al., 2022). Electrical synapses with gap junctions acting as biophysical substrates allow the bidirectional propagation of current to flow between neurons without neurotransmitters or presynaptic spikes initiating the interneuronal communication (Vaughn & Haas, 2022). Electrode-implant-based deep brain stimulation (DBS) is recognized as a safe and effective treatment, but it poses the same risks of infection and bleeding as any other surgical procedure (Vaughn & Haas, 2022). The precise wireless stimulation of deep brain neurons has the potential to treat neurological diseases, including Parkinson's disease, multiple sclerosis, epilepsy, and tremors noninvasively (Brittain & Cagnan, 2018; Hoque Apu et al., 2022; Smith et al., 2023). Neurodegenerative diseases are defined as such by their abnormal or inhibited neural signaling. Neurodegenerative diseases can occur in multiple regions of the brain and cause damage to parts of the cerebrum, cerebellum, and brainstem. Parkinson's disease leads to motor function deficits due to the degeneration of neurons responsible for the synthesis of dopamine, while Alzheimer disease is characterized by the neuronal degeneration due to abnormally folded beta-amyloid protein deposits in the brain, sometimes referred to as plaque, which first build up in regions of the brain responsible for memory and cognition (Brittain & Cagnan, 2018; Hoque Apu et al., 2022; Maldonado & Alsayouri, 2023; Nguyen et al., 2021). Demyelinating diseases, like multiple sclerosis, are caused by damage to the myelin sheath that insulates the axons of nerve cells (Fig. 14.10f), resulting in misfiring of signals (Nguyen et al., 2021). This misfiring of signals can result in sensory and/or motor impairments, depending on the location of the damage to the myelin sheath and underlying neurons. Not only do neurodegenerative diseases greatly affect patients' quality of life, they also are extremely costly to diagnose and treat. As shown in Fig. 14.10b, the annual cost of treatment for all patients with Parkinson disease in the United States is $210 billion USD (Maldonado & Alsayouri, 2023). Brain stimulation is one method that can repair abnormal or damaged neural signals, which could have a profound effect on disease management, symptom control, and fundamental understanding of the brain and neural system (Chen et al., 2017; Hoque Apu et al., 2022; Smith et al., 2023).

Yue et al. reported a computational study detailing the use of MENs to stimulate neural activity in the deep brain (Hoque Apu et al., 2022; Maldonado & Alsayouri, 2023). The model assumed a uniform distribution of MENs in the brain and treated each MEN as a point dipole. It found that an aqueous solution of 3×10^6 particles mL^{-1} MENs with diameter of 20 nm and ME coefficient of 100 $V\ cm^{-1}\ Oe^{-1}$ could be used to stimulate neurons in the Parkinson's disease computational models to modify the electric field levels in the brain to those of a healthy brain (Yue et al., 2012). An in vivo study was conducted by Guduru et al., where mice were given an IV injection in the tail vein comprising of 30 nm $CoFe_2O_4-BaTiO_3$ core–shell MENs (Yue et al., 2012). After the administration of the core–shell MENs, a d.c. magnetic field gradient on the order of 3000 $Oe\ cm^{-1}$ was applied for 30 min to pull the MENs from the blood circulation to the brain across the BBB (Guduru et al., 2015). The presence of the MENs in the brain was confirmed via scanning electron microscopy (SEM), atomic force microscopy (AFM), and magnetic force microscopy (MFM) (Fig. 14.11) (Guduru et al., 2015). This study confirmed that MENs were able to be pulled across the BBB, opening the door to new noninvasive diagnostic and treatment options for neurodegenerative diseases, such as mood disorders.

These prior studies have sparked the recent research to discover improved treatments for the neurodegenerative diseases. It was reported by Nguyen et al. that core–shell $CoFe_2O_4-BaTiO_3$ MENs were able to wirelessly stimulate cortical neuronal activity (Guduru et al., 2015). After IV injection of the fluorescent-labeled MENs, an applied magnetic field gradient allowed for the MENs to arrive in the targeted region of the brain. After that, under a low intensity applied magnetic field (350–450 Oe), MENs were stimulated and induced cortical activity within both individual neurons, as well as extensive neural networks in live subjects (Nguyen et al., 2021). The magnetostrictive $CoFe_2O_4$ core was intrinsically coupled to the piezoelectric $BaTiO_3$ shell via lattice-shell

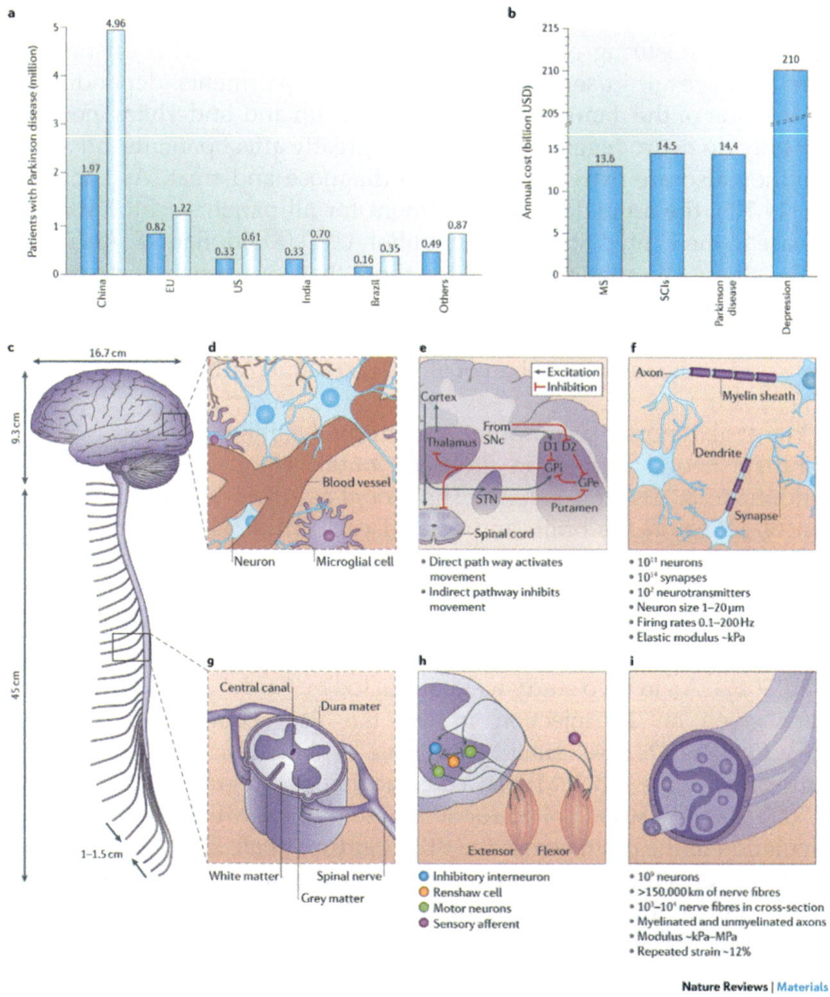

FIGURE 14.10 (a) Global population of Parkinson disease diagnoses in 2005, and corresponding projection of cases in 2030, by country; (b) annual costs for the diagnosis and treatment of all patients in the United States with multiple sclerosis, spinal cord injuries (SCIs), Parkinson disease, and depression, in billions USD; (c) dimensions of the brain and spinal cord, with (d) diagram of brain tissue composed of neurons, blood vessels, and glia, and (g) cross-sectional diagram of spinal cord; circuit diagram for (e) direct and indirect motor signaling pathways in the brain, (h) motor signaling pathway in the spinal cord; (f) diagram of neurons connected via synapses, with myelin sheath insulating axons; (i) cross-sectional diagram of peripheral nerve. D1 and D2 are dopamine receptors, GPe is the external globus pallidus, GPi is the internal globus pallidus, SNc is the substantia nigra pars compacta, and STN is the subthalamic nucleus. *Reprinted with permission from Ludewig et al. (2022).*

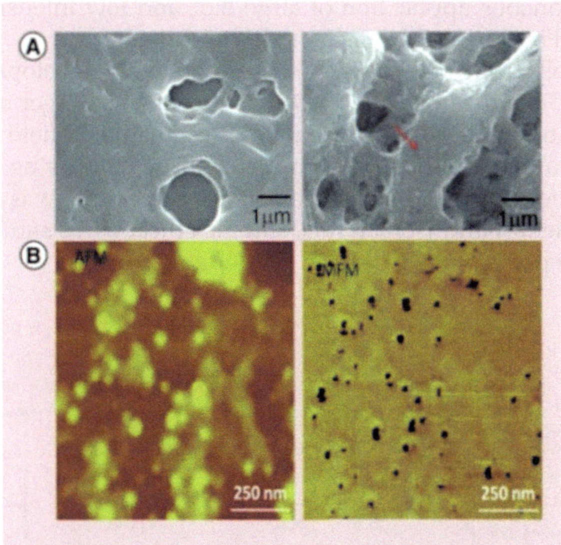

FIGURE 14.11 (a) Scanning electron microscopy (SEM) images of sagittal brain sections of two mice: (left) no magnetoelectric nanoparticles (MENs) in the brain and (right) MENs present in the brain. The red arrow points to a region of MENs. (b) (left) atomic force microscopy (AFM) and (right) magnetic force microscopy (MFM) images of a small region of MENs in the brain. *Reprinted with permission from Chen et al. (2017), © 2015 Future Medicine Ltd.*

matching. This allowed for the deformation of the magnetostrictive layer to propagate to the piezoelectric shell and generate an electric field on the order of ~1000 V/m (Nguyen et al., 2021). It was also found that the structural integrity of the MENs was maintained during the movement across the blood–brain barrier and into the localized region of the brain, and that the MENs were cleared from the brain in approximately 3 days, which could reduce potential side effects and neurotoxicity. Kozielski et al. studied deep brain stimulation (DBS) in mice with ME nanoelectrodes to electrically modify the deep brain for treatment and management of various neurodegenerative disorders (Nguyen et al., 2021). The ME nanoelectrodes were injected into the animal which avoided the need for surgical placement (Kozeilski et al., 2021). Surgical placement of devices currently utilized for treatments require wired power connections and are centimeter-sized, which increases risk for bleeding, infection, and damage to the device with activities of daily living. They demonstrated neuronal modulation in vivo as well as modulation changes in other regions of the brain connected by the basal ganglia leading to behavioral changes in the mice due to modulation in the motor cortex and limbic thalamus (Kozeilski et al., 2021). The optimal ME coefficient was found upon simultaneous application of large d.c., and low intensity a.c. magnetic fields (Fig. 14.12) (Kozeilski et al., 2021).

A remotely powered neural stimulator was developed utilizing Metglas-PVDF and Metglas-PZT ME composites by Singer et al. in a rat model (Karatum et al., 2023). An alternating magnetic field was able to power the millimeter-scale ME stimulator in freely moving rats at therapeutic frequencies (Fig. 14.13). The wireless ME device was also studied as a DBS in hemi-parkinsonian rats, and the efficacy was similar to that of previously reported wired devices (Singer et al., 2020).

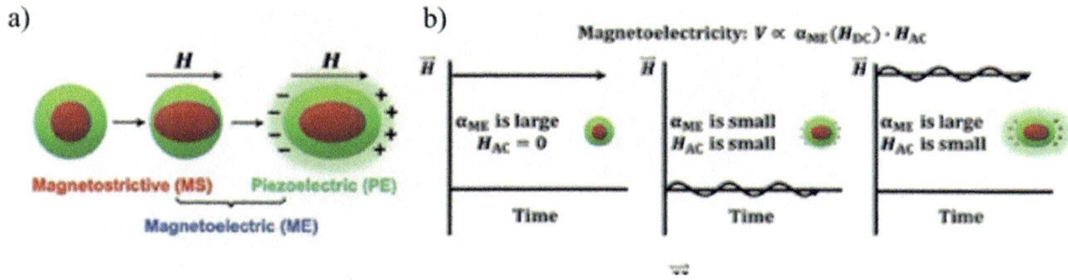

FIGURE 14.12 (a) Production of strain via ME effect of strain-coupled magnetostrictive and piezoelectric core-shell nanoparticles; (b) optimal magnetoelectric coefficient achieved from application of large d.c. magnetic field combined with low intensity a.c. magnetic field. *Reprinted with permission from Kozeilski et al. (2021).*

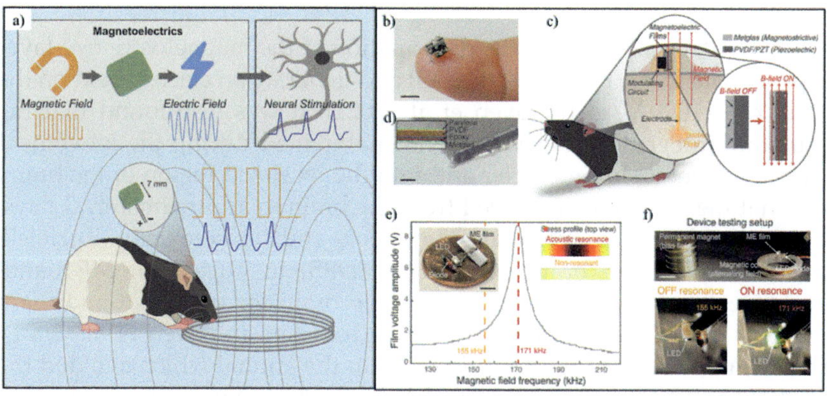

FIGURE 14.13 (a) Overview schematic for wireless deep brain stimulation (DBS) for treatment of Parkinson disease in rodents able to move freely; (b) ME device shown on fingertip for scale (5 mm scale bar); (c) freely moving rat with ME device deployed for neural stimulation, inset depicts operating principle: Strain in magnetostrictive layer is transferred to piezoelectric layer resulting in a voltage across the film; (d) cross-sectional image of cut ME film (0.2 mm scale bar); (e) resonant response curve for ME film, maximum voltage at acoustic resonance of 171 kHz; (f) device testing setup consisting of permanent magnet to apply bias field and electromagnetic coil to apply alternating magnetic field (top scale bar = 1 cm, bottom scale bar = 2 cm). *Reprinted with permission from Singer et al. (2020).*

The stimulation of peripheral nerves can be used to treat disorders that do not respond to conventional medications. However, invasive surgeries have historically been required to access the peripheral nerves and implant a bulky device (Singer et al., 2020). Recently, a study by Chen et al. reported a proof-of-concept millimeter-sized implantable to wirelessly stimulate animals' sciatic and femoral nerves in vivo (Chen et al., 2022). The ME Bio ImplanT (ME-BIT) was implanted into the hind leg of a pig near the femoral nerve and femoral artery, with an insulated stimulation wire connected to the implant placed via a catheter into the blood vessel (Fig. 14.14). A magnetic field was applied at the surface of the skin to wirelessly power the device and stimulate the femoral nerve through the femoral artery at multiple stimulation frequencies. This proof-of-concept demonstrates the feasibility of devices like the ME-BIT in humans from the large animal model data but requires more research to improve

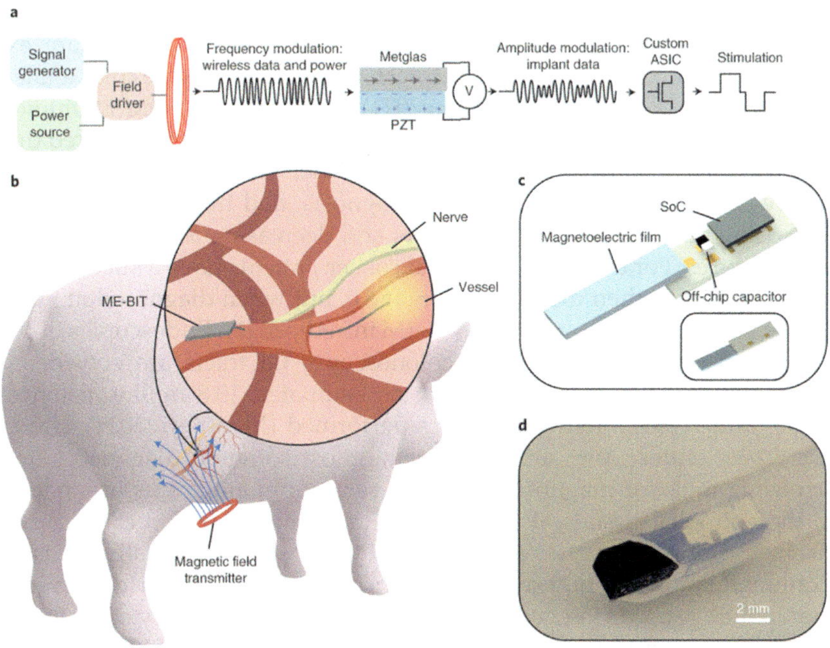

FIGURE 14.14 (a) Endovascular neural stimulator system consisting of magnetic field driver connected to a resonant coil tuned to the ME field's resonant frequency. The coil produces a frequency-dependent magnetic field, and switches between three frequencies to modulate power and data generated by ME implant; (b) ME-BIT implanted near femoral artery with stimulation lead introduced into vessel to stimulate femoral nerve and powered wirelessly via magnetic coil implanted in pig; (c) schematic of implant with external components: system on a chip (SoC), external capacitor, ME transducer; (d) photograph of packaged implant in 3D-printed capsule inside a clear sheath. *Reprinted from Chen et al. (2022).*

biocompatibility and study the durability of the capsule for long-term use and interactions between biological tissue and the applied magnetic field over time (Chen et al., 2022).

More research is required to improve the control of MENs to the region of interest and to fabricate cell-type specific activation MENs. Studies done thus far have shown proof of concept prototypes but require in-depth studies to assess the therapeutic effect of wirelessly modulated MENs in animal disease models (Chen et al., 2022). There is promise in the utilization of MENs as a contactless, targeted approach for neural stimulation for the diagnosis and treatment of a multitude of neurological conditions (Kozeilski et al., 2021).

14.4 ME composites for wireless endoscopy devices

14.4.1 Background on endoscopy devices

Gastrointestinal (GI) disease accounts for significant healthcare expenditures and deaths. In 2018 the total GI-related healthcare costs totaled $119.6 billion USD (Kozeilski et al., 2021). In 2020, there were 1.9 million new cases of colorectal cancer diagnoses and 935,000 deaths occurred due to colorectal cancer alone, not including other GI cancers and diseases (Peery et al., 2022). Early detection, diagnosis, and treatment are key in increasing survival rates in GI cancers and decreasing associated annual expenditures. For example, colorectal cancer patients have a survival rate of nearly 100% when diagnosed at Stage 0, but when diagnosed at Stage IV, the survival rate drops below 5% (Manfredi, 2021). Endoscopies of the upper (endoscopy) and lower (colonoscopy) GI tracts are considered routine outpatient procedures, with an estimated 13.8 million colonoscopies and 7.46 million endoscopies performed (Manfredi, 2021). Upper and lower endoscopies are some of the most common diagnostics and screening tools for diagnosis of a wide variety of pathologies originating in the epithelial tissues of the GI tract, including GI/colon cancers and Crohn's disease (Peery et al., 2022). They allow for minimally invasive examination of the digestive tract, and the ability to perform minor therapeutic procedures, such as foreign body removal (Peery et al., 2022), sclerosis/banding of esophageal varices (Chen et al., 2013), esophageal stenting (Poza Cordon et al., 2012), and removal of GI polyps (Richardson et al., 2018). The conventional GI endoscopy involves the insertion of a flexible tube in the human esophagus or rectum and consists of an illumination component, lens/camera system for visualization throughout the procedure with a waterjet lens cleaner, and an additional tube for the introduction of medical instruments, manipulators, or suction to remove residual stool (Akpunonu et al., 2022; Manfredi, 2021; Wallace, 2017).

Colonoscopies cause pain and discomfort to the patients during the preparation for the procedure and require trained medical staff and expensive equipment/sterilization tools (Aliqab et al., 2023). Advances in fiber optics and imaging techniques have vastly improved the visualization capabilities of conventional endoscopes. Regardless, the reach and maneuverability of tethered endoscopes is limited, as the small intestine is not accessible using traditional endoscopes (Akpunonu et al., 2022). More recently, wireless capsule endoscopies (WCEs) have been pursued to address these challenges, leveraging the advances in telemetry, miniaturized optics, and microelectronics. WCEs represent a paradigm shift in the visualization of the GI tract for clinical diagnosis. WCEs are approved by the Food and Drug Administration (FDA) in the United States for the diagnosis of various GI diseases. WCEs provide increased patient comfort and provide access to regions difficult to access with traditional endoscopes (Manfredi, 2021). An Israeli medical technology company, Given Imaging, developed a line of PillCam ingestible capsules to address the limitations of conventional endoscopies: SB series for visualization of the small bowel mucosa, COLON series for visualization of the colon, and ESO series for visualization of the esophageal mucosa (Nikolayev et al., 2016; Schneider & Feussner, 2017). WCEs were approved for use in patients by the FDA in 2001, and although much less invasive than traditional endoscopies, WCEs still possess limitations (Aliqab et al., 2023; Tettey et al., 2023). The capsule consists of an ingestible coating with an optical dome window, with LEDs situated behind to provide the required lighting, an image sensor that translates the signals acquired through a short-focus lens, then processed by a microcontroller (Schneider & Feussner, 2017; Tettey et al., 2023). Utilizing electrodes placed on the patient's abdomen connected to a radiofrequency (RF) transmitter, the information is transferred from the WCE, and the images are stored in the receiving box worn by the patient as shown in Fig. 14.15 (Akpunonu et al., 2022). The WCE still requires patient preparation, although it is typically less uncomfortable compared to traditional colonoscopy preparation. WCE is relatively safe, as there is no sedation or administration of pain medication and complications are rare. Complication include capsule retention, which was only reported to occur in 2% of all studies, battery failure, missed polyps/lesions, bowel obstructions or perforation, and aspiration of the capsule into the trachea (Schneider & Feussner, 2017). The main limitations of WCEs include passive locomotion, limiting the precise localization of any lesions or polyps, low frame rate and low image resolution, compared to conventional endoscopies, limited working time due to energy constraints with internal battery, and a lack of therapeutic options for removal of polyps, rendering it a purely diagnostic tool (Akpunonu et al., 2022).

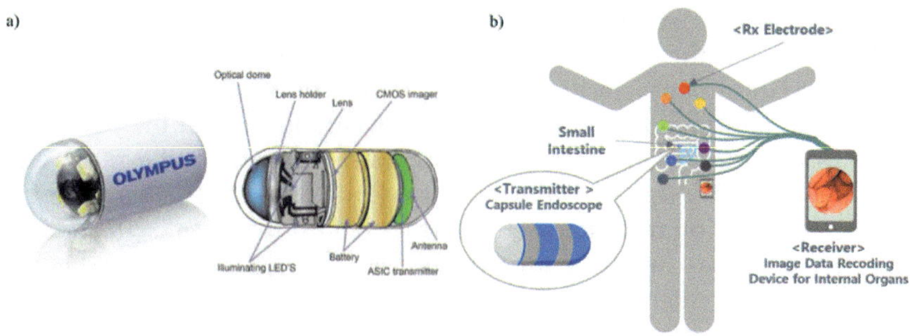

FIGURE 14.15 (a) Commercial small intestine WCE. (b) Schematic of WCE deployed in GI tract, with patient connected to electrodes and the receiver processing and recording data. *(a) Reprinted from Pradhan et al. (2021). (b) Reprinted from Park et al. (2018).*

14.4.2 ME-guided wireless capsule endoscopies

Incorporation of ME materials has the potential to address many of the above mentioned challenges. Micron-sized self-powered ME devices have been developed, utilizing an ME antenna array (Schneider & Feussner, 2017). The NanoNeuroRFID was able to harvest electromagnetic energy to power the system while sensing changing magnetic fields and communicate with an external transceiver, which could be modified for applications in wireless endoscopies (Zaeimbashi et al., 2019). An ME composite could also be utilized as a power generator to improve current WCEs. For WCEs to possess locomotion capabilities, more power is required. Rather than increase the size of the capsule for a larger battery, one could incorporate an ME composite in the WCE and utilize wireless power generation from an applied magnetic field to charge the device. Such a device would require an internal receiving coil and external transmitting coil design. Another proposed design included the transmitter coil with an ME receiver instead of the classic coil–coil system (Zaeimbashi et al., 2019). A phantom study demonstrated that this transmitter–receiver configuration was able to convert sufficient power to recharge an implant up to 20 mm away from the transmitter coil in an extremely unfavorable configuration (receiver perpendicular to transmitter), which was impossible to achieve with the coil-coil system (Rizzo, 2020). A similar inductive coupling design was demonstrated for a swimming capsule with motor-based propellers and was able to generate 1 W power and recharge the capsule in 20 min after an operating time of 10–15 min (Rizzo, 2020; Tortora et al., 2015). Another desired quality for next-generation WCEs is precise tracking of the device. Localization is critical for pathology detection and closed-loop control of the WCE. One method to achieve precise localization is an external array of sensors measuring the magnitude and direction of a field generated by a magnet-embedded WCE (Shamsudhin et al., 2017). Other methods proposed include field sensors in the WCE to measure a field generated by field generators placed outside the body, an algorithm for 6-DOF localization, and multiobject tracking. There have been proof-of-concept studies done to demonstrate these improvements with the use of ME materials, but no commercial WCEs utilize this technology (Weitschies et al., 1994).

14.5 ME device—based simultaneous wireless energy harvesting

Wireless energy harvesting technology has emerged as a promising solution to extend the lifetime of energy-constrained mobile electronics, implantable medical devices (IMDs), sensor networks, and Internet of Things (IoT) devices (Shamsudhin et al., 2017). The growth of IoT has led to the development of different types of wireless sensors that are used for monitoring, transduction, communication, and IMDs. Wireless sensors and transceivers in a typical IoT are currently powered by batteries. Power requirements for wireless sensors vary from one application to another, typically in the range of a few μW to several mW (Jiang et al., 2021; Rinne et al., 2017; Roundy et al., 2004, pp. 1—17). Due to the large number of IoT devices that are deployed in difficult to access areas, regular battery replacement is practically impossible due to complexity from inaccessibility or large volume. In all these scenarios, wireless power transfer combined with energy harvesting is a viable alternative for powering the devices. There are several mechanisms for converting ambient energy into electricity including inductive, photovoltaics, thermoelectric, magneto-thermoelectric, electromagnetic, magneto-mechano-electric conversion, piezoelectric, triboelectric, pyroelectric, and various other principles (Roundy et al., 2004, pp. 1—17). Wireless recharging technology for in-vitro and in-vivo applications has been widely investigated (Karan et al., 2024). However, wireless power transfer using radio frequencies has been observed to experience attenuation as it passes through tissue, thereby restricting the amount of power that can be transmitted to a harvesting device. Exposure to high levels of transmitted electrical (or radio) power can cause tissue damage in humans during transmission. Inductive coupling technology is highly dependent on the distance between the coils, and for it to operate efficiently, must be operated at high frequencies (MHz range). The restrictions on safety limits are particularly strict at high frequencies, which results in limitations on the amount of power that can be transmitted. The low frequency (kHz range) ME and ultrasound-based methods of energy harvesting offer higher efficiencies and power density, making them promising alternatives (Khan et al., 2020). ME transducers offer a promising solution for low-frequency magnetic field energy harvesting with high power density and safer field levels, as they result in less absorption and heating, and greater penetration through different lossy media. Alternatively, ultrasound (US) energy has low attenuation in biological tissue compared to that of electromagnetic waves, which not only reduces energy dissipation but also enables a deeper travel depth at a given power (Karan et al., 2024). Individual magnetic and ultrasound technology has difficulty in obtaining the high-power capability within the human body safety limit and angular misalignment. (Karan et al., 2024) explored a novel simultaneous energy harvesting technology through ME devices under both the ultrasound and magnetic energy in water/tissue (ex-vivo) mediums (Fig. 14.16a–d).

The magnetic field and ultrasound-induced dual generator (MUDG) produces ultrahigh rms power of ~52.1 mW with power density of ~597 mW/cm^3 at ~500 μT magnetic field and 675 mW/cm^2 ultrasound intensity. The MUDG can overcome the limitations of traditional wireless power transfer systems for human body application, providing a novel platform with the advantages of ultra-high-power density, small size, biocompatibility, and tolerance to angular misalignment (xy/yz plane).

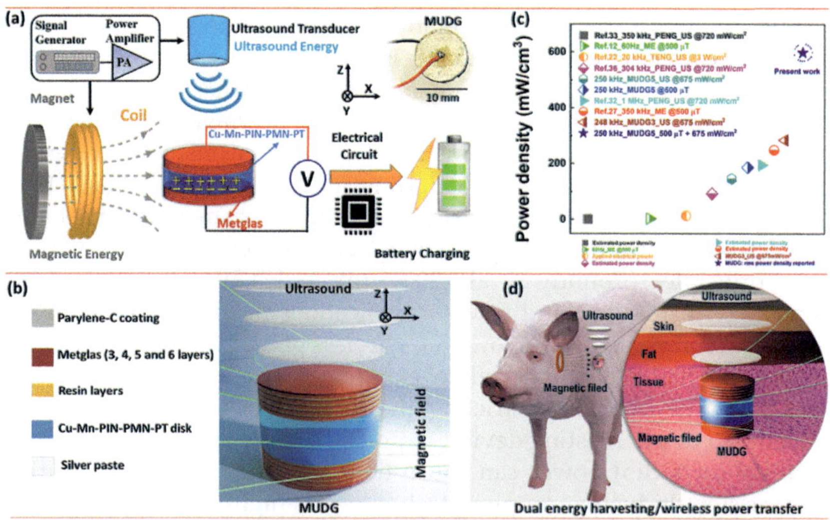

FIGURE 14.16 (a) The overview of the present study related to magnetic and ultrasound energy-based dual energy harvesting and storage in a battery (inset shows real image of MUDG). (b) Schematic representation of the MUDG device for dual (magnetic field and ultrasound) energy harvesting with description of each component. (c) The comparison of power density of MUDG5 device with the other published technologies. (d) For the proof-of-concept experiments, the MUDG is implanted inside the porcine tissue and is wirelessly powered through magnetic field and ultrasound energy simultaneously. *Reprinted from Karan et al. (2024).*

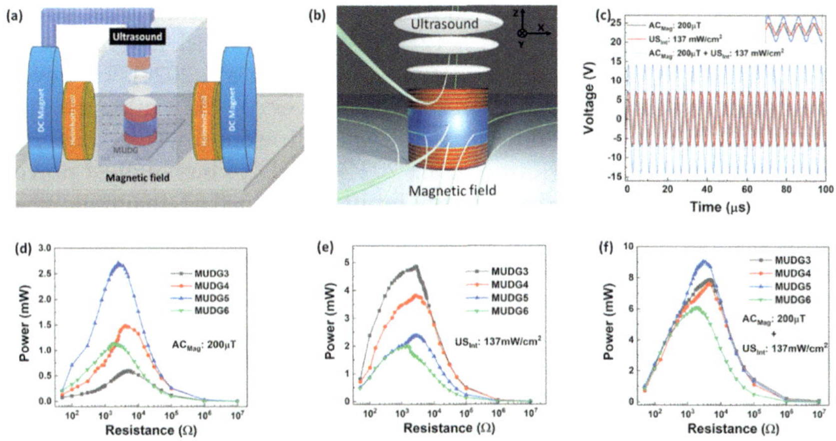

FIGURE 14.17 (a) Measurement set up for magnetic-ultrasound dual energy harvester and (b) illustration scheme for dual energy harvesting under water. (c) The generated output voltage from the MUDG devices under individual magnetic field/ultrasound technology and combined (magnetic and ultrasound) technologies (inset shows zoom graph). (d, e) Generation of output power in MUDG devices varying with external resistance under 200 μT magnetic field and 137 mW/cm^2 ultrasound intensity, respectively. (f) The generated power varying with external load resistance under simultaneous magnetic field and ultrasound energy. *Reprinted from Karan et al. (2024).*

The simultaneous energy harvesting capability of MUDG devices *has been tested* under magnetic field and ultrasound, either individually or concurrently, where a continuous sinusoidal mode with infinite cycles was utilized (Karan et al., 2024). The experimental setup for measuring dual energy harvesting of MUDG devices is demonstrated in Fig. 14.17a. The schematic illustration of dual energy harvesting in a water medium, where ultrasound is applied perpendicular to the magnetic field, is demonstrated in Fig. 14.17b (Karan et al., 2024). Water was employed as the medium to measure the performance of dual energy harvesting since it has no impact on magnetic fields and has acoustic impedance similar to that of human tissue, particularly for ultrasound. Under 200 μT magnetic field, MUDG5 produced an output voltage (Vp) of ~7.29 V, and it generated ~7.14 V under 137 mW/cm^2 ultrasound intensity (Fig. 14.17c) (Karan et al., 2024). Simultaneous measurement confirms the additive output voltage, i.e., ~14.35 V, generated from MUDG5 device under 200 μT magnetic field and 137 mW/cm^2 ultrasound intensity (Karan et al., 2024). MUDG5 exhibits output power of ~2.65 mW under 200 μT magnetic field (Fig. 14.17d) and produces best output power of ~2.38 mW under 137 mW/cm^2 ultrasound intensity (Fig. 14.17e). Fig. 14.17f confirms the addition of output power of MUDGs under simultaneous magnetic field and ultrasound intensity at similar measurement condition

(Karan et al., 2024). They reported MUDG5 device has individual power generation ability of ~15.7 and ~12.56 mW under 500 μT magnetic field and 675 mW/cm2 ultrasound intensity (Karan et al., 2024). However, simultaneous stimulation produces output power of ~52.1 mW, which is 3× and 4× greater than individual ME and ultrasound-induced energy harvesting, respectively (Karan et al., 2024). MUDG5 has highest power generation ability under dual energy input, i.e., at 500 μT magnetic field and 675 mW/cm^2 ultrasound intensity (Karan et al., 2024).

An ex-vivo study on porcine tissue was carried out to validate the implantability of MUDG5 as demonstrated for dual energy harvesting under 22 × 40 mm^2 area with different thickness (5, 10 and 15 mm) (Fig. 14.18a–c) (Karan et al., 2024). Typically, tissue with thickness between 5 and 15 mm is considered appropriate implants for wireless energy transfer. The MUDG5 device wash shown to generate a high output power of ~52.17 mW in porcine tissue under 500 μT magnetic field and 675 mW/cm^2 ultrasound intensity (Fig. 14.18d). The output power of the MUDG5 device drops to 22 mW at 15 mm thickness in porcine tissue, primarily due to the device's lower ultrasound pressure and increased damping of tissue that restricts its vibration (Fig. 14.18e) (Karan et al., 2024). If the transducer is moving in lateral direction from 5 to 15 mm, the energy harvesting performance of MUDG5 vanished due to misalignment

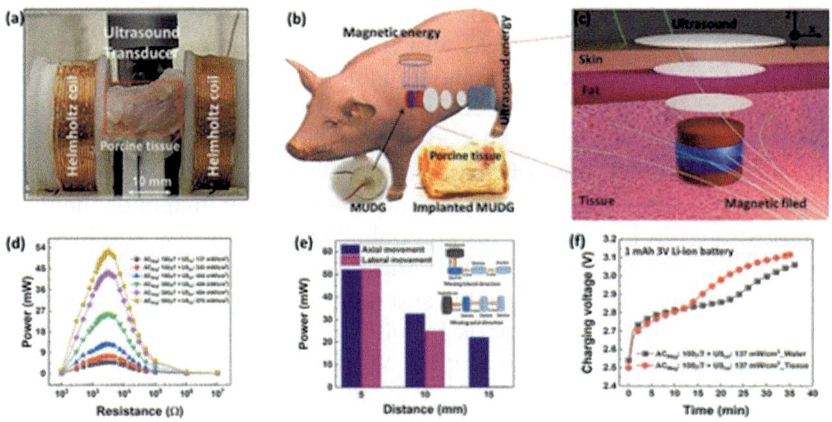

FIGURE 14.18 (a) Picture of the implanted MUDG5 device under 5 mm porcine tissue to demonstrate dual energy harvesting. (b) Schematic of the porcine ex-vivo and the possible location of implanted MUDG5 for implantable device realization. (c) Illustration of dual energy harvesting from MUDG under porcine tissue. (d) Power output of MUDG5 device under simultaneous energy harvesting varying with external load resistance. (e) Power output of MUDG5 at 5, 10, and 15 mm thick porcine tissue under 500 μT magnetic field and 675 mW/cm^2 ultrasound intensity. (f) 1 mAh 3 V Li-ion battery recharging characteristic under water and porcine tissue during simultaneous energy harvesting. *Reprinted from Karan et al. (2024).*

between the transducer and MUDG5. The MUDG5 device can recharge the 1 mAh Li-ion battery up to 3.11 V within 35 min with a 1.78 times faster rate in tissue compared to water (3.06 V within 36 min) (Fig. 14.17f). The faster charging rate in tissue medium is attributed to the more concentrated magnetic field strength compared to that in water medium (Karan et al., 2024).

Karan et al. (2024) conducted experiments using Parylene-C-coated samples to evaluate their cytotoxicity through live/dead cell staining and cell viability using MTT assays (Fig. 14.19a and b) (Karan et al., 2024). Subsequently, they examined Huh7 cell adhesion on various samples, including a control sample coated with poly-L-Lysine, and Parylene-C-coated samples both before and after exposure to oxygen plasma treatment for 1 and 5 min, over a 24-hour period. To promote improved cell adhesion, oxygen plasma treatment was applied to the Parylene-C-coated surface for 1 and 5 min to enhance the hydrophilic properties of the Parylene-C surface, which was confirmed by measuring changes in contact angles. Oxygen plasma treatment is a widely adopted technique to enhance cell adhesion without compromising material compatibility and other polymer characteristics. The oxygen plasma-treated samples (1 min) exhibited an increase in cell adhesion, displaying a uniform distribution likely due to the favorable surface roughness and hydrophilic properties, which are conducive to the growth of live cells. The fluorescence images

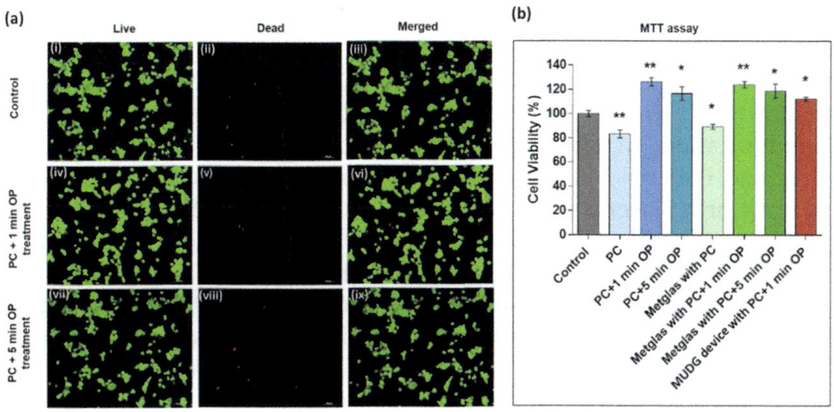

FIGURE 14.19 (a) Live/dead staining (green: live cells, red: dead cells). (a(i–iii)) fluorescent images of control (poly-L-lysine coated coverslips without PC coating) sample with live, dead, and merged view. (a(iv–ix)) Fluorescent images for Parylene-C coated and oxygen plasma treated (1 and 5 min) samples. Scale bar: 100 μm. (b) Cytotoxicity results of cultured Huh7 cells after 24 h of incubation using Parylene-C-coated samples using MTT assay. The statistics was done comparing all treatments w.r.t control. Results are mean ± SEM, $*P < 0.05$; $**P < 0.01$. PC: Parylene-C, and OP: Oxygen plasma. *Reprinted from Karan et al. (2024).*

demonstrated that the control samples (Fig. 14.19a(i–iii)), which lacked Parylene-C coating, allowed cells to grow well on the surface with a uniform distribution and limited cell death. In Fig. 14.19a, the live/dead staining results indicated that the number of live cells on the oxygen plasma-treated Parylene-C coated samples (1 and 5 min) closely resembled that of the controls. Additionally, there were very few observed dead cells, suggesting minimal cytotoxicity. The MTT assay results (Fig. 14.19b) indicate that Parylene-C coated samples (coverslips/Metglas/device) exhibited cell viability in the range of ~85%–90% suggesting limited toxicity (Karan et al., 2024). Similarly, oxygen plasma treated Parylene-C coated samples exhibited limited toxicity. All these findings suggest that Parylene-C-coated samples are noncytotoxic, promote good cell viability, and suggest Parylene-C-coated MUDG devices have great potential for future biomedical implants.

14.6 ME devices for wireless power and data transfer

Fully wireless biomedical implants at the millimeter (mm) scale with closed-loop functionality—encompassing sensing/recording and actuation/stimulation—are crucial for both basic research and clinical applications. These implants have the potential to replace pharmacological treatments and provide groundbreaking bioelectronic medicine for conditions such as epilepsy, cardiovascular diseases, gastroparesis, and diabetes. They address the need for monitoring and modulating signaling patterns in the peripheral nervous system (PNS) and various organs (Chen et al., 2022; Khan et al., 2020). For chronic use, mm-sized implants require fully wireless interrogation, including power and bidirectional communication, to avoid the complications associated with tethered wires and batteries, such as infections, limited operation, and frequent charging or replacement (Alrashdan et al., 2021; Meng & Kiani, 2017).

Several methods are currently available for the wireless interrogation of medical devices, including inductive (Karan et al., 2024), capacitive (Kiani et al., 2011), and ultrasound (US) (Erfani et al., 2018) technologies. Recently, a new approach has emerged using a low-frequency magnetic field (MF) for wireless powering, achieved through an ME transducer as the power receiver in the implant (Alrashdan et al., 2021; Chen et al., 2022; Hosur et al., 2021; Meng & Kiani, 2017). ME transducers, which operate at acoustic wave resonance, offer greater efficiency than inductive coils in converting low sub-MHz MFs into electric voltages, especially when miniaturized. An ME transducer consists of a multilayered composite with a piezoelectric material sandwiched between one or more magnetostrictive layers. A time-varying MF (H_{AC}) induces strain in the magnetostrictive layer, which then directly couples to the piezoelectric layer, resulting in an electric voltage through the direct piezoelectric effect.

Traditional technology for interrogating implants primarily uses inductive coupling, which has two major drawbacks. First, an inductive link with a millimeter-sized receiver (Rx) coil must operate at high sub-GHz frequencies (or even GHz), which limits the transmitted power due to safety restrictions. Second, this method has poor tolerance for misalignment, particularly with Rx coil orientation, resulting in reduced received power (Khan et al., 2020). Capacitive technology has seen limited acceptance due to the large size of the Rx and its short powering range. In contrast, both ultrasound (US) and ME technologies have demonstrated significant effectiveness in wirelessly interrogating millimeter-sized implants at depths of several centimeters. This success is attributed to (1) the low loss propagation of US waves and low-frequency magnetic fields (MFs) in tissue, (2) higher permissible intensities of US and low-frequency MFs within safety limits, and (3) the focusing capability of US using phased arrays (Yu et al., 2022).

A bidirectional data link is often needed to send actuation (or stimulation) control parameters to the implant (downlink) and to receive sensed (or recorded) data from the implant (uplink). The downlink is relatively straightforward, as it requires a low data rate and the data transmitter (Tx) is located externally. It typically uses amplitude, frequency, or phase shift keying by modulating the power carrier. In contrast, the uplink is more complex due to the constraints of the implant's size and power budget, as well as the need for a higher data rate. Uplink strategies include power carrier backscattering and the use of multiple frequency bands, which necessitate additional components within the implant.

Several studies have demonstrated wireless data transfer for medical implants using the ME modality (Dong et al., 2020; Hosur et al., 2023; Jin et al., 2021; Kashani & Kiani, 2023; Yu et al., 2022). For instance, Yu et al., reported the transmission of commands to an ME transducer within an implant from an external unit at 60 kbps by modulating the power carrier, while backscattering was used for uplink data at a low speed (up to 8 kbps) (Yu et al., 2022). In a study by Dong et al. (2020), a large (9 cm^2) ME antenna achieved an uplink data rate of 0.1 kbps with a power consumption of 400 mW. Using multiple transducers, one for powering and another for data communication is not ideal as it increases the implant size and overall complexity (Dong et al., 2020).

The work presented in Jin et al. (2021) introduces a communication technique known as monotonic energy transmission (MET) for transmitting information from an implant using an ME transducer to an

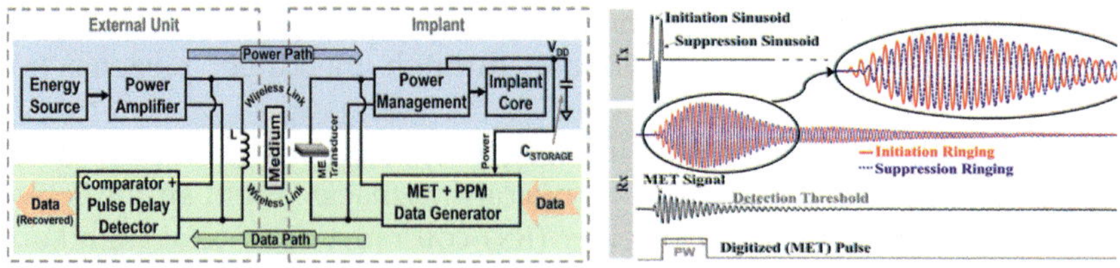

FIGURE 14.20 Generic block diagram of a wireless link for powering an implant with the ME transducer as a power receiver (power path) and wirelessly transmitting data from the implant to the external unit (uplink data path) using the ME effect. Also, the conceptual waveforms demonstrating the operation of the monotonic energy transmission (MET) technique is shown. *Reproduced with permission from Hosur et al. (2023).*

external unit (see Fig. 14.20). MET is energy-efficient, transmitting each bit by utilizing a full cycle of initiation and suppression sinusoids that are 180 degrees out of phase, as shown in Fig. 14.20. Applying a full-cycle initiation sinusoid to the ME transducer causes the external LC tank on the receiver side to ring, with the most significant ME effect observed when the sinusoid frequency is close to the transducer's resonance frequency. Using an external LC tank enhances the received voltage amplitude and signal-to-noise ratio (SNR) but also prolongs the ringing. In digital terms, ringing presence is interpreted as a bit "1" and its absence as a bit "0." To improve data rates and reduce the number of ringing cycles, a suppression sinusoid with a slightly lesser amplitude and a 180° phase shift is sent immediately after the initiation sinusoid. This suppression sinusoid cancels out the initiation ringing, while still leaving a detectable signal due to the one-cycle delay between the two sinusoids. While MET alone can transmit data, combining it with conventional pulse-position modulation (PPM) can further enhance data rates. To illustrate these concepts, a bar-shaped ME transducer measuring $6.56 \times 2.63 \times 0.67\ mm^3$ (length $\times$ width $\times$ thickness) was demonstrated (Hosur et al., 2023). This transducer features a PZT-5A piezoelectric layer sandwiched between six magnetostrictive Metglas layers—three layers on each side of the piezoelectric layer.

Fig. 14.21 displays the measurement waveforms of the MET technique using the fabricated ME transducer. The Rx coils were positioned 10 mm from the transducer, with the resonated coil having a Q factor of 16.7 at approximately 275 kHz. The received signals were amplified by a two-stage amplifier with a gain of 48 dB at 275 kHz. Without the suppression sinusoid, the received ringing from the initiation sinusoid lasted over 300 μs, but with the suppression sinusoid, the ringing was reduced to under 50 μs, improving the signal duration by more than six times. With the combined MET and PPM technique (Fig. 14.21-right), the highest achieved data rate was 220 kbps, with a bit energy of 5.4 pJ/bit.

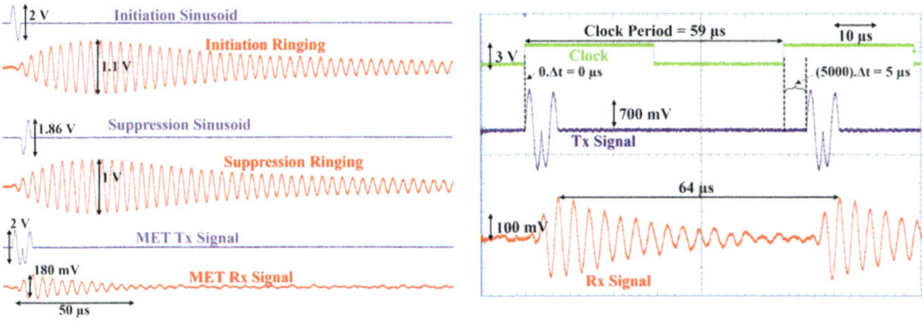

FIGURE 14.21 Left: Measured waveforms that demonstrate the MET technique using ME effect. Measured response of the initiation sinusoid, the suppression sinusoid, and their combined MET effect are all shown. Right: Measured waveforms that demonstrate the combined MET and PPM technique for low energy, high speed data transfer using ME effect. *Reproduced with permission from Hosur et al. (2023).*

Despite significant advancements, several challenges remain in wirelessly interrogating millimeter-sized implants using current technologies. First, to minimize the implant's bulkiness, the same modality is often used for both powering and uplink during sensing (Hosur et al., 2024; Piech et al., 2020). This approach can suffer from issues such as interference from the large power carrier affecting the data link or frequent interruptions in wireless powering to avoid interference. Second, nearly all interrogation methods face alignment problems, especially concerning the orientation of the implant (Hosur et al., 2024; Piech et al., 2020). Third, during the actuation/stimulation phase, which often requires high power, a single modality may not provide sufficient power while adhering to safety limits (Hosur et al., 2024; Piech et al., 2020).

The study by Hosur et al. (2023) demonstrates that the ME transducer can generate and receive both US waves and magnetic fields (MFs). The piezoelectric layer of the ME transducer can be actuated in two ways: (1) through the direct ME effect, where MFs induce mechanical stress on the piezoelectric layer, and (2) through the direct piezoelectric effect, where US pressure generates a response in the piezoelectric layer. Conversely, applying an a.c. voltage across the ME transducer can produce US pressure via the converse piezoelectric effect and MFs via the converse ME effect. Leveraging the ME transducer's dual capabilities, they have introduced MagSonic, a hybrid magnetic-ultrasonic wireless interrogation platform that utilized 1 ME transducer in the implant (see the conceptual representation in Fig. 14.22). This platform allows simultaneous power reception using one modality (MF or US) and uplink data transmission with the other, or even power reception with both modalities. MagSonic offers three key benefits over individual modalities: (1) Up to four times increased received power from both MF and US excitations, (2)

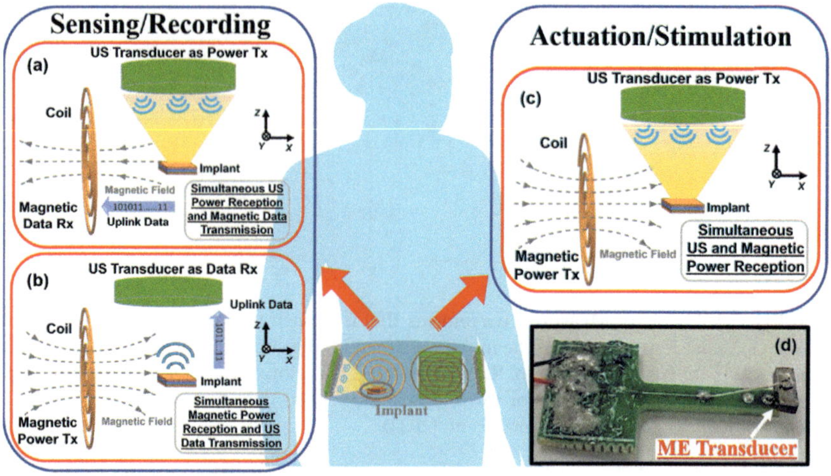

FIGURE 14.22 Conceptual representation of the MagSonic platform. (a) Continuous power reception with US while simultaneously sending uplink data signals via MF. (b) Continuous power reception with MF while simultaneously sending uplink data signals via US. (c) ME transducer's capability to simultaneously receive power from both MF and US modalities. (d) Fabricated ME transducer on a PCB with ~10 μm Parylene-C coating. *Reproduced with permission from Hosur et al. (2024).*

enhanced robustness to angular and linear misalignments due to the use of dual modalities, and (3) the ability to simultaneously receive power and communicate data using a single transducer, reducing interference between the power carrier and data link.

This study experimentally evaluated the feasibility of using a single ME transducer to simultaneously receive power via one modality and transmit uplink data via another (Hosur et al., 2024). Measurements demonstrated continuous power reception using US waves at 330 kHz while communicating uplink data through MF at 709 kHz (see Fig. 14.23a) (Hosur et al., 2024). The data pattern was applied to the ME transducer using a switching circuit, which allowed data transmission only when enabled. Another test showed simultaneous power reception via 330 kHz MF and uplink communication through 1.93 MHz US waves (see Fig. 14.23b) (Hosur et al., 2024). The wireless range was 30 mm. Using different modalities for power reception and data transmission avoid significant interference from the power carrier, which would otherwise lower the signal-to-noise ratio (SNR) and complicate data detection. Although the ME transducer generates some inherent interference while receiving power, this interference is minimal compared to the larger interference from the power transmitter. As a result, the MagSonic link effectively detects uplink data signals with high SNR despite the presence of power carrier interference.

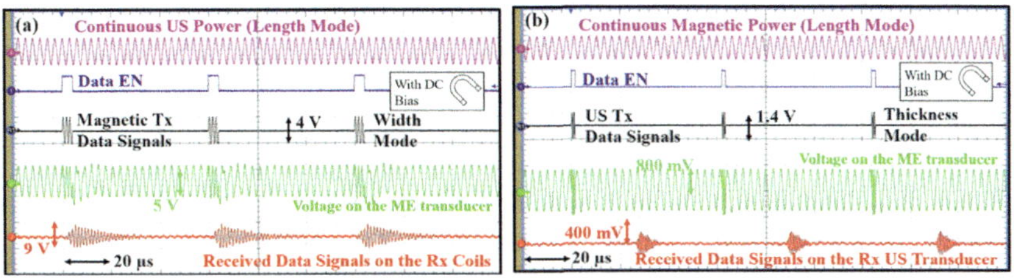

FIGURE 14.23 MagSonic operation using a single ME transducer in the receiver. (a) Simultaneous wireless powering using US and uplink communication through MF (power: length mode; data: width mode). (b) Continuous power reception through MF and simultaneous uplink communication through US (power: length mode; data: thickness mode). *Reproduced with permission from Hosur et al. (2024).*

This study also demonstrated the effectiveness of MagSonic using a single ME transducer in the receiver for increasing the power density (Hosur et al., 2024). At a 40 mm Tx-Rx distance, the ME transducer's power level (P_L) was measured under different conditions using a magnetic field (MF) of 5.16 Oe peak and US intensity of 620 mW/cm^2, both within safety limits. The individual P_L for MF and US modalities was 2.44 and 2.47 mW, respectively. When combined using the MagSonic approach (both MF and US in perpendicular planes), the P_L increased approximately fourfold to 9.72 mW at an optimal load resistance (R_L) of 2.2 kΩ. Adding a chicken breast tissue layer reduced the voltage by about 8.8%, resulting in a P_L of 8 mW at the same R_L (Hosur et al., 2024) (Fig. 14.24).

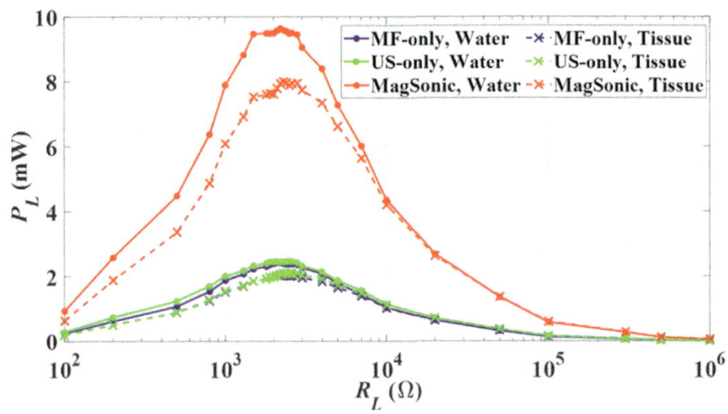

FIGURE 14.24 ME transducer PL versus RL for MF only, US only, and MagSonic (combined MF and US) excitation in aqueous medium and tissue. *Reproduced with permission from Hosur et al. (2024).*

When interrogating millimeter-sized implants, alignment between the external unit and the implant is crucial. Implants can experience angular misalignment and/or linear misalignment. ME transducers excited by magnetic fields have demonstrated some tolerance to angular misalignment (Hosur et al., 2024). Ultrasound links, especially those using phased arrays, are less sensitive to linear misalignment due to their ability to focus and steer beams to different locations. However, they are typically more sensitive to angular misalignments (Chen et al., 2022; Kashani & Kiani, 2023). The MagSonic approach, which utilizes both modalities, offers greater tolerance to a wider range of misalignments, leveraging the strengths of each modality in different planes. A comparison of various state-of-the-art millimeter-sized implants is provided in Table 14.1 (Chen et al., 2022).

The development of fully wireless, millimeter-sized biomedical implants with closed-loop functionality marks a significant advancement in both basic research and clinical applications. These implants hold the potential to revolutionize bioelectronic medicine by providing alternatives to pharmacological treatments for various conditions, including epilepsy and cardiovascular diseases. The MagSonic platform, which leverages the dual capabilities of ME transducers to operate with both magnetic fields (MF) and US modalities, represents a major breakthrough in wireless interrogation technology. By enabling simultaneous power reception and uplink communication through a single ME transducer, MagSonic overcomes the limitations of traditional modalities, offering enhanced power efficiency, robustness to misalignments, and high data rates. The experimental results demonstrate that MagSonic not only increases the delivered power density significantly but also maintains a high signal-to-noise ratio (SNR) even in the presence of power carrier interference. As such, MagSonic stands out as a highly effective and versatile solution for the wireless interrogation of mm-sized implants, paving the way for safer and more efficient bioelectronic devices that could transform patient care.

Acknowledgments

The authors acknowledge M. Paluszek, D. Avirovik, Y. Zhou, A. Chopra, S. Kundu, and R. Montague for their contribution to the chapter on this same topic in the first edition. The chapter in the second edition has been updated and new content has been added. Please refer to the first edition of the book for if there is interest in reading prior content.

TABLE 14.1 Comparison with the state-of-the-art studies for wireless interrogation of mm-sized implants.

Reference	Ibrahim et al. (2024)		Sonmezoglu et al. (2021)	Piech et al. (2020)	Ozeri and Shmilovitz (2014)	Ghanbari et al. (2019)	Chang et al. (2017)	Chen et al. (2022)	Yu et al. (2022)	Hosur et al. (2023)
Transducer surface area (mm^2)	$\pi \times 0.6^2$	$\pi \times 0.6^2$	0.75 × 0.75	0.75 × 0.75	$\pi \times 7.5^2$	0.75 × 0.75	Rx: 0.9 × 0.9 Tx: 0.55 × 0.55	4 × 3	5 × 2	5.2 × 2
Transducer Thickness (mm)	1	1	0.75	0.75	3.12	0.75	Rx: 0.5 Tx: 0.4	0.16[b]	0.28	1.6
Technology	Inductive coil	Ultrasound	Ultrasound	Ultrasound	Ultrasound	Ultrasound	Ultrasound	Magneto-electric	Magneto-electric	MagSonic
Tx-Rx distance (mm)	52	50	50	48	150	50	60	40	30	40
Medium	Tissue	Oil	Water/Tissue	Ultrasound gel	Water	Oil	Tissue	Tissue	Air/Tissue	Water/Tissue
Powering frequency (kHz)	20,000	1100	2000	1850	765	1780	950	345	335	330
Max. Delivered power (mW)	0.0056	1	<0.2	0.147	20	–	–	1.17	3.8	8
Data communication frequency (kHz)	–	–	Backscatter	Backscatter	Backscatter	1780	2500	–	335	1930 (US)

Continued

TABLE 14.1 Comparison with the state-of-the-art studies for wireless interrogation of mm-sized implants.—cont'd

Reference	Ibrahim et al. (2024)		Sonmezoglu et al. (2021)	Piech et al. (2020)	Ozeri and Shmilovitz (2014)	Ghanbari et al. (2019)	Chang et al. (2017)	Chen et al. (2022)	Yu et al. (2022)	Hosur et al. (2023)
Uplink data rate (kbps)	–	–	60	–	1.2	35	95	–	8	100
BER	–	–	10^{-5}	–	–	–	10^{-4}	–	0.32	$\leq 10^{-5}$
SPRUDT[a]	No	No	No	No	No	No	Yes[c]	No	No	Yes
High Tolerance to angular misalignment	☹	☹	☹	☹	☹	☹	☹	☹	☹	☺

Compares state-of-the-art studies using inductive, US, and ME modalities for wirelessly interrogating millimeter-sized implants. The MagSonic method demonstrates a significantly higher power level (P_L) at comparable transmission-reception (Tx-Rx) distances than other studies, even when accounting for differences such as depth, transducer surface area, and operating frequency. MagSonic also supports one of the highest uplink data rates with a bit error rate (BER) of $\leq 10^{-5}$ while simultaneously powering the implant using a single ME transducer. No other study has achieved continuous simultaneous powering and uplink communication with a single transducer. Additionally, MagSonic is the only method robust to misalignments, maintaining high power reception even at a 90 degrees angular misalignment in any axis.

[a] *SPRUDT—Simultaneous power reception and uplink data transmission.*

[b] *Thickness estimated based on given dimensions of different layers and assuming epoxy was* $\sim 10\ \mu m$.

[c] *Two separate transducers used for simultaneous power reception and uplink data transmission.*

Reproduced with permission from Hosur et al. (2024).

References

Abdelazim, H. (2017). Magneto-electric nanocarriers for drug delivery: An overview. *Journal of Drug Delivery Science and Technology, 37*, 46–50.

Akpunonu, B., Hummell, J., Akpunonu, J. D., & Ud Din, S. (2022). Capsule endoscopy in gastrointestinal disease: Evaluation, diagnosis, and treatment. *Cleveland Clinic Journal of Medicine, 89*, 200–211.

Aliqab, K., Nadeem, I., & Khan, S. R. (2023). A comprehensive review of in-body biomedical antennas: Design, challenges and applications. *Micromachines, 14*.

Alrashdan, F. T., Chen, J. C., Singer, A., Avants, B. W., Yang, K., & Robinson, J. T. (2021). Wearable wireless power systems for 'ME-BIT' magnetoelectric-powered bio implants. *Journal of Neural Engineering, 18*.

Amirov, A. (2022). Multiferroic, magnetic, and magnetoelectric nanomaterials for medical applications. In *Magnetic materials and technologies for medical applications*.

Assi, M. A., Hezmee, M. N., Haron, A. W., Sabri, M. Y., & Rajion, M. A. (2016). The detrimental effects of lead on human and animal health. *Veterinary World, 9*, 660–671.

Brittain, J. S., & Cagnan, H. (2018). Recent trends in the use of electrical neuromodulation in Parkinson's disease. *Current Behavioral Neuroscience Reports, 5*, 170–178.

Chang, T. C., Wang, M. L., Charthad, J., Weber, M. J., & Arbabian, A. (2017). 27.7 A 30.5mm3 fully packaged implantable device with duplex ultrasonic data and power links achieving 95kb/s with $<10^{-4}$ BER at 8.5cm depth. In *2017 IEEE international solid-state circuits conference (ISSCC)* (pp. 460–461).

Chen, R., Canales, A., & Anikeeva, P. (2017). Neural recording and modulation technologies. *Nature Review Materials, 2*.

Chen, J. C., Kan, P., Yu, Z., Alrashdan, F., Garcia, R., Singer, A., Lai, C. S. E., Avants, B., Crosby, S., Li, Z., Wang, B., Felicella, M. M., Robledo, A., Peterchev, A. V., Goetz, S. M., Hartgerink, J. D., Sheth, S. A., Yang, K., & Robinson, J. T. (2022). A wireless millimetric magnetoelectric implant for the endovascular stimulation of peripheral nerves. *Nature Biomedical Engineering, 6*, 706–716.

Chen, T., Wu, H. F., Shi, Q., Zhou, P. H., Chen, S. Y., Xu, M. D., Zhong, Y. S., & Yao, L. Q. (2013). Endoscopic management of impacted esophageal foreign bodies. *Diseases of the Esophagus, 26*, 799–806.

Chu, Z., Pourhosseiniasl, M., & Dong, S. (2018). Review of multi-layered magnetoelectric composite materials and devices applications. *Journal of Physics D: Applied Physics, 51*.

Dong, C., He, Y., Li, M., Tu, C., Chu, Z., Liang, X., Chen, H., Wei, Y., Zaeimbashi, M., Wang, X., Lin, H., Gao, Y., & Sun, N. X. (2020). A portable very low frequency (VLF) communication system based on acoustically actuated magnetoelectric antennas. *IEEE Antennas and Wireless Propagation Letters, 19*, 398–402.

Erfani, R., Marefat, F., Sodagar, A. M., & Mohseni, P. (2018). Modeling and experimental validation of a capacitive link for wireless power transfer to biomedical implants. *IEEE Transactions on Circuits and Systems II: Express Briefs, 65*, 923–927.

Fiocchi, S., Chiaramello, E., Marrella, A., Suarato, G., Bonato, M., Parazzini, M., & Ravazzani, P. (2022). Modeling of core-shell magneto-electric nanoparticles for biomedical applications: Effect of composition, dimension, and magnetic field features on magnetoelectric response. *PLoS One, 17*, Article e0274676.

Ghanbari, M. M., Piech, D. K., Shen, K., Faraji Alamouti, S., Yalcin, C., Johnson, B. C., Carmena, J. M., Maharbiz, M. M., & Muller, R. (2019). A sub-mm3 ultrasonic free-floating implant for multi-mote neural recording. *IEEE Journal of Solid-State Circuits, 54*, 3017–3030.

Grant, S., & Coyne, K. (2012). MRI: A guided tour. https://nationalmaglab.org/magnet-academy/read-science-stories/science-simplified/mri-a-guided-tour/.

Guduru, R., Liang, P., Hong, J., Rodzinski, A., Hadjikhani, A., Horstmyer, J., Levister, E., & Khizroev, S. (2015). Magnetoelectric 'spin' on stimulating the brain. *Nanomedicine (London), 10*, 2051–2061.

Guduru, R., Liang, P., Runowicz, C., Nair, M., Atluri, V., & Khizroev, S. (2013). Magneto-electric nanoparticles to enable field-controlled high-specificity drug delivery to eradicate ovarian cancer cells. *Scientific Reports, 3*, 2953.

Guduru, R., Liang, P., Yousef, M., Horstmyer, J., & Khizroev, S. (2018). Mapping the Brain's electric fields with Magnetoelectric nanoparticles. *Bioelectronic Medicine, 4*, 10.

Hannachi, E., & Slimani, Y. (2022). Advanced progress in magnetoelectric multiferroic composites. In *Handbook of magnetic hybrid nanoalloys and their nanocomposites.*

Hiller-Sturmhöfel, S., & Bartke, A. (1998). The endocrine system: An overview. *Alcohol Health and Research World, 22*, 153–164.

Hilton, G. (2002). Melatonin and the pineal gland. *Journal of Neuroscience Nursing, 34*, 74.

Hochberg, L. R., Serruya, M. D., Friehs, G. M., Mukand, J. A., Saleh, M., Caplan, A. H., Branner, A., Chen, D., Penn, R. D., & Donoghue, J. P. (2006). Neuronal ensemble control of prosthetic devices by a human with tetraplegia. *Nature, 442*, 164–171.

Hoque Apu, E., Nafiujjaman, M., Sandeep, S., Makela, A. V., Khaleghi, A., Vainio, S., Contag, C. H., Li, J., Balasingham, I., Kim, T., & Ashammakhi, N. (2022). Biomedical applications of multifunctional magnetoelectric nanoparticles. *Materials Chemistry Frontiers, 6*, 1368–1390.

Hosur, S., Karan, S. K., Priya, S., & Kiani, M. (2023). Short-range communication for small biomedical implants using magnetoelectric effect. In *2023 IEEE international symposium on circuits and systems (ISCAS).*

Hosur, S., Kashani, Z., Karan, S. K., Priya, S., & Kiani, M. (2024). MagSonic: Hybrid magnetic-ultrasonic wireless interrogation of millimeter-scale biomedical implants with magneto-electric transducer. *IEEE Transactions on Biomedical Circuits and Systems, 18*, 383–395.

Hosur, S., Sriramdas, R., Karan, S. K., Liu, N., Priya, S., & Kiani, M. (2021). A comprehensive study on magnetoelectric transducers for wireless power transfer using low-frequency magnetic fields. *IEEE Transactions on Biomedical Circuits and Systems, 15*, 1079–1092.

Ibrahim, M. A., Hazhirkarzar, B., & Dublin, A. B. (2024). *Gadolinium magnetic resonance imaging*. StatPearls Publishing.

Jiang, C., Li, X., Lian, S. W. M., Ying, Y., Ho, J. S., & Ping, J. (2021). Wireless technologies for energy harvesting and transmission for ambient self-powered systems. *ACS Nano, 15*, 9328–9354.

Jin, P., Fu, J., Wang, F., Zhang, Y., Wang, P., Liu, X., Jiao, Y., Li, H., Chen, Y., Ma, Y., & Feng, X. (2021). A flexible, stretchable system for simultaneous acoustic energy transfer and communication. *Science Advances, 7.*

Karan, S. K., Hosur, S., Kashani, Z., Leng, H., Vijay, A., Sriramdas, R., Wang, K., Poudel, B., Patterson, A. D., Kiani, M., & Priya, S. (2024). Magnetic field and ultrasound induced simultaneous wireless energy harvesting. *Energy and Environmental Science, 17*, 2129–2144.

Karatum, O., Han, M., Erdogan, E. T., Karamursel, S., & Nizamoglu, S. (2023). Physical mechanisms of emerging neuromodulation modalities. *Journal of Neural Engineering, 20.*

Kashani, Z., & Kiani, M. (2023). A study on ultrasonic wireless power transfer with phased array for biomedical implants. *IEEE Transactions on Biomedical Circuits and Systems, 17*, 713–724.

Kaushik, A., Jayant, R. D., Nikkhah-Moshaie, R., Bhardwaj, V., Roy, U., Huang, Z., Ruiz, A., Yndart, A., Atluri, V., EL-Hage, N., Khalili, K., & Nair, M. (2016). Magnetically guided central nervous system delivery and toxicity evaluation of magneto-electric nanocarriers. *Scientific Reports, 6*, Article 25309.

Kaushik, A., Nikkhah-Moshaie, R., Sinha, R., Bhardwaj, V., Atluri, V., Jayant, R. D., Yndart, A., Kateb, B., Pala, N., & Nair, M. (2017). Investigation of ac-magnetic field stimulated nanoelectroporation of magneto-electric nano-drug-carrier inside CNS cells. *Scientific Reports, 7*, Article 45663.

Kaushik, A., Rodriguez, J., Rothen, D., Bhardwaj, V., Jayant, R. D., Pattany, P., Fuentes, B., Chand, H., Kolishetti, N., El-Hage, N., Khalili, K., Kenyon, N. S., & Nair, M. (2019). MRI-guided, noninvasive delivery of magneto-electric drug nanocarriers to the brain in a nonhuman primate. *ACS Applied Bio Materials, 2*, 4826–4836.

Kaushik, A., Yndart, A., Atluri, V., Tiwari, S., Tomitaka, A., Gupta, P., Jayant, R. D., Alvarez-Carbonell, D., Khalili, K., & Nair, M. (2019). Magnetically guided non-invasive CRISPR-Cas9/gRNA delivery across blood-brain barrier to eradicate latent HIV-1 infection. *Scientific Reports, 9*, 3928.

Khan, S. R., Pavuluri, S. K., Cummins, G., & Desmulliez, M. P. Y. (2020). Wireless power transfer techniques for implantable medical devices: A review. *Sensors, 20*.

Kiani, M., Jow, U. M., & Ghovanloo, M. (2011). Design and optimization of a 3-coil inductive link for efficient wireless power transmission. *IEEE Transactions on Biomedical Circuits and Systems, 99*, 1.

Kolishetti, N., Vashist, A., Arias, A. Y., Atluri, V., Dhar, S., & Nair, M. (2022). Recent advances, status, and opportunities of magneto-electric nanocarriers for biomedical applications. *Molecular Aspects of Medicine, 83*, Article 101046.

Kopyl, S., Surmenev, R., Surmeneva, M., Fetisov, Y., & Kholkin, A. (2021). Magnetoelectric effect: Principles and applications in biology and medicine-a review. *Materials Today Bio, 12*, Article 100149.

Kozeilski, K. L., Jahanshahi, A., Gilbert, H. B., Yu, Y., Erin, Ö., Francisco, D., Alosaimi, F., Temel, Y., & Sitti, M. (2021). Nonresonant powering of injectable nanoelectrodes enables wireless deep brain stimulation in freely moving mice. *Science Advances, 7*.

Liang, X., Chen, H., & Sun, N. X. (2021). Magnetoelectric materials and devices. *APL Materials, 9*.

Liang, X., Matyushov, A., Hayes, P., Schell, V., Dong, C., Chen, H., He, Y., Will-Cole, A., Quandt, E., Martins, P., Mccord, J., Medarde, M., Lanceros-Mendez, S., VAN Dijken, S., Sun, N. X., & Sort, J. (2021). Roadmap on magnetoelectric materials and devices. *IEEE Transactions on Magnetics, 57*, 1–57.

Ludewig, P., Graeser, M., Forkert, N. D., Thieben, F., Randez-Garbayo, J., Rieckhoff, J., Lessmann, K., Forger, F., Szwargulski, P., Magnus, T., & Knopp, T. (2022). Magnetic particle imaging for assessment of cerebral perfusion and ischemia. *Wiley Interdisciplinary Reviews Nanomedicine and Nanobiotechnology, 14*, Article e1757.

Maldonado, K. A., & Alsayouri, K. (2023). *Physiology, brain*. StatPearls Publishing. https://www.ncbi.nlm.nih.gov/books/NBK551718/.

Manfredi, L. (2021). Endorobots for colonoscopy: Design challenges and available technologies. *Front Robot AI, 8*, Article 705454.

Meng, M., & Kiani, M. (2017). Design and optimization of ultrasonic wireless power transmission links for millimeter-sized biomedical implants. *IEEE Transactions on Biomedical Circuits and Systems, 11*, 98–107.

Mhambi, S., Fisher, D., Tchokonte, M. B. T., & Dube, A. (2021). Permeation challenges of drugs for treatment of neurological tuberculosis and HIV and the application of magneto-electric nanoparticle drug delivery systems. *Pharmaceutics, 13*.

Mitoseriu, L. (2005). Magnetoelectric phenomena in single-phase and composite systems. *Boletín de la Sociedad Española de Cerámica y Vidrio, 44*, 177–184.

Morais, J. M., Papadimitrakopoulos, F., & Burgess, D. J. (2010). Biomaterials/tissue interactions: Possible solutions to overcome foreign body response. *The AAPS Journal, 12*, 188–196.

Nagesetti, A., Rodzinski, A., Stimphil, E., Stewart, T., Khanal, C., Wang, P., Guduru, R., Liang, P., Agoulnik, I., Horstmyer, J., & Khizroev, S. (2017). Multiferroic coreshell magnetoelectric nanoparticles as NMR sensitive nanoprobes for cancer cell detection. *Scientific Reports, 7*, 1610.

Nguyen, T., Gao, J., Wang, P., Nagesetti, A., Andrews, P., Masood, S., Vriesman, Z., Liang, P., Khizroev, S., & Jin, X. (2021). In vivo wireless brain stimulation via non-invasive and targeted delivery of magnetoelectric nanoparticles. *Neurotherapeutics, 18*, 2091–2106.

NIH. (2023). *Brain basics: Know your brain*. National Institute of Health: National Institute of Neurological Disorders and Stroke. https://www.ninds.nih.gov/health-information/public-education/brain-basics/brain-basics-know-your-brain.

Nikolayev, D., Zhadobov, M., Sauleau, R., & Karban, P. (2016). Antennas for ingestible capsule telemetry. In Q. H. Abbasi, M. U. Rehman, K. Qaraqe, & A. Alomainy (Eds.), *Advances in body-centric wireless communication: Applications and state-of-the-art*. IET.

Ozeri, S., & Shmilovitz, D. (2014). Simultaneous backward data transmission and power harvesting in an ultrasonic transcutaneous energy transfer link employing acoustically dependent electric impedance modulation. *Ultrasonics, 54*, 1929–1937.

Pardo, M., & Khizroev, S. (2022). Where do we stand now regarding treatment of psychiatric and neurodegenerative disorders? Considerations in using magnetoelectric nanoparticle. *WIREs Nanomedicine and Nanobiotechnology, 14*.

Park, M., Kang, T., Lim, I., Oh, K. I., Kim, S. E., Lee, J. J., & Park, H. I. (2018). Low-power, high data-rate digital capsule endoscopy using human body communication. *Applied Sciences, 8*.

Peery, A. F., Crockett, S. D., Murphy, C. C., Jensen, E. T., Kim, H. P., Egberg, M. D., Lund, J. L., Moon, A. M., Pate, V., Barnes, E. L., Schlusser, C. L., Baron, T. H., Shaheen, N. J., & Sandler, R. S. (2022). Burden and cost of gastrointestinal, liver, and pancreatic diseases in the United States: Update 2021. *Gastroenterology, 162*, 621–644.

Piech, D. K., Johnson, B. C., Shen, K., Ghanbari, M. M., Li, K. Y., Neely, R. M., Kay, J. E., Carmena, J. M., Maharbiz, M. M., & Muller, R. (2020). A wireless millimetre-scale implantable neural stimulator with ultrasonically powered bidirectional communication. *Nature Biomedical Engineering, 4*, 207–222.

Poza Cordon, J., Froilan Torres, C., Burgos Garcia, A., Gea Rodriguez, F., & Suarez de Parga, J. M. (2012). Endoscopic management of esophageal varices. *World Journal of Gastrointestinal Endoscopy, 4*, 312–322.

Pradhan, D. K., Kumari, S., Rack, P. D., & Kumar, A. (2021). Applications of strain-coupled magnetoelectric composites. *Reference Module in Materials Science and Materials Engineering, 10*, 2072.

Priya, S., & Inman, D. J. (2009). *Energy harvesting technologies*. Springer.

Pullarao, T., Vishnunarayanan, S., Suresh, P., & Nagaraju, G. (2019). Synthesis and electrical properties of lead-free piezo electrics for magnetoelectric applications. *IOP Conference Series: Materials Science and Engineering, 577*.

Richardson, T., Naidoo, G., Rupasinghe, N., Smart, H., & Bhattacharya, S. (2018). Biodegradable stents in resistant peptic oesophageal stricture: Is it the right way to go? *Clinical Medicine Insights: Gastroenterology, 11*.

Rinne, J., Keskinen, J., Berger, P. R., Lupo, D., & Valkama, M. (2017). Viability bounds of M2M communication using energy-harvesting and passive wake-up radio. *IEEE Access, 5*, 27868–27878.

Rizzo, G. (2020). *Study of a magnetoelectric transducer to wireless power medical implants*. Ph.D., Université Paris-Saclay.

Rodzinski, A., Guduru, R., Liang, P., Hadjikhani, A., Stewart, T., Stimphil, E., Runowicz, C., Cote, R., Altman, N., Datar, R., & Khizroev, S. (2016). Targeted and controlled anticancer drug delivery and release with magnetoelectric nanoparticles. *Scientific Reports, 6*, Article 20867.

Rosenfeld, J. V., & Wong, Y. T. (2017). Neurobionics and the brain-computer interface: Current applications and future horizons. *Medical Journal of Australia, 206*, 363–368.

Roundy, S., Steingart, D., Frechette, L., Wright, P., & Rabaey, J. (2004). Power sources for wireless sensor networks. In *European workshop on wireless sensor networks*Springer.

Schmidt, C. K., Medina-Sanchez, M., Edmondson, R. J., & Schmidt, O. G. (2020). Engineering microrobots for targeted cancer therapies from a medical perspective. *Nature Communications, 11*, 5618.

Schneider, A., & Feussner, H. (2017). Diagnostic procedures. In *Biomedical engineering in gastrointestinal surgery.*

Shamsudhin, N., Zverev, V. I., Keller, H., Pane, S., Egolf, P. W., Nelson, B. J., & Tishin, A. M. (2017). Magnetically guided capsule endoscopy. *Medical Physics, 44*, e91–e111.

Siegel, R. L., Miller, K. D., Wagle, N. S., & Jemal, A. (2023). Cancer statistics, 2023. *CA: A Cancer Journal for Clinicians, 73*, 17–48.

Singer, A., Dutta, S., Lewis, E., Chen, Z., Chen, J. C., Verma, N., Avants, B., Feldman, A. K., O'malley, J., Beierlein, M., Kemere, C., & Robinson, J. T. (2020). Magnetoelectric materials for miniature, wireless neural stimulation at therapeutic frequencies. *Neuron, 107*, 631–643 e5.

Smith, I. T., Zhang, E., Yildirim, Y. A., Campos, M. A., Abdel-Mottaleb, M., Yildirim, B., Ramezani, Z., Andre, V. L., Scott-Vandeusen, A., Liang, P., & Khizroev, S. (2023). Nanomedicine and nanobiotechnology applications of magnetoelectric nanoparticles. *Wiley Interdisciplinary Reviews Nanomedicine and Nanobiotechnology, 15*, Article e1849.

Song, H., Listyawan, M. A., & Ryu, J. (2022). Core–shell magnetoelectric nanoparticles: Materials, synthesis, magnetoelectricity, and applications. *Actuators, 11*.

Sonmezoglu, S., Fineman, J. R., Maltepe, E., & Maharbiz, M. M. (2021). Monitoring deep-tissue oxygenation with a millimeter-scale ultrasonic implant. *Nature Biotechnology, 39*, 855–864.

Stimphil, E., Nagesetti, A., Guduru, R., Stewart, T., Rodzinski, A., Liang, P., & Khizroev, S. (2017). Physics considerations in targeted anticancer drug delivery by magnetoelectric nanoparticles. *Applied Physics Reviews, 4*.

Tettey, F., Parupelli, S. K., & Desai, S. (2023). A review of biomedical devices: Classification, regulatory guidelines, human factors, software as a medical device, and cybersecurity. *Biomedical Materials & Devices, 2*, 316–341.

Tortora, G., Mulana, F., Ciuti, G., Dario, P., & Menciassi, A. (2015). Inductive-based wireless power recharging system for an innovative endoscopic capsule. *Energies, 8*, 10315–10334.

Vaughn, M. J., & Haas, J. S. (2022). On the diverse functions of electrical synapses. *Frontiers in Cellular Neuroscience, 16*, Article 910015.

Vilela, M., & Hochberg, L. R. (2020). Applications of brain-computer interfaces to the control of robotic and prosthetic arms. In N. F. Ramsey, & J. D. R. Millán (Eds.), *Handbook of clinical neurology.* Elsevier B. V.

Wallace, M. B. (2017). Endoscopic removal of polyps in the gastrointestinal tract. *Gastroenterology and Hepatology, 13*.

Weitschies, W., Wedemeyer, J., Stehr, R., & Trahms, L. (1994). Magnetic markers as a noninvasive tool to monitor gastrointestinal transit. *IEEE Transactions on Biomedical Engineering, 41*.

Yu, Z., Alrashdan, F. T., Wang, W., Parker, M., Chen, X., Chen, F. Y., Woods, J., Chen, Z., Robinson, J. T., & Yang, K. (2022). Magnetoelectric backscatter communication for millimeter-sized wireless biomedical implants. In *Proceedings of the 28th annual international conference on mobile computing and networking.*

Yue, K., Guduru, R., Hong, J., Liang, P., Nair, M., & Khizroev, S. (2012). Magneto-electric nano-particles for non-invasive brain stimulation. *PLoS One, 7*, Article e44040.

Zaeimbashi, M., Lin, H., Dong, C., Liang, X., Nasrollahpour, M., Chen, H., Sun, N., Matyushov, A., He, Y., Wang, X., Tu, C., Wei, Y., Zhang, Y., Cash, S. S., Onabajo, M.,

Shrivastava, A., & Sun, N. (2019). NanoNeuroRFID: A wireless implantable device based on magnetoelectric antennas. *IEEE Journal of Electromagnetics, RF and Microwaves in Medicine and Biology, 3*, 206–215.

Zhang, W., Liang, X., Zhu, L., Zhang, X., Jin, Z., Du, Y., Tian, J., & Xue, H. (2022). Optical magnetic multimodality imaging of plectin-1-targeted imaging agent for the precise detection of orthotopic pancreatic ductal adenocarcinoma in mice. *EBioMedicine, 80*, Article 104040.

CHAPTER

15

Beyond magnetoelectrics—An outlook on emerging directions in magnetoelectrics research

Bin Luo[1], *Alexandria Ruth Will-Cole*[2], *and Nian Sun*[1]

[1] **Northeastern University, Boston, MA, United States;** [2] **Sandia National Laboratory, Albuquerque, NM, United States**

Are multiferroics dead? While at this point the underlying physics of multiferroics and magnetoelectrics is well understood, there are still many impactful research directions for multiferroics and magnetoelectrics at the intersections with other avenues of research, such as spintronics and quantum transduction. In this chapter, we summarize a brief outlook on emerging, active areas of novel magnetoelectrics research including magnetoelectric control of spintronic THz emitters, spin-to-charge conversion phenomenon, and spin-defects in diamond.

15.1 Magnetoelectric-based spintronic THz emission

The terahertz (THz) spectrum has a broadband range of 100 GHz to 30 THz and is utilized for many applications including information communication, medical imaging, and security (Ferguson & Zhang, 2002; Mittleman, 2017; Papaopannou & Beigang, 2021). Although there is a critical need for THz emitters, there are few competitive, broadband THz emission sources. Current technologies are based on photoconductive antennas, optical rectification, and magnetic spintronics (Wu et al., 2021). The typical bandwidth for photoconductive antennas based on LT-GaAs is 0.1 − 4 THz and InGaAs is 0.1 − 6 THz (Dreyhaupt et al., 2005; Globisch et al., 2017; Klatt et al., 2009, 2010). The typical bandwidth for optical

https://doi.org/10.1016/B978-0-443-18606-6.00005-8

rectification based on ZnTe is 0.1 − 4 THz and $LiNbO_3$ is 0.1 − 4 THz (Liu et al., 2004). Magnetic spintronics-based THz emitters are extremely promising due to their broad, useable bandwidth range (W/CoFeB/Pt 0.1 THz to 30 THz) (Seifert et al., 2016; Wu et al., 2021). Magnetic spintronics-based THz emitters are based on a photoexcited spin current and the inverse spin hall effect (ISHE). These are typically heterostructures comprised of ferromagnetic materials and materials with large spin-orbit-coupling (SOC), i.e., heavy metals, topological insulators, etc. (Seifert et al., 2016; Wang et al., 2018). When the heterostructure is excited by an ultrafast femtosecond (fs) laser pulse, the spin-polarized electrons in the ferromagnetic material are excited and diffuse as a spin current (j_s) into the adjacent, nonmagnetic, large SOC layer. Due to the ISHE, the diffused spin current is converted into a transverse, transient charge current (j_c), which then produces a THz-frequency electric field pulse (Papaopannou & Beigang, 2021). The process is defined by Eq. (15.1), where M is defined as the magnetization, and Θ_{SH} is the spin-Hall angle.

$$\boldsymbol{j_c} = \theta_{SH} \boldsymbol{j_s} \times \frac{\boldsymbol{M}}{|\boldsymbol{M}|} \tag{15.1}$$

This phenomenon is shown in Fig. 15.1 where the ferromagnetic layer's magnetization points in the positive x-direction, j_s is in the z-direction, and j_c is in the y-direction. Overall, this THz-emission is driven by an oscillating electric dipole due to the time varying ISHE electric current.

There are several methods to tune magnetic spintronics-based THz emitters, namely, intrinsic and extrinsic methods. Intrinsic approaches are

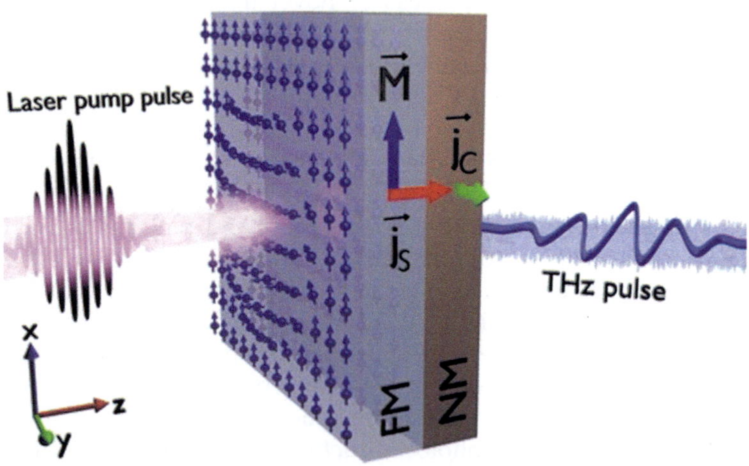

FIGURE 15.1 Mechanism schematic for magnetic spintronics-based THz emission. *This figure is reproduced from Papaopannou and Beigang (2021) under a Creative Commons Attribution 4.0 International License.*

based on careful heterostructure design, specifically the placement of the large SOC materials in the heterostructure and their respective spin-Hall angle magnitude and sign. Experimental studies have reported a correlation of the sign and amplitude of the spin Hall conductivity to the relative emitted THz amplitude and sign (Seifert et al., 2017). This means that these magnetic spintronics-based THz heterostructures can be engineered via SOC material selection and placement. Experimentally, this has been shown with W/FM/Pt heterostructure—since the W has an appreciable negative spin-Hall angle and the Pt has an appreciable positive spin-Hall angle; the trilayer emitter converts a backward- and forward-flowing spin current into a unidirectional charge current, which then enhances the emitted THz signal amplitude compared with a bilayer of W/FM or FM/Pt alone (Seifert et al., 2016, 2017). Additionally, the bandwidth of this optimized W/FM/Pt spintronic emitter far exceeds the bandwidth of traditional emitters (Dreyhaupt et al., 2005; Globisch et al., 2017; Klatt et al., 2009, 2010; Liu et al., 2004; Seifert et al., 2016, 2017; Wu et al., 2021). Extrinsic methods to tune the magnetic spintronics-based THz emitters mainly focused on polarization tuning of the emitted E-field signal through external magnetic field bias and orientation. Elliptically polarized THz-signal was experimentally generated via orientation rotation of an applied magnetic field which enabled the chirality and azimuthal angle to be effectively tuned (Kong et al., 2019). Additionally, for an applied curved magnetic field, the polarization ellipticity can be tuned from left-handed (δ^+) to right-handed (δ^-) polarization by reversing the field orientation (Kong et al., 2019).

Since magnetic-based spintronic THz emitters behave according to the ISHE that imposes an orthogonality between the magnetization and the emitted E-field polarization, tuning the magnetization state can alter the emitted E-field polarization. This has largely been explored through bulk, external magnetic field manipulation, which is not ultimately a technologically practical approach to tune the magnetization and thus the emitted E-field polarization. Several groups have recently utilized magnetoelectric coupling to enable the voltage/E-field control of the magnetic-based spintronic THz emitters (Agarwal et al., 2022; Chaurasiya et al., 2021; Cheng et al., 2021; Dreyhaupt et al., 2005; Globisch et al., 2017; Klatt et al., 2009, 2010; Kong et al., 2019; Lezier et al., 2022; Liu et al., 2004; Mittleman, 2017; Papaopannou & Beigang, 2021; Seifert et al., 2016, 2017; Wang et al., 2018; Wu et al., 2021). Cheng et al. demonstrated a 28 degrees polarization rotation and a 69% tuned THz amplitude in a $Pb(Mg_{1/3}Nb_{2/3})_{0.7}Ti_{0.3}O_3$ (PMN-PT)/$Co_{40}Fe_{40}B_{20}$/Pt with the application of 400 V (Papaopannou & Beigang, 2021). P. Agarwal et al. demonstrated 270% tuning of the THz amplitude utilizing two different magnetic materials in their heterostructures, e.g., PMN-PT/$Co_{75}Fe_{25}$/Pt and PMN-PT/CoFeB/Pt (Agarwal et al., 2022). G. Lezier et al. improved the polarization rotation to 70 degrees with a PMN-PT/W/FeCo/$TbCo_2$/FeCo/Pt-optimized ISHE heterostructure with the application of 400 V (Lezier et al., 2022). While great tuning performance of magnetic-based spintronic-based THz emitters has been demonstrated, Will-Cole et al. has demonstrated 59 degrees magnetization rotation with just a 100 V applied voltage in an ultrafast demagnetization-based THz emitter heterostructure of PMN-PT/Cu/FeGaB (Will-Cole et al., 2022; Will-Cole, Wang, et al., 2022)—this heterostructure has superior tuning performance due to the low voltage application to the magnetization tuning. Thus, there is still ample opportunity to further improve the polarization rotation control of magnetic-based spintronic THz emitters through the use of materials with high magnetostriction to improve the magnetoelectric coupling in these devices (Dong et al., 2018; Lou et al., 2007; Will-Cole et al., 2022; Will-Cole, Wang, et al., 2022). Overall compared with magnetic field control of magnetic spintronic-based THz emission, magnetoelectric control, and therefore, voltage control, is a flexible, compact, and practical approach for THz emission control. Recently, a hybrid magnon-phonon cavity structure has been proposed based on a freestanding metal/magnetic insulator (MI)/dielectric multilayer, designed to enable efficient excitation of large-amplitude THz standing exchange spin waves through prolonged resonant magnon-phonon interactions (Zhuang et al., 2024). The paramagnetic metal layer functions both as a photoacoustic transducer for generating THz acoustic phonons via femtosecond laser irradiation and as a spin–charge-current transducer for electrical detection of the acoustically excited THz magnons. Compared to traditional all-metallic THz cavities, the proposed multilayer structure minimizes eddy-current losses, enhancing performance at high frequencies, revealing the potential for designing nonlinear magnonic devices, such as magnon-based neural networks and switches, driven by acoustic waves. This work provides a pathway for advancing THz technologies and developing high-frequency, acoustically controlled magnonic devices. In the future, this device topology may enable compact, programmable THz emitter arrays.

15.2 Strain manipulation of spin-to-charge conversion

Recently, there is heightened interest in "spin-orbitronics," which exploits the interplay between charge currents and spin currents in systems with large spin-orbit-coupling, specifically the generation of pure spin currents that then can be utilized to manipulate magnetization in adjacent magnetic layers through spin-orbit torques (Kondou et al., 2016; Soumyanarayanan et al., 2016). These interfacial spin polarized states include Rashba states or topological surface states. The helical spin polarization of these states enables the conversion between spin and charge currents through the Edelstein and inverse Edelstein effects (Edelstein, 1990). The efficiency of the charge-to-spin conversion is defined by the spin Hall angle (ratio of the spin current density to the charge current density). Remarkably, the spin Hall angle in topological insulators can be tuned via electric field through coupling with a ferroelectric substrate. Q. Lu et al. demonstrated a giant voltage tunable (600%), and reversible spin Hall angle for a PMN-PT/Bi_2Se_3/NiFe/TaO_x heterostructure—the tunability here is attributed to both the strain and the charge doping at the interface between the ferroelectric substrate and the topological insulator and its corresponding effect on the Berry curvature near the Fermi level (Li, Li, Guo, & et al, 2022). Additionally, the transport properties of a magnetic topological insulator (magnetism achieved through doping) can be tuned via an electric field through coupling with a ferroelectric substrate. X.-W. Zhao et al. demonstrated the reversible and nonvolatile manipulation of the carrier density and carrier type of a Cr-doped Bi_2Se_3 film grown on a PMN-PT substrate (Zhao et al., 2018). The combination of topological insulators with ferroelectrics can enable control over the unique and exotic properties of topological insulating materials that may allow for enhanced device functionalities. The same can be said for the combination of Rashba interfaces with ferroelectricity. For instance, $SrTiO_3$ is a paraelectric at room temperature but undergoes a ferroelectric phase transition at low temperature with the application of high electric field. Thus, at low temperature, an electric field can be used to tune the spin-orbit properties of a two-dimensional electron gas (2DEG) at the $SrTiO_3$/Al interface. P. Noël et al. demonstrated the nonvolatile control of the spin−charge conversion in an $SrTiO_3$/Al/NiFe heterostructure (Noël et al., 2020). While it is interesting that the spin-charge conversion can be manipulated in $SrTiO_3$/Al heterostructure-based 2DEG system via electric field, the low-temperature requirements are not practical for potential commercial applications. Others have explored a similar phenomenon in a ferroelectric Rashba semiconductor system. At room temperature, S. Varotto et al. demonstrated nonvolatile control voltage control of spin-to-charge conversion in GeTe/Fe heterostructures. However, the ferroelectric properties of GeTe are poor, which limits the overall efficiency of such single phase Rashba ferroelectrics (Kolobov et al. n.d.; Picozzi, 2022).

15.3 Nanoscale magnetometry and hybrid quantum transducer based on NV^- centers in diamond

Negatively charged nitrogen vacancy centers (NV^- centers) are a quantum spin defects in diamond, which consist of a nitrogen atom and a nearby lattice vacancy site (Lenef & Rand, 1996). Owing to their long coherence time above 100 μs at room temperature (Ohno et al., 2012), high fidelity (Robledo et al., 2011), and easy optical initiation and readout of quantum states (Awschalom et al., 2018) tuned by magnetic excitations over a broad temperature range, NV^- centers are intriguing solid-state platforms for quantum computing, communication, information processing (Childress & Hanson, 2013; Fuchs et al., 2011; Taminiau et al., 2014), and sensitive non-invasive nanoscale magnetic sensors (de Lange et al., 2011; Rondin et al., 2014; Taylor et al., 2008). The triplet orbital ground states (^3A) of NV^- centers can be excited to the corresponding excited states (^3E) by a 532 nm pulsed green laser, which follows the spin conservation manner. On ^3A and ^3E levels, there are three spin states ($m_s = \pm 1, 0$). Due to the crystal field in diamond, the $m_s = \pm 1$ spin states are separated ~2.87 GHz away from $m_s = 0$. When a DC bias field B_{DC} is applied along the NV^- axis, the Zeeman effect can split the $m_s = \pm 1$ spin states with a frequency gap of $2\gamma B_{DC}$, where γ is the gyromagnetic ratio (Stanwix et al., 2010). This Zeeman splitting can be observed in optical detected magnetic resonance (ODMR) spectrum, where two resonance peaks at $f_1 = 2.87 - 2\gamma B_{DC}$ and $f_2 = 2.87 + 2\gamma B_{DC}$ show up. The local static magnetic field sensitivity can reach ~10^{-9} T. In addition, the NV^- electron spin resonance (ESR) transition at the spin splitting frequencies can lead to Rabi oscillations between spin states (Maletinsky et al., 2012). The AC magnetic field B_{AC} perpendicular to the NV^- axis can be detected by the oscillation frequency. The fluctuating magnetic field sensitivity can be promoted from 30 nT $Hz^{-1/2}$ (Maze et al., 2008) to 4.3 nT $Hz^{-1/2}$ (Balasubramanian et al., 2009) by isotopic engineering, enabling the NV^- center an excellent magnetometer for weak magnetic fluctuations during electric charge transport and spin wave relaxations (Du et al., 2017; Kolkowitz et al., 2015; van der Sar et al., 2015). In terms of spatial distribution, high-density NV^- center ensembles can be formed several nanometers beneath the diamond substrate surface by doping (Ohno et al., 2012) or ion-implantation (Rabeau et al., 2006), leading to a non-invasive, nanoscale proximity of NV^- center to the sample surface. This high proximity leads to a nanoscale spatial resolution of NV^- center magnetometry and even the magnetic field detection of individual spin.

The NV^- center can be coherently controlled by an AC magnetic field, electric field (Whiteley, Heremans, et al., 2019) and strain waves (Miao et al., 2019) for hybrid quantum transducer and information processing unit. Compared to the direct AC magnetic field excitation via microwave antennas, the spin wave (magnon) excitations can realize long-distance, coherent control of the NV^- centers from several hundred micron to millimeter level and enable 100 times stronger local driving field with the same driving power (Andrich et al., 2017; Kikuchi et al., 2017; Muhlherr et al., 2019). The power-efficient magnon excitation also reduces the off-resonant spurious spin wave noise at high power (Awschalom et al., 2021) and maintains long coherence time for the magnon-mediated entanglement and communication between two NV^- center qubits, enabling an on-chip quantum information processor and computation unit. NV^- center qubits can further couple to THz optic photons as mentioned previously, holding promises for microwave-to-optical quantum transducers converting GHz qubit excitations to the THz optical photon (Hensen et al., 2015) for long-distance quantum information communication to another distance quantum processing unit. The magnon-NV^- and NV^--optic-photon couplings are building blocks for quantum transduction and communication with great potential for a hybrid quantum network protocol (Barfuss et al., 2015; MacQuarrie et al., 2013; Ovartchaiyapong et al., 2014; Trifunovic et al., 2013; Wang et al., 2020).

Compared to magnons, phonons are also excellent information carriers for coherent and quantum information processing owing to their long lifetime in solid-state platforms. The small wavelength of acoustic waves makes phonons very suitable for on-chip hybrid quantum systems (Li et al., 2020). The excitation of magnons with strain waves is much more power-efficient than microwave excitation and can greatly reduce the size of magnon-mediated NV^--center-based quantum transducer. In the strong coupling region, the magnon–phonon interaction leads to the formation of magnon polaron hybridized state and enables the on-chip information transduction between magnon and phonon systems, creating a phonon-magnon-NV-photon hybrid quantum information transduction platform.

15.4 Magnetoelectric control of NV^- centers in diamond

Recently, magnetoelectric control of NV^- centers was realized by magnetoelectric coupling in CoFeB/300 um PMN-PT heterostructures (Solanki et al., 2022), as shown in Fig. 15.2. The electrical polarization of PMN-PT tuned by an external voltage, V, can generate the lattice strain transferred to the CoFeB film, which in turn, changes its magnetic anisotropy, as shown in Fig. 15.2a–d. Under a fixed external magnetic field (H_{ext}) along the [100] x-axis, the voltage increase from $V = V_{off}$ to

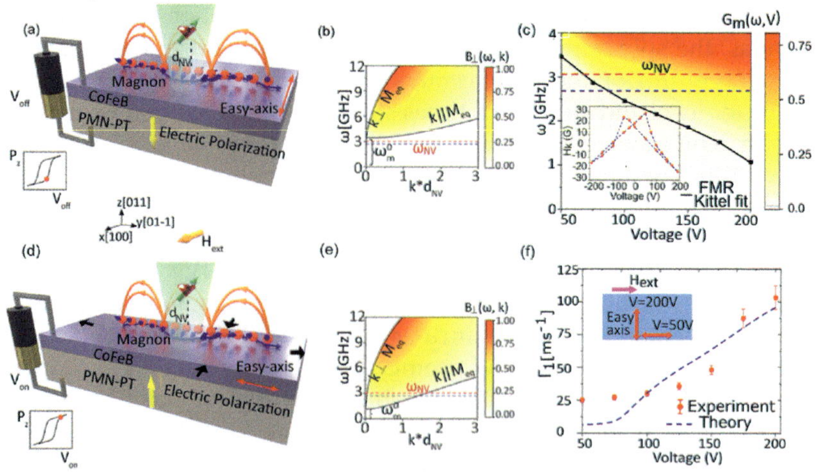

FIGURE 15.2 Magnetoelectric control of NV^- center by CoFeB/PMN-PT BAW resonator. (a,d) Schematic illustration of the quantum spin defect (QSD)-magnon hybrid sample fabricated by dispersing nanodiamonds containing ensembles of NV^- centers on a thin ferromagnetic film of CoFeB on a 300 μm thick PMN-PT ferroelectric substrate. (b,e) Maps of normalized B as a function of ω–k for both V_{on} and V_{off}. The black lines enveloping the colormap are the calculated magnon dispersion lines for bulk modes ($\boldsymbol{k}\|$M) and surface modes ($\boldsymbol{k}\perp$M). The dashed colored lines represent the NV^- electron spin resonance (ESR) lines ω_{NV}. (c) Ferromagnetic resonance (FMR) frequency (ω_m^0) (data in black lines) as a function of applied voltage extracted from the experimental results for a fixed external magnetic field $H_{ext} = 57$G along x-axis. The color map represents the calculated values of the magnetic noise spectral density $G_m(\omega, V)$ for an effective NV^- height $d_{NV} = 77$ nm. The dashed colored lines represent maximum spread of the NV^- ESR lines ω_{NV}. The inset shows the detailed measurements of magnetic anisotropy field as a function of applied voltage. (f) Measured relaxation rates Γ_1 as a function of applied voltage for a fixed $H_{ext} = 57$G along x-axis. The inset shows a schematic illustration of magnetic anisotropy field for the two different voltages for a fixed H_{ext}. The dashed line represents theoretical fit of relaxation rates Γ_1. These figures are reproduced from (Solanki et al., 2022) under a Creative Commons Attribution 4.0 International License.

$V = V_{\text{on}}$ reverses P_z and brings the magnon band into resonance with the NV^- ESR transitions, as shown in Fig. 15.2b–e. The movement of magnon bands results from the coupled electric, elastic, and magnetic orders in the multiferroic heterostructure. As the voltage is increased from 50 to 200 V, the anisotropy field H_k decreases monotonically from $H_\text{k} = 30G$ to $H_\text{k} = -22G$, indicating the flipped easy axis from x to y and the reversal of P_z, as shown in Fig. 15.2c. The lowest spin relaxation rate monotonically increases from 25.1 ± 2 $[\text{ms}]^{-1}$ at $V = V_{\text{off}}$ to 102.8 ± 9 $[\text{ms}]^{-1}$ at $V = V_{\text{on}}$, indicating a 400% tuning of the relaxation rate Γ_1. The large tunability in Γ_1 shows that magnetoelectric control can realize strong interactions between magnons and NV^- centers in a power efficient way. Further work can be done in promoting magnon–phonon interactions by using highly magneto-strictive and low damping materials such as boron- or carbon-doped FeGa and FeCo alloys (Liang et al., 2019, 2020; Schoen et al., 2016), spinel NiZnAl-ferrite (Emori et al., 2017), and magnesium aluminum ferrite oxide $MgAl_{0.5}Fe_{1.5}O_4$ (MAFO) (Li, Li, Yi, et al., 2022).

15.5 SAW driven spin wave control of NV$^-$ centers in diamond

Compared to bulk acoustic wave resonators, surface acoustic wave (SAW) resonators are more favorable for strong magnon–phonon interactions since they can highly concentrate acoustic energy on the surface to interact with spin waves. SAW resonators are also easier to fabricate owing to their planar structures. Recently, SAW-driven magnon resonance achieved local and efficient control of NV$^-$ centers, showing comparable Rabi frequencies to microwave excitation using over 1000 times more power (Andrich et al., 2017; Kikuchi et al., 2017). 20 nm Ni or Co film was placed between interdigital transducers (IDTs) with a fundamental frequency of 287 MHz (Labanowski et al., 2018). The power absorption shows a 4-fold symmetry and is maximal at $\phi = 45°$, as shown in Fig. 15.3a. The power absorption is tunable by a bias magnetic field at 45°, as shown in Fig. 15.3b. At 1429 MHz (fifth harmonic frequency of IDTs), the power absorption is higher than the fundamental and third harmonic modes. At 1429 MHz, a direct NV$^-$ photoluminescence (PL) change signal can be observed since this coincides closely with the NV$^-$'s excited state resonance frequency, as shown in Fig. 15.3c. The PL change is due to ferromagnetic resonance (FMR), since no PL was seen under a high field and close to the IDT. The NV$^-$ PL change decreases with a distance from the short edge of the film, due to the absorption of the excitation signal as it travels through the magnetic layer, as shown in Fig. 15.3d. To enhance the performance of ME SAW devices, it is critical to incorporate magnetic materials with low damping and strong magnetoelastic coupling as mentioned previously, and there needs to be surface acoustic wave energy confinement and focusing technique—this will help to realize an efficient, and local control of NV$^-$ centers through SAW-driven magnetic resonance. Gaussian focusing technique has also recently been applied in SAW devices to suppress acoustic diffraction loss and significantly enhance the load and internal Q of 22,400 at low temperatures of $\sim$30 K (Whiteley, Wolfowicz, et al., 2019). In addition, Bragg reflectors were demonstrated to help achieve strong magnon–phonon coupling with a cooperativity of 1.2 (Hatanaka et al., 2022). Implementing advanced acoustic wave focusing and confinement technique is a future research direction to

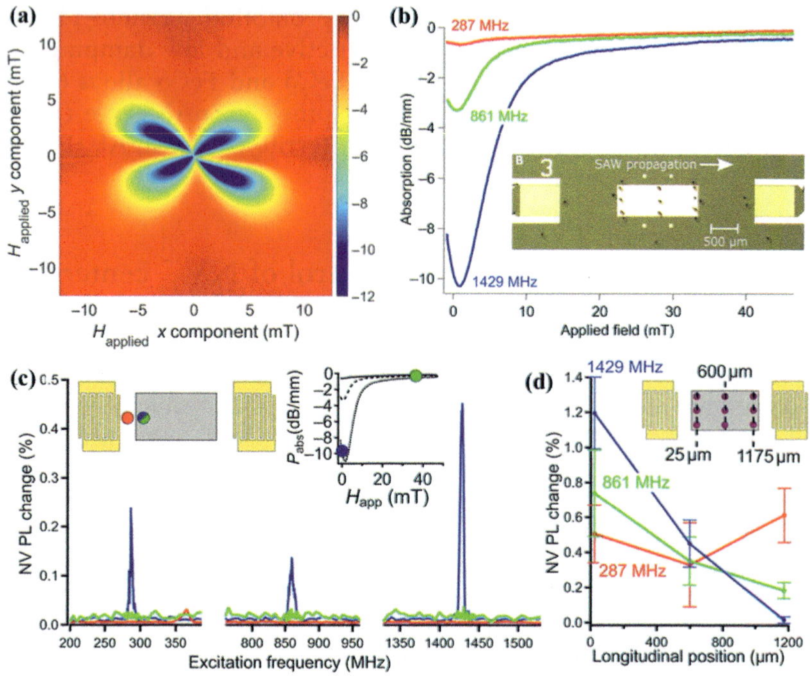

FIGURE 15.3 Energy-efficient and local control of NV^- center by SAW driven magnon resonance. (a) Plot of power absorption as a function of applied magnetic field for a 20-nm nickel acoustically driven ferromagnetic resonance (ADFMR) device at 1429 MHz. The x component of the field is taken to be parallel to the direction of SAW propagation, and the y component is in-plane and perpendicular to the direction of SAW propagation. The color bar indicates absorption in decibels per millimeter. (b) Line cut along the angle of highest absorption (45 degrees) showing a large field-dependent attenuation at 287, 861, and 1429 MHz. The insets show the photograph of IDTs, magnetoelastic film and clusters of nanodiamonds on the measured device. (c) Change in NV^- center PL normalized to the DC level for NV^- centers located off the ferromagnetic pad (*red*), and NV^- centers on the pad with zero field (*blue*) and a high (35.8 mT) applied bias field (*green*) (d) NV^- PL change in a 20-nm nickel sample as a function of longitudinal position from the edge of the ferromagnet closest to the excitation IDT at zero applied magnetic field. *These figures are reproduced from Labanowski et al. (2018) under a Creative Commons Attribution 4.0 International License.*

enable strong magnon–phonon interaction for efficient coherent control of NV^- centers and the realization of hybrid quantum transducers involving magnons, phonons, NV^- centers, and photons.

References

Agarwal, P., Huang, L., Ter Lim, S., & Singh, R. (2022). *Nature Communications, 13*, 4072.

Andrich, P., de las Casas, C., Liu, X., Bretscher, H., Berman, J., Heremans, F., Nealey, P., & Awschalom, D. (2017). Long-range spin wave mediated control of defect qubits in nanodiamonds. *Npj Quantum Inform, 3*.

Awschalom, et al. (2021). Quantum engineering with hybrid magnonic systems and materials (invited paper). *IEEE Transactions on Quantum Engineering, 2*, 1–36.

Awschalom, D., Hanson, R., Wrachtrup, J., & Zhou, B. (2018). Quantum technologies with optically interfaced solid-state spins. *Nature Photonics, 12*, 516–527.

Balasubramanian, G., Neumann, P., Twitchen, D., Markham, M., Kolesov, R., Mizuochi, N., Isoya, J., Achard, J., Beck, J., Tissler, J., Jacques, V., Hemmer, P., Jelezko, F., & Wrachtrup, J. (2009). Ultralong spin coherence time in isotopically engineered diamond. *Nature Materials, 8*, 383–387.

Barfuss, A., Teissier, J., Neu, E., Nunnenkamp, A., & Maletinsky, P. (2015). Strong mechanical driving of a single electron spi,n. *Nature Physics, 11*.

Chaurasiya, A., Li, Z., Medwal, R., Gupta, S., Mohan, J.R., Fukuma, Y., Asada, H., Chia, E.E.M., & Rawat, R.S., (2021). arXiv:2111.13823.

Cheng, H., Huang, Q., He, H., Zhao, Z., Sun, H., Wu, Q., Jiang, Z., Wang, J., Huang, H., Fu, Z., & Lu, Y. (2021). *Physical Review Applied, 16*, Article 054011.

Childress, L., & Hanson, R. (2013). Diamond NV centers for quantum computing and quantum networks *MRS Bulletin. 38*, 134–138.

de Lange, G., Riste, D., Dobrovitski, V., & Hanson, R. (2011). Single-spin magnetometry with multipulse sensing sequences. *Physical Review Letters, 106*.

Dong, C., et al. (2018). *Applied Physics Letters, 113*, Article 262401.

Dreyhaupt, A., Winnerl, S., Dekorsy, T., & Helm, M. (2005). *Applied Physics Letters, 86*, Article 12112.

Du, C., van der Sar, T., Zhou, T., Upadhyaya, P., Casola, F., Zhang, H., Onbasli, M., Ross, C., Walsworth, R., Tserkovnyak, Y., & Yacoby, A. (2017). Control and local measurement of the spin chemical potential in a magnetic insulator. *Science, 357*, 195–198.

Edelstein, V. M. (1990). *Solid State Communications, 73*, 233–235.

Emori, S., Gray, B., Jeon, H., Peoples, J., Schmitt, M., Mahalingam, K., Hill, M., McConney, M., Gray, M., Alaan, U., Bornstein, A., Shafer P, P., N'Diaye, A., Arenholz, E., Haugstad, G., Meng, K., Yang, F., Li, D., Mahat, S., ... Howe, B. (2017). Coexistence of low damping and strong magnetoelastic coupling in epitaxial spinel ferrite thin films. *Advanced Materials, 29*.

Ferguson, B., & Zhang, X.-C. (2002). Materials for terahertz science and technology. *Nature Materials, 1*, 26–33.

Fuchs, G., Burkard, G., Klimov, P., & Awschalom, D. (2011). A quantum memory intrinsic to single nitrogen-vacancy centres in diamond. *Nature Physics, 7*, 789–793.

Globisch, B., Dietz, R. J. B., Kohlhaas, R. B., Göbel, T., Schell, M., Alcer, D., Semtsiv, M., & Masselink, W. T. (2017). *Journal of Applied Physics, 121*, Article 053102.

Hatanaka, D., Asano, M., Okamoto, H., Kunihashi, Y., Sanada, H., & Yamaguchi, H. (2022). On-chip coherent transduction between magnons and acoustic phonons in cavity magnomechanics. *Physical Review Applied, 17*.

Hensen, B., Bernien, H., Dreau, A., Reiserer, A., Kalb, N., Blok, M., Ruitenberg, J., Vermeulen, R., Schouten, R., Abellan, C., Amaya, W., Pruneri, V., Mitchell, M., Markham, M., Twitchen, D., Elkouss, D., Wehner, S., Taminiau, T., & Hanson, R. (2015). Loophole-free Bell inequality violation using electron spins separated by 1.3 kilometres. *Nature, 526*, 682–686.

Kikuchi, D., Prananto, D., Hayashi, K., Laraoui, A., Mizuochi, N., Hatano, M., Saitoh, E., Kim, Y., Meriles, C., & An, T. (2017). Long-distance excitation of nitrogen-vacancy centers in diamond via surface spin waves. *Applied Physics Express, 10*.

Klatt, G., Gebs, R., Janke, C., Dekorsy, T., & Bartels, A. (2009). *Optics Express, 17*, Article 22847.

Klatt, G., Hilser, F., Qiao, W., Beck, M., Gebs, R., Bartels, A., Huska, K., Lemmer, U., Bastian, G., Johnston, M. B., et al. (2010). *Optics Express, 18*, 4939.

Kolkowitz, S., Safira, A., High, A., Devlin, R., Choi, S., Unterreithmeier, Q., Patterson, D., Zibrov, A., Manucharyan, V., Park, H., & Lukin, M. (2015). Probing Johnson noise and ballistic transport in normal metals with a single-spin qubit. *Science, 347*, 1129–1132.

Kolobov A.V., et al. *APL Materials, 2*, 066101. 202.

Kondou, K., et al. (2016). *Nature Physics, 12*, 1027–1031.

Kong, D., Wu, X., Wang, B., Nie, T., Xiao, M., Pandey, C., Gao, Y., Wen, L., Zhao, W., Ruan, C., Miao, J., Li, Y., & Wang, L. (2019). *Adv. Optical Mater., 7*, Article 1900487.

Labanowski, D., Bhallamudi, V., Guo, Q., Purser, C., McCullian, B., Hammel, P., & Salahuddin, S. (2018). Voltage-driven, local, and efficient excitation of nitrogen-vacancy centers in diamond. *Science Advances, 4.*

Lenef, A., & Rand, S. (1996). Electronic structure of the N-V center in diamond: Theory. *Physical Review B: Condensed Matter, 53*, 13441–13455.

Lezier, G., Koleják, P. P., Lampin, J.-F., Postava, K., Vanwolleghem, M., & Tiercelin, N. (2022). *Applied Physics Letters, 120*, Article 32404.

Li, Q., Li, P., Guo, Z., et al. (2022). *Nature Communications, 13*, 1650.

Li, R., Li, P., Yi, D., Riddiford, L., Chai, Y., Suzuki, Y., Ralph, D., & Nan, T. (2022). Anisotropic magnon spin transport in ultrathin spinel ferrite thin films-evidence for anisotropy in exchange stiffness. *Nano Letters, 22*, 1167–1173.

Li, Y., Zhang, W., Tyberkevych, V., Kwok, W., Hoffmann, A., & Novosad, V. (2020). Hybrid magnonics: Physics, circuits, and applications for coherent information processing. *Journal of Applied Physics, 128.*

Liang, X., Dong, C., Celestin, S., Wang, X., Chen, H., Ziemer, K., Page, M., McConney, M., Jones, J., Howe, B., & Sun, N. X. (2019). Soft magnetism, magnetostriction, and microwave properties of Fe-Ga-C alloy films. *IEEE Magnetics Letters, 10.*

Liang, X., Dong, C., Chen, H., Wang, J., Wei, Y., Zaeimbashi, M., He, Y., Matyushov, A., Sun, C., & Sun, N. (2020). A review of thin-film magnetoelastic materials for magnetoelectric applications. *Sensors, 20.*

Liu, R., Gu, C.-M., He, L.-R., Wu, S., Shen, W.-Z., Ogawa, H., & Guo, Q.-X. (2004). *Acta Physica Sinica, 53*, 1217.

Lou, J., et al. (2007). *Applied Physics Letters, 91*, Article 182504.

MacQuarrie, E., Gosavi, T., Jungwirth, N., Bhave, S., & Fuchs, G. (2013). Mechanical spin control of nitrogen-vacancy centers in diamond. *Physical Review Letters, 111.*

Maletinsky, P., Hong, S., Grinolds, M., Hausmann, B., Lukin, M., Walsworth, R., Loncar, M., & Yacoby, A. (2012). A robust scanning diamond sensor for nanoscale imaging with single nitrogen-vacancy centres. *Nature Nanotechnology, 7*, 320–324.

Maze, J., Stanwix, P., Hodges, J., Hong, S., Taylor, J., Cappellaro, P., Jiang, L., Dutt, M., Togan E, E., Zibrov, A., Yacoby, A., Walsworth, R., & Lukin, M. (2008). Nanoscale magnetic sensing with an individual electronic spin in diamond. *Nature, 455.*

Miao, K., Bourassa, A., Anderson, C., Whiteley, S., Crook, A., Bayliss, S., Wolfowicz, G., Thiering, G., Udvarhelyi, P., Ivady, V., Abe, H., Ohshima, T., Gali, A., & Awschalom, D. (2019). Electrically driven optical interferometry with spins in silicon carbide. *Science Advances, 5.*

Mittleman, D. M. (2017). Perspective: Terahertz science and technology. *Journal of Applied Physics, 122*, Article 230901.

Muhlherr, C., Shkolnikov, V., & Burkard, G. (2019). Magnetic resonance in defect spins mediated by spin waves. *Physical Review B: Condensed Matter, 99.*

Noël, P., Trier, F., Vicente Arche, L. M., et al. (2020). *Nature, 580*, 483–486.

Ohno, K., Heremans, F., Bassett, L., Myers, B., Toyli, D., Jayich, A., Palmstrom, C., & Awschalom, D. (2012). Engineering shallow spins in diamond with nitrogen delta-doping. *Applied Physics Letters, 101.*

Ovartchaiyapong, P., Lee, K., Myers, B., & Jayich, A. (2014). Dynamic strain-mediated coupling of a single diamond spin to a mechanical resonator. *Nature Communications, 5.*

Papaopannou, E. T., & Beigang, R. (2021). THz spintonic emitters: A review on achievements and future challenges. *Nanophotonics, 10*(4), 1243–1257.

Picozzi, S. (2022). *Frontiers in Physiology, 2.*

Rabeau, J., Reichart, P., Tamanyan, G., Jamieson, D., Prawer, S., Jelezko, F., Gaebel, T., Popa, I., Domhan, M., & Wrachtrup, J. (2006). Implantation of labelled single nitrogen vacancy centers in diamond using N-15. *Applied Physics Letters, 88.*

Robledo, L., Childress, L., Bernien, H., Hensen, B., Alkemade, P., & Hanson, R. (2011). High-fidelity projective read-out of a solid-state spin quantum register. *Nature, 477*, 574–578.

Rondin, L., Tetienne, J., Hingant, T., Roch, J., Maletinsky, P., & Jacques, V. (2014). Magnetometry with nitrogen-vacancy defects in diamond. *Reports on Progress in Physics, 77.*

Schoen, M., Thonig, D., Schneider, M., Silva, T., Nembach, H., Eriksson, O., Karis, O., & Shaw, J. (2016). Ultra-low magnetic damping of a metallic ferromagnet. *Nature Physics, 12*, 839–842.

Seifert, T., Jaiswal, S., Martens, U., Hannegan, J., Braun, L., Maldonado, P., Freimuth, F., Kronenberg, A., Henrizi, J., Radu, I., Beaurepaire, E., Mokrousov, Y., Oppeneer, P. M., Jourdan, M., Jakob, G., Turchinovich, D., Hayden, L. M., Wolf, M., Münzenberg, M., Kläui, M., & Kampfrath, T. (2016). *Nature Photonics, 10*, 483.

Seifert, T., Jaiswal, S., Sajadi, M., Jakob, G., Winnerl, S., Wolf, M., Kläui, M., & Kampfrath, T. (2017). *Applied Physics Letters, 110*, Article 252402.

Solanki, A. B., Bogdanov, S. I., Rahman, M. M., Rustagi, A., Dilley, N. R., Shen, T., Tong, W., Debashis, P., Chen, Z., & Appenzeller, J. (2022). Electric field control of interaction between magnons and quantum spin defects. *Physical Review Research, 4*, L012025.

Soumyanarayanan, A., Reyren, N., Fert, A., et al. (2016). *Nature, 539*, 509–517.

Stanwix, P., Pham, L., Maze, J., Le Sage, D., Yeung, T., Cappellaro, P., Hemmer, P., Yacoby, A., Lukin, M., & Walsworth, R. (2010). Coherence of nitrogen-vacancy electronic spin ensembles in diamond. *Physical Review B: Condensed Matter, 82.*

Taminiau, T., Cramer, J., van der Sar, T., Dobrovitski, V., & Hanson, R. (2014). Universal control and error correction in multi-qubit spin registers in diamond. *Nature Nanotechnology, 9*, 171–176.

Taylor, J., Cappellaro, P., Childress, L., Jiang, L., Budker, D., Hemmer, P., Yacoby, A., Walsworth, R., & Lukin, M. (2008). High-sensitivity diamond magnetometer with nanoscale resolution. *Nature Physics, 4*, 810–816.

Trifunovic, L., Pedrocchi, F., & Loss, D. (2013). Long-distance entanglement of spin qubits via ferromagnet. *Physics Reviews, X*, 3.

van der Sar, T., Casola, F., Walsworth, R., & Yacoby, A. (2015). Nanometre-scale probing of spin waves using single-electron spins. *Nature Communications, 6.*

Wang, X., Cheng, L. L., Zhu, D., Wu, Y., Chen, M., Wang, Y., Zhao, D., Boothroyd, C. B., Lam, Y. M., Zhu, J.-X., Battiato, M., Song, J. C. W., Yang, H., & Chia, E. E. M. (2018). *Advances in Materials, 30*(52), Article 1802356.

Wang, X., Xiao, Y., Liu, C., Lee-Wong, E., McLaughlin, N., Wang, H., Wu, M., Wang, H., Fullerton, E., & Du, C. (2020). Electrical control of coherent spin rotation of a single-spin qubit. *Npj Quantum Inform, 6.*

Whiteley, S., Heremans, F., Wolfowicz, G., Awschalom, D., & Holt, M. (2019). Correlating dynamic strain and photoluminescence of solid-state defects with stroboscopic x-ray diffraction microscopy. *Nature Communications, 10.*

Whiteley, S. J., Wolfowicz, G., Anderson, C. P., Bourassa, A., Ma, H., Ye, M., Koolstra, G., Satzinger, K. J., Holt, M. V., Heremans, F. J., Cleland, A. N., Schuster, D. I., Galli, G., & Awschalom, D. D. (2019). Spin–phonon interactions in silicon carbide addressed by Gaussian acoustics. *Nature Physics, 15*, 490–495.

Will-Cole, A. R., et al. (2022). *IEEE Transactions on Magnetics, 58*, 1.

Will-Cole, A. R., Wang, C., Bhattacharjee, N., Liu, Y., & Sun, N. X. (2022). *Physical Review B: Condensed Matter, 106*, Article 174401.

Wu, W., Ameyaw, C. Y., Doty, M. F., & Jungfleisch, M. B. (2021). Principles of spintronic THz emitters. *Journal of Applied Physics, 130*, Article 091101.

Zhao, X.-W., Dong, S.-N., Gao, G.-Y., et al. (2018). *Npj Quantum Materials, 3*, 52.

Zhuang, S., Zhang, X., Zhu, Y., Sun, N. X., Eom, C., Evans, P. G., & Hu, J. (2024). Hybrid magnon-phonon cavity for large-amplitude terahertz spin-wave excitation. *Physical Review Applied, 21*(044009).

Index

D

M

N

S

T

U

V

W

9780443186066